# Ultrafast Electronic and Structural Dynamics

Kiyoshi Ueda
Editor

# Ultrafast Electronic and Structural Dynamics

Springer

*Editor*
Kiyoshi Ueda
Tohoku University
Sendai, Japan

ISBN 978-981-97-2913-5 ISBN 978-981-97-2914-2 (eBook)
https://doi.org/10.1007/978-981-97-2914-2



This Springer imprint is published by the registered company Springer Nature Singapore Pte Ltd.
The registered company address is: 152 Beach Road, #21-01/04 Gateway East, Singapore 189721, Singapore

# Preface

A natural time scale of the movement of electrons in matter, or the electronic wave packet motion, is of the order of tens to hundreds attoseconds (as). 1 as corresponds to $10^{-18}$ s. One atomic unit in time is ~24 as. This is the intrinsic smallest time under which an electron takes a circle on the first Bohr orbital divided by $2\pi$. Thanks to the recent developments of laser technologies, catching electron motions in matter is no longer a dream but a tangible reality. The 2023 Nobel Prize in Physics was awarded jointly to Pierre Agostini, Ferenc Krausz, and Anne L'Huillier "for methods that generate attosecond pulses light for the study of electron dynamics in matter".

A natural time scale of the movement of atoms in matter is of the order of tens to hundreds of femtoseconds. 1 fs is $10^{-15}$ s. The fastest atomic motion is the vibration of the hydrogen molecule, and its vibrational period is ~8 fs. The 1999 Nobel Prize in Chemistry was awarded to Ahmed H. Zewail for "for showing that it is possible with rapid laser technique to see how atoms in a molecule move during a chemical reaction". This area, called femtochemistry, has been developed significantly in the last decades, thanks to the developments of X-ray free electron lasers (XFEL) and ultrafast electron diffractions (UED) that allow us to make a 3D molecular movie of atomic motions during chemical reactions and structural changes of matter.

On the other hand, time scales of the macroscopic structural changes, energy transfers, proton transfers, etc. in functional materials and macromolecules are significantly slower than the above-described time scales. These range from picoseconds (ps; 1 ps = $10^{-12}$ s) to milliseconds (ms; 1 ms = $10^{-3}$ s) because these slow processes are caused by a series of chain reactions. These slow dynamics are directly related to the functionalities of matter and living species. Here, again newly developed XFEL-based 3D molecular movie camera is opening a new route to investigate, e.g., protein's functionalities.

The present book describes time-resolved studies on atoms, molecules, condensed matters, and proteins in a wide range of time scales from attoseconds to milliseconds for the study of electronic and structural dynamics and the resulting functions of matters. We start with attosecond electron dynamics in atoms, molecules, liquids, and solids, then move to femtosecond structural dynamics in isolated molecules, molecules in solution, solids, and protein molecules, and finally picosecond to millisecond structural dynamics, energy flow, proton tunneling and functions of proteins.

In Chap. 1, D. Busto, S. Zhong, J. M. Dahlström, A. L'Huillier, and M. Gisselbrecht describe how to study electron dynamics in atoms, focusing on the timescale for the photoelectric effect in atoms. In brief, they use high-order harmonic generation (HHG) in gases irradiated by an intense 800 nm near-infrared (NIR) laser pulse to create a train of extreme ultraviolet (XUV) attosecond pulses. The XUV photons ionize atoms in the presence of a *weak* fraction of the NIR pulses used for HHG, with a variable delay, and electrons are detected with an electron spectrometer. Two-photon transitions lead to the creation of sidebands in the photoelectron spectrum, which oscillate with the delay between the XUV and the NIR fields. From these RABBIT (reconstruction of attosecond beating by interference of two-photon transitions) measurements, attosecond time delays in atomic photoionization can be extracted. Theoretical backgrounds of photoionization time delays and the RABBIT technique are described in detail in this chapter.

The HHG spectrum by itself also includes the information about the electron dynamics in matter induced by an *intense* IR pulse, which is the origin of HHG in matter. In Chap. 2, M. Krüger and N. Dudovich describe how to make use of the HHG spectra to study such attosecond electron dynamics in atoms, molecules, and solids in intense fields. They use all-optical interferometry that allows them to access phase information and thus to reconstruct the electron wave packet dynamics. They discuss two approaches, internal and external interferometry. In internal interferometry, they manipulate the quantum paths within the HHG mechanisms and resolve the phase of the electron wave packet in strong IR laser-field processes. In the external interferometry, measuring amplitude and phase information of the harmonic light field through its interference with another light field, they determine the relative spectral phase of the two beams. We can see that both approaches reveal valuable information regarding the strong-field light-matter interaction as well as the atomic, molecular, or condensed system itself.

Water is essential to life. Attosecond dynamics in water may play an important role in chemical and biological processes that dominantly take place in water. In Chap. 3, H. J. Wörner, A. Schild, D. Jelovina, I. Jordan, C. Perry, T. Luu, and Z. Yin address attosecond dynamics in liquid water employing two experimental methods. The first one is attosecond photoelectron spectroscopy. This is closely related to the RABBIT method described in Chap. 1 but employs a liquid microjet as a target, which extends the applicability of RABBIT to most liquids and solutions. In addition, they employ "a principal-component analysis" of RABBIT data, which generalizes the extraction of photoionization delays to complex systems that often possess overlapping photoelectron spectra. They show that photoemission from liquid water is delayed by 50–70 as compared to photoemission from isolated water molecules at photon energies of 20–30 eV. The origin of the delays is fully discussed and experimentally supported by a RABBIT experiment on size-selected water clusters. The second method is HHG spectroscopy described in Chap. 2 but this first demonstration of

HHG in liquid opens a new route to study attosecond dynamics also in liquid, solutes, and solvents.

As can be seen in Chaps. 1–3, HHG represents a promising attosecond coherent light source in the XUV regime, playing a key role in attosecond science. As noted, HHG is a result of the interaction between an *intense* laser pulse and matter. Especially, attosecond electron dynamics driven by an ultrashort intense laser pulse in solids is of particular interest from the viewpoint of both fundamental physics and applications. In Chap. 4, K. L. Ishikawa, Y. Shinohara, T. Sato, and T. Otobe theoretically discuss momentum-space strong-field electron dynamics in graphene, crystalline dielectrics, and semiconductors, focusing on HHG and tunneling effect in solids.

On a fundamental level, the underlying mechanism behind almost any optical phenomenon is the light-driven motion of charges on atomic dimensions in space and time, as noted above. Visualizing such dynamics, however, requires ängstrom resolution in space and attosecond resolution in time. In Chap. 5, P. Baum and Y. Morimoto describe how to reach this goal using attosecond *electron* pulses in diffraction and microscopy, together with the first proof-of-concept experiments.

Chapters 1–5 discuss attosecond electron dynamics, or coherent electronic wave packet motion in the system, where the atomic motions in the systems are assumed to be frozen in these time scales. On the other hand, Chaps. 6–11 discuss femtosecond structural dynamics due to atomic motions that occur in tens to hundreds of femtoseconds in various systems. All the structural dynamics discussed here are photoinduced dynamics. In such photoexcited systems, coherent atomic motions, or nuclear wave packet dynamics such as vibrations in molecules and phonons in solids are induced. Along these atomic motions, various types of reactions, such as isomerization and intersystem crossing in molecules, light-induced phase transitions in solids, etc. take place. During these non-adiabatic reactions, electronic and nuclear degrees of freedom cannot be described independently. The main subject here would then be how to probe the coherent motion of atoms in matter that may vary through non-adiabatic reactions. It is worth noting that, even if the coherent electronic wave packet studied in Chaps. 1–5 is created upon photoexcitation, it may dephase quickly during the atomic motions. Thus, in most cases, the implicit assumption in Chaps. 6–11 is that the photoexcitation creates mainly coherent nuclear wave packet in a particular electronic state in the systems.

In Chap. 6, M. Yamazaki, T. Endo, A. Hishikawa, and M. Takahashi describe two experimental approaches that aim to visualize the change of molecular orbital pattern of isolated molecules during a light-induced intramolecular reaction. One is to employ electron momentum spectroscopy, which is known as a versatile tool to visualize the molecular orbital in a momentum space, to photoexcited molecules ($S_2$ acetone and $S_1$ toluene), in a time-resolved manner. Another attempt is to measure ultrafast laser tunneling ionization of photoexcited molecules. The tunneling ionization comes from the outermost molecular orbital and its rate is enhanced when the electron cloud is along the laser polarization direction. Here, they employ ion momentum imaging of fragment ions produced by dissociative ionization to image the outermost molecular orbital of photoexcited NO molecules. We can see both approaches offer opportunities for investigating chemical reactions via time-resolved molecular orbital imaging.

As noted above, the advent of ultrafast pulsed XFELs with very high brightness has enabled the determination of transient molecular structures of molecules in excited states and undergoing chemical dynamics using X-ray scattering. Chapter 7 by P. M. Weber, B. Stankus, and A. Kirrander provides an introduction to X-ray scattering theory and considers several important aspects relating to the experimental implementation as well as the interpretation of the experiments. We can see here that ultrafast gas-phase X-ray scattering provides new observables to elucidate the dynamics of chemical reactions by providing complete, time-dependent molecular structures. Notably, the experiments also yield valuable information beyond molecular structure, such as the redistribution of electron density in excited states or the identification of the transition dipole moment in a molecule via the anisotropy of the scattering pattern.

In Chap. 8, J. Yang, M. Centurion, X. Wang, Th. Wolf, and M, Gühr show that also UED from isolated molecules allows insight into ultrafast quantum wave packet dynamics of molecules with sub-ängstrom spatial resolution. For the simple example of an excited state vibrational wave packet in iodine, they reconstruct the mean atomic distance as a function of delay after the optical excitation. For the case of $CF_3I$, they reconstruct the dissociative motion after UV excitation on the A-band states. They also show how the signature of non-adiabatic wave packet dynamics is visible in the diffraction data.

As seen in Chaps. 7 and 8, revealing details of coherent atomic motions during photochemical reactions on the femtosecond timescale is crucial for understanding the reaction mechanism in the photoexcited molecules. In Chap. 9, T. Katayama, T. J. Penfold, and Ch. Bressler describe how coherent atomic motions, or vibrational wave packets, are observed employing another experimental technique, i.e., time-resolved X-ray absorption near edge structure (TR XANES) spectroscopy with XFELs. TR XANES is sensitive not only to the local electronic structure changes but also to the structural dynamics in the vicinity of the selected absorbing atom. Since no long-range order information is required, it is highly suited to study disordered systems without any crystalline order, such as solvated transition metal complexes in the solution phase. Here, they present experimental studies on femtosecond vibrational modes for two prototypical transition metal complexes in solution.

In Chap. 10, M. Trigo, M. P. M. Dean, and D. A. Reis discuss structural and electronic dynamics in solids. The experimental techniques they introduce include XFEL-based time-resolved X-ray diffraction (TR XRD) and time-resolved resonant inelastic X-ray scattering (TR RIXS). As showcase examples, they first describe coherent phonon spectroscopy with TR XRD and its application to study couplings among coupled charge, spin, orbital, and lattice degrees of freedom in the photoexcited states, where the optical excitation is used to alter this delicate balance and produce properties not accessible in equilibrium. Then they show case studies in which the *intense* pump induces non-adiabatic dynamics exemplified by lattice symmetry changes. TR RIXS is also very powerful to study variations of charge, orbital, and spin degrees of freedom. Here, they describe the first TR RIXS measurement on a magnon.

Chapter 11 is the last chapter on femtosecond molecular dynamics and the first chapter on proteins. Here, A. Yabushita firstly describes details on ultrafast transient absorption spectroscopy setup developed for probing *primary* photochemical reactions of proteins. Ultrafast transient absorption signals of proteins also show modulation reflecting the real-time motion of the molecular vibration in time domain. Thus, time-gated Fourier analysis of the signal elucidates the temporal variation of the molecular vibration frequency, which elucidates molecular structural change during the primary photoreaction. Two showcase experiments are presented for ultrafast photoisomerization in bacteriorhodopsin (BR) and its mutants and ultrafast photo-dissociation in nitric oxide synthase.

As seen in Chaps. 7–11, the primary photoreactions start with coherent atomic motions in any form of matter, including proteins, but such vibrational coherences dephase in longer time scales, say, above 1 ps, as seen in these chapters. Chapters 12–15, on the other hand, focus on such long timescale reactions in the range of picoseconds to milliseconds in proteins, where coherent vibrational wave packet no longer plays a role because of dephasing. This long timescale is also very important because protein's functions appear in this time range because of chain reactions initiated by the primary reactions, such as the ones described in Chap. 11.

In Chap. 12, Y. Mizutani, S. Yamashita, and M. Mizuno describe time-resolved Raman mapping of protein energy flow that occurs in the time scale in tens of picoseconds. Their main interest is to provide a microscopic picture of energy dissipation on the atomic spatial scale. Heme proteins are ideal molecular systems to study energy flow because a photoexcited heme group is converted to an electronically ground state via ultrafast internal conversion and thus excess vibrational energy can be deposited locally at the heme site in proteins immediately after the photoexcitation. In consequence, heme acts as a very efficient energy convertor. To observe subsequent energy relaxation processes, they adopted anti-Stokes ultraviolet resonance Raman (UVRR) spectroscopy. Because of the resonance effect, UVRR spectroscopy probes Raman bands of aromatic amino acid residues with high selectivity, allowing site-selective detection of energy at the level of a single amino acid residue in a large protein molecule.

In Chap. 13, D. Zigmantas and T. Mančal introduce another powerful experimental method to study energy transfer and couplings within molecular assemblies, i.e., two-dimensional electronic spectroscopy (2DES). They review firstly the theoretical foundation for understanding of 2DES signals in molecular systems and then discuss encompassing crucial aspects of experimental implementations of 2DES. Illustrative applications of 2DES to investigate light-harvesting functions in an isolated pigment-protein complex and an intact photosynthetic unit in green sulfur bacteria reveal the extensive insights into the photophysical functions of photosynthetic machinery. The time range of revealed energy transfer processes is in the range from 50 fs to 250 ps.

In Chap. 14, A. Benabbas and P. M. Champion discuss the use of vibrational coherence and ultrafast wide dynamic-range population kinetics to probe biological molecules. They use impulsive stimulated Raman scattering to develop the method of vibrational coherence spectroscopy, which reveals both the structural and functional aspects of the difficult to detect low-frequency modes in proteins. Studies of electron tunneling in cytochrome c as well as the kinetics of the methionine–heme binding reaction are emphasized. They also examine vibrational coherence and its potential participation in the excited state proton transfer of green fluorescent proteins and conclude that coherent motion does not affect the excited state proton transfer rate that occurs on the ps timescale. For the ground-state proton back-transfer reaction, they find that (incoherent) vibrationally assisted proton tunneling is the dominant transport channel and that the tunneling rate is ~400 ps at room temperature.

The last Chap. 15 by A. M. Orville, E. Nango, S. Iwata, S. Mous, J. Standfuss, P. Nogly, M. Suga, J.-R. Shen, and M. Kubo is dedicated to time-resolved serial femtosecond X-ray crystallography (SFX) of protein molecules. XFELs offer new opportunities in structural biology, especially for samples at near-physiological conditions and for time-resolved studies that link together the analysis of structure and function within the same samples, with the crystal size requirements such that even submicron crystals can yield high-quality diffraction data. SFX is a new technique developed to exploit the fs pulses from XFELs and to use thousands of micron-size crystals or smaller. This chapter starts with an introduction to SFX by Orville. Then, Nango, Iwata, Mous, Standfuss, and Nogly describe the structure, function, and dynamics of the light-driven proton pump, bacteriorhodopsin. The time range covered here is from tens of femtoseconds (primary isomerization) to milliseconds (functions). Next, Suga and Shen summarize results from photosystem II, one of the hottest topics in time-resolved SFX. Kubo concludes the chapter with an example of a caged nitric oxide compound driving the P450nor enzyme reaction, which is an unusual member of the P450 superfamily of heme-dependent proteins.

I thank all the above-listed chapter authors for their invaluable contributions to this book.

Sendai, Japan Kiyoshi Ueda

# Contents

# Contributors

**Peter Baum** Universität Konstanz, Konstanz, Germany

**Abdelkrim Benabbas** Department of Physics and Center for Interdisciplinary Research on Complex Systems, Northeastern University, Boston, MA, USA

**Christian Bressler** European XFEL, Schenefeld, Germany;
The Hamburg Centre of Ultrafast Imaging, Hamburg, Germany;
Department of Physics, Universität Hamburg, Hamburg, Germany

**David Busto** Department of Physics, Lund University, Lund, Sweden

**Martin Centurion** Department of Physics and Astronomy, University of NE–Lincoln, Lincoln, NE, USA

**Paul M. Champion** Department of Physics and Center for Interdisciplinary Research on Complex Systems, Northeastern University, Boston, MA, USA

**Jan Marcus Dahlström** Department of Physics, Lund University, Lund, Sweden

**Mark P. M. Dean** Condensed Matter Physics and Materials Science Department, Brookhaven National Laboratory, Upton, NY, USA

**Nirit Dudovich** Weizmann Institute of Science, Department of Physics of Complex Systems, Rehovot, Israel

**Tomoyuki Endo** Kansai Institute for Photon Science, National Institutes for Quantum Science and Technology, Kyoto, Japan

**Mathieu Gisselbrecht** Department of Physics, Lund University, Lund, Sweden

**Markus Gühr** Deutsches Elektronen-Synchrotron DESY, Hamburg, Germany

**Akiyoshi Hishikawa** Department of Chemistry, Graduate School of Science, Nagoya University, Nagoya, Japan;
Research Center for Materials Science, Nagoya University, Nagoya, Japan

**Kenichi L. Ishikawa** Department of Nuclear Engineering and Management, Graduate School of Engineering, The University of Tokyo, Bunkyo-ku, Tokyo, Japan

**So Iwata** RIKEN SPring-8 Center, Hyogo, Japan;
Department of Cell Biology, Graduate School of Medicine, Kyoto University, Kyoto, Japan

**Denis Jelovina** Laboratorium für Physikalische Chemie, ETH Zürich, Zürich, Switzerland

**Inga Jordan** Laboratorium für Physikalische Chemie, ETH Zürich, Zürich, Switzerland

**Tetsuo Katayama** Japan Synchrotron Radiation Research Institute, Sayo, Hyogo, Japan;
RIKEN, Sayo, Hyogo, Japan

**Adam Kirrander** Department of Chemistry, Physical and Theoretical Chemistry Laboratory, University of Oxford, Oxford, UK

**Michael Krüger** Technion – Israel Institute of Technology, Department of Physics and Solid State Institute and Helen Diller Quantum Center, Haifa, Israel

**Minoru Kubo** Graduate School of Science, University of Hyogo, Ako, Hyogo, Japan

**Anne L'Huillier** Department of Physics, Lund University, Lund, Sweden

**Tran Trung Luu** Laboratorium für Physikalische Chemie, ETH Zürich, Zürich, Switzerland

**Tomáš Mančal** Faculty of Mathematics and Physics, Charles University, Prague, Czech Republic

**Misao Mizuno** Department of Chemistry, Graduate School of Science, Osaka University, Toyonaka, Osaka, Japan

**Yasuhisa Mizutani** Department of Chemistry, Graduate School of Science, Osaka University, Toyonaka, Osaka, Japan

**Yuya Morimoto** RIKEN Cluster for Pioneering Research and RIKEN Center for Advanced Photonics, RIKEN, Wako, Saitama, Japan

**Sandra Mous** Linac Coherent Light Source, SLAC National Accelerator Laboratory, Menlo Park, USA

**Eriko Nango** RIKEN SPring-8 Center, Hyogo, Japan;
Institute of Multidisciplinary Research for Advanced Materials, Tohoku University, Sendai, Miyagi, Japan

**Przemyslaw Nogly** Dioscuri Centre for Structural Dynamics of Receptors, Faculty of Biochemistry, Biophysics and Biotechnology, Jagiellonian University, Kraków, Poland

**Allen M. Orville** DiamondLight Source Ltd., Didcot, UK; Harwell Science and Innovation Campus, Didcot, UK

**Tomohito Otobe** Ultrafast Dynamics Group, National Institutes for Quantum and Radialogical Science and Technology (QST), Kizugawa, Kyoto, Japan

**Thomas J. Penfold** Chemistry-School of National and Environmental Sciences, Newcastle University, Newcastle Upon-Tyne, UK

**Conaill Perry** Laboratorium für Physikalische Chemie, ETH Zürich, Zürich, Switzerland

**David A. Reis** Stanford PULSE Institute and Stanford Institute for Materials and Energy Sciences (SIMES), SLAC National Accelerator Laboratory, Menlo Park, CA, USA

**Takeshi Sato** Department of Nuclear Engineering and Management, Graduate School of Engineering, The University of Tokyo, Bunkyo-ku, Tokyo, Japan

**Axel Schild** Laboratorium für Physikalische Chemie, ETH Zürich, Zürich, Switzerland

**Jian-Ren Shen** Graduate School of Environmental, Life, Natural Science and Technology, Research Institute for Interdisciplinary Science, Okayama University, Okayama, Japan

**Yasushi Shinohara** Photon Science Center, Graduate School of Engineering, The University of Tokyo, Bunkyo-ku, Tokyo, Japan

**Joerg Standfuss** Laboratory of Biomolecular Research, Biology and Chemistry Division, Paul Scherrer Institute, Villigen PSI, Switzerland

**Brian Stankus** Department of Chemistry, Western Connecticut State University, Danbury, Connecticut, USA

**Michihiro Suga** Graduate School of Environmental, Life, Natural Science and Technology, Research Institute for Interdisciplinary Science, Okayama University, Okayama, Japan

**Masahiko Takahashi** Institute of Multidisciplinary Research for Advanced Materials, Tohoku University, Sendai, Japan

**Mariano Trigo** Stanford PULSE Institute and Stanford Institute for Materials and Energy Sciences (SIMES), SLAC National Accelerator Laboratory, Menlo Park, CA, USA

**Xijie Wang** Fakultät für Physik, Universität Duisburg Essen, Essen, Germany

**Peter M. Weber** Department of Chemistry, Brown University, Providence, Rhode Island, USA

**Thomas Wolf** SLAC National Accelerator Laboraty, Menlo Park, USA

**Hans Jakob Wörner** Laboratorium für Physikalische Chemie, ETH Zürich, Zürich, Switzerland

**Atsushi Yabushita** Department of Electrophysics, National Yang Ming Chiao Tung University, Hsinchu, Taiwan;
Research Institute for Engineering, Kanagawa University, Yokohama, Japan

**Satoshi Yamashita** Department of Chemistry, Graduate School of Science, Osaka University, Toyonaka, Osaka, Japan

**Masakazu Yamazaki** Department of Chemistry, School of Science, Tokyo Institute of Technology, Tokyo, Japan

**Jie Yang** Center of Basic Molecular Science, Department of Chemistry, Tsinghua University, Beijing, China

**Zhong Yin** Laboratorium für Physikalische Chemie, ETH Zürich, Zürich, Switzerland

**Shiyang Zhong** Department of Physics, Lund University, Lund, Sweden

**Donatas Zigmantas** Chemical Physics, Lund University, Lund, Sweden

# Chapter 1
# Attosecond Dynamics of Non-resonant Atomic Photoionization

**David Busto, Shiyang Zhong, Jan Marcus Dahlström, Anne L'Huillier, and Mathieu Gisselbrecht**

**Abstract** The interaction of light with matter played a key role in the establishment of quantum mechanics, and each development of new light sources led to important advances in science, technology and society. An exciting area of physics has been opened with the new millennium, attosecond physics, where the motion of electrons can be studied at the fastest time scale ever reached with light. We review the field of non-resonant photoionization of rare gases using attosecond pulses, with an emphasis on the theoretical background of the reconstruction of attosecond beating by interference of two-photon transition (RABBIT) measurement technique. This approach allows the description of photoionization in the time domain while keeping a high spectral resolution. Examples show the status and the progress in the field during the last decade.

## 1.1 Introduction

The interaction of electromagnetic radiation with matter has considerably contributed to our understanding of the quantum nature of the microcosmos. Photoionization is one of the most fundamental processes in nature, that occurs when a high-energy photon is absorbed. Since its explanation by Einstein [1], which supported the concept of light quanta, a comprehensive quantum mechanical description of atomic photoionization has been achieved, thanks to the advent of synchrotron radiation light sources in the 1960s and the progress in photoelectron detection techniques [2]. An important aspect of photoionization studies is to characterize the partial amplitudes and the relative phases between the ionization channels. One of the goals is to perform

---

The original version of the chapter has been revised: The incorrect name has been updated. A correction to this chapter can be found at https://doi.org/10.1007/978-981-97-2914-2_16

---

D. Busto · S. Zhong · J. M. Dahlström · A. L'Huillier · M. Gisselbrecht (✉)
Department of Physics, Lund University, Box 118, 221 00 Lund, Sweden
e-mail: mathieu.gisselbrecht@sljus.lu.se

A. L'Huillier
e-mail: Anne.Lhuillier@fysik.lth.se



K. Ueda (ed.), *Ultrafast Electronic and Structural Dynamics*,
https://doi.org/10.1007/978-981-97-2914-2_1

'complete' experiments in order to capture the dynamics between different photoionization channels [3–5]. Experimentally, specific requirements have to be met [2]. The energy resolution should be high enough to disentangle the ionization channels leading to different ionic states. Angular resolution is needed to quantify the interference between paths leading to the same final ionic state, which results in anisotropic angular distributions [6, 7]. Finally, in most atomic systems, the magnetic sub-levels of the initial state are degenerate in energy. Without the detection of the outgoing electron spin [8–10] and the alignment and/or orientation of the residual ion [11], the relative contribution of these channels cannot be deduced from the measurement. The combination of laser and synchrotron radiation has led to new ways to perform 'complete' experiments [12]. In particular, resonant two-photon ionization schemes allow the manipulation of the alignment of the atoms with the light polarization, thus recovering the relative phase of different ionization channels. Hence, in-depth studies of photoemission with ever-increasing energy and angle resolution, have been performed to benchmark theoretical treatments in a variety of excited atoms [13].

The generation of high-order harmonics of a laser in a gas is one of the most recognized non-linear processes that lead to the emission of coherent radiation in the eXtreme UltraViolet (XUV) range. Discovered in the late 80's [14, 15], the unique properties of this radiation opened new ionization schemes, such as two-colour non-resonant photoionization from the ground state of rare gases [16–18], combining harmonics and infrared (IR) laser radiation. In parallel to the exploration of applications of this new radiation, world-wide efforts were devoted to understand, optimize and control the high-order harmonic generation (HHG) process [19, 20]. The exploitation of the attosecond (1 as = $10^{-18}$s) temporal structure of this XUV radiation began at the beginning of the millennium, following a series of seminal experiments [21, 22]. Since the energy spectrum of attosecond light sources lies essentially above the ionization threshold of atoms, photoionization is central to many applications.

An electron emitted after the absorption of an attosecond pulse is no longer well-defined in momentum space but is characterized by a momentum distribution, i.e. an Electron Wave-Packet (EWP). Its dispersion directly provides information on both the temporal properties of the light pulse and the potential seen by the escaping electron from the target. In particular, when applied to photoionization, the concept of scattering delays, introduced by Eisenbud, Wigner and Smith [23–25] in the 1950s–1960s, implies that when the electron propagates in the ionic potential, it acquires a delay relative to a reference wave-packet propagating in vacuum.

This information is extracted with the help of experimental techniques, such as streaking [26] and RABBIT[1] [22]. Using these techniques, photoionization time delays were measured for the first time at the beginning of the 2010's [27, 28]. The theoretical foundations for extracting dynamical information on photoionization from streaking or RABBIT experiments have been thoroughly discussed and reviewed, e.g. in [29–31]. Numerous experimental studies have been performed in different atomic systems and spectral regions, e.g. close to threshold [32–35],

[1] Various spellings of the acronym can be found in the literature. The present convention is based on the practices of the optical community.

autoionizing resonances [36–42] or Cooper minima [28, 43–45], over broad ranges [27, 46–48], including satellite transitions [49], double ionization [50] and/or Auger decay [51, 52]. Finally, in the recent years, a strong effort has been devoted to characterize the dynamics between different photoionization channels [53–56], with the aim of performing 'complete' experiments.

The development of the next generation of attosecond light sources [57] with higher flux [58, 59], repetition rate [60–62], exotic polarization state [63–66] and photon energies extending in the soft X-ray regime [67, 68] motivates the present review of attosecond atomic photoionization. We present in Sect. 1.2 the quantum mechanical description of photoionization by absorption of one and two photons. In Sect. 1.3, we discuss how the interferometric RABBIT technique, based on two-photon ionization, can be applied to determine the photoionization phase and amplitude in angle-integrated and angular-resolved experiments. Finally, in Sect. 1.4 we provide experimental details and examples showing how photoionization can be characterized in the temporal domain while keeping high spectral resolution.

## 1.2 Theoretical Background

The photoionization process provides a unique ground for fundamental studies of many-body effects and in particular electron-electron interactions. Its quantum mechanical description requires the knowledge of the atomic ground state wave function, which can be computed rather straightforwardly in centro-symmetric systems, and the final state wave function, which is the most demanding for theoretical formulation and numerical treatment. Indeed, the continuum wave function is generally expanded, using scattering theory, in partial waves, where a complete set of quantum numbers characterize both the electron and the residual ionic core. In the most general case, this fragmentation of the initial system into an ion core and photoelectron is subject, among other things, to angular momentum couplings and electron–electron interactions. Here, we will discuss, for the sake of simplicity, the case of one-photon and two-photon single ionization within the central field approximation for a single active electron.

### 1.2.1 Photoionization with Attosecond Pulses

Atoms exposed to XUV attosecond pulses get ionized. The photoionization transition amplitude can be calculated within first-order perturbation theory, due to the weak intensity of the electromagnetic field of the attosecond pulses, and can be written as

$$\mathcal{A}_{fg} = -\frac{i}{\hbar}\int dt\,\langle\varphi_f|\,T(t)\,|\varphi_g\rangle\,e^{i\Omega_{fg}t}, \tag{1.1}$$

where $\hbar$ is the reduced Planck constant, $|\varphi_f\rangle$ and $|\varphi_g\rangle$ represent the final continuum state and the ground state with respective energies $\varepsilon_f$ and $\varepsilon_g$ and $\Omega_{fg} = (\varepsilon_f - \varepsilon_g)/\hbar$ is the Bohr angular frequency. The atom–photon interaction operator, $T(t)$, is an expansion of electric and magnetic multipoles [69], which can be reduced to the non-relativistic dipole term in the spectral region spanning from 1 nm to 100 nm.[2]

Since attosecond pulses are in most cases linearly polarized [71], we consider in the following the electrical field of the attosecond pulses, $\tilde{\mathcal{E}}_{\mathrm{XUV}}(t)$, to be along the $\hat{z}$ axis. Within the dipole approximation, the interaction of the light field with an atom can be expressed in the length gauge form as

$$T(t) = e\tilde{\mathcal{E}}_{\mathrm{XUV}}(t)z. \tag{1.2}$$

#### 1.2.1.1 Electron Wave-Packet in the Continuum

Using scattering states $\varphi_{\mathbf{k}}$, where $\mathbf{k}$ is the electron wave vector as final continuum states and Eq. (1.2), Eq. (1.1) become

$$\mathcal{A}_{\mathbf{k}g} = -\frac{ie}{\hbar}\int dt\tilde{\mathcal{E}}_{\mathrm{XUV}}(t)e^{i\Omega_{\mathbf{k}g}t}\left\langle\varphi_{\mathbf{k}}\right|z\left|\varphi_g\right\rangle, \tag{1.3}$$

which, using $\mathcal{E}_{\mathrm{XUV}}$, equal to the Fourier transform of the attosecond pulse electric field, $\tilde{\mathcal{E}}_{\mathrm{XUV}}$, takes the simple form,

$$\mathcal{A}_{\mathbf{k}g} = -\frac{ie}{\hbar}\mathcal{E}_{\mathrm{XUV}}(\Omega_{kg})\left\langle\varphi_{\mathbf{k}}\right|z\left|\varphi_g\right\rangle. \tag{1.4}$$

Attosecond pulses have a large bandwidth of typically a few tens of eV. $\mathcal{A}_{\mathbf{k}g}$ is a distribution in momentum space, which represents the electron wave-packet created by absorption of the XUV attosecond pulses.

The scattering states $\varphi_{\mathbf{k}}$ (normalized in momentum) can be expanded in partial waves as [72]

$$\varphi_{\mathbf{k}}(\mathbf{r}) = \frac{1}{\sqrt{k}}\sum_{L=0}^{\infty}\sum_{M=-L}^{M=L} i^L e^{-i\eta_L(k)}Y^*_{LM}(\hat{k})R_{kL}(r)Y_{LM}(\hat{r}). \tag{1.5}$$

where $L$ and $M$ are the electron orbital angular momentum and magnetic quantum number, respectively, $\hat{r} = \mathbf{r}/r$ and $\hat{k} = \mathbf{k}/k$. The $R_{kL}$ radial wave functions are normalized as $\langle R_{kL}|R_{k'L}\rangle = \delta(\varepsilon_k - \varepsilon_{k'})$, and $Y_{LM}$ are spherical harmonics. The scattering phase $\eta_L(k)$ can be decomposed into two contributions $\eta_L = \sigma_L + \delta_L$, where $\sigma_L$ is the universal Coulomb phase shift due to the singularity at the origin and $\delta_L$ is an atom-specific phase shift originating from modifications of the short-range potential due to electron correlations. The asymptotic radial wave function of the fully

[2] See [70] for a report on the contribution of higher order terms in the soft X-ray region.

fragmented atom (ion with a free electron) is a modified plane wave [29, 73],

$$\lim_{r\to\infty} R_{kL}(r) = \sqrt{\frac{2}{\pi k}\frac{1}{r}} \sin\left[kr + \frac{Z\ln(2kr)}{k} + \eta_L(k) - \pi L/2\right], \tag{1.6}$$

where $Z$ is the charge of the ion. This expression shows that an electron propagating away from the ion core accumulates a phase shift. This phase shift, $\phi_{kL}(r)$, includes the effect of the long-range Coulomb potential through a $r$-dependent phase, $Z\ln(2kr)/k$, the contribution of the short-range potential, $\eta_L(k)$ and a centrifugal term, $-\pi L/2$, that depends on the scattering channel. Consequently, for an EWP that contains a distribution of wave vectors $\mathbf{k}$, different accumulated phases will contribute to the EWP dispersion. This EWP dispersion will thus differ from a free EWP and will entail information on the photoionization energy dependence and short-range electron correlations.

#### 1.2.1.2 Photoionization Time Delays

The concept of time delay was first introduced more than 60 years ago by Eisenbud, Wigner and Smith [23–25] in the framework of scattering theory for a short-range potential, $V(r)$. As the EWP propagates in the potential, the local momentum of the electron, given by

$$\hbar k(r) = \sqrt{2m_e[E - V(r)]}, \tag{1.7}$$

where $E$ is the electron kinetic energy, varies. This local variation of momentum, and hence of the EWP phase, leads to a global phase shift $\phi = 2\eta$ compared to an EWP propagating in free space. Consequently, the outgoing wave-packet acquires an effective group delay $\tau_S$ defined similarly to that of an optical wave-packet,

$$\tau_S = \hbar\frac{\partial\phi}{\partial E} = 2\hbar\frac{\partial\eta}{\partial E}. \tag{1.8}$$

Photoionization is a half-scattering process since the EWP only propagates through half of the potential ($r > 0$). As a result, using Eq. (1.6), the delay of the wave-packet relative to a free electron for the scattering channel characterized by the angular momentum $L$ is given by[3]

$$\begin{aligned}\tau_{\mathrm{EWP}} =&\hbar\frac{\partial}{\partial E}\left(\frac{Z\ln(2kr)}{k} + \eta_L(k) - \pi L/2\right)\\ =&\hbar\frac{\partial}{\partial E}\left(\frac{Z\ln(2kr)}{k}\right) + \hbar\frac{\partial\eta_L}{\partial E}\\ =&\tau_{\mathrm{Coul}}(k, r) + \tau_W(k, L),\end{aligned} \tag{1.9}$$

[3] Note that the difference of factor 2 compared to Eq. (1.8).

where $\tau_W(k, L)$ is called the Wigner time delay and is a finite quantity which can be interpreted in the same way as the time delay introduced by Wigner for a short range potential [Eq. (1.8)]. The term $\tau_{\text{Coul}}(k, r)$ is a universal position-dependent delay which accounts for the outgoing electron progressively slowing down because of the attractive Coulomb potential [30]. Because the range of the Coulomb interaction is infinite, it is not meaningful to compare the delay of the outgoing EWP to a free EWP, since it is not possible to define a radius after which the phase shift has converged to a stationary value. Instead, one can compare the delay between two EWPs originating from two different photoionization processes such that the contributions of $\tau_{\text{Coul}}(k, r)$ cancel out, and thereby study the difference between the Wigner time delays $\tau_W(k, L)$, which, as pointed out previously, influences the dispersion of the EWP.[4]

## 1.2.2 *Laser-Assisted Photoionization*

EWP group delays can be measured experimentally with techniques [22, 26] resembling those used in ultra-fast optics, such as cross-correlation and spectral interferometry. These techniques require, in addition to the XUV pulses, a probe laser field at a variable delay. We will consider here the weak-field regime, with only one- or two-photon transitions. This allows for a treatment of ionization within the framework of perturbation theory, up to second order.

### 1.2.2.1 Two-Photon Ionization

Without losing any generality, we use a monochromatic description of the XUV field [29]; for effects related to the finite pulse duration, see [75]. In a two-photon transition, an XUV photon with angular frequency $\Omega$ is absorbed, taking an electron from the ground state $|\varphi_g\rangle$ to an intermediate continuum state $|\varphi_\nu\rangle$, followed by further absorption or emission of a laser photon with angular frequency $\omega$ taking the electron to a final continuum state $|\varphi_{\mathbf{k}}\rangle$.[5] If both XUV and laser fields are linearly polarized along the $z$ axis, the two-photon transition matrix element can be expressed as

$$M_{\mathbf{k}g}^{(\pm)}(\Omega) = -\frac{ie^2}{\hbar} \lim_{\epsilon \to 0^+} \sum\!\!\!\!\!\!\int_{\varphi_\nu} \frac{\langle \varphi_{\mathbf{k}}|z|\varphi_\nu\rangle \langle \varphi_\nu|z|\varphi_g\rangle}{\hbar\Omega - \varepsilon_\nu + \varepsilon_g + i\epsilon}. \tag{1.10}$$

The $(\pm)$ superscript in the matrix element indicates whether the laser photon is absorbed $(+)$ or emitted $(-)$ in the second step. The sum integral in Eq. (1.10) runs

[4] See [74] for a discussion of the concept of time in quantum mechanics.

[5] It is in principle possible that the laser interacts first, taking the electron to a virtual intermediate state. However, this process is in general very improbable and we do not consider it in the following.

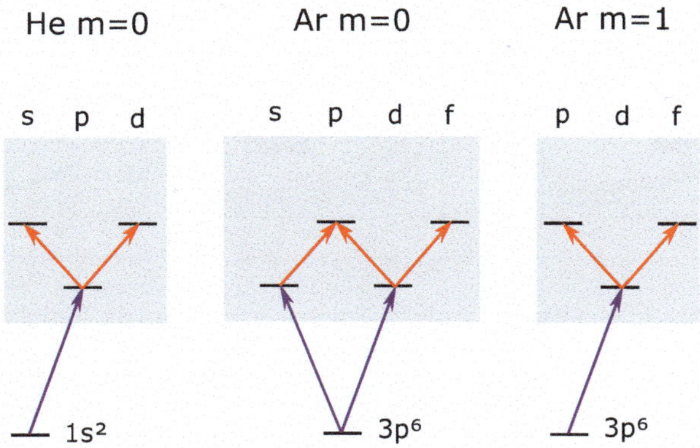


**Fig. 1.1** Example of channel-resolved two-photon transitions in He and Ar when a laser photon is absorbed. Figure taken from [77]

over both continuum and discrete states, meaning that an infinite number of quantum paths can lead to the same final state $|\varphi_{\mathbf{k}}\rangle$. The intermediate state $|\varphi_\nu\rangle$ that conserves energy, $\varepsilon_\nu = \varepsilon_g + \hbar\Omega$, contributes most significantly to the transition, while the non-resonant states are required for a correct description of the phase of the two-photon transition [29]. The use of different polarizations between the XUV and the laser fields is discussed in [76].

Expanding in partial waves the final state [see Eq. (1.5)], one can express $M^{(\pm)}_{\mathbf{k}g}$ as a sum of angular-momentum paths that build up the total two-photon transition matrix element:

$$M^{(\pm)}_{\mathbf{k}g} = \sum_{L,\lambda} M^{(\pm)}_{L\lambda m} Y_{Lm}(\theta, \phi). \tag{1.11}$$

where $\hat{k}$ is represented in spherical coordinates by the angles of emission $(\theta, \phi)$ and where the index $m$ describes the magnetic number of the initial state, $\varphi_g$[6,7], while $L = \lambda \pm 1 \geq 0$, $\lambda = \ell_0 \pm 1 \geq 0$ and $\ell_0$ are the orbital quantum numbers of the final, intermediate and initial states, respectively. Figure 1.1 shows examples of these angular-momentum paths in helium and argon, using the spectroscopic notation s, p, d and f to denote the orbital angular momentum.

Two-photon transitions involve Continuum–Continuum (CC) transitions between the intermediate and final states due to the laser field. Below, we examine the behaviour of the CC transitions in the absence of intermediate or final-state interactions.

[6] It is conserved in the two-photon process since both XUV and probe fields are linearly polarized in the same direction.

[7] If there are several initial states, e.g. with different $m$, the contributions of the EWPs created from the different initial states will add incoherently.

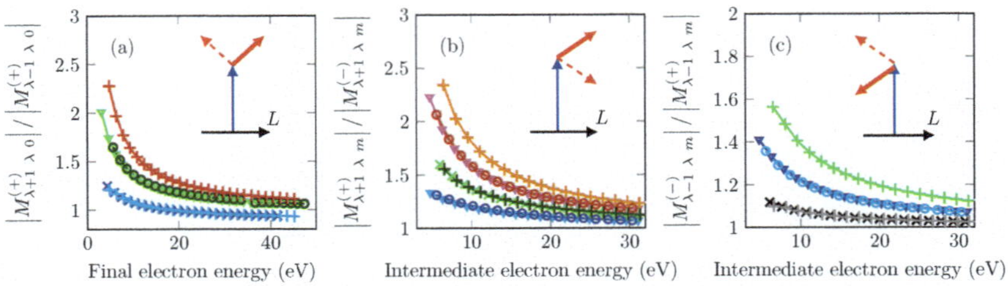


**Fig. 1.2** Propensity rules in laser-assisted photoionization in He $1s$ (×), Ne $2p$ (▽), Ar $3p$ (○) and Kr $3d$ (+). The colour of the curves indicates the angular momentum of the intermediate state in **a** and final state in **b**, **c** [shades of grey $s$, shades of blue $p$, shades of green $d$, shades of red $f$ and orange $g$]. **a** Probability ratio between increasing ($L = \lambda + 1$) and decreasing angular momentum ($L = \lambda - 1$) in the case of absorption of a photon from the intermediate state. **b** Probability ratio between absorbing and emitting a photon in the case of increasing angular momentum. **c** Probability ratio between emitting and absorbing a photon in the case of decreasing angular momentum. The insets present an energy and angular momentum diagram illustrating the propensity rule in each case. Figure taken from ref [34]. 

#### 1.2.2.2 Continuum–Continuum Transition Amplitudes—Fano's Propensity Rule

In 1985, Ugo Fano showed that for one-photon transitions, from the ground state to the continuum, the channel in which the angular momentum increases dominates over the channel in which angular momentum decreases [78]. For example, in neon, the transition $|2\text{p}\rangle \rightarrow |\epsilon d\rangle$ dominates over the transition $|2\text{p}\rangle \rightarrow |\epsilon s\rangle$. The validity of this propensity rule was investigated theoretically for CC transitions [34]. Figure 1.2(a–c) shows a comparison of the moduli of the different matrix elements in He, Ne, Ar outer shells and Kr 3d inner shell. In contrast to analytical expressions based on the Wentzel–Kramers–Brillouin (WKB) theory [29], and recently reformulated using hypergeometric functions [79], we here extract the CC transitions from two-photon transition matrix elements calculated using the Relativistic Random Phase Approximation (RRPA) for the first XUV photon absorption and an effective spherical potential for the second photon transition [80].

Figure 1.2a compares the probability of increasing or decreasing angular momentum when a photon is absorbed between two continuum states. In most cases, we find that the ratio $|M^{(+)}_{(\lambda+1)\lambda m}|/|M^{(+)}_{(\lambda-1)\lambda m}|$ is larger than one in agreement with Fano's propensity rule, which states that increasing orbital angular momentum is a more probable process.[8] Figure 1.2b, c compares the probability of absorbing or emitting a photon when angular momentum increases (b) or decreases (c). An increase of the orbital angular momentum ($L = \lambda + 1$) is more likely when a photon is absorbed while a decrease of the orbital angular momentum ($L = \lambda - 1$) is favoured when a photon is emitted. This result supports the idea that absorption and emission have time-reversal symmetry. Finally, it is worth noting that the ratios are independent of

---

[8] At high kinetic energy and low angular momentum, however, Fano's propensity rule is violated, as explained below.

the atomic species, i.e. the curves in Fig. 1.2 are universal and only depend on the intermediate angular momentum and electron kinetic energy.

The physical origin of this propensity rule can be discussed in terms of the local momentum of the electron in the continuum [see Eq. (1.7)]. The potential experienced by the electron has two contributions, $V_0(r)$, an atom-specific potential and $V_\ell(r)$ the centrifugal potential equal to

$$V_\ell(r) = \frac{\hbar^2 \ell(\ell+1)}{2m_e r^2}. \tag{1.12}$$

The strength of the transition between the intermediate and final states is highest when the difference in local momentum is the smallest. As a result, when the laser photon is absorbed, resulting in a higher kinetic energy $E$, increasing angular momentum is favoured as it increases the centrifugal potential. In the case where the laser photon is emitted, the total energy decreases thus favouring the transition to a lower angular momentum state.

Finally, when $E \gg \hbar\omega$, the matrix elements for laser-assisted absorption and emission processes become equal. In this limit, known as the soft-photon limit, the ratio of the two-photon matrix elements approaches asymptotic values that are given by ratios of the corresponding angular integrals of the dipole transitions in the continuum [81]. Due to the fact that angular integrals favour low angular momentum, a violation of Fano's propensity rule is observed at high energy for the $p \to d/s$ probability ratio, as seen in Fig. 1.2(a), blue curve, which tends to 0.8.

#### 1.2.2.3 Continuum–Continuum Transition Phases

The phase of the two-photon transition matrix element is given by

$$\arg\left(M^{(\pm)}_{L\lambda m}\right) = \pi - \frac{\pi\lambda}{2} + \eta_\lambda(\kappa^{(\pm)}) + \phi_{\rm CC}^{\lambda\to L}(k,\kappa^{(\pm)}), \tag{1.13}$$

where $\kappa^{\pm}$, $k$, with $\kappa^{(+)}<k<\kappa^{(-)}$, are the wave numbers of the electron in the intermediate and final states respectively, $\eta_\lambda(\kappa^{(\pm)}) = \eta_\lambda^{(\pm)}$ is the scattering phase in the intermediate state and $\phi_{\rm CC}^{\lambda\to L}(k,\kappa^{(\pm)}) = \phi_{\rm CC}^{(\pm)}(\lambda \to L)$ is the phase induced by the laser-driven continuum transition from $\kappa^{(\pm)}$ to $k$ in the long-range Coulomb tail, for absorption (+) and emission of a laser photon (−). In the asymptotic approximation, valid at kinetic energy above 25–30 eV, the CC phase only depends on the final and intermediate wave numbers $\phi_{\rm CC}^{(\pm)}(\lambda \to L) \sim \phi_{\rm CC}^{(\pm)}$ [29]. This approximation has proved a useful tool for understanding the delays measured in various attosecond experiments [28, 43, 82].

Fano's propensity rule for CC transitions shows that, far from resonances, the amplitude of the transitions induced by interaction with the IR depends only on the angular momentum and kinetic energy of the electron. It is independent of the atom. Figure 1.3 shows that the same is true for the phase of the CC transitions. In

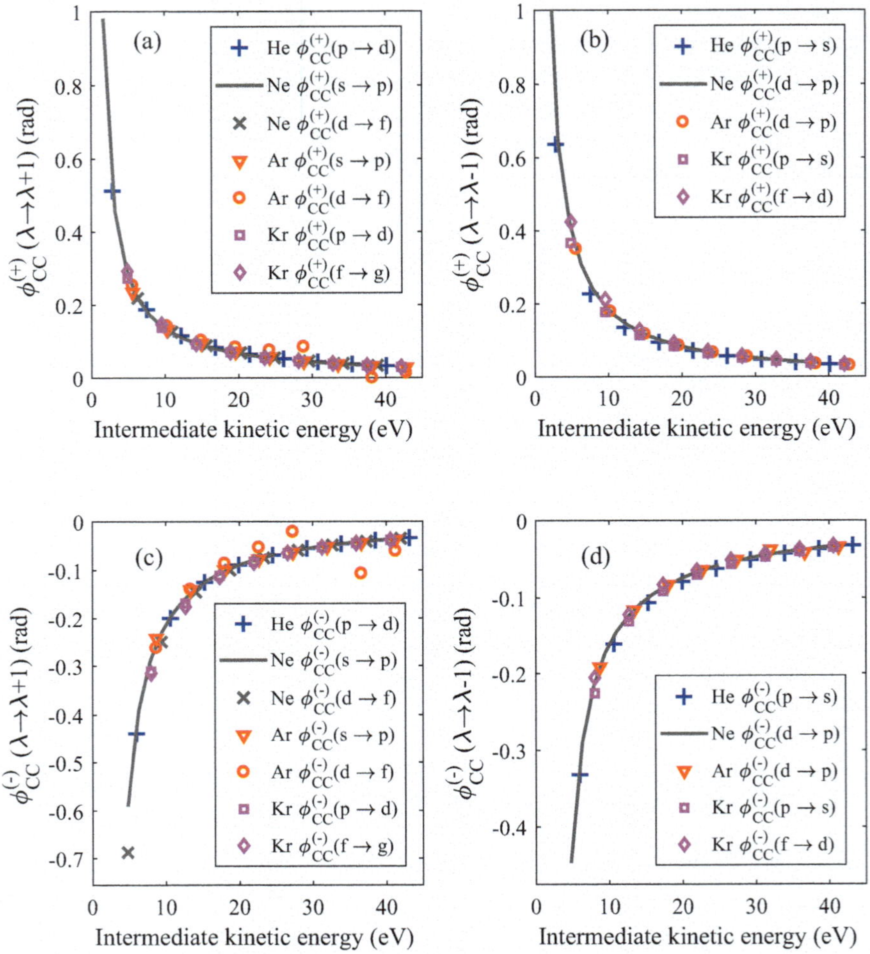


**Fig. 1.3** Angular momentum dependence of the CC phases in He 1$s$, Ne 2$p$, Ar 3$p$ and Kr 3d. **a** CC phases for absorption of an IR photon and increase of angular momentum. **b** CC phases for absorption of an IR photon and decrease of angular momentum. **c** CC phases for emission of an IR photon and increase of angular momentum. **d** CC phases for emission of an IR photon and decrease of angular momentum

fact, contrary to the amplitude of the CC transitions, the phase of these transitions does not depend on the absolute value of the electron angular momentum, but only on whether the angular momentum increases or decreases in the transition. For example, Fig. 1.3(a) shows that $\phi^{(+)}(\lambda \to L)$ is almost identical for He (p→d), Ne (s→p), Ar (d→f) and Kr (f→g) over a large range of kinetic energies, as low as 5 eV. The same is true in the case of emission of an IR photon or in the case of channels with decreasing angular momentum, see Figs. 1.3(b–d). At low kinetic energy, both in the case of

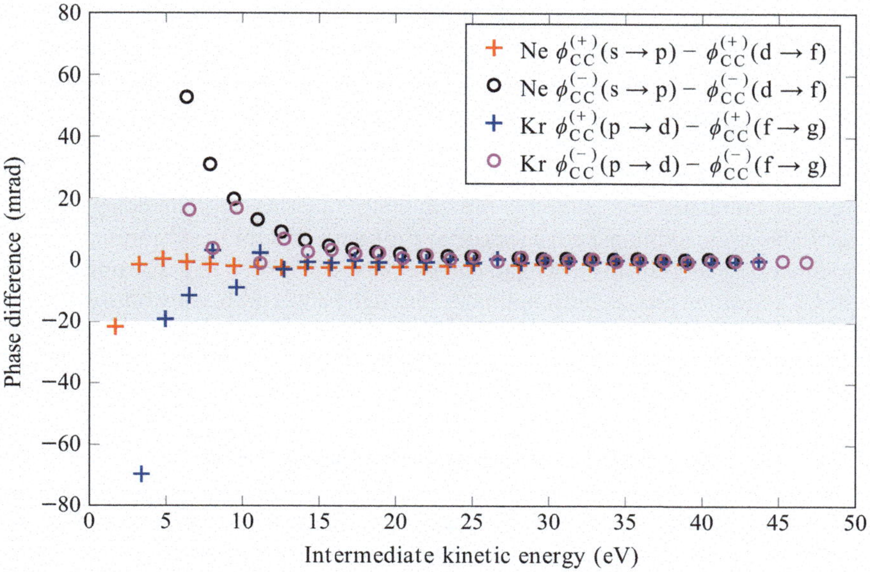


**Fig. 1.4** Deviation from the universality of the CC phase in neon and krypton. For each atom, the phase difference between two CC transitions increasing angular momentum is shown. The shaded grey area corresponds to the region for which the phase difference is smaller than 20 mrad

absorption and emission, small differences between channels with increasing and decreasing angular momentum can be observed. We note, however, that contrary to the amplitude, the CC phase is more sensitive to strong spectral features such as Cooper minima. Fig. 1.3(c) shows that in argon, a deviation from the universal behaviour can be observed in the vicinity of the 3p Cooper minimum around 30 eV, while the propensity rule for transition amplitudes is not affected [83]. Additionally, at low kinetic energy, small difference between the different channels can also be observed.

Figure 1.4 presents the phase difference between different CC transitions increasing angular momentum in neon and krypton atoms. The figure shows that at high kinetic energy (above $\approx 25$ eV), no difference is observed between the different channels. At lower kinetic energy a small difference is observable between different channels. In particular, the calculated phase difference is positive in the case of emission of an IR photon while it remains close to zero in the case of absorption of a photon. Nonetheless, the phase difference remains very small (<20 mrad) for kinetic energies down to 10 eV and it is almost the same for neon and krypton atoms. This may indicate that in the spectral region between 10 and 25 eV, the (universal) centrifugal potential starts affecting the CC phases. Below 10 eV, the difference between the channels diverges and clear differences can be observed between the different atoms, indicating the onset of atom-specific short-range effects in CC transitions.

These results are important because they show that the universality of the CC phase is not tied to the validity of the asymptotic approximation. This has also been pointed out when comparing noble gas atoms and negative ions [84]. Physically this universality can be understood in analogy to classical electrodynamics, where the interaction of a charged particle with an electrical field leads to bremsstrahlung or inverse bremsstrahlung effects. In this context, continuum–continuum transitions can be seen as stimulated one-photon bremsstrahlung. In terms of group delays the CC phase induces an additional advance (delay)[9] of the EWP for the absorption (emission) process in addition to the dynamical one-photon Wigner delay. The universality of the CC transition phases and amplitudes (through Fano's propensity rule) offers the possibility to recover information on single-photon ionization using angle-resolved interferometric two-photon measurement schemes [53].

#### 1.2.2.4 Angle-Integrated and Angle-Resolved Measurement

Finally, we introduce a set of useful notations to discuss the angle-integrated and angle-resolved two-photon ionization measurements. For each angular-momentum path, we define

$$a^{(\pm)}_{L\lambda m}(\tau) = \mathcal{E}_{\mathrm{XUV}}(\Omega)\mathcal{E}_{\mathrm{IR}}(\omega)e^{\pm i\omega\tau} M^{(\pm)}_{L\lambda m}, \tag{1.14}$$

where $\mathcal{E}_{\mathrm{IR}}$ is the amplitude of the laser field and $\tau$ the delay between the XUV and laser fields. We further introduce the quantities

$$\mathcal{A}^{(\pm)}_{L\lambda m}(\theta, \phi, \tau) = a^{(\pm)}_{L\lambda m}(\tau) Y_{Lm}(\theta, \phi), \tag{1.15}$$

$$A^{(\pm)}_{L\lambda m}(\theta, \tau) = (-1)^m a^{(\pm)}_{L\lambda m}(\tau)\sqrt{\frac{(2L+1)(L-m)!}{2(L+m)!}} P^m_L(\cos\theta), \tag{1.16}$$

where $P^m_L$ are the associated Legendre polynomials related to the spherical harmonics by

$$Y_{Lm}(\theta, \phi) = (-1)^m \sqrt{\frac{(2L+1)(L-m)!}{4\pi(L+m)!}} P^m_L(\cos\theta)e^{im\phi}. \tag{1.17}$$

The total angle-resolved signal is given by the modulus square of the sum over the transition amplitudes to all accessible final states. When the ionic state is not measured, the ionization probabilities corresponding to different final $m$ states add incoherently,

[9] This delay is defined as the *derivative* of the CC phase with energy and should not be confused with the $\tau$ delay, which depends on the *difference* between the emission and absorption CC phases, see Sect. 1.3.

$$\mathrm{S}^{(\pm)}(\theta,\phi,\tau)=\sum_{m}\left|\sum_{L,\lambda}\mathcal{A}^{(\pm)}_{L\lambda m}(\theta,\phi,\tau)\right|^2. \tag{1.18}$$

If the photoelectron signal is integrated over all azimuthal angles, the former expression reduces to

$$\mathcal{S}^{(\pm)}(\theta,\tau)=\int_0^{2\pi}\mathrm{S}^{(\pm)}(\theta,\phi,\tau)d\phi=\sum_{m}\left|\sum_{L,\lambda}A^{(\pm)}_{L\lambda m}(\theta,\tau)\right|^2. \tag{1.19}$$

Finally, due to the orthogonality of the spherical harmonics, and of the Legendre Polynomials, the total angle-integrated signal, $S^{(\pm)}$, corresponds to the incoherent sum of the transition amplitudes to all final states

$$S^{(\pm)}(\tau)=\int_0^{\pi}\mathcal{S}^{(\pm)}(\theta,\tau)\sin\theta d\theta=\sum_{L,m}\left|\sum_{\lambda}a^{(\pm)}_{L\lambda m}(\tau)\right|^2, \tag{1.20}$$

where only intermediate angular momentum paths that lead to the same final state are added coherently. Even for a single final state, when $\lambda$ can take two values, the total phase of the two-photon transition amplitude does not correspond to that of the individual channels.

## 1.3 Measuring Photoionization Time Delays with RABBIT

High-order harmonic generation in gases produces a coherent comb of odd harmonics which is used to photoionize an atom. The resulting photoelectron spectrum (PES) exhibits a series of peaks due to the comb structure for the XUV. When the XUV and IR laser fields are overlapped, additional peaks, called sidebands, appear in the PES due to absorption or emission of an IR photon between continuum states. Because two consecutive harmonics are spaced by the equivalent of two IR photons, a given sideband can be reached via two quantum paths as shown in Fig. 1.5(a): absorption of one harmonic plus absorption of one IR photon or absorption of the following harmonic plus emission of one IR photon. As a result, the signal of sideband $q$, which corresponds to a net absorption of energy equal to $q\hbar\omega$, is given by

$$\begin{aligned}\mathrm{S}_q&=\left|\mathcal{A}^{(+)}_{q-1}+\mathcal{A}^{(-)}_{q+1}\right|^2\\&=\left|\mathcal{A}^{(+)}_{q-1}\right|^2+\left|\mathcal{A}^{(-)}_{q+1}\right|^2+2\left|\mathcal{A}^{(+)}_{q-1}\right|\left|\mathcal{A}^{(-)}_{q+1}\right|\cos\left(2\omega\tau-\Delta\varphi\right),\end{aligned} \tag{1.21}$$

where $\mathcal{A}^{(\pm)}_{q\mp1}$ is the two-photon transition amplitude corresponding to absorption of harmonic $q\mp1$ and absorption (+) or emission (−) of an IR photon. This quantity

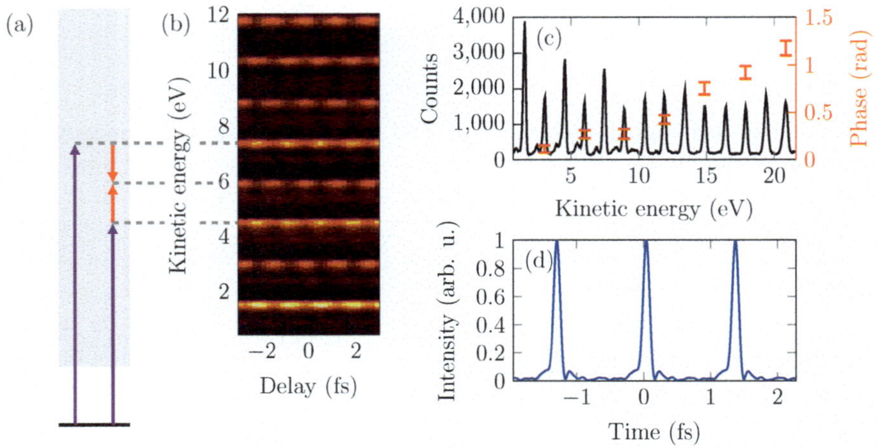


**Fig. 1.5** Characterization of an attosecond pulse train. **a** Schematic representation of the RABBIT technique. **b** Low energy part of a RABBIT scan in Ne from harmonics generated in Ne. **c** Delay integrated photoelectron spectrum and phase measured at the sidebands. **d** Reconstructed attosecond pulse train. Figures taken from [77]

is angle- and channel-dependent as we will discuss in the next sections. Here, we use a compact notation to illustrate the principle of the RABBIT technique. $\mathcal{A}_{q\mp1}^{(\pm)}$ can be expressed as

$$\mathcal{A}_{q\mp1}^{(\pm)} = \mathcal{E}_{q\mp1}(\Omega_{q\mp1})\mathcal{E}_{\text{IR}}(\omega)e^{\pm i\omega\tau}M_{\mathbf{k}g}^{(\pm)}(\Omega_{q\mp1}), \tag{1.22}$$

where $\Omega_{q\mp1} = (q \mp 1)\omega$ and $\mathcal{E}_{q\mp1}$ is the complex amplitude of the $q \mp 1$ harmonic field. The interference of the two paths to the sideband gives rise to oscillations of the sideband intensity as a function of the delay $\tau$ between the XUV and IR fields as shown in Fig. 1.5(b). The phase of these oscillations, $\Delta\varphi$, can be decomposed into two contributions, $\Delta\varphi = \Delta\varphi_{\text{XUV}} + \Delta\varphi_{\text{A}}$ which are defined as [22]:

$$\begin{aligned}\Delta\varphi_{\text{XUV}} &= \arg\left[\mathcal{E}_{q+1}(\Omega_{q+1})\right] - \arg\left[\mathcal{E}_{q-1}(\Omega_{q-1})\right]\\ \Delta\varphi_{\text{A}} &= \arg\left[M_{\mathbf{k}g}^{(-)}(\Omega_{q+1})\right] - \arg\left[M_{\mathbf{k}g}^{(+)}(\Omega_{q-1})\right].\end{aligned} \tag{1.23}$$

$\Delta\varphi_{\text{XUV}}$ originates from the fact that the attosecond pulses produced by HHG are intrinsically chirped such that consecutive harmonics have different spectral phases. Usually, this phase difference varies linearly as a function of the sideband order. This effect is called 'attochirp'. $\Delta\varphi_{\text{A}}$ is the phase difference between the two-photon transition matrix elements, called the atomic phase. By fitting the sideband oscillation with Eq. (1.21), the phase offset for the different sidebands can be extracted as shown in Fig. 1.5(c). The measured phase can be related to a time delay using a finite difference approximation of the energy derivative

$$\frac{\partial \varphi}{\partial E} \approx \frac{\Delta \varphi}{2\omega} = \tau_{\text{XUV}} + \tau_{\text{A}}, \tag{1.24}$$

where $\tau_{\text{XUV}}$ is the XUV group delay and $\tau_{\text{A}}$ is the atomic delay. In the case of ionization to a flat, featureless, continuum, the variation of the total phase $\tau_{\text{XUV}}$ is usually much larger than the atomic phase. This can be seen in Fig. 1.5(c) where the close-to-linear phase variation is due to the attochirp [85].

Historically, this interferometric technique, RABBIT, first theoretically suggested in [86], was applied to characterize experimentally the XUV radiation emitted via HHG [22, 87]. During the past decade, the main application of the RABBIT technique has shifted from the characterization of the attosecond pulse trains to the study of attosecond photoionization dynamics [88]. To access atomic time delays, two RABBIT scans are performed with one scan serving essentially as a reference for the other, for instance using a photoionization process originating from different sub-shells of the same atom or from two different atoms. Hence, the contribution of the attosecond pulses can be cancelled out by evaluating the phase difference between two scans. Accuracy and precision of the technique have been recently discussed in [89].

### *1.3.1 RABBIT Measurements for a Single Angular-Momentum Path*

To gain insight into photoionization with a RABBIT measurement, we present first how the atomic phase difference, $\Delta\varphi_{\text{A}}$, can be extracted in the case where a single two-photon ionization channel ($\ell_0 \to \lambda \to L$) dominates. Using Eqs. (1.13) and (1.23),

$$\Delta\varphi_{\text{A}} = \Delta\eta_\lambda + \Delta\phi_{\text{CC}}, \tag{1.25}$$

where $\Delta\eta_\lambda = \eta_\lambda^{(-)} - \eta_\lambda^{(+)}$ and $\Delta\phi_{\text{CC}} = \phi_{\text{CC}}^{(-)}(\lambda \to L) - \phi_{\text{CC}}^{(+)}(\lambda \to L)$. The oscillatory part of the RABBIT equation [Eq. (1.21)] can be expressed as $\cos[2\omega(\tau - \tau_{\text{XUV}} - \tau_{\text{A}})]$, where

$$\tau_{\text{A}} = \tau_{\text{W}} + \tau_{\text{CC}}, \tag{1.26}$$

is called the atomic delay. $\tau_{\text{W}} = \Delta\eta_\lambda/(2\omega)$ is the Wigner delay and $\tau_{\text{CC}} = \Delta\phi_{\text{CC}}/(2\omega)$ is the continuum–continuum delay.[10]

In Fig. 1.6(a), we test the universality of the CC delay by comparing $\tau_{\text{CC}}$ calculated using diagrammatic many-body perturbation theory for different channels in different atoms [80]. All the CC delays lie in a region of $\pm 10$ as around the $\tau_{\text{CC}}$ calculated in helium for the $s \to p \to s$ channel, as indicated by the blue shaded area. Even at low

[10] Note that the CC delay is *not* the derivative of the CC phase with energy, but rather it is proportional to the difference between emission (−) and absorption (+) CC phases (see e.g. [88]).

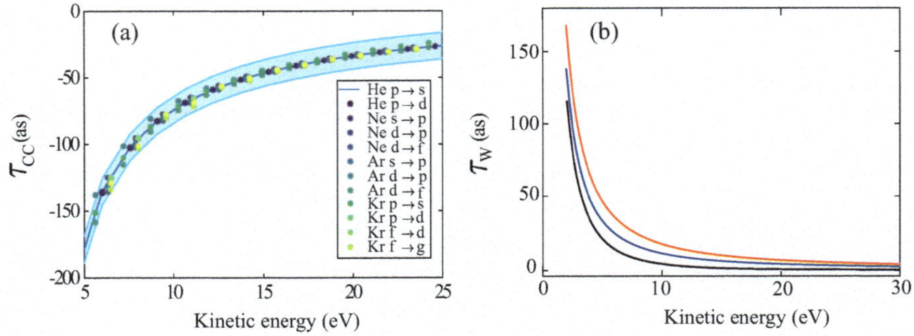


**Fig. 1.6** **a** Angular momentum dependence of $\tau_{CC}$ for the photoionization from He 1s, Ne 2p, Ar 3p and Kr 3d. The shaded area corresponds to a region of $\pm 10$ as around $\tau_{CC}(p \rightarrow s)$ calculated for helium. The calculated CC delays for the different channels lie within the shaded area for an IR wavelength of 800 nm. **b** Wigner time delays in hydrogen for $\lambda = 0$ (black), $\lambda = 1$ (blue), $\lambda = 2$ (red)

kinetic energy the spread of the $\tau_{CC}$ for the different atoms and angular momentum channels is small. For helium, the CC delay is nearly identical for the $s \rightarrow p \rightarrow s$ and $s \rightarrow p \rightarrow d$ channels. This implies that, to a good approximation, $\tau_{CC}$ can be considered universal for the analysis of the photoionization time delays. Nevertheless, the difference between the different calculated $\tau_{CC}$ may become relevant for precise and accurate measurements with error bars below 10 as at low kinetic energies. In this case, other effects such as the harmonic chirp, plasma blueshift in the generation, volume averaging and experimental fluctuations can also contribute to delay shifts with uncertainties of the same order of magnitude [89].

In the case of the hydrogen atom, the Wigner time delay is given by the energy derivative of the Coulomb phase $\sigma_L(k)$, which has a well-known analytical expression [73]. This Wigner delay contribution is present in all atoms and can be used as a reference to investigate the short-range correction to the Coulomb potential in heavier atoms. As shown in Fig. 1.6(b), close to threshold, the Coulomb time delay increases very quickly and can be subtracted to make visible the additional contribution to the Wigner time delay originating from electron correlations [52].

In general, a comprehensive interpretation of the photoionization time delays requires including two-photon processes where the order of absorption (emission) of the photons is exchanged. However, the reversed time-order process leads to corrections that are typically less than an attosecond for calculations performed in length gauge [84]. The separation made in Eq. (1.26) can break down in highly correlated processes, e.g. at the Cooper minimum for the 3$s$ orbital in argon, which has been studied using the 2-Photon 2-Colour Random Phase Approximation with Exchange (2P2C-RPAE) method [83]. Physically, this can be understood as the result of additional interactions of the atomic system with the laser probe field, e.g. when the laser field stimulates hole transitions in the ion rather than electron transitions in the continuum. Similarly, resonances in the continuum strongly affect two-photon transitions [37, 90]. Therefore, fully correlated two-photon matrix elements are sometimes required for a complete analysis of RABBIT experiments [83]. In recent years, $R$-matrix approaches have been applied to tackle complex targets beyond the capabilities of many-body perturbation theory [91, 92].

### *1.3.2 RABBIT Measurements with Multiple Angular-Momentum Paths*

When multiple channels contribute to the photoionization process, the interpretation of the measured phase, and its comparison to the theory, is more delicate. In the following, we will discuss first angle-integrated measurements and then angle-resolved measurements. Finally, we explain why measurements performed along the polarization axis are especially useful for the determination of Wigner delays from atoms using RABBIT.

#### 1.3.2.1 Angle-Integrated RABBIT Measurements

For angle-integrated measurements, contributions from different final scattering states have to be summed incoherently (see Sect. 1.2.2.4). As a result, the total delay-dependent RABBIT signal corresponds to the incoherent sum of RABBIT signal $S_{Lm}$ for each final ($L$) and initial ($m$) states,

$$S(\tau) = \sum_{L,m} S_{Lm}(\tau), \tag{1.27}$$

with

$$S_{Lm}(\tau) = \left| \sum_{\lambda} \left( a^{(+)}_{L\lambda m}(\tau) + a^{(-)}_{L\lambda m}(\tau) \right) \right|^2 . \tag{1.28}$$

where $a^{(\pm)}_{L\lambda m}$ is the two-photon transition amplitude corresponding to an angular-momentum path characterized by $(L\lambda m)$ introduced in Eq. (1.14). In helium, there are two final scattering states ($s$ and $d$) but only one intermediate one ($p$). Therefore, the RABBIT scans associated with the two final states will have the same atomic phase, $\Delta\varphi_{\mathrm{A}} = \Delta\eta_1 + \Delta\phi$, enabling the accurate extraction of the one-photon Wigner time delay.

For heavier rare gas, however, as initially discussed in [43, 82], the final state with angular momentum $L = 1$ can be reached via two different intermediate states with angular momentum $\lambda = L \pm 1$ which interfere (see Fig.1.1). In the case of a strong variation of the cross sections as a function of the energy, e.g. in the vicinity of a resonance or a Cooper minimum, the interference may have large effects on the retrieved phase.

We provide here the general method to analyse the RABBIT signal in this case. The total signal can be written as

$$\begin{aligned} S(\tau) =& 2\left|a^{(+)}_{32\pm1} + a^{(-)}_{32\pm1}\right|^2 + 2\left|a^{(+)}_{12\pm1} + a^{(-)}_{12\pm1}\right|^2 + \left|a^{(+)}_{320} + a^{(-)}_{320}\right|^2 \\ &+ \left|a^{(+)}_{120} + a^{(-)}_{120} + a^{(+)}_{100} + a^{(-)}_{100}\right|^2 . \end{aligned} \quad (1.29)$$

The first three terms oscillate in phase because they can be only reached via $d$ intermediate states and their incoherent addition will not lead to any blurring of the measured phase. However, for $m = 0$, the final $p$ state can be reached via the intermediate $s$ and $d$ states with respective scattering phases $\eta_0^{(\pm)}$ and $\eta_2^{(\pm)}$. These two paths interfere, resulting, for each two-photon ionization process, in an effective phase $\eta_{\text{int}}^{(\pm)}$, given by the equation[11]

$$\eta_{\text{int}}^{(\pm)} = \arctan\left\{\frac{|a^{(\pm)}_{120}|\sin[\eta_2^{(\pm)}] - |a^{(\pm)}_{100}|\sin[\eta_0^{(\pm)}]}{|a^{(\pm)}_{120}|\cos[\eta_2^{(\pm)}] - |a^{(\pm)}_{100}|\cos[\eta_0^{\pm}]}\right\}. \quad (1.30)$$

where the minus sign comes from the phase $\pi\lambda/2$ in Eq. (1.10). Consequently, the term associated to the final $p$ state, $m = 0$, in Eq. (1.29) does not oscillate with the same phase as the other terms. The sum of the oscillations results in a sideband oscillating at the same frequency, $2\omega$, but with a slightly different phase and a reduced contrast. The total interferometric signal measured is

$$\begin{aligned} I =& I_0 + I_2 \cos[2\omega(\tau - \tau_{\text{XUV}} - \tau_2 - \tau)] + I_{\text{int}} \cos[2\omega(\tau - \tau_{\text{XUV}} - \tau_{\text{int}} - \tau)] \\ =& I_0 + I_{\text{tot}} \cos[2\omega(\tau - \tau_{\text{XUV}} - \tau_{\text{tot}} - \tau)]. \end{aligned} \quad (1.31)$$

where $I_0$ is a constant, $I_2$ and $\tau_2$ are the oscillation amplitude and time delay of sidebands associated with all the channels via the intermediate $d$ state, $I_{\text{int}}$ and $\tau_{\text{int}}$ are those for the final $p$ state, $m = 0$. $I_{\text{tot}}$ and $\tau_{\text{tot}}$ are the intensity and delay for the total signal.

$$\begin{aligned} I_{\text{tot}} &= \left[\{I_2 + I_{\text{int}}\cos[2\omega(\tau_2 - \tau_{\text{int}})]\}^2 + \{I_{\text{int}}\sin[2\omega(\tau_2 - \tau_{\text{int}})]\}^2\right]^{1/2} \\ \tau_{\text{tot}} &= \tau_2 + \frac{1}{2\omega}\arctan\left[\frac{I_{\text{int}}\sin[2\omega(\tau_2 - \tau_{\text{int}})]}{I_2 + I_{\text{int}}\cos[2\omega(\tau_2 - \tau_{\text{int}})]}\right] \end{aligned} \quad (1.32)$$

For this reason, it is not possible to exactly retrieve Wigner time delays from angle-integrated RABBIT measurements from heavy noble gas atoms with initial $p$-orbital.[12]

---

[11] Here, we neglect any $\lambda$-dependence of $\phi^{\pm}$.

[12] It is, however, possible to extract the Wigner delay of electrons emitted along the polarization axis to a good approximation, as discussed in Sect. 1.3.2.4. In this case, the Wigner delay is defined as the energy derivative of the argument of the complex dipole given in Eq. (1.3) with the final scattering state from Eq. (1.5) that determines the superposition of $s$- and $d$-waves.

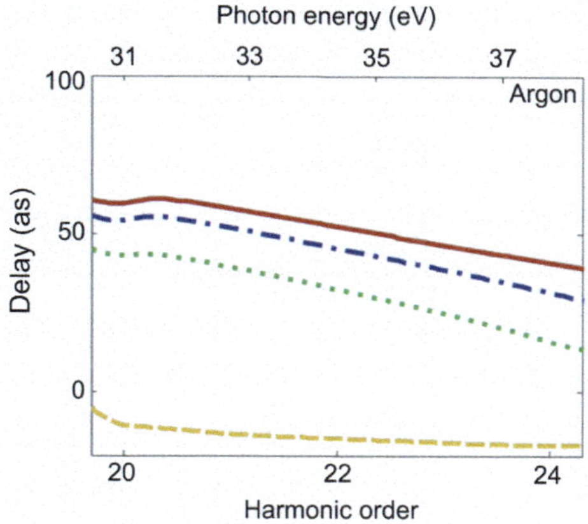


**Fig. 1.7** Theoretical time delays $\tau_2$ (red solid), $\tau_{tot}$ (blue dashed), $\tau_{int}$ (green dotted) and $\tau_0$ (yellow dashed) (see text) in argon extracted from a two-photon ionization calculation using the Random Phase Approximation with Exchange (RPAE). Figure taken from [82]. Licensed under CC-BY-3.0

Nonetheless, from Fano's propensity rule, we know that some transitions are more probable than others. For example, in Ar, the one-photon $p \to d$ channel dominates over the $p \to s$ channel at low energy below the Cooper minimum [93]. In addition, the $p \to d$ channel is possible for all magnetic quantum numbers ($m = 0, \pm 1$), while the $p \to s$ channel is only possible for $m = 0$. In this case, it is therefore reasonable to neglect the channel with lowest intermediate angular momentum, thus, allowing for extraction of the Wigner time delay for the intermediate state with the highest angular momentum. We present in Fig. 1.7 calculations of $\tau_{tot}$, $\tau_0$, $\tau_2$ and $\tau_{int}$ for a few sidebands in argon using 800-nm radiation.[13] While the $p \to s$ channel affects $\tau_{int}$, its influence on $\tau_{tot}$ is significantly reduced due to the dominating role of $p \to d$ ($m = \pm 1$) channels. Thus the delay extracted in a RABBIT measurement can still be interpreted to a good approximation as the Wigner delay of the $d$ state, with an 'error' of the order of 10 as.

#### 1.3.2.2 Angle-Resolved RABBIT Measurements

Angle-resolved RABBIT measurements are performed by detecting photoelectron signals as a function of energy, emission angle and delay. More precisely, for each delay and sideband energy, a Photoelectron Angular Distribution (PAD) is recorded. The angle and delay-dependent sideband signal is given by (see Sect. 1.2.2.4)

$$S(\theta, \tau) = \sum_m \left| \sum_{L,\lambda} A^{(+)}_{L\lambda m}(\theta, \tau) + A^{(-)}_{L\lambda m}(\theta, \tau) \right|^2 . \tag{1.33}$$

[13] In these calculations, the $\tau$, which, to a good approximation, is identical for all the channels, is not included.

This equation shows that the measured angular distribution results from the interference between different ionization channels. The asymmetry between absorption and emission, predicted by Fano's propensity rule (see Sect. 1.2.2.3), is central to interpret this interference, and is at the origin of modifications of the PAD as function of delay [32, 39, 94] as well as of anisotropic time delays [34, 77, 81]. In general, all sub-magnetic levels contribute incoherently to the angular distribution. These different contributions may have phase differences, as discussed previously in angle-integrated RABBIT.

The measured PADs can be decomposed on the basis of Legendre polynomials, as originally introduced by Cooper and Zare [6], according to

$$\mathcal{S}(\theta, \tau) = \frac{\sigma_{\text{int}}}{4\pi} \left( 1 + \sum_{n=1}^{\infty} \beta_n(\tau) P_n(\cos\theta) \right), \tag{1.34}$$

where $P_n$ are the Legendre polynomials, $\beta_n$ are the coefficients of the Legendre expansion, known as asymmetry parameters, and $\sigma_{\text{int}}$ is the total ionization cross section. For unpolarized atoms, this sum can be truncated to $n_{\text{max}} = 2N$, where $N$ is the number of photons exchanged (absorbed or emitted) with the light fields [95]. Additionally, provided that the sideband does not spectrally overlap with the main bands (due to one-photon absorption), only partial waves of the same parity contribute to the sideband, resulting in an up–down symmetry relative to the plane perpendicular to the laser polarization axis, and leading to a cancellation of the odd asymmetry parameters. Therefore, in RABBIT measurements with linearly polarized XUV and IR fields, the PAD of the sidebands can be parametrized with only two asymmetry parameters: $\beta_2$ and $\beta_4$.[14]

In the following, we discuss the case of helium where $\lambda = 1$ and $m = 0$ to gain an insight into the interference between partial waves and its relation to angle-resolved photoemission time delays. Using Eqs. (1.14) and (1.16), assuming the same amplitude for the harmonics contributing to the same sideband, and making use of the fact that $P_0 = 1$, the angle-resolved RABBIT signal is proportional to

$$\begin{aligned} \mathcal{S}(\theta, \tau) \propto \Bigg| & \overbrace{\left[ M_{010}^{(+)} e^{i\omega\tau} + M_{010}^{(-)} e^{-i\omega\tau} \right]}^{\mathcal{D}_0(\tau)} \\ & + \underbrace{\left[ M_{210}^{(+)} e^{i\omega\tau} + M_{210}^{(-)} e^{-i\omega\tau} \right]}_{\mathcal{D}_2(\tau)} \sqrt{5} P_2(\cos\theta) \Bigg|^2, \end{aligned} \tag{1.35}$$

where the Legendre polynomial $P_2(x) = \frac{1}{2}(3x^2 - 1)$. The term in the first row, $\mathcal{D}_0(\tau)$ corresponds to the 'RABBIT amplitude' associated to the $s$ final state while the term in the second row, $\mathcal{D}_2(\tau)$ is that associated to the $d$ final state. This equation shows

---

[14] A general discussion on the parameterization of photoelectron angular distributions can be found in the review [96].

that, in general, it is not possible to write $S(\theta, \tau)$ as the product of a temporal factor and an angular factor. This implies that the PAD, characterized experimentally by the asymmetry parameters $\beta_2$ and $\beta_4$, is delay-dependent.

The evolution of the RABBIT signals for the $s$ and $d$ states, $|\mathcal{D}_0|^2$ and $|\mathcal{D}_2|^2$ respectively, as a function of delay is shown in Fig. 1.8(a), where the sinusoidal oscillation over one sideband period is presented in a logarithmic scale. The maxima of the oscillations are similar, while the minimum is much smaller for the $s$ state than for the $d$ state. The contrast of the oscillations is therefore higher for the $s$ state. This reflects that absorption and emission paths to a given final angular momentum do not have the same strength due to Fano's propensity rule, and more specifically show that the asymmetry between absorption and emission increases with the final angular momentum (Fig. 1.2). Consequently, the contribution of the $s$ and $d$ waves to the total PAD varies as a function of time. In Fig. 1.8(b), we show the oscillations of the $\beta$ parameters, together with the 3-dimensional representation of the angular distributions at specific delays. In Fig. 1.8(c), we present the angle and delay resolved sideband intensity in a logarithmic scale. Both figures show the large modification of the angular distribution around the minimum of the sideband oscillation, where the ratio of the contributions of the $s$ and $d$ states changes dramatically. This effect decreases with the kinetic energy of the photoelectron.

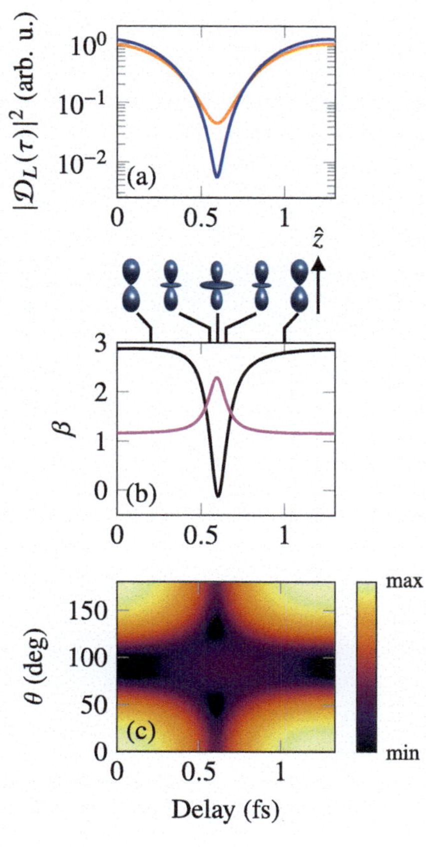


**Fig. 1.8** **a** RABBIT signals $|\mathcal{D}_L(\tau)|^2$ for the $s$ (blue) and $d$ (orange) final states. **b** Asymmetry parameters $\beta_2$ (black) and $\beta_4$ (magenta) as a function of delay. The insets on top show that 3D PAD at different delays. **c** Delay and angle-resolved sideband intensity with a logarithmic colour map. The calculations are performed for electrons with a kinetic energy $E = 6.42$ eV. Figures taken from [77]

A relevant way to examine the PAD is to disentangle the angle-resolved transition amplitudes of the absorption and emission pathways, thus highlighting the anisotropy of the time delays. We rewrite Eq. (1.35) as

$$\mathcal{S}(\theta,\tau) \propto \Bigg| \overbrace{\left[M_{010}^{(+)} + M_{210}^{(+)}\sqrt{5}P_2(\cos\theta)\right]}^{D^{(+)}(\theta)} e^{i\omega\tau} + \underbrace{\left[M_{010}^{(-)} + M_{210}^{(-)}\sqrt{5}P_2(\cos\theta)\right]}_{D^{(-)}(\theta)} e^{-i\omega\tau} \Bigg|^2, \tag{1.36}$$

where the term in the first row, $D^{(+)}(\theta)$, corresponds to the angle-resolved transition amplitude of the absorption path and the term in the second row, $D^{(-)}(\theta)$ corresponds to the emission path. Figures 1.9a, b show the square modulus and argument of $D^{(+)}$ and $D^{(-)}$, respectively.[15] The argument is displayed as a 'time delay' $\tau^{(\pm)} = \arg[D^{(\pm)}]/(2\omega)$ ($\pi$ rad is equivalent to 667 as at 800 nm). The square modulus of $D^{(+)}$ varies rapidly at $\theta = 75°$ and $105°$ and the time delay jumps by almost 600 as. On the contrary, $|D^{(-)}|^2$ varies smoothly as a function of the angle and the time delay is almost independent of the angle. As a result, the angle-dependent atomic delay $\tau_A(\theta) = \tau^{(-)}(\theta) - \tau^{(+)}(\theta)$, shown in Fig. 1.9(c), exhibits a large delay variation as a function of the angle.

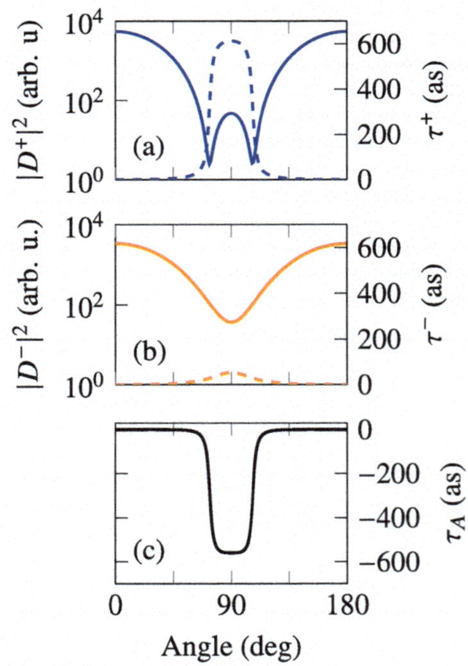


**Fig. 1.9** **a** Square modulus (solid line) and time delay (dashed line) of $D^{(+)}(\theta)$. **b** Square modulus (solid line) and time delay (dashed line) of $D^{(-)}(\theta)$. **c** Angle-resolved atomic time delay. The calculations are performed for electrons with a kinetic energy $E = 6.42$ eV. Figures taken from [77]

[15] These results take into account the exact $L$-dependence of $\phi^{(\pm)}$, which leads to a smoothing of the rapid variations of $D^{(+)}$ predicted by the asymptotic approximation [34].

To understand the physical origin of the delay jump, we use the asymptotic approximation given in Eq. (1.13), neglecting the small $L$-dependence of $\phi^{(\pm)}$, and express $D^{(\pm)}(\theta)$ as

$$D^{(\pm)}(\theta) \approx \exp[i(\pi/2 + \eta_1^{(\pm)} + \phi^{(\pm)})] \left[ |M_{010}^{(\pm)}| + |M_{210}^{(\pm)}| \sqrt{5} P_2(\cos\theta) \right], \quad (1.37)$$

where the second factor on the right-hand side is real and can be negative or positive.[16] The term corresponding to $L = 2$ is negative for $54.7° \leq \theta \leq 125.3°$. For these angles, the partial waves corresponding to $L = 0$ and $L = 2$ interfere destructively. At other angles, and in particular, along the polarization axis, both contributions are positive and add constructively. Depending on the relative weight $|M_{L10}^{(\pm)}|$ of the two scattering waves, the second factor on the right-hand side in Eq. (1.37) may or may not change of sign. In the absorption path, according to Fano's propensity rule, the $d$ wave dominates. As a result, for electrons with kinetic energy $E = 6.42$ eV, around 75° and 105°, $|M_{010}^{(\pm)}| + |M_{210}^{(\pm)}| \sqrt{5} P_2(\cos\theta)$ crosses zero. In the emission path, it is the $s$ wave that dominates. Here the second term of Eq. (1.37) never changes sign, leading to a smooth variation of $|D^{(-)}|^2$. The difference between the absorption and emission due to Fano's propensity rule is, therefore, at the origin of the angle dependence of the time delays.

The discussion on angle-resolved photoionization time delays and delay-dependent PADs has so far been done in helium atoms, where all the ionization channels contribute coherently. A similar analysis can be performed with other atoms, explaining the general angle dependence of the photoionization time delays [97]. In this case, the angle-resolved RABBIT signal can be expanded as

$$S(\theta, \tau) = \sum_m \left[ |D_m^{(+)}|^2 + |D_m^{(-)}|^2 \right] + e^{-i2\omega\tau} \sum_m D_m^{(+)*} D_m^{(-)} + c.c., \quad (1.38)$$

where $D_m^{(\pm)}$ is a generalization of $D^{(\pm)}$ previously introduced

$$D_m^{(\pm)}(\theta) = \sum_{L,\lambda} M_{L\lambda m}^{(\pm)} Y_{Lm}(\theta, 0). \quad (1.39)$$

The term $W(\theta) = \sum_m D_m^{(+)*} D_m^{(-)}$ can be parameterized as

$$W(\theta) = \frac{W_0}{4\pi} \left( 1 + \sum_{n=1}^{\infty} \tilde{\beta}_n P_n(\cos\theta) \right), \quad (1.40)$$

where $W_0$ is a complex angle-independent amplitude and $\tilde{\beta}_n$ are complex asymmetry parameters [97]. As for Eq. (1.34), for unpolarized atoms, the expansion in terms of Legendre polynomials can be truncated at $n_{\max} = 4$ and only even asymmetry

[16] In Eq. (8) of Ref. [34] the minus sign between the two corresponding terms is a typo.

parameters are non-zero in the absence of an up–down asymmetry. While the traditional asymmetry parameters are delay-dependent in the RABBIT measurements, as shown in Fig. 1.8(b), the two complex asymmetry parameters $\tilde{\beta}_2$ and $\tilde{\beta}_4$ encapsulate the amplitude and phase information of the complex interference term in RABBIT measurements and its delay dependence. This parameterization thus offers an opportunity to quantitatively analyse angle-resolved RABBIT measurement, similarly to what has been done for single-photon ionization using real asymmetry parameters [96]. The angle-resolved oscillation amplitude is then given by $|W(\theta)|$ and the angle-resolved RABBIT phase corresponds to $\arg[W(\theta)]$.

#### 1.3.2.3 From Angle-Resolved to Channel-Resolved Measurements

Angle-resolved RABBIT measurements are very sensitive to the relative amplitude and phase of the intermediate and final scattering states. However, disentangling the contributions of the bound–continuum transitions from CC transitions induced by the probe field is extremely challenging, and a proper representation of the two-photon ionization process is crucial. For instance, an angle-resolved representation can lead to a controversial interpretation of the CC delay. More precisely, the angle-resolved CC delay, defined as the difference $\tau_A(\theta) - \tau_W(\theta)$, is strongly dependent on the angle [97], and is not a universal quantity. In contrast, a channel-resolved representation is appropriate since it allows for a clear separation of the one-photon delay from that induced by the IR. It requires, however, further parametrization of the angle-resolved measurements.

Joseph et al. [98] suggest to parametrize the delay-dependent sideband PAD using

$$\mathcal{S}(\theta, \tau) = \sum_{n=0}^{2} h_{2n}(\tau) P_{2n}(\cos\theta). \tag{1.41}$$

This parameterization is very similar to that in Eq. (1.34), with an important advantage: the $h$ functions oscillate sinusoidally as a function of delay while, as shown in Fig. 1.8, the $\beta$ parameter have a more complex delay dependence. It is then possible to express the delay dependence of the $h$ coefficients as $h_{2n}(\tau) = a_{2n} + b_{2n}\cos(2\omega\tau - \phi_{2n})$. It is then clear that for each sideband, one can extract 9 observables: $a_{\{0,2,4\}}$, $b_{\{0,2,4\}}$ and $\phi_{\{0,2,4\}}$. Using Eqs. (1.14), (1.16), (1.33), and (1.41), it is possible to express these 9 observables in terms of the two-photon transition matrix elements $M^{(\pm)}_{L\lambda m}$. Interestingly, even if for individual magnetic sublevels, the expansion of Eq. (1.33) in Legendre polynomials can contain powers of $\cos\theta$ as high as $2\ell + 4$, the final angular distribution of unpolarized targets can always be described by Eq. (1.41), regardless of the value of the initial orbital quantum number $\ell$ [95]. Each of the matrix elements can then be decomposed as

$$M^{(\pm)}_{L\lambda m} = C^{m}_{L\lambda} C^{m}_{\lambda\ell} \sigma^{(\pm)}_{L\lambda\ell} e^{i[\pi - \frac{\pi\lambda}{2} + \eta^{(\pm)}_{\lambda} + \phi^{(\pm)}(\lambda \to L)]}, \tag{1.42}$$

where $C^m_{L\lambda} = \langle Lm|Y_{10}|\lambda m\rangle$ are angular coefficients and $\sigma^{(\pm)}_{L\lambda\ell}$ is the two-photon radial amplitude. In general, the number of unknowns in the problem, $\sigma^{(\pm)}_{L\lambda\ell}$, $\eta^{\pm}_{\lambda}$, $\phi^{(\pm)}(\lambda \to L)$, is larger than the number of observables [98].

In order to extract the relevant channel-resolved amplitudes and phases, different approaches have been developed. One possibility is to take advantage of the universality of the CC transitions to simplify the problem. For example, using Fano's propensity rule for CC transitions, one can relate the two-photon radial amplitudes $\sigma^{(\pm)}_{\lambda+1\lambda\ell}$ $\sigma^{(\pm)}_{\lambda-1\lambda\ell}$ of channels sharing the same intermediate state. Additionally one can also use the fact that to a large extent the CC phases only depend on the difference in angular momentum ($\pm 1$) between intermediate and final states or simply take the exact value of the CC phases obtained from numerical calculations (Fig. 1.3). Finally, it is also possible to use analytical models such as the one developed by Boll et al. [99] to get an approximate value of the angular momentum-dependent phases and relative amplitude for CC transitions.

An alternative approach is to perform RABBIT measurements with different polarizations for the XUV and IR pulses [76]. The principle is analogous to measuring dichroism in two-photon ionization.[17] By changing the polarization of the light pulses, different selection rules for the orbital angular momentum and magnetic quantum number apply, and multiple angular distributions and angle-resolved phases can be measured. Combining measurements with linear [54, 56] or circular [55, 66] polarization can provide additional information to extract the amplitude and phase of the different ionization channels.

#### 1.3.2.4 RABBIT Measurements Along the Polarization Axis

Theoretically, the problem of unresolved ion states in the experiment can be circumvented by detecting photoelectrons along the polarization axis, $\mathbf{k}_{||} = k\hat{z}$. In this direction only, the $m = 0$ electrons will be detected due to properties of spherical harmonics. Furthermore, the CC-delay in this direction is universal for noble gas atoms [84] and it coincides with the CC-delays of isolated angular momentum paths, shown in Fig. 1.6(a). In this subsection, we provide an explanation for this useful result within the asymptotic approximation [29].

In the case of an initial state $ns$, where helium is a prototypical example, with a single intermediate $p$-wave, the final state consists of two partial waves: $s$-wave and $d$-wave. Using Eqs. (1.11) and (1.13), the two-photon transition matrix element in the direction $k_{||}$ along the polarization axis $z$ can be expressed as

[17] Dichroism in photoelectron spectra is defined as the difference in intensity for two different alignments/orientations of the target atom or polarization directions of the ionizing photons [5]. This difference makes it possible to cancel out many terms from Eq. (1.38), leading to a substantial reduction in the number of unknowns.

$$M^{(\pm)}_{\mathbf{k}_{\parallel} ns} \propto [\overbrace{C^0_{01} Y_{00}(0,0)}^{\text{Final } s-\text{wave}} + \overbrace{C^0_{21} Y_{20}(0,0)}^{\text{Final } d-\text{wave}}] \times \langle\, \kappa^{(\pm)} p \mid z \mid ns \,\rangle \times \underbrace{i \exp[i(\phi^{(\pm)} + \eta_1^{(\pm)})]}_{\text{Common asymptotic phase shifts}} \tag{1.43}$$

Equation (1.43) is strictly valid within the asymptotic limit, when *both* the amplitude and phase of the radial integrals are independent of the final angular momentum.[18] Along the polarization direction, the two amplitudes are *in phase* and add up constructively. Both amplitudes are of comparable strength with the final $d$ wave being approximately twice as strong as the $s$-wave due to properties of the spherical harmonics. Since both final states share the common asymptotic phase shifts, $\eta_1^{(\pm)}$ and $\phi^{(\pm)}$, the separation of the atomic delay, as $\tau_A = \tau_W + \tau$ in Eq. (1.26), should be valid for *any* atom from an initial $s$-orbital with electrons detected along the polarization axis.

In the case of an initial $p$-orbital, where neon and argon are prototypical examples with two intermediate channels: $s$-wave and $d$-wave for $m = 0$, we again use Eqs. (1.11) and (1.13), to write the intermediate partial wave contributions separately as

$$M^{(\pm)}_{\mathbf{k}_{\parallel} np} \propto \exp[i\phi^{(\pm)}] \times \{\overbrace{-C^0_{10} Y_{10}(0,0) \langle\, \kappa^{(\pm)} s \mid z \mid np \,\rangle \exp[i\eta_0^{(\pm)}]}^{\text{Intermediate } s-\text{wave, Final } p-\text{wave}} + \underbrace{[C^0_{12} Y_{10}(0,0) + C^0_{32} Y_{30}(0,0)] \langle\, \kappa^{(\pm)} d \mid z \mid np \,\rangle \exp[i\eta_2^{(\pm)}]}_{\text{Intermediate } d-\text{wave, Final } p-\text{and } f-\text{waves}}\}, \tag{1.44}$$

where $\langle\, \kappa^{(+)} \lambda \mid z \mid np \,\rangle \approx \langle\, \kappa^{(-)} \lambda \mid z \mid np \,\rangle$ are one-photon dipole elements for photoionization to $\lambda = 0$ and 2. Because the contributions of the two intermediate waves are added coherently,[19] the amplitude ratio must be equal to that of the one-photon ionization case, for the Wigner delay to appear correctly in Eq. (1.26). Indeed, this is the case due to the following ratio of intermediate waves:

$$\frac{\overbrace{[C^0_{12} Y_{10}(0,0) + C^0_{32} Y_{30}(0,0)] C^0_{21}}^{\text{Intermediate } d-\text{wave}}}{\underbrace{C^0_{10} C^0_{01} Y_{10}(0,0)}_{\text{Intermediate } s-\text{wave}}} = 2, \tag{1.45}$$

[18] The amplitude from the radial part of the CC transition is omitted in Eq. (1.43) since it is the same for both final waves. Exact numerical results show that small corrections in phase and amplitude tend to cancel out to very good approximation.

[19] Each intermediate wave has its own scattering phase, $\eta_\lambda^{\pm}$. Note also that there is a relative minus sign between the two terms due to the intermediate angular momenta being different by two units.

which is equal to the ratio of final waves in the one-photon process

$$\frac{\overbrace{C_{21}^{0}\,Y_{20}(0,0)}^{\text{Final } d-\text{wave}}}{\underbrace{C_{01}^{0}\,Y_{00}(0,0)}_{\text{Final } s-\text{wave}}} = 2 \tag{1.46}$$

Clearly, this result is general because the atom-specific dipole matrix elements for photoionization, $\langle\, \kappa\lambda \mid z \mid np\,\rangle$ will affect the intermediate step of the two-photon process in the same way as the final state in the one-photon process. In other words, the 'intermediate' phase $\eta_{\text{int}}$ or delay $\tau_{\text{int}}$ [see Eq. (1.30) in Sect. 1.3.2.1] is the same for one- and two-photon ionization along the polarization axis, with the approximations described above. This means that the separation in Eq. (1.26), with a Wigner delay not related to a particular channel but to the final state reached by one-photon ionization, is valid from *any* atom with an initial $p$-orbital.

A similar calculation for an initial $d$-orbital gives the ratios 1.5 for the $f-$wave and $p$-wave in the final one-photon and intermediate two-photon case, respectively. By extrapolation, the separation $\tau_{\text{A}} = \tau_{\text{W}} + \tau$ should be an excellent approximation for photoionization from any atom and orbital as long of photoelectrons are detected along the polarization axis, provided that the laser field does not interact with the atomic system in any other way beyond CC transitions.

## 1.4 Experimental Methods and Results

Since the early works on photoionization time delays [27, 28], several studies, using the RABBIT technique, have been performed both with angle-integrated [33, 43–48, 50] or angle-resolved [32, 34, 35, 53, 54, 66, 100] detection. Efforts to improve the temporal precision down to a few attosecond [33], while keeping good spectral resolution to disentangle different ionization processes [45, 48], have allowed stringent tests of the theory for non-resonant photoionization. In the following, we briefly review the experimental methods and illustrate the difference between angle-integrated and angle-resolved detection on three measurements recently performed in Lund [34, 48, 53].

### *1.4.1 Methods*

There is today an increased diversity of high-order harmonics/attosecond sources driven by a variety of lasers ranging from high energy lasers at low repetition rate to high average power lasers, based upon optical parametric amplification or simply high-power oscillators. HHG sources can be vastly different, with parameters such as peak power or repetition rate varying by several orders of magnitude [101]. The generation of high-order harmonics and attosecond pulse trains requires a strong interaction between laser light and atoms, and therefore laser intensities higher than $10^{14}$ W/cm$^2$. These intensities can be achieved using Ti:Sapphire laser technology together with chirped pulse amplification [102], providing laser pulses in the near IR spectral region with typically a few tens fs pulse duration, a few mJ pulse energy at a few kHz repetition rate. These pulses can be further broadened and post-compressed to the few-cycle regime to generate isolated attosecond pulses [103]. The high repetition rate is an essential parameter for coincidence measurements [104–106]. The development of lasers based on optical parametric chirped pulse amplification allows nowadays the generation of attosecond light source with near MHz repetition rate [61, 62, 107]. Recently, the combination of post-compression techniques using hollow core fibres or multi-pass cells [108, 109], and industrial-grade ytterbium-based lasers has also proven to be a very interesting source for attosecond pulse generation due to the long term stability and high repetition rate of these laser systems [110]. Finally, lasers in the mid-IR spectral range can be used to generate attosecond pulses in the X-ray spectral range, opening a new range of applications [111–114].

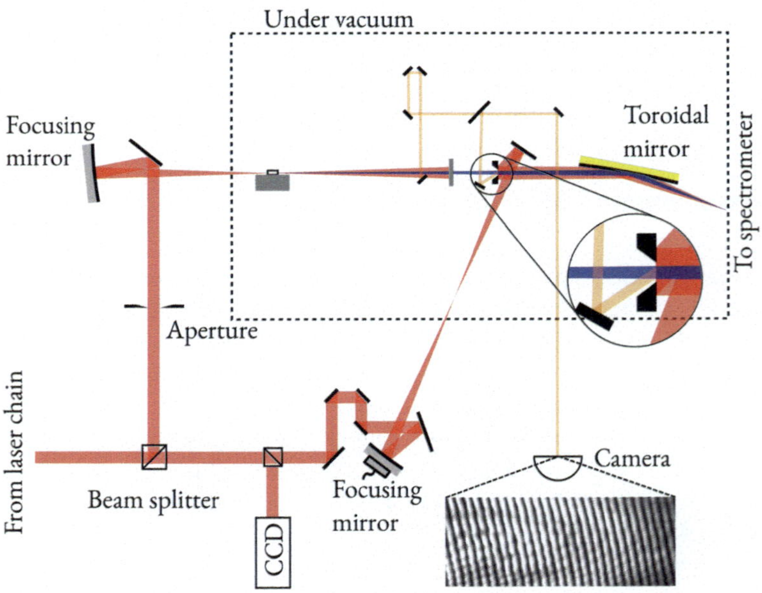


**Fig. 1.10** A schematic of the Mach–Zehnder interferometer used for RABBIT measurements in Lund. In one arm, the XUV pulses is generated and then recombined with IR pulse from the other arm. The delay is monitored and controlled with attosecond precision making small portions of the IR in both arms interfere on a camera (see beam path in orange). Figure taken from [115]

Besides an attosecond pulse train, RABBIT measurements require an XUV–IR interferometer and an optical element focusing the light into an interaction chamber containing a gas medium at low pressure and a photoelectron spectrometer. The usual approach is to split the femtosecond laser pulses in a Mach–Zehnder interferometer, in which one arm is used to generate the attosecond pulse train (APT) by focusing the laser pulses in a gas target, and the other arm is used to delay the probe pulse with attosecond precision [22, 57, 85, 116–120]. Both XUV and IR pulses are then recombined using a drilled mirror that transmits the attosecond pulses and reflects the IR probe pulse. Alternative designs based on a collinear XUV–IR interferometer have also been demonstrated [121–123]. Thin metallic filters can be used to filter out the IR that co-propagates with the APT, shape the XUV spectrum and partly compensate for the group delay of the attosecond pulses [124, 125]. The accuracy of RABBIT measurements is essentially limited by the temporal stability of the interferometer, the total number of sampling points of the sideband oscillations and the pulse intensity fluctuations [89]. Therefore, an active delay and beam-pointing stabilization is often implemented to ensure attosecond delay stability over several hours. The XUV pulse train and IR pulse are then focused in a gas jet using one or a combination of grazing incidence focusing mirrors (e.g. [118, 125]). Some setups also use a normal incidence geometry with a multi-layer focusing mirror which offers to possibility to further shape the XUV spectrum [47]. Figure 1.10 presents schematically the optical interferometer used at Lund University.

The photoelectrons resulting from the interaction of the XUV and IR pulses with the atomic gas can be measured using different spectrometers. Magnetic bottle electron spectrometers (MBES) [126–128] offer high spectral resolution over a large energy range and a collection efficiency close to 100% at the expense of angular integration. Photoelectrons generated in the interaction region are directed towards the flight tube using a strong inhomogeneous magnetic field acting as a magnetic mirror. The electrons in the flight tube are then guided towards the detectors using a weaker linear magnetic field as shown in Fig. 1.11(a). Alternatively, velocity map imaging spectrometers (VMIS) [129] or cold-target recoil ion momentum spectrometers (COLTRIMS) [130] offer angular resolution at the expense of reduced spectral resolution. Figure 1.11b shows the VMIS principle, where the electron momentum distribution after photoionization is projected on a 2D screen using a set of electrodes, with an energy resolution of the order of 1–2%. This momentum projection can then be inverted to reconstruct the 3D momentum distribution as long as the momentum distribution has a cylindrical symmetry. Several methods exist to perform such inversion [131–133] based on the Abel transform. Contrary to VMIS, COLTRIMS do not require cylindrical symmetry of the momentum distribution allowing the measurement of the sideband phase as a function of both polar and azimuthal angles.

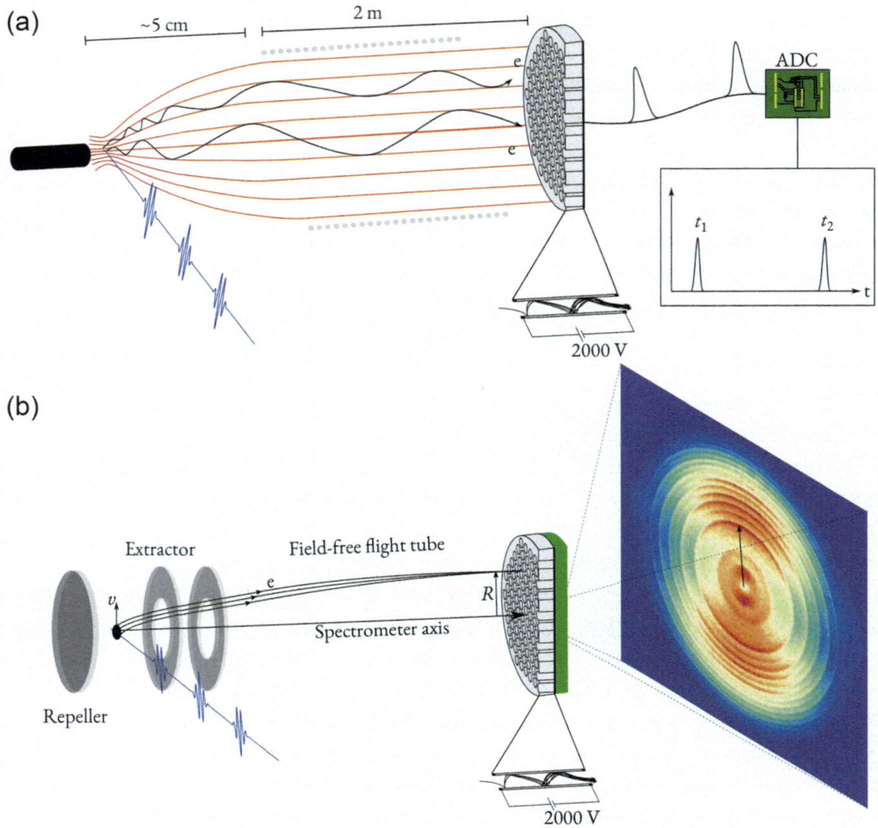


**Fig. 1.11** Schematic illustration of **a** an MBES and **b** a VMIS. In (**a**), a strong magnet is used to collect electrons emitted in all directions and to push them into the flight tube. A weak magnetic field is used to guide the electrons to the MCP detector, ensuring $\sim 100\%$ collection efficiency. In (**b**), a set of electrodes is used to project the initial momentum distribution on an MCP detector with a phosphor screen behind it. The raw image from the phosphor screen is also shown. Figures taken from [115]

### *1.4.2 Three Examples of Experimental Results*

To illustrate the theory discussed in the present article, we present three 'typical' experimental results with angle-integrated [48] and angle-resolved [34, 53] detection. In the first experiment, neon atoms are ionized in the $2s$ and $2p$ shells in the 70–100 eV energy range. The second experiment investigates the angular dependence of photoionization time delays in argon. In the third experiment, based on angle-resolved measurements in neon, the amplitude and phase of the individual angular momentum channels are extracted.

#### 1.4.2.1 Angle-Integrated Photoionization Time Delays

The first photoionization time delay measurements in atoms were performed with the attosecond streaking technique [134] by Schultze and co-workers in 2010 [27]. A time delay difference of 21 $\pm$ 5 as was measured between electrons emitted from the $2p$ ($I_p = 21.6$ eV) and $2s$ ($I_p = 48.5$ eV) shells of the neon atom. This measurement, which could not be explained theoretically in the original article, triggered a lot of attention. In the following years, despite numerous attempts to explain the experimental result, theoretical calculations consistently predicted a time delay difference smaller than that measured by Schultze et al. [135–139].

In 2017, Isinger et al. investigated the photoionization time delays between the $2s$ and $2p$ shells of neon using the angle-integrated RABBIT technique [48]. The photon energy of the attosecond pulse trains was high enough to ionize electrons from both $2s$ and $2p$ shells, thereby obtaining two RABBIT traces at different kinetic energies as illustrated in Fig. 1.12(a). The atomic time delay difference, $\tau_{2s} - \tau_{2p}$, between photoelectrons emitted from the two subshells was obtained by measuring the phase differences between sidebands of the same order in the two RABBIT traces. Using different sets of metallic filters, photoionization time delays could be measured in an energy range of 70–100 eV. Contrary to the broad photoelectron peaks obtained using the attosecond streaking technique, the discrete nature of the high-order harmonic spectrum leads to a photoelectron spectrum composed of narrow peaks. A spectrally resolved analysis called Rainbow RABBIT [38] revealed the presence of additional photoelectrons, partially overlapping with $3s$ electrons peaks, as shown in Fig. 1.12(c). These additional contributions were attributed to ionization with shake-up in which the escaping photoelectron transferred part of its energy to excite another electron to the $2p^4(^1D)3p(^2P)$ state ($I_p = 55.8$ eV) [139, 140]. Disentangling the contributions from the direct and shake-up ionization processes, excellent agreement with theoretical calculations was obtained as shown in Fig. 1.12(b). Let us emphasize that in this experiment both high temporal resolution of a few tens of attosecond and high spectral resolution, limited by the harmonic bandwith (200 meV), were achieved.

In Fig. 1.12(d), the Wigner delay, calculated along the polarization direction (see Sect. 1.3.2.4), is compared with the angle-averaged one-photon ionization time delay ($\tau_{\text{tot}}$ in Sect. 1.3.2.1). For $2p$ ionization, the two curves differ by at most 2 as, due to the dominance of $2p \rightarrow d$ transition. For $2s$ ionization, the difference is not visible, which confirms our discussion in Sects. 1.3.2.1 and 1.3.2.4 for a single intermediate channel. The absolute Wigner delay for $2s$ ionization can be extracted from the experimental data by subtracting the CC contribution $\tau_{\text{CC}}(2s)$ from the measured delay difference $\tau_{2s} - \tau_{2p}$, using the calculated $\tau_{2p}$. The $2s$ Wigner delay is approximately -3 and does not vary much in this energy region. The results shown in Fig. 1.12(b) reflect essentially the CC-time delay variation between the $2s$ and $2p$ channels, which varies with the final kinetic energy.

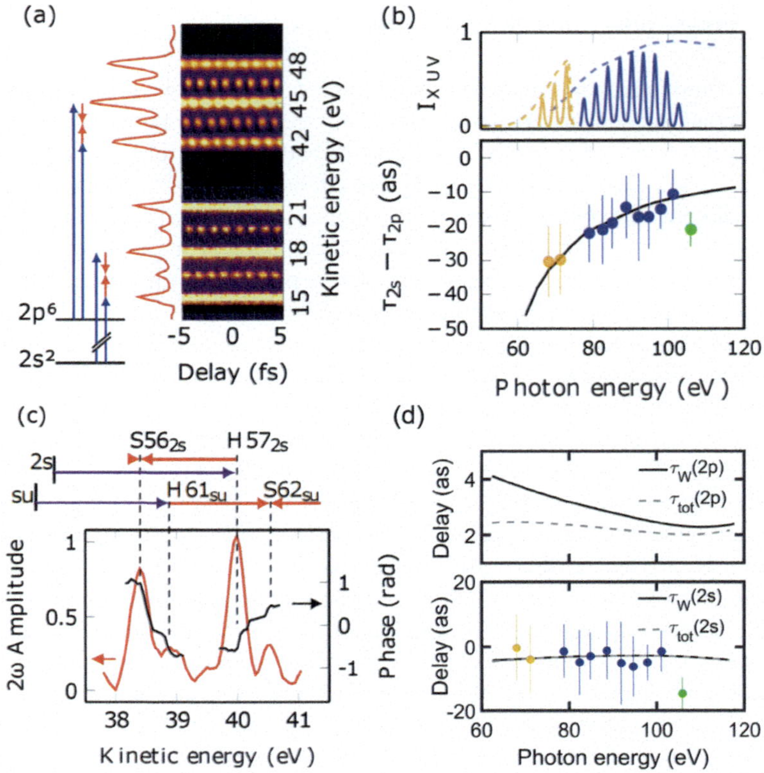


**Fig. 1.12** **a** Schematic representation of the double RABBIT measurement for photoelectrons from the $2s$ and $2p$ shells of neon. **b** Atomic time delay difference between $2s$ and $2p$ measured with different XUV spectra using Al+Zr (orange) or Zr (blue) filters. The green point is the measurement by Schultze et al. [27]. The solid black curve is the theoretical prediction. **c** Energy scheme and $2\omega$ amplitude showing the partial overlap of shake-up harmonics with $3s$ sidebands. The black curve shows the phase measured with Rainbow RABBIT. **d** Calculated Wigner delay along the direction of the light polarization $\tau_W$ (black solid lines) and angle-averaged one-photon ionization time delay $\tau_{tot}$ (dashed lines) for $2p$ (upper panel) and $2s$ (lower pannel) ionization. The experimental data is transformed to $\tau_{tot}(2s)$ by subtraction of analytical $\tau(2s)$ and simulated $\tau_{tot}(2p)$. Figure adapted from [48]. Reprinted with permission from AAAS

#### 1.4.2.2 Angle-Resolved Photoionization Time Delays

Angle-resolved RABBIT was first introduced in 2003 to characterize APTs [94]. However, it is only in 2016 that Heuser and coworkers used this technique to investigate angle-resolved photoionization time delays in He [32]. This method relies on measuring the sideband phase as a function of the emission angle of the photoelectrons. Since then, the number of theoretical [141–143] and experimental [35, 144, 145] angle-resolved studies on photoionization time delays has steadily increased.

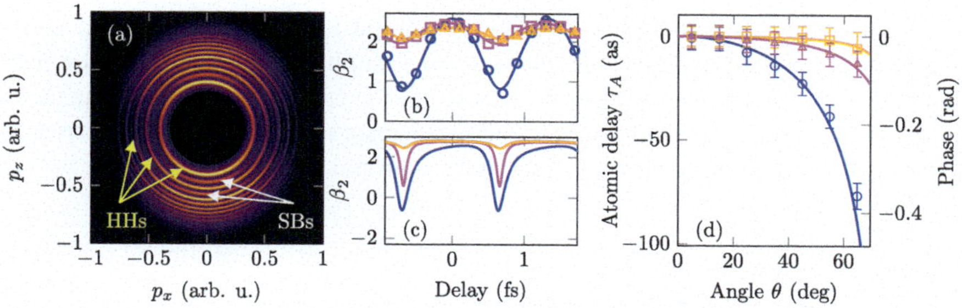


**Fig. 1.13** Experimental and theoretical results in Ar: **a** Inverted VMI image showing harmonics (HH) and sidebands (SBs). **b** Experimental and **c** theoretical variation of $\beta_2$ as a function of delay for SBs 14 (blue), 20 (magenta) and 22 (yellow) in Ar. Solid lines in (**a**) are fits to the data. **c** Experimental (circles) and theoretical (solid curves) angle dependence of the atomic delay for SBs 14, 20 and 22. The error bars correspond to the standard deviation returned from the fitting algorithm. Figure adapted from [34]. Licensed under CC-BY-4.0

Figure 1.13a shows a photoelectron momentum map obtained when Ar is photoionized by an attosecond pulse train in the presence of a weak IR field. The figure shows harmonics and sidebands for a given delay between the two pulses. When the delay is varied, the angular distribution of the sidebands changes resulting in a modification of the asymmetry parameters. Figure 1.13b presents the $\beta_2$ parameters measured as a function of the delay in [34]. The periodic modification of the angular distribution as a function of delay was used in [94] to extract the group delay of attosecond pulse trains generated in different gases. In 2018, Cirelli et al. showed that the modification of the angular distribution with delay decreases as a function of the kinetic energy of the photoelectrons, characterized by a decrease of the modulation amplitude of the $\beta_2$ oscillations [39]. The extension of Fano's propensity rule to laser-assisted photoionization in [34], as discussed in Sect. 1.3.2.2 shows that the modifications of the angular distribution with delay originates from an asymmetry between absorption and emission. The decrease of the $\beta$ oscillation amplitude with kinetic energy [Fig. 1.13(b, c)] can be explained by the fact that this asymmetry is less and less pronounced, converging towards the prediction of the soft-photon approximation [146].

Figure 1.13d presents angle-resolved atomic time delays measured in argon in [34]. The angular dependence of the photoionization time delays was first demonstrated by Heuser et al. in He [32]. The authors showed that, despite the ground state of He being spherically symmetric, the photoionization time delay strongly varies as a function of the emission angle. In addition, the angular dependence of the delays decreases with increasing kinetic energy. Since then, anisotropic photoionization time delays have been measured in other noble gases, both for resonant [39, 147, 148], and non-resonant photoionization [34, 53–55, 66, 98, 100]. The angle dependence of the photoionization time delays is another consequence of Fano's propensity rule [34] (see Sect. 1.3.2.2). The transition probabilities to the final continua have different strengths depending on whether the IR photon is absorbed or emitted so that the resulting angle-dependent phase is different for the absorption and emission paths. In particular, in helium, the transition amplitude for the absorption path changes of sign for certain angles, while it is not the case for the emission path, as shown in Sect. 1.3.2.2. The asymmetry between absorption and emission decreases with increasing kinetic energy, yielding more isotropic photoionization time delays as shown in Fig. 1.13(d).

#### 1.4.2.3 Channel-Resolved Photoionization Time Delays

The understanding of photoionization is directly coupled to the theoretical interpretation of the experimentally observed quantities. Hence, attosecond experiments combining coherent excitation of many continuum states and angular resolution have progressively explored the possibility of measuring channel-resolved photoionization time delays [53–56]. The goal is to determine the energy variation of the amplitudes and phases of all ionization channels, as discussed in Sect. 1.3.2.3.

Interestingly, unlike traditional angle-resolved photoionization experiments using synchrotron radiation, which measure the phase difference between ionization channels, experiments using attosecond light sources give access to the phase variation of the individual channels over broad energy ranges, providing direct insight into the dynamics of the ionization process for each channel. However, in general, the number of unknowns in angle-resolved RABBIT measurements is larger than the number of observable quantities. To retrieve the information about the individual channels two approaches have been followed.

The first approach consists in taking advantage of the universal properties of continuum–continuum transitions far from resonances. In this context, Peschel *et al.* [53] measured RABBIT spectrograms using a VMIS in neon. For a given delay, the photoelectron angular distribution of each sideband was fitted using Eq. (1.41). The resulting fitted coefficients, $h_i(\tau),\ i \in \{0, 2, 4\}$, oscillate sinusoidally as a function of the delay. Figure 1.14(a–c) presents this delay dependence of the $h_i$ parameters (black dots) and the result of a global fit (red solid line) for sideband 18. Each of these oscillations is characterized by an amplitude, a phase and a mean value, giving rise to a total of nine observables. These quantities, expressed in terms of two-photon matrix elements for all channels, allow, after a global fit, the determination of one-photon phases. Figure 1.14d presents the result for the one-photon scattering phases $\eta_0$ and $\eta_2$. This result shows that the scattering phase of the d partial wave varies by almost $\pi$ rad over the range 2–20 eV, while the s partial wave increases by one radian over the range 2–5 eV and then remains almost constant from 5–20 eV. For each channel, the phase could be decomposed as the sum of a universal Coulomb phase, a centrifugal phase shift and a phase due to the short-range potential. The latter one was found to be close zero for the d channel, while, for the s channel, electron correlations led to a phase offset of $1.2\pi$ [53]. The one-photon amplitudes presented in Fig. 1.14(d) show that the d channel dominates over the s one, as predicted by Fano's propensity rule for bound–continuum transitions [78]. These amplitudes were measured separately, in the absence of the dressing IR field.

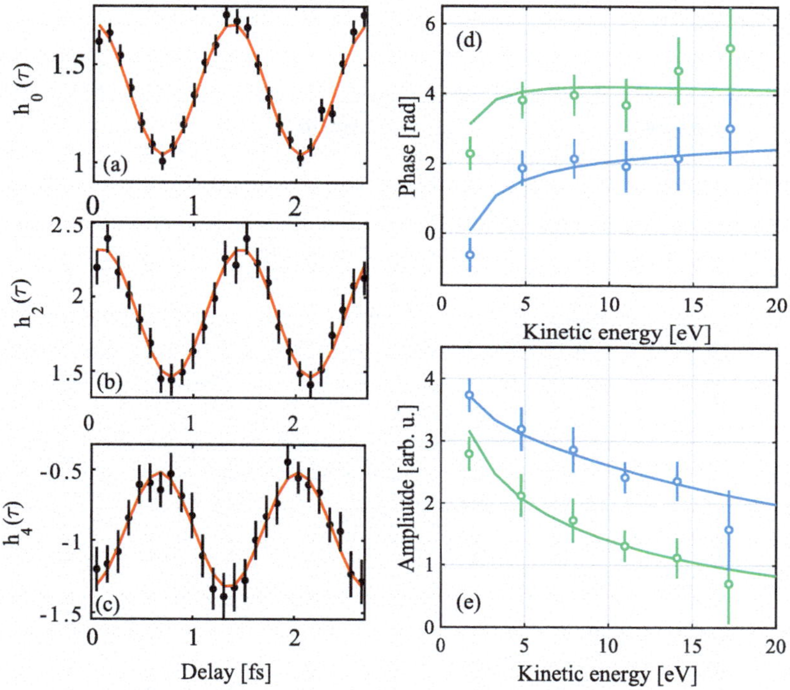


**Fig. 1.14** Experimental and theoretical results in Ne : (**a**–**c**) Delay dependence of the coefficients $h_0(\tau)$ (**a**), $h_2(\tau)$ (**b**), $h_4(\tau)$ (**c**), for sideband 18. The dots correspond to the value of the coefficients obtained at each delay by fitting the angular distribution with Eq. (1.41). The error bars indicated the standard error returned by the fitting function. The red curves are the result of the simultaneous fit of the $h_i(\tau)$) oscillations using the explicit expression of these functions in terms of the unknown quantities of the problem. (**d**–**e**) Experimental (circles) and theoretical (solid lines) one-photon scattering phases $\eta_0$ (green) and $\eta_2$ (blue) and radial amplitudes $\sigma_0$ and $\sigma_2$. The error bars correspond to the standard deviation obtained by repeating 500 times the fitting of the $h_i(\tau)$ functions with randomized initial guess within a normal distribution. Figure adapted from [53]. Licensed under CC-BY-4.0

An alternative approach has been implemented which does not require any assumption on CC-transitions [54–56]. In two-photon ionization, different ionization pathways can generally lead to the same partial wave, i.e. to the observable. As the which-way information is unknown from the measurement, the observable only encodes the interference between these pathways. By controlling the polarizations between the XUV and IR pulses, the interference can be manipulated, and angle-resolved RABBIT spectra allow partial wave analysis to recover the amplitude and phase of selected channels, giving access to the one-photon scattering phases and continuum–continuum phases of some of the channels.

## 1.5 Summary

Over the past decade, attosecond atomic photoionization in rare gas has been the subject of many studies to understand electron dynamics at the fastest relevant timescales. To circumvent experimentally the inherent loss of spectral resolution due to the ultra-short pulse duration, attosecond pulse trains can be used. The ultimate spectral resolution becomes limited to the sharpness of the harmonics, while the temporal resolution is given by the precision with which the spectral phases between harmonics are locked.

In this review, we showed that it is possible to rationalize the dynamics of the EWP emitted after the absorption of XUV light pulses within the framework of the second-order perturbation theory for the RABBIT technique. The influence of the weak dressing field can be understood for a single angular momentum path as a universal CC-delay induced on the photoelectron in the long-range Coulomb potential. Deviation of this universal behaviour is expected close to the ionization threshold for high-precision measurements ($\leq$10 as), or in the presence of resonances in the continuum. While unique angular-momentum paths are rare in practice, they do allow for a simple definition of the Wigner time delay. As soon as multiple angular momentum paths contribute, the interpretation of the experiment depends on the physical quantity measured.

With an angle-integrated spectrometer, interference due to the contribution of different one-photon intermediate states to the same two-photon final state may occur. This effect is small when there is a dominant angular momentum path and no strong variation of the cross section, allowing thus the retrieval of a quantity similar to a Wigner time delay. With an angle-resolved spectrometer, interference between the partial waves leads to the measurement of an angle-dependent delay. Channel-resolved photoionization measurements allow the separation of the one-photon delay from the CC delay. Whenever the universality of the CC-transition holds, it is in principle possible to perform a 'complete' experiment. Finally, along the polarization direction, only photoelectrons with $m = 0$ are selected, solving partially the problem of alignment/orientation of the residual ion. The measured time delay in this configuration becomes particularly interesting as it allows for accurate extraction of the Wigner delays over large energy ranges.

While a lot of progress has been achieved in the measurement of atomic photoionization time delays, many exciting lines of research are currently being investigated and/or remain to be explored. Photoionization time delays have been mostly measured from valence shells and extending these measurements to inner-shell photoionization remains an outstanding challenge. Additionally, very little is known regarding two-electron wavepackets. Studies on autoionizing resonances have shown that the concept of time delay is not valid as the dressing field affects the dynamics in a non-trivial way [31, 75]. Finally, efforts are currently underway to study negative ions as model systems for attosecond metrology since the effects of CC-transition are non-existent.

The study of photoionization of rare gases at the attosecond time scale presented in this article can serve as a benchmark for the study of more complex systems, for example, molecules[149]. Photoelectron spectroscopy measurements reveal the importance of the coupling between the electronic and nuclear degrees of freedom in autoionizing [150–152] and shape resonances [153–156] and, using coincidence technique, in dissociation [157–159]. In general, the molecular potential is not centro-symmetric and distortion of this potential due to the environment can be investigated [160–162]. In molecules, however, the effect of the CC-transition cannot fully be inferred from the atomic case due to the non-spherical potential and a high density of states in the valence region, which calls for further work.

As a concluding remark, attosecond photoionization studies inherently rely on the coherence of the light fields and the photoionization process. In recent years, a few studies have been initiated to quantify coherence in photoionization [163, 164]. Two protocols to implement photoelectron quantum state tomography, thus allowing the determination of the photoelectron density matrix, have recently been proposed [163, 165]. New approaches have been developed to perform multi-colour fields interferometric measurement using high harmonic generation [165, 166] and Free-Electron-Laser [167–169]. It can be expected that these measurements will be of increasing interest in the future, as they pave the way for the investigation of coherence and entanglement in matter.

**Acknowledgements** The authors would like to thank Eva Lindroth for sharing theoretical data and critically reviewing the work. The research was supported by the Swedish Research Council (2013-8185,2020-05200,2020-06384), the European Research Council (advanced grant QPAP 889400), the Knut and Alice Wallenberg Foundation and the Crafoord Foundation. AL is partly supported by the Wallenberg Center for Quantum Technology (WACQT) funded by the Knut and Alice Wallenberg Foundation.

## References

1. A. Einstein, Annalen der Physik **322**, 132 (1905)
2. V. Schmidt, Rep. Prog. Phys. **55**, 1483 (1992)
3. B. Bederson, Comm. Atom. Mol. Phys. **1**, 49 (1969)
4. U. Becker, J. Electron. Spectrosc. Relat. Phenom. **96**, 105 (1998)
5. H. Kleinpoppen, B. Lohmann, A.N. Grum-Grzhimailo, in *Perfect/Complete Scattering Experiments*. ed. by H. Kleinpoppen, B. Lohmann, A.N. Grum-Grzhimailo (Springer, Berlin, Heidelberg, 2013), pp.201–292
6. J.W. Cooper, R.N. Zare, J. Chem. Phys. **48**, 942 (1968)
7. J.W. Cooper, Phys. Rev. A **47**, 1841 (1993)
8. N. Cherepkov, J. Phys. B **12**, 1279 (1979)
9. U. Heinzmann, J. Phys. B **13**, 4353 (1980)
10. U. Heinzmann, J. Phys. B **13**, 4367 (1980)
11. N. Cherepkov, J. Electron Spectrosc. Relat. Phenom. **144**, 1197 (2005)
12. K. Godehusen, P. Zimmermann, A. Verweyen, A. von dem Borne, P. Wernet, B. Sonntag, Phys. Rev. A **58**, R3371 (1998)
13. F.J. Wuilleumier, M. Meyer, J. Phys. B **39**, R425 (2006)

14. A. McPherson, G. Gibson, H. Jara, U. Johann, T.S. Luk, I.A. McIntyre, K. Boyer, C.K. Rhodes, J. Opt. Soc. Am. B **4**, 595 (1987)
15. M. Ferray, A. L'Huillier, X. Li, L. Lompre, G. Mainfray, C. Manus, J. Phys. B **21**, L31 (1988)
16. P. O'Keeffe, R. Lopez-Martens, J. Mauritsson, A. Johansson, A. L'Huillier, V. Veniard, R. Taieb, A. Maquet, M. Meyer, Phys. Rev. A **69**, 051401 (2004)
17. L. Haber, B. Doughty, S. Leone, Phys. Rev. A **79**, 031401(R) (2009)
18. L. Haber, B. Doughty, S. Leone, Phys. Rev. A **84**, 013416 (2011)
19. T. Brabec, F. Krausz, Rev. Mod. Phys. **72**, 545 (2000)
20. F. Calegari, G. Sansone, M. Nisoli, in *Optical Technologies for Extreme-Ultraviolet and Soft X-ray Coherent Sources*. ed. by F. Canova, L. Poletto (Springer, Berlin, Heidelberg, 2015), pp.41–62
21. M. Hentschel, R. Kienberger, C. Spielmann, G.A. Reider, N. Milosevic, T. Brabec, P. Corkum, U. Heinzmann, M. Drescher, F. Krausz, Nature **414**, 509 (2001)
22. P.M. Paul, E.S. Toma, P. Breger, G. Mullot, F. Augé, P. Balcou, H.G. Muller, P. Agostini, Science **292**, 1689 (2001)
23. E.P. Wigner, Phys. Rev. **98**, 145 (1955)
24. L. Eisenbud, The formal properties of nuclear collisions. Ph.D. thesis, Princeton University (1948)
25. F.T. Smith, Phys. Rev. **118**, 349 (1960)
26. M. Hentschel, R. Kienberger, C. Spielmann, G.A. Reider, N. Milosevic, T. Brabec, P. Corkum, U. Heinzmann, M. Drescher, F. Krausz, Nature **414**, 509 (2001)
27. M. Schultze, M. Fieß, N. Karpowicz, J. Gagnon, M. Korbman, M. Hofstetter, S. Neppl, A.L. Cavalieri, Y. Komninos, T. Mercouris, C.A. Nicolaides, R. Pazourek, S. Nagele, J. Feist, J. Burgdörfer, A.M. Azzeer, R. Ernstorfer, R. Kienberger, U. Kleineberg, E. Goulielmakis, F. Krausz, V.S. Yakovlev, Science **328**, 1658 (2010)
28. K. Klünder, J.M. Dahlström, M. Gisselbrecht, T. Fordell, M. Swoboda, D. Guénot, P. Johnsson, J. Caillat, J. Mauritsson, A. Maquet, R. Taïeb, A. L'Huillier, Phys. Rev. Lett. **106**, 143002 (2011)
29. J.M. Dahlström, D. Guénot, K. Klünder, M. Gisselbrecht, J. Mauritsson, A. L'Huillier, A. Maquet, R. Taïeb, Chem. Phys. **414**, 53 (2013)
30. R. Pazourek, S. Nagele, J. Burgdörfer, Rev. Mod. Phys. **87**, 765 (2015)
31. L. Argenti, A. Jiménez-Galán, J. Caillat, R. Taïeb, A. Maquet, F. Martín, Phys. Rev. A **95**, 043426 (2017)
32. S. Heuser, A. Jiménez Galán, C. Cirelli, C. Marante, M. Sabbar, R. Boge, M. Lucchini, L. Gallmann, I. Ivanov, A.S. Kheifets, J.M. Dahlström, E. Lindroth, L. Argenti, F. Martín, U. Keller, Phys. Rev. A **94**, 063409 (2016)
33. I. Jordan, M. Huppert, S. Pabst, A.S. Kheifets, D. Baykusheva, H.J. Wörner, Phys. Rev. A **95**, 013404 (2017)
34. D. Busto, J. Vinbladh, S. Zhong, M. Isinger, S. Nandi, S. Maclot, P. Johnsson, M. Gisselbrecht, A. L'Huillier, E. Lindroth, J.M. Dahlström, Phys. Rev. Lett. **123**, 133201 (2019)
35. J. Fuchs, N. Douguet, S. Donsa, F. Martín, J. Burgdörfer, L. Argenti, L. Cattaneo, U. Keller, Optica **7**, 1541 (2020)
36. M. Sabbar, S. Heuser, R. Boge, M. Lucchini, T. Carette, E. Lindroth, L. Gallmann, C. Cirelli, U. Keller, Phys. Rev. Lett. **115**, 133001 (2015)
37. M. Kotur, D. Guénot, A. Jiménez-Galán, D. Kroon, E.W. Larsen, M. Louisy, S. Bengtsson, M. Miranda, J. Mauritsson, C.L. Arnold, S.E. Canton, M. Gisselbrecht, T. Carette, J.M. Dahlström, E. Lindroth, A. Maquet, L. Argenti, F. Martín, A. L'Huillier, Nat. Commun. **7**, 10566 (2016)
38. V. Gruson, L. Barreau, Á. Jiménez-Galan, F. Risoud, J. Caillat, A. Maquet, B. Carré, F. Lepetit, J.F. Hergott, T. Ruchon, L. Argenti, R. Taïeb, F. Martín, P. Salières, Science **354**, 734 (2016)
39. C. Cirelli, C. Marante, S. Heuser, C. Petersson, Á. Galán, L. Argenti, S. Zhong, D. Busto, M. Isinger, S. Nandi, S. MacLot, L. Rading, P. Johnsson, M. Gisselbrecht, M. Lucchini, L. Gallmann, J. Dahlström, E. Lindroth, A. L'Huillier, F. Martín, U. Keller, Nat. Commun. **9**, 955 (2018)

40. D. Busto, L. Barreau, M. Isinger, M. Turconi, C. Alexandridi, A. Harth, S. Zhong, R.J. Squibb, D. Kroon, S. Plogmaker, M. Miranda, Á. Jiménez-Galán, L. Argenti, C.L. Arnold, R. Feifel, F. Martín, M. Gisselbrecht, A. L'Huillier, P. Salières, J. Phys. B **51**, 044002 (2018)
41. L. Barreau, C.L.M. Petersson, M. Klinker, A. Camper, C. Marante, T. Gorman, D. Kiesewetter, L. Argenti, P. Agostini, J. González-Vázquez, P. Salières, L.F. DiMauro, F. Martín, Phys. Rev. Lett. **122**, 253203 (2019)
42. D. Busto, H. Laurell, D. Finkelstein-Shapiro, C. Alexandridi, M. Isinger, S. Nandi, R.J. Squibb, M. Turconi, S. Zhong, C.L. Arnold, R. Feifel, M. Gisselbrecht, P. Salières, T. Pullerits, F. Martín, L. Argenti, A. L'Huillier, Eur. Phys. J. D **76**, 112 (2022)
43. D. Guénot, K. Klünder, C.L. Arnold, D. Kroon, J.M. Dahlström, M. Miranda, T. Fordell, M. Gisselbrecht, P. Johnsson, J. Mauritsson, E. Lindroth, A. Maquet, R. Taïeb, A. L'Huillier, A.S. Kheifets, Phys. Rev. A **85**, 053424 (2012)
44. C. Palatchi, J.M. Dahlström, A.S. Kheifets, I.A. Ivanov, D.M. Canaday, P. Agostini, L.F. DiMauro, J. Phys. B **47**, 245003 (2014)
45. C. Alexandridi, D. Platzer, L. Barreau, D. Busto, S. Zhong, M. Turconi, L. Neoričić, H. Laurell, C.L. Arnold, A. Borot, J.F. Hergott, O. Tcherbakoff, M. Lejman, M. Gisselbrecht, E. Lindroth, A. L'Huillier, J.M. Dahlström, P. Salières, Phys. Rev. Res. **3**, L012012 (2021)
46. D. Kiesewetter, R.R. Jones, A. Camper, S.B. Schoun, P. Agostini, L.F. DiMauro, Nat. Phys. **14**, 68 (2018)
47. A. Jain, T. Gaumnitz, A. Bray, A. Kheifets, H.J. Wörner, Opt. lett. **43**, 4510 (2018)
48. M. Isinger, R.J. Squibb, D. Busto, S. Zhong, A. Harth, D. Kroon, S. Nandi, C.L. Arnold, M. Miranda, J.M. Dahlström, E. Lindroth, R. Feifel, M. Gisselbrecht, A. L'Huillier, Science **358**, 893 (2017)
49. M. Ossiander, F. Siegrist, V. Shirvanyan, R. Pazourek, A. Sommer, T. Latka, A. Guggenmos, S. Nagele, J. Feist, J. Burgdörfer, R. Kienberger, M. Schultze, Nat. Phys. **13**, 280 (2017)
50. E.P. Månsson, D. Guénot, C.L. Arnold, D. Kroon, S. Kasper, J.M. Dahlström, E. Lindroth, A.S. Kheifets, A. L'Huillier, S.L. Sorensen, M. Gisselbrecht, Nat. Phys. **10**, 207 (2014)
51. M. Drescher, M. Hentschel, R. Kienberger, M. Uiberacker, V. Yakovlev, A. Scrinzi, T. Westerwalbesloh, U. Kleineberg, U. Heinzmann, F. Krausz, Nature **419**, 803 (2002)
52. S. Zhong, J. Vinbladh, D. Busto, R.J. Squibb, M. Isinger, L. Neoričić, H. Laurell, R. Weissenbilder, C.L. Arnold, R. Feifel, J.M. Dahlström, G. Wendin, M. Gisselbrecht, E. Lindroth, A. L'Huillier, Nat. Commun. **11**, 5042 (2020)
53. J. Peschel, D. Busto, M. Plach, M. Bertolino, M. Hoflund, S. Maclot, J. Vinbladh, H. Wikmark, F. Zapata, E. Lindroth, M. Gisselbrecht, J.M. Dahlström, A. L'Huillier, P. Eng-Johnsson, Nat. Commun. **13**, 5205 (2022)
54. W. Jiang, G.S.J. Armstrong, J. Tong, Y. Xu, Z. Zuo, J. Qiang, P. Lu, D.D.A. Clarke, J. Benda, A. Fleischer, H. Ni, K. Ueda, H.W. van der Hart, A.C. Brown, X. Gong, J. Wu, Nat. Commun. **13**, 5072 (2022)
55. M. Han, J.B. Ji, T. Balčiūnas, K. Ueda, H.J. Wörner, Nat. Phys. **19**, 230 (2023)
56. W. Jiang, G.S.J. Armstrong, L. Han, Y. Xu, Z. Zuo, J. Tong, P. Lu, J.M. Dahlström, K. Ueda, A.C. Brown, H.W. van der Hart, X. Gong, J. Wu, Phys. Rev. Lett. **131**, 203201 (2023)
57. S. Kühn, M. Dumergue, S. Kahaly, S. Mondal, M. Füle, T. Csizmadia, B. Farkas, B. Major, Z. Várallyay, E. Cormier, M. Kalashnikov, F. Calegari, M. Devetta, F. Frassetto, E. Månsson, L. Poletto, S. Stagira, C. Vozzi, M. Nisoli, P. Rudawski, S. Maclot, F. Campi, H. Wikmark, C.L. Arnold, C.M. Heyl, P. Johnsson, A. L'Huillier, R. Lopez-Martens, S. Haessler, M. Bocoum, F. Boehle, A. Vernier, G. Iaquaniello, E. Skantzakis, N. Papadakis, C. Kalpouzos, P. Tzallas, F. Lépine, D. Charalambidis, K. Varjú, K. Osvay, G. Sansone, J. Phys. B **50**, 132002 (2017)
58. K. Midorikawa, Y. Nabekawa, A. Suda, Prog. Quant. Eletron. **32**, 43 (2008)
59. B. Xue, K. Midorikawa, E.J. Takahashi, Optica **9**, 360 (2022)
60. C. Guo, A. Harth, S. Carlström, Y.C. Cheng, S. Mikaelsson, E. Mårsell, C. Heyl, M. Miranda, M. Gisselbrecht, M.B. Gaarde, K.J. Schafer, A. Mikkelsen, J. Mauritsson, C.L. Arnold, A. L'Huillier, J. Phys. B **51**, 034006 (2018)
61. S. Mikaelsson, J. Vogelsang, C. Guo, I. Sytcevich, A.L. Viotti, F. Langer, Y.C. Cheng, S. Nandi, W. Jin, A. Olofsson, R. Weissenbilder, J. Mauritsson, A. L'Huillier, M. Gisselbrecht, C.L. Arnold, NanoPhotonics **10**, 117 (2020)

62. F.J. Furch, T. Witting, M. Osolodkov, F. Schell, C.P. Schulz, M.J.J. Vrakking, J. Phys. Photonics **4**, 032001 (2022)
63. P.C. Huang, C. Hernández-García, J.T. Huang, P.Y. Huang, C.H. Lu, L. Rego, D.D. Hickstein, J.L. Ellis, A. Jaron-Becker, A. Becker, S.D. Yang, C.G. Durfee, L. Plaja, H.C. Kapteyn, M.M. Murnane, A.H. Kung, M.C. Chen, Nat. Photon. **12**, 349 (2018)
64. K.M. Dorney, L. Rego, N.J. Brooks, J. San Román, C.T. Liao, J.L. Ellis, D. Zusin, C. Gentry, Q.L. Nguyen, J.M. Shaw, A. Picón, L. Plaja, H.C. Kapteyn, M.M. Murnane, C. Hernández-García, Nat. Photon. **13**, 123 (2019)
65. U. Bengs, N. Zhavoronkov, Sci. Rep. **11**, 9570 (2021)
66. M. Han, J.B. Ji, K. Ueda, H.J. Wörner, Optica **10**, 1044 (2023)
67. T. Popmintchev, M.C. Chen, D. Popmintchev, P. Arpin, S. Brown, S. Ališauskas, G. Andriukaitis, T. Balčiunas, O.D. Mücke, A. Pugzlys, A. Baltuška, B. Shim, S.E. Schrauth, A. Gaeta, C. Hernández-García, L. Plaja, A. Becker, A. Jaron-Becker, M.M. Murnane, H.C. Kapteyn, Science **336**, 1287 (2012)
68. S.L. Cousin, N.D. Palo, B. Buades, S.M. Teichmann, M. Reduzzi, M. Devetta, A. Kheifets, G. Sansone, J. Biegert, Phys. Rev. X **7**, 041030 (2017)
69. C. Cohen-Tannoudji, B. Diu, F. Laloë, *Quantum Mechanics*, vol. 2 (Wiley-Interscience, 1977)
70. O. Hemmers, R. Guillemin, D. Lindle, Rad. Phys. Chem. **70**, 123 (2004)
71. F.A. Weihe, S.K. Dutta, G. Korn, D. Du, P.H. Bucksbaum, P.L. Shkolnikov, Phys. Rev. A **51**, R3433 (1995)
72. A.F. Starace, in Fundamental Processes, in *Energetic Atomic Collisions*. ed. by H.O. Lutz, J.S. Briggs, H. Kleinpoppen (Springer, New York, NY, 1983), pp.69–110
73. L.D. Landau, E.M. Lifshitz, *Quantum Mechanics: Non-relativistic Theory*, vol. 3 (Elsevier, 2013)
74. A. Maquet, J. Caillat, R. Taïeb, J. Phys. B **47**, 204004 (2014)
75. Á. Jiménez-Galán, F. Martín, L. Argenti, Phys. Rev. A **93**, 023429 (2016)
76. J. Sörngård, J.M. Dahlström, E. Lindroth, J. Phys. B **53**, 134003 (2020)
77. D. Busto, Quantum interference effects attosecond photoionization dynamics. Ph.D. thesis, Lund University (2020). https://portal.research.lu.se/en/publications/quantum-interference-effects-in-attosecond-photoionization-dynami
78. U. Fano, Phys. Rev. A **32**, 617 (1985)
79. D.I.R. Boll, L. Martini, O.A. Fojón, Phys. Rev. A **106**, 023116 (2022)
80. J. Vinbladh, J.M. Dahlström, E. Lindroth, Atoms **10**, 80 (2022)
81. M. Bertolino, D. Busto, F. Zapata, J.M. Dahlström, Phys. Rev. Lett. **123**, 133201 (2019)
82. D. Guénot, D. Kroon, E. Balogh, E. Larsen, M. Kotur, M. Miranda, T. Fordell, P. Johnsson, J. Mauritsson, M. Gisselbrecht, K. Varju, C. Arnold, T. Carette, A.S. Kheifets, E. Lindroth, A. L'Huillier, J.M. Dahlstrom, J. Phys. B **47**, 245602 (2014)
83. J. Vinbladh, J.M. Dahlström, E. Lindroth, Phys. Rev. A **100**, 043424 (2019)
84. E. Lindroth, J.M. Dahlström, Phys. Rev. A **96**, 013420 (2017)
85. Y. Mairesse, A. de Bohan, L.J. Frasinski, H. Merdji, L.C. Dinu, P. Monchicourt, P. Breger, M. Kovačev, R. Taïeb, B. Carré, H.G. Muller, P. Agostini, P. Salières, Science **302**, 1540 (2003)
86. V. Véniard, R. Taïeb, A. Maquet, Phys. Rev. A **54**, 721 (1996)
87. A. Müller, M. Laubscher, Opt. Lett. **26**, 1915 (2001)
88. J. Dahlström, A. L'Huillier, A. Maquet, J. Phys. B **45**, 183001 (2012)
89. M. Isinger, D. Busto, S. Mikaelsson, S. Zhong, C. Guo, P. Salières, C. Arnold, A. L'Huillier, M. Gisselbrecht, Philos. T. R. Soc. A **377**, 20170475 (2019)
90. A. Jiménez-Galán, L. Argenti, F. Martín, Phys. Rev. Lett. **113**, 263001 (2014)
91. A.C. Brown, G.S. Armstrong, J. Benda, D.D. Clarke, J. Wragg, K.R. Hamilton, Z. Mašín, J.D. Gorfinkiel, H.W. van der Hart, Comput. Phys. Commun. **250**, 107062 (2020)
92. J. Benda, Z. Mašín, J.D. Gorfinkiel, Phys. Rev. A **105**, 053101 (2022)
93. J.W. Cooper, Phys. Rev. **128**, 681 (1962)
94. S.A. Aseyev, Y. Ni, L.J. Frasinski, H.G. Muller, M.J.J. Vrakking, Phys. Rev. Lett. **91**, 223902 (2003)

95. E. Arnous, S. Klarsfeld, S. Wane, Phys. Rev. A **7**, 1559 (1973)
96. K.L. Reid, Annu. Rev. Phys. Chem. **54**, 397 (2003)
97. S. Saha, J. Vinbladh, J. Sörngård, A. Ljungdahl, E. Lindroth, Phys. Rev. A **104**, 033108 (2021)
98. J. Joseph, F. Holzmeier, D. Bresteau, C. Spezzani, T. Ruchon, J.F. Hergott, O. Tcherbakoff, P. D'Oliveira, J.C. Houver, D. Dowek, J. Phys. B **53**, 184007 (2020)
99. D.I.R. Boll, L. Martini, O.A. Fojón, Phys. Rev. A **106**, 023116 (2022)
100. D. Bharti, H. Srinivas, F. Shobeiry, K.R. Hamilton, R. Moshammer, T. Pfeifer, K. Bartschat, A. Harth, Phys. Rev. A **107**, 022801 (2023)
101. C.M. Heyl, C.L. Arnold, A. Couairon, A. L'Huillier, J. Phys. B **50**, 013001 (2017)
102. D. Strickland, G. Mourou, Opt. Commun. **56**, 219 (1985)
103. E. Goulielmakis, M. Uiberacker, R. Kienberger, A. Baltuska, V. Yakovlev, A. Scrinzi, T. Westerwalbesloh, U. Kleineberg, U. Heinzmann, M. Drescher, F. Krausz, Science **305**, 1267 (2004)
104. S. Hädrich, A. Klenke, J. Rothhardt, M. Krebs, A. Hoffmann, O. Pronin, V. Pervak, J. Limpert, A. Tünnermann, Nat. Photon. **8**, 779 (2014)
105. F. Emaury, A. Diebold, C.J. Saraceno, U. Keller, Optica **2**, 980 (2015)
106. A. Harth, C. Guo, Y.C. Cheng, A. Losquin, M. Miranda, S. Mikaelsson, C.M. Heyl, O. Prochnow, J. Ahrens, U. Morgner, A. L'Huillier, C.L. Arnold, J. Opt. **20**, 014007 (2017)
107. M. Krebs, S. Hädrich, S. Demmler, J. Rothhardt, A. Zaïr, L. Chipperfield, J. Limpert, A. Tünnermann, Nat. Photon. **7**, 551 (2013)
108. T. Nagy, P. Simon, L. Veisz, Adv. Phys. X **6**, 1845795 (2021)
109. A.L. Viotti, M. Seidel, E. Escoto, S. Rajhans, W.P. Leemans, I. Hartl, C.M. Heyl, Optica **9**, 197 (2022)
110. R. Klas, A. Kirsche, M. Gebhardt, J. Buldt, H. Stark, S. Hädrich, J. Rothhardt, J. Limpert, PhotoniX **2**, 4 (2021)
111. C. Schmidt, Y. Pertot, T. Balciunas, K. Zinchenko, M. Matthews, H.J. Wörner, J.P. Wolf, Opt. Express **26**, 11834 (2018)
112. S.M. Teichmann, F. Silva, S. Cousin, M. Hemmer, J. Biegert, Nat. Commun. **7**, 11493 (2016)
113. D. Popmintchev, B.R. Galloway, M.C. Chen, F. Dollar, C.A. Mancuso, A. Hankla, L. Miaja-Avila, G. O'Neil, J.M. Shaw, G. Fan, S. Ališauskas, G. Andriukaitis, T. Balčiunas, O.D. Mücke, A. Pugzlys, A. Baltuška, H.C. Kapteyn, T. Popmintchev, M.M. Murnane, Phys. Rev. Lett. **120**, 093002 (2018)
114. J. Pupeikis, P.A. Chevreuil, N. Bigler, L. Gallmann, C.R. Phillips, U. Keller, Optica **7**, 168 (2020)
115. M. Isinger, Time-frequency analysis in attosecond spectroscopy. Ph.D. thesis, Lund University (2019). https://portal.research.lu.se/en/publications/time-frequency-analysis-in-attosecond-spectroscopy
116. M. Chini, H. Mashiko, H. Wang, S. Chen, C. Yun, S. Scott, S. Gilbertson, Z. Chang, Opt. Express **17**, 21459 (2009)
117. D. Kroon, D. Guénot, M. Kotur, E. Balogh, E.W. Larsen, C.M. Heyl, M. Miranda, M. Gisselbrecht, J. Mauritsson, P. Johnsson, K. Varjú, A. L'Huillier, C.L. Arnold, Opt. Lett. **39**, 2218 (2014)
118. M. Huppert, I. Jordan, H.J. Wörner, Rev. Sci. Instrum. **86**, 123106 (2015)
119. J. Vaughan, J. Bahder, B. Unzicker, D. Arthur, M. Tatum, T. Hart, G. Harrison, S. Burrows, P. Stringer, G.M. Laurent, Opt. Express **27**, 30989 (2019)
120. S. Luo, R. Weissenbilder, H. Laurell, M. Ammitzböll, V. Poulain, D. Busto, L. Neoričić, C. Guo, S. Zhong, D. Kroon, R.J. Squibb, R. Feifel, M. Gisselbrecht, A. L'Huillier, C.L. Arnold, Adv. Phys. X **8**, 2250105 (2023)
121. A. Zaïr, E. Mével, E. Cormier, E. Constant, J. Opt. Soc. Am. B **35**, A1105 (2018)
122. H. Ahmadi, S. Kellerer, D. Ertel, M. Moioli, M. Reduzzi, P. Maroju, A. Jäger, R. Shah, J. Lutz, F. Frassetto et al., J. Phys. Photonics **2**, 024006 (2020)
123. D. Ertel, M. Schmoll, S. Kellerer, A. Jäger, R. Weissenbilder, M. Moioli, H. Ahmadi, D. Busto, I. Makos, F. Frassetto, L. Poletto, C.D. Schröter, T. Pfeifer, R. Moshammer, G. Sansone, Rev. Sci. Instrum. **94** (2023)

124. R. López-Martens, K. Varjú, P. Johnsson, J. Mauritsson, Y. Mairesse, P. Salières, M.B. Gaarde, K.J. Schafer, A. Persson, S. Svanberg, C.G. Wahlström, A. L'Huillier, Phys. Rev. Lett. **94**, 033001 (2005)
125. E. Gustafsson, T. Ruchon, M. Swoboda, R. López-Martens, P. Balcou, A. L'Huillier, Opt. Lett. **32**, 1353 (2007)
126. P. Kruit, F.H. Read, J. Phys. E **16**, 313 (1983)
127. C.Y. Cha, G. Ganteför, W. Eberhardt, Rev. Sci. Instrum. **63**, 5661 (1992)
128. J.H. Eland, O. Vieuxmaire, T. Kinugawa, P. Lablanquie, R. Hall, F. Penent, Phys. Rev. Lett. **90**, 053003 (2003)
129. A.T.J.B. Eppink, D.H. Parker, Rev. Sci. Instrum. **68**, 3477 (1997)
130. J. Ullrich, R. Moshammer, A. Dorn, R. Dorner, L.P.H. Schmidt, H.S.B. cking, Rep. Prog. Phys. **66**, 1463 (2003)
131. K. Zhao, T. Colvin, W. Hill III., G. Zhang, Rev. Sci. Instrum. **73**, 3044 (2002)
132. M.J.J. Vrakking, Rev. Sci. Instrum. **72**, 4084 (2001)
133. G.A. Garcia, L. Nahon, I. Powis, Rev. Sci. Instrum. **75**, 4989 (2004)
134. J. Itatani, F. Quéré, G.L. Yudin, M.Y. Ivanov, F. Krausz, P.B. Corkum, Phys. Rev. Lett. **88**, 173903 (2002)
135. L. Moore, M. Lysaght, J. Parker, H. van der Hart, K. Taylor, Phys. Rev. A **84**, 061404 (2011)
136. A.S. Kheifets, I.A. Ivanov, Phys. Rev. Lett. **105**, 233002 (2010)
137. J. Dahlström, T. Carette, E. Lindroth, Phys. Rev. A **86**, 061402 (2012)
138. A. Kheifets, Phys. Rev. A **87**, 063404 (2013)
139. J. Feist, O. Zatsarinny, S. Nagele, R. Pazourek, J. Burgdörfer, X. Guan, K. Bartschat, B.I. Schneider, Phys. Rev. A **89**, 033417 (2014)
140. U. Becker, R. Wehlitz, O. Hemmers, B. Langer, A. Menzel, Phys. Rev. Lett. **63**, 1054 (1989)
141. I. Ivanov, A. Kheifets, Phys. Rev. A **96**, 013408 (2017)
142. A.W. Bray, F. Naseem, A.S. Kheifets, Phys. Rev. A **97**, 063404 (2018)
143. P. Hockett, J. Phys. B **50**, 154002 (2017)
144. S. Beaulieu, A. Comby, A. Clergerie, J. Caillat, D. Descamps, N. Dudovich, B. Fabre, R. Géneaux, F. Légaré, S. Petit et al., Science **358**, 1288 (2017)
145. D. Villeneuve, P. Hockett, M. Vrakking, H. Niikura, Science **356**, 1150 (2017)
146. A. Maquet, R. Taïeb, J. Mod. Opt. **54**, 1847 (2007)
147. A. Autuori, D. Platzer, M. Lejman, G. Gallician, L. Maëder, A. Covolo, L. Bosse, M. Dalui, D. Bresteau, J.F. Hergott, O. Tcherbakoff, H.J. Marroux, V. Loriot, F. Lépine, L. Poisson, R. Taïeb, J. Caillat, P. Salières, Sci. Adv. **8**, abl7594 (2022)
148. L. Neoričić, D. Busto, H. Laurell, R. Weissenbilder, M. Ammitzböll, S. Luo, J. Peschel, H. Wikmark, J. Lahl, S. Maclot, R.J. Squibb, S. Zhong, P. Eng-Johnsson, C.L. Arnold, R. Feifel, M. Gisselbrecht, E. Lindroth, A. L'Huillier, Front. Phys. **10**, 964586 (2022)
149. D. Baykusheva, H.J. Wörner, in *Molecular Spectroscopy and Quantum Dynamics*, ed. by R. Marquardt, M. Quack (Elsevier, 2021), pp. 113–161
150. S. Haessler, B. Fabre, J. Higuet, J. Caillat, T. Ruchon, P. Breger, B. Carré, E. Constant, A. Maquet, E. Mével, P. Salières, R. Taïeb, Y. Mairesse, Phys. Rev. A **80**, 011404 (2009)
151. L. Cattaneo, J. Vos, R.Y. Bello, A. Palacios, S. Heuser, L. Pedrelli, M. Lucchini, C. Cirelli, F. Martín, U. Keller, Nat. Phys. **14**, 733 (2018)
152. L. Cattaneo, L. Pedrelli, R.Y. Bello, A. Palacios, P.D. Keathley, F. Martín, U. Keller, Phys. Rev. Lett. **128**, 063001 (2022)
153. M. Huppert, I. Jordan, D. Baykusheva, A. von Conta, H.J. Wörner, Phys. Rev. Lett. **117**, 093001 (2016)
154. V. Loriot, A. Marciniak, S. Nandi, G. Karras, M. Hervé, E. Constant, E. Plésiat, A. Palacios, F. Martín, F. Lépine, J. Phys photonics **2**, 024003 (2020)
155. A. Kamalov, A.L. Wang, P.H. Bucksbaum, D.J. Haxton, J.P. Cryan, Phys. Rev. A **102**, 023118 (2020)
156. S. Nandi, E. Plésiat, S. Zhong, A. Palacios, D. Busto, M. Isinger, L. Neoričić, C.L. Arnold, R.J. Squibb, R. Feifel, P. Decleva, A. L'Huillier, F. Martín, M. Gisselbrecht, Sci. Adv. **6**, eaba7762 (2020)

157. A.L. Wang, V.V. Serov, A. Kamalov, P.H. Bucksbaum, A. Kheifets, J.P. Cryan, Phys. Rev. A **104**, 063119 (2021)
158. D. Ertel, D. Busto, I. Makos, M. Schmoll, J. Benda, H. Ahmadi, M. Moioli, F. Frassetto, L. Poletto, C.D. Schröter, T. Pfeifer, R. Moshammer, Z. Mašín, S. Patchkovskii, G. Sansone, Sci. Adv. **9**, eadh7747 (2023)
159. X. Gong, É. Plésiat, A. Palacios, S. Heck, F. Martín, H.J. Wörner, Nat. Commun. **14**, 4402 (2023)
160. J. Vos, L. Cattaneo, S. Patchkovskii, T. Zimmermann, C. Cirelli, M. Lucchini, A. Kheifets, A.S. Landsman, U. Keller, Science **360**, 1326 (2018)
161. S. Biswas, B. Förg, L. Ortmann, J. Schötz, W. Schweinberger, T. Zimmermann, L. Pi, D. Baykusheva, H.A. Masood, I. Liontos, A.M. Kamal, N.G. Kling, A.F. Alharbi, M. Alharbi, A.M. Azzeer, G. Hartmann, H.J. Wörner, A.S. Landsman, M.F. Kling, Nat. Phys. **16**, 778 (2020)
162. H. Ahmadi, E. Plésiat, M. Moioli, F. Frassetto, L. Poletto, P. Decleva, C.D. Schröter, T. Pfeifer, R. Moshammer, A. Palacios, F. Martín, G. Sansone, Nat. Commun. **13**, 1242 (2022)
163. C. Bourassin-Bouchet, L. Barreau, V. Gruson, J.F. Hergott, F. Quéré, P. Salières, T. Ruchon, Phys. Rev. X **10**, 031048 (2020)
164. L.M. Koll, L. Maikowski, L. Drescher, T. Witting, M.J.J. Vrakking, Phys. Rev. Lett. **128**, 043201 (2022)
165. H. Laurell, D. Finkelstein-Shapiro, C. Dittel, C. Guo, R. Demjaha, M. Ammitzböll, R. Weissenbilder, L. Neoričić, S. Luo, M. Gisselbrecht, C.L. Arnold, A. Buchleitner, T. Pullerits, A. L'Huillier, D. Busto, Phys. Rev. Res. **4**, 033220 (2022)
166. H.J.B. Marroux, A.P. Fidler, D.M. Neumark, S.R. Leone, Sci. Adv. **4**(9), eaau3783 (2018)
167. P.K. Maroju, C. Grazioli, M. Di Fraia, M. Moioli, D. Ertel, H. Ahmadi, O. Plekan, P. Finetti, E. Allaria, L. Giannessi, G. De Ninno, C. Spezzani, G. Penco, S. Spampinati, A. Demidovich, M.B. Danailov, R. Borghes, G. Kourousias, C.E. Sanches Dos Reis, F. Billé, A.A. Lutman, R.J. Squibb, R. Feifel, P. Carpeggiani, M. Reduzzi, T. Mazza, M. Meyer, S. Bengtsson, N. Ibrakovic, E.R. Simpson, J. Mauritsson, T. Csizmadia, M. Dumergue, S. Kühn, H. Nandiga Gopalakrishna, D. You, K. Ueda, M. Labeye, J.E. Bækhøj, K.J. Schafer, E.V. Gryzlova, A.N. Grum-Grzhimailo, K.C. Prince, C. Callegari, G. Sansone, Nature **578**, 386 (2020)
168. P.K. Maroju, M. Di Fraia, O. Plekan, M. Bonanomi, B. Merzuk, D. Busto, I. Makos, M. Schmoll, R. Shah, P.R. Ribič, L. Giannessi, G. De Ninno, C. Spezzani, G. Penco, A. Demidovich, M. Danailov, M. Coreno, M. Zangrando, A. Simoncig, M. Manfredda, R.J. Squibb, R. Feifel, S. Bengtsson, E.R. Simpson, T. Csizmadia, M. Dumergue, S. Kühn, K. Ueda, J. Li, K.J. Schafer, F. Frassetto, L. Poletto, K.C. Prince, J. Mauritsson, C. Callegari, G. Sansone, Nat. Photon. **17**, 200 (2023)
169. D. You, K. Ueda, E.V. Gryzlova, A.N. Grum-Grzhimailo, M.M. Popova, E.I. Staroselskaya, O. Tugs, Y. Orimo, T. Sato, K.L. Ishikawa, P.A. Carpeggiani, T. Csizmadia, M. Füle, G. Sansone, P.K. Maroju, A. D'Elia, T. Mazza, M. Meyer, C. Callegari, M. Di Fraia, O. Plekan, R. Richter, L. Giannessi, E. Allaria, G. De Ninno, M. Trovò, L. Badano, B. Diviacco, G. Gaio, D. Gauthier, N. Mirian, G. Penco, P.c.v.R. Ribič, S. Spampinati, C. Spezzani, K.C. Prince, Phys. Rev. X **10**, 031070 (2020)

# Chapter 2
# Attosecond Interferometry

**Michael Krüger and Nirit Dudovich**

**Abstract** In this chapter, we introduce all-optical attosecond interferometry using high-harmonic generation (HHG). Interferometry provides an access to phase information, enabling the reconstruction of ultrafast electron dynamics with attosecond precision. We discuss two main pathways—internal and external attosecond interferometry. In internal interferometry, the manipulation of quantum paths within the HHG mechanism enables phase-resolved studies of strong-field processes, such as field-induced tunneling. In external interferometry, the phase of the light emitted during the HHG process can be determined using optical interference in the extreme-ultraviolet regime. Both pathways have significantly progressed the state of the art of ultrafast spectroscopy, as evidenced by numerous examples described in this chapter. All-optical attosecond interferometry is applicable to a wide range of systems, such as atomic and molecular gases and condensed-matter systems. Combining the two pathways has the potential to access to hitherto elusive ultrafast multi-electron and chiral phenomena.

## 2.1 Introduction

Interferometry is one of the most basic phenomena in optics and relies on the coherence of light. It provides a powerful and straightforward approach to tackle the phase retrieval problem. While experimental detection of light enables us to resolve intensity information, the underlying phase information remains hidden. In ultrafast optics, interferometry became an essential step in resolving ultrafast dynamics. In this field, interferometry is at the heart of a wide variety of methods to characterize

M. Krüger (✉)
Technion – Israel Institute of Technology, Department of Physics and Solid State Institute and Helen Diller Quantum Center, 32000 Haifa, Israel
e-mail: krueger@technion.ac.il

N. Dudovich
Weizmann Institute of Science, Department of Physics of Complex Systems, 234 Herzl Street, 76100 Rehovot, Israel
e-mail: nirit.dudovich@weizmann.ac.il

 
K. Ueda (ed.), *Ultrafast Electronic and Structural Dynamics*,
https://doi.org/10.1007/978-981-97-2914-2_2

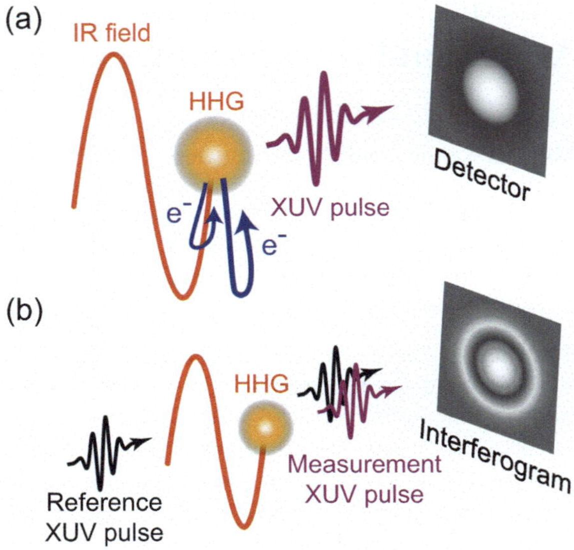


**Fig. 2.1 Schematic description of attosecond interferometry. a** Internal attosecond interferometry. A strong infrared (IR) pulse drives HHG in an atomic system. The spectral intensity of the emitted XUV light is the result of quantum-path interference. In this example, we consider the interference between two electron trajectories. **b** External attosecond interferometry. XUV light from HHG is interfered with a reference XUV beam, resulting in an optical interference at the detector

laser pulses in amplitude and phase [1]. In the attosecond time domain, the phase retrieval problem does not only apply to the optical properties of attosecond light pulses, but also to the electronic quantum dynamics on attosecond time scales. The temporal evolution of an electronic matter wave function is governed by its phase. Hence, in order to characterize ultrafast dynamics, the retrieval of the associated phase evolution is required. In the present chapter, we describe the basic principles of attosecond interferometry and its application to various phase retrieval problems in attosecond science. However, the scope of interferometric methods in attosecond science is wide. One of the most successful methods, the reconstruction of attosecond beating by interference of two-photon transitions (RABBITT, [2, 3]), is introduced in the first chapter of this book. In this chapter, we will focus on all-optical interferometry based on high-harmonic generation (HHG [4–6]). Temporal and spatial coherence is an inherent property of HHG [7, 8], allowing an interferometric retrieval of phase the information it encodes. As we will elaborate below, attosecond interferometry based on HHG has emerged as a successful method to study ultrafast quantum dynamics in atomic, molecular, and solid-state systems.

In our discussion, we briefly introduce the mechanism underlying attosecond interferometry, HHG, and the three-step model (Sect. 2.2). Then, we describe two distinct pathways of attosecond interferometry based on HHG: *internal* and *external* attosecond interferometry (see Fig. 2.1). Internal interferometry relies on quantum-path interference occurring within the three-step HHG mechanism itself (see Sect. 2.3), whereas external interferometry refers to optical interferometry of coherent extreme-ultraviolet (XUV) light beams produced by the HHG process (see Sect. 2.4).

## 2.2 High-Harmonic Generation and the Emergence of Electron Trajectories

First observed in 1987 [4, 5], HHG has become the cornerstone of attosecond science. With exceptional resourcefulness, research groups around the world were able to utilize the intrinsic attosecond light–matter interaction to produce attosecond light pulses in the XUV optical regime [2, 9–12]. While a plethora of attosecond experiments involve the downstream interaction of the attosecond pulses with matter, it has been recognized that the HHG mechanism by itself provides a spectroscopic fingerprint of the internal quantum dynamics [13, 14]. This insight has opened up HHG spectroscopy [15–18] as one of the primary approaches of attosecond time-resolved measurements.

HHG in gas-phase media is commonly described in the three-step picture [6, 19–21]. Here, under the influence of a strong laser field a bound electron is liberated by tunneling ionization, propagates in the laser field, and is driven back to the parent ion. Recollision and recombination of the electron with the ion leads to the emission of photons in the XUV regime. Importantly, the three steps—ionization, propagation, and recollision—are driven by the instantaneous electric field of the driving laser on a sub-optical cycle time scale. For instance, for a driving wavelength of 800 nm and an optical duration of 2.7 fs, we find that the time window for recollision events is a fraction of 1 fs [22]. The internal dynamics of HHG is defined mostly by the optical field, while the medium's basic properties do not play an important role. In the classical picture, the impact of the medium is limited to the ionization and recollision steps. The kinetic energy of the light-driven recolliding electron, $E_{\mathrm{rec}}$, is converted into a high-harmonic photon according to $\hbar\Omega = E_{\mathrm{rec}} + I_{\mathrm{p}}$, where $\hbar$ is the reduced Planck constant, $\Omega$ is the harmonic frequency, and $I_{\mathrm{p}}$ is the ionization potential of the target.

In a more accurate quantum-mechanical model of the HHG mechanism, developed by Lewenstein et al. [6], the emitted radiation is described by the frequency-domain dipole,

$$
\begin{aligned}
d(\Omega) \propto \int_{-\infty}^{+\infty} dt_1 \int_{-\infty}^{t_1} dt_0 \int_{-\infty}^{+\infty} dp \, \exp\left(\frac{i}{\hbar} I_{\mathrm{p}} t_0\right) d_{\mathrm{tr},x}[p - eA(t_0)]\, E(t_0) \\
\times \exp\left\{-\frac{i}{\hbar} \int_{t_0}^{t_1} dt' \frac{1}{2m} [p - eA(t')]^2\right\} \\
\times\, d^*_{\mathrm{tr},x}[p - eA(t_1)] \exp(i\Omega t_1) \exp\left(-\frac{i}{\hbar} I_{\mathrm{p}} t_1\right) + c.c. \qquad (2.1)
\end{aligned}
$$

Here we assume that the fundamental laser field $E(t)$ is linearly polarized along the $x$-axis and the dipole emission follows the fundamental's polarization direction, as does the continuum electron momentum $p$. $A(t) = -\int_{-\infty}^{t} dt' E(t)$ is the time-dependent vector potential. $e = -|e|$ and $m$ are the electron's charge and rest mass, respectively. $d_{\mathrm{tr},x}(p)$ is the dipole matrix element corresponding to transitions between the ground state and a continuum state with momentum $p$. The first line of the integrand describes the phase of the ground state wave function of the system at the time instant $t_0$, which can be interpreted as ionization time: Here a part of the wave function is leaving the vicinity of the core and acquires the momentum $p - eA(t_0)$ during the transition driven by the laser field. The second term describes the (semi-classical) evolution of the propagating part of the wave function. The third term considers the transition back to the ground state and the emission of a photon with frequency $\Omega$ at the time instant $t_1$, interpreted as the recombination or recollision time.

A natural consequence of the Lewenstein model is the emergence of semi-classical electron trajectories when applying the stationary phase approximation (SPA), also called the saddle point approximation (see, e.g., [6, 23, 24] for more details). The SPA is applicable for $\gamma \sim 1$ or smaller, where $\gamma = \sqrt{I_\mathrm{p}/2U_\mathrm{p}}$ is the Keldysh parameter and $U_\mathrm{p}$ is the ponderomotive energy in the laser field [25]. In brief, the three-fold integral can be converted into a sum over all stationary solutions $(t_{0,\mathrm{s}}, t_{1,\mathrm{s}}, p_\mathrm{s})$ to three stationary equations $\nabla_{(t_0,t_1,p)}\{S(t_{0,\mathrm{s}}, t_{1,\mathrm{s}}, p_\mathrm{s}) - \hbar\Omega t_1\} = 0$. Here $S$ denotes the quasi-classical action of the electronic wave function, given by

$$S(t_0, t_1, p) = \int_{t_0}^{t_1} dt \left\{ \frac{[p - eA(t)]^2}{2m} + I_\mathrm{p} \right\}. \tag{2.2}$$

Therefore, we retrieve three stationary equations:

$$\frac{1}{2m}[p - eA(t_0)]^2 = -I_\mathrm{p}, \tag{2.3}$$

$$\int_{t_0}^{t_1} dt' \left[p - eA(t')\right] = 0, \tag{2.4}$$

$$\frac{1}{2m}[p - eA(t_1)]^2 = \hbar\Omega - I_\mathrm{p}. \tag{2.5}$$

Again, we can see the correspondence to the three-step model: The energy conservation at the moment of ionization, the closed trajectory of the recolliding electron returning to the origin, and the energy conservation at the moment of recollision, converting the electron's kinetic energy into the photon's energy. As a consequence of the stationary phase approximation, all stationary parameters are complex solutions—a clear hallmark of the quantum nature of the interaction. Setting $I_\mathrm{p} = 0$ in the first equation, the stationary solution converges to the classical one. Calculating the final HHG dipole using the SPA involves a coherent sum over all stationary solutions $j$,

$$d(\Omega) \propto \sum_j \frac{1}{\left(t_{1,\mathrm{s}}^{(j)} - t_{0,\mathrm{s}}^{(j)}\right)^{3/2}} \exp\left[-\frac{i}{\hbar} S\left(t_{0,\mathrm{s}}^{(j)}, t_{1,\mathrm{s}}^{(j)}, p_\mathrm{s}^{(j)}\right) + i\Omega t_{1,\mathrm{s}}^{(j)}\right] \times d_{\mathrm{tr},z}^{*}\left[p_\mathrm{s}^{(j)} - eA\left(t_{1,\mathrm{s}}^{(j)}\right)\right]. \tag{2.6}$$

Here we neglect the pre-exponential factors related to the SPA over $t_0$ and $t_1$ and the ionization transition dipole moment. We only retained the pre-exponential "quantum diffusion" term, which leads to a suppression of trajectories with long travel times. The HHG dipole, calculated with the SPA, enables a physical insight into the highly complicated strong-field light–matter interaction process. The natural emergence of semi-classical electron trajectories, or equivalently "quantum orbits" or quantum paths, allows for the development of internal attosecond interferometry, as introduced in the next section.

## 2.3 Internal Attosecond Interferometry

It has been recognized in the early 2000s that HHG represents an interferometer where the amplitude and phase of the emitted XUV light encode the internal electronic dynamics of the HHG process in a fully coherent way [13]. Multiple quantum paths interfere due to their coherent nature in the HHG spectrum. Quantum-path interference in HHG can be divided into two categories: Intercycle interference denotes interference of quantum paths separated in time by half an optical cycle or more, while intracycle interference refers to interference occurring within half an optical cycle. Our discussion will focus on all-optical attosecond interferometry using HHG in atomic, molecular, and solid-state systems.

### *2.3.1 Intercycle Interference*

#### 2.3.1.1 Atomic Systems

When the HHG mechanism is driven by a multiple cycle field, an attosecond pulse train (APT) is produced, leading to the appearance of discrete harmonic orders. In the limit of a long fundamental pulse, which includes a large number of optical cycles, we can neglect the role of pulse's envelope and approximate it to be a continuous wave light field with a single frequency $\omega$. In this case, we consider the interference between two attosecond pulses generated at two consecutive half-cycles, defining a temporal interferometer. As long as the process is driven by a single color field, there is a destructive interference at even-harmonic frequencies and a constructive interference at odd harmonic frequencies. Therefore, the HHG spectrum is composed of odd-only harmonics. The manipulation of the relative phase between the two arms, represented by consecutive half-cycles, can be achieved by adding a weak second-harmonic (SH) field, polarized parallel to the strong fundamental field. The SH field perturbs the strong-field interaction, adding a small complex phase shift $\sigma$, accumulated by the electron as it interacts with the two-color field. The imaginary component $\mathrm{Im}(\sigma)$ represents the perturbation of the instantaneous tunneling probability, while the real component, $\mathrm{Re}(\sigma)$, represents the perturbation of the phase. The complex phase shift is inverted between the two half-cycles, leading to the appearance of both even and odd harmonics. For a harmonic order $N$, the modulation of the harmonic signal by the two-color field can be expressed as [26]

$$I_N(\varepsilon, \varphi) = I_N^{(0)} \begin{cases} |\cos(\varepsilon\sigma(N, \varphi))|^2, & N \text{ odd} \\ |\sin(\varepsilon\sigma(N, \varphi))|^2, & N \text{ even} \end{cases} \tag{2.7}$$

Here $\varepsilon \ll 1$ is the ratio of the SH field and the fundamental field, and $\varphi$ is the relative two-color phase—the control parameter in this scheme. The two-color interferometer was first observed and treated in the classical regime [27]. In this study, the authors assume that the Keldysh parameter is zero, hence the perturbation is purely real as well. Neglecting the role of the tunneling mechanism $\mathrm{Re}(\sigma)$ reflects the mapping between the lengths of the trajectories and the HHG spectrum. Figure 2.2a, b schematically describes the temporal interferometer. Two electron trajectories induced along the positive and negative half-cycles of the laser field define the two arms of the interferometer. The SH field unbalances the interferometer, breaking the symmetry and producing even harmonics of the fundamental field. The modulation of the even-harmonic signal as a function of the two-color delay (see Fig. 2.2c) is dictated by two fundamental properties—the ionization and recollision times. The modification of this phase with the harmonic number probes the recollision dynamics, reflecting the increase of the trajectory length and its mapping to the emitted harmonic.

Once the tunneling mechanism is taken into account, the stationary solutions, and therefore the perturbation phase $\sigma$ itself, become complex. Here $\mathrm{Im}(\sigma)$ describes the two-color perturbation of the tunneling probability. Measuring the intensity of even and odd harmonics as a function of two-color delay and applying Eq. 2.7 enable the separation between the real and imaginary components (see Fig. 2.2d). Such separation allows the reconstruction of the evolution of the instantaneous tunneling probability within the optical cycle [28] (see Fig. 2.2e).

Intercycle interference does not take place if only a single isolated attosecond pulse is produced, for instance, using an extremely short laser pulse or applying polarization gating [29]. The phase shift introduced by a perturbation field can be made visible, however, if the incident perturbation field is not collinear with respect to the fundamental laser beam that drives HHG, resulting in a small angular shift of the harmonics as a function of the two-color delay. The underlying mechanism is the deformation of the electron trajectories by the perturbation field. A recent experiment using this scheme shows that the recombination dipole moment phase can be extracted from such a measurement, elucidating the Cooper minimum in argon and the corresponding time delays [30].

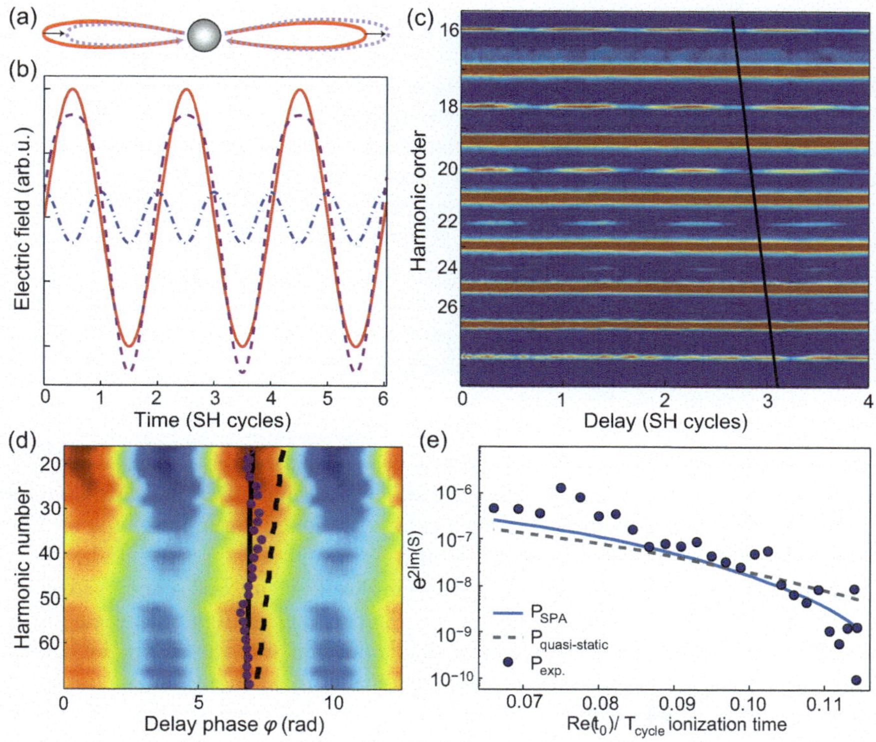


**Fig. 2.2 Two-color attosecond interferometry. a** Schematic description of symmetry breaking induced by a two-color field. The arms of the interferometer are generated by adjacent half-cycles of the light field. Two symmetric trajectories are generated by a single color field (solid line). Adding a second-harmonic field (dashed line) breaks the symmetry and results in the appearance of even harmonics. **b** Two-color field consisting of the fundamental field (solid line) and second-harmonic field (dashed-dotted line). For a relative phase of $\pi/2$, the fields add in phase to enhance the negative half-cycle, and out of phase to suppress the positive half-cycle (dashed line). Varying the phase changes the balance in the interferometer. **c** HHG spectrum of argon as a function of the two-color delay. Even harmonics appear between the odd harmonics. The two-color delay that optimizes the harmonic yield changes with the harmonic number. **d** $\mathrm{Im}(\sigma)$ as a function of harmonic number and two-color delay phase, extracted from a two-color experiment in helium. Red (blue) represents an increase (decrease) of the harmonic yield. The black line represents the SPA prediction while the dashed line represents the quasi-static barrier approximation. **e** Reconstructed sub-cycle time dependence of the instantaneous ionization probability (blue dots), quasi-static barrier approximation (gray dashed curve), and the SPA (blue curve). Figure parts **a–c** adapted with permission from [27], **d**, **e** from [28]

#### 2.3.1.2 Molecular Systems

Internal attosecond interferometry in atomic systems reveals the underlying sub-cycle recollision dynamics. Advancing from atoms to molecular systems raises fundamental questions—what is the role of the structure of the molecular orbitals within the interferometer? Can we decouple structural from temporal information?

In most molecular systems the tunneling process involves several orbitals, thus launching a "hole" wavepacket [17, 31, 32]. The dynamics of the multi-electron system is dictated by the electron rearrangement that evolves on an attosecond time scale. The different ionization channels can be considered as an internal interferometer, where tunneling serves as a beam splitter while recollision serves as a beam combiner. Over the past decade an extensive theoretical and experimental effort has been invested in studying these phenomena (see [17] and references therein), therefore their detailed review is beyond the scope of this chapter.

In the following, we focus on the case of a single ionization channel where the hole dynamics can be neglected. A natural step to take integrates the time-domain interferometer, induced by the two-color field in parallel configuration, with molecular alignment. Such an integration probes the coupling between the spatial and the dynamical degrees of freedom [33, 34]. Applying well-established schemes to small molecular systems such as $N_2$ molecules, we can control the alignment angle of the molecule with respect to the fundamental field's polarization. Resolving the phases of the oscillations of the harmonic signal for each alignment angle shows a universal behavior—the two-color response of the harmonics is independent of the alignment angle. While rotating the alignment angle leads to significant structural effects in HHG, the dynamical properties as probed by the two-color measurement remain invariant. These observations show the universality of recollision picture which, in the absence of hole dynamics, can be decoupled from the molecular system.

The structure of molecular systems can be probed by manipulating the *spatial* properties of the interfering quantum paths. Such manipulation has been applied on either the molecular system itself or the electron trajectories that define the interferometer arms. The first approach has been demonstrated with oriented polar molecules. When the HHG mechanism is driven by a single color field, electron trajectories induced during two consecutive half-cycle probe the polar molecule from two opposite orientations, forming a spatial interferometer [35, 36]. Since the symmetry between the two half-cycles is broken by the molecular system, even harmonics appear, reflecting the differences in the complex dipole moment associated with the recollision events. Advanced studies applied this approach to study orientation mechanisms [37], laser-induced electronic structure [38], and attosecond-scale charge transfer [39].

The above scheme is based on the following fundamental requirement—the symmetry of the interaction is broken by the molecular system itself. In addition to the molecular alignment degree of freedom, a more generalized scheme can be applied to molecular HHG by manipulating the spatial properties of the electron trajectories. Adding an SH field, now orthogonally polarized to the fundamental field, enables us to shape the 2D properties of the electron trajectories [41, 42], manipulating the

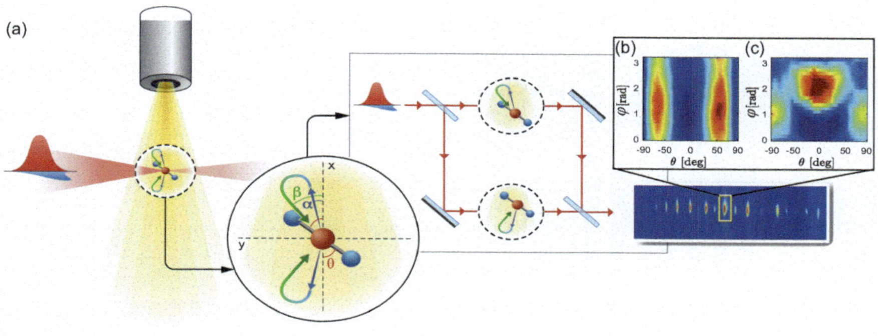


**Fig. 2.3 Decoupling structural and temporal information in a molecular interferometer. a** Strong-field interaction between a two-color, orthogonally polarized field and a molecular system can be viewed as a two-arm interferometer induced on a single molecule level. The interferometer is defined by two electron trajectories induced during the two half-cycles of the laser field. The interferometer arms are defined by the two steps of the interactions, tunneling and recombination, controlled by the ionization angle $\alpha$ and the recollision angle $\beta$, respectively. **b**, **c** The interference pattern is mapped into the polarization state of the HHG spectrum. Experimentally measured polarization components of harmonic 16, polarized parallel **b** and perpendicular **c** to the fundamental field, as a function of the interferometer knobs: the alignment angle $\theta$ and the two-color delay $\varphi$. Figure adapted with permission from [40]

angle of ionization and the angle of recollision. Breaking the symmetry between the positive and negative half-cycles of the laser field defines a spatial interferometer (see Fig. 2.3), where its two arms are associated with two recollision events induced by the positive and negative half-cycles [40]. These trajectories are inverted along the fundamental field polarization, probing the molecular system at two opposite angles of return $\pm\beta$. The interference is encoded in the polarization state of the HHG spectrum, probing the complex ionization and recollision dipole moments.

#### 2.3.1.3 Solids

High-harmonic generation inside a bulk dielectric solid was first demonstrated by Ghimire et al. in 2011 [43], opening up a new field of research in attosecond science (see reviews [44, 45]). As a coherent process, solid-state HHG maps the internal strong-field-driven electron dynamics into high harmonics. A long-standing question that still puzzles the scientific community is the nature of the underlying HHG mechanism. Two basic processes can be distinguished, intraband acceleration of electrons and interband electron–hole recollisions [46]. Both have analogies in atomic HHG: The first process is similar to Brunel harmonics [47], while the second process is reminiscent of the recollision-based three-step HHG mechanism.

Vampa et al. employed the two-color interferometry scheme applied to the gas phase in order to demonstrate the recollision picture in solid-state HHG (ZnO crystal, 3.8$\mu$m laser pulses). The authors measured the modulation of even and odd harmonics

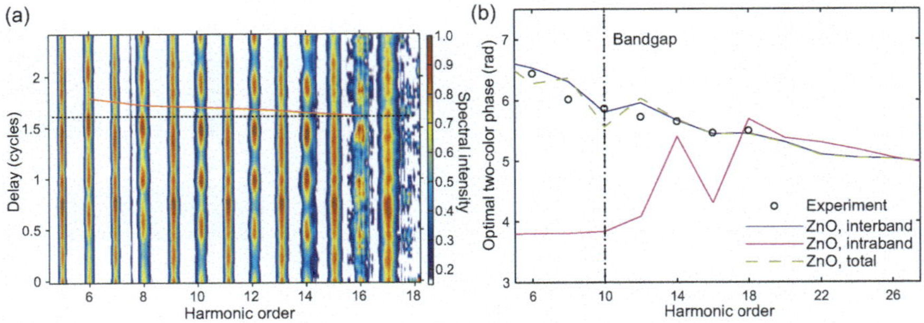


**Fig. 2.4 Two-color interferometry in solid-state HHG. a** Measured high-harmonic spectra in ZnO versus the two-color delay. The phase of the even-harmonic modulation depends on the harmonic order, as evidenced by the red solid line relative to the dashed black reference line. The red solid line links the minima in the modulation of the even-harmonic intensity. Each harmonic order is normalized separately. The delay of the second harmonic is defined in cycles of the second harmonic. **b** The phase of the modulation of the even harmonics is extracted from the experiment (black circles) and compared to the simulated intraband (purple line) and interband (blue line) phase. The simulated phase for their combined emission (intraband plus interband, yellow dashed line) agrees with the interband emission. Figure adapted with permission from [48]

in the two-color field and determined the optimal two-color phase that maximizes the yield of each even-harmonic (see Fig. 2.4). The slope of the optimal phases (Fig. 2.4b) agrees well with an interband model calculation, providing a direct probe of the internal dynamics.

Two-color interferometry has been used to observe dynamical band structure effects in MgO where the strong field leads to an effective closure of higher lying bandgaps between conduction bands, which manifests itself in the behavior of the two-color oscillation phase as a function of laser intensity and crystal orientation [49].

## 2.3.2 *Intracycle Interference*

When we zoom into the optical cycle we realize that quantum-path interference can be induced within half an optical cycle—so-called intracycle interference. In the following, we will discuss various experimental schemes, based on intracycle interference.

### 2.3.2.1 Interference of Long and Short Trajectories

In HHG, two distinct trajectory classes exist, known as the short and long trajectories. While for short trajectories the return energy increases with the trajectory length, leading to a positive chirp, for long trajectories the energy reduces with the return time, leading to a negative chirp. The phase associated to each quantum path $j$ of

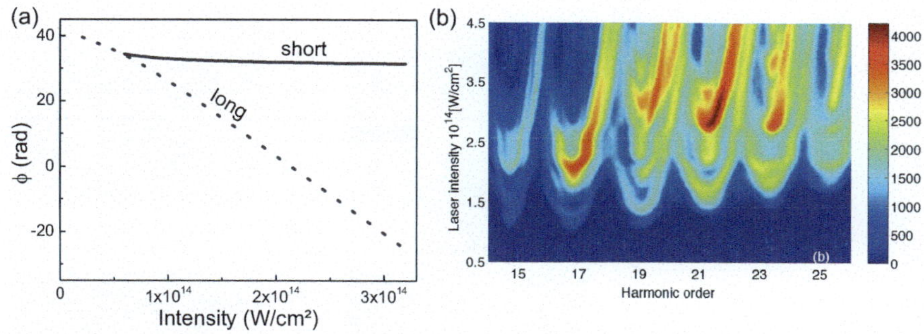


**Fig. 2.5** **Intracycle interference of short and long trajectories. a** Spectral phase $\phi$ as a function of the laser intensity, calculated from the SPA for Ar, harmonic 19. The long trajectory phase (dashed) and short trajectory phase (solid) exhibit a different intensity scaling, causing a rapid change of their phase difference. **b** Measured laser intensity dependence of the harmonic spectrum of argon. Using off-axis spatial filtering, the rapidly varying intracycle interference pattern can be detected on the spectrograph [51]. Figure part **a** adapted with permission from [50], part **b** from [51]

harmonic order $N$ is described as

$$\phi_N^{(j)} = \arg\left[d^{(j)}(N\omega)\right]. \tag{2.8}$$

This phase is dictated mostly by the electron's excursion time [23, 50]. The two families of trajectories interfere in the HHG spectrum, with a relative phase $\phi_N^{(l)} - \phi_N^{(s)}$ (see Fig. 2.5a).

Their spatio-temporal interference is mapped into the spatio-spectral interference in the far field, measured on a spatially resolving spectrograph [51]. Controlling the fundamental field's intensity manipulates the trajectory length and therefore their relative phase. Such control, combined with careful spatio-spectral detection, enabled the first observation of such quantum-path interference on the attosecond time scale (see Fig. 2.5b). The first observation was followed by more advanced studies, demonstrating that these quantum-path interferences are indeed an intrinsic phenomenon of HHG and can be observed for various generation media [52]. At later stages, this scheme was applied to probe molecular dynamics (see, e.g., [53]). A non-collinear beam configuration for HHG enables the isolation of the quantum-path interference of short and long trajectories from other effects [54]. In a recent work, the quantum-path interference was studied in aligned $N_2$ and $CO_2$ molecules, revealing signatures of a shape resonance and multichannel interference, respectively [55].

#### 2.3.2.2 Spectral Caustics

A unique class of intracycle interferometers is observed at points where the SPA fails to describe the recollision dynamics. When the semi-classical action becomes stationary to higher orders, the model exhibits singularities, leading to the appearance

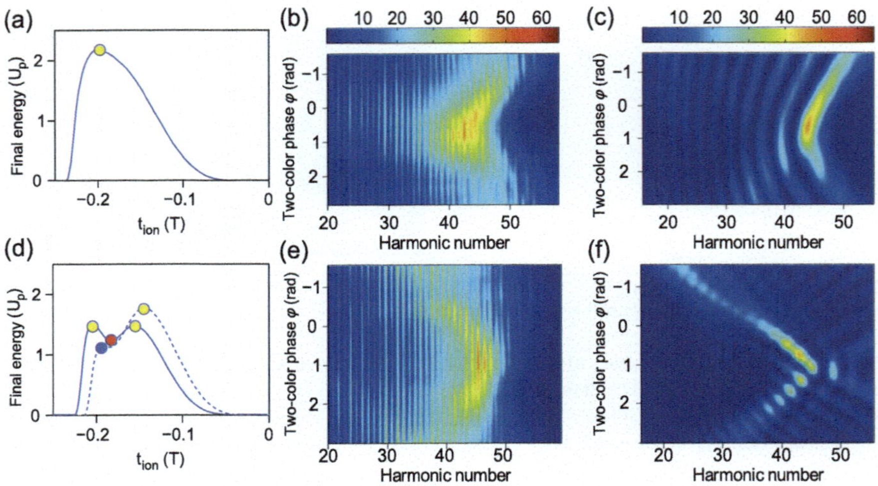


**Fig. 2.6 Spectral caustics due to intracycle interference. a, d** Final kinetic energy of the recolliding electron as function of ionization time $t_{ion}$ for specific two-color field ratios $\epsilon$ and phases $\varphi$. **b, e** Measured HHG spectra as a function of $\varphi$ for specific values of $\epsilon$. **c, f** Theoretical "swallowtail" diffraction patterns. **a–c** $\epsilon \sim 0.3$ leads to the swallowtail standard diffraction pattern because there is one singular point, a local maximum. **d–f** $\epsilon \sim 0.75$ supports two different characteristic scenarios (see **d**), controlled by $\varphi$. Remarkable qualitative agreement is observed: the curvature of enhancement changes its direction, facing the lower end of the spectrum in **b** and the higher end in **e**. The color code is consistent for comparison. Figure adapted with permission from [56]

of spectral caustics. These caustics describe the interference of electron trajectories that coalesce together, generating a bright narrow spectral feature. Catastrophe theory enables a universal description of these singularities, providing an analytical solution near each singular point, while maintaining the simple classical description of the mechanism. The most common singularity is observed at the HHG cut-off frequency, describing the coalescence between the short and long trajectory branches. Higher order singularities can be engineered using two-color [56] or three-color [57] configurations. The additional fields, which are not perturbative, lead to the emergence of more electron trajectories and singularities within each half-cycle (see Fig. 2.6).

Recently, spectral caustics have been observed in solid-state HHG in MgO [58]. HHG in MgO at infrared wavelengths is governed by the interband emission process. Here electron–hole recombination leads to HHG, in close analogy to gas-phase experiments. The emission of XUV photons is dictated by the energy difference between conduction and valence bands, hence information about the band structure in the Brillouin zone is mapped into the HHG spectrum. The strong laser field creates an electron–hole wavepacket and drives it along the polarization direction. At extrema, for example, around the lattice momentum $k = 0$, or at other places where the electron–hole velocity approaches zero, many trajectories coalesce together, leading to the emergence of singularities and spectral caustics.

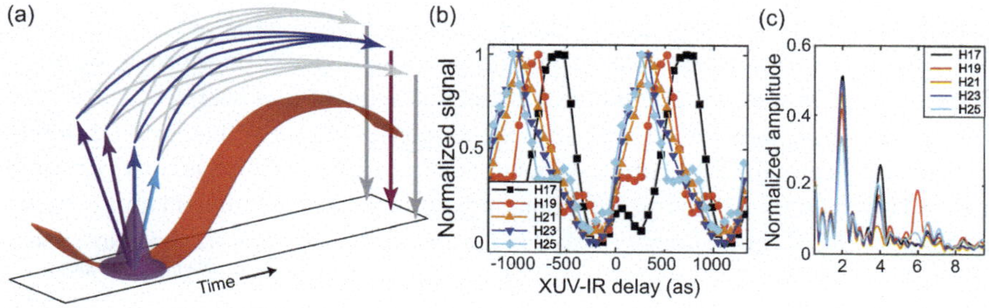


**Fig. 2.7** **XUV-initiated HHG.** **a** Illustration of XUV-initiated HHG. Absorption of XUV photons of an attosecond pulse (violet) gives rise to a number of ionization channels (colorful arrows). An IR field (red) leads to recollision trajectories (blue and gray curves), resulting in the emission of high harmonics (purple and gray arrows). Depending on the relative phase of the ionization channels, destructive (gray curves) or constructive interference (blue curves) of the trajectories is observed. **b** Oscillations of the HHG signal with the XUV-IR delay due to quantum-path interference. **c** Fourier spectrum of the signal oscillations. The observation of Fourier peaks at $4\omega$ and $6\omega$ is consistent with four ionization channels. Figure adapted with permission from [64]

#### 2.3.2.3 Intracycle Interference in XUV-Initiated HHG

XUV-initiated HHG presents an advanced scheme in HHG spectroscopy, where the tunneling ionization step is replaced with photoionization by an XUV field [59–62]. This scheme offers potential applications in ultrafast spectroscopy of hole-decay phenomena in molecules [63]. When XUV-initiated HHG is driven with discrete harmonics, each incoming harmonic gives rise to an ionization channel (see Fig. 2.7a) for an illustration). The IR field, whose intensity is not strong enough to produce tunneling ionization, acts as the driving field for recolliding electron trajectories. The resulting harmonics are subject to interference of multiple quantum paths, corresponding to each ionization channel. Their relative phases are controlled by the delay between the XUV and IR fields. Varying this delay leads to oscillations in the interference signal (see Fig. 2.7b). A Fourier transformation of the signal reveals frequency components at $2\omega$, $4\omega$ and $6\omega$, which is consistent with the number of ionization channels in the experiment (see Fig. 2.7c). The authors use the phase information encoded in the oscillations in order to reconstruct the temporal evolution of the birth of the photoelectron wavepacket in the presence of a strong IR field.

## 2.4 External Attosecond Interferometry

In this section, we introduce the conceptionally simple and versatile experimental approach of external attosecond interferometry. The idea is straightforward—measuring amplitude and phase information of the harmonic light field through its interference with another light field. The interference signal of two monochromatic waves of frequency $\Omega$ can then be described by $I(\Delta t) = E_1^2 + E_2^2 + 2E_1E_2$

$\cos[\Omega\Delta t - (\phi_2 - \phi_1)]$, where $E_{1,2}$ and $\phi_{1,2}$ are the waves' amplitudes and phases, respectively, and $\Delta t$ is the interferometric delay. Using the interferometric trace $I(\Delta t)$, we can determine the relative spectral phase of the two beams.

The main advantage of external attosecond interferometry is that it does not rely on the modeling of the HHG process, but is an *ex situ* approach that is fully optical in nature. However, its realization is challenging due to the lack of efficient transmission optics in the XUV spectral region. Mainly the absence of efficient transmissive beam splitters impedes the implementation of Mach–Zehnder-type optical interferometers. Two main pathways have been pursued: The first pathway relies on splitting an XUV beam using reflective optics, for example, split mirrors. The second pathway exploits the coherent nature of HHG. Instead of splitting the XUV beam, two or more HHG sources are driven coherently by the same laser beam. The resulting XUV light is temporally and spatially overlapped on the detector, creating an interference pattern that encodes the desired phase information. This approach has been implemented in various embodiments that we will discuss further below.

### 2.4.1 Interferometric Auto-correlation of Attosecond Pulses

With the lack of efficient transmissive beam splitters, reflective beam splitters present an important alternative. This approach has been pursued first in the so-called "atto-correlator" [65–67], where attosecond pulse trains (APTs) from an HHG source were reflected off a harmonic separator split mirror consisting of a pair of flat Si mirror segments (see Fig. 2.8a). One of the segments can be moved with high precision. Imaging the beam with a spectrograph yields a vertical interference fringe pattern due to the diffraction at edges of the harmonic separator. Since two copies of the same field are interfered with each other, an interferometric auto-correlation is measured.

The diffraction pattern observed for each individual harmonic oscillates according to the temporal delay with its own fundamental frequency. When switching the spectrograph's imaging mode to the zeroth-order diffraction peak, the harmonics are now superimposed onto each other. Scanning the delay leads to the first-order interferometric auto-correlation trace (Fig. 2.8b), enabling Fourier-transform spectroscopy. The Fourier transform of the collective oscillation signal corresponds to the spectral intensity of the APT. The symmetry properties of HHG are captured by the $\pi$-phase flip observed for each subsequent "pulse". However, the atto-correlator does not reveal any information about the CEP of the pulses or any dispersion effects including the actual attosecond pulse duration. While a determination of the CEP remains elusive, the latter can be measured with second-order (nonlinear) auto-correlation, for example, using electronic two-photon processes in atomic and molecular gas media in the XUV [68–72].

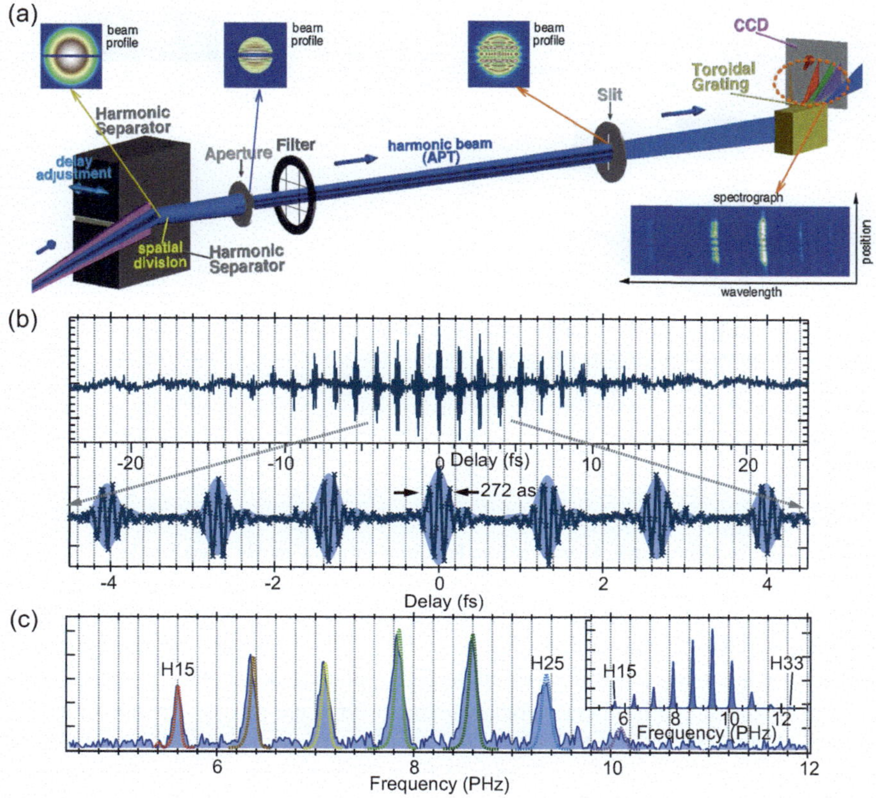


**Fig. 2.8** **Atto-correlator.** **a** Sketch of the experimental setup of the atto-correlator. An APT is split into two components by a harmonic separator mirror pair. After several filtering stages, spatial interference fringes are detected in an XUV grating spectrograph. Moving one of the mirrors changes the optical delay in the interferometer. **b** Resulting interferometric auto-correlation trace. Measuring the zeroth-order beam in the spectrograph in a delay scan experiment yields the auto-correlation trace of the APT. **c** The Fourier transform of the auto-correlation trace (blue) yields the spectrum of the APT, in good agreement with the dispersively measured APT spectrum (dashed curves and inset). Figure adapted with permission from [65]

## 2.4.2 *Two-Source Attosecond Interferometry*

HHG is a coherent process where the amplitude and phase of the driving laser pulse are imprinted on the generated XUV light. As a consequence, when HHG is driven at two spatially separated sources, their interference can be observed on a detector. Their relative phase depends on the properties of the two driving pulses, such as polarization, intensity, or wavelength, and the generation media, such as the gas species and its initial quantum state. Two-source interferometry hence enables direct access to phase information of the HHG process and the interacting medium.

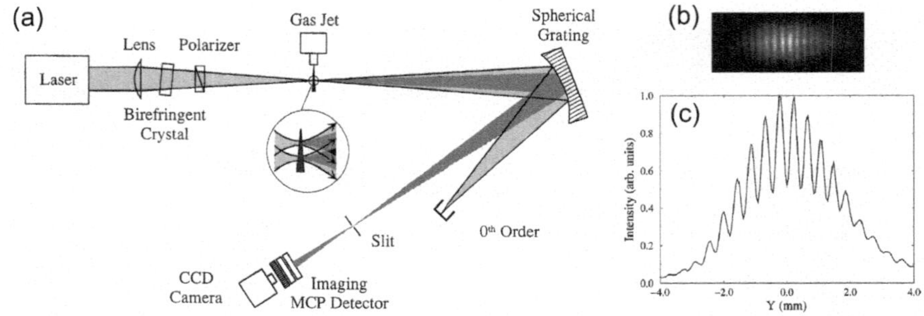


**Fig. 2.9 First realization of two-source XUV interferometry.** **a** Sketch of the experimental setup. The driving laser beam is split into two equally intense components using a birefringent crystal and polarizer combination. Two laser foci separated by about 0.15 mm lead to HHG at two sources. The resulting XUV beam is spectrally dispersed and imaged on a spectrograph. **b** The resulting far-field fringe pattern. **c** The fringe pattern after vertical integration. Figure adapted with permission from [7]

#### 2.4.2.1 High-Harmonic Generation Interferometry with Two Foci

In two-source interferometry, the beam splitter is usually applied to the driving pulse that leads to HHG, and can therefore be conveniently performed in the optical or infrared domain using transmissive optics. The implementation of the interferometer is straightforward and has been widely pursued ever since its first demonstration by Zerne et al. in 1997 [7]. Figure 2.9a shows the experimental setup. A Nd:YAG laser beam is split into two slightly displaced beam components using a combination of a $BaB_2O_4$ crystal and a polarizer, inducing two separate HHG sources. In the focal plane, two twin foci with a distance of about $150\mu$m are created. Imaging the resulting beam in the far field using a spectrograph yields vertical interference fringes, as a result of the interference of the two fully phase-locked sources. The experiment provides the first demonstration of the transfer of the coherent properties of the fundamental field into the HHG process. The authors also introduced a strong imbalance in the intensity of the two foci, leading to a decrease of the visibility of the interference pattern. This effect is due to the fact that the interferometer is sensitive to the intensity scaling of the spectral amplitude of the HHG process. Also a phase shift is expected since the phase scales with intensity (cf. Eq. 2.8), but experimental limitations did not permit its detection.

Further experimental studies followed, establishing two-source interferometry as a viable spectroscopic approach. Bellini et al. determined the coherence time of XUV light generated in HHG and found a striking difference between the short and long trajectory components [8]. Two-source interferometry in the two-foci configuration has also been used to study the physical properties of objects in the beam path, for instance, a plasma plume, by measuring the induced XUV phase shifts [73–75]. More advanced schemes demonstrated two-source Fourier-transform spectroscopy [76–79]. In studies of molecular dynamics, two-source interferometry has resolved the

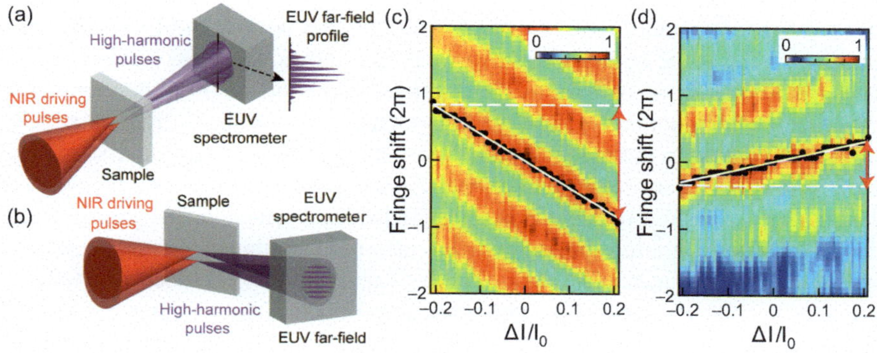


**Fig. 2.10 Two-source interferometry of solid-state HHG.** **a** Sketch of the transmission geometry HHG experiment. XUV light is generated by two parallel light beams inside a MgO crystal, leading to interference fringes in the far field. **b** Sketch of the reflection geometry HHG experiment. Here the XUV light is created in reflection. **c** Intensity-induced fringe shift of H15 in MgO[110] in transmission geometry. $\Delta I$ is the difference between the driving intensities and $I_0 \sim 22\,\mathrm{TW\,cm^{-2}}$. **d** The same for the reflection geometry. Adapted with permission from [87]

dependence of the XUV phase on the molecular alignment angle ([31], see also [80]). This measurement revealed hole dynamics in the carbon dioxide molecule. The state of the art in two-foci interferometry in terms of stability is using a spatial light modulator to shape a single beam into generating two foci with a digitally controlled time delay, resulting in a precision in the zeptosecond time domain [81–83].

Two-foci HHG has enabled precision studies of HHG dynamics in atoms [84], molecules [85, 86], and recently also in solids [87]. In the latter work, the authors show that the intensity scaling of solid-state HHG depends on the geometry of the experiment (see Fig. 2.10). In a two-foci configuration, the first source driven by an IR beam serves as a reference, whereas the intensity in the second source is varied. The change in intensity leads to a phase shift and a corresponding fringe shift in the far field. While the transmission geometry shows an anomalous phase slope due to propagation effects (Fig. 2.10c), the intensity scaling in the reflection geometry follows the theoretical prediction (Fig. 2.10d).

#### 2.4.2.2 Transient-Grating Spectroscopy

Transient-grating spectroscopy can be viewed as an extended version of two-source interferometry in a two-foci configuration. In this scheme, two crossed beams interfere at the focal plane, initiating rotational, electronic, or dissociation dynamics in a molecular gas. These beams create horizontal planes of excited molecules that alternate with planes of unexcited molecules, forming the transient grating. A third delayed beam, which serves as the probe, leads to HHG. In this scheme, harmonics are produced from a modulated medium, where the phase and amplitude of the excitation are encoded in the phases and amplitudes of the harmonics and projected into a diffraction pattern in the far field. An important advantage of the transient-grating scheme is the ability to resolve weak excitations with zero background in the far-field diffraction pattern, as the diffracted signal arises from the modulated excitation only. Importantly, as in the two-source scenario, the diffraction pattern provides an access to the spectral phase information. This scheme was first applied to perform HHG spectroscopy of rotational dynamics [88]. Advanced studies applied this approach to resolve molecular dissociation [89] and conical intersection dynamics [90]. Transient-grating spectroscopy has also been combined with an additional two-source interferometer serving as a probe, enabling vectorial spectroscopy [91]. Recently, a transient-grating-type interferometer formed by two circularly polarized IR beams with opposite handedness combined with XUV polarimetry has enabled the extraction of intensity-dependent dipole phases of noble gas atoms [92].

#### 2.4.2.3 Collinear XUV Interferometry

In collinear interferometry, the interfering beams share a common path before reaching the detector. The simplest collinear XUV interferometer that avoids the two-foci configuration is HHG performed in a gas mixture. Every atomic or molecular system in the mixture coherently contributes to the harmonic signal, resulting in the interference between the different emitters on the detector. Kanai et al. performed the first gas mixture experiment with helium and neon and measured HHG spectra for the individual gases and their mixture [93]. The authors demonstrated areas of constructive and destructive interference in the HHG spectrum of the mixture. By measuring the interference phase, the excursion time of the three-step electron trajectories, $t_1 - t_0$, could be determined. The gas mixture scheme has also been applied to molecular systems, resolving the phase associated with the molecular recombination dipole momentum [94, 95].

The spectral interference in gas mixtures can be used for characterizing attosecond pulses [96]. Measuring HHG from two individual gases and their mixture enables reconstruction of the spectral amplitude and phase of the attosecond pulses from these gases with the help of double-blind holography. The results from this simple experimental scheme are in reasonable agreement with a well-established, but more complex approach, frequency-resolved optical gating for complete reconstruction of attosecond bursts (FROG-CRAB [97]).

The drawback of interferometry with gas mixtures is the fact that the interferometer arms and their delay cannot be controlled independently. A straightforward way to enable such a control is the use of two time-delayed copies of an IR pulse in order to drive HHG in a single gas medium. For example, this scheme has been applied to HHG in order to corroborate the attosecond temporal confinement of the generated XUV light [98]. Recently, such a two-pulse interferometer has been applied to study and control quantum-mechanical entanglement of a molecular ion and a photoelectron [99, 100].

An alternative approach applies a single laser pulse and two consecutive jets with a variable distance. In this scheme, two spatially separated gas sources are lined up

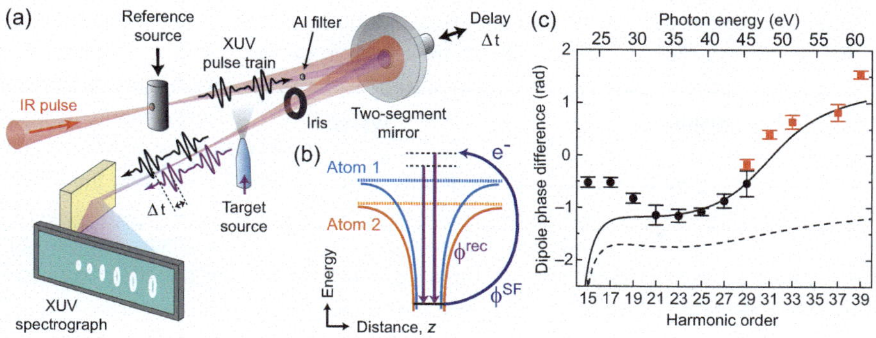


**Fig. 2.11 Collinear two-source interferometry. a** Experimental setup. An IR pulse is focused into the first gas source, generating the reference XUV pulse train (black). An aluminum foil transmits the XUV pulse train while the inner part of the IR pulse is blocked. Both beams are refocused into the second gas source by a two-segment mirror in order to generate the target XUV pulse train (purple) by the remaining IR beam. The delay $\Delta t$ between both XUV pulse trains is controlled by moving the inner segment of the mirror. A spectrograph resolves their interference. **b** Schematic depiction of the HHG process for two different target atoms. A laser field ionizes an atom and accelerates the photoelectron back to its parent ion (blue arrow). After recollision and recombination, XUV light is emitted (purple arrows). The phase of the emitted light is given by the sum of the electron wavepacket phase $\phi_{\mathrm{SF}}$ and the phase of the recombination dipole $\phi_{\mathrm{rec}}$. The differential interferometric measurement yields the difference in $\phi_{\mathrm{rec}}$ between the two species. **c** Experimental photo-recombination dipole phase difference of neon and argon atoms as function of photon energy and harmonic order. The black- and red-filled squares show the experimental result from two measurement series with different reference gas. The solid and dashed lines represent numerically calculated dipole phase differences without and with averaging over the scattering angle, respectively. Figure adapted with permission from [107]

along the propagation axis of the IR driver beam. Originally devised for achieving quasi-phase matching and strong enhancement of HHG [101–103], two gas jets are placed in consecutive order within a focused Gaussian beam. Controlling the relative phase is achieved by changing the relative distance of the jets. The Gouy phase, a geometrical phase shift in a Gaussian beam, affects the local phase of the IR pulse as $\phi_{\mathrm{Gouy}}(z) = -\tan^{-1}(z/z_{\mathrm{R}})$, where $z$ is the longitudinal coordinate and $z_{\mathrm{R}}$ is the Rayleigh length. If two jets are located at coordinates $z_1$ and $z_2$, this causes a phase difference $\Delta\phi_{\mathrm{Gouy}}$ between the IR drivers. This phase difference is imprinted on the high harmonics where controlling the distance of the jets controls the arms of the interferometer. The resulting phase shift of each harmonic is given by $N \times \Delta\phi_{\mathrm{Gouy}}$, where $N$ is the harmonic number. This scheme has been demonstrated in several works [104–106]. Its disadvantage lies in the fact that scanning the distance between the two jets slightly affects the local IR intensity, leading to modifications of the HHG spectra on top of the optical interference of the HHG sources.

In an alternative approach, these limitations have been overcome, demonstrating external attosecond interferometry with two HHG sources separated by about 1.5 m [107]. The two sources are driven by the same IR driver pulse in a collinear scheme (see Fig. 2.11a). The time delay between the IR pulse and the XUV light from the first source is controlled with the help of a two-segment mirror. The IR pulse drives HHG in the second source, resulting in XUV interference on the detector. Importantly, this collinear scheme enables full manipulation of the properties of the two HHG processes, for instance, different gas species, different driver pulse intensities, or different light polarizations [108]. In Ref. [107], the authors directly compared the recombination dipole phase of different atomic species using the fact that the phase of the HHG light is essentially given by the sum of the electron wavepacket phase $\phi_{SF}$ and the phase of the recombination dipole $\phi_{rec}$ (cf. Eq. 2.6, see also [109]). Since $\phi_{SF}$ is given by the HHG process and can be determined from the SPA, $\phi_{rec}$ can be accessed directly (see Fig. 2.11b). In contrast, photo-electron-based techniques like RABBITT or streaking gain access only to the group delay, which is the first derivative of the spectral phase. Figure 2.11c shows the result of an interferometric study of neon and argon HHG. The experiment directly resolves the phase jump around the Cooper minimum in argon, demonstrating the spectroscopic potential of external attosecond interferometry.

#### 2.4.2.4 Integrating Internal and External Attosecond Interferometry

The ability to directly measure spectral phases in the XUV regime using external attosecond interferometry can be integrated with internal interferometry where a secondary field perturbs the HHG process by imprinting a small phase shift on electron trajectories. In a first realization, Kneller et al. demonstrate this integration by combining collinear XUV interferometry with internal two-color interferometry induced by a perturbative SH field (see Fig. 2.12a). The odd-numbered harmonics of the reference XUV source interfere with those of the target XUV source, where the latter is perturbed by the two-color delay. With this, the interferogram has now three dimensions—the photon energy (harmonic order from the HHG process), the XUV-XUV delay (external interferometer), and the IR-SH delay (internal interferometer), allowing for the retrieval of rich phase information. The observed interferogram enabled the authors to isolate the evolution of the tunneling barrier during the optical cycle. Due to this, tunneling electron wavepackets are gaining energy in the classically forbidden region (nonadiabatic tunneling) and are exiting the barrier closer to the origin (shift $\Delta x_{exit}$), acquiring an additional phase shift $\Delta\Phi_{tunnel}$ with respect to the static case (see Fig. 2.12b and **c**). The external XUV-XUV interferometer resolves this phase and follows its sub-cycle evolution.

## 2.5 Summary and Future Perspectives

In summary, this chapter reviews a broad range of approaches, where the complex phase information encoded in sub-cycle light–matter interactions is resolved via attosecond interferometry. These approaches can be classified into two main branches—internal and external interferometry. While internal interferometry is

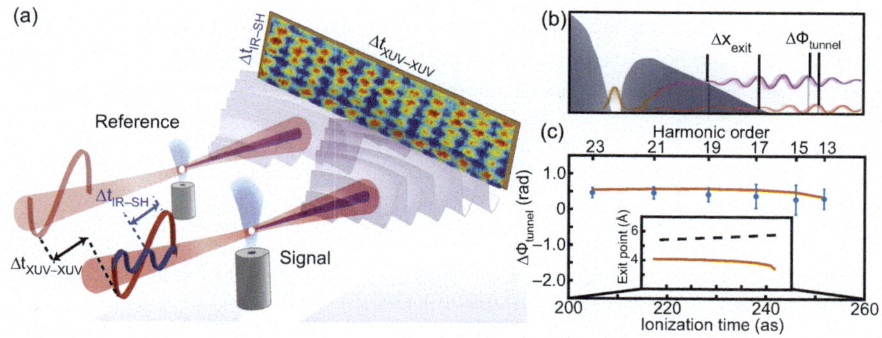


**Fig. 2.12 Integration of external and internal attosecond interferometry.** **a** Schematic description of the experiment. **b** Illustration of the effects of nonadiabatic tunneling of an electron wavepacket (violet) compared to a wavepacket tunneling through a static barrier (orange). **c** Phase shift $\Delta\Phi_{\text{tunnel}}$ as a function of ionization time (blue: reconstructed from the experiment, orange: quantum-path model). Inset: The full quantum-path model tunnel exit points (orange) deviates strongly from the static case (black). Figure adapted with permission from [110]

based on the engineering of quantum-path interference, external interferometry is based on the accurate control of XUV optical path interference. Both branches reveal valuable information regarding the strong-field light–matter interaction as well as the atomic, molecular, or condensed system itself. In future studies, we expect more experiments where the two branches are integrated with each other. Internal interferometry will probe the instantaneous sub-cycle evolution of strong-field light–matter interaction while external XUV interferometry will resolve the complete optical information encoded in the attosecond pulses. This scheme holds the potential of resolving the sub-cycle evolution of phase accumulated during fundamental phenomena such as field-induced tunneling, hole dynamics, resonance excitation, or multi-electron correlations. The experiments described in this chapter focus on the study of simple atomic, molecular, or solid-state systems.

Looking forward, attosecond interferometry opens up new perspectives in the observation of fundamental phenomena at the forefront of attosecond science, such as multi-electron dynamics in solid-state systems and dynamical chiral processes. In solids, the phase information will provide a direct access into the sub-cycle evolution of the band structure, multiple-band currents, or even sub-cycle phase transitions. In chiral systems, attosecond interferometers will resolve the sub-cycle evolution of chiral phenomena and shed new light on laser-induced chirality.

Interferometers have been instrumental in quantum photonics where non-classical light states are generated and characterized. Since quantum optics is commonly associated with small photon numbers, it is not obvious that HHG with its intense driving fields also enables interesting quantum optical effects. A recent experiment shows that through a post-selection method the HHG driver field is converted to a non-classical Schrödinger cat state [111, 112]. Several more theory works [113–116] have appeared that explore the non-classical nature of HHG, including the fact that harmonic light can be squeezed. Integrating internal and external attosecond interferometry will help to extract the properties of the generated non-classical XUV light and elucidate the complex internal dynamics.

## References


1. J.C. Diels, W. Rudolph, *Ultrashort Laser Pulse Phenomena*, 2nd edn. (Academic Press, San Diego, 2006). https://doi.org/10.1016/B978-0-12-215493-5.X5000-9
2. P.M. Paul, E.S. Toma, P. Breger, G. Mullot, F. Augé, P. Balcou, H.G. Muller, P. Agostini, Science **292**(5522), 1689 (2001). https://doi.org/10.1126/science.1059413
3. H. Muller, Appl. Phys. B **74**(1), s17 (2002). https://doi.org/10.1007/s00340-002-0894-8
4. M. Ferray, A. L'Huillier, X.F. Li, L.A. Lompre, G. Mainfray, C. Manus, J. Phys. B: At. Mol. Opt. Phys. **21**(3), L31 (1988). https://doi.org/10.1088/0953-4075/21/3/001
5. A. McPherson, G. Gibson, H. Jara, U. Johann, T.S. Luk, I.A. McIntyre, K. Boyer, C.K. Rhodes, J. Opt. Soc. Am. B **4**, 595 (1987). https://doi.org/10.1364/JOSAB.4.000595
6. M. Lewenstein, P. Balcou, M.Y. Ivanov, A. L'Huillier, P.B. Corkum, Phys. Rev. A **49**(3), 2117 (1994). https://doi.org/10.1103/PhysRevA.49.2117
7. R. Zerne, C. Altucci, M. Bellini, M.B. Gaarde, T.W. Hänsch, A. L'Huillier, C. Lyngå, C.G. Wahlström, Phys. Rev. Lett. **79**, 1006 (1997). https://doi.org/10.1103/PhysRevLett.79.1006
8. M. Bellini, C. Lyngå, A. Tozzi, M.B. Gaarde, T.W. Hänsch, A. L'Huillier, C.G. Wahlström, Phys. Rev. Lett. **81**, 297 (1998). https://doi.org/10.1103/PhysRevLett.81.297
9. M. Hentschel, R. Kienberger, C. Spielmann, G.A. Reider, T. Brabec, P. Corkum, U. Heinzmann, M. Drescher, F. Krausz, Nature **414**, 509 (2001). https://doi.org/10.1038/35107000
10. R. Kienberger, E. Goulielmakis, M. Uiberacker, A. Baltuška, V. Yakovlev, F. Bammer, A. Scrinzi, T. Westerwalbesloh, U. Kleineberg, U. Heinzmann, M. Drescher, F. Krausz, Nature **427**, 817 (2004). https://doi.org/10.1038/nature02277
11. G. Sansone, E. Benedetti, F. Calegari, C. Vozzi, L. Avaldi, R. Flammini, L. Poletto, P. Villoresi, C. Altucci, R. Velotta, S. Stagira, S.D. Silvestri, M. Nisoli, Science **314**, 443 (2006). https://doi.org/10.1126/science.1132838
12. T. Gaumnitz, A. Jain, Y. Pertot, M. Huppert, I. Jordan, F. Ardana-Lamas, H.J. Wörner, Opt. Express **25**(22), 27506 (2017). https://doi.org/10.1364/OE.25.027506
13. H. Niikura, F. Légaré, R. Hasbani, A.D. Bandrauk, M.Y. Ivanov, D.M. Villeneuve, P.B. Corkum, Nature **417**, 917 (2002). https://doi.org/10.1038/nature00787
14. P.B. Corkum, F. Krausz, Nat. Phys. **3**(6), 381 (2007). https://doi.org/10.1038/nphys620
15. S. Haessler, J. Caillat, P. Salières, J. Phys. B: At. Mol. Opt. Phys. **44**(20), 203001 (2011). https://doi.org/10.1088/0953-4075/44/20/203001
16. P. Salières, A. Maquet, S. Haessler, J. Caillat, R. Taïeb, Rep. Prog. Phys. **75**(6), 062401 (2012). https://doi.org/10.1088/0034-4885/75/6/062401
17. B.D. Bruner, Z. Mašín, M. Negro, F. Morales, D. Brambila, M. Devetta, D. Faccialà, A.G. Harvey, M. Ivanov, Y. Mairesse, S. Patchkovskii, V. Serbinenko, H. Soifer, S. Stagira, C. Vozzi, N. Dudovich, O. Smirnova, Faraday Discuss. **194**, 369 (2016). https://doi.org/10.1039/C6FD00130K
18. P. Peng, C. Marceau, D.M. Villeneuve, Nat. Rev. Phys. **1**, 144 (2019). https://doi.org/10.1038/s42254-018-0015-1
19. P.B. Corkum, Phys. Rev. Lett. **71**(13), 1994 (1993). https://doi.org/10.1103/PhysRevLett.71.1994
20. K.C. Kulander, K.J. Schafer, J.L. Krause, in *Super-Intense Laser-Atom Physics (SILAP) III*, ed. by B. Piraux, A. L'Huillier, K. Rzażewski (Plenum Press (New York), 1993), p. 95
21. K.J. Schafer, B. Yang, L.F. DiMauro, K.C. Kulander, Phys. Rev. Lett. **70**, 1599 (1993). https://doi.org/10.1103/PhysRevLett.70.1599

22. P.B. Corkum, N.H. Burnett, M.Y. Ivanov, Opt. Lett. **19**(22), 1870 (1994). https://doi.org/10.1364/OL.19.001870
23. P. Salières, B. Carré, L. Le Déroff, F. Grasbon, G.G. Paulus, H. Walther, R. Kopold, W. Becker, D.B. Milošević, A. Sanpera, M. Lewenstein, Science **292**(5518), 902 (2001). https://doi.org/10.1126/science.108836
24. O. Smirnova, M. Ivanov, *Multielectron High Harmonic Generation: Simple Man on a Complex Plane* (Wiley, Ltd., 2014), Chap. 7, pp. 201–256. https://doi.org/10.1002/9783527677689.ch7
25. L.V. Keldysh, Sov. Phys. JETP **20**(5), 1307 (1965)
26. J.M. Dahlström, A. L'Huillier, J. Mauritsson, J. Phys. B: At. Mol. Opt. Phys. **44**(9), 095602 (2011). https://doi.org/10.1088/0953-4075/44/9/095602
27. N. Dudovich, O. Smirnova, J. Levesque, Y. Mairesse, M.Y. Ivanov, D. Villeneuve, P.B. Corkum, Nat. Phys. **2**(11), 781 (2006). https://doi.org/10.1038/nphys434
28. O. Pedatzur, G. Orenstein, V. Serbinenko, H. Soifer, B.D. Bruner, A.J. Uzan, D.S. Brambila, A.G. Harvey, L. Torlina, F. Morales, O. Smirnova, N. Dudovich, Nat. Phys. **11**, 815 (2015). https://doi.org/10.1038/nphys3436
29. O. Tcherbakoff, E. Mével, D. Descamps, J. Plumridge, E. Constant, Phys. Rev. A **68**, 043804 (2003). https://doi.org/10.1103/PhysRevA.68.043804
30. C. Zhang, G. Brown, D.H. Ko, P.B. Corkum, Ultrafast Sci. **3**, 0034 (2023). https://doi.org/10.34133/ultrafastscience.0034
31. O. Smirnova, Y. Mairesse, S. Patchkovskii, N. Dudovich, D. Villeneuve, P. Corkum, M.Y. Ivanov, Nature **460**(7258), 972 (2009). https://doi.org/10.1038/nature08253
32. S. Haessler, J. Caillat, W. Boutu, C. Giovanetti-Teixeira, T. Ruchon, T. Auguste, Z. Diveki, P. Breger, A. Maquet, B. Carré, R. Taïeb, P. Salières, Nat. Phys. **6**, 200 (2010). https://doi.org/10.1038/nphys1511
33. M. Spanner, J.B. Bertrand, D.M. Villeneuve, Phys. Rev. A **94**, 023825 (2016). https://doi.org/10.1103/PhysRevA.94.023825
34. G. Orenstein, O. Pedatzur, A.J. Uzan, B.D. Bruner, Y. Mairesse, N. Dudovich, Phys. Rev. A **95**, 051401 (2017). https://doi.org/10.1103/PhysRevA.95.051401
35. E. Frumker, N. Kajumba, J.B. Bertrand, H.J. Wörner, C.T. Hebeisen, P. Hockett, M. Spanner, S. Patchkovskii, G.G. Paulus, D.M. Villeneuve, A. Naumov, P.B. Corkum, Phys. Rev. Lett. **109**, 233904 (2012). https://doi.org/10.1103/PhysRevLett.109.233904
36. P.M. Kraus, A. Rupenyan, H.J. Wörner, Phys. Rev. Lett. **109**, 233903 (2012). https://doi.org/10.1103/PhysRevLett.109.233903
37. P.M. Kraus, D. Baykusheva, H.J. Wörner, Phys. Rev. Lett. **113**, 023001 (2014). https://doi.org/10.1103/PhysRevLett.113.023001
38. P.M. Kraus, O.I. Tolstikhin, D. Baykusheva, A. Rupenyan, J. Schneider, C.Z. Bisgaard, T. Morishita, F. Jensen, L.B. Madsen, H.J. Wörner, Nat. Commun. **6**, 7039 (2015). https://doi.org/10.1038/ncomms8039
39. P.M. Kraus, B. Mignolet, D. Baykusheva, A. Rupenyan, L. Horný, E.F. Penka, G. Grassi, O.I. Tolstikhin, J. Schneider, F. Jensen, L.B. Madsen, A.D. Bandrauk, F. Remacle, H.J. Wörner, Science **350**(6262), 790 (2015). https://doi.org/10.1126/science.aab2160
40. A.J. Uzan, H. Soifer, O. Pedatzur, A. Clergerie, S. Larroque, Z. Masin, B.D. Bruner, B. Pons, M. Ivanov, O. Smirnova, N. Dudovich, Nat. Photon. **14**, 188 (2020). https://doi.org/10.1038/s41566-019-0584-2
41. D. Shafir, Y. Mairesse, D. Villeneuve, P. Corkum, N. Dudovich, Nat. Phys. **5**(6), 412 (2009). https://doi.org/10.1038/nphys1251
42. D. Shafir, H. Soifer, B.D. Bruner, M. Dagan, Y. Mairesse, S. Patchkovskii, M.Y. Ivanov, O. Smirnova, N. Dudovich, Nature **485**, 343 (2012). https://doi.org/10.1038/nature11025
43. S. Ghimire, A.D. DiChiara, E. Sistrunk, P. Agostini, L.F. DiMauro, D.A. Reis, Nat. Phys. **7**, 138 (2011). https://doi.org/10.1038/nphys1847
44. G. Vampa, T. Brabec, J. Phys. B: At. Mol. Opt. Phys. **50**(8), 083001 (2017). https://doi.org/10.1088/1361-6455/aa528d
45. S. Ghimire, D.A. Reis, Nat. Phys. **15**, 10 (2019). https://doi.org/10.1038/s41567-018-0315-5

46. D. Golde, T. Meier, S.W. Koch, Phys. Rev. B **77**, 075330 (2008). https://doi.org/10.1103/PhysRevB.77.075330
47. F. Brunel, J. Opt. Soc. Am. B **7**(4), 521 (1990). https://doi.org/10.1364/JOSAB.7.000521
48. G. Vampa, T.J. Hammond, N. Thiré, B.E. Schmidt, F. Légaré, C.R. McDonald, T. Brabec, P.B. Corkum, Nature **522**, 462 (2015). https://doi.org/10.1038/nature14517
49. A.J. Uzan-Narovlansky, Álvaro Jiménez-Galán, G. Orenstein, R.E.F. Silva, T. Arusi-Parpar, S. Shames, B.D. Bruner, B. Yan, O. Smirnova, M. Ivanov, N. Dudovich, Nat. Photon. **16**, 428 (2022). https://doi.org/10.1038/s41566-022-01010-1
50. K. Varjù, Y. Mairesse, B. Carré, M.B. Gaarde, P. Johnsson, S. Kazamias, R. López-Martens, J. Mauritsson, K.J. Schafer, P. Balcou, A. L'Huillier, P. Salières, J. Mod. Opt. **52**(2–3), 379 (2005). https://doi.org/10.1080/09500340412331301542
51. A. Zaïr, M. Holler, A. Guandalini, F. Schapper, J. Biegert, L. Gallmann, U. Keller, A.S. Wyatt, A. Monmayrant, I.A. Walmsley, E. Cormier, T. Auguste, J.P. Caumes, P. Salières, Phys. Rev. Lett. **100**, 143902 (2008). https://doi.org/10.1103/PhysRevLett.100.143902
52. M. Holler, A. Zaïr, F. Schapper, T. Auguste, E. Cormier, A. Wyatt, A. Monmayrant, I.A. Walmsley, L. Gallmann, P. Salières, U. Keller, Opt. Express **17**(7), 5716 (2009). https://doi.org/10.1364/OE.17.005716
53. A. Zaïr, T. Siegel, S. Sukiasyan, F. Risoud, L. Brugnera, C. Hutchison, Z. Diveki, T. Auguste, J.W. Tisch, P. Salières, M.Y. Ivanov, J.P. Marangos, Chem. Phys. **414**, 184 (2013). https://doi.org/10.1016/j.chemphys.2012.12.022
54. K.T. Kim, C. Zhang, A.D. Shiner, S.E. Kirkwood, E. Frumker, G. Gariepy, A. Naumov, D.M. Villeneuve, P.B. Corkum, Nat. Phys. **9**, 159 (2013). https://doi.org/10.1038/nphys2525
55. A. Camper, A. Ferré, V. Blanchet, D. Descamps, N. Lin, S. Petit, R. Lucchese, P. Salières, T. Ruchon, Y. Mairesse, Phys. Rev. Lett. **130**, 083201 (2023). https://doi.org/10.1103/PhysRevLett.130.083201
56. O. Raz, O. Pedatzur, B.D. Bruner, N. Dudovich, Nat. Photon. **6**, 170 (2012). https://doi.org/10.1038/nphoton.2011.353
57. S. Haessler, T. Balčiunas, G. Fan, G. Andriukaitis, A. Pugžlys, A. Baltuška, T. Witting, R. Squibb, A. Zaïr, J.W.G. Tisch, J.P. Marangos, L.E. Chipperfield, Phys. Rev. X **4**, 021028 (2014). https://doi.org/10.1103/PhysRevX.4.021028
58. A.J. Uzan, G. Orenstein, Á. Jiménez-Galán, C. McDonald, R.E.F. Silva, B.D. Bruner, N. Klimkin, V. Blanchet, T. Arusi-Parpar, M. Krüger, A.N. Rubtsov, O. Smirnova, M. Ivanov, B. Yan, T. Brabec, N. Dudovich, Nat. Photon. **14**, 183 (2020). https://doi.org/10.1038/s41566-019-0574-4
59. K.J. Schafer, M.B. Gaarde, A. Heinrich, J. Biegert, U. Keller, Phys. Rev. Lett. **92**, 023003 (2004). https://doi.org/10.1103/PhysRevLett.92.023003
60. M.B. Gaarde, K.J. Schafer, A. Heinrich, J. Biegert, U. Keller, Phys. Rev. A **72**, 013411 (2005). https://doi.org/10.1103/PhysRevA.72.013411
61. J. Biegert, A. Heinrich, C.P. Hauri, W. Kornelis, P. Schlup, M.P. Anscombe, M.B. Gaarde, K.J. Schafer, U. Keller, J. Mod. Opt. **53**, 87 (2006). https://doi.org/10.1080/09500340500167669
62. G. Gademann, F. Kelkensberg, W.K. Siu, P. Johnsson, M.B. Gaarde, K.J. Schafer, M.J.J. Vrakking, New J. Phys. **13**(3), 033002 (2011). https://doi.org/10.1088/1367-2630/13/3/033002
63. J. Leeuwenburgh, B. Cooper, V. Averbukh, J.P. Marangos, M.Y. Ivanov, Phys. Rev. Lett. **111**, 123002 (2013). https://doi.org/10.1103/PhysRevLett.111.123002
64. D. Azoury, M. Krüger, G. Orenstein, H.R. Larsson, S. Bauch, B.D. Bruner, N. Dudovich, Nat. Commun. **8**, 1453 (2017). https://doi.org/10.1038/s41467-017-01723-w
65. Y. Nabekawa, T. Shimizu, Y. Furukawa, E.J. Takahashi, K. Midorikawa, Phys. Rev. Lett. **102**, 213904 (2009). https://doi.org/10.1103/PhysRevLett.102.213904
66. Y. Nabekawa, T. Shimizu, Y. Furukawa, E.J. Takahashi, K. Midorikawa, Chem. Phys. **414**, 20 (2013). https://doi.org/10.1016/j.chemphys.2012.01.003
67. M.C. Chen, C. Mancuso, C. Hernández-García, F. Dollar, B. Galloway, D. Popmintchev, P.C. Huang, B. Walker, L. Plaja, A.A. Jaroń-Becker, A. Becker, M.M. Murnane, H.C. Kapteyn, T. Popmintchev, Proc. Nat. Acad. Sci. USA **111**(23), E2361 (2014). https://doi.org/10.1073/pnas.1407421111

68. P. Tzallas, D. Charalambidis, N.A. Papadogiannis, K. Witte, G.D. Tsakiris, Nature **426**, 267 (2003). https://doi.org/10.1038/nature02091
69. Y. Nabekawa, T. Shimizu, T. Okino, K. Furusawa, H. Hasegawa, K. Yamanouchi, K. Midorikawa, Phys. Rev. Lett. **96**, 083901 (2006). https://doi.org/10.1103/PhysRevLett.96.083901
70. Y. Nabekawa, T. Shimizu, T. Okino, K. Furusawa, H. Hasegawa, K. Yamanouchi, K. Midorikawa, Phys. Rev. Lett. **97**, 153904 (2006). https://doi.org/10.1103/PhysRevLett.97.153904
71. Y. Nomura, R. Hörlein, P. Tzallas, B. Dromey, S. Rykovanov, Z. Major, J. Osterhoff, S. Karsch, L. Veisz, M. Zepf, D. Charalambidis, F. Krausz, G.D. Tsakiris, Nat. Phys. **5**, 124 (2009). https://doi.org/10.1038/nphys1155
72. E.J. Takahashi, P. Lan, O.D. Mücke, Y. Nabekawa, K. Midorikawa, Nat. Commun. **4**, 2691 (2013). https://doi.org/10.1038/ncomms3691
73. P. Salières, L. Le Déroff, T. Auguste, P. Monot, P. d'Oliveira, D. Campo, J.F. Hergott, H. Merdji, B. Carré, Phys. Rev. Lett. **83**, 5483 (1999). https://doi.org/10.1103/PhysRevLett.83.5483
74. D. Descamps, C. Lyngå, J. Norin, A. L'Huillier, C.G. Wahlström, J.F. Hergott, H. Merdji, P. Salières, M. Bellini, T.W. Hänsch, Opt. Lett. **25**(2), 135 (2000). https://doi.org/10.1364/OL.25.000135
75. J.F. Hergott, T. Auguste, P. Salières, L.L. Déroff, P. Monot, P. d'Oliveira, D. Campo, H. Merdji, B. Carré, J. Opt. Soc. Am. B **20**(1), 171 (2003). https://doi.org/10.1364/JOSAB.20.000171
76. M. Kovačev, S.V. Fomichev, E. Priori, Y. Mairesse, H. Merdji, P. Monchicourt, P. Breger, J. Norin, A. Persson, A. L'Huillier, C.G. Wahlström, B. Carré, P. Salières, Phys. Rev. Lett. **95**, 223903 (2005). https://doi.org/10.1103/PhysRevLett.95.223903
77. Y. Meng, C. Zhang, C. Marceau, A.Y. Naumov, P.B. Corkum, D.M. Villeneuve, Opt. Express **23**(22), 28960 (2015). https://doi.org/10.1364/OE.23.028960
78. G.S.M. Jansen, D. Rudolf, L. Freisem, K.S.E. Eikema, S. Witte, Optica **3**(10), 1122 (2016). https://doi.org/10.1364/OPTICA.3.001122
79. G.S.M. Jansen, X. Liu, K.S.E. Eikema, S. Witte, Opt. Lett. **44**(15), 3625 (2019). https://doi.org/10.1364/OL.44.003625
80. W. Boutu, S. Haessler, H. Merdji, P. Breger, G. Waters, M. Stankiewicz, L.J. Frasinski, R. Taïeb, J. Caillat, A. Maquet, P. Monchicourt, B. Carré, P. Salières, Nat. Phys. **4**, 545 (2008). https://doi.org/10.1038/nphys964
81. J. Tross, G. Kolliopoulos, C.A. Trallero-Herrero, Opt. Express **27**(16), 22960 (2019). https://doi.org/10.1364/OE.27.022960
82. G.R. Harrison, T. Saule, B. Davis, C.A. Trallero-Herrero, Appl. Opt. **61**(30), 8873 (2022). https://doi.org/10.1364/AO.472926
83. G.R. Harrison, T. Saule, R.E. Goetz, G.N. Gibson, A.T. Le, C.A. Trallero-Herrero. Generation and control of non-local quantum equivalent extreme ultraviolet photons (2023). https://doi.org/10.48550/arXiv.2305.17263
84. C. Corsi, A. Pirri, E. Sali, A. Tortora, M. Bellini, Phys. Rev. Lett. **97**, 023901 (2006). https://doi.org/10.1103/PhysRevLett.97.023901
85. X. Zhou, D. Yoshitomi, Y.H. Kobayashi, K.J. Torizuka, Opt. Express **16**(10), 7055 (2008)
86. A. Camper, T. Ruchon, D. Gauthier, O. Gobert, P. Salières, B. Carré, T. Auguste, Phys. Rev. A **89**(4), 043843 (2014). https://doi.org/10.1103/PhysRevA.89.043843
87. J. Lu, E.F. Cunningham, Y.S. You, D.A. Reis, S. Ghimire, Nat. Photon. **13**, 96 (2019). https://doi.org/10.1038/s41566-018-0326-x
88. Y. Mairesse, D. Zeidler, N. Dudovich, M. Spanner, J. Levesque, D.M. Villeneuve, P.B. Corkum, Phys. Rev. Lett. **100**, 143903 (2008). https://doi.org/10.1103/PhysRevLett.100.143903
89. H.J. Wörner, J.B. Bertrand, D.V. Kartashov, P.B. Corkum, D.M. Villeneuve, Nature **466**(7306), 604 (2010). https://doi.org/10.1038/nature09185
90. H.J. Wörner, J.B. Bertrand, B. Fabre, J. Higuet, H. Ruf, A. Dubrouil, S. Patchkovskii, M. Spanner, Y. Mairesse, V. Blanchet, E. Mével, E. Constant, P.B. Corkum, D.M. Villeneuve, Science **334**(6053), 208 (2011). https://doi.org/10.1126/science.1208664

91. A. Camper, A. Ferré, V. Blanchet, F. Burgy, D. Descamps, S. Petit, T. Ruchon, Y. Mairesse, Opt. Lett. **40**(22), 5387 (2015). https://doi.org/10.1364/OL.40.005387
92. K.Y. Chang, L.C. Huang, K. Asaga, M.S. Tsai, L. Rego, P.C. Huang, H. Mashiko, K. Oguri, C. Hernández-García, M.C. Chen, Optica **8**(4), 484 (2021). https://doi.org/10.1364/OPTICA.413531
93. T. Kanai, E.J. Takahashi, Y. Nabekawa, K. Midorikawa, Phys. Rev. Lett. **98**(15), 153904 (2007). https://doi.org/10.1103/PhysRevLett.98.153904
94. N. Wagner, X. Zhou, R. Lock, W. Li, A. Wüest, M. Murnane, H. Kapteyn, Phys. Rev. A **76**, 061403 (2007). https://doi.org/10.1103/PhysRevA.76.061403
95. B.K. McFarland, J.P. Farrell, P.H. Bucksbaum, M. Gühr, Phys. Rev. A **80**, 033412 (2009). https://doi.org/10.1103/PhysRevA.80.033412
96. O. Pedatzur, A. Trabattoni, B. Leshem, H. Shalmoni, M.C. Castrovilli, M. Galli, M. Lucchini, E. Månsson, F. Frassetto, L. Poletto, B. Nadler, O. Raz, M. Nisoli, F. Calegari, D. Oron, N. Dudovich, Nat. Photon. **13**, 91 (2019). https://doi.org/10.1038/s41566-018-0308-z
97. Y. Mairesse, F. Quéré, Phys. Rev. A **71**, 011401 (2005). https://doi.org/10.1103/PhysRevA.71.011401
98. N.A. Papadogiannis, B. Witzel, C. Kalpouzos, D. Charalambidis, Phys. Rev. Lett. **83**, 4289 (1999). https://doi.org/10.1103/PhysRevLett.83.4289
99. L.M. Koll, L. Maikowski, L. Drescher, M.J. Vrakking, T. Witting, Opt. Express **30**(5), 7082 (2022). https://doi.org/10.1364/OE.452018
100. L.M. Koll, L. Maikowski, L. Drescher, T. Witting, M.J. Vrakking, Phys. Rev. Lett. **128**(4), 043201 (2022). https://doi.org/10.1103/PhysRevLett.128.043201
101. J. Seres, V.S. Yakovlev, E. Seres, C. Streli, P. Wobrauschek, C. Spielmann, F. Krausz, Nat. Phys. **3**, 878 (2007). https://doi.org/10.1038/nphys775
102. A. Pirri, C. Corsi, M. Bellini, Phys. Rev. A **78**, 011801 (2008). https://doi.org/10.1103/PhysRevA.78.011801
103. A. Willner, F. Tavella, M. Yeung, T. Dzelzainis, C. Kamperidis, M. Bakarezos, D. Adams, M. Schulz, R. Riedel, M.C. Hoffmann, W. Hu, J. Rossbach, M. Drescher, N.A. Papadogiannis, M. Tatarakis, B. Dromey, M. Zepf, Phys. Rev. Lett. **107**, 175002 (2011). https://doi.org/10.1103/PhysRevLett.107.175002
104. D.E. Laban, A.J. Palmer, W.C. Wallace, N.S. Gaffney, R.P.M.J.W. Notermans, T.T.J. Clevis, M.G. Pullen, D. Jiang, H.M. Quiney, I.V. Litvinyuk, D. Kielpinski, R.T. Sang, Phys. Rev. Lett. **109**, 263902 (2012). https://doi.org/10.1103/PhysRevLett.109.263902
105. M.M. Mang, D.T. Lloyd, P.N. Anderson, D. Treacher, A.S. Wyatt, S.M. Hooker, I.A. Walmsley, K. O'Keeffe, Opt. Lett. **43**(21), 5275 (2018). https://doi.org/10.1364/OL.43.005275
106. M.H. Mustary, D.E. Laban, J.B.O. Wood, A.J. Palmer, J. Holdsworth, I.V. Litvinyuk, R.T. Sang, J. Phys. B: At. Mol. Opt. Phys. **51**(9), 094006 (2018). https://doi.org/10.1088/1361-6455/aaba40
107. D. Azoury, O. Kneller, S. Rozen, B.D. Bruner, A. Clergerie, Y. Mairesse, B. Fabre, B. Pons, N. Dudovich, M. Krüger, Nat. Photon. **13**(1), 54 (2019). https://doi.org/10.1038/s41566-018-0303-4
108. D. Azoury, O. Kneller, M. Krüger, B.D. Bruner, O. Cohen, Y. Mairesse, N. Dudovich, Nat. Photon. **13**, 198 (2019). https://doi.org/10.1038/s41566-019-0350-5
109. A.T. Le, T. Morishita, C.D. Lin, Phys. Rev. A **78**, 023814 (2008). https://doi.org/10.1103/PhysRevA.78.023814
110. O. Kneller, D. Azoury, Y. Federman, M. Krüger, A.J. Uzan, G. Orenstein, B.D. Bruner, O. Smirnova, S. Patchkovskii, M. Ivanov, N. Dudovich, Nat. Photon. **16**, 304 (2022). https://doi.org/10.1038/s41566-022-00955-7
111. M. Lewenstein, M.F. Ciappina, E. Pisanty, J. Rivera-Dean, P. Stammer, T. Lamprou, P. Tzallas, Nat. Phys. **17**, 1104 (2021). https://doi.org/10.1038/s41567-021-01317-w
112. U. Bhattacharya, T. Lamprou, A.S. Maxwell, A. Ordóñez, E. Pisanty, J. Rivera-Dean, P. Stammer, M.F. Ciappina, M. Lewenstein, P. Tzallas, Rep. Prog. Phys. **86**(9), 094401 (2023). https://doi.org/10.1088/1361-6633/acea31

113. A. Gorlach, O. Neufeld, N. Rivera, O. Cohen, I. Kaminer, Nat. Commun. **11**, 4598 (2020). https://doi.org/10.1038/s41467-020-18218-w
114. A. Gorlach, M.E. Tzur, M. Birk, M. Krüger, N. Rivera, O. Cohen, I. Kaminer, Nat. Phys. **19**, 1689 (2023). https://doi.org/10.1038/s41567-023-02127-y
115. A. Pizzi, A. Gorlach, N. Rivera, A. Nunnenkamp, I. Kaminer, Nat. Phys. **19**, 551 (2023). https://doi.org/10.1038/s41567-022-01910-7
116. M.E. Tzur, M. Birk, A. Gorlach, M. Krüger, I. Kaminer, O. Cohen, Nat. Photon. **17**, 501 (2023). https://doi.org/10.1038/s41566-023-01209-w

# Chapter 3
# Attosecond Dynamics in Liquids

**Hans Jakob Wörner, Axel Schild, Denis Jelovina, Inga Jordan, Conaill Perry, Tran Trung Luu, and Zhong Yin**

**Abstract** Attosecond science is well developed for atoms and promising results have been obtained for molecules and solids. Here, we review the first steps in developing attosecond time-resolved measurements in liquids. These advances provide access to time-domain studies of electronic dynamics in the natural environment of chemical reactions and biological processes. We concentrate on two techniques that are representative of the two main branches of attosecond science: pump-probe measurements using attosecond pulses and high-harmonic spectroscopy (HHS). In the first part, we discuss attosecond photoelectron spectroscopy with cylindrical microjets and its application to measure time delays between liquid and gaseous water. We present the experimental techniques, the new data-analysis methods, and the experimental results. We describe in detail the conceptual and theoretical framework required to describe attosecond chronoscopy in liquids at a quantum-mechanical level. This includes photoionization delays, scattering delays, as well as a coherent description of electron transport and (laser-assisted) photoemission and scattering. As a consequence, we show that attosecond chronoscopy of liquids is, in general, sensitive to both types of delays, as well as the electron mean-free paths. Through detailed modeling, involving state-of-the-art quantum scattering and Monte-Carlo trajectory methods, we show that the photoionization delays dominate in attosecond chronoscopy of liquid water at photon energies of 20–30 eV. This conclusion is supported by a near-quantitative agreement between experiment and theory. In the second part, we introduce HHS of liquids based on flat microjets. These results represent the first observation of high-harmonic generation (HHG) in liquids extending well beyond the visible into the extreme-ultraviolet regime. We show that the cut-off energy scales almost linearly with the peak electric field of the driver and that the yield of all harmonics scales non-perturbatively. We also discuss the ellipticity dependence of the liquid-phase harmonics, which is systematically broadened compared to the gas phase. We introduce a strongly driven few-band model as a zero-order approximation of HHG in liquids and demonstrate the sensitivity of HHG to the electronic structure of liquids. Finally, we discuss more advanced approaches

H. J. Wörner (✉) · A. Schild · D. Jelovina · I. Jordan · C. Perry · T. T. Luu · Z. Yin
Laboratorium für Physikalische Chemie, ETH Zürich, 8093 Zürich, Switzerland
e-mail: hwoerner@ethz.ch

 
K. Ueda (ed.), *Ultrafast Electronic and Structural Dynamics*,
https://doi.org/10.1007/978-981-97-2914-2_3

for modeling liquid-phase HHG. In the conclusion, we present an outlook on future studies of attosecond dynamics in liquids.

## 3.1 Introduction

Attosecond science has the potential to address fundamental open questions in chemical and physical sciences. By directly accessing the electronic dynamics of matter on a time scale that naturally freezes any structural dynamics, attosecond spectroscopy targets the most fundamental electronic processes that define the properties of matter. The methods of attosecond science have been well established and most broadly demonstrated on atoms [1–4]. Very promising results have also been obtained on molecules [5–10] and solids [11–16]. The field of attosecond molecular dynamics has recently been reviewed in Ref. [17].

Attosecond dynamics in liquids have so far not been accessible to time-resolved experiments. However, they are expected to play a particularly important role in chemical and biological processes that dominantly take place in the liquid phase. The primary processes underlying charge and energy transfer in the liquid phase take place on few-femtosecond to attosecond time scales [18] and have eluded direct experimental access to date. These processes include, e.g., the ionization of liquid water, electron scattering during its transport through the liquid phase, as well as intermolecular Coulombic decay (ICD) [19] and electron-transfer-mediated decay (ETMD) [20], to name only a few examples.

Among these, the ionization of liquid water plays a particularly important role since it initiates the dominant pathway of radiation damage in living matter (see Ref. [21] for a recent review). The ionization of liquid water is inherently an attosecond process and therefore offers a target of primary relevance for attosecond science. However, the primary ionization event itself is not the only attosecond process involved in the ionization of liquid water. The subsequent electron scattering during its transport through the liquid phase also takes place on a sub-femtosecond time scale. It controls the energy deposition during electron scattering and thereby the chemistry initiated by ionizing radiation in liquid water. Important new insights into the early femtosecond time scale of water ionization have very recently been obtained using both one-photon ionization [22] and strong-field ionization [23].

In this book chapter, we discuss two recently demonstrated experimental approaches to attosecond science in the liquid phase: attosecond photoelectron spectroscopy [24] and high-harmonic spectroscopy [25]. We present the experimental methodology and discuss the first experimental results, along with the developed conceptual framework and the related theoretical methods.

In the first part, we describe the principles and methods of attosecond photoelectron spectroscopy of liquids [24]. This technique has been applied to measure the time delay between photoemission from liquid and gaseous water. Photoionization delays from isolated molecules in the gas phase have only recently been measured [9]. These measurements have been interpreted through the development of a general theoretical and computational methodology [26]. Here, we use photoemission from gaseous water molecules as a reference against which photoemission from liquid water is clocked. We find that photoemission from liquid water is *delayed* by ~50–70 attoseconds compared to photoemission from isolated water molecules at photon energies of ~20-30 eV. We show through detailed modeling that includes quantum-scattering calculations on water clusters of variable size, as well as Monte-Carlo trajectory simulations benchmarked against the time-dependent Schrödinger equation, that the measured delays are dominated by the local solvation structure in liquid water. Electron scattering during transport, in contrast, makes a negligible contribution to the measured delays. This conclusion has very recently been confirmed in a very different type of experiments that measured photoionization delays between size-resolved water clusters and water monomers and found that the photoionization delays steadily increase from the monomer to the tetramer or pentamer, but do not significantly increase towards larger clusters. This scaling of the time delays has been related to the spatial extension of the electron hole created in the photoionization process [27]. The relative time delays measured between pentamers and hexamers with respect to monomers have been found to agree with those measured for liquid water with respect to water monomers [28].

In the second part, we describe the experimental methods and first results on non-perturbative high-harmonic generation in liquids [25]. This recent discovery represented a paradigm change in attosecond science because a series of previous experimental works on water droplets [29–31] concluded that high-harmonic generation was impossible at the natural density of liquid water. Previous observations of harmonic generation from liquids were indeed limited to low-order harmonics in the visible range [32] and to the coherent-wake-emission regime [33], where high-harmonic generation takes place in a plasma created by the ultra-intense laser pulse (peak intensity $> 10^{18}$ W/cm$^2$) rather than in the original, unperturbed target. The main observations from our work [25] are a linear scaling of the highest-emitted energy with the electric field, a pronounced sensitivity of the observed spectra on the electronic structure of the liquids, and a broadened ellipticity dependence of the high-harmonic yield compared to gas-phase HHG. High-harmonic spectroscopy of liquids offers a complementary and remarkably different approach to the same fundamental problems discussed above. It provides access to the attosecond dynamics of the ionization step. It is sensitive to the dynamics of electron transport through the liquid phase, and thereby to electron scattering dynamics [34, 35]. Finally, it might also be sensitive to the spatial shape of the created electron hole through the recombination step. Liquid-phase high-harmonic spectroscopy therefore offers several promising prospects for exploring the electronic structure and dynamics of liquids, solutes, and solvents.

In the outlook that concludes this book chapter, we discuss possible future avenues of attosecond science in liquids. These include time-resolved studies of ICD and ETMD, which are the prototype elementary processes of charge and energy transfer in living matter, but have unknown time scales.

## 3.2 Attosecond Photoelectron Spectroscopy of Liquid Water

The most direct way to probe electron dynamics in liquid water is to perform a time-resolved measurement on the attosecond time scale. For this purpose, we have chosen attosecond photoelectron spectroscopy. It consists in creating electrons in the conduction band of liquid water through ionization with an attosecond pulse and probing their dynamics through interaction with a temporally delayed near-infrared (NIR) pulse before detecting them through photoelectron spectroscopy.

For the first experiment, described herein, we have used attosecond interferometry, which combines an extreme ultraviolet (XUV) attosecond pulse train (APT) with a femtosecond NIR laser pulse (see Fig. 3.1). This method has previously been applied to measure photoionization delays in atoms [3] and molecules [9]. The particular advantage of attosecond interferometry over, e.g., attosecond streaking is the high spectral resolution that is achieved [36], in addition to the high temporal resolution that is common to both methods. Whereas the spectral resolution in attosecond streaking is limited by the temporal duration of the attosecond pulse through the time-bandwidth product, the spectral resolution in attosecond interferometry is not. Instead, the resolution is limited by the spectral width of the individual harmonic orders, which are limited by the length of the APT and its chirp over the (femtosecond) time scale of the APT envelope. The high spectral resolution is a crucial aspect of the present experiments because it enables the spectral distinction of photoelectrons originating from the gas and liquid phases. These give rise, respectively, to the narrow and broad spectral features illustrated in Fig. 3.1.

Experiments based on attosecond interferometry or attosecond streaking directly measure time delays in photoionization, but all such measurements are inherently relative at present. It is therefore essential to identify a suitable reference against which these delays can be measured. In the present experiment, we turn the presence of the evaporating gas-phase molecules, usually perceived as a nuisance in liquid-microjet photoelectron spectroscopy, into an advantage. By measuring photoionization time delays of liquid water relative to those of isolated water molecules, we eliminate the "trivial" single-molecule contributions to the photoionization delays. The word "trivial" is consciously put in quotation marks because the measurement and interpretation of molecular photoionization delays is itself a highly active and intriguing area of research [9, 26, 27, 37–42]. However, the goal of the present work goes much beyond this, by aiming at identifying the specific contributions of the liquid phase to the attosecond photoemission delays. These include two types of effects, i.e., the modification of the electronic structure and therefore the scattering potential of an isolated water molecule caused by solvation, as well as electron scattering dynamics during its transport through the liquid phase.

In the following sections, we discuss the experimental techniques, which include the realization of liquid-phase attosecond interferometry and the methods developed to retrieve photoionization delays from overlapping photoelectron spectra. We also discuss in detail the novel theoretical methods, which comprise the integra-

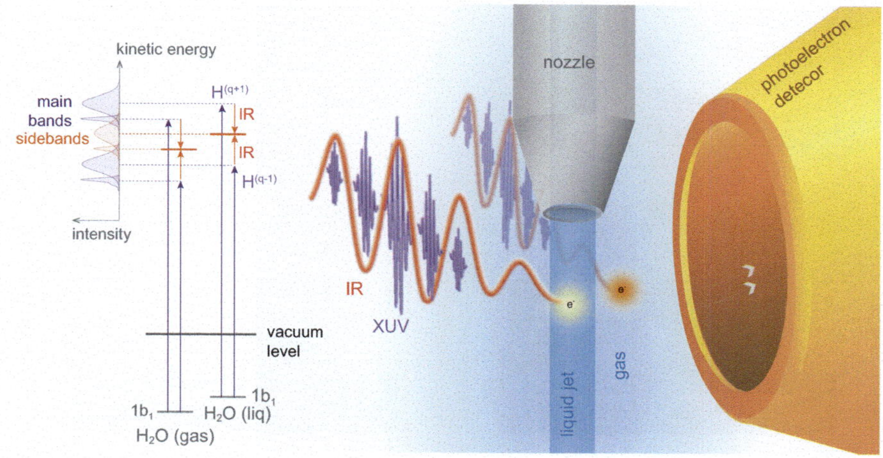


**Fig. 3.1 Attosecond time-resolved photoelectron spectroscopy of liquid water.** A spectrally filtered attosecond pulse train, composed of a few high-harmonic orders ($H^{(q-1)}$,$H^{(q+1)}$, etc.) superimposed with a near-infrared femtosecond laser pulse interacts with a microjet of liquid water. Photoelectrons are simultaneously emitted from the liquid and the surrounding gas phase. The resulting photoelectron spectra are measured as a function of the time delay between the overlapping pulses. Adapted from Ref. [24]

tion of laser-assisted scattering in the formalism of attosecond interferometry, the development of a general three-dimensional Monte-Carlo code for simulating such experiments and the identification of condensation effects on the photoionization delays from water clusters. This extensive development of experimental and theoretical methods has culminated in the quantitative interpretation of the measured photoemission time delays of liquid water [24].

### *3.2.1 Experimental Methods and Results*

The experimental setup consists of an attosecond beamline delivering XUV APTs and a femtosecond NIR pulse, a liquid microjet delivering a micrometer-thin stream of liquid into a high-vacuum chamber, and a photoelectron spectrometer. In this work, we have used the attosecond beamline described in [43], which uses nested white-light and monochromatic interferometers to actively stabilize the time delay between NIR and XUV pulses to extreme accuracy. High-harmonic generation from a NIR laser pulse centered at 800 nm in an argon gas cell is used to generate the APT. The created APT is spectrally filtered by thin metallic foils to reduce its spectral bandwidth and thereby the spectral overlap in the photoelectron spectra. Specifically, we have used Sn filters to isolate the harmonic orders 11, 13, and 15, or Ti filters to isolate the harmonic orders 17, 19, and 21. A particular advantage of these metal filters is their sharp spectral truncation at the high-energy side of the spectrum, which

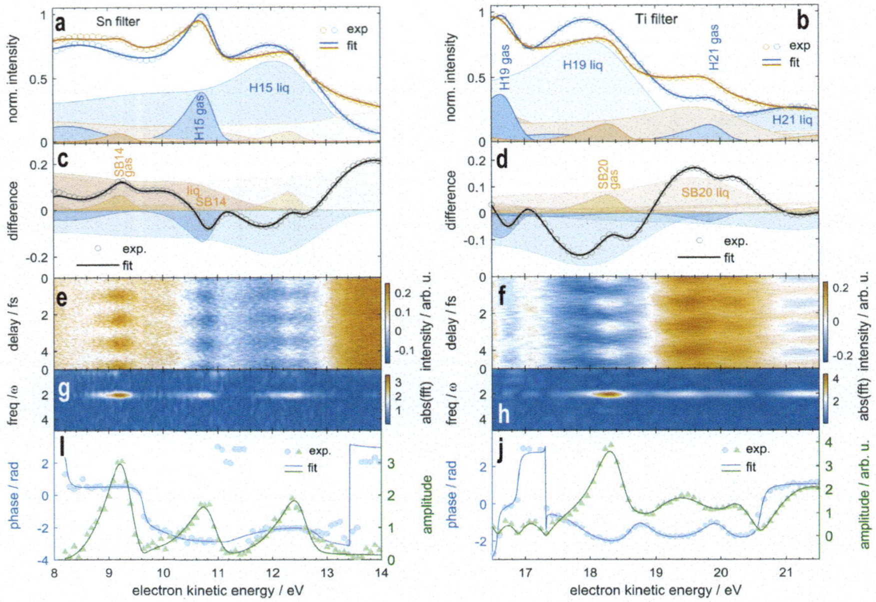


**Fig. 3.2 Attosecond photoelectron spectra of liquid and gaseous water**. Data acquired with a Sn-filtered APT (left) or a Ti-filtered APT (right). **a,b** Photoelectron spectra in the absence (blue) and presence (orange) of the IR field with their principal-component fit (full lines) and decomposition (filled curves). **c,d** Difference spectra (circles), principle-component fits (line) and decomposition (filled curves) into side bands (orange) and depletion (blue). **e,f** Difference spectra as a function of the APT-IR time delay. **g,h** Fourier-transform power spectrum of c. **i,j** Amplitude and phase of the $2\omega$ component of the Fourier transform. Adapted from Ref. [24]

makes the remaining spectral overlap manageable (see Fig. 3.2). The liquid microjet is formed by expanding high-purity liquid water (with the addition of NaCl to a concentration of 50 mM) through a $\sim$20 $\mu$m inner-diameter capillary with a high-performance liquid-chromatography pump. The photoelectron spectrometer is a 1-m long magnetic-bottle time-of-flight spectrometer previously described in Ref. [44].

The experimental results are shown in Fig. 3.2, where the left (right) columns show the data acquired with the Sn (Ti) filter. Correspondingly, the data on the left-hand side is mainly composed of photoelectrons spectra generated by H13 and H15, whereas the data on the right-hand side mainly originates from H19 and H21. These data have been analyzed through a principal-component analysis (PCA) based on photoelectron spectra of gaseous and liquid water that were measured with a monochromatized high-harmonic light source [45] and the same photoelectron spectrometer [44]. This PCA provides a unique decomposition of the measured spectra into its constituents. The panel (a) in Fig. 3.2 shows photoelectron spectra acquired with the XUV-APT only (blue) or with XUV-APT and NIR pulses (orange). The panel (b) shows the difference spectra, obtained by subtracting the XUV-only spectra from the XUV+NIR spectra. In the panels (a) and (b), the circles represent the experimental data and the full lines of the PCA fit. The contributing principal components are shown as filled areas, whereby the XUV-only contributions are shown in blue and the sideband contributions are shown in orange. The perfect agreement between the PCA fit and the experimental data shows that the measured spectra have been successfully decomposed into the contributing photoelectron spectra from liquid and gaseous water of the relevant high-harmonic orders.

The panel (c) shows the difference spectra (from panel b) as a function of the time delay between the overlapping XUV-APT and NIR pulses. Positive signals (in orange) are dominated by sideband contributions and negative signals (in blue) are dominated by the depletion of the XUV-APT photoelectron spectra. The Fourier transform of the data in panel (c) along the time-delay axis is shown in panel (d) and the phase and amplitude of the Fourier transform, integrated over the dominant $2\omega$ frequency component is shown in panel (e). These complex-valued Fourier transforms have been analyzed within the PCA with minimal assumptions, i.e., the entire photoelectron bands corresponding to the highest-occupied molecular orbital (HOMO) of the gas or liquid contributions are assigned a characteristic phase ($\phi_{\text{gas}}$ or $\phi_{\text{liq}}$). This minimal assumption is sufficient to reproduce the complete electron-kinetic-energy-dependent complex-valued Fourier transforms, as illustrated in panel (e). As a consequence, it is possible to determine the photoemission time delays between the HOMO bands of the liquid and gas phases:

$$\Delta\tau = \tau_{\text{liq}} - \tau_{\text{gas}} = \frac{\phi_{\text{liq}} - \phi_{\text{gas}}}{2\omega}. \tag{3.1}$$

We find $\tau_{\text{liq}} - \tau_{\text{gas}} = 69 \pm 20$ as at a photon energy of 21.7 eV (SB14) and $\tau_{\text{liq}} - \tau_{\text{gas}} = 48 \pm 16$ as at a photon energy of 31.0 eV (SB20).

In addition to robust delays, the PCA also enables the determination of reliable modulation depths of the sideband intensities. This additional observable of attosecond interferometry has received very little attention to date. In an idealized situation where the amplitude of the two-photon-ionization pathways leading to the same sideband state have the same amplitude, the modulation contrast will be 100%. The experimentally observed modulation contrast will in general be smaller than 100% for various possible reasons. In the gas phase, these include (i) different amplitudes of the two-photon-ionization pathways, (ii) contributions of several two-photon-ionization pathways that do not interfere, e.g., because they correspond to different initial eigenstates of they leave the system in different final states, (iii) contributions of ionization channels (e.g., emission angle or target orientation) corresponding to different time delays, (iv) fluctuations in XUV-IR path-length difference, etc. In condensed-phase attosecond interferometry, additional effects include (v) contributions from different local environments around the photoionized entity and (vi) decoherence of the photoelectron wave packet during transport through the condensed phase.

Given the many possible contributions to a finite modulation contrast, it is thus important to find a meaningful reference. In the present case, the best reference is our previously published measurement of molecular photoionization delays in water vapor [9], which was carried out with the same apparatus, under the same experimental conditions as the liquid-phase experiments. This experiment revealed a modulation contrast of HOMO SB14 indistinguishable from 100 % within the experimental signal-to-noise ratio (see Fig. 4.3 in [46]).

Using the PCA discussed above, we have determined a relative modulation depth of the liquid—compared to the gas-phase signal $M_r = M_{liq}/M_{gas}$ of $0.17 \pm 0.03$ at a photon energy of 21.7 eV and $0.45 \pm 0.06$ at a photon energy of 31.0 eV. The determination of relative modulation depths has the advantage that it eliminates the contributions (i)–(iv), at least as far as single-molecule contributions and the experimental imperfections are concerned. Since a deviation from a perfect modulation contrast was not observable in water vapor [46], the finite modulation depths reported above can mainly be attributed to the effects (v) and (vi). As further discussed in Sect. 3.2.3, contribution (v) is significant and contribution (vi) might also be important.

## 3.2.2 *Concepts and Theoretical Methods*

In this section, a detailed theoretical treatment of attosecond photoelectron spectroscopy (APES) in liquids is given. Although a significant amount of work on APES of solids has previously been published (see, e.g., [47–50]), a comprehensive treatment that includes both photoionization delays and scattering delays has been missing until our recent work [51]. Motivated by experimental results [11–13, 52–55], most previous theoretical treatments were designed to describe experiments on metallic samples. The description of APES in liquids fundamentally differs from the situation encountered in metals. This is due to the different penetration depths of the XUV and NIR fields. Whereas the XUV fields penetrate more deeply into metals than the NIR, the situation is reversed in liquids, which are practically transparent to the NIR radiation. Since the temporal information originates from the interaction of the XUV-induced photoelectron wave packet with the NIR radiation, APES on metals is mainly sensitive to the transport time of the photoelectron wave packet from the point of ionization to the metal-vacuum interface. In APES of liquids, the situation is reversed and the NIR field is present throughout the medium. Therefore APES in liquid is sensitive to other aspects of attosecond photoionization dynamics. The following sections present a detailed conceptual and theoretical analysis of this situation.

### 3.2.2.1 Time Delays in Photoionization and Scattering

Time delays in quantum scattering were first analyzed by Wigner and Smith [56, 57]. They studied conventional scattering events, corresponding to the collision of two particles. Photoionization can be understood as a half collision and can be described with a very similar quantum-mechanical formalism. Consequently, it is not surprising that photoionization delays are closely related to scattering delays, both conceptually and in terms of their magnitudes [50, 58].

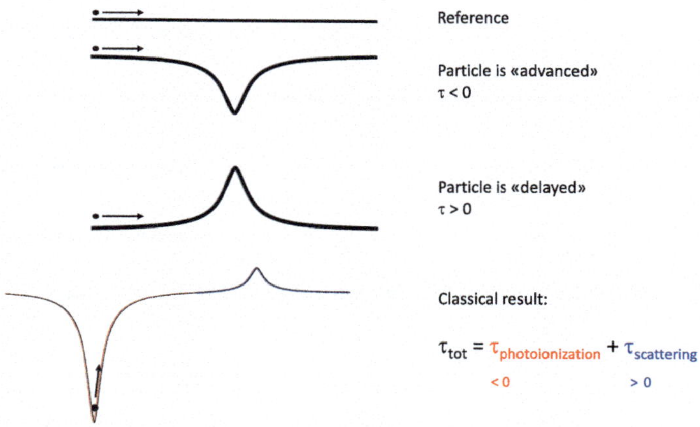


**Fig. 3.3 Time delays in photoionization and scattering in the classical limit.** The time delays are defined as the differences in the arrival times of the classical particles that have propagated through different potentials with that of a "reference" particle propagating through potential-free space with the same final (asymptotic) velocity

Both scattering and photoionization delays typically lie in the attosecond domain. They have therefore become accessible in the time domain only recently [3, 59]. Attosecond science has so far focused on the measurement of time delays in photoionization because they are directly accessible to APES. As a consequence, time delays in (conventional) scattering, i.e., full collisions have so far not been discussed. Here, we show that they can play a significant role in condensed-phase APES. More importantly, we show that laser-assisted electron scattering is a general phenomenon that needs to be included in a comprehensive treatment of condensed-phase APES.

We initiate our discussion with a classical analysis that provides a transparent basic understanding of the problem. Figure 3.3 illustrates the definition of these classical delays. It is easy to understand that an attractive (repulsive) potential results in a negative (positive) delay because the particle is locally accelerated (decelerated) compared to the potential-free propagation before returning to the same asymptotic velocity. As a consequence the time required to cover the same horizontal distance is smaller (larger) compared to potential-free propagation, which results in a negative (positive) delay, as illustrated. In these classical considerations, it is also easy to understand that such classical scattering delays are additive. As long as the potentials do not overlap, the delay contributed by a particular potential is only a property of the local potential and does not depend on the previous trajectory of the particle.

The same considerations hold true for a classical treatment of photoionization delays. This situation is illustrated at the bottom of Fig. 3.3. The classical definition of a photoionization delay corresponds to a situation where the particle starts from the bottom of an attractive potential with a finite velocity. This velocity must be chosen such that the final (asymptotic) velocity of the particle is the same as for the reference particle. This classical photoionization delay will be negative for the same reason as the classical scattering delay for an attractive potential is negative. The classical photoionization delay is equal to half of the classical scattering delay for the same potential.

Continuing on this analogy, it is easy to show that the total classical delay for one photoionization and one scattering event is simply the sum of the two classical delays, as long as the potentials do not overlap. This situation is illustrated at the bottom of Fig. 3.3.

#### 3.2.2.2 Non-local Attosecond Interferometry in One Dimension

The translation of the results from the previous section to observables of attosecond spectroscopy requires at least two modifications. First, electrons are not classical objects and should therefore be described quantum-mechanically. When this is not possible, the classical approximation has to be justified, as we further discuss below. Second, the situation analyzed in classical terms in the previous section would correspond to a time-of-flight measurement with attosecond resolution, which is impractical. Instead, attosecond spectroscopy measures such time delays through two-color photoionization. Whereas the relation of photoionization delays measured with such techniques with the (quantum-mechanical) Wigner delays has been established [50, 58], the treatment of full-scattering events itself and full-scattering in combination with photoionization has only been addressed in our recent publication [51].

In this section, we present a quantum-mechanical analysis of attosecond time delays caused by photoionization and scattering as measured by attosecond interferometry. We note that these results are not specific to attosecond interferometry, but also apply to attosecond streaking, which will be the subject of future publications. Starting from the numerical solution of the time-dependent Schrödinger equation, we show that the obtained delays remarkably deviate from the classical expectation. Building on analytical descriptions of laser-assisted photoemission (LAPE) and laser-assisted electron scattering (LAES), we develop a complete analytical description of the quantum-mechanical problem at hand. This has led to the discovery of a novel phenomenon that we call "non-local" attosecond interferometry. In contrast to the traditional understanding of attosecond interferometry, where the interaction of the photoelectron wave packet with the XUV and NIR radiation takes place in a common spatial volume, the non-local pathways involve interactions at spatially separated regions. Specifically, the XUV absorption is localized to the position of the initially bound electronic wave function, whereas the NIR interaction takes place at the position where scattering occurs. These "non-local" XUV/NIR interactions lead to the interesting result that the total delays measured in the presence of photoionization and scattering are, in general, sensitive to both photoionization and scattering delays, and to the distances traveled between XUV and NIR interactions, i.e., to the mean-free paths of electron scattering.

We solve the one-electron time-dependent Schrödinger equation (TDSE) in one dimension using a model potential consisting of an attractive and a repulsive potential well (see Fig. 3.4). Details of this calculation are given in Ref. [51]. The system is initiated in the lowest bound electronic state of the attractive potential. The TDSE

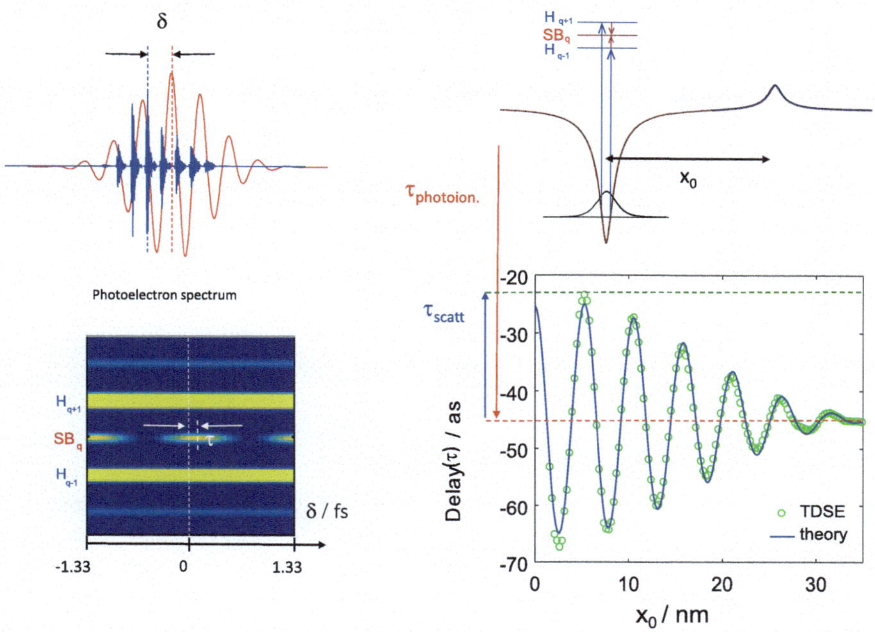


**Fig. 3.4 Time delays in photoionization and scattering as measured by attosecond interferometry**. This figure shows (schematically) the employed XUV and NIR fields (top left), the used model potential (top right), the calculated photoelectron spectrum as a function of the time delay $\delta$ (bottom left) and the calculated delay $\tau$ as a function of the separation $x_0$ between the attractive and repulsive potentials (bottom right)

is solved in the presence of an XUV field consisting of two harmonic orders ($H_{q-1}$ and $H_{q+1}$) and a NIR field. The photoelectron spectrum is calculated by projecting the total wave function at the end of the propagation onto the continuum eigenstates of the model system. This gives rise to a photoelectron spectrum as shown on the bottom left of Fig. 3.4. The calculated photoelectron spectra reveal an oscillation of the intensity of $SB_q$ as a function of the delay $\delta$ between the XUV and NIR pulses. In the calculation, we actually varied the carrier-envelope phase of the NIR pulse instead of varying the time delay $\delta$. This is equivalent for the present purpose but avoids unwanted effects caused by the envelope functions of the two fields. The temporal shift of the $SB_q$ maximum with respect to $\delta = 0$ defines the measured time delay $\tau$.

Solving the TDSE under the described conditions for different separations $x_0$ between the attractive and repulsive potentials provides the result shown in the bottom-right part of Fig. 3.4. The total delay displays an oscillatory dependence on $x_0$. This is in remarkable contrast to the naive classical expectation discussed in the previous section. For our model system, the photoionization delay amounts to -44 as, as indicated by the red dashed line, and the scattering delay to +21 as. Within our classical considerations this would lead to a total delay of -23 as, independent

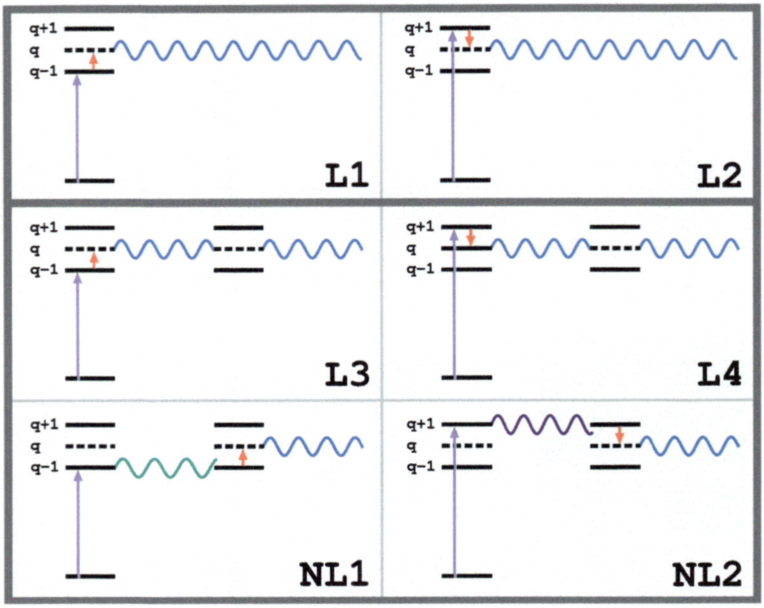


**Fig. 3.5** Schematic depiction of the four "local" (L1, L2, L3, L4) and the two "non-local" paths (NL1, NL2) contributing mainly to the attosecond interference signal if one scattering event is possible after ionization

of $x_0$ as indicated by the dashed green line. The result of the TDSE (green circles) departs markedly from this classical prediction. The calculated delay $\tau$ is found to oscillate around the value of the photoionization delay with a maximal amplitude given by the scattering delay. More precisely, the total delay $\tau$ can be expressed as a sum of the photoionization delay $\tau_{\text{PI}}$ and a non-local delay $\tau_{\text{nl}}$,

$$\tau = \tau_{\text{PI}} + \tau_{\text{nl}}, \tag{3.2}$$

where $\tau_{\text{nl}}$ oscillates with the distance $x_0$.

This result significantly contrasts with the previous understanding of attosecond interferometry on isolated particles, where only local XUV and NIR interactions are important. In such a framework, the observed oscillations cannot be explained. They could, however, be quantitatively explained in Ref. [51] by developing an analytical formalism that includes both local and non-local pathways.

Figure 3.5 shows the relevant paths that an electron can take so that it ends up in the sideband $q$. The paths L1 and L2 are "local" paths that are the basis of APES in gas-phase molecules, the paths L3 and L4 are paths where interaction with the assisting IR field is still at the ionization site but the outgoing electron wave packet undergoes scattering, and NL1 and NL2 are "non-local" paths where photon exchange happens at the scattering center. In words, those paths are as follows:

- L1: Absorption of an XUV photon of frequency $(q-1)\omega$ and absorption of one IR photon at the ionization site,
- L2: absorption of an XUV photon of frequency $(q+1)\omega$ and emission of one IR photon at the ionization site,
- L3: absorption of an XUV photon of frequency $(q-1)\omega$ and absorption of one IR photon at the ionization site, and scattering at the perturber without photon exchange,
- L4: absorption of an XUV photon of frequency $(q+1)\omega$ and emission of one IR photon at the ionization site, and scattering at the perturber without photon exchange,
- NL1: absorption of an XUV photon of frequency $(q-1)\omega$ at the ionization site and absorption of one IR photon at the perturber, and
- NL2: Absorption of an XUV photon of frequency $(q+1)\omega$ at the ionization site and emission of one IR photon at the perturber.

While L1 and L2 are the two paths that explain APES for gas-phase molecules, all six paths are needed to reproduce the delay of Fig. 3.4. Each of these paths contributes to the second-order amplitude (i.e., the amplitude for two-photon transitions) $f$, i.e.,

$$f = f_{\mathrm{L1}} + f_{\mathrm{L2}} + f_{\mathrm{L3}} + f_{\mathrm{L4}} + f_{\mathrm{NL1}} + f_{\mathrm{NL2}}, \tag{3.3}$$

and the sideband intensity is then $I \propto |f|^2$. The amplitudes are

$$f_{\mathrm{L1}} = F_q^{q-1} e^{ik_q x_0} \tag{3.4}$$
$$f_{\mathrm{L2}} = F_q^{q+1} e^{ik_q x_0} \tag{3.5}$$
$$f_{\mathrm{L3}} = F_q^{q-1} e^{ik_q x_0} f_q^q \tag{3.6}$$
$$f_{\mathrm{L4}} = F_q^{q+1} e^{ik_q x_0} f_q^q \tag{3.7}$$
$$f_{\mathrm{NL1}} = F_{q-1}^{q-1} e^{ik_{q-1} x_0} f_q^{q-1} \tag{3.8}$$
$$f_{\mathrm{NL2}} = F_{q+1}^{q+1} e^{ik_{q+1} x_0} f_q^{q+1}. \tag{3.9}$$

Here, $F_n^m$ indicates the amplitude of transition from state $m$ to state $n$ at the photoionization site and $f_n^m$ indicates the amplitude of transition from state $m$ to state $n$ at the perturber. An additional phase factor is due to propagation to the scattering site with momentum $k_q$ (for L1, L2, L3, and L4), with momentum $k_{q-1}$ (NL1), or with momentum $k_{q+1}$ (NL2).

Numerical values for the amplitudes $F_n^m$ and $f_n^m$ can be obtained within what is known as strong-field approximation, soft-photon approximation, or Kroll-Watson theory, and which is referred to as laser-assisted photoelectric effect (LAPE) and laser-assisted electron scattering (LAES), respectively [60–62]. For LAPE, it is assumed that the initial state is a bound state, the final state is that of a laser-dressed free electron in the IR field, the ionizing XUV pulse can be treated as perturbation, and the IR field is approximately a continuous wave. Similar approximations are made for LAES, with the initial and final states being laser-dressed states of the free electron in the IR field and the scattering potential is the perturbation. These theories provide the amplitudes

$$F_q^{q+n} = e^{in(\frac{\pi}{2}+\omega\Delta t)} J_n\left(\frac{k_q F_0^{\mathrm{IR}}}{\omega^2}\right) f_{q+n}^{\mathrm{PI}} \tag{3.10}$$

$$f_q^{q+n} = e^{in(\frac{\pi}{2}+\omega\Delta t)} J_n\left(\frac{(k_q - k_{q-n}) F_0^{\mathrm{IR}}}{\omega^2}\right) f_{q+n,q}^{\mathrm{ES}} \tag{3.11}$$

where $F_0^{\mathrm{IR}}$ is the maximum field strength of the IR pulse, $f_{q+n}^{\mathrm{PI}}$ is the field-free photoionization amplitude, and $f_{q+n,q}^{\mathrm{ES}}$ is the field-free elastic scattering amplitude. $J_n$ are the Bessel function of the first kind.

The side band intensity in this multi-path model is calculated as

$$I \propto |f|^2 = A\cos\left(2\omega(\Delta t + \tau_{\mathrm{PI}} + \tau_{\mathrm{nl}})\right), \tag{3.12}$$

where the photoionization delay is given by the phase differences of the LAPE amplitudes,

$$\tau_{\mathrm{PI}} = \frac{\arg(F_q^{q+1}) - \arg(F_q^{q-1})}{2\omega}. \tag{3.13}$$

With some algebra, the non-local delay can be approximated as

$$\tau_{\mathrm{nl}} \approx \frac{1}{\omega}\frac{A_{\mathrm{nl}}}{A_{\mathrm{l}}} \sin\left(\frac{1}{2}\left(k^+ x_{\mathrm{s}} + \phi^+\right)\right) \times \sin\left(\frac{1}{2}\left(k^- x_{\mathrm{s}} + \phi^-\right)\right), \tag{3.14}$$

where $x_{\mathrm{s}}$ represents the distance between the potentials responsible for LAPE and LAES processes, $A_{\mathrm{l}}$ and $A_{\mathrm{nl}}$ correspond to the amplitudes of these two processes, respectively, i.e.,

$$A_{\mathrm{l}} \approx |f_{\mathrm{L1}} + f_{\mathrm{L3}}| \approx |f_{\mathrm{L2}} + f_{\mathrm{L4}}| \tag{3.15}$$

$$A_{\mathrm{nl}} \approx |f_{\mathrm{NL1}}| \approx |f_{\mathrm{NL2}}|, \tag{3.16}$$

and the phase shifts of the two sine functions are[1]

$$\phi^+ = \arg(f_q^{q+1}) - \arg(f_q^{q-1}) \tag{3.17}$$

$$\phi^- = \arg(f_q^{q+1}) - 2\arg(1 + f_q^q) + \arg(f_q^{q-1}). \tag{3.18}$$

[1] The 1 in $1 + f_q^q$ comes from combination of the paths L1 and L3 or L2 and L4.

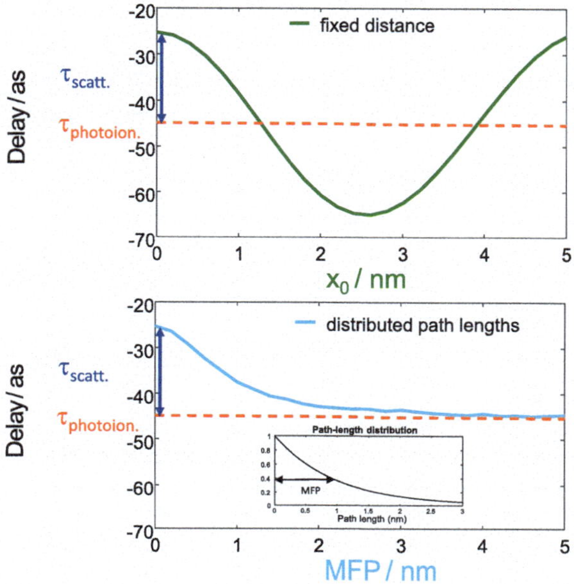


**Fig. 3.6 Effect of path-length distribution.** The top panel illustrates the case of a fixed distance $x_0$ between photoionization and scattering, whereas the bottom panel represents the case where the path lengths follow an exponential distribution. In the bottom panel, the delay is given as a function of the average path length (MFP)

The period of oscillation with $x_s$ is characterized by the two wave numbers

$$k^+ = k_{q+1} - k_{q-1} \tag{3.19}$$

$$k^- = k_{q+1} - 2k_q + k_{q-1}. \tag{3.20}$$

Clearly, the non-local delay (3.14) exhibits a beat pattern, i.e., a superposition of two waves with slightly different oscillation frequencies. In this case, the two frequencies giving rise to the beat pattern are $k_{q+1} - k_q$ and $k_q - k_{q-1}$, whose sum is $k^+$ (the fast oscillation with $x_s$) and whose difference is $k^-$ (the slow oscillation with $x_s$). We also note that due to the phase shifts the non-local delay starts close to its maximum and the oscillation is cosine-like.

The first step in translating these results from a one-dimensional model system to the condensed phase consists of including the effect of path-length distributions. The distance traveled by an electron between photoionization and the first collision or between two collisions is, in general, a distributed quantity, the average of which is defined as the mean-free path. For simplicity, we first consider only one scattering event and we assume an exponential distribution of path lengths, which is the standard assumption in condensed-phase electron-transport simulations (see, e.g., Ref. [21]).

Figure 3.6 shows the effect of the path-length distribution on the calculated delay. The top panel shows the results obtained when the distance between photoionization and scattering has a fixed value $x_0$, whereas the bottom panel shows the case where the distance $x_s$ is a distributed quantity following an exponential distribution, as a function of the average path-length MFP. We find that the oscillatory behavior from the fixed-distance case is turned into a monotonic decay from $\tau_{\text{photoion}} + \tau_{\text{scatt}}$ to

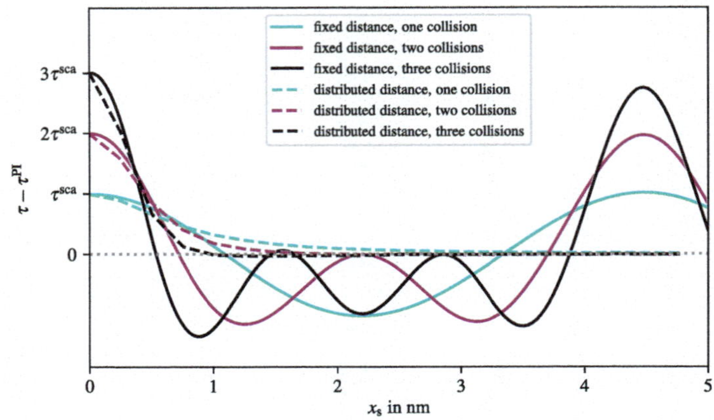


**Fig. 3.7 Effect of multiple collisions.** Calculated scattering-induced delay for the case of a finite number of collisions separated by fixed distances $x_0$ (full lines) or path lengths that follow an exponential distribution of average $x_0$ (dashed lines)

$\tau_{\mathrm{photoion}}$. This result has an intuitive explanation. When the MFP is long, specifically MFP$\geq L$, where $L = 2/k_+$, the contribution of scattering cancels because of the oscillatory nature of the delay dependence on the path length. When MFP$\ll L$ the delays behave additively, i.e., the classical situation is recovered.

In a second step, we now include the treatment of multiple collisions. Figure 3.7 shows the results obtained for $n = 1 - 4$ collisions. In the case of fixed distances (full lines), the total delay can be expressed as

$$\tau_{\mathrm{nl}} \approx \tau_{\mathrm{sca}} \sum_{j=1}^{n} \cos(nkx_{\mathrm{s}}), \tag{3.21}$$

where $\tau_{\mathrm{scatt}}$ is the scattering delay. Thus, at $x_{\mathrm{s}} = 0$, the non-local delay is ca. $\tau_{\mathrm{photoion}} + n\tau_{\mathrm{scatt}}$. In the case of distributed path lengths (dashed lines), a monotonic decay from the same values at $x_{\mathrm{s}} = 0$ to an asymptotic value of 0 is observed.

These results are in line with the same intuitive explanation given above. For long MFPs the contribution of scattering vanishes, such that the calculated delay is equal to the photoionization delay. For very short MFPs ($\ll L$), the delays again behave additively, i.e., the classical limit is reached.

### 3.2.2.3 Scattering Delays and Differential Scattering Cross Sections of Water

The previous section has established the conceptual foundations of attosecond interferometry in condensed matter on the basis of simple model potentials. The purpose

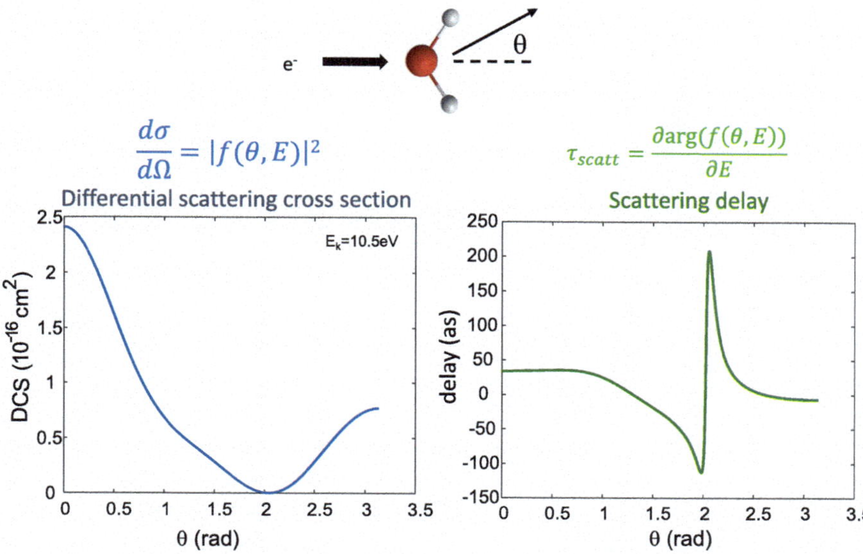


**Fig. 3.8 Calculations of DCS and scattering delay for water monomer**. These calculations were done for an electron kinetic energy of 10 eV using the commercial R-matrix code Quantemol, in which the calculation of the DCS is implemented

of this section is to present the results of state-of-the-art quantum scattering calculations on water clusters as a computationally tractable model of liquid water.

Beginning with an isolated water molecule, we calculate the complex scattering factor for the fully elastic scattering process of a free electron with a water molecule in its rovibronic ground state. In our calculations, we use orientational averaging of the scattering target. As a consequence, the scattering problem has cylindrical symmetry, such that the scattering factor $f(\theta, E)$ depends only on the polar angle $\theta$ (see Fig. 3.8) and the electron kinetic energy $E$. This scattering factor defines both the differential scattering cross section (DCS) $\frac{d\sigma}{d\Omega}$ and the scattering delay $\tau_{\mathrm{scatt}}$ as defined in Fig. 3.8.

The DCS has a global maximum at $\theta = 0$, corresponding to forward scattering and a local maximum at $\theta = \pi$, corresponding to back scattering. The scattering delay in the forward direction is positive, corresponding to the naive classical expectation for a repulsive interaction. The scattering delay displays a rapid variation around $\theta \approx 2$ rad, i.e., at the same angle at which the DCS displays a local minimum. A detailed analysis of these calculations shows that the minimum in the DCS originates from a destructive interference of partial waves, which also leads to a rapid variation of the scattering delay.

Whereas these calculations are representative of electron scattering in the gas phase, and their quantitative accuracy has been verified [63], they may not be representative of electron scattering in liquid water. Condensation is indeed known to modify the electronic structure of molecules, particularly in the case of water because

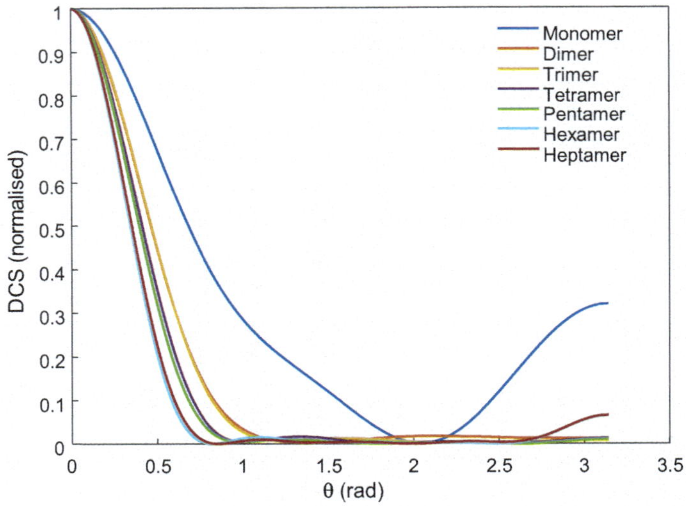


**Fig. 3.9 Effect of condensation on the DCS of electron scattering.** Calculated DCS for electron scattering with water clusters of the indicated size using Quantemol. The equilibrium structures of the most stable isomers of the water clusters of each size have been taken from [64]. All DCS have been normalized at $\theta = 0$ for comparison

of strong hydrogen bonding. We have therefore studied the evolution of the DCS with the size of the water cluster. The results are shown in Fig. 3.9.

The striking observation from these calculations is the rapid convergence of the DCS with increasing cluster size. The DCS evolves from a double-maximum structure with a relatively broad forward-scattering peak towards a quasi-Gaussian DCS without backscattering. The width of the quasi-Gaussian peak converges very rapidly with cluster size.

The rapid convergence of the DCS with the size of water clusters suggests that the highly accurate quantum-scattering calculations presented in this section can be used to model the description of electron scattering in liquid water. In addition to providing the crucial scattering factors required to model attosecond interferometry, they additionally enable us to determine both elastic and inelastic mean-free path for electron scattering in liquid water [65], as discussed in the following section. Similar results have recently been reported using ePolyScat, whereby a very similar convergence of the DCS has been found [66].

#### 3.2.2.4 Elastic and Inelastic Mean-Free Paths for Electron Scattering in Liquid Water

Given the DCS for electron scattering at a certain kinetic energy, we are now in the position to determine the elastic and inelastic mean-free paths (EMFP and IMFP) [65]. For this purpose, we turn to two liquid-jet measurements of the photoioniza-

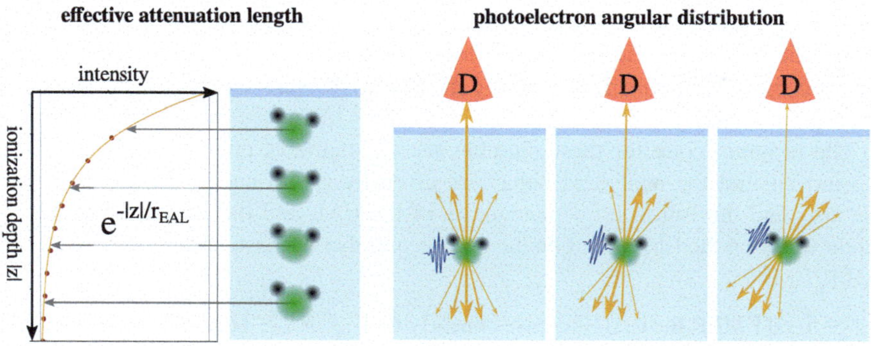


**Fig. 3.10** Left: Schematic depiction of the effective attenuation length (EAL) for ionization in a liquid. The signal of photoelectrons with kinetic energy corresponding to direct ionization decreases exponentially with the depth of the ionization site due to inelastic scattering. The EAL is the width parameter of the exponential function. Right: Concept of the measurement of the photoelectron angular distribution (PAD) of a liquid: Molecules are ionized with a beam and the photoelectron signal is measured with a detector above the surface that has a relatively small detection angle (indicated as "D"). Variation of the beam polarization axis yields the PAD of the liquid

tion of water: A measurement of the effective attenuation length (EAL) [67] and a measurement of the gas-phase and liquid-phase photoelectron angular distributions (PAD) [68] for oxygen 1s photoionization of water molecules. The underlying ideas of the experiments and of our simulations are depicted in Figure 3.10. For ionization at a depth $|z|$ below the surface the measured photoelectron signal at the kinetic energy corresponding to direct photoionization behaves as

$$S(z) \propto e^{-|z|/r_{\mathrm{EAL}}}, \tag{3.22}$$

where $r_{\mathrm{EAL}}$ is the EAL. The loss of signal with increasing $|z|$ is due to inelastic scattering, and the EAL is also a lower bound for the IMFP. We can simulate the experiment with our Monte-Carlo trajectory code by using the field-free quantities and by simply measuring the number of electrons exiting the liquid depending on the assumed ionization depth.

The PAD can, for achiral molecules, be described with an energy-dependent asymmetry parameter $\beta$ as

$$\mathrm{PAD}(\theta) \propto 1 + \beta P_2(\cos\theta), \tag{3.23}$$

where $\theta$ is the polar angle and $P_2$ is the Legendre polynomial of second order. $\beta$-parameters were measured in [68] for gas-phase and for liquid water by rotating the polarization direction of the ionizing beam relative to the detector. In particular, the detector is not moved relative to the surface of the liquid jet, as indicated in Figure 3.10. We can simulate also this experiment with our Monte-Carlo trajectory code by rotating the PAD for ionization, which we take to be that of gas-phase water, as oxygen 1s-electrons are considered and we do not expect the corresponding orbitals to be influenced by solvation, except for a small shift of the binding energy. The detected signal is obtained by counting the electrons which exit the surface under a small polar angle.

The program code for these simulations, CLstunfti, is publicly available [69]. We indeed find the analytical behaviors given by (3.22) and (3.23) numerically. We also find the following dependencies of the EAL and the $\beta$-parameter $\beta_{\text{liq}}$ of liquid water on the EMFP and on the average number of elastic scatterings $\langle N_{\text{ela}} \rangle =$ IMFP/EMFP:

- For fixed EMFP, the EAL increases linearly with $\sqrt{\langle N_{\text{ela}} \rangle}$. This is because the EAL is essentially the root-mean-square translation distance of a random walk [70].
- For fixed $\langle N_{\text{ela}} \rangle$, the EAL increases linearly with the EMFP. This happens because, for fixed $\langle N_{\text{ela}} \rangle$, both the EMFP and the IMFP are scaled by the same amount and hence essentially all trajectories are also just scaled accordingly.
- For fixed $\langle N_{\text{ela}} \rangle$, $\beta_{\text{liq}}$ does not depend on the EMFP. This independence is also a result of the discussed scaling, as only trajectories are scaled but angles are invariant to changing both EMFP and IMFP by the same amount.
- For fixed EMFP, $\beta_{\text{liq}}$ decreases to zero with increasing $\langle N_{\text{ela}} \rangle$, i.e., the PAD becomes closer and closer to the isotropic PAD with each additional collision.

From these dependencies follows that a set of parameters ($r_{\text{EAL}}$, $\beta_{\text{liq}}$) corresponds to a unique set of mean-free paths (EMFP, IMFP). We use this fact to find the EMFP and IMFP needed for our simulations. For sideband 14 we obtain an EMFP of 0.6 nm and $\langle N_{\text{ela}} \rangle \approx 7$, and for sideband 20 we obtain a slightly longer EMFP of 0.8 nm and $\langle N_{\text{ela}} \rangle \approx 6$. The values for the EMFP and IMFP from 10 eV to 300 eV determined with this approach can be found in [65]. A further refined scattering model, which incorporates the extrapolation of the EMFP and IMFP values reported in Ref. [65], can be found in Ref. [66].

#### 3.2.2.5 Theoretical Results for Liquid Water

In this section, we combine the results described in the preceding subsections to a comprehensive model of APES in liquids. Starting from the discussion of the one-dimensional model system introduced in Sect. 3.2.2.2, we generalize these results to three dimensions and include the output from *ab-initio* quantum scattering calculations on realistic models of liquid water. For this purpose, we need the elastic and inelastic mean-free paths for electron transport in liquid water. These were determined based on two independent experiments on liquid-water microjets that we have inverted within our new formalism described in the previous section. The result is a fully self-consistent description of liquid-phase attosecond spectroscopy that incorporates the newly discovered non-local processes, as well as a state-of-the-art description of time delays in photoionization and scattering.

As a first step towards interpreting the experimentally measured time delays, we return to the one-dimensional model system discussed in Sect. 3.2.2.2. The attractive

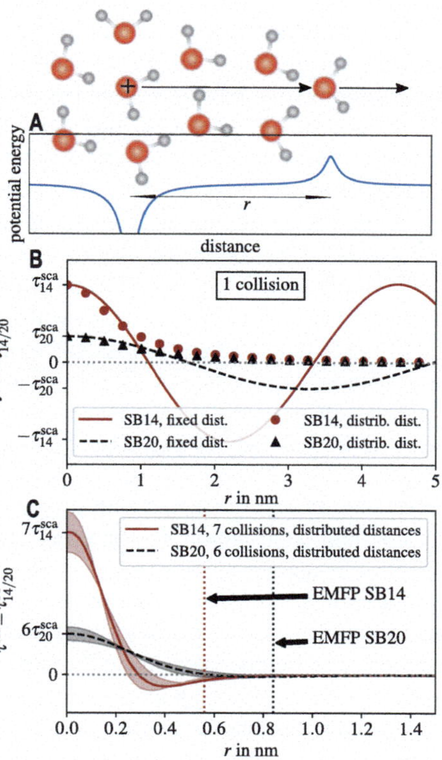


**Fig. 3.11 Contributions of photoionization and scattering to the measured delays.** Schematic representation of the potentials used in the TDSE calculations (top) modeling the photoionization and scattering as attractive and repulsive potentials, respectively. The resulting delays ($\tau_{\text{tot}}$) for the case of a single collision (middle) oscillate between $\tau_{\text{SB}}^{\text{PI}} + \tau_{\text{SB}}^{\text{sca}}$ and $\tau_{\text{SB}}^{\text{PI}} - \tau_{\text{SB}}^{\text{sca}}$ as a function of the distance $r$ (lines). In the case of an exponential path-length distribution with average $r$, a monotonic decay is obtained (symbols). In the case of $n$ elastic collisions, sampled according to an exponential path-length distribution with average $r$, the delay decays from $\tau_{\text{SB}}^{\text{PI}} + n\tau_{\text{SB}}^{\text{sca}}$ to $\tau_{\text{SB}}^{\text{PI}}$ (bottom). Adapted from Ref. [24]

potential is now chosen to mimic the ionized water molecule, whereas the repulsive potential represents electron scattering with a neutral water molecule, as illustrated in Fig. 3.11a. We additionally choose the photon energies to be those used in the experiment, i.e., harmonic orders 13/15 and 19/21 of an 800-nm driving field. Panel (b) shows the calculated delays as a function of the distance $r$ between the potentials representing photoionization and scattering. The lines show the results of calculations done with fixed distances, whereas the symbols correspond to calculations assuming an exponential distribution of path lengths. These results confirm the conclusions of Sect. 3.2.2.2. Specifically, in the case of distributed distances, the total delay decays monotonically from $\tau_q^{\text{PI}} + \tau_q^{\text{sca}}$ to $\tau_q^{\text{PI}}$ with increasing mean-free path.

With the availability of the EMFP and IMFP for liquid water at the relevant electron-kinetic energies from Sect. 3.2.2.4, we can now address the role of scattering delays in attosecond spectroscopy of liquid water within the one-dimensional model. The ratio IMFP/EMFP defines the number of elastic collisions that take place (on average) before one inelastic collision occurs. This ratio amounts to $\sim$7 in the case of SB14 and $\sim$6 in the case of SB20. Panel (c) therefore shows the delays obtained from one-dimensional simulations assuming these numbers of collisions, separated by distributed path lengths, as a function of the average path length (or EMFP) $r$. Although the variation of the calculated delay with $r$ is no longer strictly monotonic, as a consequence of the more rapid variations of the delay with $r$ in the case of multiple collisions (see Ref. [51] for details), the rapid convergence of both delays to $\tau_q^{\mathrm{PI}}$ is still obvious. Most importantly, when using the EMFP values determined in Sect. 3.2.2.4, we find that both delays have practically converged to $\tau_q^{\mathrm{PI}}$. In other words, the one-dimensional calculations already suggest that the contributions of electron scattering during transport in liquid water are negligible under our experimental conditions. Although these results were obtained with simple model potentials, they already show that any possible contribution of scattering delays will be strongly reduced as a consequence of the nature of the non-local processes.

Before attempting an interpretation of the experimental results, we need to generalize this formalism to three dimensions and include a realistic description of photoionization and scattering in liquid water. Since the fully quantum-mechanical time-dependent description of electron transport in liquid water is computationally intractable, we opted for a hybrid quantum-classical approach. The analysis of the TDSE calculations in Sect. 3.2.2.2 has shown that a quantum-mechanical description of photoionization and scattering within the Kroll-Watson formalism, combined with a classical-trajectory representation of electron transport, produces results of quantitative accuracy for electrons with kinetic energies $E_{\mathrm{k}} \gtrsim 10$ eV. Independent earlier work has shown that electron-transport simulations in liquid water based on a classical-trajectory Monte-Carlo description also reach quantitative agreement with a fully quantum simulation of electron transport, provided that the kinetic energy is larger than $\sim$10 eV [71]. We therefore based our description of liquid-phase attosecond spectroscopy on the same approach as Sect. 3.2.2.2, i.e., a quantum-mechanical description of photoionization, scattering and their laser-assisted counterparts and a classical-trajectory Monte Carlo description of the electron transport steps.

This is done as follows:

- We assume a flat interface between the liquid and the vacuum at $z = 0$ (called the "surface"), where $z < 0$ is inside the liquid. Many trajectories are initialized at a randomly selected starting depth $z_0 < 0$. Those trajectories, taken together, represent the photoelectron wave packet originating from one ionization event, thus each trajectory has a phase.
- Half of these trajectories are initialized in state $n = q - 1$ and the other half in state $n = q + 1$, where a "state" corresponds to the kinetic energy $n\hbar\omega - E_{\mathrm{b}}$.
- The generalization of formula (3.10) to three dimensions reads

$$F_q^{q+n} = e^{in(\frac{\pi}{2}+\omega\Delta t)} J_n\left(\frac{\mathbf{k}_q \cdot \mathbf{F}_0^{\mathrm{IR}}}{\omega^2}\right) f_{q+n}^{\mathrm{PI}}, \tag{3.24}$$

which describes the amplitude of changing from state $q+n$ to state $q$. We know the photoelectron angular distribution $|f_{q+n}^{\mathrm{PI}}(\theta)|^2$ as a function of the polar angle $\theta$ from photoionization calculations described in Sect. 3.9 and we do not need the phase arg $f_{q+n}^{\mathrm{PI}}(\theta)$, because we are only interested in the non-local contribution $\tau_{\mathrm{nl}}$ to the total measured delay $\tau = \tau_{\mathrm{PI}} + \tau_{\mathrm{nl}}$, cf. (3.2)[2]
Equation (3.24) can be used to model the interaction of the electron with the assisting IR field during photoionization. This is done by randomly selecting trajectories to change to states $q-2$, $q$, or $q+2$, according to the $|F_{q-2}^{q-1}|$ (emission), $|F_{q-1}^{q-1}|$ (no photon exchange), and $|F_q^{q-1}|$ (absorption) if the trajectory is initially in state $q-1$, and according to the amplitudes $|F_{q+2}^{q+1}|$ (absorption), $|F_{q+1}^{q+1}|$ (no photon exchange), and $|F_q^{q+1}|$ (emission) if the trajectory is initially in state $q+1$.[3] If a state change happens, the corresponding trajectories obtain a phase as given by (3.24) (without the unknown phase due to $f_{q+n}^{\mathrm{PI}}$). We note that as the laser field polarization is always assumed to be in $z$-direction, $F_q^{q+n}$ depends only on the offset $\Delta t$ between ionizing XUV and assisting IR pulse, the ionization direction, and the initial/final states.
- In the end, only trajectories reaching the surface in state $q$ contribute to the sideband of interest and hence are the only trajectories included in the determination of $\tau_{\mathrm{nl}}$. The probability of ending up in state $q$ is negligible if the trajectories end up once in $q \pm 2$, hence those trajectories are discarded.
- We use mean-free paths $r_{\mathrm{MFP}}$ based on the general definition

$$P(r) = \frac{1}{r_{\mathrm{MFP}}} e^{-r/r_{\mathrm{MFP}}} \tag{3.25}$$

to model random scattering in the medium. Each trajectory is attributed a randomly chosen maximal path to travel according to an IMFP $r_{\mathrm{IMFP}}$. This IMFP reflects (electronically) inelastic scattering such that the scattered electrons loose enough energy not to be detected at the kinetic energy of the considered sideband. If a trajectory reaches this maximum path length before reaching the surface, it is discarded.[4]

---

[2] We use two sets of angles: The "lab" frame with polar angle $\theta$ and azimuthal angle $\phi$, where $\theta$ is the angle relative to the $z$-axis define via the surface normal, and the relative frame with polar angle $\vartheta$ and azimuthal angle (never explicitly used here), where $\vartheta$ is the angle relative to the direction of motion of the electron/trajectory. For the simulations described here the polarization of the laser pulses is assumed to be in $z$-direction, hence there is no difference between lab frame and relative frame for ionization. This is different for scattering.

[3] The probability to stay in the respective states is obtained from $|F_{q-1}^{q-1}|$ and $|F_{q+1}^{q+1}|$. For the relevant parameters, absorption or emission of two IR photons is irrelevant.

[4] There is another way to implement the IMFP which is computationally more efficient if many IMFP values should be tested: No maximum path length is set but the contributions of the trajectories

- The trajectories are moved in the direction chosen randomly according to $|f^{\mathrm{PI}}_{q+n}(\theta)|^2$ and for a distance $r$ chosen randomly according to (3.25) with EMFP $r_{\mathrm{EMFP}}$. At this distance, elastic scattering is assumed to happen. A phase $k_n r$ is added to the phase of the trajectory, where $k_n$ is the momentum that the electron has in its current state.
- The generalization of formula (3.11) for laser-assisted scattering to three dimensions is

$$f_q^{q+n} = e^{in(\frac{\pi}{2}+\omega\Delta t)} J_n\left(\frac{(\mathbf{k}_q - \mathbf{k}_{q-n}) \cdot \mathbf{F}_0^{\mathrm{IR}}}{\omega^2}\right) f^{\mathrm{ES}}_{q+n,q} \tag{3.26}$$

where $f^{\mathrm{ES}}_{q+n,q}(\vartheta)$ is the field-free scattering amplitude depending on the polar angle $\vartheta$ relative to the current direction of motion, $\mathbf{k}_{q-n}$ is the momentum vector of the incoming electron, and $\mathbf{k}_q$ is the momentum vector of the outgoing electron. For each elastic scattering, a probability to change states as well as new directions for the motion of the electrons is chosen according to the amplitudes $f_q^{q+n}$. If a state change happens, the corresponding phase is added to the phase of the trajectory.
- From $|f_q^{q+n}|^2$ new directions for the outgoing electron trajectories are determined and the process of elastic scattering is repeated.[5]
- Trajectories are stopped if they end in states $q \pm 2$, if their path is longer than the previously sampled corresponding maximum path, or if they reach the surface at $z = 0$.
- Trajectories reaching the surface in state $q$ which originate from the same starting point are added coherently. The contribution to the total second-order amplitude originating from a given point is calculated by coherent summation over all trajectories, which reach the detector with total acquired phases $\gamma_j$:

$$f_k^{\mathrm{W}}(\Delta t) = \sum_j^{n_k^{\mathrm{det}}} f^{\mathrm{t}}_{j,k} = \sum_j^{n_k^{\mathrm{det}}} e^{\gamma_j(\Delta t, z_k^{\mathrm{ini}})}. \tag{3.27}$$

- Simulations are run for a large number of initial positions $z_j^{\mathrm{ini}}$. As it is assumed that electrons originating from different initial positions $z_j^{\mathrm{ini}}$ in the liquid do not interfere, these contributions are summed incoherently,[6] so that the total signal for a given value of the offset $\Delta t$ is

reaching the surface in state $q$ to the total signal are weighted as $e^{-r_{\mathrm{tot}}/r_{\mathrm{MFP}}}$, where $r_{\mathrm{tot}}$ is the total path that the respective trajectory traveled until reaching the surface.

[5] We note that $f_q^{q+n}$ depends on three angles, for example, on the polar angle of the incoming electron trajectory $\theta_{\mathrm{in}}$ and the polar and azimuthal angles $\theta_{\mathrm{ou}}$, $\phi_{\mathrm{ou}}$ of the outgoing electron trajectory in the lab frame, as well as on the offset $\Delta t$ and the initial and final state of the electron. If the laser field polarization axis was not aligned with the $z$-axis of the lab frame (defined by the surface normal), $f_q^{q+n}$ would also depend on the azimuthal angle $\phi_{\mathrm{in}}$ of the incoming electron trajectory.

[6] There is, however, no significant change to the result if the summation is done coherently.

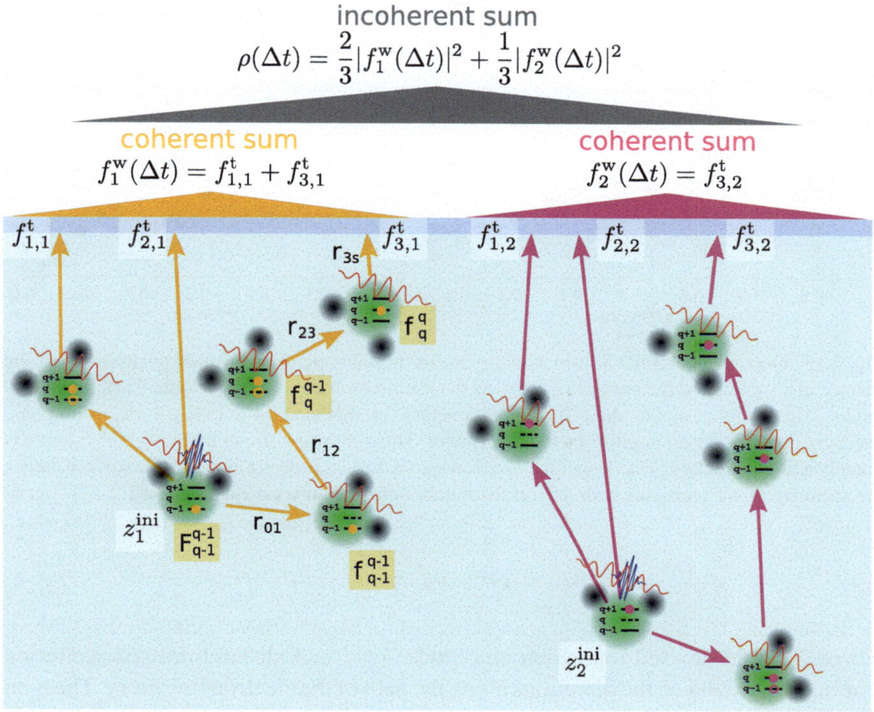


**Fig. 3.12 Three-dimensional model of attosecond interferometry in the condensed phase** Sketch of possible trajectories and how they are detected, i.e., how the intensity in the sideband $q$ is calculated as a function of the offset $\Delta t$ between XUV and IR pulse (the experimental control parameter). On the left side, three trajectories are shown which originate from a depth $z_1^{\mathrm{ini}}$, two of which end up in state $q$ ($f_{1,1}^{\mathrm{t}}$ and $f_{3,1}^{\mathrm{t}}$; for the latter, also path lengths $r_{jk}$ and amplitudes of the individual processes are given). The coherent sum of the corresponding amplitudes provides the contribution from $z_1^{\mathrm{ini}}$. On the right side, three trajectories originating from $z_2^{\mathrm{ini}}$, where only one trajectory ends in state $q$ ($f_{3,2}^{\mathrm{t}}$). Contributions for different $z_j^{\mathrm{ini}}$ are summed incoherently to obtain the sideband intensity $\rho(\Delta t)$

$$\rho(\Delta t) = \sum_k \frac{n_k^{\mathrm{det}}}{n^{\mathrm{tot}}} |f_k^{\mathrm{w}}(\Delta t)|^2. \tag{3.28}$$

Figure 3.12 illustrates the simulation procedure schematically. In the figure, two ionization sites (at $z_1^{\mathrm{ini}}$ and $z_2^{\mathrm{ini}}$) are depicted, from which three trajectories each are started. Let us first concentrate on one trajectory, i.e., the one with amplitude $f_{3,1}^{\mathrm{t}}$. This trajectory starts in state $q-1$, scatters once at a distance $r_{01}$ without photon exchange, scatters a second time at a distance $r_{12}$ and absorbs a photon to change to state $q$, and scatters one last time without photon exchange after a distance $r_{23}$ before it reaches the surface in state $q$ and is thus counted to the signal. The amplitude for the trajectory is thus

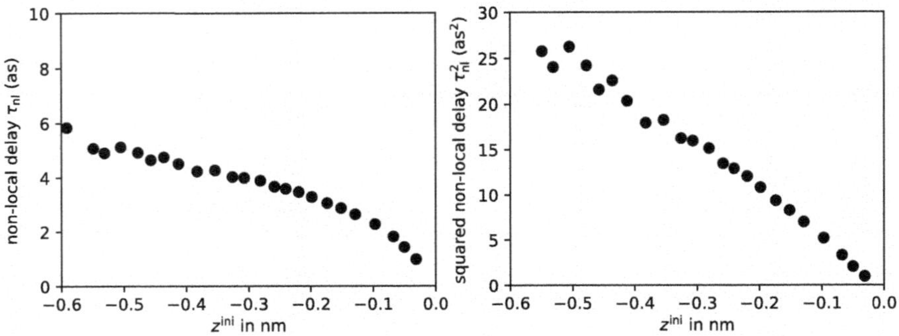


**Fig. 3.13** Dependence of the non-local delay $\tau_{\text{nl}}$ for liquid water on the initial position $z^{\text{ini}}$ of the ionized electron below the surface ($z = 0$), as determined for the classical Monte-Carlo trajectory model. This simulation was done for the parameters for sideband 14 (ca. 10 eV electron kinetic energy) of an attosecond interferometry experiment. The right panel shows that $\tau_{\text{nl}}(z^{\text{ini}})^2$ is approximately a linear function. The deeper the electron starts, the less trajectories reach the surface, hence the accuracy of $\tau_{\text{nl}}$ decreases with $|z^{\text{ini}}|$ if the number of trajectories is not increased

$$f_{3,1}^{\text{t}} = F_{q-1}^{q-1}\, e^{ik_{q-1}r_{01}}\, f_{q-1}^{q-1}\, e^{ik_{q-1}r_{12}}\, f_{q-1}^{q}\, e^{ik_q r_{23}}\, f_q^q\, e^{ik_q r_{3s}}, \tag{3.29}$$

where the field-dressed ionization amplitude $F_{q-1}^{q-1}$ and all field-dressed scattering amplitudes depend on the directions along the path of the electron trajectory. The final signal obtained from the six trajectories is obtained as given in the figure. We note that the actual simulation uses millions to billions of trajectories for each ionization site, hundreds to thousands of ionization sites for convergence of $\tau_{\text{nl}}$, and most of the trajectories do not end in state $q$ as the transition probabilities are relatively small.

Using the input parameters described in Sects. 3.2.2.3–3.2.2.4, we can simulate the contribution of the non-local delay to the total delay of APES for liquid water. We find that the non-local delay is almost independent of the average number of elastic collisions $\langle N_{\text{ela}} \rangle \in [5, 14]$ and decays with an increasing EMFP $\in [0.1, 1.3]$nm. A decrease of $\tau_{\text{nl}}$ with increasing EMFP is expected from the one-dimensional model, as we see a decay of $\tau_{\text{nl}}$ with the mean-free path, cf. 3.2.2.2. The independence of $\tau_{\text{nl}}$ on $\langle N_{\text{ela}} \rangle$ in the tested range is surprising but, as shown below, $\tau_{\text{nl}}$ is also rather small. This independence may originate from a suppression of coherence by random scattering, or from the dominance of relatively short trajectories originating from close to the surface.

Figure 3.13 shows the dependence of the non-local delay (for EMFP=0.55 nm and $\langle N_{\text{ela}} \rangle = 11$) on the initial position of the trajectories below the surface. We can see that $\tau_{\text{nl}}$ becomes larger with starting depth, but is only ca. 5 as for an initial depth of 0.5 nm. As the contribution to the total signal also decays exponentially with starting depth, this graph already suggests that $\tau_{\text{nl}}$ is small compared to the experimentally measured value of the delay. We find indeed that according to our simulations, the contribution of $\tau_{\text{nl}}$ to the total delay is only about 2 as.

Thus, we conclude that the non-local delay is negligible in our attosecond experiments on liquid water. Nevertheless, there are situations where the non-local delay may play a role, for example, if the EMFP is significantly shorter than in the present case or if the considered medium has long-range order, such as, e.g., a crystal. For these cases, the non-local delays may become measurable, which would provide additional information regarding the scattering delays and mean-free paths.

### 3.2.3 *Interpretation*

We now return to the interpretation of the experimental results described in Sect. 3.2.1. These measurements have yielded time delays between photoemission from liquid and gaseous water of $69\pm20$ as at 21.7 eV photon energy and $48\pm16$ as at 31.0 eV photon energy. In addition, relative modulation contrasts of $0.17\pm0.03$ and $0.45\pm0.06$ have been determined at the mentioned photon energies. In Sect. 3.2.2, we have introduced the concepts relevant to interpret these experimental results and quantitative methods to model such experiments. These calculations have shown that attosecond spectroscopy of liquids is, in general, sensitive to photoionization delays, scattering delays, and mean-free paths. In the case of liquid water at the relatively low photon energies studied so far, the contributions of electron scattering and transport are predicted to be negligible. This leaves one main possible source for the measured delays, i.e., the photoionization step.

To evaluate the possible contribution from the photoionization step to the measured delay, we follow the same strategy as in Sects. 3.2.2.3–3.2.2.4, i.e., we use accurate quantum-scattering calculations on water clusters as a model for liquid water. Our calculations are based on the formalism described in Ref. [26], which was developed to interpret the first measurements of photoionization delays in isolated molecules [9]. Whereas the experimental measurements can at present only access relative time delays, e.g., between gaseous and liquid water, the calculations yield absolute time delays. We will therefore discuss the absolute photoionization time delays of water clusters and isolated water molecules. Finally, we will compare the relative time delays to the experimental results.

There are several reasons why photoionization time delays of water molecules can be expected to change under the effect of condensation. First, the valence electronic structure of water molecules changes with the addition of neighboring molecules. This is a consequence of the relatively strong dipole moment of the water molecule, which causes the formation of strong hydrogen bonds between water molecules. The consequence is a significant change in the electronic structure, which becomes visible in the hybridization of the molecular valence orbitals. In addition to the local change of the electronic wave function, condensation also causes a partial delocalization of the one-electron wavefunctions (orbitals) over more than one molecule. These effects concern the initial bound electronic wave function. They will all have an influence on the photoionization delay. Second, the final continuum wave function is also affected by condensation, probably even more than the initial state. The continuum wavefunction for bulk liquid water is very complex. It consists of a complicated conduction-band part inside the liquid bulk and a simpler part in the vacuum region

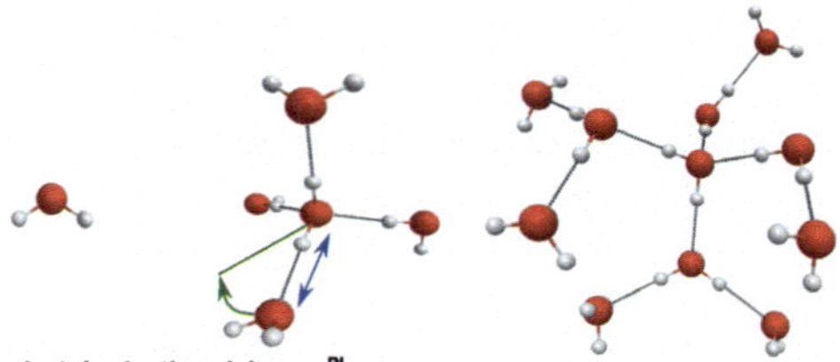

photoionization delays $\tau^{PI}$

| photon energy (side-band order) | $H_2O$ | $(H_2O)_5$ | $(H_2O)_{11}$ | theory $\tau^{PI}_{(H_2O)_{11}} - \tau^{PI}_{H_2O}$ | experiment $\tau_{liq} - \tau_{gas}$ |
|---|---|---|---|---|---|
| 21.7 eV (SB14) | 49 as | 97 as | 110 as | 61 as | 69±20 as |
| 31.0 eV (SB20) | 18 as | 29 as | 48 as | 30 as | 49±16 as |

| | $(H_2O)_5$ stretched | $(H_2O)_5$ bent | $(H_2O)_{11}$ stretched | $(H_2O)_{11}$ bent | theory stretched | theory bent |
|---|---|---|---|---|---|---|
| 21.7 eV (SB14) | 71 as | 107 as | 111 as | 157 as | 62 as | 108 as |
| 31.0 eV (SB20) | 32 as | 36 as | 47 as | 41 as | 29 as | 23 as |

**Fig. 3.14 Effect of solvation and hydrogen bonding on photoionization delays**. Calculated photoionization delays for $H_2O$, $(H_2O)_5$ and $(H_2O)_{11}$, representing water molecules with zero, one complete or one complete and one partial solvation shells, respectively. The bottom row indicates delays obtained by stretching (blue arrow) or bending (green arrow) one hydrogen bond in the clusters. Adapted from Ref. [24]

outside the liquid. However, all that is needed to accurately describe photoionization is the continuum wave function over the spatial region where the initial-state wave function has a significant amplitude. Therefore, our calculations on water clusters can be expected to converge to the results for liquid water when the water clusters become sufficiently large as to describe the relevant spatial extension of the initial-state wave function. Both effects discussed in this paragraph are naturally included in our quantum-scattering calculations.

Figure 3.14 shows the calculated photoionization delays of water clusters of increasing size at the two photon energies relevant for this work. These calculations were done with ePolyScat [72, 73] using the methods described in Ref. [26]. The first two lines of the table show the photoionization delays of isolated water molecules, a water molecule with a complete first solvation shell (pentamer) and with a partial second solvation shell in which all dangling hydrogen bonds of the first shell have been coordinated (undecamer). Interestingly, the photoionization delays systematically increase with increasing size of the water cluster. Whereas the calculations at the lower photon energy show evidence of convergence with cluster size, the convergence is less obvious at the higher photon energy. Since converged quantum-scattering calculations on $(H_2O)_{11}$ reach the limit of current computational methods ($\sim$180 CPU days for one calculation), these calculations could not yet be extended to larger water clusters. However, the relative photoionization delays between the water clusters and the water monomer are in rewarding agreement with the experimental results. Whereas the calculated relative delay of 61 as at 21.7 eV agrees with the experiment within the error bar, the relative delay of 30 as at 31.0 eV is close to the error interval of the experiment.

In addition to reaching near-quantitative agreement with the experiment, the results in the two bottom lines of Fig. 3.14 shed light on the other experimental observable introduced in this work: the modulation contrast in attosecond interferometry. Whereas the calculations shown in the upper two lines assumed a tetrahedral coordination geometry of each water molecule with O-O distances fixed to 2.8 Å, the local solvation environment is known to be partially distorted in liquid water and subject to fluctuations on picosecond time scales. To simulate the effect of the structural distortions of the hydrogen bonds, we have stretched or bent one of the hydrogen bonds by 0.7 Å or 50°, respectively, following the distortions used in Ref. [74]. In the case of the pentamer, we have moved one of the water molecules, and in the case of the undecamer, we have moved a group of 3 water molecules while keeping its internal geometry unchanged. The effects of these local distortions of the hydrogen-bond structure are remarkable. The delays calculated at 21.7 eV display a pronounced sensitivity to the structure with variations reaching -16 as in the pentamer and +47 as in the undecamer. The corresponding changes at 31.0 eV are much smaller and do not exceed $\pm 7$ as. These different sensitivities of the delays to the local solvation structure are consistent with the very different modulation contrasts observed at the two energies, i.e., $0.17 \pm 0.03$ and $0.45 \pm 0.06$, respectively. This provides an important first indication that the origin of the observed modulation depths might be the distribution of local solvation structures in liquid water. It does not exclude, of course, that other contributions, such as electron decoherence during transport through the liquid phase, also contribute to the finite modulation depths.

These results complete the consistent picture that we have built up in this book chapter. Whereas our calculations predicted that the contributions of electron scattering and transport were small, such that only the photoionization delays would play a role, the independent calculations of these photoionization delays on water clusters have confirmed this prediction, reaching a near-quantitative agreement with the experimental delays. Most recently, attosecond electron-ion-coincidence spectroscopy on size-resolved water clusters has further confirmed that the increase of the time delays saturates around 4-5 molecules and that the photoionization delays relative to water monomer quantitatively agree with those measured for liquid water [27, 28]. The remarkable sensitivity of the delays to the local solvation structure moreover offers interesting perspectives for extending such measurements to other photon energies, other liquids, and solvated species. The investigation of local solvation structures around different types of solutes, e.g., anions vs. cations, is particularly promising, as well as the investigation of other attosecond time-scale processes in the liquid phase.

## 3.3 High-Harmonic Spectroscopy of Liquids

High-harmonic generation (HHG) in gases has been extensively studied over the last three decades. Apart from its quintessential significance as a coherent light source, it has also been developed into a unique spectroscopic technique. High-harmonic spectra indeed contain a wealth of information about the structure and dynamics of the medium from which they are emitted. In the gas phase, HHG can be understood as a process that involves strong-field ionization, electron propagation in the continuum, and photorecombination. As a consequence, high-harmonic spectra indeed contain information about the electronic structure of the medium which is encoded in the orientation-dependence of the strong-field-ionization and photorecombination dipole matrix elements. They also contain dynamical information as a consequence of the unique mapping from the transit time of the electron in the continuum to the emitted photon energy, when the contributions of the "short" electron trajectories are recorded. The applications of high-harmonic spectroscopy (HHS) have led to many important results, such as the tomography of molecular orbitals [75], the observation of structural and electronic dynamics on attosecond time scales [6, 76], the identification of two-center interference minima [77, 78], the observation of Cooper minima [79], the determination of ionization and recombination times in HHG [80], and the measurement and laser control of attosecond charge migration [8]. The field of HHS has recently been reviewed in Ref. [82].

High-harmonic generation in solids is a much younger and highly dynamic research field. Although harmonic generation up to the $7^{\text{th}}$ order of a MIR driver in ZnSe has already been observed in 2001 [83], the rapid development of HHS of solids has started with the observation of high-order harmonic generation in solids in 2011 [84], the analysis of inter- and intraband contributions to HHG [15, 85, 86] and the observation of extreme-ultraviolet HHG from solids [14]. HHS of solids has developed very rapidly, including impressive developments with terahertz drivers [87, 88], observations of crystal-structure effects [89], and the measurement of the Berry curvature of solids [16]. The field of solid-state HHS has recently been reviewed in Ref. [90].

In remarkable contrast to these developments, high-harmonic generation from liquids has barely been developed so far. Although relatively low-order harmonic generation from liquids in the visible domain has been observed in 2009 [32], it took until 2018 before HHG from bulk liquids was reported [25]. This situation is certainly to some extent the consequence of the technological challenges associated with HHG from liquids. Early efforts to observe liquid-phase HHG indeed date back to 2003, when HHG from liquid-water droplets was attempted [29]. Instead of observing the expected coherent HHG emission, the authors only observed incoherent emission from a plasma formed by multiple ionization of water molecules. Coherent XUV emission was only observed following the action of a pump pulse preceding the HHG pulse that caused a hydrodynamic expansion of the droplets [29]. These results were confirmed in later experiments and combined with a model of the droplet expansion [30, 31], which led to the conclusion that HHG was not observed at the density of liquid water but only appeared at densities that were significantly lower. The results reported in Ref. [25] contrast with this interpretation by showing that coherent HHG does occur at the density of liquid water. Finally, high-harmonic emission from a plasma created through the interaction of a laser with a liquid microjet has been observed in Ref. [33]. These results were obtained in the so-called coherent-wake-emission regime, which is reached for extremely high laser intensities ($I > 10^{18}$W/cm$^2$). In this regime, HHG takes place from laser-induced plasma dynamics, which are largely insensitive to the properties of the target used to create the plasma. In this section, we discuss the methods, results, and interpretations of HHG from bulk liquids, which form the basis of liquid-phase HHS, and introduce a new type of targets for strong-field science and the development of new high-harmonic light sources.

### 3.3.1 Experimental Methods

The first successful observation of extreme-ultraviolet HHG from bulk liquids was realized with the experimental setup illustrated in Fig. 3.15 [25]. The key enabling technology for this experiment was the flatjet technique [91, 92]. The flatjet is created by colliding two cylindrical microjets under an impact angle of 48°. In the present experiment, cylindrical jets with a diameter of 50 $\mu$m were used, resulting in a flatjet with a thickness of ~1-2 $\mu$m. Since the thickness of the flatjet scales quadratically with the diameter of the colliding jets under reasonable assumptions [91], the thickness can easily be reduced, which however comes at the cost of a smaller cross section of the flatjet. High-harmonic generation is realized by focusing a short-wave infrared (SWIR) femtosecond pulse centered at 1.5 $\mu$m onto the flatjet in normal incidence, as shown in Fig. 3.15a. The high-harmonic emission is detected by a flat-field spectrometer consisting of an entrance slit, a concave variable-line-spacing grating and a microchannel-plate detector backed with a phosphor screen and a charge-coupled device camera. A photograph of the flatjet under operating HHG conditions is shown in Fig. 3.15b. The green light originates from scattering of the third-harmonic radiation. Figure 3.15c shows the simultaneously observed high-harmonic spectra emitted from bulk liquid ethanol and the surrounding gas-phase ethanol. The two spectra were independently normalized to their maximal intensity. The comparison of the two spectra immediately reveals several characteristic differences, i.e., the liquid-phase harmonics have (i) a much lower cut-off, (ii) a larger divergence, and (iii) a rapidly decreasing intensity distribution compared to the gas-phase harmonics.

The most important aspect of liquid-phase HHG is the separation of the signals from the liquid and gas phases. The operation of a liquid jet in vacuum necessarily entails the presence of a surrounding gas phase created by evaporation from the liquid jet. The challenge of separating HHG from the gas and liquid phases is elegantly solved by the natural shape of the flatjet, as illustrated in Fig. 3.16. Since the thickness of the flatjet decreases from top to bottom across the first sheet, the interfaces are not parallel to each other but form a small angle with respect to each other. This leads to two consecutive refractions of the fundamental driving field upon entrance and exit of the liquid phase. The XUV radiation generated within the liquid jet is not (significantly) refracted when it exits the liquid because the index of refraction change is negligible in the XUV. This leads to a natural separation of liquid- and gas-phase HHG on the detector. The observed gas-phase HHG originates entirely

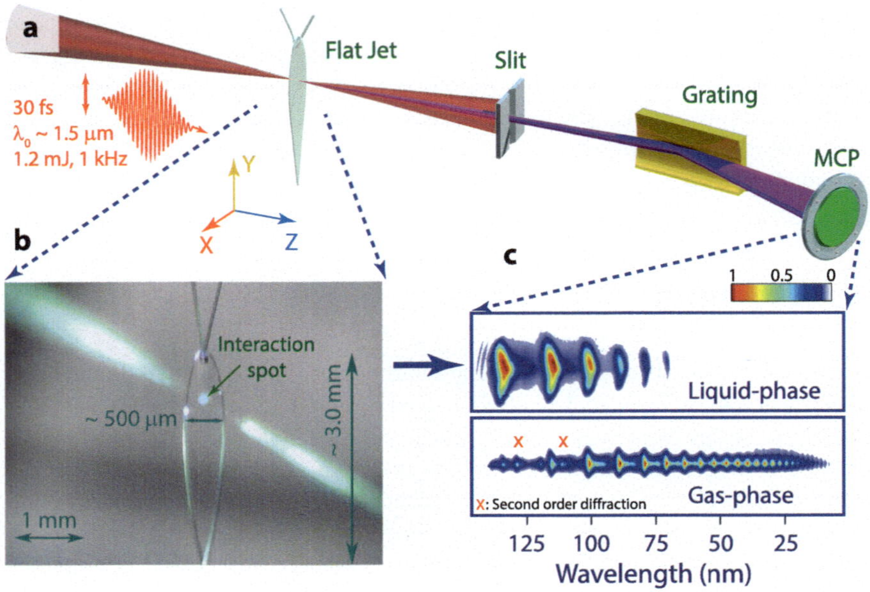


**Fig. 3.15 Observation of HHG from liquids.** **a** schematic representation of the experimental setup without the vacuum chambers, **b** photograph of the flatjet under operating conditions, **c** simultaneously recorded high-harmonic spectra of liquid and gaseous ethanol. Adapted from Ref. [25]

from the gas phase located behind the flatjet because the high-harmonic radiation created in front of the jet is completely absorbed, given the typical absorption lengths of ~10 nm at 20 eV [93]. This natural separation of HHG has several immediate benefits, which include the possibility to perform HHS experiments on the gas and liquid phases simultaneously and the separation of generated high harmonics from the driving fields, which could become relevant for high-average-power attosecond [94] or XUV-frequency-comb [95] experiments.

One of the most fundamental characteristics of HHG is the scaling of its cut-off energy with the intensity of the driving field. In gases, the cut-off scales quadratically with the peak electric field of the driver. This can intuitively be understood as the signature of the maximal kinetic energy that the continuum electron can acquire from the driving field. The experimental results obtained on liquid-phase HHG are shown in Fig. 3.17. They reveal a quasi-linear scaling ($H_{\text{cut-off}} \propto E^{1.2}$) of the cut-off photon energy with the peak electric field of the driver. This result points at a fundamentally different mechanism of HHG in liquids compared to gases. The obtained result is similar to observations made in solids, where a quasi-linear scaling of the cut-off energy was also observed [84]. This cut-off scaling is common to both types of mechanisms discussed for solids, i.e., intraband currents (Bloch oscillations) and interband polarization (generalized recollision) [14, 15, 86].

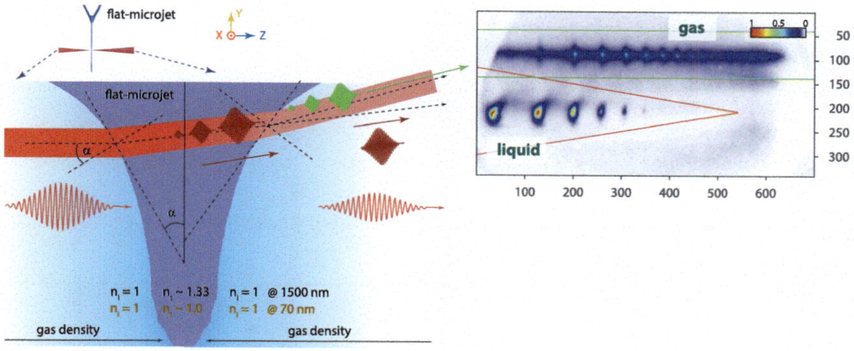


**Fig. 3.16 Separation of HHG from the liquid and gas phases**. A schematic representation of the optical path of the fundamental beam across the liquid jet is shown on the left. The curvature of the jet surfaces has been exaggerated for clarity. A spatio-spectrally resolved far-field image of the high-harmonic spectra simultaneously emitted from the liquid and gas phases is shown on the right. Adapted from Ref. [25]

We next turn to the scaling of the high-harmonic yield with the driving-field intensity. Figure 3.17 shows the dependence of the yield of H13 and H21, generated from a 1.5 $\mu$m driver in liquid ethanol on the peak electric-field strength. An electric field strength of 1 V/Å corresponds to a peak intensity of $2.65\times10^{13}$ W/cm$^2$. The dashed lines of the same color indicate the corresponding perturbative scaling laws. All observed harmonic orders thus follow non-perturbative scaling laws with a deviation from the perturbative scaling that strongly increases with harmonic order. These results demonstrate the strongly non-perturbative character of the observed HHG from liquids.

Figure 3.18 compares the ellipticity dependence of the HHG yield from liquid- and gas-phase ethanol. The ellipticity dependence in general contains important information about the properties of the medium and the strong-field-driven electron dynamics. We find that the ellipticity dependence of all harmonic orders emitted from liquid ethanol is clearly broadened compared to the gas-phase emission. In the gas phase, the ellipticity dependence narrows down with increasing harmonic order, which is a signature of the laser-driven electron dynamics in the continuum: the longer the trajectory of the electron is, the more its trajectory is influenced by the laser field, which results in a larger sensitivity to ellipticity. The same observation applies to the liquid-phase ellipticity dependence, which also narrows down with increasing harmonic order. This observation suggests that a trajectory-based understanding of high-harmonic generation in liquids might be adequate. Figure 3.18c compares the ellipticity dependence of H13 from water and several alcohols to that of gas-phase ethanol. The ellipticity dependencies of all liquids are very similar and considerably broader than those of the gas phase. The broadening of the ellipticity dependence in the liquid phase can have several origins, which include (i) differences in the continuum-electron propagation, (ii) electron scattering in liquids, and (iii) a different spatial extension of the electron-hole wavefunctions. Explanation (i)

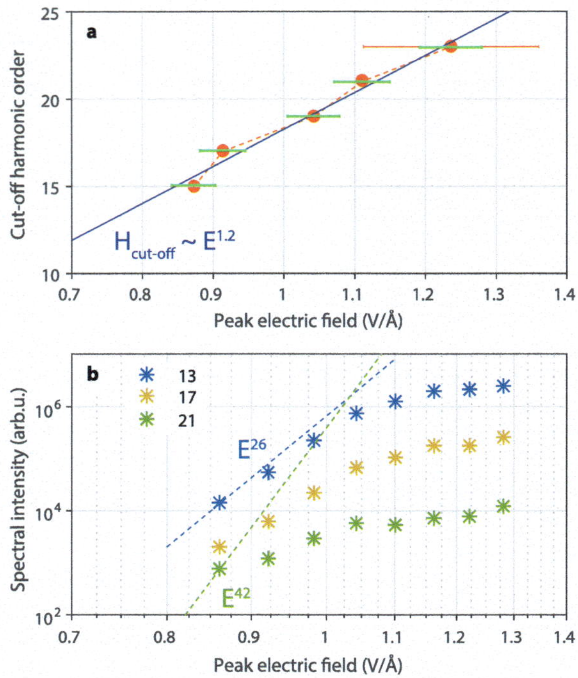


**Fig. 3.17 Scaling of the liquid-phase HHG cut-off and harmonic yields.** Scaling of the cut-off harmonic order **a** and the yield of selected harmonic orders **b** with the peak electric field of the driver (~30 fs, centered at 1.5 $\mu$m) in liquid ethanol. Adapted from Ref. [25]

is insufficient to rationalize the large observed effects because the differences in the ionization energies between the isolated and condensed molecules are on the order of 1 eV and the electron mass is not noticeably reduced for electron propagation in the liquids. Explanation (ii) is more difficult to quantify because electron elastic mean-free paths are not available for alcohols and only recently became available in liquid water (see Sect. 3.2.2.4 and Ref. [65]). In Ref. [25], we have used the mean-free paths determined for amorphous ice [96], for a lack or reliable data on liquid water. On this basis, we have assumed that electron scattering was negligible because the total propagation length of the electron trajectory emitting the cut-off harmonic (20 eV photon energy, emitted by a 1.5 $\mu$m driver with a peak electric field of 1.5 V/Å) amounts to ~3 nm, which is shorter than the mean-free paths including all types of collisions (~4 nm) at the corresponding kinetic energy. However, using the elastic mean-free path determined in our recent work (see Sect. 3.2.2.4), which amounts to ~0.6 nm, we now conclude that elastic scattering could contribute, at least to some extent, to the observed broadening of the ellipticity dependence. An important reason for our EMFP being much shorter than in ice is that it is based on a physical DCS, which was lacking in the analysis of the amorphous-ice data [96]. The importance of electron scattering in liquid-phase HHG has indeed been confirmed very recently in a study that showed that the cut-off of HHG in liquids is independent of the driving wavelength beyond a threshold intensity [34]. The ellipticity dependence of the cut-off [34] and the pulse-duration independence of the cut-off [35] additionally confirmed that electron scattering indeed plays a crucial factor in liquid-phase HHG, which can even be used to retrieve effective electron mean-free paths. Finally, effect iii) definitely contributes to, and possibly dominates, the observed broadening of the ellipticity dependence. In the gas phase, the continuum electron wave packet has to return to the parent molecule for high-harmonic emission to take place. This causes the width of the ellipticity dependence to reflect the spatial extension of the ionized orbital(s). It is known from many calculations (see, e.g., Refs. [97]) that the outermost valence orbitals of liquid water (and alcohols) are delocalized over several molecules, with a similar extent as in small water clusters [27]. The delocalization over the first solvation shell is strong, but further delocalization is rather weak as a consequence of disorder in liquids. This delocalization is illustrated in Fig. 3.18a, which shows the highest-occupied molecular orbital of a water pentamer, characteristic of a water molecule with one complete solvation shell. This significant delocalization of the electron hole leads to a broadening of the ellipticity dependence, which will contribute significantly to the observed effects. A quantitative verification of the relative importance of the discussed effects will become possible in the near future by adapting the methodological developments described in Sect. 3.2.2 and the recent work described in Refs. [34, 35] to liquid-phase HHG, as briefly discussed below (Sect. 3.3.2).

### *3.3.2 Concepts and Theoretical Methods*

The theoretical description of HHG in liquids meets with several challenges. First, liquids have a density comparable to solids, such that the typical extensions of continuum-electron trajectories known from gas-phase HHG would correspond to the electron wave packet encountering several neighboring molecules on its trajectory. Consequently, the methods developed to understand HHG in solids might be the better starting point. However, and this is the second challenge, liquids are intrinsically disordered, which prevents the rigorous application of a momentum-space description of HHG, which has enabled rapid progress in the understanding of HHG in solids. Nevertheless, the absence of long-range order does not completely exclude the application of momentum-space methods because liquids possess short-range order. As long as the spatial extension of the continuum-electron trajectory is comparable to the length scale of this short-range order, the latter can be expected to play a role in HHG.

Based on these considerations, we have chosen the semiconductor Bloch equations (SBE) as a starting point for modeling liquid-phase HHG. The electronic structures of liquids are indeed commonly understood as those of large-bandgap semiconductors [98]. The presence of local order, which is enhanced by the existence of strong hydrogen bonds, additionally justifies the use of an effective band structure [97]. The advantage of the SBE is that they naturally include both interband and intraband contributions to HHG, such that additional assumptions regarding the mechanisms at play are not needed [86]. The solution of the SBE with a realistic band structure, such

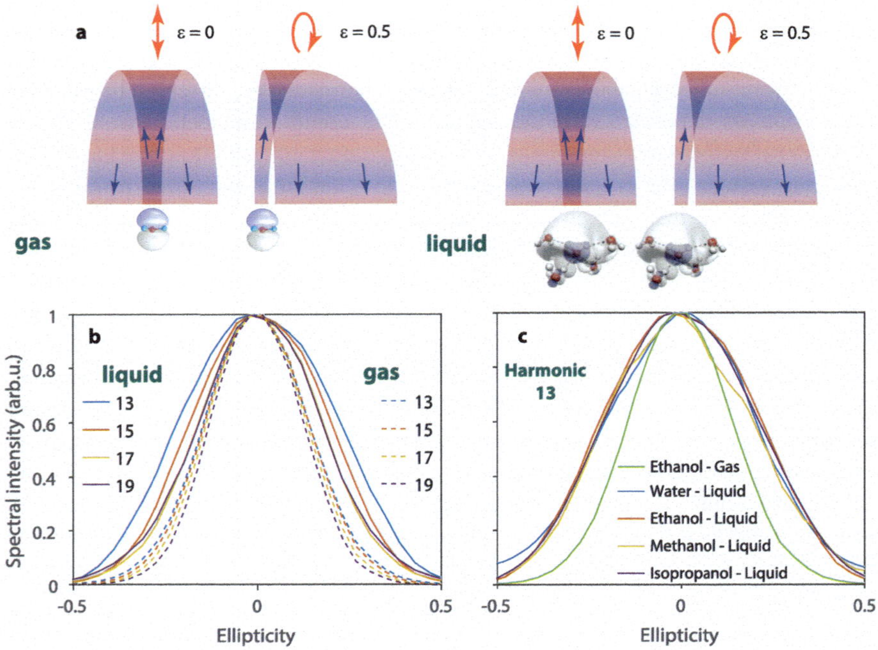


**Fig. 3.18 Ellipticity dependence of liquid- versus gas-phase HHG.** **a** Schematic representation of the effect of ellipticity on the propagation of the electron wave packet. **b** Dependence of the yield of selected harmonics on the ellipticity of the driving field in liquid and gaseous ethanol. **c** Comparison of the ellipticity dependence of H13 from different liquids to gaseous ethanol. All measurements were recorded with a ~30 fs driver, centered at 1.5 $\mu$m. Adapted from Ref. [25]

as that derived from a calculation of a slab of, e.g., 128 molecules however quickly reaches the limit of computational feasibility, in addition to the limits imposed by the accuracy of density-functional-theory calculations.

We have therefore chosen, as a zero-order approximation, an approach that circumvents all of these challenges. We have used the density of occupied and unoccupied states, measured by state-of-the-art X-ray spectroscopies, to derive a highly simplified model band structure that reflects the known densities of states, arising from the three occupied outer-most valence bands and the two lowest-lying unoccupied conduction bands. The main goals of this approach were to identify (i) the sensitivity of the calculated HHG spectra to the electronic structure of the liquids, i.e., the band gap and the width of the bands, (ii) the number of contributing bands, and (iii) the capability of such a simple model to explain the characteristic differences in the HHG spectra of liquid water as compared to the alcohols.

Figure 3.19a shows the densities of states (DOS) of liquid water and ethanol as measured by X-ray spectroscopy [99–102]. In the case of water, these consist of three outer-valence bands, which are labeled according to the symmetries of the orbitals of the isolated water molecule ($1b_1$, $3a_1$ and $1b_2$). The unoccupied states

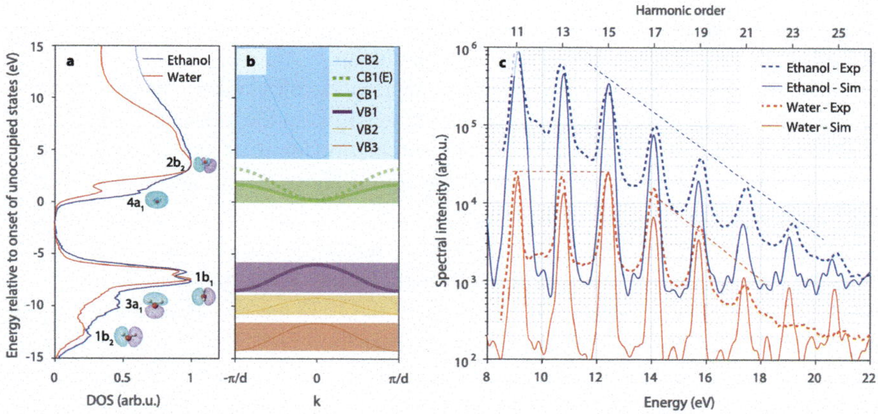


**Fig. 3.19 Sensitivity of HHG to the electronic structure of liquids. a** Densities of states of liquid water and ethanol obtained from X-ray spectroscopy, **b** model "band structure" chosen to reflect these densities of states, **c** measured HHG spectra of liquid water and ethanol using a ~30 fs laser pulse centered at 1.5 $\mu$m and spectra calculated by solving the SBE for the model "band structure" shown in **b**. Adapted from Ref. [25]

consist of two conduction bands, that are labeled according to the symmetries of the unoccupied orbitals of the isolated water molecule ($4a_1$ and $2b_2$). The DOS of ethanol is comparable to that of water, with an important difference being the absence of a local maximum at the position of the $4a_1$ band of liquid water. Based on these DOS and the results of detailed "band-structure" calculations [97], we have derived the highly simplified model band structure shown in Fig. 3.19b. The difference in the DOS of liquid water and ethanol is accounted for by choosing a larger width for the lowest conduction band of liquid ethanol compared to liquid water.

The solution of the SBE using these model band structures yields high-harmonic spectra that are in very good agreement with the observed spectra (full and dashed lines, respectively, in Fig. 3.19). Most importantly, these calculations fully reproduce the main characteristics of the high-harmonic spectra, i.e., the monotonic decrease in the spectrum of ethanol and the plateau observed at the lowest three harmonic orders in liquid water, followed by a cut-off. We have verified that the calculated spectra do not change significantly when the two lower-lying valence bands and the highest-lying conduction band are removed from the calculation. Therefore the calculated high-harmonic spectra are mainly sensitive to the properties of the highest-lying valence and the lowest-lying conduction band.

Within our model, the characteristic differences between the HHG spectra of water and ethanol therefore exclusively originate from the different widths of the conduction bands, for which a cosinusoidal shape was assumed. The larger width in the case of ethanol causes the monotonic decay of the HHG spectrum, whereas the narrower width in the case of water causes the appearance of the plateau and cut-off. An additional characteristic of the HHG spectrum of liquid water is the appearance of a local minimum at H13 (10.7 eV). This local minimum is also reproduced in the

SBE calculations. As we showed in Fig. 7 of Ref. [25], the position of this minimum shifts in energy according to the size of the bandgap.

The comparison between our measurements and calculations therefore shows that high-harmonic spectroscopy of liquids is sensitive to the electronic structure of liquids, particularly to the properties of the conduction band, as well as the bandgap.

In the future, the methods for describing liquid-phase HHG can be improved in different ways. The most accurate computationally tractable approach would consist in running real-time, real-space time-dependent density-functional-theory (TDDFT) calculations on slabs of liquids sampled from molecular-dynamics calculations. This approach, which has recently been reported in Ref. [103], has the advantage of requiring minimal assumptions beyond those inherent to the TDDFT approach. A complementary approach relies on TDDFT calculations on water clusters using structures from liquid-water simulations [104]. The TDDFT-based approaches have the disadvantage that the physical insight into the mechanisms at play remains limited. Therefore, it might be desirable to develop complementary approaches, which offer additional insights. One such approach would consist in Monte-Carlo trajectory calculations of the strong-field driven electron dynamics. These calculations could rely on the principles described in Sect. 3.2.2 and Ref. [66] and thereby incorporate elastic and inelastic electron scattering, as well as laser-assisted electron scattering. The additionally required strong-field-ionization and photorecombination matrix elements could be obtained by extending the methods established for isolated molecules to the condensed phase by using the cluster approach described in Sect. 3.2.2. This method would offer additional insights into liquid-phase HHG because the role of electron scattering, orbital delocalization, and many other effects could be disentangled. A first step in this direction has recently been taken with a very simple scattering model, in which scattering was assumed to eliminate the contributions of all electron trajectories extending beyond an effective mean-free path [34].

The results presented in this section set the foundations for the development of liquid-phase high-harmonic spectroscopy. In the future, the detailed mechanism of liquid-phase HHG will be studied with attosecond temporal resolution by using both *in-situ* methods such as two-color HHG [15, 80] and attosecond transient absorption [105, 106] based on water-window high-harmonic sources [107, 108] or *ex-situ* methods, such as attosecond photoelectron spectroscopy [24, 109]. Of particular interest are the capabilities of HHS to resolve attosecond electron dynamics in the liquid phase, such as electron-liquid scattering dynamics and the spatial characteristics of electron-hole dynamics in liquids. Targets of primary interest will be water itself, but also aqueous solutions, which offer the possibility to study solvated species in their natural environment, and other liquids.

## 3.4 Conclusions and Outlook

In this chapter, we have described two novel experimental approaches that enable attosecond time-resolved experiments to be performed on liquid samples. The crucial innovations have been the development of attosecond photoelectron spectroscopy with cylindrical liquid microjets and the demonstration of high-harmonic spectroscopy with flat microjets.

In the first case, a general novel approach to analyze and interpret overlapping attosecond photoelectron spectra has been developed. This technique generalizes attosecond photoelectron spectroscopy not only to the liquid phase, but actually to any complex sample. On the conceptual side, a general theoretical framework for condensed-phase attosecond photoelectron spectroscopy has been developed that includes, for the first time, the treatment of photoionization and scattering delays, as well as a coherent treatment of all processes involved. This framework has been validated first by benchmarking against the time-dependent Schrödinger equation and finally by reaching quantitative agreement with attosecond time delays of liquid water.

In the second case, a general approach to high-harmonic spectroscopy in liquids has been established. This result represents a change of paradigms in a field where XUV high-harmonic generation from liquids was previously assumed to be impossible. Our novel experimental approach has enabled the first observation of XUV HHG from liquids, the unequivocal separation and simultaneous measurement of liquid- and gas-phase HHG and their detailed comparison. We have found a linear scaling of the high-harmonic cut-off energy with the peak electric field of the driver and a highly non-perturbative scaling of the HHG yield. We have found a systematically and substantially broadened dependence of the HHG yields on the ellipticity of the driving field, compared to the gas phase. Finally, based on the solution of the semiconductor Bloch equations of a strongly driven model band system, we have found a pronounced sensitivity of the HHG spectra on the electronic structure of liquids, particularly the properties of the conduction band and the band gap.

These developments establish two major avenues for developing attosecond time-resolved spectroscopy of liquids and solutions. They open a myriad of future possibilities to explore the role of electronic dynamics in solvated atoms, ions, molecules, and nanoparticles in their natural environment. Of particular interest for such work are the prototypical processes of charge and energy transfer. We will briefly discuss two examples of such processes. Intermolecular Coulombic decay [19] in liquid water is the prototype for energy transfer in the liquid phase and a source of slow electrons in liquid water. ICD occurs when the inner-valence ($2a_1$) or the core ($1a_1$) orbitals of water (clusters or liquid) are ionized. ICD has indeed been observed in water dimer [110] and water clusters [111, 112], but most recently also in bulk liquid water [113]. Time-resolved experiments on ICD in water are yet to be demonstrated, but are well underway in our laboratory. Recent theoretical work predicts a time scale of 12-52 fs for ICD in small water clusters [112]. However, a different type of calculation predicts ICD lifetimes of 3.6-4.6 fs for isoelectronic $(HF)_3$ clusters [114], suggesting that the ICD lifetimes in water (clusters) could also be much shorter. Electron-transfer-mediated decay [20] is a prototype of ultrafast charge transfer reactions in liquids. ETMD occurs when the inner-valence or core hole created by ionization is filled by an electron from a neighboring particle. The energy made available in this process then serves to eject a second electron from the particle that provided the first electron to fill the hole (ETMD(2)) or from a third particle (ETMD(3)). ETMD has only recently been observed in the liquid phase [115]. Its time scale is presently unknown.

These two examples illustrate the possible future of liquid-phase attosecond science. They represent two relatively simple and very fundamental mechanisms of electronic dynamics in liquids. They are representative of a broad variety of electronic processes that play a role in most chemical reactions and biological transformations. Looking forward, the methods described in this chapter will provide access to the study of many important, but poorly understood electronic processes in the liquid phase.

**Acknowledgements** We gratefully acknowledge the contributions of many co-workers and collaborators who have contributed to this work over several years. This work was financially supported by an ERC starting grant (project no. 307270-ATTOSCOPE), an ERC consolidator grant (project no. 772797-ATTOLIQ) and the Swiss National Science Foundation (SNSF) via the National Center of Competence in Research Molecular Ultrafast Science and Technology and projects no. 200021_159875, 200021_172946 and 200021_204928. A. S. is grateful for financial support from an Ambizione grant of the SNSF. D. J. thanks the FP-RESOMUS fellowship program.

## References

1. M. Drescher, M. Hentschel, R. Kienberger, M. Uiberacker, V. Yakolev, A. Scrinzi, T. Westerwalbesloh, U. Kleineberg, F. Krausz, Nature **419**, 803 (2002)
2. E. Goulielmakis, Z.H. Loh, A. Wirth, R. Santra, N. Rohringer, V.S. Yakovlev, S. Zherebtsov, T. Pfeifer, A.M. Azzeer, M.F. Kling, S.R. Leone, F. Krausz, Nature **466**(7307), 739 (2010). http://dx.doi.org/10.1038/nature09212
3. K. Klünder, J.M. Dahlström, M. Gisselbrecht, T. Fordell, M. Swoboda, D. Guénot, P. Johnsson, J. Caillat, J. Mauritsson, A. Maquet, Taïeb, A. L'Huillier, Phys. Rev. Lett. **106**, 143002 (2011). https://doi.org/10.1103/PhysRevLett.106.143002. http://link.aps.org/doi/10.1103/PhysRevLett.106.143002
4. C. Ott, A. Kaldun, P. Raith, K. Meyer, M. Laux, J. Evers, C.H. Keitel, C.H. Greene, T. Pfeifer, Science **340**(6133), 716 (2013). https://doi.org/10.1126/science.1234407. http://www.sciencemag.org/content/340/6133/716.abstract
5. H. Niikura, F. Légaré, R. Hasbani, A.D. Bandrauk, M.Y. Ivanov, D.M. Villeneuve, P.B. Corkum, Nature **417**, 917 (2002)
6. S. Baker, J.S. Robinson, C.A. Haworth, H. Teng, R.A. Smith, C.C. Chirila, M. Lein, J.W.G. Tisch, J.P. Marangos, Science **312**, 424 (2006)
7. F. Calegari, D. Ayuso, A. Trabattoni, L. Belshaw, S. De Camillis, S. Anumula, F. Frassetto, L. Poletto, A. Palacios, P. Decleva et al., Science **346**(6207), 336 (2014)
8. P.M. Kraus, B. Mignolet, D. Baykusheva, A. Rupenyan, L. Horný, E.F. Penka, G. Grassi, O.I. Tolstikhin, J. Schneider, F. Jensen, L.B. Madsen, A.D. Bandrauk, F. Remacle, H.J. Wörner, Science **350**, 790 (2015). http://www.sciencemag.org/content/350/6262/790.abstract
9. M. Huppert, I. Jordan, D. Baykusheva, A. von Conta, H.J. Wörner. Phys. Rev. Lett. **117**(093001) (2016). https://doi.org/10.1103/PhysRevLett.117.093001
10. D.T. Matselyukh, V. Despré, N.V. Golubev, A.I. Kuleff, H.J. Wörner, Nat. Phys. **18**(10), 1206 (2022). https://doi.org/10.1038/s41567-022-01690-0
11. A.L. Cavalieri, N. Muller, T. Uphues, V.S. Yakovlev, A. Baltuska, B. Horvath, B. Schmidt, L. Blumel, R. Holzwarth, S. Hendel, M. Drescher, U. Kleineberg, P.M. Echenique, R. Kienberger, F. Krausz, U. Heinzmann, Nature **449**(7165), 1029 (2007). http://dx.doi.org/10.1038/nature06229

12. S. Neppl, R. Ernstorfer, A.L. Cavalieri, C. Lemell, G. Wachter, E. Magerl, E.M. Bothschafter, M. Jobst, M. Hofstetter, U. Kleineberg, J.V. Barth, D. Menzel, J. Burgdorfer, P. Feulner, F. Krausz, R. Kienberger, Nature **517**(7534), 342 (2015). http://dx.doi.org/10.1038/nature14094
13. Z. Tao, C. Chen, T. Szilvási, M. Keller, M. Mavrikakis, H. Kapteyn, M. Murnane, Science **353**(6294), 62 (2016)
14. T.T. Luu, M. Garg, S.Y. Kruchinin, A. Moulet, M.T. Hassan, E. Goulielmakis, Nature **521**(7553), 498 (2015). http://dx.doi.org/10.1038/nature14456
15. G. Vampa, T.J. Hammond, N. Thire, B.E. Schmidt, F. Legare, C.R. McDonald, T. Brabec, P.B. Corkum, Nature **522**(7557), 462 (2015). http://dx.doi.org/10.1038/nature14517
16. T.T. Luu, H.J. Wörner, Nat. Commun. **9**(1), 916 (2018). https://doi.org/10.1038/s41467-018-03397-4
17. D. Baykusheva, H.J. Wörner, *Molecular Spectroscopy and Quantum Dynamics* (Elsevier, 2020), Chap. Attosecond Molecular Spectroscopy and Dynamics. https://arxiv.org/abs/2002.02111
18. V. May, O. Kühn, *Charge and Energy Transfer Dynamics in Molecular Systems* (Wiley, 2011)
19. L.S. Cederbaum, J. Zobeley, F. Tarantelli, Phys. Rev. Lett. **79**, 4778 (1997). https://doi.org/10.1103/PhysRevLett.79.4778
20. J. Zobeley, R. Santra, L. Cederbaum, J. Chem. Phys. **115**(11), 5076 (2001). https://doi.org/10.1063/1.1395555. http://dx.doi.org/10.1063/1.1395555
21. H. Nikjoo, D. Emfietzoglou, T. Liamsuwan, R. Taleei, D. Liljequist, S. Uehara, Rep. Progr. Phys. **79**(11), 116601 (2016)
22. V. Svoboda, R. Michiels, A.C. LaForge, F. Stienkemeier, P. Slavíček, H.J. Wörner, Sci. Adv. **6**(3), eaaz0385 (2020). https://doi.org/10.1126/sciadv.aaz0385
23. Z.H. Loh, G. Doumy, C. Arnold, L. Kjellsson, S. Southworth, A. Al Haddad, Y. Kumagai, M.F. Tu, P. Ho, A. March, et al., Science **367**(6474), 179 (2020)
24. I. Jordan, M. Huppert, D. Rattenbacher, M. Peper, D. Jelovina, C. Perry, A. von Conta, A. Schild, H.J. Wörner, Science **369**(6506), 974 (2020). https://doi.org/10.1126/science.abb0979. https://science.sciencemag.org/content/369/6506/974
25. T.T. Luu, Z. Yin, A. Jain, T. Gaumnitz, Y. Pertot, J. Ma, H.J. Wörner, Nat. Commun. **9**(1), 3723 (2018). https://doi.org/10.1038/s41467-018-06040-4
26. D. Baykusheva, H.J. Wörner, J. Chem. Phys. **146**(12), 124306 (2017). https://doi.org/10.1063/1.4977933. http://dx.doi.org/10.1063/1.4977933
27. X. Gong, S. Heck, D. Jelovina, C. Perry, K. Zinchenko, R. Lucchese, H.J. Wörner, Nature **609**(7927), 507 (2022). https://doi.org/10.1038/s41586-022-05039-8
28. X. Gong, I. Jordan, M. Huppert, S. Heck, D. Baykusheva, D. Jelovina, A. Schild, H.J. Wörner, Chimia **76**(6), 520 (2022). https://doi.org/10.2533/chimia.2022.520
29. A. Flettner, T. Pfeifer, D. Walter, C. Winterfeldt, C. Spielmann, G. Gerber, Appl. Phys. B **77**(8), 747 (2003)
30. H.G. Kurz, D.S. Steingrube, D. Ristau, M. Lein, U. Morgner, M. Kovačev, Phys. Rev. A **87**(6), 063811 (2013)
31. H.G. Kurz, M. Kretschmar, T. Binhammer, T. Nagy, D. Ristau, M. Lein, U. Morgner, M. Kovačev, Phys. Rev. X **6**(3), 031029 (2016)
32. A.D. DiChiara, E. Sistrunk, T.A. Miller, P. Agostini, L.F. DiMauro, Optics Express **17**(23), 20959 (2009)
33. P. Heissler, E. Lugovoy, R. Hörlein, L. Waldecker, J. Wenz, M. Heigoldt, K. Khrennikov, S. Karsch, F. Krausz, B. Abel, G.D. Tsakiris, New J. Phys. **16**(11), 113045 (2014). https://doi.org/10.1088/1367-2630/16/11/113045
34. A. Mondal, O. Neufeld, Z. Yin, Z. Nourbakhsh, V. Svoboda, A. Rubio, N. Tancogne-Dejean, H. Wörner, Nat. Phys. (2023). https://doi.org/10.1038/s41567-023-02214-0
35. A. Mondal, B. Waser, T. Balciunas, O. Neufeld, Z. Yin, N. Tancogne-Dejean, A. Rubio, H. Wörner, Optics Express (2023). https://doi.org/10.1364/OE.496686
36. M. Isinger, R. Squibb, D. Busto, S. Zhong, A. Harth, D. Kroon, S. Nandi, C. Arnold, M. Miranda, J.M. Dahlström et al., Science **358**(6365), 893 (2017)

37. J. Vos, L. Cattaneo, S. Patchkovskii, T. Zimmermann, C. Cirelli, M. Lucchini, A. Kheifets, A.S. Landsman, U. Keller, Science **360**(6395), 1326 (2018). https://doi.org/10.1126/science.aao4731
38. L. Cattaneo, J. Vos, R.Y. Bello, A. Palacios, S. Heuser, L. Pedrelli, M. Lucchini, C. Cirelli, F. Martín, U. Keller, Nat. Phys. **14**(7), 733 (2018). https://doi.org/10.1038/s41567-018-0103-2
39. S. Biswas, B. Förg, J. Schötz, W. Schweinberger, L. Ortmann, T. Zimmermann, L.W. Pi, D. Baykusheva, H. Masood, I. Liontos, K.A. M., K.N. G., A.F. Alharbi, M. Alharbi, A.M. Azzeer, G. Hartmann, H.J. Wörner, A.S. Landsman, M.F. Kling, Nat. Phys. **16**, 778 (2020)
40. S. Heck, D. Baykusheva, M. Han, J.B. Ji, C. Perry, X. Gong, H.J. Wörner, Sci. Adv. **7**(49), eabj8121 (2021). https://doi.org/10.1126/sciadv.abj8121
41. S. Heck, M. Han, D. Jelovina, J.B. Ji, C. Perry, X. Gong, R. Lucchese, K. Ueda, H.J. Wörner, Phys. Rev. Lett. **129**(13), 133002 (2022). https://doi.org/10.1103/PhysRevLett.129.133002
42. X. Gong, É. Plésiat, A. Palacios, S. Heck, F. Martín, H.J. Wörner, Nat. Commun. **14**(1), 4402 (2023). https://doi.org/10.1038/s41467-023-40120-4
43. M. Huppert, I. Jordan, H.J. Wörner, Rev. Sci. Instrum. **86**(12), 123106 (2015). https://doi.org/10.1063/1.4937623. http://scitation.aip.org/content/aip/journal/rsi/86/12/10.1063/1.4937623
44. I. Jordan, M. Huppert, M.A. Brown, J.A. van Bokhoven, H.J. Wörner, Rev. Sci. Instrum. **86**(12), 123905 (2015). https://doi.org/10.1063/1.4938175. http://scitation.aip.org/content/aip/journal/rsi/86/12/10.1063/1.4938175
45. A. von Conta, M. Huppert, H.J. Wörner, Rev. Sci. Instrum. **87**(7), 073102 (2016). https://doi.org/10.1063/1.4955263. http://scitation.aip.org/content/aip/journal/rsi/87/7/10.1063/1.4955263
46. M. Huppert, Actively-stabilized attosecond beamline and its application to attosecond dynamics in atoms, molecules and liquids. Ph.D. thesis, ETH Zürich (2016)
47. C.H. Zhang, U. Thumm, Phys. Rev. Lett. **102**(12), 123601 (2009)
48. Q. Liao, U. Thumm, Phys. Rev. Lett. **112**(2), 023602 (2014)
49. A.G. Borisov, D. Sánchez-Portal, A.K. Kazansky, P.M. Echenique, Phys. Rev. B **87**, 121110 (2013). https://doi.org/10.1103/PhysRevB.87.121110. https://link.aps.org/doi/10.1103/PhysRevB.87.121110
50. R. Pazourek, S. Nagele, J. Burgdörfer, Rev. Mod. Phys. **87**, 765 (2015). https://doi.org/10.1103/RevModPhys.87.765. http://link.aps.org/doi/10.1103/RevModPhys.87.765
51. D. Rattenbacher, I. Jordan, A. Schild, H.J. Wörner, Phys. Rev. A **97**(6), 063415 (2018). https://doi.org/10.1103/PhysRevA.97.063415
52. S. Neppl, R. Ernstorfer, E.M. Bothschafter, A.L. Cavalieri, D. Menzel, J.V. Barth, F. Krausz, R. Kienberger, P. Feulner, Phys. Rev. Lett. **109**(8), 087401 (2012). http://link.aps.org/doi/10.1103/PhysRevLett.109.087401
53. W. Okell, T. Witting, D. Fabris, C. Arrell, J. Hengster, S. Ibrahimkutty, A. Seiler, M. Barthelmess, S. Stankov, D. Lei et al., Optica **2**(4), 383 (2015)
54. R. Locher, L. Castiglioni, M. Lucchini, M. Greif, L. Gallmann, J. Osterwalder, M. Hengsberger, U. Keller, Optica **2**(5), 405 (2015). http://www.osapublishing.org/optica/abstract.cfm?URI=optica-2-5-405
55. F. Siek, S. Neb, P. Bartz, M. Hensen, C. Strüber, S. Fiechter, M. Torrent-Sucarrat, V.M. Silkin, E.E. Krasovskii, N.M. Kabachnik et al., Science **357**(6357), 1274 (2017)
56. E.P. Wigner, Phys. Rev. A **98**, 145 (1955)
57. F.T. Smith, Phys. Rev. **118**(1), 349 (1960). http://link.aps.org/doi/10.1103/PhysRev.118.349
58. J.M. Dahlström, A. L'Huillier, A. Maquet, J. Phys. B: Atomic, Molecul. Opt. Phys. **45**(18), 183001 (2012). http://stacks.iop.org/0953-4075/45/i=18/a=183001
59. M. Schultze, M. Fiess, N. Karpowicz, J. Gagnon, M. Korbman, M. Hofstetter, S. Neppl, A.L. Cavalieri, Y. Komninos, T. Mercouris, C.A. Nicolaides, R. Pazourek, S. Nägele, J. Feist, J. Burgdörfer, A.M. Azzeer, R. Ernstorfer, R. Kienberger, U. Kleineberg, E. Goulielmakis, F. Krausz, V.S. Yakovlev, Science **328**(5986), 1658 (2010). http://www.sciencemag.org/cgi/content/abstract/328/5986/1658

60. N.M. Kroll, K.M. Watson, Phys. Rev. A **8**, 804 (1973). https://doi.org/10.1103/PhysRevA.8.804. https://link.aps.org/doi/10.1103/PhysRevA.8.804
61. L.B. Madsen, Am. J. Phys. **73**(1), 57 (2005). https://doi.org/10.1119/1.1796791. http://dx.doi.org/10.1119/1.1796791
62. Álvaro Jiménez Galán and Luca Argenti and Fernando Martín, New J. Phys. **15**(11), 113009 (2013). http://stacks.iop.org/1367-2630/15/i=11/a=113009
63. C. Perry, M. Peper, A. Schild, H.J. Wörner, to be published (2020)
64. B. Temelso, K.A. Archer, G.C. Shields, J. Phys. Chem. A **115**(43), 12034 (2011)
65. A. Schild, M. Peper, C. Perry, D. Rattenbacher, H.J. Wörner, J. Phys. Chem. Lett. **11**, 1128 (2020). https://doi.org/10.1021/acs.jpclett.9b02910
66. T. Gadeyne, P. Zhang, A. Schild, H.J. Wörner, Chem. Sci. **13**(6), 1675 (2022). https://doi.org/10.1039/D1SC06741
67. Y.I. Suzuki, K. Nishizawa, N. Kurahashi, T. Suzuki, Phys. Rev. E **90**(1), 010302 (2014). http://link.aps.org/doi/10.1103/PhysRevE.90.010302
68. S. Thürmer, M. Ončák, N. Ottosson, R. Seidel, U. Hergenhahn, S.E. Bradforth, P. Slavíček, B. Winter, Nat. Chem. **5**(7), 590 (2013). http://dx.doi.org/10.1038/nchem.1680
69. https://gitlab.com/axelschild/CLstunfti
70. J.P. Sethna, *Statistical Mechanics: Entropy, Order Parameters, and Complexity* (Oxford Univeristy, Oxford University Press, 2006)
71. D. Liljequist, Radiat. Phys. Chem. **77**(7), 835 (2008)
72. F.A. Gianturco, R.R. Lucchese, N. Sanna, J. Chem. Phys. **100**(9), 6464 (1994). http://link.aip.org/link/?JCP/100/6464/1
73. A.P.P. Natalense, R.R. Lucchese, J. Chem. Phys. **111**(12), 5344 (1999). http://link.aip.org/link/?JCP/111/5344/1
74. P. Wernet, D. Nordlund, U. Bergmann, M. Cavalleri, M. Odelius, H. Ogasawara, L.A. Näslund, T.K. Hirsch, L. Ojamäe, P. Glatzel, Science **304**(5673), 995 (2004)
75. J. Itatani, J. Levesque, D. Zeidler, H. Niikura, H. Pépin, J.C. Kieffer, P.B. Corkum, D.M. Villeneuve, Nature **432**, 867 (2004)
76. O. Smirnova, S. Patchkovskii, Y. Mairesse, N. Dudovich, D. Villeneuve, P. Corkum, M.Y. Ivanov, Phys. Rev. Lett. **102**(6), 063601 (2009). http://link.aps.org/abstract/PRL/v102/e063601
77. C. Vozzi, F. Calegari, E. Benedetti, J.P. Caumes, G. Sansone, S. Stagira, M. Nisoli, R. Torres, E. Heesel, N. Kajumba, J.P. Marangos, C. Altucci, R. Velotta, Phys. Rev. Lett. **95**(15), 153902 (2005). https://doi.org/10.1103/PhysRevLett.95.153902. http://link.aps.org/abstract/PRL/v95/e153902
78. C. Vozzi, M. Negro, F. Calegari, G. Sansone, M. Nisoli, S. De Silvestri, S. Stagira, Nat. Phys. **7**(10), 822 (2011). http://dx.doi.org/10.1038/nphys2029
79. H.J. Wörner, H. Niikura, J.B. Bertrand, P.B. Corkum, D.M. Villeneuve, Phys. Rev. Lett. **102**(10), 103901 (2009). http://link.aps.org/abstract/PRL/v102/e103901
80. D. Shafir, H. Soifer, B.D. Bruner, M. Dagan, Y. Mairesse, S. Patchkovskii, M.Y. Ivanov, O. Smirnova, N. Dudovich, Nature **485**(7398), 343 (2012). http://dx.doi.org/10.1038/nature11025
81. P.M. Kraus, O.I. Tolstikhin, D. Baykusheva, A. Rupenyan, J. Schneider, C.Z. Bisgaard, T. Morishita, F. Jensen, L.B. Madsen, H.J. Wörner, Nat. Comm. **6**, 7039 (2015). https://doi.org/10.1038/ncomms8039
82. P.M. Kraus, H.J. Wörner, Angewandte Chemie Int. Ed. **57**(19), 5228 (2018). https://doi.org/10.1002/anie.201702759
83. A.H. Chin, O.G. Calderón, J. Kono, Phys. Rev. Lett. **86**, 3292 (2001). https://doi.org/10.1103/PhysRevLett.86.3292. https://link.aps.org/doi/10.1103/PhysRevLett.86.3292
84. S. Ghimire, A.D. DiChiara, E. Sistrunk, P. Agostini, L.F. DiMauro, D.A. Reis, Nat. Phys. **7**(2), 138 (2011)
85. G. Vampa, C. McDonald, G. Orlando, D. Klug, P. Corkum, T. Brabec, Phys. Rev. Lett. **113**(7), 073901 (2014)

86. T.T. Luu, H.J. Wörner, Phys. Rev. B **94**(11), 115164 (2016). https://doi.org/10.1103/PhysRevB.94.115164
87. O. Schubert, M. Hohenleutner, F. Langer, B. Urbanek, C. Lange, U. Huttner, D. Golde, T. Meier, M. Kira, S.W. Koch et al., Nat. Photon. **8**(2), 119 (2014)
88. M. Hohenleutner, F. Langer, O. Schubert, M. Knorr, U. Huttner, S.W. Koch, M. Kira, R. Huber, Nature **523**(7562), 572 (2015)
89. Y.S. You, D.A. Reis, S. Ghimire, Nat. Phys. **13**(4), 345 (2017)
90. S. Ghimire, D.A. Reis, Nat. Phys. **15**(1), 10 (2019)
91. D. Hasson, R.E. Peck, AIChE J. **10**(5), 752 (1964)
92. M. Ekimova, W. Quevedo, M. Faubel, P. Wernet, E.T.J. Nibbering, Struct. Dyn. **2**(5), 054301 (2015). https://doi.org/10.1063/1.4928715. http://scitation.aip.org/content/aca/journal/sdy/2/5/10.1063/1.4928715
93. H. Hayashi, N. Hiraoka, J. Phys. Chem. B **119**(17), 5609 (2015). (PMID: 25835527). https://doi.org/10.1021/acs.jpcb.5b01567
94. D. Hammerland, P. Zhang, S. Kühn, P. Jojart, I. Seres, V. Zuba, Z. Varallyay, D. Charalambidis, K. Osvay, T.T. Luu, H.J. Wörner, J. Phys. B **52**(23), 23LT01 (2019). http://dx.doi.org/10.1088/1361-6455/ab486c
95. C. Corder, P. Zhao, J. Bakalis, X. Li, M.D. Kershis, A.R. Muraca, M.G. White, T.K. Allison, Struct. Dyn. **5**(5), 054301 (2018)
96. M. Michaud, A. Wen, L. Sanche, Radiat. Res. **159**(1), 3 (2003)
97. D. Prendergast, J.C. Grossman, G. Galli, J. Chem. Phys. **123**(1), 014501 (2005)
98. P. Cabral do Couto, S. Estácio, B. Costa Cabral, J. Chem. Phys. **123**(5), 054510 (2005)
99. T. Tokushima, Y. Harada, O. Takahashi, Y. Senba, H. Ohashi, L.G. Pettersson, A. Nilsson, S. Shin, Chem. Phys. Lett. **460**(4–6), 387 (2008)
100. O. Fuchs, M. Zharnikov, L. Weinhardt, M. Blum, M. Weigand, Y. Zubavichus, M. Bär, F. Maier, J. Denlinger, C. Heske et al., Phys. Rev. Lett. **100**(2), 027801 (2008)
101. S. Schreck, A. Pietzsch, K. Kunnus, B. Kennedy, W. Quevedo, P.S. Miedema, P. Wernet, A. Föhlisch, Struct. Dyn. **1**(5), 054901 (2014)
102. S. Schreck, P. Wernet, J. Chem. Phys. **145**(10), 104502 (2016)
103. Z. Nourbakhsh, O. Neufeld, N. Tancogne-Dejean, A. Rubio, arXiv preprint arXiv:2212.04177 (2022)
104. O. Neufeld, Z. Nourbakhsh, N. Tancogne-Dejean, A. Rubio, J. Chem. Theory Comput. **18**(7), 4117 (2022)
105. A.D. Smith, T. Balčiunas, Y.P. Chang, C. Schmidt, K. Zinchenko, F.B. Nunes, E. Rossi, V. Svoboda, Z. Yin, J.P. Wolf, H.J. Wörner, J. Phys. Chem. Lett. **11**(6), 1981 (2020). https://doi.org/10.1021/acs.jpclett.9b03559
106. Z. Yin, Y.P. Chang, T. Balčiūnas, Y. Shakya, A. Djorovic, G. Geoffrey, G. Fazio, R. Santra, L. Inhester, J.P. Wolf, H. Wörner, Nature **619**, 749 (2023). https://doi.org/10.1038/s41586-023-06182-6
107. Y. Pertot, C. Schmidt, M. Matthews, A. Chauvet, M. Huppert, V. Svoboda, A. von Conta, A. Tehlar, D. Baykusheva, J.P. Wolf, H.J. Wörner, Science **355**, 264 (2017). https://doi.org/10.1126/science.aah6114
108. K.S. Zinchenko, F. Ardana-Lamas, I. Seidu, S.P. Neville, J. van der Veen, V.U. Lanfaloni, M.S. Schuurman, H.J. Wörner, Science **371**(6528), 489 (2021). https://doi.org/10.1126/science.abf1656. https://science.sciencemag.org/content/371/6528/489
109. P.M. Paul, E.S. Toma, P. Breger, G. Mullot, F. Augé, P. Balcou, H.G. Muller, P. Agostini, Science **292**, 1689 (2001)
110. T. Jahnke, H. Sann, T. Havermeier, K. Kreidi, C. Stuck, M. Meckel, M. Schöffler, N. Neumann, R. Wallauer, S. Voss, A. Czasch, O. Jagutzki, A. Malakzadeh, F. Afaneh, T. Weber, H. Schmidt-Bocking, R. Dorner, Nat. Phys. **6**(2), 139 (2010). http://dx.doi.org/10.1038/nphys1498
111. M. Mucke, M. Braune, S. Barth, M. Forstel, T. Lischke, V. Ulrich, T. Arion, U. Becker, A. Bradshaw, U. Hergenhahn, Nat. Phys. **6**(2), 143 (2010). http://dx.doi.org/10.1038/nphys1500
112. C. Richter, D. Hollas, C.M. Saak, M. Förstel, T. Miteva, M. Mucke, O. Björneholm, N. Sisourat, P. Slaviček, U. Hergenhahn, Nat. Commun. **9**(1), 4988 (2018). https://doi.org/10.1038/s41467-018-07501-6

113. P. Zhang, C. Perry, T.T. Luu, D. Matselyukh, H.J. Wörner, Phys. Rev. Lett. **128**(13), 133001 (2022). https://doi.org/10.1103/PhysRevLett.128.133001
114. A. Ghosh, S. Pal, N. Vaval, Molecul. Phys. **112**(5–6), 669 (2014)
115. I. Unger, R. Seidel, S. Thürmer, M.N. Pohl, E.F. Aziz, L.S. Cederbaum, E. Muchová, P. Slaviček, B. Winter, N.V. Kryzhevoi, Nat. Chem. **9**(7), 708 (2017)

# Chapter 4
# Strong-Field Electron Dynamics in Solids

**Kenichi L. Ishikawa, Yasushi Shinohara, Takeshi Sato, and Tomohito Otobe**

**Abstract** Solid-state materials have recently emerged as a new stage of strong-field physics and attosecond science. The mechanism of the electron dynamics driven by an ultrashort intense laser pulse is under intensive discussion. Here we theoretically discuss momentum-space strong-field electron dynamics in graphene and crystalline dielectrics and semiconductors. First, within massless Dirac fermion and tight-binding models for graphene, we rigorously derive intraband displacement and interband transition, which form the basis for understanding solid-state strong-field physics including high-harmonic generation (HHG). Then, based on the time-dependent Schrödinger equation for a one-dimensional model crystal, we introduce a simple, multiband, momentum-space three-step model that incorporates intraband displacement, interband tunneling, and recombination with a valence band hole. We also analyze how the model is modified by electron-hole interaction. Finally, actual three-dimensional materials are investigated. We present a time-dependent density-matrix method whose results for HHG are compared with experimental measurement results. Moreover, we describe the dynamical Franz-Keldysh effect in femtosecond time resolution, i.e., the time-dependent modulation of a dielectric function under an intense laser field, using a real-time time-dependent density functional theory.

K. L. Ishikawa (✉) · T. Sato
Department of Nuclear Engineering and Management, Graduate School of Engineering, The University of Tokyo, 7-3-1 Hongo, Bunkyo-ku, Tokyo 113-8656, Japan
e-mail: ishiken@n.t.u-tokyo.ac.jp

T. Sato
e-mail: sato@atto.t.u-tokyo.ac.jp

Y. Shinohara
Photon Science Center, Graduate School of Engineering, The University of Tokyo, 7-3-1 Hongo, Bunkyo-ku, Tokyo 113-8656, Japan
e-mail: shinohara@atto.t.u-tokyo.ac.jp

T. Otobe
Ultrafast Dynamics Group, National Institutes for Quantum and Radialogical Science and Technology (QST), 8-1-7, Umemidai, Kizugawa, Kyoto 619-0215, Japan
e-mail: otobe.tomohito@qst.go.jp

 
K. Ueda (ed.), *Ultrafast Electronic and Structural Dynamics*,
https://doi.org/10.1007/978-981-97-2914-2_4

## 4.1 Introduction

The Physics Nobel Prize in 2018 was awarded for groundbreaking inventions in the field of laser physics to Arthur Ashkin, Gérard Mourou, and Donna Strickland. Among them, the prize motivation for G. Mourou and D. Strickland was "for their method of generating high-intensity, ultra-short optical pulses." Ultrashort (typically femtosecond), intense laser pulses enabled by their invention, chirped-pulse amplification (CPA) [1], have become an important tool in scientific research as well as industrial applications. In *Scientific Background* [2], the Nobel Committee for Physics at the Royal Swedish Academy of Sciences names CPA technology's major applications, among which the first is *strong-field physics and attosecond science* and the third *high-intensity lasers in industry and medicine*.

Atoms and molecules in the gas phase irradiated by a high-intensity femtosecond laser pulse exhibits highly nonlinear behavior such as above-threshold ionization, tunneling ionization, non-sequential double ionization, and high-harmonic generation (HHG) [3, 4]. The strong-field physics is a field that studies these *strong-field phenomena*. HHG, in particular, represents a highly successful avenue toward an attosecond coherent light source in the extreme-ultraviolet and soft x-ray spectral ranges [5–7], which has opened new research possibilities including attosecond science [8–11] to observe and manipulate ultrafast electron dynamics. The gas-phase strong-field phenomena can be consistently explained by the so-called three-step model [12, 13], in which an electron is first ejected by tunneling ionization by the strong field, then accelerated classically by an oscillating laser field, and finally recombines or recollides with the parent ion.

Thanks to the advent of high-intensity mid-infrared to terahertz radiation sources, solid-state materials have recently emerged as a new stage of strong-field physics and attosecond science [14]. In particular, many experimental observations of HHG from solids have been reported since the first discovery by Ghimire et al. [15–28], revealing unique aspects of solid-state HHG such as linear scaling of cutoff energy with field strength [15, 17] and multiple plateau structure [20, 26]. In contrast to the gas-phase case, the mechanism underlying the strong-field electron dynamics in solids has turned out to be complex and depend on experimental conditions. Among factors specific to solids are,

- Co-presence of intraband and interband transitions
- electron-hole interaction (e.g., exciton formation) and electron correlation (e.g., carrier scattering and excitonic molecule formation)
- dependence on crystal orientation and laser polarization due to crystal anisotropy.

Such complexity and diversity will make strong-field electron dynamics in solids offer even richer information on band structure and ultrafast dynamic electron correlation. Strong-field physics and attosecond science have enabled detailed analysis on intense laser interaction with matter. The extension of its frontier from atomic and molecular systems to solid-state materials will further advance industrial and medical applications of high-intensity lasers such as laser material processing.

In this Chapter, we present theories on momentum-space electron dynamics in graphene and crystalline dielectrics and semiconductors subject to intense laser fields. First, we discuss graphene within massless Dirac fermion and tight-binding models, where we rigorously derive intraband and interband transitions, forming the basis for understanding solid-state strong-field physics (Sect. 4.2). Then, based on the time-dependent Schrödinger equation for a one-dimensional model crystal within a single-electron approximation, we introduce a simple, *multiband*, momentum-space three-step model that incorporates intraband displacement, interband tunneling, and recombination with a valence band hole. We further analyze electron-hole interaction effects, using the time-dependent Hartree-Fock calculations (Sect. 4.3). Finally, actual three-dimensional materials are investigated. We present a time-dependent density-matrix method useful to quantitatively understand and explain experimental results (Sect. 4.4). Moreover, we describe the dynamical Franz-Keldysh effect in femtosecond time resolution, i.e., the time-dependent modulation of a dielectric function under an intense laser field, using a real-time time-dependent density functional theory (Sect. 4.5). Summary is given in Sect. 4.6. Hartree atomic units are used throughout unless otherwise stated.

## 4.2 Graphene

It is instructive to examine the laser-driven coherent electron dynamics in graphene [29–32], for which the intraband and interband transitions, key to understand solid-state HHG, can be rigorously and simply derived. Let us consider a single-electron response in the monolayer graphene placed in the $xy$ plane subject to normal incidence of a laser pulse with its electric field $\mathbf{E}(t)$ and vector potential $\mathbf{A}(t) = -\int \mathbf{E}(\mathbf{t})dt$ being in the graphene plane.

### *4.2.1 Graphene Bloch Equations (GBEs)*

The two-component wave function $\psi(t)$ of the electron with an initial wave vector of $\mathbf{k}$ and canonical momentum $\mathbf{p} = \hbar\mathbf{k}$ is governed by the following time-dependent Schrödinger equation (TDSE):

$$i\hbar\frac{\partial}{\partial t}\psi(t) = H(t)\psi(t), \tag{4.1}$$

with the time-dependent Hamiltonian $H(t)$,

$$H(t) = \begin{pmatrix} 0 & h(t) \\ h(t)^* & 0 \end{pmatrix} \qquad h(t) = \epsilon(t)e^{-i\theta(t)}, \tag{4.2}$$

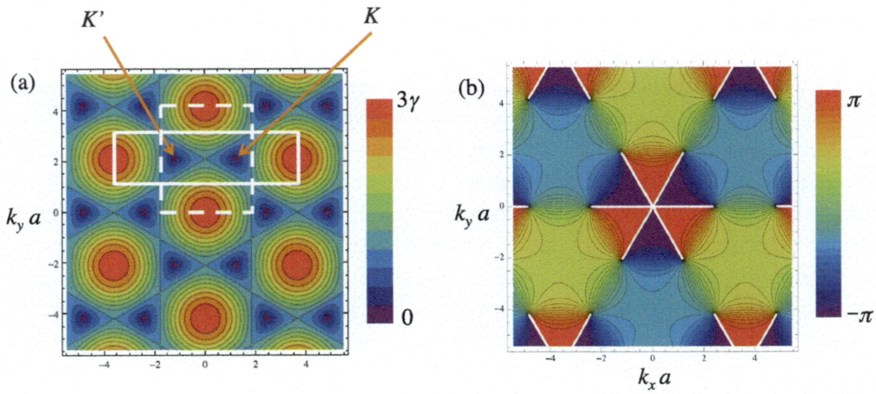


**Fig. 4.1** (color online) Contour and false color plots of **a** $\epsilon_{\mathbf{k}}$ and **b** the principal value $-\mathrm{Arg}\, h_{\mathbf{k}}$, defined in the range $[-\pi, \pi)$ [38]. The thick solid and dashed white squares in panel **a** outline examples of simulational Brillouin zone, i.e., the area of integration in Eq. (4.16). The Dirac points $K$ and $K'$ are located at $\left(\pm\frac{2\pi}{3\sqrt{3}}, \frac{2\pi}{3}\right) = (\pm 1.2092, 2.0944)$. In panel **b**, the value jumps between $\pi$ and $-\pi$ on solid white lines. Reprinted from Ref. [38] with permission under the Creative Commons Attribution 3.0 license

where $\epsilon(t) = |h(t)|$ and $\theta(t) = -\arg h(t)$.

Within the framework of the tight-binding (TB) model of nearest-neighbor interactions [33, 34],

$$h(t) = -\gamma \sum_{\alpha=1}^{3} e^{i\boldsymbol{\kappa}\cdot\boldsymbol{\delta}_\alpha}, \tag{4.3}$$

where $\gamma \approx 2.5 - 2.8\,\mathrm{eV}$ denotes the hopping energy, and $\boldsymbol{\delta}_1 = a(0, 1)$ and $\boldsymbol{\delta}_{2,3} = \frac{a}{2}(\pm\sqrt{3}, -1)$ the locations of nearest neighbors separated by distance $a \approx 1.42\,\text{Å}$. $\boldsymbol{\kappa} = \boldsymbol{\pi}/\hbar$ is the wave vector corresponding to the kinetic momentum $\boldsymbol{\pi}(t) = \mathbf{p} + e\mathbf{A}(t)$ with $e(> 0)$ being the elementary charge. $\boldsymbol{\kappa}$ and $\boldsymbol{\pi}$ vary with time and describe the laser-driven intraband displacement. $\epsilon$ denotes the magnitude of the energy eigenvalue in the absence of the field whose value $\epsilon_{\mathbf{k}}$ for $\mathbf{k}$ is given by [Fig. 4.1a],

$$\epsilon_{\mathbf{k}} = \gamma\sqrt{3 + 2\cos\sqrt{3}k_x a + 4\cos\frac{\sqrt{3}k_x a}{2}\cos\frac{3k_y a}{2}}. \tag{4.4}$$

If we resort to the massless Dirac fermion (MDF) picture [33] applicable near the Dirac point and take the Dirac point as the origin of $\boldsymbol{\pi}$ (Fig. 4.2), $h(t)$ and $\epsilon(t)$ are simplified to

$$h(t) = v_F(\pi_x - i\pi_y) = v_F[(p_x + eA_x) - i(p_y + eA_y)], \tag{4.5}$$

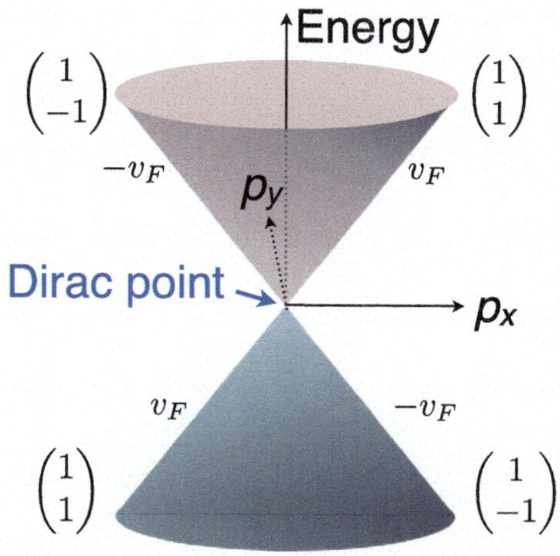


**Fig. 4.2** Electronic band structure of monolayer graphene near the Dirac point (Dirac cone) within the massless Dirac fermion picture. Reprinted with permission from Ref. [35]. Copyright 2010 by American Physical Society

and

$$\epsilon(t) = v_F|\boldsymbol{\pi}(t)| = v_F\hbar|\boldsymbol{\kappa}(t)|, \tag{4.6}$$

respectively, with $v_F = \frac{3\gamma a}{2\hbar} \approx c/300$. It should be noted that TDSE Eq. (4.1) has a form similar to the Dirac equation but is different from the original Dirac equation in that the fermion mass is zero, which leads to the two-component, instead of four-component, wave function. We also notice that $\theta$ becomes the directional angle of $\boldsymbol{\kappa}$: $\kappa_x = |\boldsymbol{\kappa}|\cos\theta$ and $\kappa_y = |\boldsymbol{\kappa}|\sin\theta$ (around $\mathbf{K}$), $-|\boldsymbol{\kappa}|\sin\theta$ (around $\mathbf{K}'$).

In the field-free case, whether we may use the TB or MDF pictures, the TDSE has the following two solutions:

$$\psi(t) = \frac{1}{\sqrt{2}}\exp\left(\mp i\frac{\epsilon}{\hbar}t\right)\begin{pmatrix} e^{-\frac{i}{2}\theta} \\ \pm e^{\frac{i}{2}\theta} \end{pmatrix} \tag{4.7}$$

with their energy eigenvalues are $\pm\epsilon$. The upper sign refers to the upper band (electron band), and the lower sign to the lower band (hole band).

Let us now turn on the laser pulse and express the wave function as a superposition,

$$\psi(t) = c_+(t)\psi_+(t) + c_-(t)\psi_-(t), \tag{4.8}$$

of the instantaneous upper and lower band states (Volkov states),

$$\psi_\pm(t) = \frac{1}{\sqrt{2}}\exp[\mp i\Omega(t)]\begin{pmatrix} e^{-\frac{i}{2}\theta(t)} \\ \pm e^{\frac{i}{2}\theta(t)} \end{pmatrix} \tag{4.9}$$

with the instantaneous temporal phase or dynamical phase $\Omega(t)$ defined as

$$\Omega(t) = \int \frac{\epsilon(t)}{\hbar}dt. \tag{4.10}$$

We find that Eq. (4.8) is indeed the exact solution of the TDSE Eq. (4.1) if the expansion coefficients $c_\pm(t)$ satify the equations of motion:

$$\dot{c}_\pm(t) = \frac{i}{2}\dot{\theta}(t)c_\mp(t)e^{\pm 2i\Omega(t)}. \tag{4.11}$$

Introducing the population difference $n = |c_+|^2 - |c_-|^2$ between the two band and the interband coherence $\rho = c_+c_-^*$, we can transform Eq. (4.11) into the graphene Bloch equations [35]:

$$\dot{n} = -i\,\dot{\theta}(t)\rho(t)\,e^{-2i\Omega(t)} + \text{c.c.}, \tag{4.12}$$

$$\dot{\rho} = -\frac{i}{2}\dot{\theta}(t)n(t)e^{2i\Omega(t)}. \tag{4.13}$$

To take account of the Fermi distribution at finite temperature $T$, we solve Eqs. (4.12) and (4.13) under initial conditions $n = F(p) - F(-p)$ and $\rho = 0$, where $F(p) = \{1 + \exp[(\epsilon(p) - \mu)/k_BT]\}^{-1}$ is the Fermi-Dirac function, where $\mu$ and $k_B$ denote the chemical potential and Boltzmann constant, respectively.

Through $\Omega(t)$ and $\theta(t)$, Eq. (4.9) incorporates the field-induced intraband dynamics (transition) of the electron that changes its kinetic momentum $\mathbf{p} + e\mathbf{A}(t)$ following the acceleration theorem. On the other hand, Eqs. (4.11)–(4.13) indicate that the electron undergoes interband transitions while retaining coherence. Thus, the GBEs are physically more transparent than the TDSE.

If we defined $\theta(t)$ by the principal value $\theta(t) = -\text{Arg}\,h(t)$ with $\text{Arg}\,z \in (-\pi, \pi]$, as plotted in Fig. 4.1b, $\theta(t)$ would undergo $2\pi$ jumps on the white lines linking Dirac points. When we follow the electron dynamics based on the GBEs, instead, we need to define $\theta(t) = -\arg h(t)$, with $\arg z = \text{Arg}\,z + 2\pi n$ ($n$ is an integer), in such a way that it varies continuously along the path of $\boldsymbol{\kappa}(t)$. It should be noticed that if $\boldsymbol{\kappa}(t)$ takes a $\mathbf{k}$-space trajectory surrounding a Dirac point, $\psi_\pm(t)$ acquires a geometrical phase of $\pi$, in addition to the dynamical phase $\Omega(t)$. Thus, Berry's phase [36, 37] is incorporated in the GBEs.

It should be noticed that the whole dynamics within the MDF model is invariant under multiplication of quantities of energy dimension, $\hbar\omega$, $v_F p$, $v_F eA$, $\mu$, $k_BT$, and $\hbar/t$, by a common factor. In the weak-field limit, one can derive the universal conductivity $e^2/4\hbar$ from Eqs. (4.17)–(4.19) [38].

An example of carrier occupation distribution calculated within the MDF model is visualized in Fig. 4.3 for the case where $T = 0$, $\mu = v_F eA_0/5$, and the vector potential $A(t)$ is assumed to be a sine pulse with a Gaussian intensity envelope whose full-width-at-half-maximum width corresponds to two optical cycles, and peak amplitude $A_0$ satisfies $\hbar\omega/v_F eA_0 = 9.46 \times 10^{-3}$ [Fig. 4.3a]. Figure 4.3b and c shows carrier occupation distribution at moments $A$ and $B$ marked in Fig. 4.3a. The green circle represents the electrons originally in the upper band around the Dirac point (note that $\mu > 0$), which undergoes intraband displacement. In addition, outside the green circle, part of the electrons initially in the lower band transfers to the upper band through interband transitions. An electron with a given initial

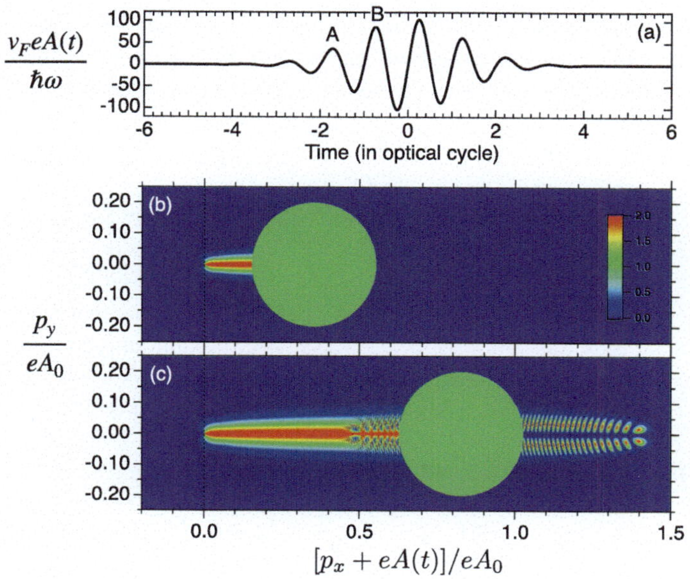


**Fig. 4.3** **a** Normalized vector potential $v_F eA(t)/\hbar\omega$ of the incident optical pulse. **b** and **c** Carrier occupation distribution calculated for the moment labeled as $A$ and $B$, respectively, in panel **a**. Reprinted with permission from Ref. [35]. Copyright 2010 by American Physical Society

momentum may transfer to the other band each time it passes near the Dirac point in the oscillating laser field, thus split into different quantum pathways to reach the conduction band. Their interference is clearly seen in Fig. 4.3c. Such an interference effect has been experimentally observed and controlled by Higuchi et al. [39].

### 4.2.2 Electric Current and Harmonic Generation

The single-electron electric current is given by $-e\mathbf{j}_e$, where $\mathbf{j}_e$ is defined as

$$\mathbf{j}_e = \psi^\dagger(t)\frac{\partial H}{\partial \boldsymbol{\pi}}\psi(t). \tag{4.14}$$

For the case of the TB model, using $\rho(t)$ and $n(t)$, we obtain the following explicit form:

$$\mathbf{j}_e = \frac{\gamma}{\hbar}\sum_{\alpha=1}^{3}\left[n\,\sin\left(\boldsymbol{\kappa}\cdot\boldsymbol{\delta}_\alpha+\theta\right) - i\left\{\rho\, e^{-i2\Omega}\cos\left(\boldsymbol{\kappa}\cdot\boldsymbol{\delta}_\alpha+\theta\right) - \text{c.c.}\right\}\right]\boldsymbol{\delta}_\alpha. \tag{4.15}$$

Then, to obtain the macroscopic electric current $\mathbf{J}(t)$ generated by the laser field, we calculate the carrier current $\mathbf{j}_c$ by replacing $n$ with the carrier occupation $n+1$ in Eq. (4.15) then integrate $-e\mathbf{j}_c$ over the honey-comb lattice Brillouin zone [Fig. 4.1a] as

$$\mathbf{J}(t) = -\frac{g_s e}{(2\pi)^2} \int_{\mathrm{BZ}} \mathbf{j}_c(t) d\mathbf{k}, \tag{4.16}$$

where $g_s = 2$ denotes the spin-degeneracy factor.

The expressions in the MDF picture are simpler. Each component of the carrier current is written as

$$j_{c,x} = v_F \left[ (n+1)\cos\theta + i \sin\theta \{\rho e^{-2i\Omega} - \text{c.c.}\} \right], \tag{4.17}$$

$$j_{c,y} = v_F \left[ (n+1)\sin\theta - i \cos\theta \{\rho e^{-2i\Omega} - \text{c.c.}\} \right]. \tag{4.18}$$

Then, the macroscopic electric current $\mathbf{J}(t)$ is given by

$$\mathbf{J}(t) = -\frac{g_s g_v e}{(2\pi\hbar)^2} \int \mathbf{j}_c(t) d\mathbf{p} = -\frac{g_s g_v e}{(2\pi)^2} \int \mathbf{j}_c(t) d\mathbf{k}, \tag{4.19}$$

with $g_v = 2$ being the valley-degeneracy factor.

One can calculate the intensity spectrum $I(\omega)$ of harmonic generation using the Fourier transform $\hat{\mathbf{J}}(\omega)$ of $\mathbf{J}(t)$ by

$$I(\omega) \propto |\omega \hat{\mathbf{J}}(\omega)|^2. \tag{4.20}$$

In Eqs. (4.17) and (4.18), we can identify the contribution to harmonic generation from the oscillating interband polarization $\rho(t)e^{-2i\Omega(t)}$ and the intraband transition $\theta(t)$ as well as the temporal variation in population $n(t)$. Among them, the population variation contributes only to the first term, and the interband polarization only to the second, whereas both terms contain the contribution from the intraband transition.

In Fig. 4.4, we plot the harmonic spectrum for the case of Fig. 4.3. Curiously, the harmonic intensity (red thick solid line) is reduced compared with the case of the pure intraband dynamics (thick dotted line) [35], obtained by switching off interband transitions in the calculation. Thus, in spite of the small contribution from the interband polarization itself (thin solid lines in Fig. 4.4), i.e., the second term of Eq. (4.17), the interband dynamics strongly modify the optical response of graphene, relaxing nonlinearity. Nevertheless, harmonic generation of up to the thirteenth order can be seen, revealing high nonlineality; high-harmonic generation from graphene has been experimentally observed [40, 41].

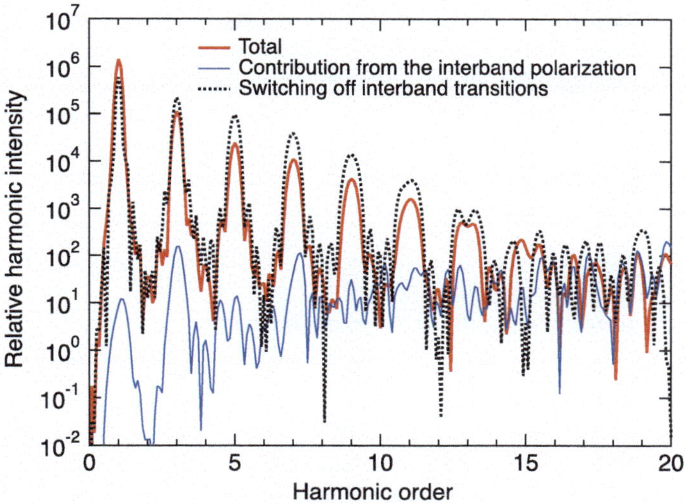


**Fig. 4.4** Harmonic intensity spectra for the case of Fig. 4.3. Thick solid line: total spectrum calculated with Eq. (4.20), thin solid line: contribution from the interband polarization, thick dotted line: calculated by switching off the interband transitions. Reprinted with permission from Ref. [35]. Copyright 2010 by American Physical Society

## 4.3 Solid-State Three-Step Model

What characterizes high-harmonic generation is that its spectrum consists of a *plateau* where the harmonic intensity is nearly constant over many orders and a sharp *cutoff*. The gas-phase HHG can be intuitively and even quantitatively captured by the so-called three-step model [12, 13], in which an electron is first ejected by tunneling ionization by the strong field, then accelerated classically by an oscillating laser field, and finally radiatively recombines with the parent ion emitting a harmonic photon. The cutoff energy, i.e., the maximal harmonic photon energy $E_c$ is given by

$$E_c = I_p + 3.17U_p, \tag{4.21}$$

where $I_p$ is the ionization potential of the target atom or molecule, and $U_p[\text{eV}] = E_0^2/4\omega^2 = 9.337 \times 10^{-14}\ I\ [\text{W/cm}^2]\ (\lambda\ [\mu\text{m}])^2$ the ponderomotive energy, with $E_0$, $I$, $\omega$, and $\lambda$ being the strength, intensity, angular frequency, and wavelength of the driving field, respectively. Hence, the cutoff energy is roughly proportional to the square of the laser electric field strength.

High-harmonic generation from solid-state materials is, on the other hand, quite distinct from its gas-phase counterpart, exhibiting unique aspects such as linear scaling of cutoff energy with field strength [15, 17] and multiple plateau formation [20, 26], to name only a few. The comprehensive mechanism underlying solid-state HHG is under active investigation; both real-space, as in the gas-phase case, and momentum-space pictures are on the market. In this Section, using a one-dimensional (1D) model periodic crystal, we present a solid-state momentum-space three-step model [42–45] that considers electron dynamics across multiple bands, incorporating field-induced intraband displacement, interband tunneling, and recombination with the valence-band (VB) hole, suitable to discuss harmonic generation from interband polarization. We first describe an independent-electron picture (Sect. 4.3.1) [42] and then discuss electron-hole interaction effects [45].

### 4.3.1 Independent-Electron Approximation

We consider the electron dynamics in a 1D model crystal along linear laser polarization, assuming that VBs are initially fully occupied across the whole Brillouin zone (BZ), as is usually the case for wide-band-gap semiconductors. Let us calculate harmonic spectra, based on the effective TDSE for each independent electron,

$$i\frac{\partial}{\partial t}\psi_{nk}(x,t) = \hat{H}(t)\psi_{nk}(x,t) = \left\{\frac{1}{2}\left[\hat{p} + A(t)\right]^2 + V(x)\right\}\psi_{nk}(x,t), \tag{4.22}$$

within the velocity gauge for the electron initially located in band $n$ with a crystal momentum $k$, where $\hat{p} = -i\frac{\partial}{\partial x}$, $V(x)$ denotes the periodic single-electron effective potential with lattice constant $a$, i.e., $V(x+a) = V(x)$. As the initial state of the time-dependent wave function $\psi_{nk}(x,t)$, we take the Bloch function $\phi_{nk}$, i.e., the eigenstate of the field-free Hamiltonian with the energy eigenvalue $\varepsilon_{nk}$. We use the Mathieu-type potential [46] given by

$$V(x) = -V_0\left[1 + \cos(2\pi x/a)\right], \tag{4.23}$$

with $V_0 = 0.37$ and $a = 8$, which supports a band structure (Fig. 4.5) with minimum band gap 4.2 eV at $k = 0$, while the first and second CBs approach each other at the Bragg plane ($k = \pm\frac{\pi}{a}$). We assume that the two VBs ($n = 0, 1$ in Fig. 4.5) are initially filled across the whole BZ.

Rather than expand the wave functions with basis functions, we resort to direct numerical integration of the TDSE Eq. (4.22) in real space. One of the advantages of the velocity gauge is that the Hamiltonian retains lattice periodicity even under the action of the laser pulse. As a consequence, the initial crystal momentum $k$ is always a good quantum number, and, thus, we can solve the TDSE for each $k$ independently. Using Bloch's theorem, the wave function $\psi_{nk}(x,t)$ can be factorized as

$$\psi_{nk}(x,t) = e^{ikx}u_{nk}(x,t) \tag{4.24}$$

where $u_{nk}(x,t)$ satisfies $u_{nk}(x+a,t) = u_{nk}(x,t)$. The substitution of Eq. (4.24) into Eq. (4.22) leads the equation of motion for $u_{nk}(x,t)$,

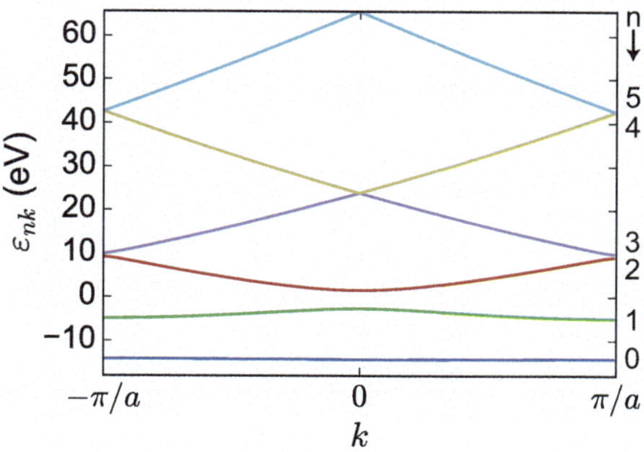


**Fig. 4.5** Two valence bands ($n = 0, 1$) and first four conduction bands ($n = 2, \ldots, 5$) of the field-free Hamiltonian. The integers on the right axis are the band indices $n$. Reprinted with permission from Ref. [42]. Copyright 2017 by American Physical Society

$$i\frac{\partial}{\partial t}u_{nk}(x,t) = \left\{\frac{1}{2}\left[\hat{p}+k+A(t)\right]^2 + V(x)\right\}u_{nk}(x,t). \tag{4.25}$$

This is to be solved only within the unit cell $x \in [0, a]$, which brings substantial computational-cost reduction. It is interesting to notice that $k + A(t)$ in Eq. (4.25) automatically accounts for the intraband dynamics [47, 48] and that Eq. (4.25) couples different bands. For a given pair of $(n, k)$, we numerically integrate the equation of motion (4.25), using the finite-difference method with the grid spacing 0.53 a.u., time step size $2.67 \times 10^{-4}$ fs $= 1.10 \times 10^{-2}$ a.u., and the number $N$ of $k$-points 141.

We calculate the expectation value of velocity to obtain the contribution to the field-induced current from each $(n, k)$,

$$j_{nk}(t) = \langle\psi_{nk}(t)|\,\hat{p} + A(t)\,|\psi_{nk}(t)\rangle = \int_0^a u_{nk}^*(x,t)\left[\hat{p}+k+A(t)\right]u_{nk}(x,t)dx. \tag{4.26}$$

Then, we obtain the total current by summing $j_{nk}$ over the initial band indices $n(= 1, 2)$ and initial crystal momenta $k$,

$$j(t) = \frac{1}{Na}\sum_{nk} j_{nk}(t). \tag{4.27}$$

It should be remembered that $n$ and $k$ denote the band index and crystal momentum, respectively, that the electron initially occupies. The harmonic spectrum is calculated as the modulus square of the Fourier transform of $j(t)$. Before applying the Fourier transform, we multiply $j(t)$ by a mask function $W(t) = \sin^4(t/\tau)$ of the same form as the field envelop in order to suppress the current remaining after the pulse.

We specifically consider a laser electric field of a form $E(t) = E_0 \sin^4(t/\tau)\sin[\omega(t - \pi\tau/2)]$ for $t \in [0, \pi\tau]$ and $E(t) = 0$ for $t \notin [0, \pi\tau]$, where $E_0$, $\tau$ denote the peak electric field amplitude and a measure of pulse width, respectively (Fig. 4.6). Figure 4.7a shows the high-harmonic spectra for a central

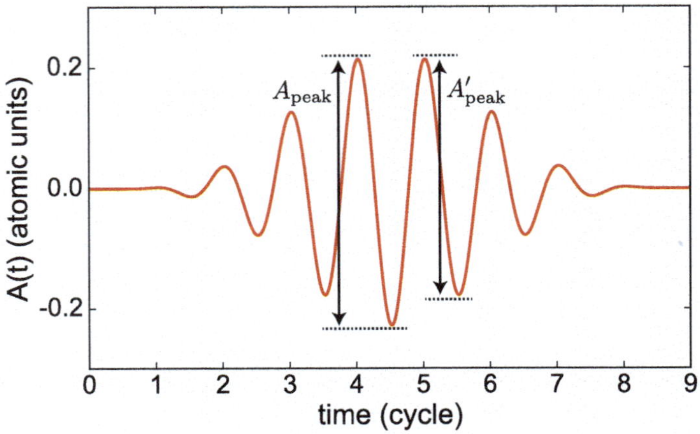


**Fig. 4.6** The waveform of the vector potential $A(t)$ of the laser pulse with $E_0 = 1.65$ V/nm and $\tau = 99.66$ fs. The maximum and the second maximum peak-to-valley amplitude $A_{\rm peak}$ and $A'_{\rm peak}$ are defined as depicted in the figure. Reprinted with permission from Ref. [42]. Copyright 2017 by American Physical Society

wavelength 3200 nm, $\tau = 96.66$ fs corresponding to a full-width-at-half-maximum (FWHM) duration of 48 fs, and several field amplitudes. We immediately notice that, whereas the spectrum for $E_0 = 0.87$ V/nm has a single plateau and cutoff, those for $E_0 = 1.65$ and 2.11 V/nm have two additional plateaus of lower intensity. Moreover, the transition from the single- to multiple-plateau structure takes place not gradually but suddenly [Fig. 4.7b]; while the cutoff energy increases smoothly and quasi-linearly with $E_0$ up to $\approx 1.4$ V/nm, second and third plateaus suddenly appear there, and the cutoff jumps up from 15 eV to 45 eV. This result qualitatively reproduces the previously reported unique features of solid-state HHG [20, 46, 49]. Also, another cutoff jump is seen at $E_0 \approx 2.8$ V/nm.

We define the maximum peak-to-valley amplitude $A_{\rm peak}$ of $A(t)$ (see Fig. 4.6) and put it on the top axis of Fig. 4.7b. Then, we find that the jump-up positions interestingly satisfy the condition that $A_{\rm peak} = \frac{\pi}{a} = 0.393$ a.u. and $\frac{2\pi}{a} = 0.786$ a.u. [vertical white dashed lines in Fig. 4.7b]. Note that $A_{\rm peak}$ characterizes the largest crystal momentum gain in the intraband dynamics and that $\frac{\pi}{a}$ is the distance from the $\Gamma$ point to the first-BZ edge (Fig. 4.5).

The following simple model [42, 44, 50, 51] explains the above findings as well as cutoff positions and the time-frequency structure of HHG: Its essential ingredients are summarized as follows:

(i) *Tunneling ionization*: each electron is tunnel ionized to an upper band predominantly at the minimum band gap to a first approximation, e.g., from band 1 to 2 at $k = 0$ and from 2 to 3 at the BZ edge in the present model crystal.
(ii) *Intraband acceleration*: the electron is displaced in the momentum space (laser-driven intraband dynamics), following the acceleration theorem [47, 48])

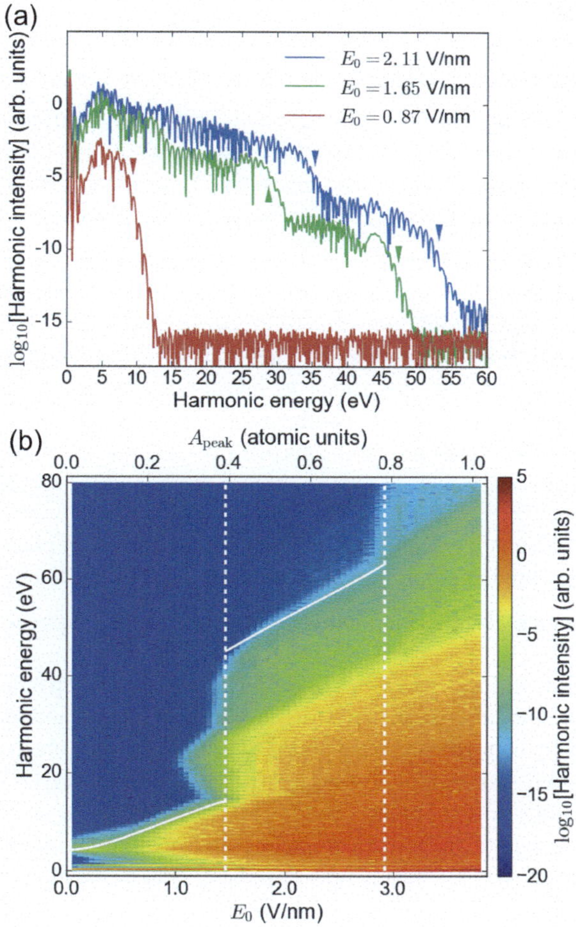


**Fig. 4.7** **a** High-harmonic spectra for $E_0 = 0.87$ V/nm (red (lower) line), $E_0 = 1.65$ V/nm (green (middle) line), and $E_0 = 2.11$ V/nm (blue (upper) line). Arrowheads indicate the positions given by $\Delta\varepsilon_{21}(A_{\text{peak}})$ for $E_0 = 0.87$ V/nm (red), and Eqs. (4.30) and $\Delta\varepsilon_{41}\left(\frac{\pi}{a} - A_{\text{peak}}\right)$ for $E_0 = 1.65$ (green) and 2.11 (blue) V/nm. **b** False-color representation of the harmonic spectra as functions of $E_0$. $A_{\text{peak}}$ corresponding to $E_0$ is shown on the top axis in the atomic unit. The two vertical white dashed lines represent $A_{\text{peak}} = \pi/a$ and $\frac{2\pi}{a}$. The two white solid lines represent the cutoff energy positions given by $\Delta\varepsilon_{21}(A_{\text{peak}})$ for $0 < A_{\text{peak}} < \frac{\pi}{a}$, and Eq. (4.30) for $\frac{\pi}{a} < A_{\text{peak}} < \frac{2\pi}{a}$. Reprinted with permission from Ref. [42]. Copyright 2017 by American Physical Society

$k(t) = k_0 + A(t)$ with $k_0$ being the initial crystal momentum. The resulting oscillating current leads to photoemission (*intraband* contribution to HHG).

(iii) *Interband recombination*: the electron emits a photon when it undergoes an interband transition to the initial band, i.e., recombination with the valence-band hole (*interband* contribution to HHG). The photon energy is given by the particle-hole energy

$$\Delta\varepsilon_{n(t)n_0}(k(t)) = \varepsilon_{n(t)k(t)} - \varepsilon_{n_0k(t)}, \tag{4.28}$$

between the band $n(t)$ where the electron is located at $t$ and the initial band $n_0$.

Comprising tunneling ionization, acceleration, and recombination, this model can be viewed as a solid-state, momentum-space counterpart of the familiar coordinate-space three-step model [12, 13] of gas-phase HHG. Nevertheless, there are important differences:

- All the electrons in the VB undergo the intraband acceleration (ii) together [52, 53] even before the first tunneling. Thus, VB electrons starting from not only $k_0 = 0$ but also any arbitrary initial momenta $k_0$ are considered.[1]
- Electrons can climb up to higher and higher bands by repeating (i) and (ii).
- Not only (iii) but also (ii) contribute to harmonic generation, while harmonic photons are emitted only upon recombination in the gas phase [17, 46, 50, 54, 55]. Thus, there are intraband and interband contributions to solid-state HHG. They can be rigorously derived for the case of graphene [35, 38], as discussed in Sec. 4.2.
- (iii) can take place at any time, in principle, while, in the gas phase, the electron can recombine with the parent ion only at the position of the latter.

This electron dynamics is conceptually similar to that in graphene [35, 38, 39] (Sect. 4.2).

In analogy to the trajectory analysis of the gas-phase three-step model, explaining the cutoff law and the time-frequency structure, we can understand many aspects of solid-state HHG by using the above-mentioned recipes to trace electron trajectories in the band diagram. An example for $A_{\text{peak}} = 0.44 > \frac{\pi}{a}$ is displayed in Fig. 4.8.

Electrons initially in the valence band are accelerated (①), and excited to the CB at $k = 0$ at $t = t_0$ (②). Vertical tunneling and recombination being assumed, once a waveform of $A(t)$ is given [Fig. 4.8a], the crystal momentum history is fully described as

$$k(t) = k_0 + A(t) = A(t) - A(t_0), \tag{4.29}$$

regardless of the band where the electron resides [Fig. 4.8b]. It should especially be noticed that $|k(t)| < A_{\text{peak}}$. Thus, if $A_{\text{peak}} < \frac{\pi}{a}$, the electron cannot reach the BZ edge but oscillates in the first CB without further excitation. Therefore, the emitted photon energy is given by $\Delta\varepsilon_{21}[k(t)]$ as a function of recombination time $t$, and the cutoff energy is given by $\Delta\varepsilon_{21}(A_{\text{peak}})$, which agrees with the position represented by the white solid line in Fig. 4.7b.

Now that $A_{\text{peak}} > \frac{\pi}{a}$ in Fig. 4.8, after excitation to the first CB (① - ②), part of electrons can be accelerated to reach the BZ edge (③), and open a channel to climb up to the upper CB (④) within a half cycle. The promoted electrons then undergo intraband displacement to the reversed direction in the second CB ($n = 3$)

---

[1] This does not violate the Pauli exclusion principle, since all the electrons in the VB move uniformly together [52, 53], and thus, no $(n, k)$ point is occupied simultaneously by more than one electron at any time.

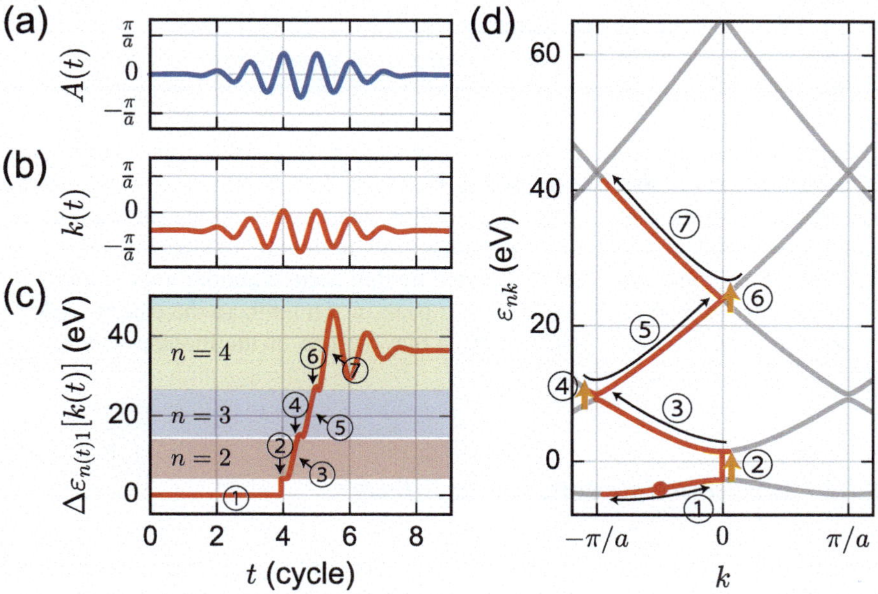


**Fig. 4.8** Momentum-space trajectory of an electron excited from a VB ($n = 1$) to the first CB ($n = 2$) at $t = 4T$ ($T$ denotes an optical cycle), drawn based on the solid-state three-step model when $E_0 = 1.65$ V/nm or $A_{\text{peak}} = 0.44 > \frac{\pi}{a}$, for which $k_0 = -0.49 \times \frac{\pi}{a}$. **aa** waveform of $A(t)$ **b** instantaneous crystal momentum $k(t)$ **c** temporal evolution of the particle-hole energy, i.e., emitted photon energy **d** pictorial representation of the momentum-space electron trajectory in the band diagram. Reprinted with permission from Ref. [42]. Copyright 2017 by American Physical Society

in the next half cycle, enabling photon emission of higher energy (⑤). This simple pictorial analysis neatly explains why multiple plateaus suddenly appear at $A_{\text{peak}} \approx \frac{\pi}{a}$ [Fig. 4.7b]. Electrons can experience interband transitions not only precisely at the minimum band gaps but also in their vicinities. This is the origin of some high-energy components, which appear even before $A_{\text{peak}}$ reaches $\frac{\pi}{a}$ in Fig. 4.7b, from $E_0 \sim 1.1$ V/nm.

Each time the electrons reach the minimum energy gap to next CB every half cycle, they either undergo further interband excitation (⑤ - ⑦) or pass through it. They can climb up to the third CB ($n = 4$) if $A'_{\text{peak}} < \frac{\pi}{a}$, where $A'_{\text{peak}}$ denotes the second maximum peak-to-valley amplitude (Fig. 4.6), and the fourth CB ($n = 5$) if $A'_{\text{peak}} > \frac{\pi}{a}$ at $t \approx 5.5T$ with $T$ being the optical cycle. From this scenario, we can estimate the maximum energy gain as

$$E_c = \begin{cases} \Delta\varepsilon_{41}(A'_{\text{peak}}) & (A'_{\text{peak}} < \frac{\pi}{a}) \\ \Delta\varepsilon_{51}(A'_{\text{peak}}) & (\frac{\pi}{a} < A'_{\text{peak}}), \end{cases} \tag{4.30}$$

which reproduces the highest harmonic energy in Fig. 4.7b. Thanks to the band-climbing process,[2] the highest cutoff energy can exceed the value expected in the gas phase for the same laser parameters and ionization potential (band gap energy in the solid case) [20].

Electrons that start from $k_0 \sim 0$ are excited when $A(t) \approx 0$, i.e., at an extremum of $E(t)$, promoting tunneling transition. However, they cannot reach the BZ edge and are confined in the first CB if $A_{\text{peak}} < \frac{2\pi}{a}$. Consequently, their contributions are limited to the range below $E_{31}$. In contrast, the harmonic components above $E_{31}$ including the highest cutoff are dominated by the electrons initially far from the $\Gamma$ point and first excited in the vicinity of a peak of $A(t)$, where the electric field is weak, thus with smaller probability. This may be one of the reasons why higher plateaus are weaker in intensity.

## 4.3.2 Electron-Hole Interaction Effects

Sections 4.2 and 4.3.1 as well as most of the works investigating the mechanisms of solid-state HHG have used independent-electron approximation. On the other hand, multielectron effects in the strong-field regime is largely unexplored. Let us focus on the role of the electron-hole interaction (EHI), which forms excitions in the linear response regime, in this Subsection, based on the time-dependent Hartree-Fock (TDHF) calculation [45].

We again consider a 1D model crystal along laser polarization. A 1D system, which has a strong electron-hole correlation [56], is suitable for the investigation of EHI. We solve a set of the spin-restricted TDHF equation,

$$i\frac{\partial}{\partial t}\psi_{nk_0}(x,t) = \hat{h}(t)\psi_{nk_0}(x,t) = \left[\frac{1}{2}[\hat{p} + A(t)]^2 + U(x) + \hat{w}[\rho(t)]\right]\psi_{nk_0}(x,t), \tag{4.31}$$

in the velocity gauge, where $U(x)$ denotes the periodic potential from the crystal nuclei, $\rho(t)$ the density matrix,

$$\rho(x,x',t) = 2\sum_{n\in\text{VB},\,k_0}\psi_{nk_0}(x,t)\,\psi_{nk_0}(x',t)^*, \tag{4.32}$$

and the operator $\hat{w}[\rho]$, composed of the Coulomb and exchange terms, describes the contribution from the interelectronic Coulomb interactions, reflecting the dynamics of the other electrons within a mean-field treatment. As the initial state of $\psi_{nk_0}(t)$, we take the VB Bloch function $\phi_{nk_0}$, obtained as the self-consistent eigenstate of the field-free Hartree-Fock Hamiltonian with the energy eigenvalue $\varepsilon_{bk_0}$. We calculate

[2] This somewhat reminds us of *Donkey Kong*, an arcade game released by Nintendo (https://en.wikipedia.org/wiki/Donkey_Kong_(video_game)).

the HHG spectrum as the modulus square of the Fourier transform of the induced current $j(t) = 2\sum_{n\in \mathrm{VB},\, k_0} \langle \psi_{nk_0}(t) |\, \hat{p} + A(t) \,| \psi_{nk_0}(t) \rangle$.

Let us compare the TDHF equation Eq. (4.31) with the independent-electron TDSE Eq. (4.22). Aside from the exchange terms not included in the latter, the effective potential $V(x)$ is considered to include the Coulomb terms formed by the initial state, in addition to $U(x)$. Therefore, to mimic the independent-electron treatment, we also perform simulations using the *frozen* TDHF Hamiltonian

$$\hat{h}_f(t) = [\hat{p} + A(t)]^2/2 + U(x) + \hat{w}[\rho_0], \tag{4.33}$$

with $\rho_0(x, x') = e^{-iA(t)\cdot x}\rho(0)e^{iA(t)\cdot x'}$, where electrons move independently in the potential constructed by the ground state Bloch functions. The factors $e^{-iA(t)\cdot x}$ and $e^{iA(t)\cdot x'}$ are required in the velocity gauge. The difference $\hat{w}[\delta\hat{\rho}(t)]$ with $\delta\hat{\rho}(t) = \hat{\rho}(t) - \hat{\rho}_0$ between the *full* TDHF Hamiltonian $\hat{h}(t)$ and frozen Hamiltonian $\hat{h}_f(t)$ takes account of EHI.

Specifically, our system is a 1D model hydrogen chain insulator with a lattice constant of $a = 3.6$ a.u., composed of a series of hydrogen dimers whose bond length is 1.6 a.u.. We use a soft-Coulomb potential $v(x, x') = [(x - x')^2 + 1]^{-1/2}$ for both electron-nucleus and electron-electron interactions. Figure 4.10 shows the band structure, the set of the energy eigenvalues $\varepsilon_{nk_0}$, with a gap energy of 9.5 eV. The lowest band or VB is initially fully occupied. The laser field is assumed to be $E(t) = E_0 \sin^2(t/\tau)\sin(\omega t)$ with $\tau = 702.3$ (5 cycle), $\hbar\omega = 0.387$ eV. We numerically integrate the full and frozen TDHF equations, using the finite-difference method with the grid spacing 0.24 atomic units, time step size $4.4 \times 10^{-3}$ atomic units, and the number of $k$ points 201.

Figure 4.9 displays the calculated harmonic spectra as functions of the field amplitude $E_0$ and the corresponding $A_{\mathrm{peak}}$ [inset of Fig. 4.10a] In the case of the *frozen* TDHF [Fig. 4.9a], i.e., within the independent-electron approximation, we can well understand the appearance of multiple plateaus at $A_{\mathrm{peak}} = \frac{\pi}{a} = 0.87$ and the cutoff positions on the basis of the solid-state momentum-space three-step model [42–44] discussed in Sect. 4.3.1. A typical trajectory is depicted in Fig. 4.10a for $A_{\mathrm{peak}} < \frac{\pi}{a}$, for which no excited electrons in the first conduction band (CB) can reach the next MBG ($k = 0$), and they only oscillate in the first CB, which forms a single plateau in the high-harmonic spectra.

The full TDHF results, with EHI turned on, are shown in Fig. 4.9b. We find two distinct features. First, at low intensity [inset in Fig. 4.9b], there is an exciton peak at 3.8 eV below the gap energy, which indicates that the TDHF simulations capture EHI appropriately. Note that TDDFT at present cannot reproduce excitons, which is based on the simple adiabatic local-density approximation in practical implementations without any nonlocal exchange-like term [57].[3] Second and more remarkably, the

[3] The excitonic physics seems to be guaranteed by the mixture of the exchange term, called hybrid-functional, within the TDDFT framework [65–67]. However, it is not fully investigated that the potential of the hybrid-functional for electron excitation of the extended systems due to few applications.

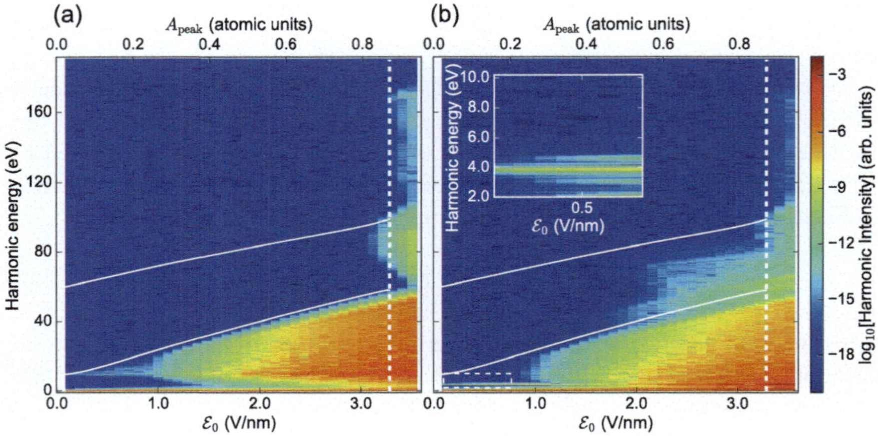


**Fig. 4.9** Harmonic spectra as functions of the field amplitude $\mathcal{E}_0$ (bottom axis) and corresponding $A_{\text{peak}}$ (top axis) obtained from **a** frozen TDHF and **b** full TDHF simulations. The white dashed vertical lines denote $A_{\text{peak}} = \frac{\pi}{a} = 0.87$, which characterizes the position where the multiple plateaus appear according to the solid-state three-step model [42]. Two white solid lines are the energy differences between CBs and VB as function of $A_{\text{peak}}$, i.e., $\varepsilon_{10}(\frac{\pi}{a} - A_{\text{peak}})$ (lower) and $\varepsilon_{20}(A_{\text{peak}})$ (higher). Inset: close-up of the low-field region represented by a dashed rectangle in **b**. Reprinted with permission from Ref. [45]. Copyright 2018 by American Physical Society

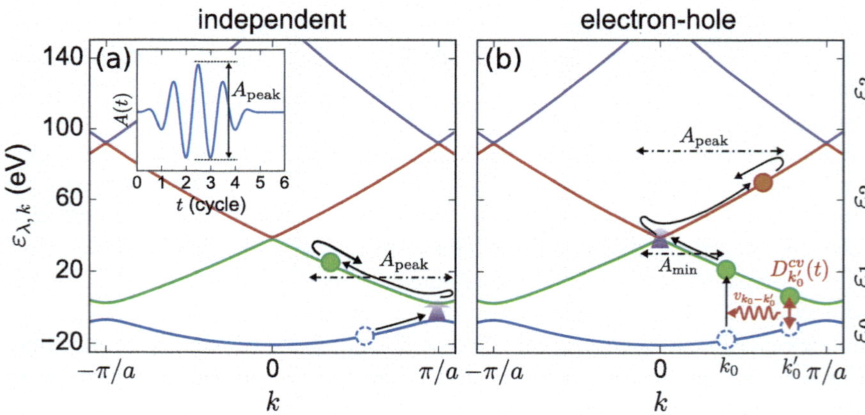


**Fig. 4.10** Pictorial representation of momentum-space electron dynamics **a** within the independent-electron approximation and **b** involving hauling-up excitation. The inset in **a** shows the waveform of the vector potential used in TDHF and *frozen* TDHF simulations and the definition of $A_{\text{peak}}$. The single VB and first three CBs are shown for a 1D model hydrogen chain insulator (see text). The band index $n$ is labeled as 0, 1, 2, ... from the bottom. Reprinted with permission from Ref. [45]. Copyright 2018 by American Physical Society

second plateau already appears at $A_{\text{peak}} \sim 0.5$, much smaller than $\frac{\pi}{a}$. Thus, EHI qualitatively alters HHG spectra.

In order to understand the microscopic mechanism underlying the latter feature, let us expand the orbital functions $\psi_{bk_0}(x,t)$ with Houston states $e^{-iA(t)x}\phi_{nk(t)}(x)$ [48], the instantaneous eigenstates of $\hat{h}_f(t)$ with eigenvalues $\varepsilon_{nk(t)}$, as

$$\psi_{nk_0}(x,t) = \sum_m \alpha^m_{nk_0}(t) e^{-i\int_0^t \varepsilon_{mk(t')}dt'} e^{-iA(t)x} \phi_{mk(t)}(x), \tag{4.34}$$

where $k(t) = k_0 + A(t)$ is the instantaneous crystal momentum incorporating intraband dynamics. Since the system under consideration has a single VB, we drop the initial band index $n$ hereafter. Substituting Eq. (4.34) into Eq. (4.31), we obtain equations of motion for complex amplitudes $\alpha^m_{k_0}(t)$ expressing interband dynamics,

$$i\frac{d}{dt}\alpha^m_{k_0}(t) = \sum_n \alpha^n_{k_0}(t) e^{i\int_0^t \Delta\varepsilon_{mn}[k(t')]dt'} \left( E(t) d^{mn}_{k(t)} + \left\langle \tilde{\phi}_{mk_0} \middle| \hat{w}[\delta\rho(t)] \middle| \tilde{\phi}_{nk_0} \right\rangle \right), \tag{4.35}$$

where $d^{mn}_k = i\langle u_{km}|\nabla_k u_{kn}\rangle$ with $u_{km}(x)$ being the lattice periodic part of the initial Bloch state, i.e., $\phi_{km}(x) = e^{ikx}u_{km}(x)$, and $\tilde{\phi}_{nk_0}(x,t) = e^{-iA(t)x}\phi_{nk(t)}(x)$. The first term comes from the *frozen* TDHF Hamiltonian, and thus describes the independent electron dynamics. The second term, on the other hand, stems from EHI $\hat{w}[\delta\rho(t)]$. After some approximation and algebraic manipulations [45], we get

$$i\frac{d}{dt}\alpha^m_{k_0}(t) = \sum_n \alpha^n_{k_0}(t) e^{i\int_0^t \Delta\varepsilon_{mn}[k(t')]dt'} \left[ E(t) d^{mn}_{k(t)} - \sum_{q\in\text{BZ}} \bar{v}(-q) D^{mn}_{k(t)+q}(t) \right], \tag{4.36}$$

where $\bar{v}(q)$ denotes the spatial Fourier transform of the interelectronic soft-Coulomb potential, and $D^{mn}_{k(t)}$ the time-dependent interband polarization between $m$ and $n$ at $k(t)$:

$$D^{mn}_{k(t)}(t) = \alpha^m_{k_0}(t)\alpha^{n*}_{k_0}(t) e^{-i\int_0^t \Delta\varepsilon_{mn}[k(t')]dt'}. \tag{4.37}$$

Since the population of CBs turns out to be small ($\lesssim 10^{-3}$) [45], we introduce approximations $\alpha^0_{k_0}(t) \approx 1$ and $\alpha^{m\geq 1}_{k_0}(t) \approx 0$ [58]. Then Eq. (4.36) for the first CB ($m = 1$) becomes

$$i\frac{d}{dt}\alpha^1_{k_0}(t) \approx e^{i\int_0^t \varepsilon_{10}[k(t')]dt'} \left[ E(t) d^{10}_{k(t)} - \sum_q \bar{v}(-q) D^{10}_{k(t)+q}(t) \right], \tag{4.38}$$

for the excitation dynamics of a VB electron with an initial crystal momentum $k_0$. The second term due to EHI indicates that interband or electron-hole polarization at a remote crystal momentum $k(t) + q$,

$$D^{10}_{k(t)+q}(t) = \alpha^{1}_{k_0+q}(t) e^{-i\int_0^t \varepsilon_{10}[k(t')+q]dt'}, \tag{4.39}$$

can induce quasi-resonant excitation when $\varepsilon_{10}[k(t)] \approx \varepsilon_{10}[k(t)+q]$. Therefore, even if a VB electron starting from $k_0$ does not reach MBG through intraband displacement, it can be excited to the first CB once another electron initially at $k_0 + q$ reaches MBG and tunnels to the CB [Fig. 4.10b]. It should be noticed that neither the first nor second terms directly change the crystal momentum, thus, the instantaneous crystal momentum is always given by $k(t) = k_0 + A(t)$, in whichever band the electron actually is.

This *hauling-up* effect provides a shortcut for VB electrons to climb up to the second CB, which leads to the formation of the second plateau even if $A_{\text{peak}} < \frac{\pi}{a}$. The electrons initially at $k_0 \in [-\max(A(t)), -\min(A(t))]$ pass by $k = 0$, i.e., MBG between the first and second CB. Thus, if these VB electrons are excited to the first CB via the hauling-up effect, then they can climb up to the second CB by tunneling at $k = 0$, eventually forming the second plateau via recombination with the VB hole. Note that they cannot reach MBG at $k = \pm\frac{\pi}{a}$ between the second and third CB. Therefore, the cutoff energy is expected to be given by $\varepsilon_{20}(A_{\text{peak}})$. This prediction is in good agreement with the cutoff energy obtained from the TDHF simulation at $0.5 \lesssim A_{\text{peak}} \leq \frac{\pi}{a} = 0.87$ [the upper white line in Fig. 4.9b].

## 4.4 Time-Dependent Density-Matrix Method Combined with First-Principles Calculation for Three-Dimensional Crystals

In material science, density-functional theory (DFT) is one of the *de facto* standards for materials at electronic ground state, owing to a good balance between accuracy and computational cost. Time-dependent density-functional theory (TDDFT) is one of the most feasible theories to describe electron excitation under an intense laser field from first-principles [59, 60]. While TDDFT shows accurate results, its calculation cost is relatively expensive, e.g., a few hundred core-hour or longer for well-converged results of a laser parameter. This calculation cost is still high for modern supercomputers when we need to investigate the optical response over a wide parameter region for the laser pulse, such as photon energies, field strengths, polarization properties, and so on. We develop an alternative theoretical framework based on a first-principles theory with cheaper calculation cost, called time-dependent density-matrix (TD-DM) method.

A one-body density-matrix (DM) is the key degree of freedom in TD-DM. The equation of motion is von Neumann equation:

$$i\frac{d\rho}{dt} = [h(t), \rho] + \left(i\frac{d\rho}{dt}\right)_{\text{coll}} \tag{4.40}$$

where $h$ and $(id\rho/dt)_{\text{coll}}$ are one-body Hamiltonian and collision term. This is a standard approach in the nonlinear optics [61]. Our TD-DM can be a first-principles theoretical framework by choosing a representation that orbitals derived from self-consistent DFT. The matrix elements of DM, the Hamiltonian, and the collision term are expressed by the orbitals $\phi_{b\mathbf{k}}$, where $b$ and $\mathbf{k}$ are indices for a band and Brillouin zone, respectively. When the collision term is neglected, this theoretical framework is equivalent to independent electron dynamics for DFT one-body Hamiltonian. This framework is regarded as a generalization of the one-dimensional theories in the precedent sections to a spatially three-dimensional system with the one-body potential $V(\mathbf{r})$ obtained by the DFT calculation. Our TD-DM includes the many-body effect of the electrons reflected in the constitution of the electronic structure through DFT calculation. In other words, the dynamical correction of the many-body effect is not included in our TD-DM, except for effect provided by the collision term. The computational cost of this framework is substantially reduced by using a basis set, compared to grid-basis TDDFT [62, 63]. An additional advantage of this framework is flexible modifications on the top of DFT, *e.g.* bandgap correction and phenomenological relaxations via the scattering term. This theoretical framework is close to the de fact standard theory, semiconductor Bloch equation (SBE) [64]. The difference lies in the length gauge and the electron-hole attraction term.

The key observable of our TD-DM is induced current density evaluated by the expectation of the velocity operator $\mathbf{v}(t) = \mathbf{p} + \mathbf{A}(t)$ divided by the volume of cell $V_{\text{cell}} : \mathbf{J}(t) = -\text{tr}\,(\rho\mathbf{v})\,/\,V_{\text{cell}}$. To obtain emitted photon intensity, taking the absolute value of Fourier transform of the acceleration density, the temporal derivative of the current density.

We show our TD-DM performance by an application to a high-harmonic generation from GaSe crystal [27]. In the experiment, the GaSe sample is exposed to linearly polarized light whose polarization is on the basal plane, in which the wavelength and the pulse duration are 4.96 $\mu$m and 200 fs. Angle dependences of the driving field with respect to crystal orientation are measured for two orthogonal polarization directions of the emitted photon, parallel, and perpendicular components to the driving field. The results are shown in Fig. 4.11a–d, together with the theoretical counterparts. Our TD-DM reproduces almost all features in the experimental spectra, almost isotropic angle dependence for the parallel component of odd-order harmonics, a 60-degree period of even-order harmonics, and a 30-degree period for the perpendicular components of odd-order harmonics.

This direct comparison between the theoretical simulation and the experiment is hardly achieved if we employ TDDFT because of the tough computational cost. To achieve an expected feature for the polarization direction dependence, very dense Brillouin zone sampling, $64 \times 64 \times 12$, was mandatory. Besides, we must perform multiple calculations for different angles of the field polarization. Computations to draw Fig. 4.11a, b require 80 thousands core-hour. Typically TDDFT requires tens to a hundred times more. The core-hour estimation for TDDFT is possible with modern supercomputers in principle but unrealistic for daily use of the supercomputers. Our TD-DM is a lightweight simulation option to perform a more comprehensive investigation of strong-field phenomena in solids, keeping the nonempirical nature.

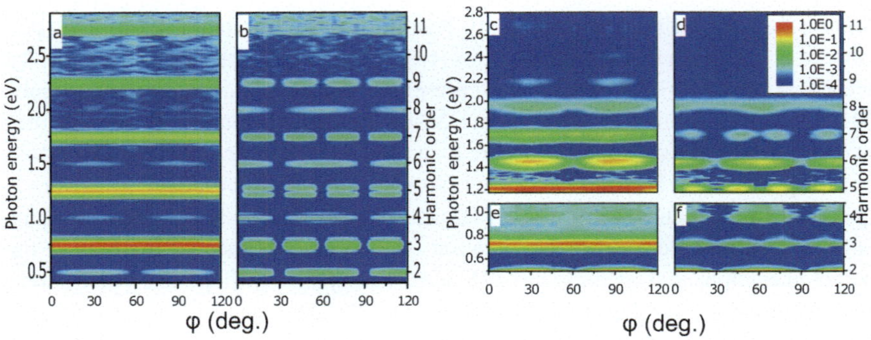


**Fig. 4.11** Harmonic spectra from GaSe crystal. Parallel **a** and perpendicular **b** components from TD-DM simulation. **c** and **d** are the same as **a** and **b** but experimental results. Reprinted with permission from Ref. [27]. Copyright 2018 by American Physical Society

The polarization-resolved analysis showed that crystal symmetry is reflected in the HHG spectra even for the non-perturbative regime beyond the susceptibility-based argument for the second- and third-order harmonics. Part of the fingerprint of HHG for the symmetry can be understood by the intraband current model with a time-independent carrier population [15, 24] capturing band-structure anisotropy of crystals. While this intraband current model gives us a clear-cutting and simple description of the symmetric aspects, a qualitative judgment of the intraband current is severe because of many assumptions to proceed with the model calculations. We made a scheme to decompose the current density, in our TD-DM, into intraband and interband contributions, like the intraband current and the interband polarization in the SBE.

We define the intraband component of the current density as a partial sum of the trace over only the diagonal contribution of DM represented by the instantaneous eigenfunction of the time-dependent Hamiltonian $h(t)\varphi_{\beta k}^{(t)} = \varepsilon_{\beta k}^{(t)}\varphi_{\beta k}^{(t)}$ as

$$\mathbf{J}(t) = \mathbf{J}_{\text{intra}}(t) + \mathbf{J}_{\text{inter}}(t), \tag{4.41}$$

$$\mathbf{J}_{\text{intra}}(t) = -\frac{1}{V_{\text{cell}}} \sum_{\beta\mathbf{k}} \left\langle \varphi_{\beta k}^{(t)} \left| \rho \right| \varphi_{\beta k}^{(t)} \right\rangle \left\langle \varphi_{\beta k}^{(t)} \left| \mathbf{v} \right| \varphi_{\beta k}^{(t)} \right\rangle, \tag{4.42}$$

$$\mathbf{J}_{\text{inter}}(t) = -\frac{1}{V_{\text{cell}}} \sum_{\beta\gamma(\beta\neq\gamma)\mathbf{k}} \left\langle \varphi_{\beta k}^{(t)} \left| \rho \right| \varphi_{\gamma k}^{(t)} \right\rangle \left\langle \varphi_{\gamma k}^{(t)} \left| \mathbf{v} \right| \varphi_{\beta k}^{(t)} \right\rangle \tag{4.43}$$

where the interband contribution is obtained as the rest of the total current subtracted by the intraband contribution or the partial sum of the trace over the off-diagonal component of the DM. The superscript parenthesis $t$ of the variables, $\bullet^{(t)}$, represent that the object parametrically depends on the time. The intraband contribution defined here is the sum over product between the population evaluated with the instantaneous basis and the group velocity of the band structure because the velocity expectation

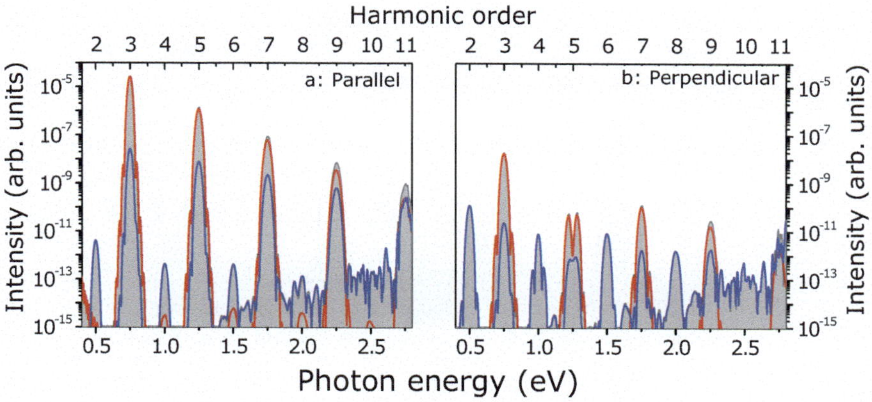


**Fig. 4.12** Decompositions of harmonic spectra into intraband and interband contributions, for parallel (h) and perpendicular (i) components. Total (gray shadow), intraband (red solid), and interband (blue solid) components are shown. Reprinted with permission from Ref. [27]. Copyright 2018 by American Physical Society

evaluated with the instantaneous Hamiltonian is equivalent to the group velocity with the vector potential $\left\langle \varphi_{\beta k}^{(t)} \left| \mathbf{v}(t) \right| \varphi_{\beta k}^{(t)} \right\rangle = \left( \partial \varepsilon_{\beta \mathbf{k}} / \partial \mathbf{k} \right)_{\mathbf{k}+\mathbf{A}(t)}$. This formula is nicely related to the intraband current in SBE. This definition of the intraband current is a generalization of the simplest intraband current model, such that the time-dependent population obtained from the microscopic theory rather than just constant. The harmonic spectra of the two contributions are shown in Fig. 4.12. In the lower odd-order harmonics, the intraband contribution dominates the total yield of the harmonics. The interband contributions increase with increasing the photon energy toward 2 eV, which is the bandgap of the GaSe. The two contributions become comparable for the ninth and eleventh harmonics.

One of the most striking results is that all even-order harmonics are exclusively from the interband contributions. This fact invokes that band structure with spin-independent Hamiltonian is always spatially symmetric. The intraband current is expected not to produce even-order harmonics because of the symmetric band structure. A significance of this investigation is that the expected results are demonstrated by a microscopic quantum mechanical simulation based on a first-principles theory. The exclusion of the even-harmonic for the intraband current supports our definition of the intraband and interband contributions to the total current density.

The intraband current can be further decomposed into band-resolved contributions by taking the partial sum of Eq. (4.42) over a part of bands. This analysis allows us to investigate which bands, associated with atomic orbital nature, mainly produce the harmonics for a specific situation. We apply this analysis to HHG from the $CsPbCl_3$ perovskite [28].

We show the band-resolved intraband contribution to HHG in Fig. 4.13. We employ a pulsed electric field that has 0.62 eV photon energy, 1.0 V/nm field strength,

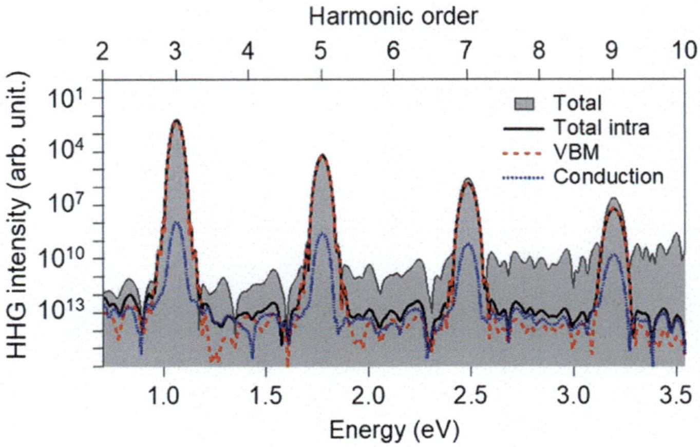


**Fig. 4.13** Decompositions of harmonic spectra into band-resolved intraband and interband contributions. Total (gray shadow), total intraband (black solid), VBM intraband (red dashed), and conduction intraband (blue solid) components are shown. Reprinted from Ref. [28] with permission under the Creative Commons Attribution (CC BY) license

160 fs full width at half maximum pulse duration. Reflecting on the inversion symmetry of the crystal, only odd-order harmonics appear in the spectrum. The total intraband current dominates the power spectrum of the total current density. For a perovskite containing halide and lead ions, the topmost valence bands have the characteristic function for the optical absorption and are frequently argued as valence band maximum (VBM). VBM of the halide-lead perovskite is composed of p-nature halide and s-nature leads orbitals. According to the band-resolved intraband current analysis, we define VBM and conduction intraband currents as partial sums of Eq. (4.42) over the VBM and conduction bands:

$$\mathbf{J}_{\mathrm{VBM}}(t) = -\frac{1}{V_{\mathrm{cell}}} \sum_{\beta(\in \mathrm{VBM})\mathbf{k}} \left\langle \varphi_{\beta k}^{(t)} \left| \rho \right| \varphi_{\beta k}^{(t)} \right\rangle \left\langle \varphi_{\beta k}^{(t)} \left| \mathbf{v} \right| \varphi_{\beta k}^{(t)} \right\rangle, \tag{4.44}$$

$$\mathbf{J}_{\mathrm{conduction}}(t) = -\frac{1}{V_{\mathrm{cell}}} \sum_{\beta(\in \mathrm{conduction})\mathbf{k}} \left\langle \varphi_{\beta k}^{(t)} \left| \rho \right| \varphi_{\beta k}^{(t)} \right\rangle \left\langle \varphi_{\beta k}^{(t)} \left| \mathbf{v} \right| \varphi_{\beta k}^{(t)} \right\rangle. \tag{4.45}$$

The contributions of the two intraband currents are shown as a red dashed line in Fig. 4.13. The intensity of the total intraband current is well dominated by the VBM components. The conduction band does not affect almost anything for the spectrum. The conduction band is frequently regarded as a source of the intraband band harmonic generation within the simplest intraband current model, because of more dispersive band curves than valence ones. This explicit decomposition raises a counter-intuitive point that VBM intraband current due to the hole motion mainly produces the HHG, at least for $CsPbCl_3$ perovskite.

## 4.5 Dynamical Franz-Keldysh Effect

In recent years, it has become possible to generate attosecond pulsed light using high-order harmonics, which are nonlinear interactions between the gas phase and ultrashort pulse lasers. Currently, the pulse width is reduced to tens of attoseconds. Since the 2010s, changes in optical properties of solids shorter than the electric field period of light have been reported.

Phenomena that occur inside a solid in a laser field can be broadly classified according to the presence or absence of electronic excitation. Since the recombination of electron-hole pairs is generally on a long time scale of the order of pico-second, electronic excitation causes a change in physical properties that is not recovered in the attosecond time scale. Many phenomena including the electron excitation such as the saturable absorption due to occupation of the conduction band, and generation of higher harmonics in solids have been attracting interests. On the other hand, it is considered that the phenomenon that recovers quickly after passing through the laser does not contribute to electronic excitation.

A strong laser can be treat as an oscillating electric field. The change in the dielectric function without electronic excitation by an electrostatic field is known as the Franz-Keldysh effect (FKE). FKE is photon absorption via the tunnel effect and can be considered as a long wavelength or strong laser field limit. The dynamical Franz-Keldysh effect (DFKE) is an extension of FKE to periodic oscillating electric fields [68]. DFKE has been investigated theoretically and experimentally for the time-averaged modulation which corresponds to the blueshift of the band gap by the ponderomotive energy [68–70]. Time-resolved observation of the DFKE is reported by Novelli et al., employing the THz light [71]. We have recently proposed the time-resolved analysis for dynamical Franz-Keldysh effect (Tr-DFKE). Analytical theory and first-principles simulations have revealed that Tr-DFKE is a modulation faster than the oscillation of the laser electric field. Recently, the sub-cycle modulation has been confirmed experimentally. In this section, we would like to introduce our recent works on the construction of DFKE analytical theoretical formulas and first-principles calculations.

### 4.5.1 *Time-Resolved Spectroscopy*

We would like to clarify time-resolved spectroscopy before moving on to specific processe. The optical property of a material is described by susceptibility $\chi$, which connects the polarization $P(t)$ to a given electric field E $(t)$ as $P_i(t) = \sum_j \int_{-\infty}^{\infty} dt' \chi_{ij}(t-t')E_j(t')$. Here indexes $i$ and $j$ indicate the components $(x, y, z)$ of the vector. It should be noted that the $\chi$ is the function of the relative time, $t - t'$. The time-dependence of $\chi$ indicates the time-invariance of the optical properties. If the system depends on the time, we should reconsider the $\chi$ as the function of two independent time,

$$\chi(t-t') \rightarrow \chi'(t,t'). \tag{4.46}$$

In the same way, dielectric function $\varepsilon$ and optical conductivity $\sigma$ become the function of $t$ and $t'$, $\varepsilon_{ij}(t,t') = \delta_{ij}\delta(t-t') + 4\pi\chi'_{ij}(t,t')$, $\chi'_{ij}(t,t') = \int_{-\infty}^{\infty} dt"\Theta(t-t")\sigma_{ij}(t",t')$, where $\Theta(t-t")$ is the Heaviside function.

The optical properties is observed as the modulation of the probe pulse whose peak intensity is at the time $T_p$. The detected susceptivity is the function of the frequency $\omega$ and the $T_p$, $\chi(T_p,\omega)$. If we assume the probe pulse as $f_p\delta(t-T_p)$, the polarization becomes $P(t) = f_p\chi'(t,T_p)$. In this step, we assume that the $\chi'(t,T_p)$ has only diagonal part. The susceptivity in the frequency-domain $\chi(T_p,\omega)$ can be defined from the $\chi'(t,T_p)$, $\chi(T_p,\omega) = \int dte^{i\omega t}P(t)/\int dte^{i\omega t}f_p\delta(t-T_p)$.

### 4.5.2 Analytical Theory by Houston Function

In this section, we would like to show the derivation of the analytical formula employing the model Hamiltonian for the spatially periodic system,

$$i\hbar\frac{\partial u_{n,\mathbf{k}}(\mathbf{r},t)}{\partial t} = \left(\frac{(\mathbf{p}+\hbar\mathbf{k}+\frac{e}{c}\mathbf{A}(t))^2}{2m} + V(\mathbf{r})\right)u_{n,\mathbf{k}}(\mathbf{r},t). \tag{4.47}$$

The time-dependent wave function $u_{n,\mathbf{k}}(\mathbf{r},t)$ can be expressed by the Houston function[72],

$$w_{n,\mathbf{k}}(\mathbf{r},t) = u^G_{n,\mathbf{k}+\frac{e}{c}\mathbf{A}(t)}(\mathbf{r})\exp\left[-\frac{i}{\hbar}\int^t dt'\epsilon^G_{n,\mathbf{k}+\frac{e}{c}\mathbf{A}(t')}\right], \tag{4.48}$$

as $u_{n,\mathbf{k}}(\mathbf{r},t) = \sum_i C_i(t)w_{i,\mathbf{k}}(\mathbf{r},t)$. Here, $\epsilon$ is the eigenenergy of the electron, $\mathbf{k}$ is the Bloch wavevector, $n$ in the band index, and $G$ indicates the ground state.

To simplify the system, we assume the parabolic two-band system defined as

$$\epsilon_{c,\mathbf{k}} - \epsilon_{v,\mathbf{k}} = \epsilon_g + \frac{\hbar^2|\mathbf{k}|^2}{2\mu}, \tag{4.49}$$

where $\epsilon_g$ is the band gap, $\mu$ is the reduced mass, and $v(c)$ presents valence (conduction) band. The Houston function can be expanded by the $e^{-il\Omega t}$

$$w_{v(c),\mathbf{k}}(\mathbf{r},t) = \sum_l W^l_{v(c),\mathbf{k}}(\mathbf{r},t)e^{-il\Omega t}, \tag{4.50}$$

with continuous wave$\mathbf{A}(t) = \mathbf{A_0}\cos\Omega t$ [73, 74]. Here, $W^l_{v(c),\mathbf{k}}(\mathbf{r},t)$ is the $l-$th order coeffient. Since Eq. (4.50) corresponds to the expansion in to the dressed states

(Floquet states) [75], the transient absorption can be understood as the response of the dressed states at time $T_p$.

The electronic current ($\mathbf{J}(t)$) is important to consider the optical response. From the Fourier transformation of the current, we can estimate the conductivity as $\sigma(\omega) = \frac{\tilde{J}(\omega)}{\tilde{E}(\omega)}$, where $\tilde{J}(\omega)$ is the Fourier component of current, and $\tilde{E}(\omega)$ is the applied field. From Eq. (4.47), the current is expressed as

$$\mathbf{J}(t) = -\frac{n_e e^2}{mc}\mathbf{A}(t) - \frac{e}{2m}\sum_{n,\mathbf{k}}\left[\frac{1}{\Omega_{cell}}\int_{\Omega_{cell}} d\mathbf{r} u^*_{n,\mathbf{k}}(\mathbf{r},t)\mathbf{p}u_{n,\mathbf{k}}(\mathbf{r},t) + C.C.\right] \tag{4.51}$$

where $n_e$ is the electron density, $\Omega_{cell}$ is the volume of the unit cell.

We can derive the transient dielectric function with the usual linear response treatment under the elliptically polarized light, $\mathbf{A}(t) = A_0(\eta \sin\Omega t, 0, \cos\Omega t)$ $(0 \le \eta \le 1)$, as

$$\begin{aligned}\varepsilon_E(T_p,\omega) = 1 &- \frac{4\pi e^2}{m\omega^2}n_e - \frac{2e^2|p_{cv}|^2\mu^{3/2}}{\sqrt{2}m^2\pi} \\ &\times \int_0^\infty \sqrt{\epsilon_k}d\epsilon_k \int_{-1}^{1} d\cos\theta \int_0^{2\pi} d\phi \sum_{l_1,l_2,\zeta_1,\zeta_2} \\ &\times \frac{(-1)^{\zeta_1-\zeta_2}e^{i2\zeta_2\Omega T}}{\omega+2\zeta_2\Omega}\tilde{J}_{l_1}(\alpha,\beta)\tilde{J}_{l_1-2\zeta_1}(\alpha,\beta) \\ &\times \left[\frac{J_{l_2}(\gamma)J_{l_2+2(\zeta_1-\zeta_2)}(\gamma)}{\omega-(\epsilon_g+\epsilon_k+U_E+(l_1+l_2-2\zeta_2)\Omega)}\right. \\ &\left.- \frac{J_{l_2}(\gamma)J_{l_2+2(\zeta_1+\zeta_2)}(\gamma)}{\omega+(\epsilon_g+\epsilon_k+U_E+(l_1+l_2+2\zeta_2)\Omega)}\right]\end{aligned} \tag{4.52}$$

[74]. Here, $\alpha, \beta, \gamma$ are defined as

$$\alpha = \frac{ekA_0}{c\mu\Omega}\cos\theta, \tag{4.53}$$

$$\beta = \frac{e^2A_0^2}{8c^2\mu\Omega}(1-\eta^2), \tag{4.54}$$

$$\gamma = \eta\frac{ekA_0}{c\mu\Omega}\sin\theta\cos\phi, \tag{4.55}$$

respectively, $U_E$ is the ponderomotive energy, $J_l$ is the $l$-th order Bessel function, $\tilde{J}_l(a,b)$ is the generalized Bessel function [76],

$$\tilde{J}_l(a,b) = \sum_m J_{l-2m}(a)J_m(b), \tag{4.56}$$

and $p_{cv}$ is the transition moment between valence and conduction band. The $\theta$ ($\phi$) is the angle between $\mathbf{k}$ and $z-$ ($x-$) axis, $\mathbf{k} = k(\sin\theta\cos\phi, \sin\theta\sin\phi, \cos\theta)$.

Transient spectroscopy using ultrashort pulses observes not only the absorption by the dressed states , but also the phase difference between them. Therefore, the oscillation of $\varepsilon_E(T_p, \omega)$ in probe time $T_p$ is derived from the energy difference between the dressed states. The oscillation period of $\varepsilon_E(T_p, \omega)$ is an even multiple of the frequency of the pump light due to the symmetry of the system.

### 4.5.3 Time-Dependent Density Functional Theory

A theory describing the ground state of a multielectron system is density functional theory (DFT). Electronic states are obtained by solving the Kohn-Sham equation, which is the basic equation of DFT[77]. On the other hand, the electron dynamics under the laser fields can be described by the time-dependent Kohn-Sham (TDKS) equation

$$i\hbar\frac{\partial}{\partial t}u_{n,\mathbf{k}}(t) = \hat{H}(t)u_{n,\mathbf{k}}(t) \tag{4.57}$$

$$\hat{H}(t) = \frac{1}{2m}\left(\mathbf{p} + \hbar\mathbf{k} + \frac{e}{c}\mathbf{A}(t)\right)^2 + V_{ion} + V_H + V_{xc}, \tag{4.58}$$

based on the time-dependent density functional theory (TDDFT) [78]. Here, $\mathbf{A}(t)$ is the vector potential, $V_{ion}$ is the coulomb potential from ions, $V_H$ is the Hartree potential, and $V_{xc}$ is the exchange-correlation potential. TDDFT is a good approach to describe the nonlinear attosecond electron dynamics [79–83].

We approximate the time-evolution of the wave function by the fourth-order Taylor expansion time-evolution operator [84, 85]. The wave functions and the Hamiltonian are discretized with three-dimensional grid [86]. The core electrons are neglected by employing the norm-conserve pseudopotential [87, 88].

### 4.5.4 Tr-DFKE in Diamond

We present the numerical results for a diamond by the TDDFT. We assume a cubic unit cell including eight carbon atoms. We descretize the unit cell with $24^3$ grid points, and the K-space with $22^3$ grid points. We assume the adiabatic approximation, and employed a local density approximation for the exchange-correlation potential [89]. The pump laser pulse is defined as the quasi-continuous wave,

$$\mathbf{E}(t) = \hat{e}E_0F_P(t)\sin\Omega t, \tag{4.59}$$

with frequency of $\Omega = 0.4$ eV. The $F_P(t)$ is the envelope function to apply the laser field adiabatically. The probe laser is the gaussian type pulse,

$$\mathbf{E}_p(t) = \hat{e}_p f_p \sin(\omega_p t) \exp\left(-\frac{(t - T_p)^2}{\zeta^2}\right), \tag{4.60}$$

where $T_p$ is the probe time. We assume the pulse duration $\zeta = 0.7$ fs, peak field intensity $f_p = 2.7 \times 10^{-3}$ MV/cm, and frequency of $\omega_p = 5.5$ eV. The intensity of probe field is weak so as to linear response is dominant. The frequency $\omega_p$ corresponds to the optical band gap in out calculation.

We apply the electric fields of $E(t)$ (dashed line) and $E_p(t)$ (solid line) as shown in Fig. 4.14a. The field intensity of the pump light is 20 MV / cm, and the time of the probe $T_p$ is 13 fs. The polarization of the pump (probe) field is parallel to [1,0,0] ([0.0.1]). A solid line in Fig. 4.14b indicates an electronic current density $J_p(t)$ induced by the $E_p(t)$. We use atomic unit (a.u.) in all calculation. A dashed line in Fig. 4.14c indicates the imaginary part of the dielectric function Im[$\varepsilon$], which is calculated from the Fourier component. the $J_p(t)$ and $E_p(t)$. For comparison, Im[$\varepsilon$] without the pump is indicated by a dotted line, and the difference between them, $\Delta$Im[$\epsilon(\mathrm{T_p}, \omega)$], is indicated by a solid line. We can see the absorption below, and transparency above the band gap.

Figure 4.15 shows the probe time dependence of $\Delta$Im[$\varepsilon(\mathrm{T_p}, \omega)$] and its dependence on peak intensity of the pump field as a function of probe time $T_p$ (fs) and frequency $\omega$. In the case of $E_0 = 5$ MV/cm (Fig. 4.15b), the $\Delta$Im[$\varepsilon(\mathrm{T_p}, \omega)$] show the maximum at the time when the pump light field is zero around the optical band gap. The peak of the $\Delta$Im[$\varepsilon(\mathrm{T_p}, \omega)$] at each $\omega$ shifts backward as the $\omega$ decreases from the band gap. The peak of the $\Delta$Im[$\varepsilon(\mathrm{T_p}, \omega)$] shifts backward as the pump field intensity increases. In the case of $E_0 = 50$ MV/cm (Fig. 4.15e), whereas the peak coincides with the peak of the pump field intensity, the $\omega$ dependent shift becomes weak. The coincidence between the pump field and the $\Delta$Im[$\varepsilon(\mathrm{T_p}, \omega)$] in Fig. 4.15e indicates the adiabatic response of the electrons.

Let us now compare the TDDFT and analytical model calculations. Figure 4.16 shows the results by the model calculation [73, 74]. We present the time depencen of $\Delta$Im[$\varepsilon(T_p, \omega)$] in left panels, and time-averaged modulation in right panels. Whereas the peak of $\Delta$Im[$\varepsilon(T_p, \omega)$] coincident with the minimum of the pump field intensity, it shifts to the maximum of the pump field intensity as the field intensity increases. The $\omega$ dependence is also qualitatively agree with Fig. 4.15. The agreement of model calculation with the TDDFT results indicates that the Tr-DFKE can be understood by the response of the dressed states. We also present in Fig. 4.16e–h the modulation by the static electric field assuming the FKE. The interesting point is that the DFKE and FKE show similar behavior as the pump laser field increases.

We would like to see the ellipticity ($\eta$) dependence in the next step. Figure 4.17 shows the $\eta$ dependence in TDDFT (left panels) and model calculation (right panels). Fig. 4.17b and e is the circularly polarization, (c) and (f) is elliptic polarization with $\eta = \sqrt{2}$, and (d) and (g) is linear polarization case. We present only the positive modulation of Im[$\varepsilon(T_p, \omega)$] in Fig. 4.17. As the reference, the pump light field with

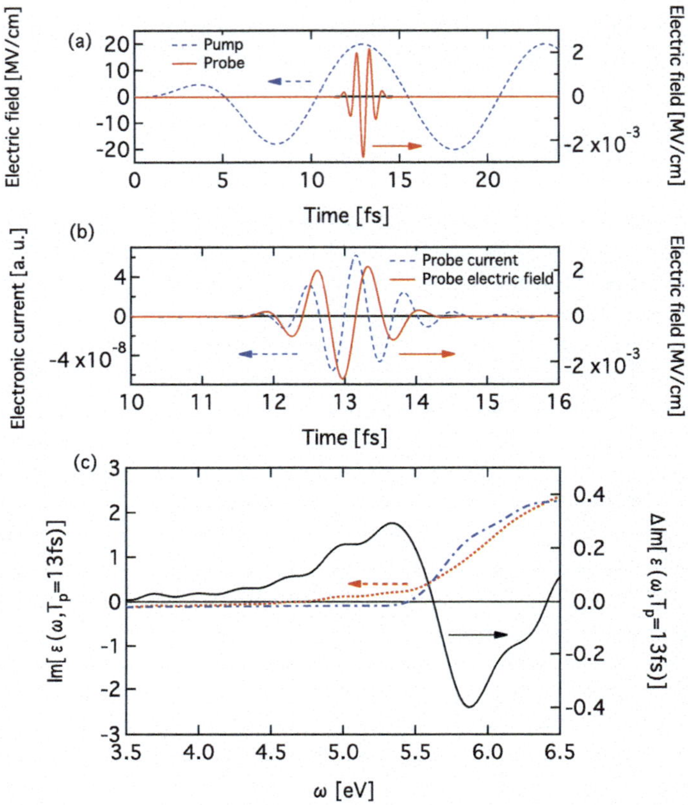


**Fig. 4.14** **a** Pump (dashed line) and probe (solid line) electric fields are shown. **b** The electronic current (solid line) induced by the probe electric field (dashed line). **c** The imaginary part of the dielectric function in the presence of the pump field, $\mathrm{Im}[\varepsilon(T_p = 13fs, \omega)]$ (dashed line), and in the absence of the pump field, $\mathrm{Im}[\varepsilon(\omega)]$ (dot line). Solid line shows the difference, $\mathrm{Im}[\varepsilon(T_p = 13fs, \omega)] - \mathrm{Im}[\varepsilon(\omega)]$. Reprinted with permission from Ref. [73]. Copyright 2016 by American Physical Society

linear polarization is shown in Fig. 4.17a. The maximum field intensity is set to 10 MV/cm for all calculations. The time-dependence of $\Delta\mathrm{Im}[\varepsilon(T_p, \omega)]$ becomes weak as the ellipticity increases in both case. It should be noted that the time-dependence disappears in circularly polarization. In our model, we assume parabolic bands which is isotropic system. The diamond is also relatively isotropic system. Therefore, in the circularly polarization, the electron cannot distinguish the oscillation of the field in average.

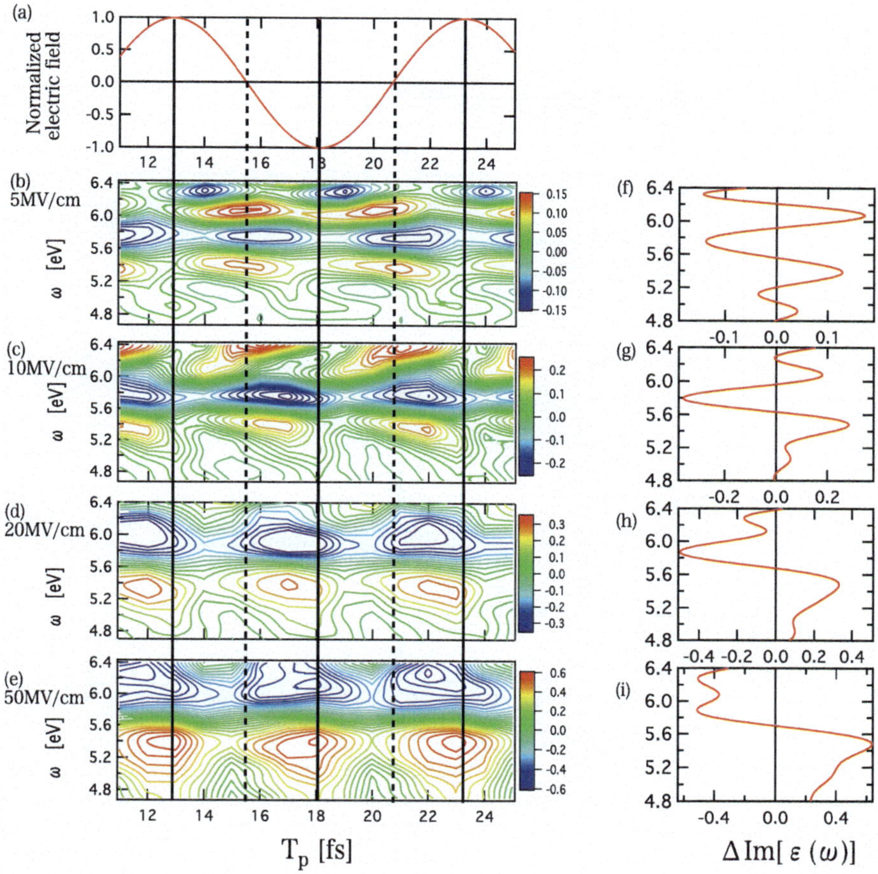


**Fig. 4.15** Contour plots of $\Delta\text{Im}[\varepsilon(T_p, \omega)]$ under the pump field of the intensity of **b** 5, **c** 10, **d** 20, and **e** 50 MV/cm. **f–i** present the time average of the $\Delta\text{Im}[\varepsilon(T_p, \omega)]$. The normalized pump field is shown in **a**. The vertical solid lines indicate the time when the pump field is maximum, and the dashed lines indicate the minimum. Reprinted with permission from Ref. [73]. Copyright 2016 by American Physical Society

## 4.6 Summary

We have compiled recent development in theoretical and numerical modeling of strong-field electron dynamics in solids. First, we have the concept of introduced intraband and interband transitions, forming the basis for discussion in the momentum space, through the rigorous derivation for graphene. Then, we have extended it to the multiband, momentum-space three-step model. The electron-hole interaction effects can also be incorporated in this model. Moreover, we have presented the TD-DM and TDDFT methods for actual three-dimensional materials, whose predictions can be quantitatively compared with experimental results.

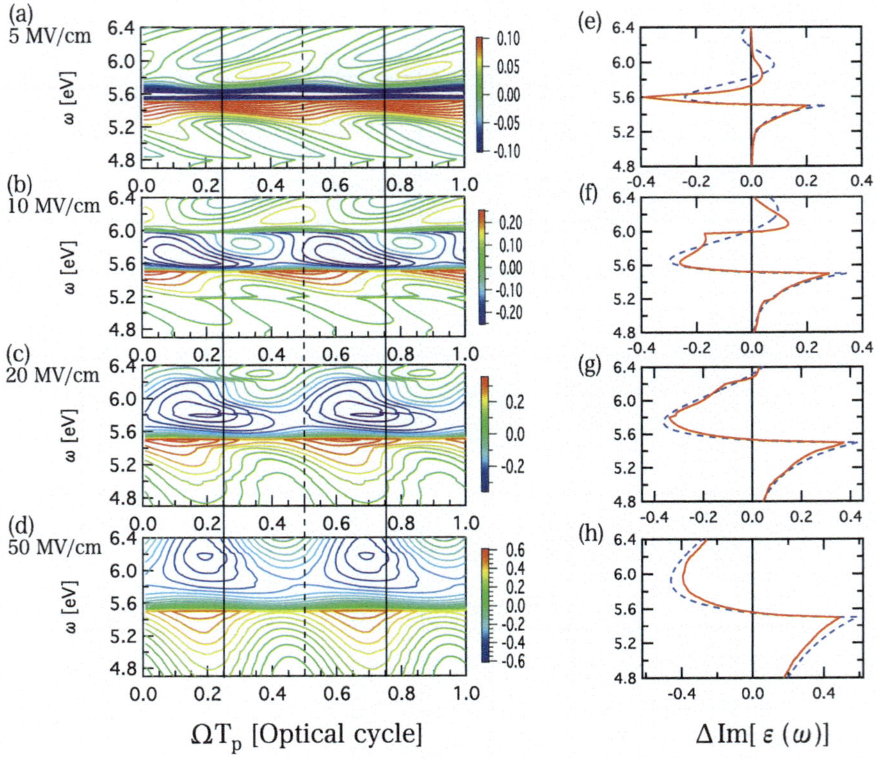


**Fig. 4.16** Contour plots of $\Delta\mathrm{Im}[\varepsilon(T_p, \omega)]$ in the two-band model for the pump field intensities of **a** $E_0 = 5$, **b** 10, **c** 20, and **d** 50 MV/cm. The horizon axis is the phase defined by $\Omega T_p$. **e**–**h** indicates the time average of the $\Delta\mathrm{Im}[\varepsilon(T_p, \omega)]$ (red solid lines) and the modulation assuming usual FKE. Reprinted with permission from Ref. [73]. Copyright 2016 by American Physical Society

However, much more theoretical and experimental investigations are yet to be done, in order to reach comprehensive understanding of the electron dynamics in various solid-state materials subject to intense laser pulses. It is expected that further accumulation of knowledge from different perspectives, such as suitable use and eventual unification of real-space and momentum-space pictures, effects of electron correlation, impurity, relaxation, and decoherence, and high-field phenomena in topological insulators and quantum materials, will lead to discovery of novel phenomena in various functional materials as well as innovative applications of high-intensity lasers.

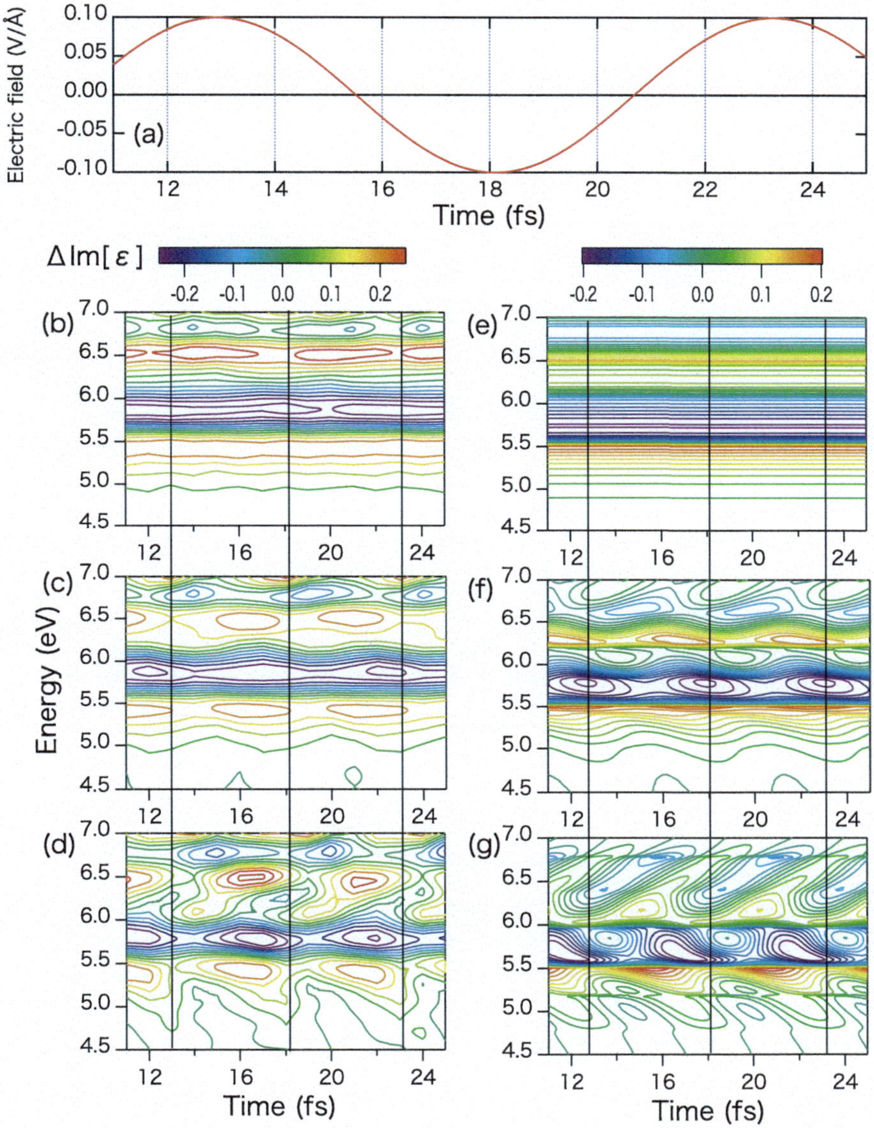


**Fig. 4.17** $\Delta\text{Im}[\varepsilon(T_p, \omega)]$ under circularly (**b** and **e**), elliptically (**c** and **f**), and linearly (**d** and **g**) polarized lasers. **b–d** presents the numerical results from TDDFT, and **e–g** is the results from Eqs. (4.52). Panel **a** plots the electric field of the pump laser against the probe time, the range of which corresponds to that for the real-time TDDFT simulation. The vertical solid lines indicate the maximum of the pump field intensity. Reprinted with permission from Ref. [74]. Copyright 2016 by American Physical Society

**Acknowledgements** This research was supported in part by a Grant-in-Aid for Scientific Research (Grants No. 23104708, No. 26390076, No. 17K05070, No. 18H03891, No. 18K14145, No. 19H02623, and No. 19H00869) from the Ministry of Education, Culture, Sports, Science and Technology (MEXT) of Japan and also by the Photon Frontier Network Program of MEXT. This research was also partially supported by the Center of Innovation Program (Grant No. JPMJCE1313) from the Japan Science and Technology Agency (JST), JST CREST (Grant No. JPMJCR15N1), MEXT Quantum Leap Flagship Program (Grant No. JPMXS0118067246), the Research and Education Consortium for Innovation of Advanced Integrated Science by JST, and the Exploratory Challenge on Post-K Computer from MEXT. The computation in this work was done using the facilities of the Supercomputer Center, the Institute for Solid State Physics, The University of Tokyo, and also using the K computer provided by the RIKEN Advanced Institute for Computational Science through the HPCI System Research project (Project ID: hp160260, hp170235, and hp180174).

## References

1. D. Strickland, G. Mourou, Opt. Commun. **56**, 219 (1985)
2. https://www.nobelprize.org/uploads/2018/10/advanced-physicsprize2018.pdf
3. M. Protopapas, C.H. Keitel, P.L. Knight, Rep. Prog. Phys. **60**, 389 (1997)
4. T. Brabec, F. Krausz, Rev. Mod. Phys. **72**, 545 (2000)
5. T. Popmintchev, M.C. Chen, D. Popmintchev, P. Arpin, S. Brown, S. Alisauskas, G. Andriukaitis, T. Balciunas, O.D. Mucke, A. Pugzlys, A. Baltuška, B. Shim, S.E. Schrauth, A. Gaeta, C. Hernandez-Garcia, L. Plaja, A. Becker, A. Jaroń-Becker, M.M. Murnane, H.C. Kapteyn, Science **336**, 1287 (2012)
6. Z. Chang, *Fundamentals of Attosecond Optics* (CRC Press, Boca Raton, 2011)
7. L. Plaja, R. Torres, A. Zaïr (eds.), *Attosecond Physics, Springer Series in Optical Sciences*, vol. 177 (Springer, Berlin, 2013)
8. P. Salières, A. Maquet, S. Haessler, J. Caillat, R. Taïeb, Rep. Prog. Phys. **75**, 062401 (2012)
9. P. Agostini, L.F. DiMauro, Rep. Prog. Phys. **67**, 813 (2004)
10. F. Krausz, M. Ivanov, Rev. Mod. Phys. **81**, 163 (2009)
11. L. Gallmann, C. Cirelli, U. Keller, Ann. Rev. Phys. Chem. **63**, 447 (2013)
12. P.B. Corkum, Phys. Rev. Lett. **71**, 1994 (1993)
13. K.C. Kulander, K.J. Schafer, J.L. Krause, in *Super-Intense Laser-Atom Physics*, NATO ASI, Series B, vol. 316, ed. by B. Piraux, A. L'Huillier, K. Rzążewski (Plenum Press, New York, 1993), p. 95
14. S. Ghimire, D.A. Reis, Nat. Phys. **15**, 10 (2019)
15. S. Ghimire, A.D. Di Chiara, E. Sistrunk, P. Agostini, L.F. Di Mauro, D.A. Reis, Nat. Phys. **7**, 138 (2011)
16. O. Schubert, M. Hohenleutner, F. Langer, B. Urbanek, C. Lange, U. Huttner, D. Golde, T. Meier, M. Kira, S.W. Koch, R. Huber, Nat. Photon. **8**, 119 (2014)
17. T.T. Luu, M. Garg, S.Y. Kruchinin, A. Moulet, M.T. Hassan, E. Goulielmakis, Nature **521**, 498 (2015)
18. G. Vampa, T.J. Hammond, N. Thiré, B.E. Schmidt, F. Légaré, C.R. McDonald, T. Brabec, P.B. Corkum, Nature **522**, 462 (2015)
19. M. Hohenleutner, F. Langer, O. Schubert, M. Knorr, U. Huttner, S.W. Koch, M. Kira, R. Huber, Nature **523**, 572 (2015)
20. G. Ndabashimiye, S. Ghimire, M. Wu, D.A. Browne, K.J. Schafer, M.B. Gaarde, D.A. Reis, Nature **534**, 520 (2016)
21. S. Han, H. Kim, Y.W. Kim, Y.-J. Kim, S. Kim, I.-Y. Park, S.-W. Kim, Nat. Commun. **7**, 13105 (2016)

22. M. Garg, M. Zhan, T.T. Luu, H. Lakhotia, T. Klostermann, A. Guggenmos, E. Goulielmakis, Nature **538**, 359 (2016)
23. H. Liu, Y. Li, Y.S. You, S. Ghimire, T.F. Heinz, D.A. Reis, Nat. Phys. **13**, 262 (2017)
24. Y.S. You, D.A. Reis, S. Ghimire, Nat. Phys. **13**, 345 (2017)
25. F. Langer, M. Hohenleutner, U. Huttner, S. Koch, M. Kira, R. Huber, Nat. Photon. **11**, 227 (2017)
26. Y.S. You, M. Wu, Y. Yin, A. Chew, X. Ren, S. Gholam-Mirzaei, D.A. Browne, M. Chini, Z. Chang, K.J. Schafer, M.B. Gaarde, S. Ghimire, Opt. Lett. **42**, 1816 (2017)
27. K. Kaneshima, Y. Shinohara, K. Takeuchi, N. Ishii, K. Imasaka, T. Kaji, S. Ashihara, K.L. Ishikawa, J. Itatani, Phys. Rev. Lett. **120**, 243903 (2018)
28. H. Hirori, P. Xia, Y. Shinohara, T. Otobe, Y. Sanari, H. Tahara, N. Ishi, J. Itatani, K.L. Ishikawa, T. Aharen, M. Ozaki, A. Wakamiya, Y. Kanemitsu, APL Mater. **7**, 041107 (2019)
29. K.S. Novoselov, A.K. Geim, S.V. Morozov, D. Jiang, Y. Zhang, S.V. Dubonos, I.V. Grigorieva, A.A. Firsov, Science **306**, 666 (2004)
30. C. Berger, Z. Song, T. Li, X. Li, A.Y. Ogbazghi, R. Feng, Z. Dai, A.N. Marchenkov, E.H. Conrad, P.N. First, W.A. de Heer, J. Phys. Chem. B **108**, 19912 (2004)
31. A.K. Geim, K.S. Novoselov, Nat. Mater. **6**, 183 (2007)
32. A.K. Geim, Science **324**, 1530 (2009)
33. A.H. Castro Neto, F. Guinea, N.M.R. Peres, K.S. Novoselov, A.K. Geim, Rev. Mod. Phys. **81**, 109 (2009)
34. V.P. Gusynin, S.G. Sharapov, J.P. Carbotte, Int. J. Mod. Phys. B **21**, 4611 (2007)
35. K.L. Ishikawa, Phys. Rev. B **82**, 201402(R) (2010)
36. M.V. Berry, Proc. R. Soc. A **392**, 45 (1984)
37. Y. Zhang, Y.-W. Tan, H.L. Stomer, P. Kim, Nature **438**, 201 (2005)
38. K.L. Ishikawa, New J. Phys. **15**, 055021 (2013)
39. T. Higuchi, C. Heide, K. Ullmann, H.B. Weber, P. Hommelhoff, Nature **550**, 224 (2017)
40. P. Bowlan, E. Martinez-Moreno, K. Reimann, T. Elsaesser, M. Woerner, Phys. Rev. B **89**, 041408(R) (2014)
41. H.A. Hafez, S. Kovalev, J.-C. Deinert, Z. Mics, B. Green, N. Awari, M. Chen, S. Germanskiy, U. Lehnert, J. Teichert, Z. Wang, K.-J. Tielrooij, Z. Liu, Z. Chen, A. Narita, K. Müllen, M. Bonn, M. Gensch, D. Turchinovich, Nature **561**, 507 (2018)
42. T. Ikemachi, Y. Shinohara, T. Sato, J. Yumoto, M. Kuwata-Gonokami, K.L. Ishikawa, Phys. Rev. A **95**, 043416 (2017)
43. M. Wu, D.A. Browne, K.J. Schafer, M.B. Gaarde, Phys. Rev. A **94**, 063403 (2016)
44. T.-Y. Du, X.-B. Bian, Opt. Express **25**, 151 (2017)
45. T. Ikemachi, Y. Shinohara, T. Sato, J. Yumoto, M. Kuwata-Gonokami, K.L. Ishikawa, Phys. Rev. A **98**, 023415 (2018)
46. M. Wu, S. Ghimire, D.A. Reis, K.J. Schafer, M.B. Gaarde, Phys. Rev. A **91**, 043839 (2015)
47. C. Kittel, *Quantum Theory of Solids*, 2nd edn. (Wiley, New York, 1987), pp.190–193
48. J.B. Krieger, G.J. Iafrate, Phys. Rev. B **33**, 5494 (1986)
49. C.R. McDonald, G. Vampa, P.B. Corkum, T. Brabec, Phys. Rev. A **92**, 033845 (2015)
50. G. Vampa, C.R. McDonald, G. Orlando, D.D. Klug, P.B. Corkum, T. Brabec, Phys. Rev. Lett. **113**, 073901 (2014)
51. G. Vampa, C.R. McDonald, G. Orlando, P.B. Corkum, T. Brabec, Phys. Rev. B **91**, 064302 (2015)
52. C. Kittel, *Introduction to Solid State Physics*, 8th edn. (Wiley, New York, 2004), p.197
53. N. Ashcroft, N. Mermin, *Solid State Physics* (Brooks Cole, Boston, 1976), pp.221–225
54. S. Ghimire, A.D. DiChiara, E. Sistrunk, G. Ndabashimiye, U.B. Szafruga, A. Mohammad, P. Agostini, L.F. DiMauro, D.A. Reis, Phys. Rev. A **85**, 043836 (2012)
55. P.G. Hawkins, M.Y. Ivanov, Phys. Rev. A **87**, 063842 (2013)
56. H. Haug, S. Koch, *Quantum Theory of the Optical and Electronic Properties of Semiconductors* (World Scientific, Singapore, 2009)
57. G. Onida, L. Reining, A. Rubio, Rev. Mod. Phys. **74**, 601 (2002)
58. C.R. McDonald, G. Vampa, P.B. Corkum, T. Brabec, Phys. Rev. Lett. **118**, 173601 (2017)

59. T. Otobe, J. Appl. Phys. **111**, 093112 (2012)
60. N. Klemke, N. Tancogne-Dejean, G.M. Rossi, Y. Yang, F. Scheiba, R.E. Mainz, G. Di Sciacca, A. Rubio, F.X. Kärtner, O.D. Mücke, Nat. Commun. **10**, 1319 (2019)
61. R.W. Boyd, *Nonlinear Optics*, 4th edn. (Academic Press, 2020)
62. M. Noda, S.A. Sato, Y. Hirokawa, M. Uemoto, T. Takeuchi, S. Yamada, A. Yamada, Y. Shinohara, M. Yamaguchi, K. Iida, I. Floss, T. Otobe, K.-M. Lee, K. Ishimura, T. Boku, G.F. Bertsch, K. Nobusada, K. Yabana, Comp. Phys. Comm. **235**, 356 (2018)
63. X. Andrade, D.A. Strubbe, U. De Giovannini, A.H. Larsen, M.J.T. Oliveira, J. Alberdi-Rodriguez, A. Varas, I. Theophilou, N. Helbig, M. Verstraete, L. Stella, F. Nogueira, A. Aspuru-Guzik, A. Castro, M.A.L. Marques, A. Rubio, Phys. Chem. Chem. Phys. **17**, 31371 (2015)
64. M. Lindberg, S.W. Koch, Phys. Rev. B **38**, 3342 (1988)
65. S.A. Sato, Y. Taniguchi, Y. Shinohara, K. Yabana, J. Chem. Phys. **143**, 224116 (2015)
66. J. Paier, M. Marsman, G. Kresse, Phys. Rev. B **78**, 121201 (2008)
67. C.D. Penmmaraju, Comput. Mater. Sci. **18**, e00348 (2019)
68. A.-P. Jauho, K. Johnsen, Phys. Rev. Lett. **76**, 4576 (1996)
69. K. Johnsen, A.-P. Jauho, Phys. Rev. B **57**, 8860 (1998)
70. K.B. Nordstrom, K. Johnsen, S.J. Allen, A.-P. Jauho, B. Birnir, J. Kono, T. Noda, H. Akiyama, H. Sakaki, Phys. Rev. Lett. **81**, 457 (1998)
71. F. Novelli, D. Fousti, F. Giusti, F. Parmigiani, M. Hoffmann, Sci. Rep. **3**, 1227 (2013)
72. W.V. Houston, Phys. Rev. **51**, 184 (1940)
73. T. Otobe, Y. Shinohara, S.A. Sato, K. Yabana, Phys. Rev. B **93**, 045124 (2016)
74. T. Otobe, Phys. Rev. B **94**, 165152 (2016)
75. Y. Mizumoto, Y. Kayanuma, A. Srivastava, J. Kono, A.H. Chin, Phys. Rev. B **74**, 045216 (2006)
76. H.R. Reiss, V.P. Krainov, J. Phys. A: Math. Gen. **36**, 5575 (2003)
77. W. Kohn, L.J. Sham, Phys. Rev. **140**, A1133 (1965)
78. E. Runge, E.K.U. Gross, Phys. Rev. Lett. **52**, 997 (1984)
79. A. Schiffrin, T. Paasch-Colberg, N. Karpowicz, V. Apalkov, D. Gerster, S. Mühlbrandt, M. Korbman, J. Reichert, M. Schultze, S. Holzner et al., Nature **493**, 70 (2013)
80. A. Sommer, E.M. Bothschafter, S.A. Sato, C. Jakubeit, T. Latka, O. Razskazovskaya, H. Fattahi, M. Jobst, W. Schweinberger, V. Shirvanyan et al., Nature **534**, 86 (2016)
81. M. Schultze, K. Ramasesha, C.D. Pemmaraju, S.A. Sato, D. Whitmore, A. Gandman, J.S. Prell, L.J. Borja, D. Prendergast, K. Yabana et al., Science **346**, 1348 (2014)
82. M. Lucchini, S. Sato, J. Herrmann, A. Ludwig, M. Volkov, L. Kasmi, Y. Shinohara, K. Yabana, L. Gallmann, U. Keller, Science **353**, 916 (2016)
83. G. Wachter, C. Lemell, J. Burgdörfer, S.A. Sato, X.-M. Tong, K. Yabana, Phys. Rev. Lett. **113**, 087401 (2014)
84. K. Yabana, G.F. Bertsch, Phys. Rev. B **54**, 4484 (1996)
85. G.F. Bertsch, J.-I. Iwata, A. Rubio, K. Yabana, Phys. Rev. B **62**, 7998 (2000)
86. J.R. Chelikowsky, N. Troullier, K. Wu, Y. Saad, Phys. Rev. B **50**, 11355 (1994)
87. N. Troullier, J.L. Martins, Phys. Rev. B **43**, 1993 (1991)
88. L. Kleinman, D.M. Bylander, Phys. Rev. Lett. **48**, 1425 (1982)
89. J.P. Perdew, A. Zunger, Phys. Rev. B **23**, 5048 (1981)

# Chapter 5
# Attosecond Space–Time Imaging with Electron Microscopy and Diffraction

**Peter Baum and Yuya Morimoto**

**Abstract** The first step of most light-matter interactions is a field-driven motion of electron density in and around the atoms of a material. Fully visualizing such dynamics and its consequences for the macroscopic functionality of a material, therefore, requires atomic resolution in space and sub-light-cycle resolution in time. Here, we review our latest progress with attosecond space–time imaging by using attosecond electron pulses in diffraction and microscopy. We start with a brief review of recent technological advancements for the temporal compression of ultrashort electron pulses with radio-frequency waves, terahertz pulses and now optical field cycles. We then report on the light-wave control of electron beams at thin metallic or dielectric membranes, which form the basis of attosecond electron imaging in our laboratory. We report the first demonstrations of attosecond electron diffraction and microscopy in proof-of-principle experiment that reveal an upper limit for the delays associated with electron-crystal scattering and visualize the oscillations and propagation of a traveling light wave on a nanometer-thick membrane. These unprecedented space–time resolutions provided by attosecond electron microscopy and diffraction now enable to capture the dynamics of electrons inside of atoms, molecules, crystals or nanostructures as a function of space and time.

**Keywords** Attosecond electron pulse · Attosecond electron imaging · Attosecond microscopy · Attosecond electron diffraction · Space–time imaging

P. Baum (✉)
Universität Konstanz, 78464 Konstanz, Germany
e-mail: peter.baum@uni-konstanz.de

Y. Morimoto (✉)
RIKEN Cluster for Pioneering Research and RIKEN Center for Advanced Photonics, RIKEN, 2-1 Hirosawa, Wako, Saitama 351-0198, Japan
e-mail: yuya.morimoto@riken.jp

 
K. Ueda (ed.), *Ultrafast Electronic and Structural Dynamics*,
https://doi.org/10.1007/978-981-97-2914-2_5

## 5.1 Introduction

When a light wave interacts with a material, its bound or free electrons are driven around by the electric and magnetic field cycles and subsequently return to their original positions via the restoring forces provided by the structural environment. Light-driven electron dynamics is therefore often complicated and gives rise to a wealth of optical phenomena with fundamental and technical relevance, for example the refractive index, birefringence, dispersion, optical parametric amplification, frequency-doubling, self-phase modulation, multi-photon absorption, higher-harmonic generation and many more. Textbooks treat perturbative linear and nonlinear optics via a time-dependent polarization density $P(t) = \epsilon_0(\chi^{(1)}E(t) + \chi^{(2)}E^2(t) + \chi^{(3)}E^3(t) + \cdots)$, where $\chi^{(n)}$ is the $n$-th order susceptibility tensor, and $E(t)$ is the applied electric field. In the non-perturbative and strong-field regimes, $P(t)$ depends on the cycle-level history of $E(t)$ in time, for example via field-induced intraband currents, dynamical metallization, core excitons, sub-cycle energy transfer, dynamical electron localization to d-orbitals, or seed-induced changes of the band structure; see references in [1]. Although many such nonlinear-optical and strong-field phenomena in atoms, molecules, solids and nanostructures are currently rather well understood on basis of fields and polarizations, we lack a comprehensive understanding of the atomistic origin of the macroscopic quantities $P(t)$ and $\chi^{(n)}$ and how the structure of a material causes the effective nonlinear-optical phenomena that we observe. In our laboratory, we, therefore, aim at visualizing the spatiotemporal dynamics of bound or free electrons in space and time at the fundamental resolutions associated with their natural dynamics, typically picometres and attoseconds [2]. This quest requires novel measurement approaches, and our choice is ultrafast electron microscopy and diffraction with light-cycle controlled single-electron wave functions [3–5] because we can, in this way, unite the sub-atomic spatial resolution and imaging flexibility of freely propagating electrons with the temporal accuracy and stability provided by the cycles of light. Here, we report the basic concepts, explain the technological basis, discuss the first proof-of-principle experiments and conclude with an outlook of the perspectives that we see for our approach in the future.

## 5.2 Light-Wave Control of Free Electrons

The electron microscope or an electron diffraction apparatus is a very versatile instrument for characterizing the atomic structure and composition of complex materials of almost any morphology or shape. At tens of keV to MeV electron energies, the de Broglie wavelength is picometres, and electron microscopy and diffraction, therefore, routinely provide nanometer to picometer spatial resolution of atoms [6, 7] and the electric fields that hold them together [8–10]. However, a comparable temporal resolution at the atomic units of time is difficult to achieve because electrons carry an electrical charge that leads in short pulses to space charge effects. Furthermore, electrons also carry a rest mass that causes wave packet dispersion in vacuum [3]. Although picosecond and femtosecond electron pulses can be produced by laser-triggered photoemission from a cathode and subsequent acceleration by an electrostatic voltage [11, 12], the shortest pulse duration achievable with this scheme is limited by Coulomb repulsion and dispersion effects to ~100 fs even for the most compact gun designs at highest acceleration field [11, 13, 14]. Note that researchers in ultrafast science usually apply a full-width-at-half-maximum definition for pulse durations and time resolutions, while accelerator physicists often use a root-mean-square definition that appears to be >2 times shorter. When an electron pulse contains multiple electrons, ideally so many as to record structural dynamics in a single shot [12], Coulomb repulsion between the electrons becomes a dominant contribution and the temporal apparatus function reaches hundreds of femtoseconds.

In order to achieve electron pulses that are shorter than tens of femtoseconds, the fundamental time scale of phonons and molecular vibrations, an additional pulse compression technique is required. In general, researchers use time-dependent electric fields for such purpose (compare Fig. 5.1). A rapidly oscillating electric field in a sub-wavelength resonator decelerates the leading parts of an electron pulse and accelerates the rear part. Higher-energy parts of the electron pulses now catch up with the lower-energy parts, compressing the pulses in time. After the temporal focus, the pulses over-disperse and become long again. In microwave cavities, terahertz compressors or on bunching schemes driven by the optical cycles of light, the energy bandwidth of the electron pulse necessarily increases by a few eV, but the overall bandwidth remains below a fraction of $10^{-4}$ to $10^{-5}$ of the central energy. Such a degree of monochromaticity is sufficient for atomic-scale diffraction and sub-nanometer real-space imaging [15, 16].

Initially, researchers applied microwaves in mm-sized enhancement cavities and electron pulses as short as ~20 fs (FWHM) have been achieved [17–20]. An intrinsic technical problem here is temporal jitter between electrons and the microwave field, typically tens of femtoseconds even with state-of-the-art feedback [19, 21]. A demonstration of our laboratory was in 2016 the use of THz waves for the compression of

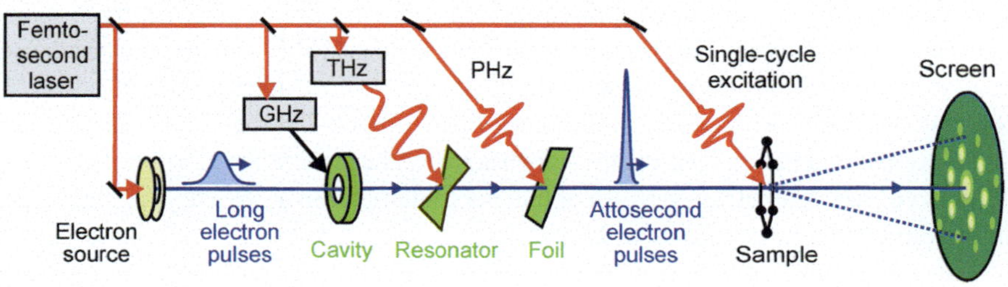


**Fig. 5.1** Overview of different concepts for electron pulse compression. Long electron pulses (blue) can either be compressed by radio-frequency radiation in microwave cavities, by all-optical means in a terahertz-driven resonator or—topic of this work—by the optical field cycles of visible or near-infrared light at a dielectric or metallic planar membrane. The compressed electron pulses may be useful for time-resolved imaging of ultrafast light-matter interaction via attosecond electron microscopy or diffraction. Reprinted from [2], Copyright 2017, with permission from Elsevier

freely propagating electron pulses, drawing on earlier simulations [22–25], cycle-averaging experiments [26] and studies at low energy [27]. THz-driven electron pulse control is now widely used in accelerator physics for diagnostics [28], and researchers also aim at all-optical electron pulse formation driven and controlled entirely by THz fields. However, the attosecond regime of pulse duration has not been achieved in this way, mainly because of temporal aberrations by the nonlinear optical cycles of the terahertz wave and by the too long temporal focus distances that are realistic at the typical peak intensities of terahertz pulses [29].

## 5.3 Attosecond Control of Electron Beams at a Membrane

Even shorter electron pulses with a duration of attoseconds are achieved via compression with optical light of PHz frequency [5, 30–34]. Interestingly, it is possible to produce in this way electron pulses that are shorter than any other process in the experiment, including the pulses from the driving lasers [4], because we apply here a light-cycle-driven time-dependent energy modulation and not an energy-dependent time modulation (i.e. constant bandwidth) like in many optical prism or grating compressors. Quantum mechanical calculation shows that the shortest electron pulse duration is given by $\tau_{\min} \approx 4\pi\hbar L_{\text{focus}}/\gamma^3 m_e v_e^3 T_{\text{comp}} \approx 4\hbar/\Delta W_{\max}$, where $\hbar$ is reduced Planck constant, $\gamma$ is Lorentz factor, $m_e$ is the rest mass of electron, $v_e$ is the speed of electron before compression, $T_{\text{comp}}$ is the cycle period of compression field, and $\Delta W_{\max}$ is the amount of energy modulation in full width [35, 36]. $L_{\text{focus}}$ is the distance where the electron pulse duration becomes shortest measured from the compression stage and is given by $L_{\text{focus}} = v_e^2 T_{\text{comp}}/2\pi\Delta v_{\max}$, where $\Delta v_{\max}$ is the peak amplitude in the velocity modulation. These equations suggest that stronger and faster modulation fields can provide shorter electron pulses at shorter temporal focal distance.

The practical implementation of such an extreme compression, its underlying physics and further details of this light-wave control are discussed in the next sections, before reporting two proof-of-principle applications in diffraction and microscopy [5]. Related to our results, generation of attosecond electron pulses has also been reported by using an inverse free-electron-laser process [30], a two-color ponderomotive potential [33] or optical near-fields around nanostructures [31, 32, 34]. In contrast, we use free-standing membranes of metals or dielectrics [14, 36] as the modulation elements. In contrast to nanostructures, these membranes offer the possibility to compress an electron beam of extended diameter in a velocity-matched way and without spatiotemporal distortions [37].

When an electron interacts with a laser field, the momentum of electron is modulated in time, given by the Lorentz force, $\Delta \boldsymbol{p}(t) = -e\int_{-\infty}^{t} \boldsymbol{E}(\boldsymbol{x}, t)dt - e\int_{-\infty}^{t} \boldsymbol{v}_e \times \boldsymbol{B}(\boldsymbol{x}, t)dt$, where $e$ is the unit charge, $\boldsymbol{E}(\boldsymbol{x}, t)$ and $\boldsymbol{B}(\boldsymbol{x}, t)$ are electric and magnetic fields at position $\boldsymbol{x}(t)$ and time $t$.. When the electron-laser interaction occurs in free space, the net momentum shift after the interaction ($t \to +\infty$) always averages out to zero, except for much weaker and cycle-averaging ponderomotive effects. The situation is completely different when a nanometer-thick foil is present in the optical focus region (see Fig. 5.2). For a metallic foil, the incident electromagnetic field is almost fully reflected and for a dielectric material, there occur thin-film interference and a phase shift of the transmitted wave due to the foil's refractive index. As a result, in both cases, we obtain electromagnetic fields with different amplitudes and phases on each side of the membrane. This break of symmetry can be used to impart a net momentum that remains there after the interaction [36]. In a simplified picture, the electrons pass through the membrane within a time that is shorter than half an optical cycle. They feel an abrupt change of the field's amplitude and phase (see Fig. 5.2) within sub-cycle times and therefore acquire after passage through the laser focus a non-zero final momentum from the field. In a way, the laser-electron-membrane interaction is a three-body problem in which the membrane material enables efficient momentum exchange between the electrons and the laser photons. For a laser wave with multiple field cycles (see Fig. 5.3), the amount of momentum shift becomes a sinusoidal function of the electron arrival time at the membrane with respect to the optical cycles. The amplitude is proportional to the electric field strength. After further propagation in free space, such a field-cycle-driven periodic acceleration and deceleration of the electrons modulate the temporal shape of the electron pulses into a train of attosecond pulses [38]. If the membrane angles and laser polarization are chosen such that there is rather a periodic transversal momentum change, we obtain a time-dependent sideways deflection of the electrons as a function of time, useful for directly characterizing attosecond electron pulses in the time domain by streaking [5]. A typical attosecond electron microcopy or diffraction experiment therefore involves two electron-laser interactions at a finite distance, like for example depicted in Fig. 5.3, where we show the experimental setup that has enabled first attosecond electron microscopy or diffraction experiments (see Sects. 5.5 and 5.6).

Before we report these applications, it is instructive to explain some key dependencies of the laser-membrane-electron interaction on wavelength, field strength, angles and polarization [36]. We obtain eight essential dependencies as summarized below. (a) Like in microwave compression cavities, the role of the magnetic field is significant, although the frequencies are about a million times higher. Magnetic fields can never accelerate or decelerate the electrons and therefore do not contribute to longitudinal compression, but they can induce substantial and relevant transversal momentum shifts that can tilt the electrons pulses to have a non-normal propagation direction [37] or contribute to the timing and amplitude of sideways streaking for pulse characterization [4, 36]. Explicit consideration of the magnetic fields is therefore crucial for design of optimized geometries. (b) Only the *p*-component of the laser field can induce any substantial longitudinal and transversal momentum changes. For almost any membrane material and angle combination, there is an almost complete cancellation of electric and magnetic field contributions for the *s*-polarization component of the laser light [36]. Only membranes of extended thickness, microstructures resonators [4] or dedicated waveguides [39] break this symmetry by the onset of localized mode concentration or quasi-phase matching in the longitudinal direction. (c) The longitudinal and transversal momentum shifts always happen simultaneously and in synchrony to the optical cycles of the laser wave at almost any angle combinations between electrons, laser and membrane [36]. (d) Practical pulse compression

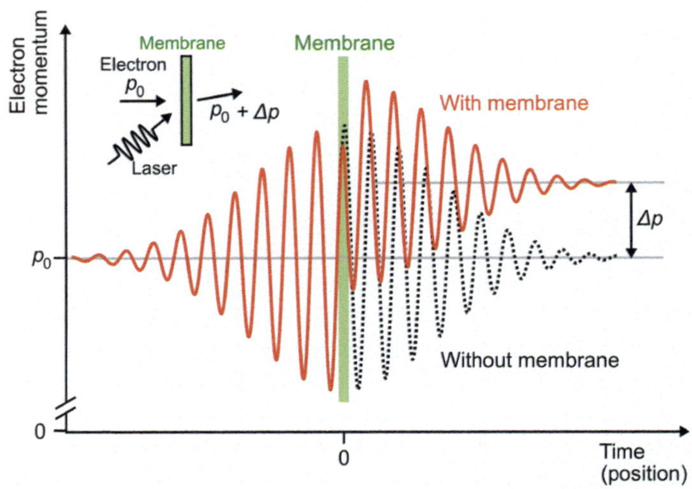


**Fig. 5.2** Mechanism of laser-electron momentum transfer with the help of a planar dielectric, metallic and absorbing membrane. Without a membrane, the momentum (black dotted curve) returns to the initial value after interaction with a laser field. In contrast, the electron acquires in the presence of the membrane a finite momentum (red curve) due to the electromagnetic fields with different amplitudes and phases on each side of the membrane. Reprinted figure with permission from [36] Copyright 2018 by the American Physical Society

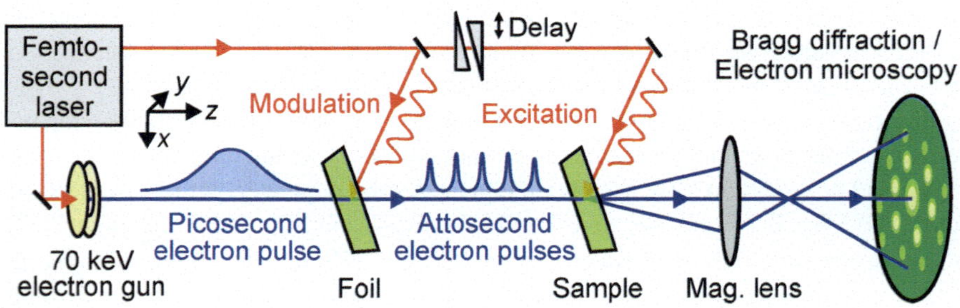


**Fig. 5.3** Experimental setup for attosecond imaging with freely propagating electron pulses. Femtosecond or picosecond electron pulses (blue) from a laser-driven source (yellow) are intersected with the optical cycles of a laser beam (red) at a dielectric membrane (green). The resulting time-dependent modulations of forward momentum convert the electron beam into a sequence of attosecond pulses (blue). Those pulses are locked in time to the optical cycles of a second laser wave (red) that can serve for pulse metrology or for sample excitation. A magnetic lens system (grey) can be set up to either produce diffraction data or real-space images with magnification. Reprinted from [5]

and streaking require consideration of extended beam diameters, such as unavoidable for electron beams of finite emittance. There are special conditions between the membrane angle, the angle of incidence of the electron and the laser polarization that match the surface velocity of the electron incidence to the surface velocity of the optical laser cycles along the compression membrane, thereby ensuring that every lateral point of an extended electron beam experiences the same physics in time [40]. Membranes operated at proper angles are therefore ideal for compressing electron beams of finite emittance or reaching ultimate pulse duration without any spatiotemporal distortions. (e) At the special angles satisfying this electron-light velocity matching, the transversal momentum shift is always zero, regardless of the type of membrane. Extended laser and electron beams are, therefore, accelerated/decelerated or temporally compressed without any sideways deflection [37]. (f) Irrespective of the type of membrane material (metallic, dielectric, absorbing), the net momentum change is proportional to the laser field amplitude and synchronized in time to the optical cycle, albeit with a constant phase shift that depends on the membrane material. In the case of dielectric membranes, the amplitude of the momentum change is almost directly related to the amount of phase shift of the transmitted laser fields. In other words, the momentum change originates from the 'missing' or 'additional' time that the electrons spend in the optical cycles while being in the membrane. (g) At a constant laser intensity, longer wavelengths usually give stronger momentum modulation with metallic and absorbing foils. In contrast, the wavelength dependence is weak with dielectric membranes [36]. (8) At given laser wavelength, the maximum amount of momentum modulation that a membrane can provide depends predominantly on its optical damage threshold. Dielectric membranes excel here due to their low linear and nonlinear absorption [36]. For example, a silicon membrane at a field strength just slightly below damage threshold [41] for near-infrared picosecond laser pulses allows to achieve electron pulses as short as ~15 as at a focal distance of 0.1 mm with 70 keV electron beam. Even shorter pulses can be generated at higher electron energies and longer laser wavelengths [35].

## 5.4 Generation and Detection of Attosecond Electron Pulses

In this section, we report the concepts and details on our recent experimental demonstration of first direct time-domain streaking measurements of attosecond electron pulses [5]. We refer to Fig. 5.3 and its sketch of the experiment. A regenerative Yb:YAG thin-disk amplifier generates ~900-fs pulses at 1030 nm central wavelength. A pulsed electron beam at an energy of 70 keV is generated through two-photon photo-emission of second-harmonic light at a gold photocathode [42] and subsequent acceleration by a static acceleration voltage. The de Broglie wavelength of the electrons is 4.5 pm. The electron beam is collimated by a magnetic lens that is aligned to avoid temporal distortions [43, 44] and subsequently guided to two free-standing dielectric membranes for pulse compression and streaking characterization, respectively [5]. The incident electron pulse duration is about 1 ps (full width at half maximum, FWHM) with <1 eV energy bandwidth [4]. The electrons transmitted through the foils are detected by a single-electron detector (F816, TVIPS GmbH) located at 1.3 m distance from the foils.

We employ two types of the free-standing foils; 50 nm of $Si_3N_4$ for temporal modulation and 60 nm of Si for streaking characterization. Both membranes are available as electron microscopy support grids and uncomplicated to obtain. They are placed in parallel to each other at a distance of 4 mm. The laser beam for attosecond control is stretched to 1.7 ps by a grating sequence in order to match its duration to the length of the electron pulses for achieving a temporally homogeneous interaction. The laser beam is then split in two beams by an optical interferometer and both parts are focused on the two dielectric membranes by a common lens. The optical focus of each beam is placed slightly before the foils in order to achieve a homogeneous electromagnetic field that covers the entire electron beam diameter and to avoid different Gouy phases on the two foils [5, 36]. The first beam on the $Si_3N_4$ membrane temporally modulates the longitudinal momentum of the electron beam (see above) and free-space propagation converts it into a train of attosecond electron pulses. The laser power and thereby the compression field strength is adjusted to move the temporal focus to the location of the second membrane. There, the geometry is chosen such that the electrons are periodically streaked in sideways direction (see above), like in a cathode-ray oscilloscope but at the frequency of light. Due to the high damage threshold of dielectric membranes, we achieve a streaking speed as high as ~0.2 mrad/fs. The time-dependent streaking projects the temporal shape of the electron pulse into a spatial profile on the screen. Given the temporal synchrony between the attosecond electron pulses from the compression interaction with the optical cycles of the characterization laser [45], each electron pulse in the sequence sees the same streaking dynamics and the experiment therefore measured an average temporal profile of all attosecond electron pulses in the train [5].

Figure 5.4 shows the observed streaking results as a function of the electron-streaking delay. The localized spot on the detector moves up and down along the laser polarization direction as a function of the sub-femtosecond delay between the streaking field cycles and the compressed electron pulses. This oscillation directly demonstrates the successful compression of electron pulses to attosecond durations, for every laser shot in the form of a pulse train or burst in synchrony to the optical cycles. We obtain a mean pulse duration through a global fitting of the entire 2D data of Fig. 5.4c by assuming a Gaussian temporal shape. At a field amplitude of 60 MV/m for the compression, we obtain a shortest electron pulse duration of 820 as (FWHM) above a ~30% uncompressed background, which originates from the non-converging half-cycles of the modulation field. These results mark the first generation of attosecond electron pulses in a table-top diffraction apparatus, complementing related results from the group of C. Ropers with a transmission electron microscope [31] and from the group of P. Hommelhoff with ponderomotive compression [33].

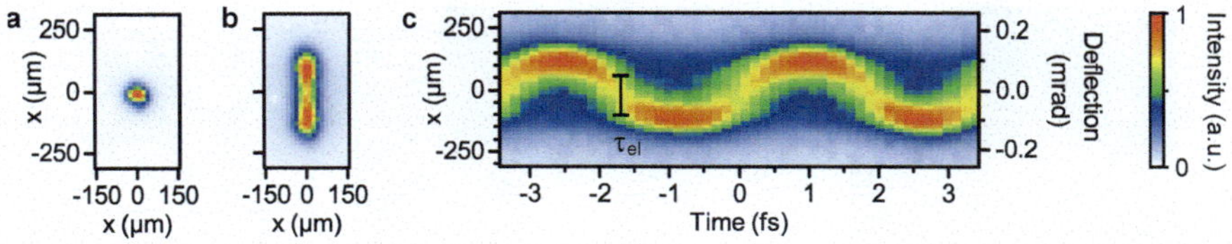


**Fig. 5.4** Results on attosecond electron pulses. **a** Beam profile of the long-pulse electron beam without any modulation. **b** Beam profile in case of sideways streaking but no temporal compression (compare Fig. 5.3). The measured elongation and double-peak structure of the spot indicate the rapid change of transverse momentum by the characterization laser (second in Fig. 5.3). **c** Time-resolved deflection data. The measured change of the sideways deflection as a function of the delay between the electron pulses and the cycles of the streaking laser reveals the presence of attosecond electron pulses at the diffraction or microscopy target. Reprinted from [5]

## 5.5 Attosecond Electron Diffraction

Theoretical considerations have indicated the possibility to measure spatiotemporal charge carrier dynamics in atoms [46], molecules [46], condensed matter [47] or nanostructures [48] by using attosecond electron diffraction or microscopy. Here, we report our first two proof-of-concept experiments [5] that indicate the feasibility of such endeavors. The first result is attosecond electron diffraction from a crystalline membrane [5]. The purpose is to obtain an upper limit on the scattering time that it takes to convert an incoming electron into one that travels into a Bragg spot. Figure 5.5a shows the electron diffraction pattern of single-crystalline Si membrane at the orientation of $[-1/\sqrt{2}, \sqrt{2}, 1/\sqrt{2}]$, as recorded with the attosecond electron pulses. A large set of diffraction spots with Miller indices up to (606) and (404) are clearly visible in the raw data, demonstrating atomic spatial resolution. The geometry of the laser-electron interaction is chosen such that whole diffraction pattern is streaked in a sideways direction as a function of time. While the streaking of the direct beam is given by integration over its entire trajectory, the Bragg electrons are only generated within the crystal and therefore potentially with a delay that may originate from the scattering or the interferences of multiple-scattering pathways. Figure 5.5e shows the results of the measured streaking signals of the Bragg spots (0, 0, 0) and (1, −1, 3) and an evaluation of the two traces' potential attosecond delays using a center-of-mass analysis. The results of such measurements for eight different Bragg spots are summarized in Fig. 5.5f. We find that the scattering delay of the electrons by the nanometer-thick crystalline membrane is not in a discernible way depending on the Miller index. The average delay amounts to $4.6 \pm 5.4$ as or less. We conclude that Bragg spot electrons are delayed by crystal diffraction by less than tens of attoseconds, even in the regime of substantial multiple-scattering dynamics in our highly crystalline material. Estimations indicate that few-attosecond delays should originate from multiple scattering and sub-attosecond delays are associated with the atomic scattering process itself. Given the limited signal-to-noise of our very first attosecond diffraction experiment, we expect that future experiments will reveal those delays and therefore access the physics of electron-atom interaction directly

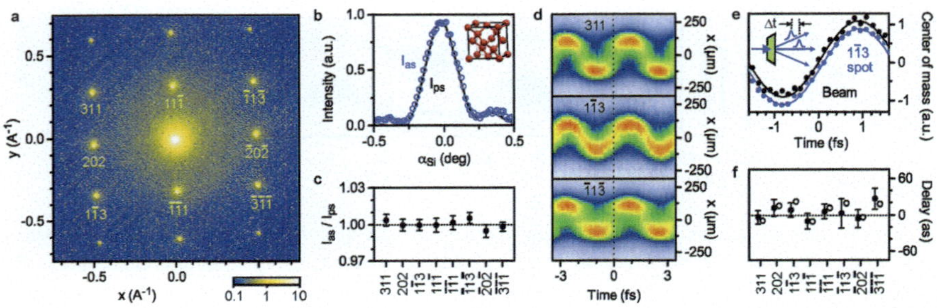

**Fig. 5.5** Attosecond electron diffraction results. **a** Transmission diffraction pattern of a single-crystalline silicon foil, taken with the attosecond electron pulses. Bragg spots up to 6th order are visible in the raw data, limited by absorption at high angles by beamline components. **b** Rocking curve of the (1, −1, 3) spot. **c** Intensity changes of different Bragg spots measured with the attosecond pulses in comparison to data taken with uncompressed picosecond electron pulses. **d** Streaking deflectogram data of three selected Bragg spots. **e** Centre-of-mass analysis of the deflectogram of the (1, −1, 3) Bragg spot and the direct beam, slightly displaced for visibility. **f** Observed attosecond-level delay of different Bragg spots with respect to the direct beam. Reprinted from [5]

in the time domain. Also, a slightly improved signal-to-noise ratio or a roughly 10 times stronger excitation field will be sufficient to perform diffractive imaging of field-driven carrier dynamics, in order to visualize the light-driven motion of valence electrons in materials such as graphene in real-space and real-time [47].

## 5.6 Attosecond Electron Microscopy

In a second proof-of-concept application of our attosecond electron pulses, we demonstrate attosecond electron microscopy of a traveling electromagnetic waveform. We, hereby, advance ultrafast waveform electron microscopy [48] from the regime of THz-driven resonator analysis to optical frequencies and attosecond time resolution. Figure 5.6d–e illustrate the experimental geometry. We record transmission electron microscopy images of a nanometer-thick dielectric membrane that is excited by an optical laser beam that impinges from a slightly off-angle direction. In this geometry, we generate a traveling optical wave on the nanometer-thick foil in which the optical cycles move along the surface at superluminal phase velocity. The attosecond electron pulses pass through the foil as a collimated beam. The transit time is faster than half an optical cycle period and the electrons, therefore, see approximately time-frozen electromagnetic fields that deflect them in sideways direction as a function of space and time. Figure 5.6a, b show static and delay-dependent transmission images of the silicon foil as measured with our attosecond electron pulses, respectively. The entire pattern moves rapidly along the horizontal direction when varying the attosecond delay between the optical excitation field and the attosecond electron pulses because the intensity variations at each delay time are directly related to the local and instantaneous optical fields' amplitudes and phases. The parts with

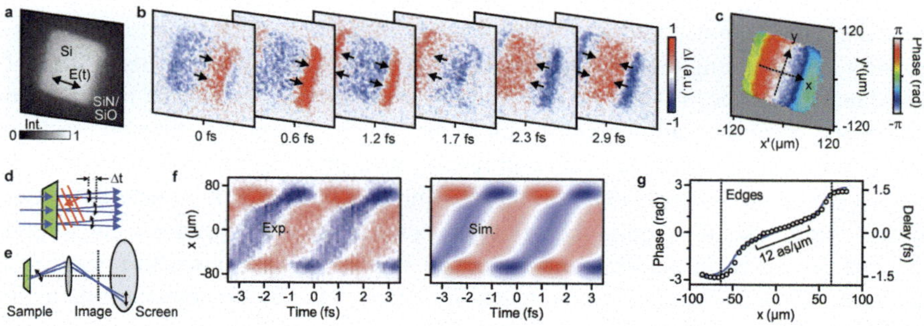


**Fig. 5.6** **a** Transmission electron microscopy image of a 60-nm-thick silicon membrane. **b** Measured intensity changes as a function of attosecond-level time delay. **c** Phase map of the observed intensity oscillations. **d** Illustration of how a traveling-wave excitation produces converging (red part in **b**) and diverging (blue part in **b**) electron trajectories. **e** Schematic of the imaging geometry with differential phase contrast. **f** Position-dependent intensity changes as a function of delay time. **g** Phase and delay of the wave across the sample (dots) in comparison with a simulation (blue). The optical traveling wave advances in space–time by 12 as/μm. Reprinted from [5]

higher signal intensity (red) indicate such locations where the instantaneous electromagnetic deflection angles are convergent. In contrast, the less intense parts (blue) represent a diverging deflection. The data, hence, directly reveal the propagation dynamics of the optical wave in space and time, resolved on attosecond time scales. More complex materials or nanostructure arrays, such as commonly applied for metamaterials [49, 50], can therefore be investigated on the basis of the electromagnetic field cycles in space and time, thus revealing the origin of their performance on the level of the optical cycles of light.

## 5.7 Outlook: Attosecond Electron Imaging

Based on the reported advances and breach of ultrafast electron microscopy and diffraction into the regime of attosecond dynamics [5], we here convey some further ideas how to utilize the reported combination of atomic spatial resolution, given by the picometre de Broglie wavelength of high-energy electrons beams, with the attosecond timing accuracy provided by the cycles of light for future novel imaging applications. We first note that attosecond electron pulses are simple to generate, versatile to apply and useful in practice as a potentially valuable alternative to high-harmonic photon sources for understanding basic light-matter interaction by visualization in space and time [51, 52]. With the help of dielectric or metallic membranes in proper geometry, attosecond electron pulses can rather easily be implemented into almost any existing table-top electron diffraction experiments but also into modern scanning/transmission electron microscopes at a point in the column where the beam is collimated, in order to mitigate the need for near-field structures that have so far been the only demonstrated way of generating attosecond electron pulses in transmission and scanning electron microscopy [31, 34]. If the temporal distortions of the magnetic lenses [43] can be resolved [44], it might be possible to integrate an attosecond compression interaction before the twin lens into a collimated electron beam, in order to disentangle the compression and imaging interactions from each other. One could also speculate on the possibility to use a continuous-wave laser in combination with a continuous electron beam to create a beam of attosecond electron pulses at an unprecedented average brightness and signal-to-noise [53]; we have, in the meantime, demonstrated this idea [54, 55]. Dielectric or metallic modulation membranes and the reported proofs of feasibility of attosecond electron diffraction and microscopy of condensed-matter specimen open in our opinion up a novel direction of research, in which fundamental light-matter interaction and applied nano-plasmonics are investigated on their natural dimensions in space and time [5, 55].

## References

1. Y. Morimoto, Y. Shinohara, M. Tani, B.-H. Chen, K.L. Ishikawa, P. Baum, Asymmetric single-cycle control of valence electron motion in polar chemical bonds. Optica **8**, 382–387 (2021). https://doi.org/10.1364/OPTICA.414213
2. P. Baum, F. Krausz, Capturing atomic-scale carrier dynamics with electrons. Chem. Phys. Lett. **683**, 57–61 (2017). https://doi.org/10.1016/j.cplett.2017.03.073
3. M. Aidelsburger, F.O. Kirchner, F. Krausz, P. Baum, Single-electron pulses for ultrafast diffraction. Proc. Natl. Acad. Sci. **107**, 19714–19719 (2010). https://doi.org/10.1073/pnas.1010165107
4. C. Kealhofer, W. Schneider, D. Ehberger, A. Ryabov, F. Krausz, P. Baum, All-optical control and metrology of electron pulses. Science **352**, 429–433 (2016). https://doi.org/10.1126/science.aae0003
5. Y. Morimoto, P. Baum, Diffraction and microscopy with attosecond electron pulse trains. Nat. Phys. **14**, 252–256 (2018). https://doi.org/10.1038/s41567-017-0007-6
6. P.E. Batson, N. Dellby, O.L. Krivanek, Sub-ångstrom resolution using aberration corrected electron optics. Nature **418**, 617–620 (2002). https://doi.org/10.1038/nature00972
7. R. Erni, M.D. Rossell, C. Kisielowski, U. Dahmen, Atomic-resolution imaging with a sub-50-pm electron probe. Phys. Rev. Lett. **102**, 096101 (2009). https://doi.org/10.1103/PhysRevLett.102.096101
8. N. Shibata, S.D. Findlay, Y. Kohno, H. Sawada, Y. Kondo, Y. Ikuhara, Differential phase-contrast microscopy at atomic resolution. Nat. Phys. **8**, 611–615 (2012). https://doi.org/10.1038/nphys2337
9. Y. Jiang, Z. Chen, Y. Han, P. Deb, H. Gao, S. Xie, P. Purohit, M.W. Tate, J. Park, S.M. Gruner, V. Elser, D.A. Muller, Electron ptychography of 2D materials to deep sub-ångström resolution. Nature **559**, 343–349 (2018). https://doi.org/10.1038/s41586-018-0298-5
10. K. Müller, F.F. Krause, A. Béché, M. Schowalter, V. Galioit, S. Löffler, J. Verbeeck, J. Zweck, P. Schattschneider, A. Rosenauer, Atomic electric fields revealed by a quantum mechanical approach to electron picodiffraction. Nat. Commun. **5**, 5653 (2014). https://doi.org/10.1038/ncomms6653
11. A.H. Zewail, Four-dimensional electron microscopy. Science **328**, 187–193 (2010). https://doi.org/10.1126/science.1166135
12. R.J.D. Miller, Femtosecond crystallography with ultrabright electrons and x-rays: capturing chemistry in action. Science **343**, 1108–1116 (2014). https://doi.org/10.1126/science.1248488

13. A. Feist, K.E. Echternkamp, J. Schauss, S.V. Yalunin, S. Schäfer, C. Ropers, Quantum coherent optical phase modulation in an ultrafast transmission electron microscope. Nature **521**, 200–203 (2015). https://doi.org/10.1038/nature14463
14. F.O. Kirchner, A. Gliserin, F. Krausz, P. Baum, Laser streaking of free electrons at 25 keV. Nat. Photonics **8**, 52–57 (2013). https://doi.org/10.1038/nphoton.2013.315
15. A. Feist, N. Bach, N. Rubiano da Silva, T. Danz, M. Möller, K.E. Priebe, T. Domröse, J.G. Gatzmann, S. Rost, J. Schauss, S. Strauch, R. Bormann, M. Sivis, S. Schäfer, C. Ropers, Ultrafast transmission electron microscopy using a laser-driven field emitter: femtosecond resolution with a high coherence electron beam. Ultramicroscopy **176**, 63–73 (2017). https://doi.org/10.1016/j.ultramic.2016.12.005
16. N. Bach, T. Domröse, A. Feist, T. Rittmann, S. Strauch, C. Ropers, S. Schäfer, Coulomb interactions in high-coherence femtosecond electron pulses from tip emitters. Struct. Dyn. **6**, 014301 (2019). https://doi.org/10.1063/1.5066093
17. R.P. Chatelain, V.R. Morrison, C. Godbout, B.J. Siwick, Ultrafast electron diffraction with radio-frequency compressed electron pulses. Appl. Phys. Lett. **101**, 081901 (2012). https://doi.org/10.1063/1.4747155
18. J. Maxson, D. Cesar, G. Calmasini, A. Ody, P. Musumeci, D. Alesini, Direct measurement of sub-10 fs relativistic electron beams with ultralow emittance. Phys. Rev. Lett. **118**, 154802 (2017). https://doi.org/10.1103/PhysRevLett.118.154802
19. A. Gliserin, M. Walbran, F. Krausz, P. Baum, Sub-phonon-period compression of electron pulses for atomic diffraction. Nat. Commun. **6**, 8723 (2015). https://doi.org/10.1038/ncomms9723
20. T. Van Oudheusden, P.L.E.M. Pasmans, S.B. Van Der Geer, M.J. De Loos, M.J. Van Der Wiel, O.J. Luiten, Compression of subrelativistic space-charge-dominated electron bunches for single-shot femtosecond electron diffraction. Phys. Rev. Lett. **105**, 1–4 (2010). https://doi.org/10.1103/PhysRevLett.105.264801
21. H.W. Kim, N.A. Vinokurov, I.H. Baek, K.Y. Oang, M.H. Kim, Y.C. Kim, K.-H. Jang, K. Lee, S.H. Park, S. Park, J. Shin, J. Kim, F. Rotermund, S. Cho, T. Feurer, Y.U. Jeong, Towards jitter-free ultrafast electron diffraction technology. Nat. Photonics 6–11 (2019). https://doi.org/10.1038/s41566-019-0566-4
22. S.D. Vartak, N.M. Lawandy, Breaking the femtosecond barrier: a method for generating attosecond pulses of electrons and photons. Opt. Commun. **120**, 184–188 (1995). https://doi.org/10.1016/0030-4018(95)00427-A
23. R.B. Yoder, J.B. Rosenzweig, Side-coupled slab-symmetric structure for high-gradient acceleration using terahertz power. Phys. Rev. Spec. Top. - Accel. Beams **8**, 111301 (2005). https://doi.org/10.1103/PhysRevSTAB.8.111301
24. L.J. Wong, A. Fallahi, F.X. Kärtner, Compact electron acceleration and bunch compression in THz waveguides. Opt. Express **21**, 9792 (2013). https://doi.org/10.1364/OE.21.009792
25. J. Fabiańska, G. Kassier, T. Feurer, Split ring resonator based THz-driven electron streak camera featuring femtosecond resolution. Sci. Rep. **4**, 5645 (2015). https://doi.org/10.1038/srep05645
26. E.A. Nanni, W.R. Huang, K. Hong, K. Ravi, A. Fallahi, G. Moriena, R.J. Dwayne Miller, F.X. Kärtner, Terahertz-driven linear electron acceleration. Nat. Commun. **6**, 8486 (2015). https://doi.org/10.1038/ncomms9486
27. L. Wimmer, G. Herink, D.R. Solli, S.V. Yalunin, K.E. Echternkamp, C. Ropers, Terahertz control of nanotip photoemission. Nat. Phys. **10**, 432–436 (2014). https://doi.org/10.1038/nphys2974
28. L. Zhao et al., Terahertz oscilloscope for recording time information of ultrashort electron beams. Phys. Rev. Lett. **122**, 144801 (2019). https://doi.org/10.1103/PhysRevLett.122.144801
29. D. Ehberger, K.J. Mohler, T. Vasileiadis, R. Ernstorfer, L. Waldecker, P. Baum, Terahertz compression of electron pulses at a planar mirror membrane. Phys. Rev. Appl. **11**, 024034 (2019). https://doi.org/10.1103/PhysRevApplied.11.024034
30. C.M.S. Sears, E. Colby, R. Ischebeck, C. Mcguinness, J. Nelson, R. Noble, R.H. Siemann, J. Spencer, D. Walz, T. Plettner, R.L. Byer, Production and characterization of attosecond electron bunch trains. Phys. Rev. Spec. Top. - Accel. Beams**061301**, 1–6 (2008)https://doi.org/10.1103/PhysRevSTAB.11.061301

31. K.E. Priebe, C. Rathje, S.V. Yalunin, T. Hohage, A. Feist, S. Schäfer, C. Ropers, Attosecond electron pulse trains and quantum state reconstruction in ultrafast transmission electron microscopy. Nat. Photonics **11**, 793–797 (2017). https://doi.org/10.1038/s41566-017-0045-8
32. D.S. Black, U. Niedermayer, Y. Miao, Z. Zhao, O. Solgaard, R.L. Byer, K.J. Leedle, Net acceleration and direct measurement of attosecond electron pulses in a silicon dielectric laser accelerator. Phys. Rev. Lett. **123**, 264802 (2019). https://doi.org/10.1103/PhysRevLett.123.264802
33. M. Kozák, N. Schönenberger, P. Hommelhoff, Ponderomotive generation and detection of attosecond free-electron pulse trains. Phys. Rev. Lett. **120**, 103203 (2018). https://doi.org/10.1103/PhysRevLett.120.103203
34. N. Schönenberger, A. Mittelbach, P. Yousefi, J. McNeur, U. Niedermayer, P. Hommelhoff, Generation and characterization of attosecond microbunched electron pulse trains via dielectric laser acceleration. Phys. Rev. Lett. **123**, 264803 (2019). https://doi.org/10.1103/PhysRevLett.123.264803
35. P. Baum, Quantum dynamics of attosecond electron pulse compression. J. Appl. Phys. **122**, 223105 (2017). https://doi.org/10.1063/1.5006864
36. Y. Morimoto, P. Baum, Attosecond control of electron beams at dielectric and absorbing membranes. Phys. Rev. A **97**, 033815 (2018). https://doi.org/10.1103/PhysRevA.97.033815
37. D. Ehberger, A. Ryabov, P. Baum, Tilted electron pulses. Phys. Rev. Lett. **121**, 094801 (2018). https://doi.org/10.1103/PhysRevLett.121.094801
38. P. Baum, A.H. Zewail, Attosecond electron pulses for 4D diffraction and microscopy. Proc. Natl. Acad. Sci. **104**, 18409–18414 (2007). https://doi.org/10.1073/pnas.0709019104
39. D. Zhang, A. Fallahi, M. Hemmer, X. Wu, M. Fakhari, Y. Hua, H. Cankaya, A.-L. Calendron, L.E. Zapata, N.H. Matlis, F.X. Kärtner, Segmented terahertz electron accelerator and manipulator (STEAM). Nat. Photonics **12**, 336–342 (2018). https://doi.org/10.1038/s41566-018-0138-z
40. F.O. Kirchner, A. Gliserin, F. Krausz, P. Baum, Laser streaking of free electrons at 25 keV. Nat. Photonics 3–8 (2013). https://doi.org/10.1038/nphoton.2013.315
41. Y. Morimoto, I. Roland, S. Rennesson, F. Semond, P. Boucaud, P. Baum, Laser damage of free-standing nanometer membranes. J. Appl. Phys. **122**, 215303 (2017). https://doi.org/10.1063/1.5004081
42. L. Kasmi, D. Kreier, M. Bradler, E. Riedle, P. Baum, Femtosecond single-electron pulses generated by two-photon photoemission close to the work function. New J. Phys. **17**, 033008 (2015). https://doi.org/10.1088/1367-2630/17/3/033008
43. C. Weninger, P. Baum, Temporal distortions in magnetic lenses. Ultramicroscopy **113**, 145–151 (2012). https://doi.org/10.1016/j.ultramic.2011.11.018
44. D. Kreier, P. Baum, Avoiding temporal distortions in tilted pulses. Opt. Lett. **37**, 2373 (2012). https://doi.org/10.1364/OL.37.002373
45. P. Baum, A.H. Zewail, Breaking resolution limits in ultrafast electron diffraction and microscopy. Proc. Natl. Acad. Sci. **103**, 16105–16110 (2006). https://doi.org/10.1073/pnas.0607451103
46. H. Shao, A.F. Starace, Detecting electron motion in atoms and molecules. Phys. Rev. Lett. **105**, 263201 (2010). https://doi.org/10.1103/PhysRevLett.105.263201
47. V.S. Yakovlev, M.I. Stockman, F. Krausz, P. Baum, Atomic-scale diffractive imaging of subcycle electron dynamics in condensed matter. Sci. Rep. **5**, 14581 (2015). https://doi.org/10.1038/srep14581
48. A. Ryabov, P. Baum, Electron microscopy of electromagnetic waveforms. Science **353**, 374–377 (2016). https://doi.org/10.1126/science.aaf8589
49. A.V. Kildishev, A. Boltasseva, V.M. Shalaev, Planar photonics with metasurfaces. Science **339**, 1232009–1232009 (2013). https://doi.org/10.1126/science.1232009
50. M. Khorasaninejad, F. Capasso, Metalenses: Versatile multifunctional photonic components. Science **358**, eaam8100 (2017). https://doi.org/10.1126/science.aam8100
51. Y. Morimoto, P. Baum, Single-cycle optical control of beam electrons. Phys. Rev. Lett. **125**, 193202 (2020). https://doi.org/10.1103/PhysRevLett.125.193202

52. Y. Morimoto, Attosecond electron-beam technology: a review of recent progress. Microscopy **72**, 2–17 (2023). https://doi.org/10.1093/jmicro/dfac054
53. P. Baum, Vorrichtung zur Beobachtung mit Ladungsteilchen, Elektronenmikroskop sowie Verfahren zur zeitaufgelösten Beobachtung. Patent DE 10 2016 012 724 (2016)
54. A. Ryabov, J.W. Thurner, D. Nabben, M.V. Tsarev, P. Baum, Attosecond metrology in a continuous-beam transmission electron microscope. Sci. Adv. **6**, eabb1393 (2020). https://doi.org/10.1126/sciadv.abb1393
55. D. Nabben, J. Kuttruff, L. Stolz, A. Ryabov, P. Baum, Attosecond electron microscopy of sub-cycle optical dynamics. Nature **619**, 63–67 (2023). https://doi.org/10.1038/s41586-023-06074-9

# Chapter 6
# Towards Time-Resolved Molecular Orbital Imaging

**Masakazu Yamazaki, Tomoyuki Endo, Akiyoshi Hishikawa, and Masahiko Takahashi**

**Abstract** Understanding the mechanism of a chemical reaction is essential for controlling selectivity and yields of products or designing a novel molecular function, which is one of the ultimate goals of chemistry. Since a chemical reaction can be defined as nuclear dynamics driven by the change in electron motion, time-resolved molecular orbital imaging would open the door not only to gain a deeper insight into molecular dynamics but also to advance and extend frontiers of science and technology. In this chapter, two experimental techniques that aim to visualize the changing molecular orbital pattern during a chemical reaction are described in detail. One is the attempt to tackle the issue in momentum space, and the other is the attempt to do the same based on laser tunneling ionization. It is demonstrated that these two techniques are each applicable to short-lived excited states, thereby both offering opportunities for investigating the driving force behind chemical reaction.

**Keywords** Molecular orbital · Electron Compton scattering · Tunneling ionization

M. Yamazaki
Department of Chemistry, School of Science, Tokyo Institute of Technology, Tokyo, Japan

T. Endo
Kansai Institute for Photon Science, National Institutes for Quantum Science and Technology, Kyoto, Japan

A. Hishikawa
Department of Chemistry, Graduate School of Science, Nagoya University, Nagoya, Japan

Research Center for Materials Science, Nagoya University, Nagoya, Japan

M. Takahashi (✉)
Institute of Multidisciplinary Research for Advanced Materials, Tohoku University, Sendai, Japan
e-mail: masahiko@tohoku.ac.jp



K. Ueda (ed.), *Ultrafast Electronic and Structural Dynamics*,
https://doi.org/10.1007/978-981-97-2914-2_6

## 6.1 Introduction

When a molecule in its ground state absorbs a photon of light of the appropriate wavelength, the molecule is raised to an electronically excited state. Such molecular excited states are usually short-lived, as they may subsequently undergo, within a short period of time, deactivation reactions involving isomerization, dissociation, radiative, nonradiative, and energy-transfer processes. Furthermore, since molecules in excited states are more reactive than those in their ground state, a lot and variety of chemical reactions in which molecular excited states participate can be found in literature. For these reasons, the study of excited-state molecular dynamics is a topic of growing interest owing to the fundamental importance as well as potential application in many areas [1].

The excited-state molecular dynamics or unimolecular chemical reaction can be regarded as a successive series of rearrangements of atoms in molecules. Since the molecular structure is continuously changing during the chemical reaction that often involves the breaking and formation of chemical bonds, several approaches to observe such nuclear dynamics in real time have been proposed and demonstrated in the last decades. From the most fundamental point of view, however, the nuclear dynamics is driven by the change of electron motion, so real-time observation of electron dynamics is essential to reaching one of the goals of chemistry, that is, complete understanding, control and design of a chemical reaction. Furthermore, if the frontier orbital theory [2] is applied to the electron dynamics, spatial pattern of the highest occupied molecular orbital (HOMO) of the transient, evolving system would serve particularly as the key to understanding of the direction and underlying mechanism of the chemical reaction. In this respect, it should be noted, on one hand, that there have not yet been any experimental techniques that enable one to look at molecular orbitals changing rapidly during chemical reaction. For instance, although the widely-used time-resolved photoelectron spectroscopy provides accurate information about binding energy values of the molecular orbitals of a transient system [3, 4], spatial patterns of the orbitals are beyond the reach of this experimental technique. On the other hand, quantum chemistry theory can, in principle, predict the direction and underlying mechanism of a chemical reaction as well as the spatial patterns of the orbitals by numerically integrating the time-dependent Schrödinger equation. However, such full quantum dynamics calculations are not practically feasible for systems having more than three atoms, due not only to computational cost problem but also to the difficulty in constructing the high-dimensional and coupled global potential energy surfaces. In order to overcome these theoretical impediments, the trajectory surface hopping method [5, 6] has been developed, but there is at present a severe limitation in the number of degrees of freedom that it can handle.

Under the above-mentioned situations, one may desire to have an experimental technique that makes it possible to look at molecular orbitals changing rapidly during a chemical reaction. The key for this challenge is to develop time-resolved versions of the existing molecular orbital imaging techniques such as (*e*, 2*e*) electron momentum spectroscopy [7–13], laser tunneling imaging [14–24], and laser high-order harmonics generation [25–28]. Indeed, such attempts have already been undertaken along that line, and they are the subject of this chapter. Here, two kinds of attempts are described. One is the method to tackle the issue in momentum space and the other is the method to do the same based on laser tunneling ionization. Their experimental principle and current status are discussed in Sects. 6.2 and 6.3, respectively. Finally, this chapter is concluded in Sect. 6.4.

## 6.2 Molecular Orbital Imaging in Momentum Space

### *6.2.1 (e, 2e) Electron Momentum Spectroscopy*

Over the last five decades, electron momentum spectroscopy (EMS), also known as binary (e, 2e) spectroscopy, has been developed as a powerful means to provide unique and versatile information on the electronic structure and electron correlation effects in matter [7–13]. It is a kinematically complete high-energy electron-impact ionization experiment with large momentum transfer and large energy loss, in which both the inelastically scattered and ejected electrons are detected in coincidence. The (e, 2e) reaction that ionizes the target molecule M can be described by

$$\begin{array}{ccccccccc} \mathrm{M} & + & \mathrm{e}_0^- & \rightarrow & \mathrm{e}_1^- & + & \mathrm{e}_2^- & + & \mathrm{M}^+ \\ & & (E_0, \boldsymbol{p}_0) & & (E_1, \boldsymbol{p}_1) & & (E_2, \boldsymbol{p}_2) & & (E_{\mathrm{recoil}}, \boldsymbol{q}) \end{array} \tag{6.1}$$

Here the $E_j$'s and $\boldsymbol{p}_j$'s ($j = 0$, 1, and 2) are the kinetic energies and momenta of the incident, inelastically scattered, and ejected electrons, respectively. Similarly, $E_{\mathrm{recoil}}$ and $\boldsymbol{q}$ represent the recoil energy and recoil momentum of the residual ion $\mathrm{M}^+$. Since the thermal energy of M as well as $E_{\mathrm{recoil}}$ is negligibly small compared with any of $E_j$'s in the (e, 2e) reaction under consideration, one has the following conservation laws of energy and momentum,

$$E_{\mathrm{bind}} = E_0 - E_1 - E_2 \tag{6.2}$$

$$\boldsymbol{q} = \boldsymbol{p}_0 - \boldsymbol{p}_1 - \boldsymbol{p}_2. \tag{6.3}$$

Here, $E_{\mathrm{bind}}$ is the electron binding energy (ionization energy). Furthermore, since the experiment is performed under the high-energy Bethe ridge conditions [11, 13], where the collision kinematics most nearly corresponds to a collision of two free electrons with the residual ion $\mathrm{M}^+$ acting as a spectator, the momentum of the target electron $\boldsymbol{p}$, before ionization, is equal in magnitude but opposite in sign to $\boldsymbol{q}$,

$$\boldsymbol{p} = -\boldsymbol{q} = \boldsymbol{p}_1 + \boldsymbol{p}_2 - \boldsymbol{p}_0. \tag{6.4}$$

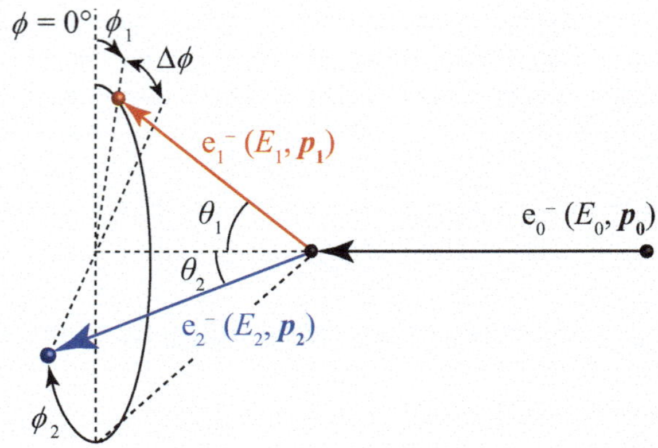


**Fig. 6.1** Symmetric noncoplanar geometry for the study of the EMS reaction

The EMS signal can thus be measured as a function of $E_{\text{bind}}$ and $\boldsymbol{p}$. Figure 6.1 shows the widely used symmetric noncoplanar geometry, in which two outgoing electrons having equal energies ($E_1 = E_2$) and making equal scattering polar angles ($\theta_1 = \theta_2 = 45°$) with respect to the incident electron beam axis are detected in coincidence. In this kinematic scheme, the magnitude of the target electron momentum $\boldsymbol{p}$ is given by

$$|\boldsymbol{p}| = \sqrt{(p_0 - \sqrt{2}p_1)^2 + (\sqrt{2}p_1 \sin(\Delta\phi/2))^2} \tag{6.5}$$

with $\Delta\phi$ being the out-of-plane azimuthal angle difference between the two outgoing electrons detected.

The most widely used scattering model for EMS is the plane-wave impulse approximation (PWIA) with the weak-coupling approximation [11, 12] that describes EMS cross-sections as

$$\frac{d^3\sigma_{\text{PWIA}}}{dE_1 d\Omega_1 d\Omega_2} \propto S_\alpha \frac{1}{4\pi} \int d\Omega |\psi_\alpha(\boldsymbol{p})|^2. \tag{6.6}$$

Here, $\psi_\alpha(\boldsymbol{p})$ is the momentum-space representation of the quasiparticle or Dyson orbital which can be defined as

$$\sqrt{S_\alpha}\psi_\alpha(\boldsymbol{p}) = \left\langle \boldsymbol{p}\Psi_f^{N-1} \,\middle|\, \Psi_i^N \right\rangle \tag{6.7}$$

Here, $\Psi_i^N$ and $\Psi_f^{N-1}$ are the $N$-electron initial neutral and the $(N-1)$-electron final ion wave functions of the target, and $S_\alpha$ is a quantity called pole strength, also known as the spectroscopic factor. $(1/4\pi) \int d\Omega$ is the spherical averaging due to random orientation of gaseous molecular targets.

In this way, EMS has a unique ability to measure momentum densities of each target electron with different binding energies or to look at individual molecular orbitals in momentum space. Since the position- and momentum-space molecular orbital functions are uniquely related by the Dirac-Fourier transform, an expansion of a molecular orbital in position space corresponds to a contraction of it in momentum space. This is the material reason why EMS has proven sensitive to the behavior of the outer, loosely bound electrons that are of central importance in chemical properties such as bonding, reactivity, and molecular recognition.

### 6.2.2 *Time-Resolved Electron Momentum Spectroscopy*

EMS has a history since 1970s. Nevertheless, EMS experiments had long been limited to studies on stable targets in their electronic ground state, except the early pioneering experiment of Zheng et al. on excited sodium atoms prepared using a cw ring dye laser [29]. This is due partly to the difficulty of performing EMS experiments that require the coincident detection of the two outgoing electrons produced by high-energy electron-impact ionization at large momentum transfer, where the (e, 2e) cross-section is generally quite small compared with those at small momentum transfers. In addition, an ultrashort-pulsed incident electron beam, which is the primary requirement of conducting EMS experiments on short-lived transient species, may suffer from the space charge effect; the more intensity an ultrashort-pulsed incident electron beam has, the more space charge effect may significantly broaden not only the temporal width but also the energy spread of the electron packet. In this respect, however, highly encouraging was continuing efforts for progress of the EMS experimental techniques [7–13], one of which has eventually improved the instrumental sensitivity as much as possible [30, 31] by covering almost completely the available azimuthal angle range for the symmetric noncoplanar EMS reaction (Fig. 6.1). Under these circumstances, time-resolved EMS (TR-EMS) has been proposed in 2008 [13], and the first-generation apparatuses [32–34] have been developed recently.

Figure 6.2 shows a schematic picture of the experimental setup employed in the first TR-EMS apparatus [32]. The 800 nm output of a 5-kHz femtosecond laser (<120 fs, 0.8 mJ) is split into a pump path and an electron-generation path. 90% of the output is used to yield with an optical parametric amplifier the pump laser pulse (195 nm, ~120 fs, 0.8 μJ), which is subsequently used to excite target molecules in an effusive gas beam after the 5-kHz repetition rate being halved by an optical chopper. On the other hand, 10% of the 800 nm output is frequency tripled (267 nm, <10 μJ) and directed toward a back-illuminated photocathode in order to produce electron pulses via the photoelectric effect. The photocathode is made of a silver film of 40 nm thickness, which is negatively biased to accelerate the electron pulses up to 1.2 keV. The resulting ultrashort pulsed electron beam (~1 ps, ~50 pA) is then used to induce EMS events which are recorded by an EMS spectrometer with an exceptionally large spherical analyzer (mean radius of 220 mm). The energy- and momentum-dispersive multichannel measurement is achieved based on the well-known property of a spherical analyzer of maintaining the azimuthal angles of the analyzed electrons while dispersing them according to their kinetic energies. The delay time $t_{\mathrm{delay}}$ between the arrival of the pump laser pulse and the probe electron

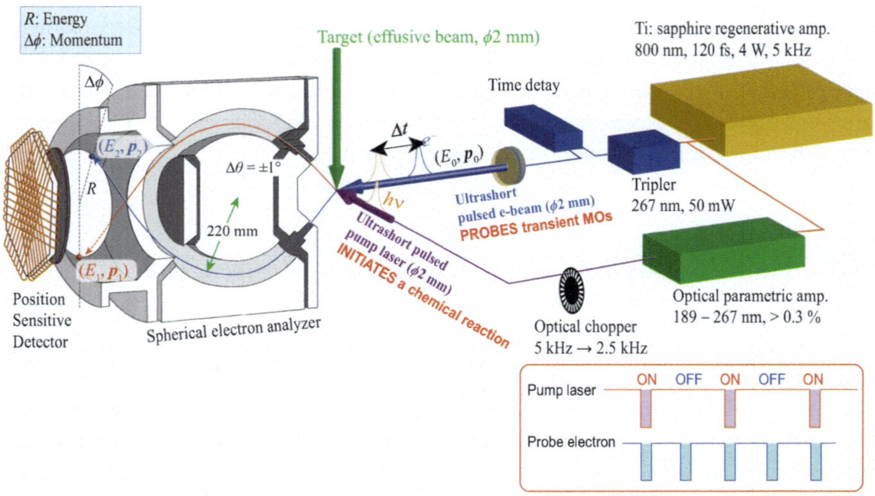


**Fig. 6.2** A schematic representation of time-resolved electron momentum spectroscopy apparatus [32]

pulse is controlled with a computer-driven translation stage. Here, in order to have a workable signal count rate, diameters of the target gas beam, the pump laser, and the probe electron beam are all set to be about 2 mm, at the expense of the experimental resolutions for the energy (5 eV FWHM), momentum (0.4 a.u. at $\Delta\phi = 0°$), and time ($\pm$35 ps). Note that the time resolution is almost entirely dominated by the group velocity mismatch between the pump laser and probe electron pulses [35]. Note also that since the 5 kHz repetition rate is halved only for the pump laser, the TR-EMS apparatus concurrently produces two kinds of EMS data sets. One is data that are measured with the pump laser (laser-on), the other is reference data that are measured without the pump laser (laser-off). The TR-EMS data are then obtained by subtracting the laser-off spectrum with an appropriate weight factor from the laser-on spectrum.

### 6.2.3 *TR-EMS Study on the* $S_2$ *Acetone*

The first application of TR-EMS to a molecular excited state has been made for the deuterated acetone molecule (acetone-*d*6), $(CD_3)_2CO$, in its second excited singlet $S_2(n, 3s)$ state [36] with a lifetime of 13.5 ps [37]. Figure 6.3a compares the laser-off binding energy spectrum with an associated theoretical spectrum for the acetone-*d*6 ground state. It can be seen that although the employed instrumental energy resolution of 5 eV FWHM does not allow the spectral peaks to be identified clearly, the laser-off spectrum is on the whole well reproduced by the associated theoretical spectrum over the entire binding energy range covered.

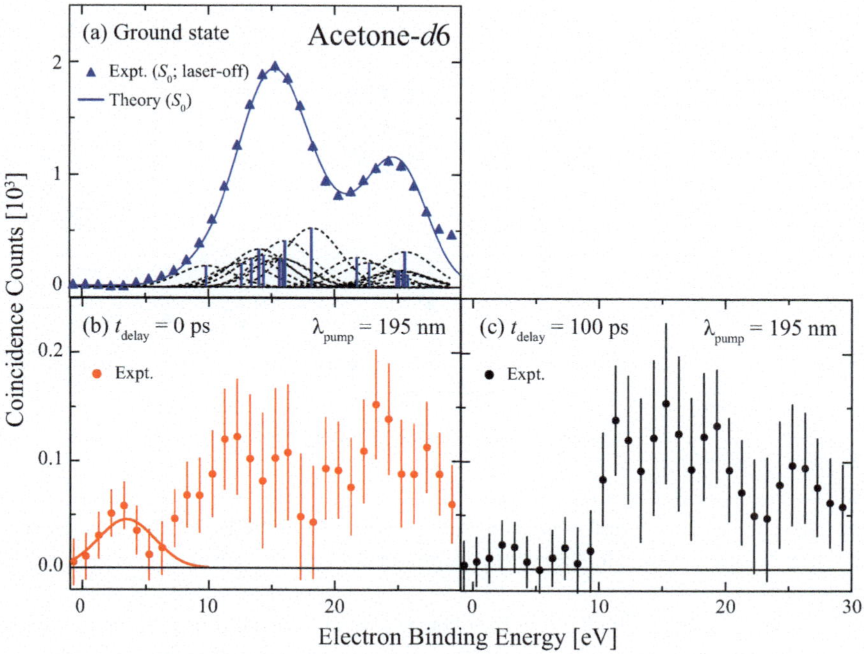


**Fig. 6.3** Experimental **a** laser-off EMS binding energy spectrum and time-resolved EMS binding energy spectra of the acetone-$d6$ $S_2$ state obtained at $t_{\text{delay}}$ of **b** 0 ps and **c** 100 ps. The broken and solid lines for **a** represent associated theoretical calculations. The solid line for **b** represents a band due to ionization from the outermost 3$s$ Rydberg orbital. Adapted with permission from Ref. [36]. Copyright 2015 by American Physical Society

Figure 6.3b shows a TR-EMS binding energy spectrum of the acetone-$d6$ $S_2$ state at 195 nm, measured at $t_{\text{delay}} = 0$ ps. It is evident from Fig. 6.3b that a band appears at around $E_{\text{bind}} = 3.5$ eV which is undoubtedly assigned, from the energy conservation, to ionization from the HOMO (outermost 3$s$ Rydberg orbital) of the acetone-$d6$ $S_2$ excited state. Furthermore, the 3.5 eV band disappears when $t_{\text{delay}}$ is changed from 0 to 100 ps, as can be seen from Fig. 6.3c. This is indeed in accordance with the fact that the acetone-$d6$ $S_2$ state has a lifetime of 13.5 ps, and it undergoes the three-body ultrafast dissociation process to eventually produce $2CD_3$ and CO [37], indicating that the $t_{\text{delay}} = 100$ ps spectrum is largely governed by the reaction products.

Further evidence for the successful measurement is given by examining the electron momentum distribution for the 3.5 eV band. Figure 6.4 shows spherically averaged electron-momentum-density distribution for the 3 s Rydberg orbital, which was constructed by plotting the number of true coincidence events that formed the 3.5 eV band as a function of the electron momentum. Also included in the figure are associated theoretical distributions calculated for an empty 3 s Kohn–Sham orbital (DFT_ 3s Ryd.) of the acetone-$d6$ $S_0$ state or a 3$s$ orbital of the $S_2$ state optimized with the

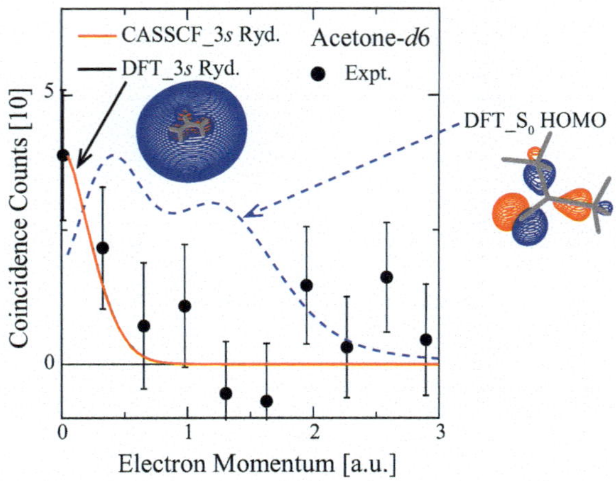


**Fig. 6.4** Comparison of spherically averaged electron momentum-density distributions between experiment and theory for the 3*s* Rydberg orbital of the acetone-*d*6 $S_2$ state. Adapted with permission from Ref. [36]. Copyright 2015 by American Physical Society

complete-active-space self-consistent-field method (CASSCF_3s Ryd.). A theoretical distribution for the HOMO of the $S_0$ state (DFT_$S_0$ HOMO) is also included as a reference.

It can be seen from Fig. 6.4 that the experimental result exhibits a maximum at the momentum origin and its intensity drops off rapidly with the increase in the electron momentum. This behavior can be understood by considering in momentum space the nature of the 3*s* Rydberg orbital. Since the angular part of a wavefunction is invariant under the Dirac-Fourier transform, a certain molecular orbital in position space has similar shape in momentum space. On the other hand, the radial part of a wave function is largely affected by the Dirac-Fourier transformation; high density at large *r* leads to high density at small *p* and vice versa. Thus, the diffuse 3 s Rydberg orbital in position space possesses the sharply peaked momentum distribution. The associated theoretical calculations support this observation. This achievement can thus be recognized as having demonstrated that EMS measurements of short-lived molecular excited states are certainly feasible [38, 39].

### *6.2.4 TR-EMS Study on the $S_1$ Toluene*

The first molecular orbital imaging experiment for a molecular excited state discussed in Sect. 6.2.3 [36] is not enough to show the potential capability of TR-EMS; its observation was limited only to the energetically well-separated HOMO of the acetone $S_2$ state. TR-EMS should be able to observe spatial distributions, in momentum space, of not only the HOMO but also all other, more tightly bound orbitals of a molecular excited state. The limitation in the first TR-EMS experiment [36] is due to the fact that the acetone $S_2$ state decays faster than the employed time resolution ($\pm$35 ps). Thus, the experiment on the higher-binding-energy region shown in Fig. 6.3b had to be affected by its subsequent decay process. The potential capability of TR-EMS to observe individual orbitals of a molecular excited state has then been demonstrated by changing its target to the toluene molecule in the $S_1(\pi, \pi^*)$ state at 267 nm [40]. Since the toluene $S_1$ state has a lifetime of 86 ns [41], much longer than the time resolution of ±35 ps, a TR-EMS experiment without any contributions of the subsequent intramolecular relaxation processes can be made while the whole valence electronic structure must be observed in the binding energy spectrum.

Figure 6.5a, b shows TR-EMS binding energy spectra obtained for the toluene $S_0$ (ground) and $S_1$ (excited) states, respectively. The former is the laser-off spectrum, while the latter was obtained at $t_{\mathrm{delay}} = 0$ ps. Also included in the figures are associated theoretical spectra calculated by using the symmetry adapted cluster configuration-interaction (SAC-CI) method [42]. It can be seen from Fig. 6.5a that the employed instrumental energy resolution of 5 eV FWHM only allows the spectral peaks to be identified as two broad bands centered at around 14 and 24 eV, which are a group of peaks due to the outer-valence and inner-valence ionization, respectively. It is also seen that the experiment is on the whole well reproduced by the theoretical spectrum over the entire binding energy range that the SAC-CI calculations covered.

As for the $S_1$ spectrum in Fig. 6.5b, there are two additional features. One is the appearance of a very weak band at a lower binding-energy region centered at 4–5 eV which is assigned to the ionization from the excited $\pi^*$ orbitals of the toluene $S_1$ state. The weak band intensity originates mainly in the electron occupation number. Indeed, the SAC-CI wave function of the toluene $S_1$ state can be approximated as a linear combination of two electronic configurations, $0.70 \times [(3\pi)^{-1}(1\pi^*)^1] + 0.63$

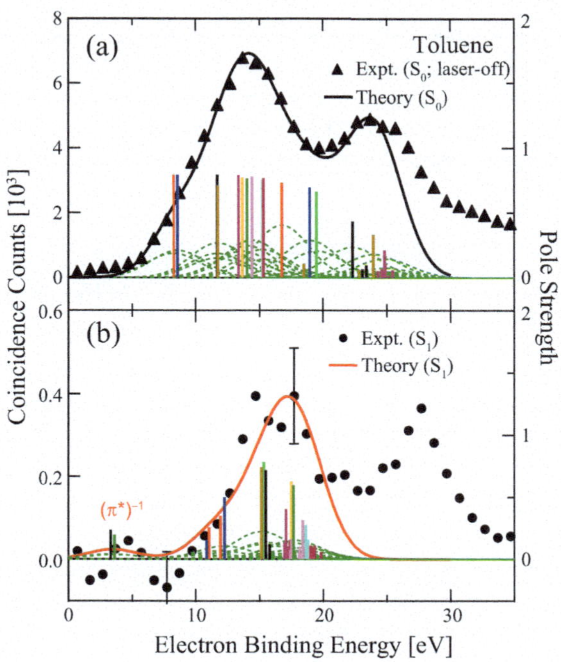


**Fig. 6.5** Comparisons of binding energy spectra between the experiments and SACCI calculations for the **a** $S_0$ and **b** $S_1$ states of toluene. The dashed lines show the contribution of each transition and the solid line is their sum. Vertical bars represent pole strength (>0.05). Reprinted with permission from Ref. [40]. Copyright 2016 by American Physical Society

$\times$ $[(2\pi)^{-1}(2\pi^*)^1]$ in which the two $1\pi^*$ and $2\pi^*$ excited orbitals are almost equally occupied by one electron. Thus, the energetically close-lying transitions from the $S_1$ state to the ground ($D_0[(3\pi)^{-1}]$) and first excited ($D_1[(2\pi)^{-1}]$) states of the toluene cation appear with small pole strength values.

Another feature of the $S_1$ spectrum in Fig. 6.5b is a shift of both the outer- and inner-valence bands towards higher energy by about 3 eV compared to those at ~14 and ~24 eV in the $S_0$ spectrum. This observation can be understood by extending the ionization propensity rule of EMS to molecular excited states. Namely, only one-electron processes are allowed, but two-electron and other multi-electron processes are all forbidden. For instance, as shown in Fig. 6.6, the ionization to the $D_2[(15\sigma)^{-1}]$ state is forbidden from the $S_1$ state, as it is a two-electron process, whilst it is allowed from the $S_0$ state. On the contrary, transitions from the $S_1$ state to higher-energy ionic states of two-hole-one-particle ($2h1p$) configurations such as $[(15\sigma)^{-1}(3\pi)^{-1}(1\pi^*)^1]$ and $[(15\sigma)^{-1}(2\pi)^{-1}(2\pi^*)^1]$ are allowed, although those from the $S_0$ state are forbidden. The same argument can be made for all other, more tightly bound molecular orbitals. It was found from the SAC-CI calculations that most of the allowed $2h1p$ configurations have the $1\pi^*$ or $2\pi^*$ orbital occupied by an electron. Thus, the outer- and inner-valence ionization bands of the $S_1$ state are expected to shift towards higher energy approximately by the difference in the $\pi$–$\pi^*$ excitation energy between the $S_0$ and $D_0$ states, i.e., $E_{\pi\pi*}^{D_0} - E_{\pi\pi*}^{S_0}$, respectively. The calculated value for $E_{\pi\pi*}^{D_0} - E_{\pi\pi*}^{S_0}$ is 2.9 eV, and this is in good agreement with the experimental value of ~3 eV.

Note that the critical role of $2h1p$ configurations is an indication of the inherent capability of TR-EMS of being able to observe spatial distributions, in momentum space, of not only the HOMO but also all other, more tightly bound orbitals of a molecular excited state. This is because in primary ionization to such a $2h1p$ configuration the Dyson orbital is always of either the fully occupied orbitals or the singly occupied orbital of lower energy, which could be separately observed if the energy resolution were improved to the desired extent.

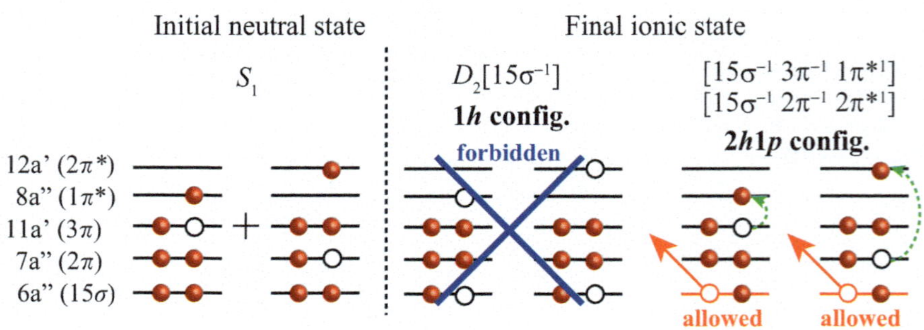


**Fig. 6.6** EMS ionization schemes for transitions from the toluene $S_1$ state to several ionic states within the frozen orbital approximation, showing their occupation with electrons (closed circles) or holes (open circles). Adapted with permission from Ref. [40]. Copyright 2016 by American Physical Society

## 6.3 Molecular Orbital Imaging by Ultrafast Laser Tunneling

Thanks to recent developments in laser technology, it is now possible to generate ultrashort intense laser fields from a table-top laser system. Due to the large electric fields, molecules in intense laser fields show a variety of characteristic features that are hardly observable in a weak field regime. In this section, we describe an approach to visualize electron distributions in molecules by utilizing one of such processes, laser tunneling ionization. The application to electronically ground and excited molecules is presented with illustrative examples on a small molecule, nitric oxide (NO) [23, 24].

### *6.3.1 Laser Tunneling Ionization*

When molecules are exposed to intense laser fields (typically ~$10^{14}$ W/cm$^2$), the electron binding potential is deformed due to the large electric fields as shown in Fig. 6.7a. The electron tunnels through a potential barrier thus formed into continuum states. This ionization process is called "laser tunneling ionization". The tunneling electron comes mainly from the outermost molecular orbital, typically HOMO, which has a thinner potential barrier than lower lying molecular orbitals such as HOMO-1 and HOMO-2. The rate of laser tunneling ionization depends on the shape of molecular orbital along the laser polarization direction. When the lobes of a molecular orbital face the laser polarization direction, the tunneling ionization is expected to be enhanced because of the large electron densities along the laser polarization direction. Since HOMO of NO has the $\pi$ symmetry, the tunneling ionization is expected to become prominent when the molecular axis is oriented at 45° with respect to the laser polarization direction as shown in Fig. 6.7b. On the other hand, when the laser electric fields are applied along the nodes, the tunneling ionization should be suppressed due to small electron densities. This corresponds to the parallel orientation of the NO molecular axis (Fig. 6.7c). Thus, the angular dependence of laser tunneling ionization reflects the electron distribution in the molecular frame, or the shape of the molecular orbital. This feature is examined by various theoretical approaches [43–47], and the roles of various factors in tunneling ionization such as the permanent dipole moment, distortion of molecular orbital, and contributions from the lower lying orbitals are discussed.

There are several experimental approaches to obtain the angular dependence of ionization probability. For aligned or oriented molecules, it can be obtained as a function of the polarization direction of laser fields in the laboratory frame [14–16]. Alternatively, electron–ion coincidence momentum measurements can be used for randomly oriented molecules, to record molecular frame photoelectron angular distribution in circularly polarized laser fields [17–19]. Laser high-order harmonics

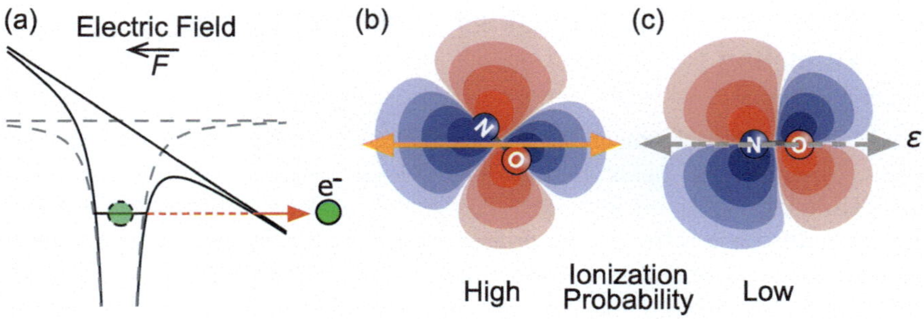


**Fig. 6.7** **a** Schematic of laser tunneling ionization. Field-free electron binding potential (grey broken line) is deformed by applying external electric field $F$. Electron tunnels through the potential barrier (black solid line). **b, c** Orientation dependence of tunneling ionization probability of NO with respect to the laser polarization direction $\varepsilon$. Tunneling ionization rate of **b** 45° oriented configuration is higher than that of **c** parallel configuration due to $\pi$ symmetry of the HOMO of NO

generation by recombination of tunneling electron to the parent ion is another interesting approach capable of determining phase of molecular orbitals [25–28]. Here, we employ ion momentum imaging of fragment ions produced by dissociative ionization or Coulomb explosion [20–24]. As illustrated in Fig. 6.7b, c, tunneling ionization of a molecular orbital determines the alignment or orientation of the resultant molecular ions. Thus, it is possible to capture the orbital shape from the fragment angular distribution under the axial recoil approximation. The simple experimental setup allowed us to explore the application of laser tunneling ionization imaging to molecular photoexcitation.

### *6.3.2 Tunneling Ionization Imaging of NO 2π orbital*

First, we discuss tunneling ionization imaging on NO in the $X^2\Pi$ ground state, where the electric configuration is $(1\sigma)^2\ (2\sigma)^2\ (3\sigma)^2\ (4\sigma)^2\ (5\sigma)^2\ (1\pi)^4\ (2\pi)^1$. The experimental setup is shown in Fig. 6.8. Briefly, a linearly polarized intense laser pulse is focused onto a molecular beam of NO introduced into an ultrahigh vacuum chamber (residual gas pressure $\leq 10^{-7}$ Pa). Ions generated by dissociative ionization of NO, $NO \rightarrow NO^+ + e^- \rightarrow N^+ + O + e^-$, are detected by a position-sensitive detector (PSD) [48]. The three-dimensional momentum vector $\mathbf{p} = (p_x, p_y, p_z)$ of each ion is calculated using the arrival time ($t$) and position ($x$, $y$) at the detector.

The $N^+$ momentum image in one-color intense laser fields (45 fs, 800 nm, $1 \times 10^{14}$ W/cm$^2$) is shown in Fig. 6.9a. The $N^+$ momentum image shows a clear butterfly-like pattern peaked at 40 and 140° with respect to the laser polarization direction, as expected from the shape of the $2\pi$ HOMO (see Fig. 6.7). The molecular dissociation subsequent to the tunneling ionization ejects the fragment ion with an offset angle against the laser polarization direction.

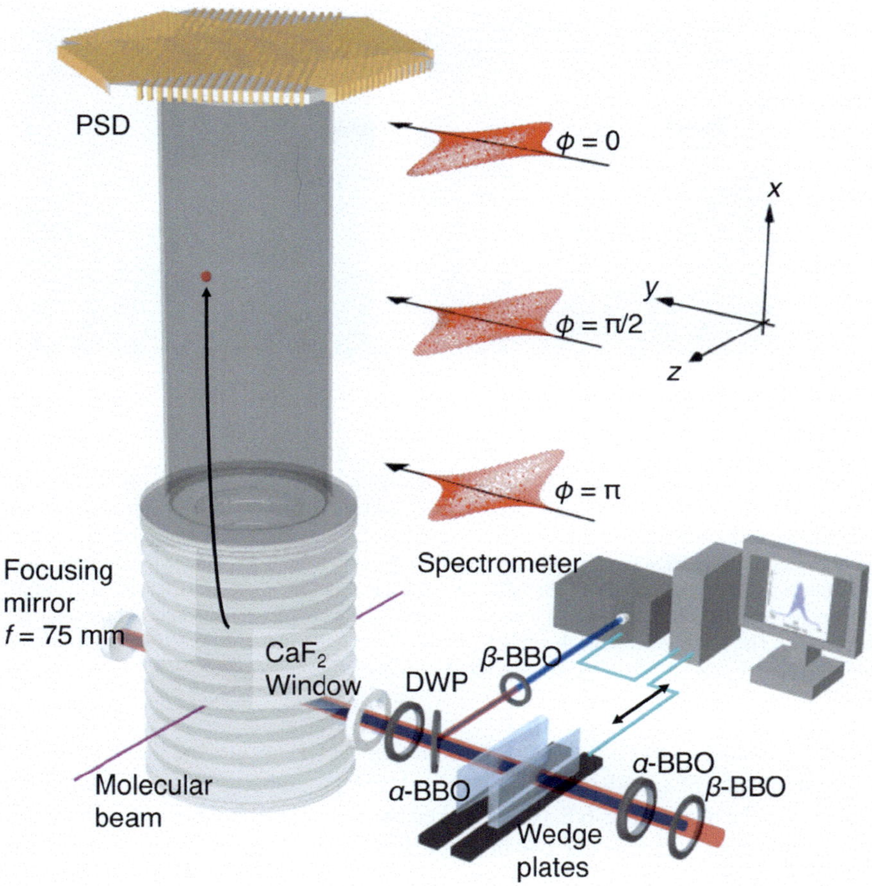


**Fig. 6.8** Schematic of the experimental setup with phase-locked two-color laser pulses. The fundamental femtosecond laser pulse (45 fs, 800 nm, 1 kHz) and the second harmonics generated by a $\beta$-barium borate (BBO) crystal are co-linearly introduced to the three-dimensional ion momentum imaging spectrometer. The time delay between $\omega$ and $2\omega$ pulses is controlled by birefringent $\alpha$-BBO crystal and a pair of fused silica wedges. The relative phase is locked by a feedback loop utilizing the $2\omega$–$2\omega$ interference spectrum [50, 51]. The fragment ions produced from the interaction region are guided by a uniform electric field (61 V/cm) to reach a PSD through a field-free region. Temporal profiles of the two-color laser electric fields for the relative phases of 0, $\pi/2$, and $\pi$ are shown in the subset. DWP: dual-wavelength wave plate; PSD: position-sensitive detector. Adapted with permission from Ref. [24]. Copyright 2019 by American Physical Society

Shown in Fig. 6.9b is the $N^+$ image obtained with asymmetric two-color laser fields [24], generated by a coherent superposition of the fundamental and second harmonic laser pulses [49–52]. The two-color laser electric field may be expressed as

$$F(t) = F_\omega(t)\cos(\omega t) + F_{2\omega}(t)\cos(2\omega t + \phi), \tag{6.8}$$

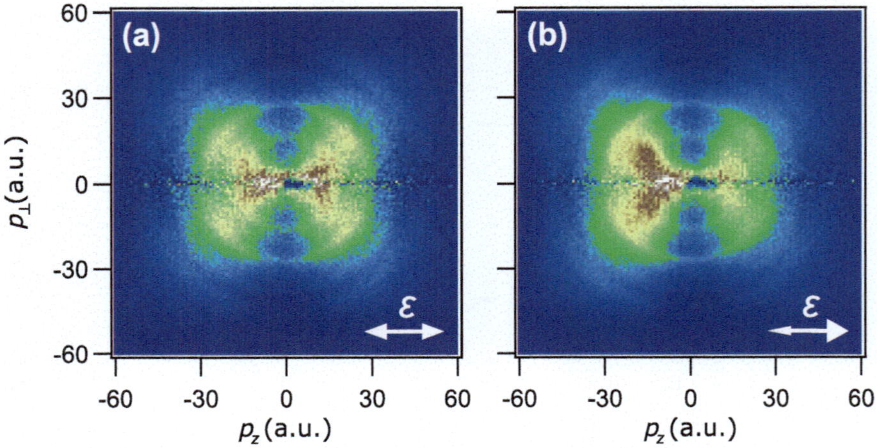


**Fig. 6.9** Two-dimensional momentum images of the $N^+$ fragment ions produced in **a** one-color intense laser fields (45 fs, 800 nm, $1 \times 10^{14}$ W/cm$^2$) and **b** phase-locked two-color intense laser fields (45 fs, 800 nm + 400 nm, $1 \times 10^{14}$ W/cm$^2$, $I_{2\omega}/I_\omega = 0.04$, $\phi = 0.1\pi$). The momentum map represents a thin slice (with a slice width of ±0.5 a.u.) of the three-dimensional momentum distribution in the plane containing the laser polarization direction denoted as $\varepsilon$. The momentum components parallel and perpendicular to the laser polarization direction are denoted with $p_z$ and $p_\perp$, respectively. The arrow represents the direction of the larger amplitude. Adapted with permission from Ref. [24]. Copyright 2019 by American Physical Society

where $F_\omega(t)$ and $F_{2\omega}(t)$ are the envelope function of laser electric fields with the carrier frequency of $\omega$ and $2\omega$, and $\phi$ is the relative phase between two laser pulses. As shown in the inset of Fig. 6.8, phase-locked two-color laser fields have asymmetric electric field amplitudes, which can be controlled by changing the relative phase and the ratio between the intensity of each field, $I_{2\omega}/I_\omega$.

The $N^+$ momentum image obtained with phase-locked two-color intense laser fields (45 fs, 800 nm + 400 nm, $1 \times 10^{14}$ W/cm$^2$, $I_{2\omega}/I_\omega = 0.04$, $\phi = 0.1\pi$ shows larger yields on the side of smaller electric field amplitude, while preserving the butterfly-like pattern as observed with the fundamental fields alone. The preferential ejection of $N^+$ fragments to the smaller amplitude side suggests that the tunneling ionization occurs more efficiently from the N atom side [56], whose electron density is larger than that on the O atom side as shown in Fig. 6.7.

The obtained ion images show concentric patterns associated with different dissociation pathways (see Fig. 6.10). Figure 6.11 shows fragment angular distributions for the $B^1\Pi$ component with the highest momentum (|**p**| ~ 50 a.u.) [57]. The fragment distribution strongly depends on the shape of laser electric fields, and the fragment asymmetry is maximum at $\phi$ ~ 0 or $\pi$ when the laser field amplitude has the largest asymmetry.

To understand the origin of the phase dependence, theoretical calculations on tunneling ionization yields are carried out. According to the many electron weak-field asymptotic theory (ME-WFAT) [58], the tunneling ionization rate in a static

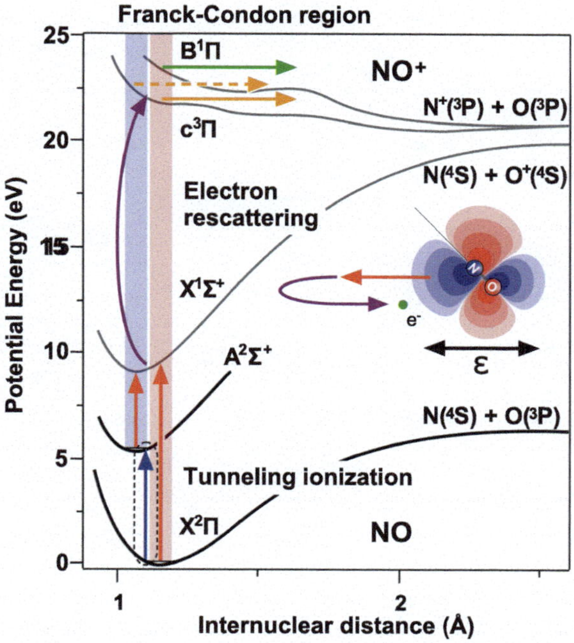


**Fig. 6.10** Potential energy curves of the selected electronic states of NO and $NO^+$ [53–55]. Arrows schematically represent the dissociation pathways. Adapted with permission from Ref. [23]. Copyright 2016 by American Physical Society

electric field can be expressed as

$$G(\beta, F) = \left\{ |G_{00}(\beta)|^2 + \frac{F}{2\varkappa^2}\left[ \left| G_{01}^{(+)}(\beta) \right|^2 + \left| G_{01}^{(-)}(\beta) \right|^2 \right] \right\} W_{00}(F), \tag{6.9}$$

where $\beta$ is the angle between the molecular axis and the electric field, $F$ is the amplitude of the electric field, $\varkappa = (2I_{\text{p}})^{1/2}$ with $I_{\text{p}}$ being the ionization potential. The structure factors $G_{00}$ and $G_{01}$ consist of factors determined by the molecular orbital of interest and the permanent dipole difference between neutral and ionic states. The field factor $W_{00}$ is written as

$$W_{00}(F) = \frac{\varkappa}{2}\left( \frac{4\varkappa^2}{F} \right)^{2/-1} \exp\left( -\frac{2\varkappa^3}{3F} \right). \tag{6.10}$$

Here, the structure factors are obtained from the HOMO within the Hartree–Fock approximation using the X2DHF code in place of the Dyson orbital, which is good approximation for small diatomic molecules. The tunneling ionization yields are obtained from the calculated rates by time integration over the laser pulse. At $\phi = 0.1\pi$, the theoretical result exhibits a clear asymmetric butterfly-like structure with

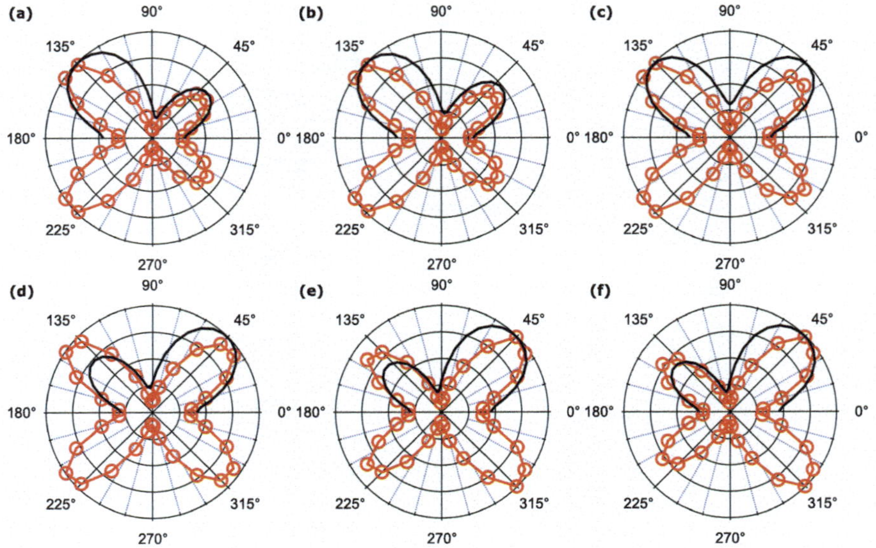


**Fig. 6.11** The angular distributions of the $B^1\Pi$ dissociation components for different relative phases $\phi$, **a** $0.1\pi$, **b** $0.3\pi$, **c** $0.5\pi$, **d** $0.7\pi$, **e** $0.9\pi$ and **f** $1.1\pi$. The laser polarization direction is along the horizontal axis. Experimental results show clear dependence on the relative phase (red circles). Theoretical results obtained by ME-WFAT are shown in the first and second quadrants (solid line) for comparison. Adapted with permission from Ref. [24]. Copyright 2019 by American Physical Society

the larger peak at ~135° and the smaller peak at ~45°, as shown in Fig. 6.11a, in agreement with the experimental results.

The theoretical calculation reproduces the experimental phase dependence of the angular distribution, which shows a gradual variation of the asymmetry along the laser polarization direction, with the peak angles remaining essentially the same as shown in Fig. 6.11. The obtained results show that the fragment angular distribution is essentially governed by the shape of the outermost molecular orbital. It is worth noting that a counter intuitive angular dependence has been observed for OCS [15], where the large permanent dipole moment modifies effective ionization potentials depending on the molecular orientation. In the present case of NO, the experimental angular distribution has slightly narrower widths with peak angles slightly shifted towards the polarization axis compared with the theoretical results. These deviations could be attributed to the multi-orbital effects [59, 60] discussed to explain the angular distributions of the tunneling ionization of CO. In addition, the electron recollisional excitation involved in the dissociation process could also contribute. Since the molecular potential of $NO^+$ is anisotropic, the cross-section of electron impact excitation, $NO^+ + e^- \rightarrow N^+ + O + e^-$, is expected to vary with the orientation of the molecular ion, which results in the modification of the angular distribution of the fragment ions.

### 6.3.3 Tunneling Ionization Imaging of Molecular Photoexcitation

Presented here is tunneling ionization imaging in the first electronically excited state ($A^2\Sigma^+$) of NO, excited by a deep-UV (DUV) pulse of a wavelength of 226 nm. The electric configuration of the excited state is $\ldots(5\sigma)^2\ (1\pi)^4\ (2\pi)^0\ (3s\sigma)^1$, where an electron in the $2\pi$ orbital is promoted to the $3s\sigma$ orbital upon the photoexcitation (Fig. 6.12a). Here, 8-fs few-cycle intense laser pulses are used as probe pulses. The pump pulses are obtained by an optical parametric amplifier and its central wavelength is tuned to the A–X (0, 0) transition of NO. Since the amount of the $A^2\Sigma^+$ state is estimated to be only 0.5% of the $X^2\Pi$ state, an optical chopper is introduced to block the pump pulse in every other shot. The net signal from the excited state is obtained by subtraction of the pump alone and probe alone signals from the pump-probe signal.

The ion momentum images of $N^+$ fragments generated from the ground and excited states in the few-cycle intense laser fields (8 fs, $1.1 \times 10^{14}$ W/cm$^2$) are

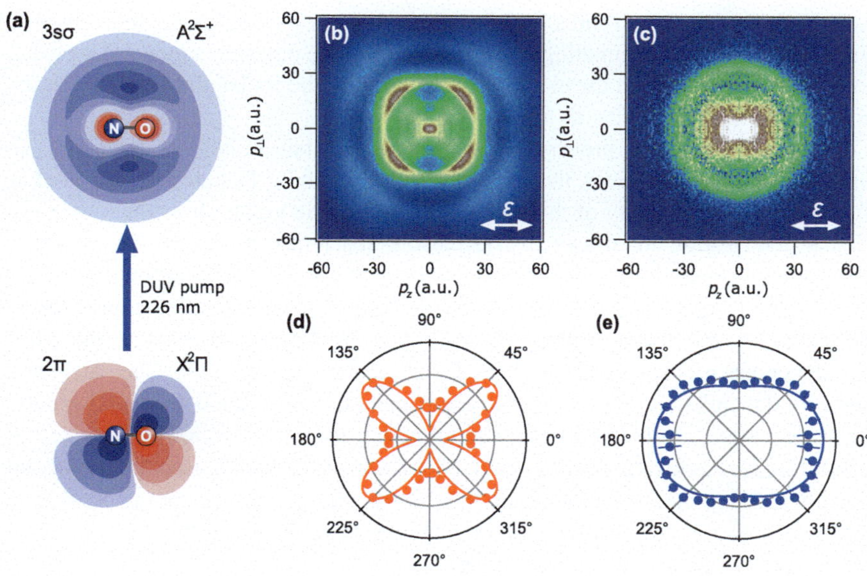


**Fig. 6.12** **a** The highest occupied molecular orbitals, $2\pi$ and $3s\sigma$, in the $X^2\Pi$ and $A^2\Sigma^+$ states. Momentum images of the $N^+$ fragment ions produced by dissociative ionization starting from **b** the $X^2\Pi$ state and **c** the $A^2\Sigma^+$ state in few-cycle intense laser fields (8 fs, $1.1 \times 10^{14}$ W/cm$^2$). Symmetries with respect to the $p_z$ and $p_\perp$ axes are utilized to reduce the statistical uncertainty. The probe NIR laser polarization direction is denoted with $\varepsilon$. Polar plots of the fragment angular distributions obtained for **d** the $X^2\Pi$ and **e** $A^2\Sigma^+$ initial states. The distributions are evaluated on the $c^3\Pi$ dissociation components. Solid lines are theoretical tunneling ionization yields calculated by the ME-WFAT under the adiabatic approximation. Adapted with permission from Ref. [23]. Copyright 2016 by American Physical Society

shown in Fig. 6.12b, c. To improve the data statistics, the ion momentum images are symmetrized along the $p_z$ and $p_\perp$ axes. For the ground state, clear peaks are observed at 45° with respect to the laser polarization direction as discussed in the previous section. Two distinct components appear at $|\mathbf{p}| = 25$ and 50 a.u., which are corresponding to the Franck–Condon dissociation via $c^3\Pi$ and $B^1\Pi$ states (see Fig. 6.10). Compared with the ion momentum images obtained with the 45 fs laser fields (Fig. 6.9), the image shows simple and sharp distributions due to suppression of post-ionization interaction with laser fields [23, 24].

Results for the excited state are shown in Fig. 6.12b, exhibiting a broad circular distribution around $|\mathbf{p}| = 32$ a.u. The increase of momentum is attributed to the shift of the Franck–Condon window to the steeper part of the $c^3\Pi$ potential, associated with the change in the equilibrium internuclear distance from 1.15 Å in $X^2\Pi$ state to 1.06 Å in $A^2\Sigma^+$ state [61]. The ion momentum image drastically changes by photoexcitation from $2\pi$ to $3s\sigma$ orbitals, showing that the fragment anisotropy reflects the changes in the shape of the outermost molecular orbital. The polar plots of the angular distribution of the $N^+$ fragments produced via the $c^3\Pi$ state are shown in Fig. 6.12d, e for the ground and excited states, respectively. Theoretical ionization yields calculated based on ME-WFAT show good agreement with the experimental results both for the ground and excited states. This presents that the fragment angular distribution reflects the tunneling ionization rates of the outermost molecular orbital and follows the change in the electron distribution in the molecular frame upon the photoexcitation from the $2\pi$ to $3s\sigma$ orbitals.

Finally, the excitation process to the dissociative state in $NO^+$ is discussed. To clarify the excitation process, the fragment yields via the $c^3\Pi$ state are measured as a function of the ellipticity of the probe laser pulse. The yields show the maximum at linear polarization and drastically decrease to reach near zero value with circular polarization for both the ground and excited states (Fig. 6.13). This ellipticity dependence suggests that electron rescattering associated with tunneling ionization is involved in the excitation to the $c^3\Pi$ state. For quantitative discussion, the ellipticity dependence is calculated by WFAT and adiabatic theory [23]. The calculated ellipticity dependence reproduced the experimental results for both of the ground and excited states, thus confirming the tunneling ionization followed by electron rescattering is responsible for dissociative ionization under the present experimental conditions. Since the ground state of $NO^+$ is singlet, electron impact excitation to triplet states may serve as a filter to favor tunneling rather than multiphoton ionization from the ground and excited states of NO.

In this section, our recent studies on molecular orbital imaging by laser tunneling ionization have been reviewed. The angular distribution of $N^+$ fragments produced by the dissociative ionization of NO, $NO \rightarrow NO^+ + e^- \rightarrow N^+ + O + e^-$ reflects changes in electron distribution in the molecular frame upon the photoexcitation of an electron, which can be considered as a model chemical reaction. Thanks to the ultrashort duration of intense laser pulses, the laser tunneling ionization provides a promising means towards the tracking of ultrafast electron dynamics during chemical reaction processes.

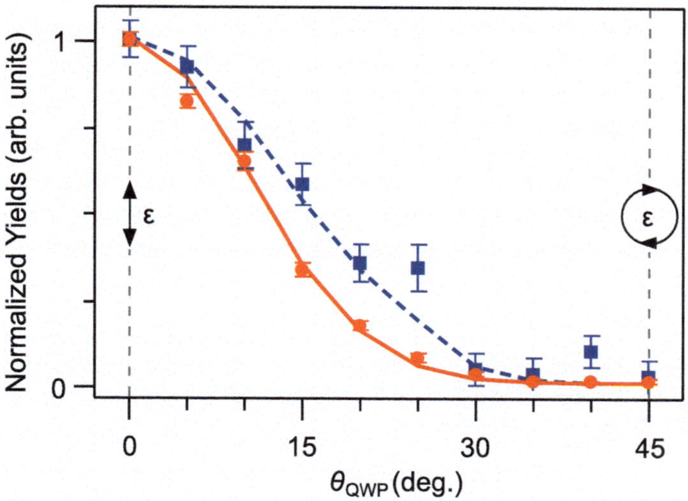


**Fig. 6.13** Relative yields of the $c^3\Pi$ dissociation components as a function of the angle of a quarter-wave plate $\theta_{QWP}$ with the $X^2\Pi$ (red circles) and $A^2\Sigma^+$ (blue squares) states of NO as the initial state. The solid line is a result of the least-square fitting to the experimental data for the ground state, while the dashed line represents a theoretical prediction for the excited state. Adapted with permission from Ref. [23]. Copyright 2016 by American Physical Society

## 6.4 Summary and Outlook

Two types of time-resolved molecular orbital imaging techniques have been discussed in this chapter, both of which would open the door to the investigation of the driving force behind chemical reaction. One is TR-EMS and the other is the laser tunneling ionization imaging. It is shown that TR-EMS has the distinctive feature in its potential capability to observe not only the HOMO but also all other, more tightly bound orbitals of a transient, evolving system during chemical reaction, with the rigorous orbital selection based on the energy conservation law. It is the advantage of TR-EMS that the well-established knowledge about the electron-molecule collision dynamics for traditional EMS can be applied in a straightforward fashion to the time-resolved measurements of chemical reaction in a laser-field-free condition. However, there is ample room for improvements, mainly in data statistics and energy and time resolution. These have to be improved by introducing the ever-developing technologies that provide more intense laser, electron and molecular beams [33].

The ultrafast laser tunneling imaging technique can offer a unique approach to visualize the dynamics of outermost electrons of molecules, with a high temporal resolution in tens of femtoseconds or shorter. On the other hand, it is necessary to take into account various factors contributing laser tunneling ionization, such as dipole moment and multielectron effects, for proper interpretation of obtained images. Further investigation of laser tunneling ionization process needs to be performed for general applications.

The qualitative difference between TR-EMS and the laser tunneling ionization imaging is that the former looks at a molecular orbital in momentum space while the latter in position space. On the other hand, the common strong point of the two techniques is that they are both sensitive to the behavior of electron moving around the outer region of a transient, evolving system, far from the nuclei, which governs its reactivity. These similarity and dissimilarity of the two techniques ensure that future work on why the atoms are dancing in such a way, which is at the heart of chemical reaction dynamics, will be tackled from multiple perspectives, when their paths are crossed.

**Acknowledgements** MY and MT thank the technical staff of the machine shop at the Institute of Multidisciplinary Research for Advanced Materials and of the equipment development center at the Institute for Molecular Science for their expertise and skills in development of the TR-EMS apparatus. MT acknowledges financial support by JSPS KAKENHI Grant Numbers 20225001, 25248002 and 17F17338, and a support by MEXT Five-star Alliance Program. MY acknowledges financial support by JSPS KAKENHI Grant Numbers 21750005 and 15H03762, and by the Sumitomo Foundation and the Yamada Science Foundation. TE and AH are grateful to Prof. Oleg. I. Tolstikhin (MIPT, Russia) and Prof. Toru Morishita (UEC, Japan) for theoretical calculations on laser tunneling ionization imaging of NO. AH acknowledges financial support by JSPS KAKENHI Grant Numbers 19H00887 and 16H04029, MEXT Quantum Leap Flagship Program (MEXT Q-LEAP) Grant Number JPMXS0118068681 and by World Research Unit (B-1) of Reaction Infography (R-ing) at Nagoya University. TE acknowledges Grant-in-Aid for JSPS fellows (25 2966).

## References

1. R.C. Evans, P. Douglas, H.D. Burrows (eds.), *Applied Photochemistry* (Springer, New York, 2013)
2. K. Fukui, Role of frontier orbitals in chemical reactions. Science **218**, 747–754 (1982). https://doi.org/10.1126/science.218.4574.747
3. A. Stolow, A.E. Bragg, D.M. Neumark, Femtosecond time-resolved photoelectron spectroscopy. Chem. Rev. **104**, 1719–1757 (2004). https://doi.org/10.1021/cr020683w
4. T. Suzuki, Femtosecond time-resolved photoelectron imaging. Annu. Rev. Phys. Chem. **57**, 555–592 (2006). https://doi.org/10.1146/annurev.physchem.57.032905.104601
5. J.C. Tully, Molecular dynamics with electronic transitions. J. Chem. Phys. **93**, 1061–1071 (1990). https://doi.org/10.1063/1.459170
6. C.-Y. Zhu, H. Nakamura, The two-state linear curve crossing problems revisited. II. Analytical approximations for the Stokes constant and scattering matrix: The Landau–Zener case. J. Chem. Phys. **97**, 8497–8514 (1992).https://doi.org/10.1063/1.463368
7. I.E. McCarthy, E. Weigold, (e, 2e) spectroscopy. Phys. Rep. **27**, 275–371 (1976)
8. C.E. Brion, Looking at orbitals in the laboratory: the experimental investigation of molecular wavefunctions and binding energies by electron momentum spectroscopy. Int. J. Quantum Chem. **29**, 1397–1428 (1986). https://doi.org/10.1002/qua.560290534
9. K.T. Leung, in *Theoretical Models of Chemical Bonding, Part 3*. ed. by Z.B. Maksic (Springer, Berlin, 1991), p.339
10. M.A. Coplan, J.H. Moor, J.P. Doering, (e, 2e) spectroscopy. Rev. Mod. Phys. **66**, 985–1014 (1994). https://doi.org/10.1103/RevModPhys.66.985
11. E. Weigold, I.E. McCarthy, *Electron Momentum Spectroscopy* (Kluwer Academic/Plenum, New York, 1999)

12. V.G. Neudachin, Y.V. Popov, Y.F. Smirnov, Electron momentum spectroscopy of atoms, molecules and thin films. Phys. Usp. **42**, 1017–1044 (1999). https://doi.org/10.1070/PU1999v042n10ABEH000492
13. M. Takahashi, Looking at molecular orbitals in three-dimensional form: from dream to reality. Bull. Chem. Soc. Jpn **82**, 751–777 (2009). https://doi.org/10.1246/bcsj.82.751
14. D. Pavičić, K.F. Lee, D.M. Rayner, P.B. Corkum, D.M. Villeneuve, Direct measurement of the angular dependence of ionization for $N_2$, $O_2$, and $CO_2$ in intense laser fields. Phys. Rev. Lett. **98**, 243001 (2007). https://doi.org/10.1103/PhysRevLett.98.243001
15. L. Holmegaard, J.L. Hansen, L. Kalhøj, S. Louise Kragh, H. Stapelfeldt, F. Filsinger, J. Küpper, G. Meijer, D. Dimitrovski, M. Abu-samha, C.P.J. Martiny, L. Bojer Madsen, Photoelectron angular distributions from strong-field ionization of oriented molecules. Nat. Phys. **6**, 428–432 (2010). https://doi.org/10.1038/nphys1666
16. J.L. Hansen, L. Holmegaard, L. Kalhøj, S.L. Kragh, H. Stapelfeldt, F. Filsinger, G. Meijer, J. Küpper, D. Dimitrovski, M. Abu-samha, C.P.J. Martiny, L.B. Madsen, Ionization of one- and three-dimensionally-oriented asymmetric-top molecules by intense circularly polarized femtosecond laser pulses. Phys. Rev. A **83**, 023406 (2011). https://doi.org/10.1103/PhysRevA.83.023406
17. A. Staudte, S. Patchkovskii, D. Pavičić, H. Akagi, O. Smirnova, D. Zeidler, M. Meckel, D.M. Villeneuve, R. Dörner, M.Y. Ivanov, P.B. Corkum, Angular tunneling ionization probability of fixed-in-space $H_2$ molecules in intense laser pulses. Phys. Rev. Lett. **102**, 033004 (2009). https://doi.org/10.1103/PhysRevLett.102.033004
18. H. Akagi, T. Otobe, A. Staudte, A. Shiner, F. Turner, R. Dörner, D.M. Villeneuve, P.B. Corkum, Laser tunnel ionization from multiple orbitals in HCl. Science **325**, 1364–1367 (2009). https://doi.org/10.1126/science.1175253
19. J. Wu, L.P.H. Schmidt, M. Kunitski, M. Meckel, S. Voss, H. Sann, H. Kim, T. Jahnke, A. Czasch, R. Dörner, Multiorbital tunneling ionization of the CO molecule. Phys. Rev. Lett. **108**, 183001 (2012). https://doi.org/10.1103/PhysRevLett.108.183001
20. A.S. Alnaser, S. Voss, X.M. Tong, C.M. Maharjan, P. Ranitovic, B. Ulrich, T. Osipov, B. Shan, Z. Chang, C.L. Cocke, Effects of molecular structure on ion disintegration patterns in ionization of $O_2$ and $N_2$ by short laser pulses. Phys. Rev. Lett. **93**, 113003 (2004). https://doi.org/10.1103/PhysRevLett.93.113003
21. A.S. Alnaser, C.M. Maharjan, X.M. Tong, B. Ulrich, P. Ranitovic, B. Shan, Z. Chang, C.D. Lin, C.L. Cocke, I.V. Litvinyuk, Effects of orbital symmetries in dissociative ionization of molecules by few-cycle laser pulses. Phys. Rev. A **71**, 031403(R) (2005). https://doi.org/10.1103/PhysRevA.71.031403
22. I. Znakovskaya, P. von den Hoff, S. Zherebtsov, A. Wirth, O. Herrwerth, M.J.J. Vrakking, R. de Vivie-Riedle, M.F. Kling, Attosecond control of electron dynamics in carbon monoxide. Phys. Rev. Lett. **103**, 103002 (2009). https://doi.org/10.1103/PhysRevLett.103.103002
23. T. Endo, A. Matsuda, M. Fushitani, T. Yasuike, O.I. Tolstikhin, T. Morishita, A. Hishikawa, Imaging electronic excitation of NO by ultrafast laser tunneling ionization. Phys. Rev. Lett. **116**, 163002 (2016). https://doi.org/10.1103/PhysRevLett.116.163002
24. T. Endo, H. Fujise, H. Hasegawa, A. Matsuda, M. Fushitani, O.I. Tolstikhin, T. Morishita, A. Hishikawa, Angle dependence of dissociative tunneling ionization of NO in asymmetric two-color intense laser fields. Phys. Rev. A **100**, 053422 (2019). https://doi.org/10.1103/PhysRevA.100.053422
25. J. Itatani, J. Levesque, D. Zeidler, H. Niikura, H. Pépin, J.C. Kieffer, P.B. Corkum, D.M. Villeneuve, Tomographic imaging of molecular orbitals. Nature **432**, 867–871 (2004). https://doi.org/10.1038/nature03183
26. H.J. Wörner, J.B. Bertrand, D.V. Kartashov, P.B. Corkum, D.M. Villeneuve, Following a chemical reaction using high-harmonic interferometry. Nature **466**, 604–607 (2010). https://doi.org/10.1038/nature09185
27. P.M. Kraus, H.J. Wörner, Time-resolved high-harmonic spectroscopy of valence electron dynamics. Chem. Phys. **414**, 32–44 (2013). https://doi.org/10.1016/j.chemphys.2012.01.013

28. W. Li, X. Zhou, R. Lock, S. Patchkovskii, A. Stolow, H.C. Kapteyn, M.M. Murnane, Time-resolved dynamics in $N_2O_4$ probed using high harmonic generation. Science **322**, 1207–1211 (2008). https://doi.org/10.1126/science.1163077
29. Y. Zheng, I.E. McCarthy, E. Weigold, D. Zhang, Direct observation of the momentum-density profile of excited and oriented sodium atoms. Phys. Rev. Lett. **64**, 1358–1360 (1990). https://doi.org/10.1103/PhysRevLett.64.1358
30. M. Yamazaki, H. Satoh, M. Ueda, D.B. Jones, Y. Asano, N. Watanabe, A. Czasch, O. Jagutzki, M. Takahashi, A highly sensitive electron momentum spectrometer incorporating a multiparticle imaging detector. Meas. Sci. Technol. **22**, 075602-1–075602-13 (2011). https://doi.org/10.1088/0957-0233/22/7/075602
31. Q.G. Tian, K.D. Wang, X. Shan, X.J. Chen, A high-sensitivity angle and energy dispersive multichannel electron momentum spectrometer with $2\pi$ angle range. Rev. Sci. Instrum. **82**, 033110-1–033110-7 (2011). https://doi.org/10.1063/1.3568744
32. M. Yamazaki, Y. Kasai, K. Oishi, H. Nakazawa, M. Takahashi, Development of an (e, 2e) electron momentum spectroscopy apparatus using an ultrashort pulsed electron gun. Rev. Sci. Instrum. **84**, 063105-1–063105-10 (2013). https://doi.org/10.1063/1.4809792
33. M. Yamazaki, Y. Kasai, K. Oishi, H. Nakazawa, M. Takahashi, Development of time-resolved (e, 2e) electron momentum spectroscopy: a tool for visualizing the motion of electrons during a chemical reaction. J. Phys. Conf. Ser. **488**, 012056-1–012056-6 (2014). https://doi.org/10.1088/1742-6596/488/1/012056
34. Y. Tang, X. Shan, Z. Liu, S. Niu, E. Wang, X. Chen, Development of an electron momentum spectrometer for time-resolved experiments employing nanosecond pulsed electron beam. Rev. Sci. Instrum. **89**, 033101-1–033101-8 (2018). https://doi.org/10.1063/1.5018665
35. J.C. Williamson, A.H. Zewail, Ultrafast electron diffraction. Velocity mismatch and temporal resolution in crossed-beam experiments. Chem. Phys. Lett. **209**, 10–16 (1993). https://doi.org/10.1016/0009-2614(93)87193-7
36. M. Yamazaki, K. Oishi, H. Nakazawa, C.Y. Zhu, M. Takahashi, Molecular orbital imaging of the acetone $S_2$ excited state using time-resolved (e, 2e) electron momentum spectroscopy. Phys. Rev. Lett. **114**, 103005-1–103005-5 (2015). https://doi.org/10.1103/PhysRevLett.114.103005
37. J.C. Owrutsky, A.P. Baronavski, Ultrafast photodissociation dynamics of the $S_1$ and $S_2$ states of acetone. J. Chem. Phys. **110**, 11206–11213 (1999). https://doi.org/10.1063/1.478003
38. E. Gibney, Nature **519**, 392 (2015). https://doi.org/10.1038/519392d
39. T. Wogan, Physics **8**, 23 (2015). https://doi.org/10.1103/Physics.8.23
40. M. Yamazaki, Y. Tang, M. Takahashi, Ionization propensity and electron momentum distribution of the toluene $S_1$ excited state studied by time-resolved binary (e, 2e) spectroscopy. Phys. Rev. A **94**, 052509-1–052509-5 (2016). https://doi.org/10.1103/PhysRevA.94.052509
41. C.G. Hickman, J.R. Gascooke, W.D. Lawrance, The $S_1$–$S_0$ ($^1B_2$–$^1A_1$) transition of jet-cooled toluene: excitation and dispersed fluorescence spectra, fluorescence lifetimes, and intramolecular vibrational energy redistribution. J. Chem. Phys. **104**, 4887–4901 (1996). https://doi.org/10.1063/1.471122
42. H. Nakatsuji, Description of two- and many-electron processes by the SAC-CI method. Chem. Phys. Lett. **177**, 331–337 (1991). https://doi.org/10.1016/0009-2614(91)85040-4
43. X.M. Tong, Z.X. Zhao, C.D. Lin, Theory of molecular tunneling ionization. Phys. Rev. A **66**, 033402 (2002). https://doi.org/10.1103/PhysRevA.66.033402
44. O.I. Tolstikhin, T. Morishita, L.B. Madsen, Theory of tunneling ionization of molecules: weak-field asymptotics including dipole effects. Phys. Rev. A **84**, 053423 (2011). https://doi.org/10.1103/PhysRevA.84.053423
45. S. Petretti, Y.V. Vanne, A. Saenz, A. Castro, P. Decleva, Alignment-dependent ionization of $N_2$, $O_2$, and $CO_2$ in intense laser fields. Phys. Rev. Lett. **104**, 223001 (2010). https://doi.org/10.1103/PhysRevLett.104.223001
46. T. Otobe, K. Yabana, J.-I. Iwata, First-principles calculations for the tunnel ionization rate of atoms and molecules. Phys. Rev. A **69**, 053404 (2004). https://doi.org/10.1103/PhysRevA.69.053404

47. B. Zhang, J. Yuan, Z. Zhao, Dynamic core polarization in strong-field ionization of CO molecules. Phys. Rev. Lett. **111**, 163001 (2013). https://doi.org/10.1103/PhysRevLett.111.163001
48. A. Hishikawa, E.J. Takahashi, A. Matsuda, Electronic and nuclear responses of fixed-in-space $H_2S$ to ultrashort intense laser fields. Phys. Rev. Lett. **97**, 243002 (2006). https://doi.org/10.1103/PhysRevLett.97.243002
49. H. Ohmura, N. Saito, T. Morishita, Molecular tunneling ionization of the carbonyl sulfide molecule by double-frequency phase-controlled laser fields. Phys. Rev. A **89**, 013405 (2014). https://doi.org/10.1103/PhysRevA.89.013405
50. T. Endo, H. Fujise, A. Matsuda, M. Fushitani, H. Kono, A. Hishikawa, Coincidence momentum imaging of asymmetric Coulomb explosion of $CO_2$ in phase-locked two-color intense laser fields. J. Electron Spectrosc. Relat. Phenom. **207**, 50–54 (2016). https://doi.org/10.1016/j.elspec.2015.12.010
51. T. Endo, H. Fujise, Y. Kawachi, A. Ishihara, A. Matsuda, M. Fushitani, H. Kono, A. Hishikawa, Selective bond breaking of $CO_2$ in phase-locked two-color intense laser fields: laser field intensity dependence. Phys. Chem. Chem. Phys. **19**, 3550–3556 (2017). https://doi.org/10.1039/C6CP07471E
52. D. Ray, F. He, S. De, W. Cao, H. Mashiko, P. Ranitovic, K.P. Singh, I. Znakovskaya, U. Thumm, G.G. Paulus, M.F. Kling, I.V. Litvinyuk, C.L. Cocke, Ion-energy dependence of asymmetric dissociation of $D_2$ by a two-color laser field. Phys. Rev. Lett. **103**, 223201 (2009). https://doi.org/10.1103/PhysRevLett.103.223201
53. F.R. Gilmore, Potential energy curves for $N_2$, NO, $O_2$ and corresponding ions. J. Quant. Spectrosc. Radiat. Transfer **5**, 369–389 (1965). https://doi.org/10.1016/0022-4073(65)90072-5
54. D.L. Albritton, A.L. Schmeltekopf, R.N. Zare, Potential energy curves for $NO^+$. J. Chem. Phys. **71**, 3271 (1979). https://doi.org/10.1063/1.438757
55. H. Partridge, S.R. Langhoff, C.W. Bauschlicher Jr., Theoretical study of the spectroscopy of $NO^+$. J. Chem. Phys. **93**, 7179 (1990). https://doi.org/10.1063/1.459716
56. H. Li, D. Ray, S. De, I. Znakovskaya, W. Cao, G. Laurent, Z. Wang, M.F. Kling, A.T. Le, C.L. Cocke, Orientation dependence of the ionization of CO and NO in an intense femtosecond two-color laser field. Phys. Rev. A **84**, 043429 (2011). https://doi.org/10.1103/PhysRevA.84.043429
57. A. Lafosse, M. Lebech, J.C. Brenot, P.M. Guyon, O. Jagutzki, L. Spielberger, M. Vervloet, J.C. Houver, D. Dowek, Vector correlations in dissociative photoionization of diatomic molecules in the VUV range: strong anisotropies in electron emission from spatially oriented NO molecules. Phys. Rev. Lett. **84**, 5987 (2000). https://doi.org/10.1103/PhysRevLett.84.5987
58. O.I. Tolstikhin, L.B. Madsen, T. Morishita, Weak-field asymptotic theory of tunneling ionization in many-electron atomic and molecular systems. Phys. Rev. A **89**, 013421 (2014). https://doi.org/10.1103/PhysRevA.89.013421
59. S. Ohmura, T. Kato, T. Oyamada, S. Koseki, H. Ohmura, H. Kono, A single-electron picture based on the multiconfiguration time-dependent Hartree-Fock method: application to the anisotropic ionization and subsequent high-harmonic generation of the CO molecule. J. Phys. B **51**, 034001 (2018). https://doi.org/10.1088/1361-6455/aa9e45
60. V.P. Majety, A. Scrinzi, Static field ionization rates for multi-electron atoms and small molecules. J. Phys. B **48**, 245603 (2015). https://doi.org/10.1088/0953-4075/48/24/245603
61. G. Herzberg, *Spectra of Diatomic Molecules*, 2nd edn. (Van Nostrand, New York, 1950), p.558

# Chapter 7
# Ultrafast X-Ray Scattering: New Views of Chemical Reaction Dynamics

**Peter M. Weber, Brian Stankus, and Adam Kirrander**

**Abstract** The advent of ultrafast pulsed X-ray free-electron lasers with very high brightness has enabled the determination of transient molecular structures of small and medium-sized organic molecules in excited states and undergoing chemical dynamics using X-ray scattering. This chapter provides an introduction into X-ray scattering theory and considers several important aspects relating to the experimental implementation. Ultrafast gas phase X-ray scattering is shown to provide new observables to elucidate the dynamics of chemical reactions by providing complete, time-dependent molecular structures. Consideration of correlations between structural parameters is important for molecules far from their equilibrium, and the changes in electron density distributions of molecules upon optical excitation need to be considered in the analysis. Future technological developments are expected to lead to further important advances.

**Keywords** X-ray scattering · Ultrafast dynamics · Excited molecular states · Transient molecular structures

## 7.1 Introduction

The determination of molecular structures is a foundational achievement of twentieth century chemistry. Started as a curiosity-driven inquiry in chemistry, molecular structures now form the bedrock of many fields including molecular biology,

P. M. Weber (✉)
Department of Chemistry, Brown University, Providence, Rhode Island 02912, USA
e-mail: peter_weber@brown.edu

B. Stankus
Department of Chemistry, Western Connecticut State University, Danbury, Connecticut 06810, USA

A. Kirrander
Department of Chemistry, Physical and Theoretical Chemistry Laboratory, University of Oxford, South Parks Road, Oxford, UK



K. Ueda (ed.), *Ultrafast Electronic and Structural Dynamics*,
https://doi.org/10.1007/978-981-97-2914-2_7

pharmacology, and material science. Applications in medicine, drug design, information storage, communication devices and many other areas have immeasurable benefit for modern society. Instrumental advances at the beginning of the twenty-first century have now made it possible to determine molecular structures in excited states. Coupled with a time resolution deep into the femtosecond regime, we can now measure time-evolving molecular structures of chemical systems during vibrational motions or chemical reactions. The measurement of molecular structures far from equilibrium will advance many fields, possibly including the selective control of chemical reactions, which might lead to the synthesis of yet unknown materials with useful molecular properties and/or the prevention of undesirable reactions. Just as knowledge of ground state molecular structures has been invaluable to the development of computational methods that now find widespread application in the molecular sciences, data on excited state and transient molecular structures can serve as benchmarks for the refinement of quantum chemical methods.

For the determination of static molecular structures in classical chemistry, X-ray diffraction is an essential tool. X-ray and electron scattering patterns are Fourier transform projections of the molecular electron and charge density distributions, respectively. Both X-ray and electron scattering are sensitive to the nuclear coordinates and the electron density distributions. Electron scattering results from electrostatic interaction of the electrons in the electron beam with the nuclei and the electrons of the molecule. X-ray scattering arises from the interaction of a molecule's electrons with the electromagnetic field, and therefore only depends on electron density distributions. Since the atomic core electrons are tightly centered on the nuclei, X-ray scattering also determines nuclear geometries. Thus, both scattering experiments can yield the molecular geometry, as well as the associated electron density distributions. Unlike in time-resolved spectroscopies, where the Heisenberg uncertainty relation $\Delta E \cdot \Delta t \geq \hbar/2$ imposes a fundamental limitation, scattering experiments measure spatial structures that can be measured with temporal resolution without being restricted by an uncertainty relation. The prospect of measuring nuclear geometries and electron density distributions, i.e. chemical bonding, with ultrafast time resolution makes time-resolved scattering an attractive field to develop. The high sensitivity of scattering experiments makes it possible to measure molecular systems even in dilute vapors.

## 7.2 Elements of Scattering Theory

### *7.2.1 X-ray Scattering*

Scattering of hard X-rays can be treated in the perturbative regime [1] using the first Born and the Waller Hartree approximations [1, 2]. The detectors used in ultrafast scattering are generally not energy-resolved, which means that we are mainly concerned with total scattering. The differential cross-section for total scattering is

[3]

$$\frac{d\sigma}{d\Omega} = r_0 |\boldsymbol{e}_0 \cdot \boldsymbol{e}_1|^2 S(\boldsymbol{q}), \tag{7.1}$$

where $r_0 = e^2/m_e c^2$ is the so-called classical electron radius ($e$ signifies the charge and $m_e$ the mass of an electron, $c$ the speed of light). The polarization factor $|\boldsymbol{e}_0 \cdot \boldsymbol{e}_1|^2$ accounts for the polarization of the incoming X-rays via,

$$|\boldsymbol{e}_0 \cdot \boldsymbol{e}_1|^2 = \begin{cases} 1 & \text{vertical} \\ \cos 2\theta & \text{horizontal,} \\ \frac{(1+\cos^2 2\theta)}{2} & \text{unpolarized} \end{cases} \tag{7.2}$$

for unpolarized sources or sources polarized in the vertical or horizontal scattering planes. The dynamic structure factor is given by

$$S(\boldsymbol{q}) = \sum_{\beta} \left| \langle \psi_\beta | \hat{L} | \psi_\alpha \rangle \right|^2 = \langle \psi_\alpha | \hat{L}^\dagger \hat{L} | \psi_\alpha \rangle, \tag{7.3}$$

where $|\psi_\beta\rangle$ and $|\psi_\alpha\rangle$ are the final and initial electronic states and the sum includes all states. The second equality constitutes a powerful result that makes it possible to calculate the total scattering for a particular electronic state (in this instance $|\psi_\alpha\rangle$) without reference to any other states [3]. The scattering vector $\boldsymbol{q}$ (a. k. a. the momentum transfer vector) is given by the difference between the incident and the scattered wave vectors,

$$\boldsymbol{q} = \boldsymbol{k}_0 - \boldsymbol{k}_1, \tag{7.4}$$

where $k_0 = |\boldsymbol{k}_0| = \frac{2\pi}{\lambda} = \frac{2\pi E}{hc}$ with $\lambda$ and $E$ the wavelength and energy, respectively, of the incoming X-rays. The scattering angle $2\theta$ is related to the magnitude of the scattering vector via $q = |\boldsymbol{q}| = |\boldsymbol{k}_0 - \boldsymbol{k}_1| = 2k_0 \sin\theta$. Finally, the scattering operator in Eq. 7.3 is given by,

$$\hat{L} = \sum_{j=1}^{N_e} e^{i\boldsymbol{q}\boldsymbol{r}_j}, \tag{7.5}$$

where the sum runs over all the $N_e$ electrons with coordinates $\boldsymbol{r}_j$. One may decompose the contributions to the dynamic structure factor into elastic and inelastic components,

$$S(\boldsymbol{q}) = |F(\boldsymbol{q})|^2 + S_{\text{inel}}(\boldsymbol{q}), \tag{7.6}$$

where the elastic component $|F(\boldsymbol{q})|^2$ is proportional to the form factor,

$$F(\boldsymbol{q}) = \langle \psi_\alpha | \widehat{L} | \psi_\alpha \rangle = \int \rho_\alpha^{(N_e)}(\boldsymbol{r}) e^{i\boldsymbol{q}\boldsymbol{r}} d\boldsymbol{r} \tag{7.7}$$

with $\rho_\alpha^{(N_e)}(\boldsymbol{r})$ the electron density. The inelastic component, $S_{\text{inel}}(\boldsymbol{q})$, is what remains to make up the total scattering and accounts for transitions to all states $\alpha \neq \beta$. Quite often, rotationally averaged signals are considered [4, 5], especially in the gas phase, in which case only the amplitude $q$ matters.

### 7.2.2 Independent Atom Model

The independent atom model (IAM), originally proposed by Debye [6], is widely used. It approximates the electron density as a sum of isotropic isolated-atom electron densities centered at the positions of the nuclei [3, 7], which makes it possible to use conveniently tabulated [8] atomic form factors, $f_i(q)$, and inelastic corrections, $S_{\text{inel},i}^{IAM}(q)$, to express the rotationally averaged total scattering as,

$$S_{IAM}(q) = \sum_{i=1}^{N_{at}} \sum_{j=1}^{N_{at}} f_i(q) f_j(q) \frac{\sin(qR_{ij})}{qR_{ij}} + \sum_{i=1}^{N_{at}} S_{\text{inel},i}^{IAM}(q) \tag{7.8}$$

where the indices $i$ and $j$ run over the $N_{at}$ atoms with $R_{ij}$ the distance between pairs of atoms. As previously in Eq. 7.6, the first term is the elastic component, while the second is the inelastic component. For IAM, the inelastic component is considered independent of the molecular geometry. The shortcomings of the IAM approximation are well-documented [3, 5, 9–11], but it remains a reasonable approximation for many molecular systems, especially in the electronic ground state, and constitutes an effective way to model the most prominent features of molecular scattering patterns.

### 7.2.3 Comparison to Electron Scattering

Most molecular structures known today were determined with either X-ray or electron diffraction. Electron diffraction is closely related to X-ray scattering, but arises from the Coulomb interactions between charged particles, meaning that the electrons scatter from both target electrons and nuclei [12–14]. In practice, the Thomson cross-section above is replaced by the Rutherford cross-section and an additional $q^{-4}$ damping appears in the resulting scattering intensity [15–18]. Electron form factors can be computed from the Fourier transform of the atomic potentials [8, 15, 19, 20] or found as approximate analytical expressions [21, 22]. It is most convenient to use the Mott-Bethe formula, which relates electron scattering form factors to X-ray atomic scattering factors,

$$f_e(q) = \frac{2m_e e^2}{\hbar^2} \cdot \frac{Z_i - f_i(q)}{q^2} \tag{7.9}$$

with $\hbar$ the reduced Planck's constant and $Z_i$ the atomic number of each atom. Note that we follow the X-ray convention and use the symbol $q$ rather than $s$ for the momentum transfer. Equation 7.9 suggests that outside of the constants and the $\frac{1}{q^2}$ dependence of the Rutherford scattering amplitude, electron and X-ray scattering have similar information content in that both depend on the X-ray form factor. Yet they are also complementary: the X-ray form factors decrease with increasing scattering angle while the electron form factors increase, Fig. 7.1. The increasing electron form factors partially compensate for the loss of electron scattering signal arising from the rapidly decaying $\frac{1}{q^4}$ term of Rutherford scattering.

Several important conclusions follow from this discussion: First, the cross-section for electron scattering is dramatically larger than that for X-rays. This is the reason why early studies explored ultrafast electron diffraction (UED) to study excited state structures and dynamics [24–31]. The first UED measurement of the ring-opening reaction of 1,3-cyclohexadiene (CHD) was conducted almost 20 years ago [32, 33]. Even so, at the time the experiments were difficult because space-charge interactions between the electrons, which are closely confined within an ultrashort electron pulse, make it challenging to achieve the excellent signal-to-noise ratio demanded by the rapid ($1/q^4$) dependence of the scattering signal on the momentum transfer vector $q$. Tremendous progress results from using relativistic electrons [34, 35] and pulse compression techniques [36, 37]. New MeV ultrafast electron diffraction instruments have yielded important advances, starting with the dynamics of $I_2$ molecules [38].

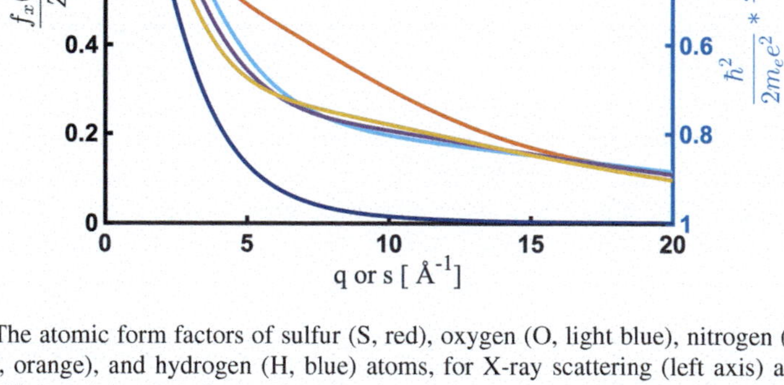


**Fig. 7.1** The atomic form factors of sulfur (S, red), oxygen (O, light blue), nitrogen (N, purple), carbon (C, orange), and hydrogen (H, blue) atoms, for X-ray scattering (left axis) and electron scattering (right axis), respectively. The form factors are scaled by the atomic number $Z$ and the electron form factors are in addition scaled by the Rutherford cross section pre-factors. From Ref. [23]

More recent experiments include the applicability to larger systems such as CHD [39]. Electron diffraction with ultrafast time resolution continues to be a valuable complement to the X-ray scattering. In a current review article, we delineate the accomplishments and promise of UED [40].

The larger cross-section for electron scattering is valuable, but it only goes so far. The tremendous brightness of modern X-ray free electron lasers (XFEL) beams has now turned the table: while the electron scattering cross section is larger by a factor of $10^5$, there are $3 \cdot 10^7$ more photons in an X-ray pulse than there are electrons in the electron pulse. Consequently, the scattering signal is about 300 times larger for X-rays. Since this comparison is on a pulse-by-pulse basis, the difference will become even more dramatic as high repetition rate XFEL technology advances.

## 7.3 Experimental Implementation of Pump-Probe X-Ray Scattering

The experimental scheme for the ultrafast gas-phase X-ray scattering experiments is conceptually straightforward. An optical pump laser pulse is focused onto the gaseous target and initiates the photochemistry, Fig. 7.2. It is followed in time by an X-ray probe pulse that produces a scattering image on an area detector. The scattering signal is recorded for variable time delays to monitor the dynamical structure of the molecules. In its implementation, the experiment requires a wide array of considerations affecting the final observed signal. Important design features relevant for the initial study of the ring-opening of 1,3-cyclohexadiene [41] were described in a detailed method paper in 2016 [42]. Since then, the experiment has been improved and implemented at the Coherent X-ray Imaging (CXI) endstation of the Linac Coherent Light Source (LCLS).

Gas-phase X-ray scattering experiments have been enabled by the rapid advancement in capabilities of XFELs such as LCLS. The high photon energies coupled with exceptional brightness (see Sect. 7.2.3) yield high signal-to-noise ratios that allow for recovery of detailed structural information. In addition to the pulse characteristics of LCLS, the CXI endstation offers several other distinct advantages. Firstly, the interaction and detection regions are fully in-vacuum, which helps to eliminate unwanted background scattering signals from air. In addition, CXI houses multiple differentially pumped sample chambers, which allows for additional measurements to be made downstream of the initial CSPAD detector. Both of these features are important for improving the signal-to-noise ratio of the measured scattering patterns. Of course, many other aspects of the experimental design, some of which are discussed below, need careful optimization to ensure stable and reliable signals.

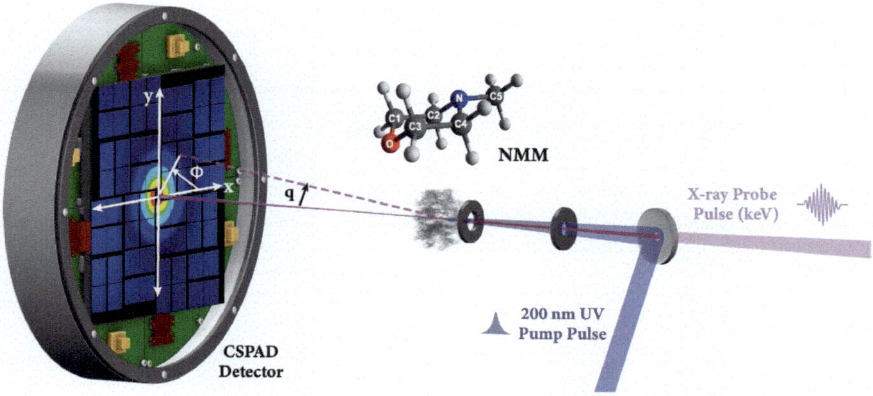


**Fig. 7.2** A typical pump-probe experimental setup for ultrafast X-ray scattering. A photochemical reaction is initiated, by an optical pump pulse and X-ray scattering patterns are recorded as a function of time delay between the laser and the X-ray pulses. The scattering signals are recorded as a function of the scattering vector $q$ and the azimuthal angle Φ

### *7.3.1 Scattering Cell Design*

The scattering cell shown in Fig. 7.3 is a 2.4 mm pathlength stainless steel gas cell, with a 3.2 mm inner diameter inlet tube for the gaseous target molecule. The cell was constructed with a 250 μm platinum entrance aperture on the upstream side and a beryllium exit window with a 250 μm aperture on the downstream side. This avoids obstruction of the primary X-ray beam and allows for sufficient gas flow between X-ray pulses (see below).

In order to prevent Bragg scattering from the primary X-ray beam on the cell entrance aperture, an upstream blocking aperture is used (see Fig. 7.3b). The upstream blocking aperture has a 200 μm diameter opening, which allows the vast majority of the primary X-ray beam (nominally focused to 30 μm FWHM) to pass through while blocking the low-intensity "edges" of the incident beam. This inevitably causes Bragg scattering from the platinum metal, which is subsequently blocked from entering the interaction region by the scattering cell itself. The entrance aperture, which has a 250 μm diameter opening, then allows the primary X-ray beam to pass cleanly to the interaction region. The diffuse scatter from the blocking aperture that enters the cell through the entrance aperture is weak and occurs at such a small scattering angle that it is not detected in the experiment.

The cell was designed to allow scattering at angles up to ~60° to exit the interaction region unobstructed, with the exception of the beryllium exit window. A very thin (100 μm thick) disk of beryllium was chosen as the exit window material because it is nearly transparent to X-rays (98.8% transmission at 9.5 keV). There is a very small dependence of the transmission on the scattering angle due to the changing path length. This effect, which is on the order of 0.5% in the absolute scattering signal, is taken into account during the data analysis.

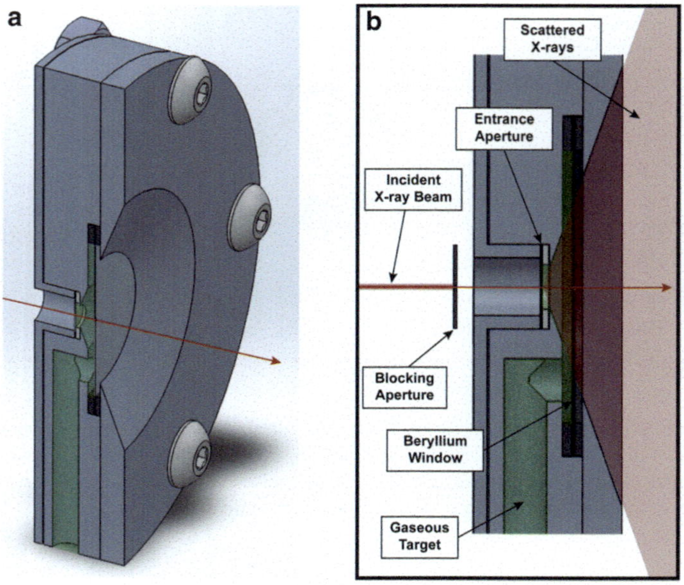


**Fig. 7.3** **a** Full cross-section of the scattering cell. The X-ray beam propagation axis is indicated in red, and the gaseous target is represented as a green-shaded region. **b** A close-up cross-section of the interaction region, with relevant components labeled. From Ref. [43]

The relatively small 2.4 mm path length of the interaction region was chosen for two reasons. Firstly, the short interaction length controls the Beer-Lambert attenuation of the UV pump pulse as it propagates through the sample. During the pump-probe experiments, it is necessary to have a significant number of excited molecules (as this is the signal being measured) while avoiding an excitation probability of >10% at any point in the interaction region to avoid multiphoton absorption. To attain both of these conditions, a near-constant excitation probability of less than 10% is desired. To achieve this, the Beer-Lambert attenuation is offset by weakly focusing the pump beam at the downstream end of the cell as detailed in a previous report [42]. The short interaction length, in concert with careful control of the gas pressure and UV intensity, helps to ensure that the desired balance is achieved.

The other benefit of a short interaction length is an improved resolution of the scattering angle. With any finite interaction length, there is an inherent limit on scattering angle resolution caused by scattering from molecules at the upstream and downstream ends of the interaction region reaching the same point on the detector. This effect is dependent on the radial distance of the detection point from the beam propagation axis. At 9.5 keV X-ray energy and an 86 mm sample-to-detector distance, the $q$ resolution is ~0.06 $\text{Å}^{-1}$ or better over the range of detection.

The sizes of the entrance and exit apertures (both 250 μm) were also carefully chosen to not only allow the pump and probe pulses to pass through but also to allow sufficient flow for proper sample turnover between X-ray pulses. Given that the sample cell is placed inside a vacuum chamber with 2000 L/s of turbomolecular pumping, we can use the approximation that the pressure outside the cell is negligible relative to the pressure inside. Thus, we can calculate the flow rate $q_{\text{cell}}$ out of the cell according to

$$q_{\text{cell}} = P_{\text{cell}} \cdot (A_{\text{entrance}} + A_{\text{exit}}) \cdot \sqrt{\frac{k_B T}{2\pi m}} \tag{7.10}$$

where $P_{\text{cell}}$ is the pressure inside the cell, $A_{\text{entrance}}$ and $A_{\text{exit}}$ are the areas of the respective apertures, $k_B$ is the Boltzmann constant, $T$ is the temperature, and $m$ is the molecular mass [44]. Using, for example, 7 Torr of N-methyl morpholine at 22 °C, $q_{\text{cell}} = 0.0427$ Torr · L/s, which means that the cell (with a volume of ~0.4 cm$^3$) will turn over every ~65 ms, or about every eighth X-ray pulse.

In the experiment, the interaction region is a very small portion of the total cell volume. So, it may be more informative to consider the motion of individual molecules as opposed to the collective flow of the ensemble. Thus, we also consider the mean distance $\langle x \rangle$ traveled between X-ray shots,

$$\langle x \rangle = \sqrt{\frac{2\lambda v}{3f}} \tag{7.11}$$

where $\lambda$ is the mean free path, $v$ is the average thermal particle velocity, and $f$ is the repetition rate of the experiment [44]. Using 7 Torr of N-methyl morpholine at 22 °C with the LCLS operating at 120 Hz, we find $\langle x \rangle \approx 825$ μm. Given that the diameter of the X-ray spot is only 30 μm FWHM, the probability of scattering off of the same molecule with multiple X-ray pulses is negligible.

### *7.3.2 Calibration of the Detector Geometry*

In order to properly calibrate the measured absolute scattering signals, it is necessary to consider the physical geometry of the detector relative to the interaction region. Scattering patterns calculated from the Independent Atom Model (IAM) represent the scattering per unit area as a function of $q$ at a fixed distance $R$ between the scattering medium and the point of detection. The IAM allows molecular scattering patterns to be predicted using the atomic form factors, the X-ray wavelength, and the interatomic distances, according to Eq. 7.8.

In the experiments, a planar detector is positioned perpendicular to the primary X-ray beam axis. Thus, the distance $R$ depends on the position on the detector, and geometric correction factors must be applied for direct comparison to the IAM. The measured intensity is divided by a $\cos(2\theta)^2$ factor to correct for the $R$ dependence as a function of $2\theta$. The measured intensity is also divided by an additional $\cos(2\theta)$ factor to normalize for the effective area of pixels at different points along the detector. Combined, the measured scattered intensity is divided by $\cos(2\theta)^3$ to compare with calculated IAM patterns [42].

Once the proper correction factors are applied, the measured scattering signal can be compared to the signal predicted from Eq. 7.8 to determine the precise orientation of the detector relative to the interaction region. The detector used is a planar 2.3-megapixel Cornell-SLAC Pixel Array Detector (CSPAD) [45] with a known internal pixel geometry. A least-squares optimization is performed between the IAM image generated from a calculated optimized molecular geometry and the experimentally measured ground-state scattering pattern. The optimization outputs five geometrical parameters. The parameters $x_0$, $y_0$, and $z_0$ are the absolute geometrical coordinates of the center of the detector relative to the interaction region, assuming that the detector plane is perpendicular to the X-ray beam. The fitting also optimizes $\phi_0$, the azimuthal angle of the detector relative to the X-ray polarization, and an overall intensity scaling factor $I_0$. By performing this optimization with ground-state scattering patterns, for which optimized geometries can be calculated reliably, we ensure that the measured excited-state patterns are also properly calibrated.

### *7.3.3 Calibration of Timing and Intensity Jitter*

Using a free-electron laser such as LCLS as the X-ray source has many advantages (such as short pulse duration and high photon flux) but also creates distinct challenges. The LCLS has significant shot-to-shot fluctuations in pulse arrival time, pulse intensity, and spatial orientation. In order to achieve a high signal-to-noise ratio, one must account for these fluctuations.

In time-resolved experiments, the relative timing of the pump and probe pulses is controlled via a motorized delay stage. In addition, we also monitor the timing jitter via a spectrally encoded cross-correlator that has been described in detail elsewhere [46]. Briefly, a chirped white light continuum is directed through a thin silicon nitride film, and then dispersed onto a CCD camera as a reference spectrum. Then, the chirped white light is crossed with the X-ray pulse on the film. The X-ray pulse changes the index of refraction of the silicon nitride, causing a drop-in transmission. The measured spectrum is then subtracted from the reference spectrum, resulting in a sharp decrease in intensity at a given point in the measured frequency spectrum. After calibrating the position of the edge as a function of X-ray pulse arrival time, this so-called '*time tool*' provides a shot-to-shot measure of the timing jitter.

In the experiments employing the scattering cell described in Sect. 7.3.1, the fluctuations in X-ray intensity incident on the sample arise from two sources: the fluctuations in total pulse-to-pulse X-ray intensity; and the spatial pointing instability of the X-ray beam, which affects transmission through the blocking and entrance apertures. In order to simultaneously correct for both effects, the transmitted X-ray intensity through the sample is monitored with a photodiode downstream of the CSPAD. The single-shot X-ray scattering patterns are then corrected for the photodiode value prior to averaging. It should be noted that this method assumes that the transmitted X-ray intensity is independent of the molecular dynamics. This is a reasonable approximation, as the total probability of an X-ray photon being scattered is ~ $10^{-13}$ for dilute molecular vapors. The total transmitted intensity is therefore a very good representation of the total incident intensity and can be used for calibration.

## 7.4 New Observables for Chemical Dynamics and Reactions

Probing molecular dynamics with scattering techniques provides unique views that are often complementary to spectroscopic methods. Consequently, time-resolved scattering can yield new insights about molecular dynamics and kinetics that might be difficult or even impossible to obtain with optical pump-probe spectroscopy. Even though very few time-resolved X-ray scattering experiments have been performed to date, they already have resulted in very important results, some of which are described in the following.

### *7.4.1 Chemical Reaction Dynamics*

The first time-resolved X-ray study of chemical reaction dynamics was published in 2015 and explored the chemical ring-opening reaction dynamics of gas phase 1,3-cyclohexadiene (CHD) [41]. Excitation at 266 nm places the molecules on the surface of the 1B electronic state, Fig. 7.4 [47]. A rapid structural evolution via two conical intersections (CI) transforms the molecule to the open 1,3-hexatriene (HT) form. The ring-opening reaction of CHD is a good test system for structural dynamics studies because the reaction is perceived to be ballistic [48]: optical excitation creates a well-defined wavepacket that propagates rapidly through the electronic surfaces. Since all molecules in the gas phase sample undergo the same dynamics, the scattering patterns of the molecular ensemble can be equated to the structural dynamics of individual molecules.

The time-dependent scattering pattern is shown in Fig. 7.5. The time resolution of the experiment, as determined from the analysis of the scattering data, was about 80 fs. To determine the time evolution of the molecular structure, trajectory calculations using the multiconfigurational Ehrenfest method were performed with potential energies and nonadiabatic couplings obtained on-the-fly from SA3-CAS(6,4)-SCF/cc-pVDZ ab initio electronic structure calculations. Three electronic states are included: the ground state, the optically accessed 1B state, and the 2A state that is implicated in the Woodward-Hoffman mechanism of the electrocyclic reaction. Each trajectory constitutes a molecular reaction path with a complete but distinct set of interatomic distances. To determine the combination of trajectories that best describes the chemical reaction, the experimental scattering patterns were compared to the scattering patterns calculated from the computed trajectories. Optimization

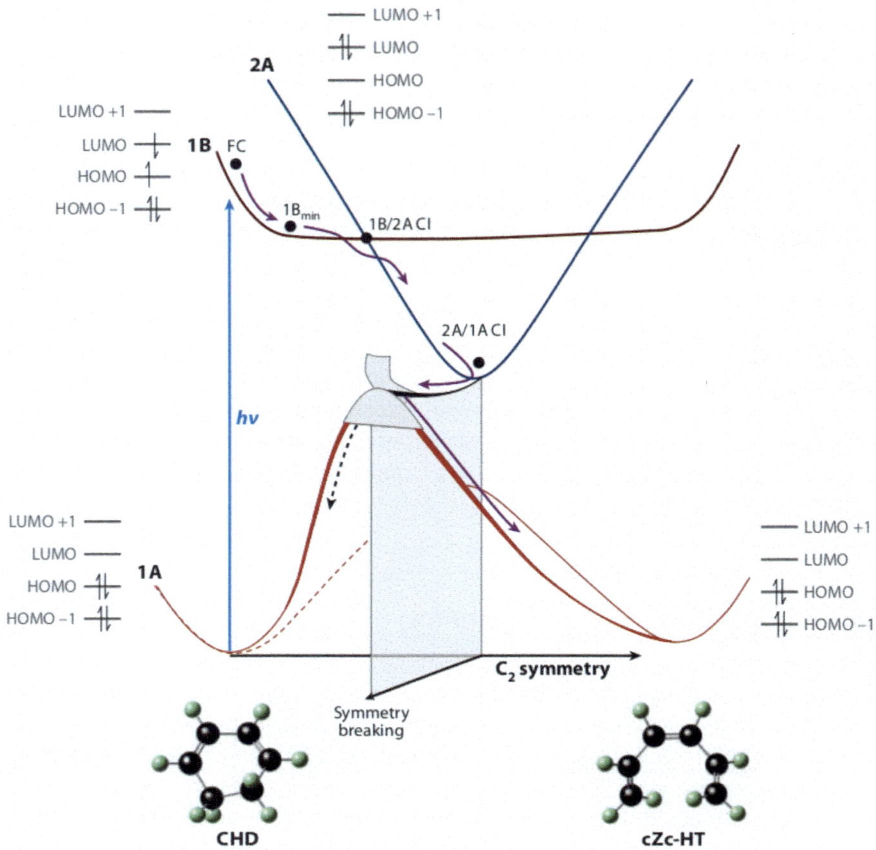


**Fig. 7.4** The potential energy surfaces of the CHD/HT system. Optical excitation of CHD leads to the 1B electronic surface. Rapid crossing to other electronic states drives the system toward the HT isomeric form in its ground electronic state. From Ref. [47]

converged on a small number of trajectories with four of them having a combined weight of approximately 80%. These trajectories therefore suffice to represent the experimentally observed data, and their graphical representation resulted in 'molecular movies' that show the time-evolving molecular structures. Interestingly, recent simulations by Polyak et al. [49] predict a branching ratio between the HT and CHD products closely aligned with that implied by the experiments. Some of the complexity inherent in the analysis of structures of polyatomic molecules can be avoided by investigating the structural dynamics of diatomic molecules. Glownia et al. measured the time-dependent X-ray scattering signals of diatomic $I_2$, Fig. 7.6 [50]. Excitation at 520 nm lifts the system to the B state, from where crossing to the B' level can lead to molecular dissociation. The signatures of the dissociation are clearly visible in the plot of the experimental excited state charge distribution versus time, although extracting the charge distribution in real space remains challenging [51–53].

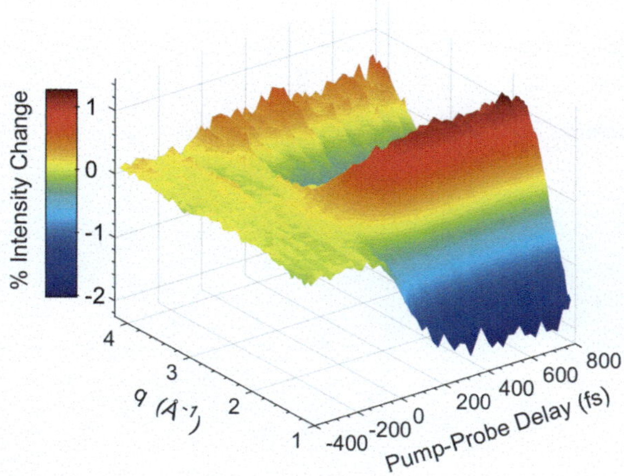


**Fig. 7.5** The time-dependent pump probe scattering signal of 1,3-cyclohexadiene upon excitation at 266 nm. Shown is the percent difference change in the scattering signal upon optical excitation, [100*(laser on-laser off)/laser off], as a function of pump-probe delay time. The scattering signals change on the order of 1%–2% as indicated by the color bar. From Ref. [41]

Dephasing of the vibrational wavepacket leads to a decline in the time-dependent oscillatory signal. The linear polarization of the laser pulse causes an anisotropy of the scattering signal that is readily observed in the experiment and which can be separated from the isotropic component by fitting to a Legendre polynomial.

## 7.4.2 Excited State Structures

Further development of the time-resolved X-ray scattering technique, both experimentally and theoretically, has more recently enabled the determination of polyatomic molecular structures in electronically excited states. In a 2019 study, the time-dependent molecular structure of N-methyl morpholine (NMM) was measured following excitation to a molecular Rydberg state [54], revealing the dephasing of coherent molecular vibrations on a picosecond time scale. Previous studies of NMM showed that the molecule is promoted to the $3p_z$ molecular Rydberg state following excitation at 200 nm [55, 56], which launches a coherent structural oscillation. This oscillation was seen to persist even following internal conversion to the 3s state on a sub-picosecond time scale [55].

Since the structural motions in the excited state are coherent, the distribution of molecular structures in the excited-state ensemble varies approximately normally [54] around a single representative structure. It is this feature that allowed for the development of a novel analysis of time-resolved X-ray scattering patterns that yields complete, time-dependent molecular structures with unprecedented precision.

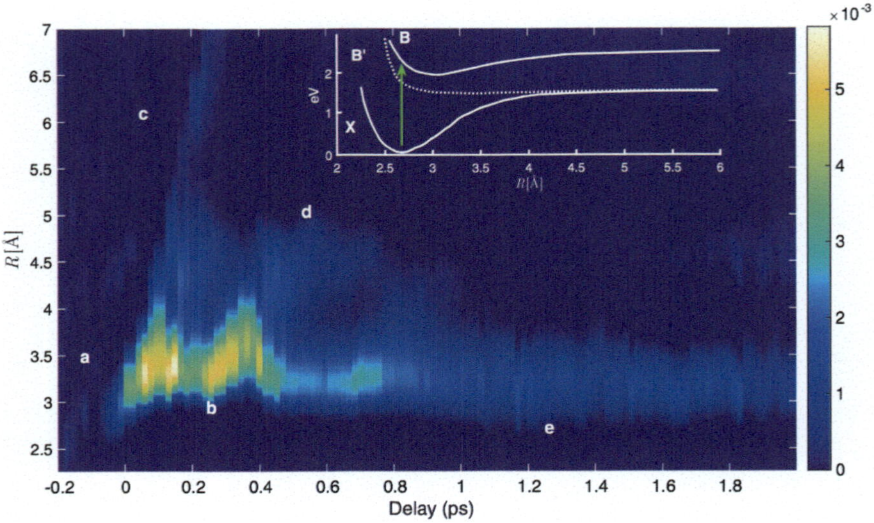


**Fig. 7.6** The pump-probe scattering signal of iodine, $I_2$, following excitation at 520 nm, which excites the molecules from the X to the B surface, inset. Shown is the radial charge distribution obtained from the isotropic component of the scattering signal. From Ref. [50]

The analysis begins with calculation of a large set of surface-hopping trajectories propagated on the ground and excited electronic states, with the appropriate amount of excess kinetic energy included. While each trajectory is unique, all trajectories sample physically reasonable and energetically accessible regions of the structural phase space. From the trajectories, approximately one million individual molecular geometries were extracted, from which a corresponding set of scattering patterns were calculated.

Using this pool of calculated scattering patterns, the least-squares fitting errors of each pattern compared to the experimentally measured signal are calculated at each time point. Given that the least-squares fitting error was observed to vary approximately normally as a function of each individual structural parameter, the best-fit values of all structural parameters are obtained independently at each time point. This results in a full set of molecular structure parameters (i.e. the molecular structure) at each time point, which are then viewed in succession to reveal a detailed picture of the time-dependent molecular motions. Selected representative structural parameters of NMM as a function of time are shown in Fig. 7.7.

### 7.4.3 Correlated Structures Far from Equilibrium

While scattering experiments measure the electron and charge density distributions in molecules, the concept of '*molecular structure*' itself merits careful thought. In

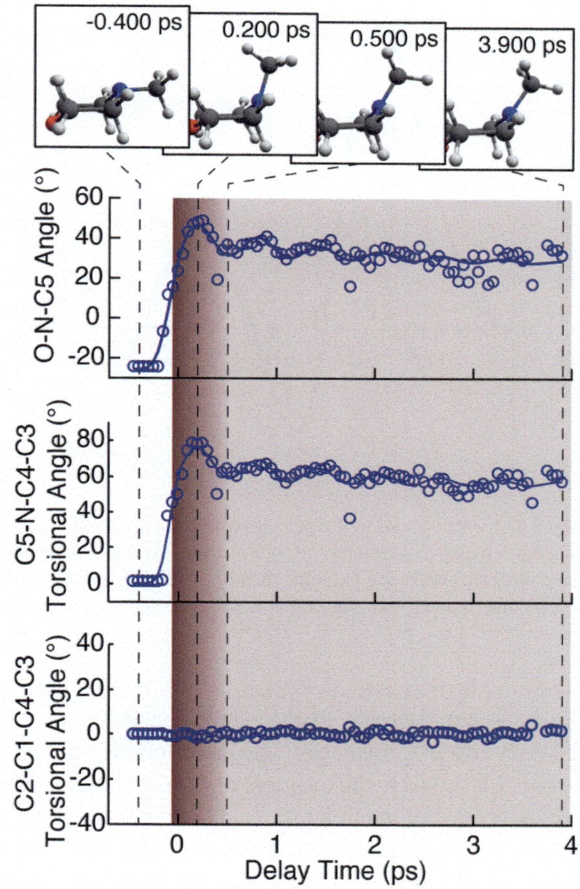


**Fig. 7.7** Time dependence of selected structural parameters of NMM following Rydberg excitation. The O–N–C5 angle (top), the C5–N–C4–C3 torsional angle (middle) and the C2–C1–C4–C3 torsional angle (bottom), extracted from the structural determination, are shown along with their respective $1\sigma$ error bars. The dynamic fits to the respective vibrational motions of the O–N–C5 angle and C5–N–C4–C3 torsional angle are also shown as solid lines. The approximate lifetime of the initially excited $3p_z$ Rydberg state (determined from separate photoelectron measurements) is shown as a dark red shaded region, which corresponds to the $3s$ state when the colour is lighter. Example molecular structures for selected time points are also shown. From Ref. [54]

reacting systems, the chemical dynamics oftentimes starts out as a wavepacket with a well-defined maximum that resembles a classical system. But during a reaction, there is a large amount of energy in play, which typically is distributed into a hot bath as the molecule approaches to a thermal state. The effect of this energy redistribution is readily seen in the experiments on $I_2$ and NMM discussed above, Figs. 7.6 and 7.7. Hot molecular systems may have large amplitude vibrations that can, depending on the structural rigidity of the molecule, lead to geometries that are far from equilibrium. To describe their scattering signals, distance shifts, anharmonicities, and structural correlations need to be considered [57]. The anharmonicity of the interatomic or normal mode potentials causes average bond distances to be longer than the equilibrium distance. A vibrationally hot molecule therefore has a different scattering pattern than a cold molecule, Fig. 7.8.

Depending on the molecular geometry, interatomic distances may shrink or increase with vibrational excitation. The inset in Fig. 7.8 illustrates the effect on

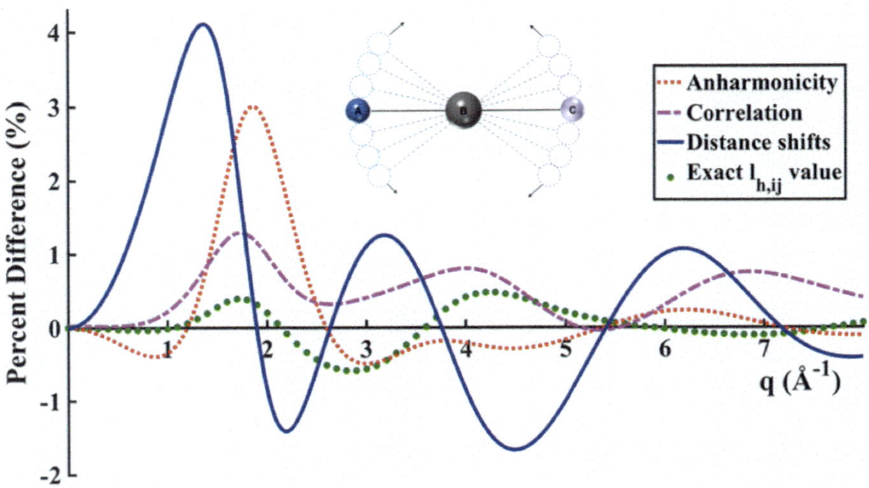


**Fig. 7.8** Contributions to the percent difference scattering signals of the CHD model system, after excitation with 6 eV photons and subsequent relaxation into thermal (2870 K) vibrations. Shown are the contributions arising from the distance shifts, anharmonicity, correlations between atom–atom pair distances and exact vibrational amplitudes $l_{h,ij}$. From Ref. [57]

the example of a triatomic system, ABC. Vibrational excitation of the bending vibration causes the average distance between atoms A and C to be less than the sum of the A-B and B-C distances. The effect on the scattering pattern can be substantial, as Fig. 7.8 shows in the example of the CHD molecule. Depending on the scattering vector, it may be much larger than the more intuitive changes arising from complicated structure distributions associated with complex potential energy surfaces far from the equilibrium structure ('exact vibrational amplitudes $l_{h,ij}$').

The Debye formula, given in Eq. 7.8, expresses the scattering signal from interferences due to the distance between pairs of atoms. In an $N_{at}$-atom molecule, there are $\frac{1}{2}N_{at}(N_{at}-1)$ such interatomic distances yet there are only $3N_{at}-6$ normal mode coordinates. For any molecule with more than 4 atoms, there are therefore fewer normal modes than interatomic distances. Consequently, there must be redundant parameters, implying that correlations between interatomic distances need to be considered for vibrationally very hot molecules. Figure 7.8 illustrates that the effect can be considerable, although not quite as large as the contributions from the distance shifts and the anharmonicity. A complete analysis, considering all the described effects, can simulate scattering patterns that are in excellent agreement with experimentally measured patterns [57]. Importantly, the analysis of the NMM excited state structure described above inherently takes structural correlations, as well as the other effects, into consideration.

### 7.4.4 Chemical Kinetics

X-ray scattering experiments have recently also proven to be powerful tools for investigating excited-state intermediates during kinetic processes. In contrast to dynamic experiments, which probe the molecular structure of a reacting molecule on the timescale of the nuclear motions, kinetic experiments seek to measure the relative populations of reactant and product species as a function of time. In the case of long-lived intermediates, the measured scattering signals include the characteristic pattern of the intermediates, and so their population can also be monitored.

Recently, X-ray scattering was used to investigate the photodissociation of a prototypical tertiary amine, trimethylamine (TMA) [58]. Previous photoionization and molecular beam studies showed that when TMA is optically excited near 200 nm, the $3p_z$ molecular Rydberg state is prepared, which quickly undergoes internal conversion to $3p_x$ and $3p_y$, which subsequently decay into the 3s Rydberg state on a picosecond time scale [59, 60]. In addition, Forde et al. measured the product distributions following excitation at 193 nm, which identified two significant photodissociation channels: the dominant one involving a stepwise reaction to form N-methylmethanimine (NMMA), and a secondary channel forming dimethylamine (DMA) and methyl radicals [61, 62]. Left unknown, however, were the kinetic timescales for product formation and the molecular geometries of the product species.

In the study by Ruddock et al. [58], TMA was excited by a 200 nm optical pulse and probed via hard X-ray scattering. The scattering patterns revealed two distinct photodissociation pathways: a minor fast dissociation pathway from the 3p state, and a dominant slower dissociation which proceeds via internal conversion to the 3s state. Both pathways were found to yield dimethylamine and methyl radicals, with no direct evidence for the formation of NMMA. In addition, the scattering signals at long delay times showed a clear signature of the DMA radical (see Fig. 7.9), which allowed for extraction of a detailed molecular structure of this transient.

More recently, the same technique was used to investigate the kinetics of cyclohexadiene (CHD) following excitation at 200 nm to the 3p molecular Rydberg state [63]. In addition to the ultrafast experiments studying CHD excited near 267 nm as previously discussed, the molecular response of CHD to deep UV excitation near 200 nm has also been the subject of several investigations. Ultrafast photoelectron experiments by Bühler et al. [64] revealed that the initially excited Rydberg state quickly decays, but the experiments showed no evidence of ring opening. In contrast, Sekikawa and coworkers [65, 66] used two-photon excitation with 400 nm pulses to investigate the same region of the absorption spectrum, and suggested that the decay of the initially excited state-initiated ring opening on a sub-picosecond time scale. Given the lack of consensus on the molecular structure following decay of the 3p state, a more definitive determination of the time-evolving molecular structure was deemed necessary.

In 2019, Ruddock et al. [63] used time-resolved X-ray scattering to probe the molecular structure of CHD following excitation with 200 nm optical pulses. The difference scattering signals were measured from the early femtosecond time scale

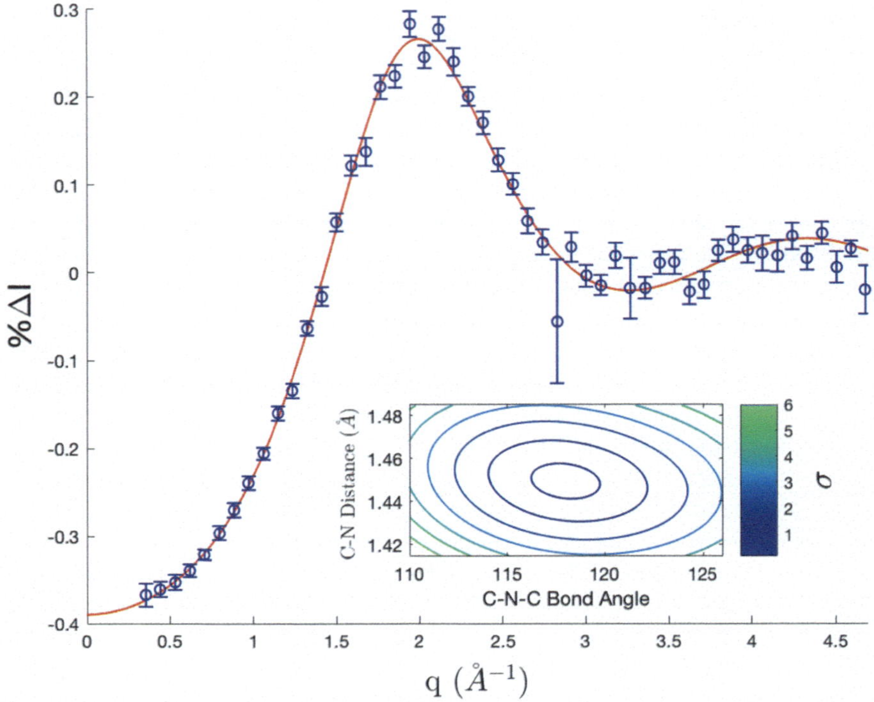


**Fig. 7.9** Difference scattering patterns of DMA transient as obtained from the experimental data at long time delays (blue dots) and computed using an independent atom model (red line). The red line extrapolates to $-0.389$ at $q \rightarrow 0$. Inset: residuals from the fit for different values of the bond length and bond angle. Contour lines are at 0.2, 1, 2, 3, 4, 5, and 6 $\sigma$. From Ref. [58]

out to 1 ns (see Fig. 7.10), providing a full picture of the reaction progress. Analysis of the measured scattering patterns showed that the initially excited $3p_x$ and $3p_y$ Rydberg states quickly find their way to the electronic ground state, where initially 76% of the molecules are in the closed-ring CHD form, with the remainder having undergone rapid ring-opening to form the hexatriene (HT) product. Then, the hot ground-state CHD undergoes a thermal reaction to form the HT product on a 174 ps time scale [63] (see Fig. 7.11). In addition to determining the kinetic parameters for this reaction, further analysis of the product scattering pattern revealed a thermal distribution of rotameric forms in the HT product. In addition to being a powerful tool for studying ultrafast reactions, time-resolved X-ray scattering has thus proven to be a valuable technique for determining reaction rates, revealing multiple reaction pathways, and precisely measuring product molecular structures in kinetic reactions.

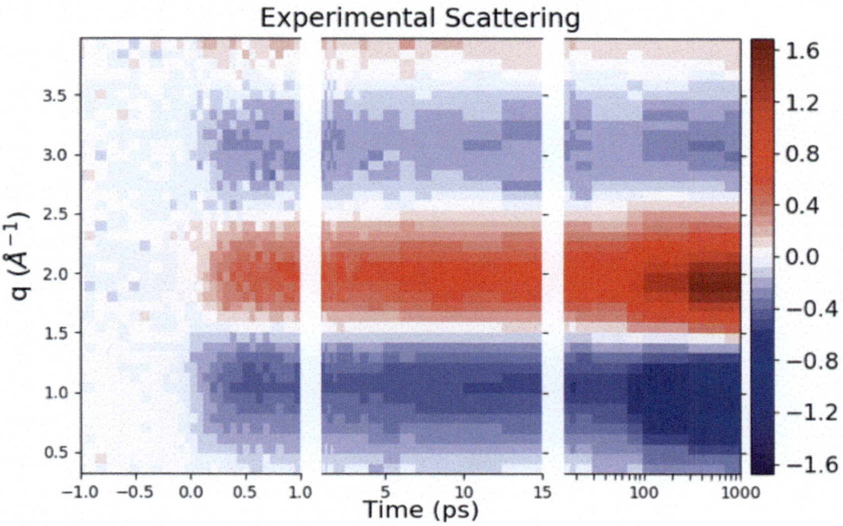


**Fig. 7.10** The isotropic component of the time-dependent experimental percent difference scattering signal of CHD as a function of time. Plotted is the percent difference in scattering intensity (color bar) induced by the laser pulse as a function of the absolute value of the scattering vector, q, and the pump-probe time delay. Because of the large range of experimental delay times the panels are divided into three time segments: one from −1 to 1 ps to show the ultrafast temporal response to the pump laser pulse; the times from 1 to 15 ps showing the initial ground-state population; and the time range from 15 ps to 1 ns (on a log scale) giving the increase in HT population as the molecules equilibrate on the ground state potential energy surface. From Ref. [63]

## *7.4.5 Counting Electrons*

For small scattering angles, the scattered amplitudes from different electrons in a molecule add coherently. As the scattering vector approaches zero, the X-ray scattering signal of a molecule with $N$ electrons is proportional to $N^2$, a result of the coherent addition of amplitudes [58],

$$S(q \rightarrow 0) = N_e^2. \tag{7.12}$$

Free molecules in the gas phase are at random separations, so that scattered amplitudes from different molecules add up incoherently. When absorption of a photon fragments a molecule into components with electron counts $N_\alpha$, with $N_e = \sum_\alpha N_\alpha$, the scattering signal close to $q = 0$ is proportional to the sum of the fragment electron counts squared,

$$S(q \rightarrow 0) = \sum\nolimits_\alpha N_\alpha^2. \tag{7.13}$$

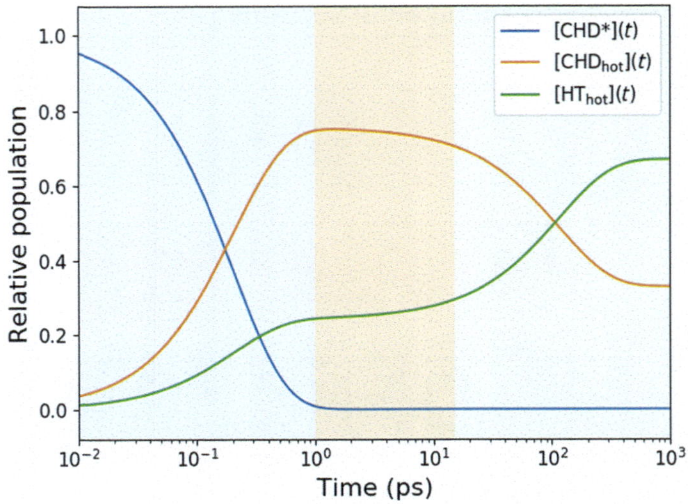


**Fig. 7.11** Relative populations of the three components of the photoreaction on a logarithmic timescale: Electronically excited CHD*, the hot ground-state $CHD_{hot}$, and the hot ground-state $HT_{hot}$. The background colors indicate the different time regimes of the kinetics: Blue is dominated by the decay of CHD*, orange is dominated by the hot CHD, and green indicates the equilibration to 67% $HT_{hot}$ and 33% $CHD_{hot}$. From Ref. [63]

The sum of the squares is always smaller than the square of the sum, and consequently any dissociation is associated with a reduction in the X-ray scattering signal for very small scattering vectors. Experimentally, this was observed on the dissociation of trimethyl amine (TMA) to methyl radicals and dimethyl amine (DMA). Figure 7.12 shows the scattering signals at various delay times. In the sub-picosecond regime, the molecules are bonded, as witnessed by the convergence of the percent difference scattering signal toward zero with small $q$, even though the optical excitation at 200 nm changes the scattering signals at larger $q$. Since TMA has 34 electrons, the scattering signal extrapolated to $q \to 0$ is proportional to $34^2 = 1156$ for the undissociated molecules.

As the fragmentation proceeds with increasing time delays, the percent difference scattering signal approaches a lower value. The methyl radical has 9 electrons, and DMA has 25, so that the sum of the scattering signals becomes proportional to $25^2 + 9^2 = 706$. Therefore, once the reaction to DMA and $CH_3$ is complete, the $q \to 0$ percent difference signal is expected to reduce by $100(706 - 1156)/1156 = -38.9\%$ compared to the TMA molecule. This is exactly what is observed in the experiment, shown in Fig. 7.12.

The quantitative measurement of the scattering signals at very low scattering angles can, therefore, provide an accurate determination of the electron counts of the fragments involved. This can be very useful for the identification of fragmentation pathways and kinetic reaction schemes. In the case of TMA, it led to the discovery of a fast reaction channel that had previously not been identified.

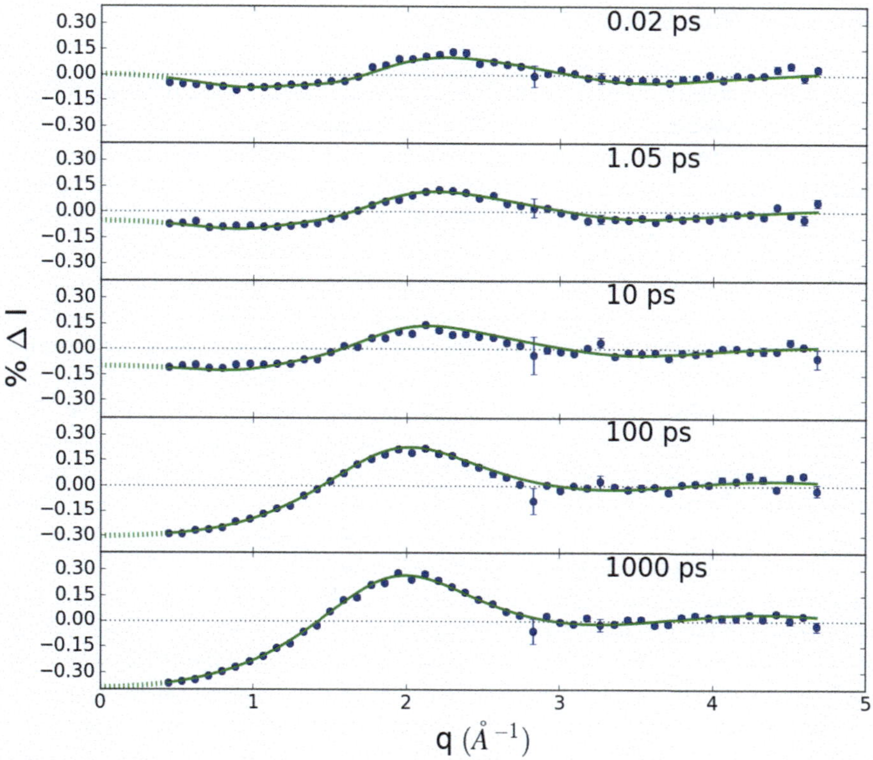


**Fig. 7.12** Experimental pump-probe X-ray scattering percent difference signals (blue dots) of TMA at select delay times as indicated. The signals are scaled by the experimentally determined percent excitation of 1.83%. The green solid lines are fits and dashed lines are extrapolated for $q \to 0$. A dissociation reaction leads to a characteristic decrease of the signal at small values of $q$. From Ref. [58]

A persistent question in pump-probe scattering experiments relates to the identity of the initially excited state. For any optical excitation that creates a sizeable population in an excited state, there is the possibility that absorption of additional photons leads to yet higher electronic states. For many experiments with electronic excitation in the UV part of the spectrum, absorption of a second photon would lead to the ionization of the molecule. A molecule with $N_e$ electrons would become a cation with $N_e - 1$ electrons, so that the small-angle scattering signal would be reduced from $N_e^2$ to $(N_e - 1)^2 + 1^2$. For many prototypical chemical reaction dynamics systems, the effect is in the range of several percent, so that a careful measurement of the pump-probe scattering signals can experimentally determine whether a significant fraction of the molecules have absorbed a second photon.

### *7.4.6 Electron Density Distributions*

As previous examples illustrate, scattering can reveal structure and structural changes in molecules. In the context of X-ray scattering, this is because a significant portion of the signal tracks the positions of nuclei, predominantly via scattering from the tightly bound core electrons. However, scattering occurs from *all* electrons and the valence electrons also contribute. These can be significantly distorted by chemical bonding. The discrepancy between a nucleus-centered electron distribution and actual observations was first observed in high-resolution X-ray crystallography, in particular for light elements such as hydrogen. In those situations, an expansion of the electron density using a multipole model was found to provide a more accurate description of the experimentally observed scattering patterns and the associated electron density [67]. Ultrafast dynamics involves excited electronic states, which have electron distributions distinct from the ground electronic state. The initial change in electronic structure due to vertical excitation by the pump pulse provides one of the cleanest examples of this. An early theoretical study by Ben-Nun et al. investigated the change in elastic X-ray scattering due to excitation of a lithium atom from the $^2S$ electronic ground state to the first excited $^2P$ state [68]. Comparing the calculated signal for the ground state versus the excited state, shown in Fig. 7.13, the effect of the redistribution of electron density upon excitation is seen clearly in the scattering pattern. More recently, Northey et al. considered the changes in the elastic scattering upon vertical excitation in molecular systems [9]. In Fig. 7.14, the calculated elastic scattering from the excited $S_1$ state in the molecule 1,3-cyclohexadiene is shown. The magnitude of the changes in the scattering pattern is comparatively small, as seen in the lower panel, which shows the difference in scattering between the $S_1$ excited and the $S_0$ ground state. This relates to the comparatively minor changes in electron density, in first approximation on the order of one electron moving between, e.g. the highest occupied and the lowest unoccupied molecular orbital. However, although the changes are small in absolute terms, as a relative percent difference change they can be significant [9, 10], with the change at specific pixels in the current example reaching 40% for an aligned molecule.

Experimental detection of changes in electronic structure via X-ray scattering requires high-quality data. Early experiments could quite adequately be interpreted using the independent atom model (IAM), which only accounts for the electron density of a molecule in an approximate manner and cannot describe excited electronic states, as discussed in Sect. 7.2. The recent experiment by Stankus et al. [54] is indicative of the advances that have been made since. In that experiment, excited state structural dynamics in the molecule NMM was measured (see Sect. 7.4.2). Interpretation of the data required that the IAM was augmented by a correction from electronic structure theory that accounted for the change in electron density due to the molecular dynamics evolving on diffuse excited $3p_z$ and 3s Rydberg states. However, the measurement was insufficient to positively identify the electronic state along the lines of previous theoretical predictions in atoms [69, 70]. As Fig. 7.15 shows, the correction due to electron density redistribution is small compared to the

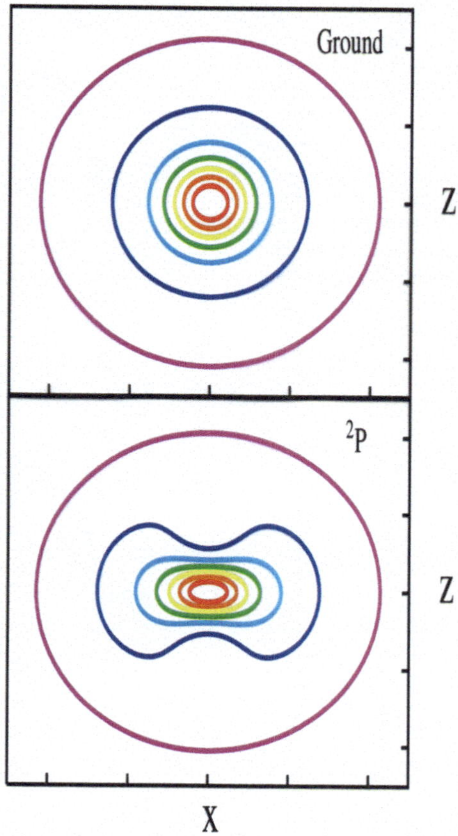


**Fig. 7.13** Calculated elastic X-ray scattering for Li atoms in the ground state (upper panel) and when excited by a linearly polarized photon to the $^2$P state (lower panel). The change in the electron density distribution upon excitation leads to a change in the X-ray scattering intensity. From Ref. [68]

changes in the signal resulting from changes in the molecular geometry, but is not negligible. Going forward, it is likely to become increasingly important to use accurate models that account for the electronic structure when interpreting experimental data. This work is well underway. Codes that calculate elastic scattering from ground and excited states have been developed [9, 10] and can handle larger molecules [71]. The codes can also compute scattering due to specific inelastic transitions [13] as well as total scattering [3]. To be practical for the interpretation of large volumes of experimental data, these methods have to be computationally efficient [4, 5].

Looking further ahead, it does appear likely that ultrafast X-ray scattering experiments will eventually succeed to track not only the initial change in electron density due to excitation by the optical pump pulse, which would be an important result in itself, but also subsequent transitions between electronic states due to nonadiabatic or spin–orbit coupling as the molecule relaxes towards the product states. Such experiments would close the gap between ultrafast structural dynamics and existing ultrafast spectroscopies. A particularly attractive aspect of such experiments would be the ability to track structural and electronic changes in a single experiment. However,

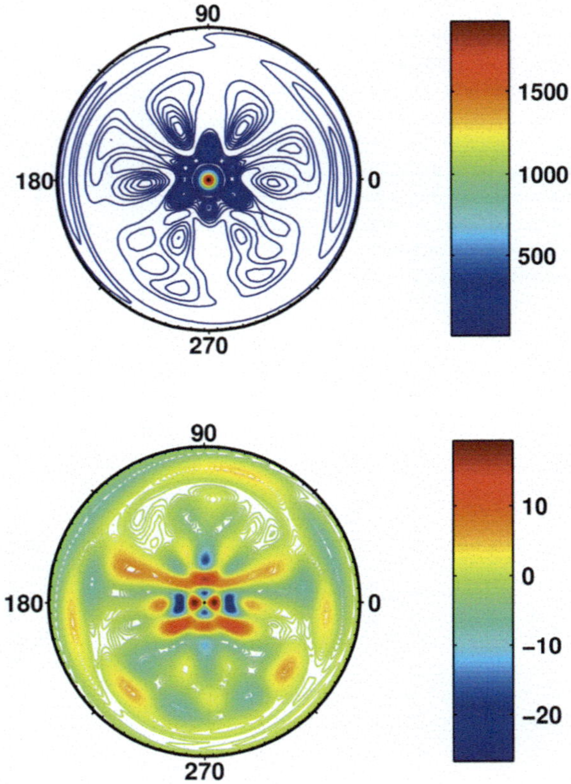


**Fig. 7.14** Calculated elastic scattering from vertically excited electronic S1 state in perfectly aligned 1,3-cyclohexadiene molecules. (Upper panel) Elastic scattering from the S1 state in the ground state geometry. (Lower panel) The difference between the ground S0 and the first excited S1 state scattering. From Ref. [9]

challenges remain. Apart from requiring experimental signals of very high quality and methods to predict the scattering for various electronic states, one would have to tackle the problem of separating contributions to the scattering signal from changes in electronic structure, changes in population, and molecular geometry. This is a nontrivial task. The most likely route forward in the immediate future is therefore experiments whose interpretation relies on a combination of scattering and spectroscopic data [72, 73]. There is also interesting overlap with quantum crystallography [74, 75], which grapples with similar challenges in the context of high-quality static X-ray diffraction data from ground state molecules

### *7.4.7 Transition Dipoles*

Most optical pump, X-ray probe experiments utilize linearly polarized optical laser radiation. The interaction of the linearly polarized light with the randomly oriented molecules in the gas phase selects out those molecules whose transition dipole

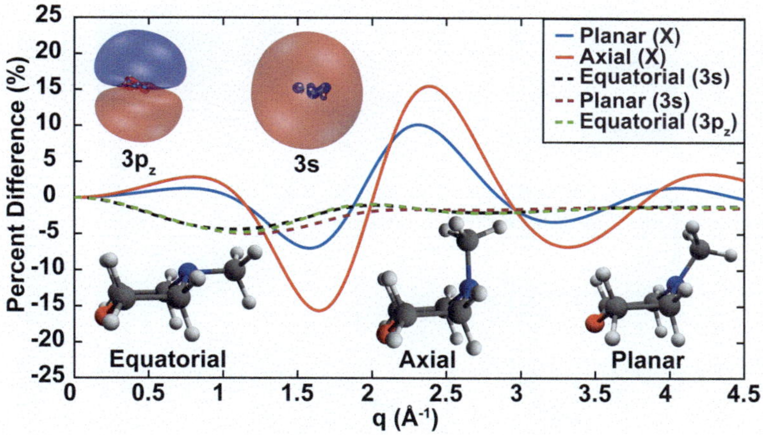


**Fig. 7.15** The calculated difference in rotationally averaged scattering caused by nuclear and electronic structure changes as a function of momentum transfer $q$, assuming 100% excitation of the sample. The three relevant molecular conformations and the orbital plots for the 3$s$ and 3$p_z$ states are included as insets. Different molecular geometries give rise to significant changes in the scattering signals, solid lines. Those arising from the rearranging electronic structures, dashed lines, are smaller, but not negligible. From Ref. [54]

moments align with the laser polarization. This induces an anisotropy that is readily detected in the scattering experiments.

Figure 7.16 illustrates this in the example of the diatomic $I_2$ system [50]. Excitation with a linearly polarized laser pulse at 520 nm leads to a mix of optically excited and ground state molecules. The distribution of both is anisotropic, as the laser excitation has removed molecules from the ground state and created the excited state population. Since the scattering signatures of ground and excited state molecules are different, the overall scattering signal reflects the anisotropy of the molecular ensemble. The anisotropy of the scattering signals turns out to be a very useful tool to determine the excited state prepared in the optical excitation. The alignment of the optical transition dipole moment, as reflected in the scattering patterns, provides important information on the symmetries of the involved states, and can therefore help identify the initially prepared state that is the subject to the observed chemical dynamics. Figure 7.17 illustrates the concept in the NMM molecule [56]. Excitation in the 200 nm region might excite any of the 3p Rydberg states, or a mixture thereof. But since the sublevels of 3p align differently with respect to the frame of the molecule, the resulting pump-probe scattering patterns would be different. A comparison to the experiment, Fig. 7.18, unambiguously determines that the optical excitation is predominantly to the 3p$_z$ level, which, therefore, is the starting point of the ensuing chemical dynamics.

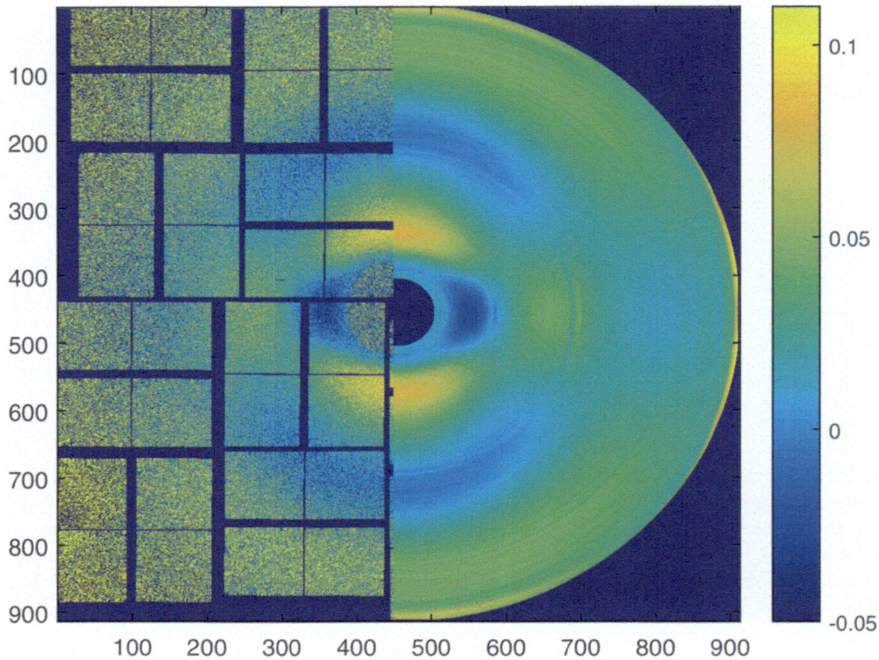


**Fig. 7.16** Left: Half of the megapixel array detector showing the fractional deviation from the mean scattering signal upon excitation of I2 to the B state using a linearly polarized laser pulse. Right: A Legendre polynomial fit. From Ref. [50]

## 7.5 Outlook

Scattering experiments are beginning to contribute a new perspective on the dynamics of molecular reactions. The tremendous intensity of X-ray free electron lasers, paired with the ability to deliver pulses with durations rivaling those of optical lasers, has made the investigation of chemical dynamics of free molecules possible using time-resolved X-ray scattering. Important insights have already been obtained, as described in this chapter. Without doubt, new XFEL technologies will continue to drive advances that will lead to a deeper understanding of the basic functioning of molecules and their transformations. Most time-resolved X-ray scattering experiments to date have been performed with X-ray photon energies below 10 keV, limited by the range of the fundamental beam of the LCLS light source. Upcoming upgrades will extend the range to include higher photon energies. This will be an important step because it will enable the measurement of larger scattering vectors, where the signatures of the detailed shapes of wavepackets are expected to be found [76]. These features occur on small real-space length scales for prototypical reactions such as the ballistic motions of CHD. Consequently, they will appear at higher scattering vectors in the scattering signals, Fig. 7.19. Directly observing the shape and time evolution

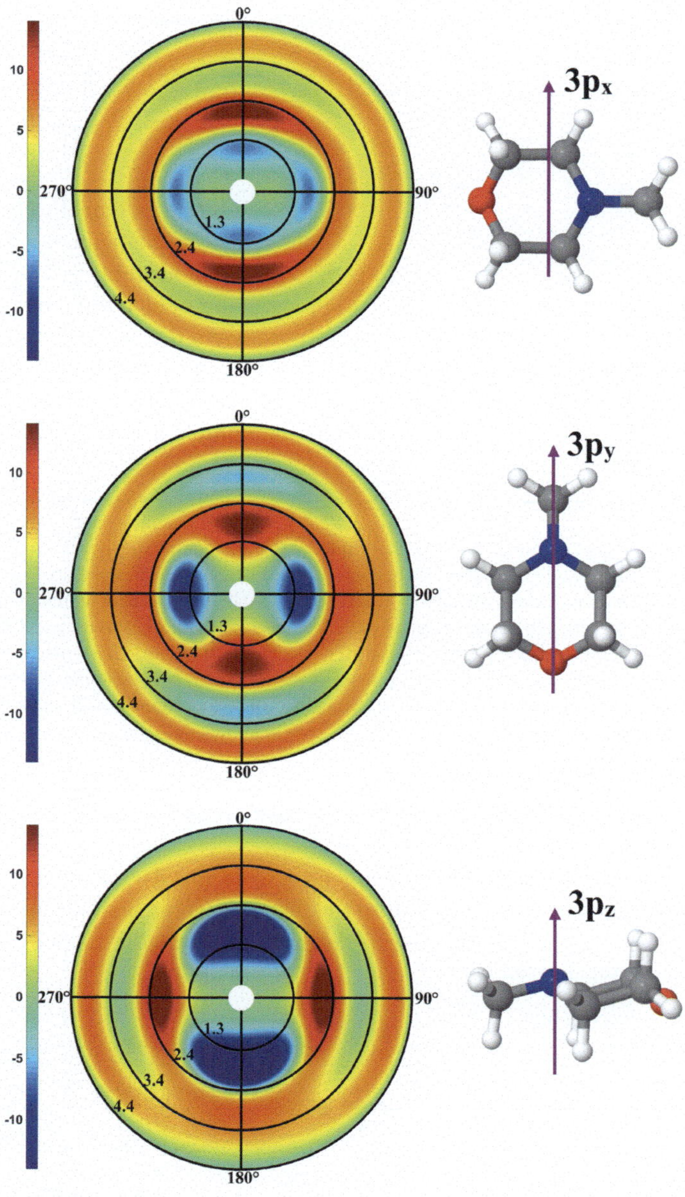


**Fig. 7.17** Simulated percent difference scattering patterns for N-methyl morpholine molecules, left column, excited to the $3p_x$, $3p_y$, and the $3p_z$ electronic Rydberg states, with the orientation of the transition dipole moment (TDM) relative the molecule shown in the right column. A $\cos^2(\theta)$ distribution with respect to the laser polarization axis is assumed and orientations due to rotation about the laser polarization axis are averaged out. In the right column, the orientation of the TDM in the molecular frame is indicated using a purple arrow, as calculated from MRCI(2,5)/6–311 + G(d). From Ref. [56]

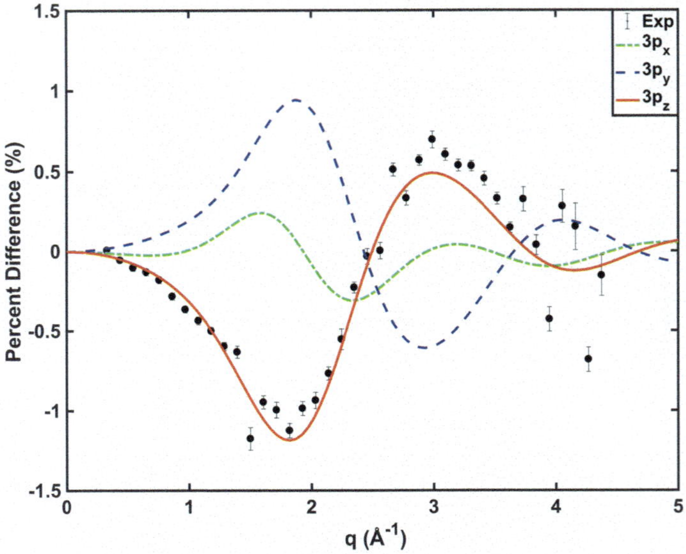


**Fig. 7.18** The anisotropic scattering signal of NMM, derived from experimental results at a pump-probe delay time of t = 150 fs, black symbols. The three theoretical signals are derived from calculations. The theoretical curves are scaled with respect to the excitation fraction, and dephasing of the $\cos^2(\theta)$ distribution after time zero is taken into consideration. Comparison of the experimental data with the theoretical curves allows for the determination of the orientation of the optical transition dipole moment. From Ref. [56]

of dynamical wavepackets would be an important advance in our exploration, and eventual control, of chemical reactions.

The manner in which scattering investigations of chemical dynamics complement spectroscopy is an important motivation for further development of the scattering methods. Intriguingly, while ultrafast spectroscopy has been developed for several

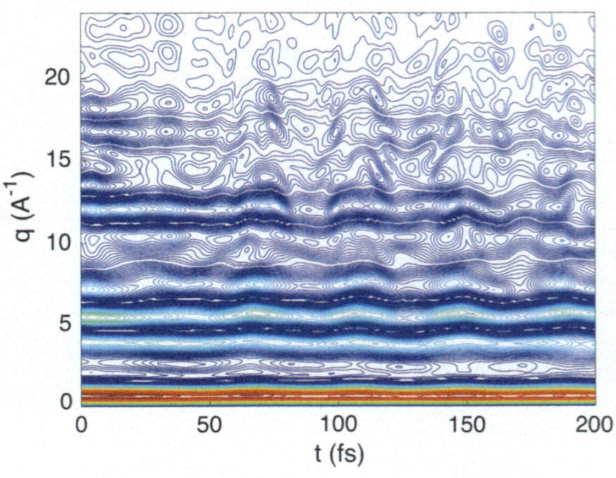


**Fig. 7.19** Simulated contour plots of the isotropic, modified (elastic) scattering intensity for the ring-opening reaction of 1,3-cyclohexadiene (CHD), according to Ref. [76]. The time-evolving wavepacket causes intricate patterns at large scattering vectors

decades and has become very powerful, it may well be ultrafast scattering that ultimately delivers more insight. As discussed in Sect. 7.4.6, we may eventually be able to assign electronic states during the dynamics based on analysis of the actual electron distribution, and theoretical predictions of coherent effects in scattering indicate that it may be possible to access information on electronic transitions, transient electron dynamics, and nuclear coherences [77–80]. Furthermore, all time-resolved spectroscopic experiments are limited by the Heisenberg uncertainty relation, which imposes a fundamental limit on measuring the energies of spectroscopic transitions and time simultaneously. A scattering experiment ultimately yields positions, in the form of charge or electron density distributions. Since the Heisenberg uncertainty relation imposes no fundamental limit on the simultaneous measurement of positions and time, it may be that, once the technologies are fully developed, scattering experiments may give us the most detailed views of chemical dynamics.

## References

1. W. Schülke, Electron Dynamics by Inelastic X-ray Scattering, 1st edn. (Oxford Science Publications, 2007)
2. I. Waller, D.R. Hartree, On the intensity of total scattering of x-rays. Proc. R. Soc. Lond. Ser. A **124**, 119–142 (1929). https://doi.org/10.1098/rspa.1929.0101
3. A. Moreno Carrascosa, H.W. Yong, D. Crittenden, P. M. Weber, A. Kirrander, Ab-initio calculation of total x-ray scattering from molecules. J. Chem. Theory Comp. **15**, 2836–2846 (2019). https://doi.org/10.1021/acs.jctc.9b00056
4. R.M. Parrish, T.J. Martínez, Ab Initio computation of rotationally-averaged pump-probe x-ray and electron diffraction signals. J. Chem. Theory Comp. **15**, 1523–1537 (2019). https://doi.org/10.1021/acs.jctc.8b01051
5. N. Zotev, A. Moreno Carrascosa, M. Simmermacher, A. Kirrander, Excited electronic states in total isotropic scattering from molecules. J. Chem. Theory Comp. (2019). https://doi.org/10.1021/acs.jctc.9b00670
6. P. Debye, Zerstreuung von Röntgenstrahlen. Ann. Phys. **351**, 809 (1915). https://doi.org/10.1002/andp.19153510606
7. A. Kirrander, K. Saita, D. Shalashilin, Ultrafast x-ray scattering from molecules. J. Chem. Theory Comput. **12**, 957–967 (2016). https://doi.org/10.1021/acs.jctc.5b01042
8. E. Prince, *International Tables for Crystallography, Vol. C, Mathematical, Physical and Chemical Tables* (Kluwer Academic Publishers, 2004)
9. T. Northey, N. Zotev, A. Kirrander, Ab initio diffraction from molecules. J. Chem. Theory Comp. **10**(11), 4911–4920 (2014). https://doi.org/10.1021/ct500096r
10. T. Northey, A. Moreno Carrascosa, S. Schäfer, A. Kirrander, Elastic x-ray scattering from state-selected molecules. J. Chem. Phys. **145**, 154304 (2016). https://doi.org/10.1063/1.4962256
11. A. Moreno Carrascosa, T. Northey, A. Kirrander, Imaging rotations and vibrations in polyatomic molecules with x-ray scattering. Phys. Chem. Chem. Phys. **19**, 7853–7863 (2017). https://doi.org/10.1039/C6CP06793J
12. M. Inokuti, Inelastic collisions of fast charged particles with atoms and molecules—the Bethe theory revisited. Rev. Mod. Phys. **43**(3), 297–347 (1971). https://doi.org/10.1103/RevModPhys.43.297
13. A. Moreno Carrascosa, A. Kirrander, Inelastic ab initio scattering. Phys. Chem. Chem. Phys. **19**, 19545–19553 (2017). https://doi.org/10.1039/C7CP02054F

14. M. Stefanou, K. Saita, D.V. Shalashilin, A. Kirrander, Comparison of ultrafast electron and x-ray diffraction—a computational study. Chem. Phys. Lett. **683**, 300–305 (2017). https://doi.org/10.1016/j.cplett.2017.03.007
15. F.M. Rudakov, J.B. Hastings, D.H. Dowell, J.F. Schmerge, P.M. Weber, in *AIP Conference Proceedings*, vol. 845 (2006), p. 1287https://doi.org/10.1063/1.2263560
16. F. Salvat, Elastic scattering of fast electrons and positrons by atoms. Phys. Rev. A **43**, 578 (1991). https://doi.org/10.1103/PhysRevA.43.578
17. J.B. Hastings, F.M. Rudakov, D.H. Dowell, J.F. Schmerge, J.D. Cardoza, J.M. Castro, S.M. Gierman, H. Loos, P.M. Weber, Ultrafast time-resolved electron diffraction with megavolt electron beams. App. Phys. Lett. **89**, 184109 (2006). https://doi.org/10.1063/1.2372697
18. L.O. Brockway, Electron diffraction by gas molecules. Rev. Mod. Phys. **8**, 231 (1936). https://doi.org/10.1103/RevModPhys.8.231
19. L.-M. Peng, J.M. Cowley, Errors arising from numerical use of the Mott formula in electron image simulation. Acta Cryst. **A44**, 1–6 (1988). https://doi.org/10.1107/S0108767387006858
20. P.A. Doyle, P.S. Turner, Relativistic Hartree-Fock x-ray and electron scattering factors. Acta Cryst. **A24**, 390–399 (1968). https://doi.org/10.1107/S0567739468000756
21. V. Vand, P.F. Eiland, R. Pepinsky, Analytical representation of atomic scattering factors. Acta Cryst. **10**, 303–306 (1957). https://doi.org/10.1107/S0365110X57000882
22. G.H. Smith, R.E. Burge, The analytical representation of atomic scattering amplitudes for electrons. Acta Cryst. **15**, 182–186 (1962). https://doi.org/10.1107/S0365110X62000481
23. L. Ma, H. Yong, J.D. Geiser, A. Moreno Carrascosa, N. Goff, P.M. Weber, Ultrafast x-ray and electron scattering of free molecules: a comparative evaluation. Struct. Dynam. **7**, 034102 (2020). PMCID: PMC7316516. https://doi.org/10.1063/4.0000010
24. J.D. Geiser, P.M. Weber, in *Proceedings, SPIE Conference on Time Resolved Electron and X-Ray Diffraction*, San Diego, vol. 2521 (July 1995), pp. 136–144. https://doi.org/10.1117/12.218345
25. P.M. Weber, S.D. Carpenter, T. Lucza, in *Proceedings, SPIE Conference on Time Resolved Electron and X-Ray Diffraction*, San Diego, vol. 2521 (July 1995), pp. 23–30https://doi.org/10.1117/12.218364
26. J. Thompson, P.M. Weber, P.J. Estrup, in *Proceedings, SPIE Conference on Time Resolved Electron and X-Ray Diffraction*, San Diego, vol. 2521 (July 1995), pp. 113–122.https://doi.org/10.1117/12.218343
27. M. Ben-Nun, T.J. Martinez, P.M. Weber, K.R. Wilson, Direct imaging of excited electronic states using diffraction techniques: theoretical considerations. Chem. Phys. Lett. **262**, 405–414 (1996). https://doi.org/10.1016/0009-2614(96)01108-6
28. J.D. Geiser, P.M. Weber, Pump-probe diffraction imaging of vibrational wave functions. J. Chem. Phys. **108**, 8004–8011 (1998). https://doi.org/10.1063/1.476239
29. S. Ryu, P.M. Weber, R.M. Stratt, The diffraction signatures of individual vibrational modes in polyatomic molecules. J. Chem. Phys. **112**, 1260–1270 (2000). https://doi.org/10.1063/1.480678
30. S. Ryu, R.M. Stratt, P.M. Weber, Diffraction signals of aligned molecules in the gas phase: tetrazine in intense laser fields. J. Phys. Chem. A **107**, 6622–6629 (2003). https://doi.org/10.1021/jp0304632
31. S. Ryu, R.M. Stratt, K.K. Baeck, P.M. Weber, Electron diffraction of molecules in specific quantum states: a theoretical study of vibronically excited s-tetrazine. J. Phys. Chem. A **108**, 1189–1199 (2004). https://doi.org/10.1021/jp031061x
32. R.C. Dudek, P.M. Weber, Ultrafast diffraction imaging of the electrocyclic ring-opening reaction of 1,3-cyclohexadiene. J. Phys. Chem. A **105**, 4167–4171 (2001). https://doi.org/10.1021/jp010122t
33. P.M. Weber, R.C. Dudek, S. Ryu, R.M. Stratt, in *Femtochemistry and Femtobiology: Ultrafast Events in Molecular Science,* ed. by M. Martin, J.T. Hynes (Elsevier, 2004), p. 19
34. J.B. Hastings, F.M. Rudakov, D.H. Dowell, J.F. Schmerge, J. Cardoza, J.M. Castro, S.M. Gierman, H. Loos, P.M. Weber, Ultrafast time-resolved electron diffraction with megavolt electron beams. Appl. Phys. Lett. **89**, 184109 (2006). https://doi.org/10.1063/1.2372697

35. F.M. Rudakov, J.B. Hastings, D.H. Dowell, J.F. Schmerge, P.M. Weber, Megavolt electron beams for ultrafast time-resolved electron diffraction. AIP Conf. Proc. **845**, 1287 (2006). https://doi.org/10.1063/1.2263560
36. R.J.D. Miller, Femtosecond crystallography with ultrabright electrons and x-rays: capturing chemistry in action. Science **343**, 1108 (2014). https://doi.org/10.1126/science.1248488
37. R.J.D. Miller, Mapping atomic motions with ultrabright electrons: the chemists' Gedanken experiment enters the lab frame. Annu. Rev. Phys. Chem. **65**, 583–604 (2014). https://doi.org/10.1146/annurev-physchem-040412-110117
38. J. Yang, M. Guehr, X. Shen, R. Li, T. Vecchione, R. Coffee, J. Corbett, A. Fry, N. Hartmann, C. Hast, K. Hegazy, K. Jobe, I. Makasyuk, J. Robinson, M.S. Robinson, S. Vetter, S. Weathersby, C. Yoneda, X. Wang, M. Centurion, Diffractive imaging of coherent nuclear motion in isolated molecules. Phys. Rev. Lett. **117**, 153002 (2016). https://doi.org/10.1103/PhysRevLett.117.153002
39. T.J.A. Wolf, D.M. Sanchez, J. Yang, R.M. Parrish, J.P.F. Nunes, M. Centurion, R. Coffee, J.P. Cryan, M. Gühr, K. Hegazy, A. Kirrander, R.K. Li, J. Ruddock, X. Shen, T. Veccione, S.P. Weathersby, P.M. Weber, K. Wilkin, H. Yong, Q. Zheng, X.J. Wang, M.P. Minitti, T.J. Martínez, The photochemical ring-opening of 1,3-cyclohexadiene imaged by ultrafast electron diffraction. Nat. Chem. **11**, 504–509 (2019). https://doi.org/10.1038/s41557-019-0252-7
40. A. A. Ischenko, P. M. Weber, R. J. D Miller, Capturing Chemistry in Action with Electrons: Realization of Atomically Resolved Reaction Dynamics, Chemical Reviews, *submitted*.
41. M.P. Minitti et al., Imaging molecular motion: femtosecond x-ray scattering of an electrocyclic chemical reaction. Phys. Rev. Lett. **114**, 1–5 (2015). https://doi.org/10.1103/PhysRevLett.114.255501
42. J.M. Budarz, M.P. Minitti, B. Stankus, A. Kirrander, J.B. Hastings, P.M. Weber, Observation of femtosecond molecular dynamics via pump-probe gas phase x-ray scattering. J. Phys. B At. Mol. Opt. Phys. **49**, 34001 (2016). https://doi.org/10.1088/0953-4075/49/3/034001
43. B. Stankus, H. Yong, J. Ruddock, L. Ma, A. Moreno Carrascosa, N. Goff, S. Boutet, X. Xu, N. Zotev, A. Kirrander, M.P. Minitti, P.M. Weber, Advances in ultrafast gas-phase x-ray scattering. J. Phys. B. Atomic Molecul. Optic. Phys. 53, 23 (2020). https://doi.org/10.1088/1361-6455/abbfea
44. J.F. O'Hanlon, *A User's Guide to Vacuum Technology* (Wiley, 2005)
45. H.T. Philipp, M. Hromalik, M. Tate, L. Koerner, S.M. Gruner, Pixel array detector for x-ray free electron laser experiments. Nucl. Instrum. Methods Phys. Res. Sect. A: Accel. Spectrometers, Detect. Assoc. Equip. **649**, 67–69 (2011). https://doi.org/10.1016/j.nima.2010.11.189
46. M.R. Bionta, et al., Spectral encoding method for measuring the relative arrival time between x-ray/optical pulses. Rev. Sci. Instrum. **85** (2014). https://doi.org/10.1063/1.4893657
47. S. Deb, P.M. Weber, The ultrafast pathway of photon-induced electrocyclic ring opening reactions: the case of 1,3-cyclohexadiene. Annu. Rev. Phys. Chem. **62**, 19–39 (2011). https://doi.org/10.1146/annurev.physchem.012809.103350
48. W. Fuß, W.E. Schmid, S.A. Trushin, Time-resolved dissociative intense-laser field ionization for probing dynamics: femtosecond photochemical ring opening of 1,3-cyclohexadiene. J. Chem. Phys. **112**, 8347 (2000). https://doi.org/10.1063/1.481478
49. I. Polyak, L. Hutton, R. Crespo-Otero, M. Barbatti, P.J. Knowles, Ultrafast photoinduced dynamics of 1,3-cyclohexadiene using XMS-CASPT2 surface hopping. J. Chem. Theory Comp. **15**(7), 3929–3940 (2019). https://doi.org/10.1021/acs.jctc.9b00396
50. J.M. Glownia, A. Natan, J.P. Cryan, R. Hartsock, M. Kozina, M.P. Minitti, S. Nelson, J. Robinson, T. Sato, T. van Driel, G. Welch, C. Weninger, D. Zhu, P.H. Bucksbaum, Self-referenced coherent diffraction x-ray movie of Ångstrom- and femtosecond-scale atomic motion. Phys. Rev. Lett. **117**, 153003 (2016). https://doi.org/10.1103/PhysRevLett.117.153003
51. K. Bennett, M. Kowalewski, S. Mukamel, Comment on "Self-Referenced Coherent Diffraction X-Ray Movie of Ångstrom- and Femtosecond-Scale Atomic Motion." Phys. Rev. Lett. **119**, 069301 (2017). https://doi.org/10.1103/PhysRevLett.119.069301
52. J.M. Glownia, A. Natan, J.P. Cryan, R. Hartsock, M. Kozina, M.P. Minitti, S. Nelson, J. Robinson, T. Sato, T. van Driel, G. Welch, C. Weninger, D. Zhu, P.H. Bucksbaum, Glownia

et al. Reply: Phys. Rev. Lett. **119**, 069302 (2017). https://doi.org/10.1103/PhysRevLett.119.069302
53. M.R. Ware, J.M. Glownia, A. Natan, J.P. Cryan, P.H. Bucksbaum, On the limits of observing motion in time-resolved x-ray scattering. Phil. Trans. R. Soc. A **377**, 20170477 (2019). https://doi.org/10.1098/rsta.2017.0477
54. B. Stankus, H. Yong, N. Zotev, J.M. Ruddock, D. Bellshaw, T.J. Lane, M. Liang, S. Boutet, S. Carbajo, J.S. Robinson, W. Du, N. Goff, Y. Chang, J.E. Koglin, M.P. Minitti, A. Kirrander, P.M. Weber, Ultrafast x-ray scattering reveals vibrational coherence following Rydberg excitation. Nat. Chem. **11**, 716–721 (2019). https://doi.org/10.1038/s41557-019-0291-0
55. Y. Zhang, H. Jónsson, P.M. Weber, Coherence in nonradiative transitions: internal conversion in Rydberg-excited N-methyl and N-ethyl morpholine. Phys. Chem. Chem. Phys. **19**, 26403–26411 (2017). https://doi.org/10.1039/C7CP05244H
56. H. Yong et al., Determining orientations of optical transition dipole moments using ultrafast x-ray scattering. J. Phys. Chem. Lett. **9**, 6556–6562 (2018). https://doi.org/10.1021/acs.jpclett.8b02773
57. H. Yong, J.M. Ruddock, B. Stankus, L. Ma, W. Du, N. Goff, Y. Chang, N. Zotev, D. Bellshaw, S. Boutet, S. Carbajo, J.E. Koglin, M. Liang, J.S. Robinson, A. Kirrander, M.P. Minitti, P.M. Weber, Scattering off molecules far from equilibrium. J. Chem. Phys. **151**, 084301 (2019). https://doi.org/10.1063/1.5111979
58. J.M. Ruddock, N. Zotev, B. Stankus, H. Yong, D. Bellshaw, S. Boutet, T.J. Lane, M. Liang, S. Carbajo, W. Du, A. Kirrander, M.P. Minitti, P.M. Weber, Simplicity beneath complexity: counting molecular electrons reveals transients and kinetics of photodissociation reactions. Angew. Chem. Int. Ed. **58**, 6371–6375 (2019). https://doi.org/10.1002/anie.201902228
59. J.D. Cardoza, F.M. Rudakov, P.M. Weber, Electronic spectroscopy and ultrafast energy relaxation pathways in the lowest rydberg states of trimethylamine. J. Phys. Chem. **112**, 10736–10743 (2008). https://doi.org/10.1021/jp8041236
60. J.D. Cardoza, P.M. Weber, Resolved: electronic states underneath broad absorptions. J. Chem. Phys. **127**, 036101–036102 (2007). https://doi.org/10.1063/1.2751189
61. N.R. Forde, M.L. Morton, S.L. Curry, S.J. Wrenn, L.J. Butler, Photodissociating trimethylamine at 193 nm to probe dynamics at a conical intersection and to calibrate detection efficiency of radical products. J. Chem. Phys. **111**, 4558–4568 (1999). https://doi.org/10.1063/1.479217
62. N.R. Forde, L.J. Butler, B. Ruscic, O. Sorkhabi, F. Qi, A. Suits, Characterization of nitrogen-containing radical products from the photodissociation of trimethylamine using photoionization detection. J. Chem. Phys. **113**, 3088–3097 (2000). https://doi.org/10.1063/1.1286973
63. J.M. Ruddock, H. Yong, B. Stankus, W. Du, N. Goff, Y. Chang, A. Odate, A.M. Carrascosa, D. Bellshaw, N. Zotev, M. Liang, S. Carbajo, J. Koglin, J.S. Robinson, S. Boutet, A. Kirrander, M.P. Minitti, P.M. Weber, A deep UV trigger for ground-state ring-opening dynamics of 1,3-cyclohexadiene. Sci. Adv. **5**, eaax6625 (2019). https://doi.org/10.1126/sciadv.aax6625
64. C.C. Bühler, M.P. Minitti, S. Deb, J. Bao, P.M. Weber, Ultrafast dynamics of 1,3-cyclohexadiene in highly excited states. J. At. Mol. Opt. Phys. **6** (2011). https://doi.org/10.1155/2011/637593
65. R. Iikubo, T. Sekikawa, Y. Harabuchi, T. Taketsugu, Structural dynamics of photochemical reactions probed by time-resolved photoelectron spectroscopy using high harmonic pulses. Faraday Disc. **194**, 147–160 (2016). https://doi.org/10.1039/C6FD00063K
66. K. Kaneshima, Y. Ninota, T. Sekikawa, Time-resolved high-harmonic spectroscopy of ultrafast photoisomerization dynamics. Opt. Express **26**, 31039–31054 (2018). https://doi.org/10.1364/OE.26.031039
67. R.F. Stewart, J. Bentley, B. Goodman, Generalized x-ray scattering factors in diatomic molecules. J. Chem. Phys. **63**(9), 3786–3793 (1975). https://doi.org/10.1063/1.431871
68. M. Ben-Nun, T.J. Martinez, P.M. Weber, K.R. Wilson, Direct imaging of excited electronic states using diffraction techniques: theoretical considerations. Chem. Phys. Lett. **262**(3), 405–414 (1996). https://doi.org/10.1016/0009-2614(96)01108-6
69. J.L. Krause, K.J. Schafer, M. Ben-Nun, K.R. Wilson, Creating and detecting shaped rydberg wave packets. Phys. Rev. Lett. **79**(25), 4978 (1997). https://doi.org/10.1103/PhysRevLett.79.4978

70. A. Kirrander, X-ray diffraction assisted spectroscopy of Rydberg states. J. Chem. Phys. **137**, 154310 (2012). https://doi.org/10.1063/1.4757913
71. T. Northey, A. Kirrander, Ab-Initio fragment method for molecular x-ray diffraction. J. Phys. Chem. A **123**, 3395–3406 (2019). https://doi.org/10.1021/acs.jpca.9b00621
72. C. Pemberton, Y. Zhang, K. Saita, A. Kirrander, P.M. Weber, From the (1B) spectroscopic state to the photochemical product of the ultrafast ring-opening of 1,3-cyclohexadiene: a spectral observation of the complete reaction path. J. Phys. Chem. A **119**, 8832–8845 (2015). https://doi.org/10.1021/acs.jpca.5b05672
73. M. Tudorovskaya, R.S. Minns, A. Kirrander, Effects of probe energy and competing pathways on time-resolved photoelectron spectroscopy: the ring-opening of 1,3-cyclohexadiene. Phys. Chem. Chem. Phys. **20**, 17714 (2018). https://doi.org/10.1039/C8CP02397B
74. C. Gatti, P. Macchi (eds.), *Modern Charge-Density Analysis* (Springer, 2012). https://doi.org/10.1007/978-90-481-3836-4
75. S. Grabowsky, A. Genoni, H.-B. Bürgi, Quantum crystallography. Chem. Sci. **8**(6), 4159–4176 (2017). https://doi.org/10.1039/C6SC05504D
76. A. Kirrander, P.M. Weber, Fundamental limits on meaningful spatial resolution in ultrafast x-ray diffraction. Appl. Sci. **7**, 534 (2017). https://doi.org/10.3390/app7060534
77. G. Dixit, O. Vendrell, R. Santra, Imaging electronic quantum motion with light. Proc. Natl. Acad. Sci. **109**(29), 11636–11640 (2012). https://doi.org/10.1073/pnas.1202226109
78. K. Bennett, M. Kowalewski, J.R. Rouxel, S. Mukamel, Monitoring molecular nonadiabatic dynamics with femtosecond x-ray diffraction. Proc. Natl. Acad. Sci. **115**(26), 6538–6547 (2018). https://doi.org/10.1073/pnas.1805335115
79. M. Simmermacher, N.E. Henriksen, K.B. Møller, A. Moreno Carrascosa, A. Kirrander, Electronic coherence in ultrafast x-ray scattering from molecular wavepackets. Phys. Rev. Lett. **122**, 073003 (2019). https://doi.org/10.1103/PhysRevLett.122.073003
80. M. Simmermacher, A. Moreno Carrascosa, N. Henriksen, K.B. Møller, A. Kirrander, Theory of ultrafast x-ray scattering by molecules in the gas phase. J. Chem. Phys. **151**, 174302 (2019). https://doi.org/10.1063/1.5110040

# Chapter 8
# Photochemical Reactions in the Gas Phase Studied by Ultrafast Electron Diffraction

**Jie Yang, Martin Centurion, Xijie Wang, Thomas Wolf, and Markus Gühr**

**Abstract** Time-resolved ultrafast electron diffraction (UED) from isolated molecules allows insight into ultrafast quantum wavepacket dynamics of molecules with sub-Angstrom spatial resolution. For the simple example of an excited state vibrational wavepacket in iodine, we can reconstruct the mean atomic distance as a function of delay after the optical excitation. For the case of $CF_3I$, we can reconstruct the dissociative motion after UV excitation on the A-band states. Furthermore, we also show how the signature of nonadiabatic wavepacket dynamics after two-photon excitation is visible in the diffraction data.

J. Yang
Center of Basic Molecular Science, Department of Chemistry, Tsinghua University, Beijing 100084, China
e-mail: jieyang1@tsinghua.edu.cn

M. Centurion
Department of Physics and Astronomy, University of NE–Lincoln, 855 N 16th Street, Lincoln, NE 68588, USA
e-mail: martin.centurion@unl.edu

X. Wang
Fakultät für Physik, Universität Duisburg Essen, Lotharstr. 1, 47057 Essen, Germany
e-mail: xijie.wang@uni-due.de

T. Wolf
SLAC National Accelerator Laboraty, 2575 Sand Hill Road, Menlo Park 94025, USA
e-mail: thomas.wolf@slac.stanford.edu

M. Gühr (✉)
Deutsches Elektronen-Synchrotron DESY, Notkestr. 85, 22607 Hamburg, Germany
e-mail: markus.guehr@desy.de



K. Ueda (ed.), *Ultrafast Electronic and Structural Dynamics*,
https://doi.org/10.1007/978-981-97-2914-2_8

## 8.1 Introduction

The first electron diffraction experiment can be dated back to 1927 [1], where the diffraction pattern of a Nickle crystal was measured and served as a direct experimental evidence of Louis De Brogile's matter wave theory. For gas molecules, electron diffraction has been used as a structural tool since the 1930s [2, 3]. Through the development of many decades, the structural determination precision by gas electron diffraction (GED) can often reach a remarkable level of sub-picometer [4].

A natural extension of the traditional time-averaged electron diffraction is time-resolved diffraction, where the change of molecular structure can be recorded, preferably as a chemical reaction takes place. For photoexcited molecules in the gas phase, this would serve as an important complementary tool to the widely used pump-probe spectroscopy, because of its direct sensitivity to changes in the molecular geometry. The natural timescale for atom rearrangement during an excited state chemical reaction is on the order of ~100 fs [5]. For this reason, time resolution is a key parameter in time-resolved diffraction experiments.

In the following, we give a short account of past work for time-resolved electron diffraction in the gas phase. The first milestone in this field was a study of the dissociation of $CF_3I$ molecules by multiphoton IR excitation observed by Ischenko et al. in 1983 with a time resolution of 1 $\mu$s [6]. The first picosecond electron scattering experiment on solid phase was reported by Williamson and Mourou in 1982 [7], with an electron pulse duration of ~100 ps. Ahmed H. Zewail's group pioneered the field of ultrafast electron diffraction (UED) on gas molecules in the 1990s, in which the excited state molecular dynamics was studied using a laser-pump-electron-probe scheme with a time resolution of a few picoseconds [8–14]. Such a time resolution was not sufficiently fast to directly monitor atomic motion during chemical reactions but was sufficient to reveal the structure of photoproducts and a number of reaction intermediates. Williamson et al. revealed the structure of the $CF_3$ fragment after $CF_3I$ dissociation [15], and the geometry of the photoproduct of $CH_2I_2$ dissociation [16]. Ihee et al. studied the photodissociation of an important organometallic compound $Fe(CO)_5$ by two-photon excitation with a 310 nm UV laser and observed photoproducts $Fe(CO)_2$, Fe(CO) and Fe [12]. In a later experiment, Ihee et al. studied the two-photon excitation of $Fe(CO)_5$ and provided a transient structural measurement of the reaction intermediate $Fe(CO)_4$ [17]. Studies on $CpCo(CO)_2$ identified cyclopentadienyl as a primary photoproduct, indicating a complete dissociation of both the Co atom and the carbonyl radical [10]. Ihee, Cao et al. resolved the two-step dissociation of $C_2F_4I_2$, identified the intermediate structure to be classical rather than bridged [9, 18, 19]. Dudek et al. [20], Ihee et al. [19], and Ruan et al. [21] studied the photochemical ring-opening of cyclohexadiene (CHD), and provided insight to the structure of the ring-opened hot ground state, mostly an elongation of the C–C bonds. Ruan et al. also studied the ring-opening reaction of cycloheptatriene (CHT) [21], where they were able to resolve the vibrational energy re-distribution within 5 ps after photoexcitation. Lobastov et al. [22] studied the so-called "channel three" dynamics in pyridine with 266 nm excitation, where they concluded a ring-opening pathway. Srinivasan et al. [23] compared the 266 nm excitation of pyridine, picoline, and lutidine, where they found that only pyridine and picoline lead to a ring-opened photoproduct.

At the same time as gas phase ultrafast electron diffraction made this progress, UED studies on condensed phase samples with sub-picosecond resolution were pioneered by R. J. Dwayne Miller's group [24–29]. Because of the high scattering cross sections for electrons, the solid samples need to be kept extremely thin. This in turn minimizes time-smearing due to different group velocities of exciting lasers and probing electron pulses [28].

In addition to directly monitoring photophysical and photochemical reactions through UED, the ~1 ps time resolution provides an opportunity to directly resolve the 3-D structure of gas molecules when combined with either photoselection in the excited state [30] or impulsive laser alignment techniques [31–34]. The mathematical foundation of retrieving 3-D molecular structure using UED from aligned molecules in the gas phase was first introduced by Baskin and Zewail [35, 36]. Hoshina et al. [37] reported the first experimental realization of an anisotropic diffraction pattern with 10 ns resolution using adiabatic alignment and a pulsed electron diffraction apparatus. Reckenthaeler et al. [30] reported an anisotropic diffraction pattern of $C_2F_4I_2$ with few ps time resolution, where they used the anisotropy of the molecular ensemble that was generated upon photoexcitation with a linearly polarized laser pulse. Hensley et al. [31] reported the first experimental 3-D structural retrieval from an anisotropic diffraction pattern, in which an intense infrared pulse was used to impulsively align $CF_3I$ molecules. Yang et al. [32] reported the alignment and structural change of $CS_2$ molecules in an intense laser field.

In recent years, technological breakthroughs led to a temporal resolution down to ~100 fs in UED experiments by several methods: accelerating electrons to relativistic energies [38–40], using an RF cavity to actively compress the electron beam [41–44], or utilizing a compact setup where the distance between the electron gun and the sample is minimized [27, 28, 45, 46]. For studying excited state dynamics of molecules in the gas phase, gas-phase ultrafast electron diffraction (GUED) utilizing a relativistic electron beam has made great progresses. Yang et al. reported the capture of impulsive alignment and rotational revivals of $N_2$ molecules with 230 fs FWHM time resolution [47], the direct observation of coherent vibration in the diatomic molecule iodine [48], and the capture of photodissociation and nonadiabatic dynamics in $CF_3I$ valence and Rydberg states [49]. Wolf et al. reported an unprecedentedly clear view of the ring-opening dynamics in CHD molecules [50], and Wilkin et al. reported the transient structure in $C_2F_4I_2$ immediately after the dissociation of an iodine atom, where they have clarified that the bridged intermediate structure has never formed, even on the timescale of a single vibrational period [51].

Following the lasing of the world's first hard X-ray free electron lasers (XFELs) in 2010 [52], X-ray diffraction has emerged as another powerful tool to study time-resolved structural dynamics on gas, liquid, and solid samples with femtosecond time resolution and ångström spatial resolution. For samples in the gas phase, Küpper et al. have observed an anisotropic diffraction pattern in a strongly aligned 2,5-diiodobenzonitrile molecular ensemble [53]. Minitti et al. reported the capture of CHD ring-opening [54], and P. M. Weber's group has published a series of papers recently on this topic, including the determination of the transition dipole moment direction from anisotropic diffraction patterns [55], understanding the information content of small-angle X-ray scattering [56], observation of vibrational coherence in *N*-methylmorpholine after a Rydberg excitation [57], and discerned ground-state ring-opening in CHD after a Rydberg excitation using a 200 nm pump laser [58].

In this chapter, we will discuss recent progresses in utilizing relativistic electron pulses to study excited state molecular dynamics in the gas phase. In Sect. 8.2, we provide a theoretical background for GUED simulations and data treatment. In Sect. 8.3, we present the cases of iodine and trifluoroiodomethane, for which we resolved nuclear motion in space on an ultrafast timescale. In Sect. 8.4, we give a conclusion and an outlook on the future development of GUED.

## 8.2 Diffraction Theory

In this section, we will introduce the theory for electron diffraction from a gas ensemble of molecules and present the methodology for both simulating diffraction patterns and retrieving real-space structural information for both GED and GUED.

### *8.2.1 Static Electron Diffraction from a Gas Ensemble*

#### 8.2.1.1 Theory of Electron Diffraction

Electron diffraction from molecules can be viewed as the scattering of an incident plane wave by the Coulomb potential of protons and electrons within the target molecule. For the diffraction of high energy (>keV) electrons by gas molecules, the first-order Born approximation assumes the incident electron only scatters once within the Coulomb potential of the molecule. Under the first-order Born approximation, the scattering intensity can be written as the sum of elastic and inelastic scattering [3, 59]

$$I(\vec{s}) = I_{\text{elastic}}(\vec{s}) + I_{\text{inelastic}}(\vec{s}) \tag{8.1}$$

Since the inelastic scattering contains very little information on the geometrical structure of the molecule, in this chapter we'll limit the discussion on elastic scattering only. The elastic scattering can be written as

$$I_{\text{elastic}} = |\tilde{f}(\vec{s})|^2 \tag{8.2}$$

The scattering amplitude $\tilde{f}(\vec{s})$ is given as

$$\tilde{f}(\vec{s}) = \frac{m}{2\pi\hbar^2}\iiint V(\vec{r})\exp(i\vec{s}\cdot\vec{r})d^3r \tag{8.3}$$

where $m$ is the electron mass, $\hbar$ is the reduced Planck constant, $\vec{s}=\vec{k}-\vec{k_0}$ is the momentum transfer, and $\vec{k}$ and $\vec{k_0}$ are the wave vectors for scattered and incident electron, respectively. $V(\vec{r})$ is the Coulomb potential of the target molecule. The integral is performed over the real space.

A commonly used approximation to calculate the elastic scattering patterns is the independent-atom-model (IAM), where the scattering from a molecule is treated as a coherent sum of scattering from independent atoms, and the effect of chemical bonds is ignored [3]. The breakdown of the IAM has been investigated intensively in the second half of the last century [60–66]. It was found that the scattering contribution from bonding electrons is mostly limited to a small momentum transfer range, and it is a common practice to ignore the range $s < 1$–$2$ Å$^{-1}$ for structural retrieval of molecules on the ground state. For GUED experiments, most existing works have found that the IAM is a sufficiently accurate approximation with currently achievable signal to noise [8, 48]. Therefore, in this chapter we will limit ourselves to the IAM framework.

For an individual atom, the Coulomb potential is isotropic and can be written as

$$V(r) = -\frac{Ze^2}{r} + e^2\iiint \frac{\rho(\vec{r}-\vec{r\prime})}{|\vec{r}-\vec{r\prime}|}d^3r\prime \tag{8.4}$$

where Z is the atomic number and $\rho(\vec{r})$ is the electron density. Inserting (8.4) into (8.3) one can derive

$$f(s) = \frac{2me^2}{\hbar^2}\frac{Z-F(s)}{s^2} \tag{8.5}$$

and

$$F(s) = 4\pi\int_0^\infty \rho(r)\frac{\sin(sr)}{sr}r^2dr \tag{8.6}$$

The atomic form factor $f(s)$ for electron scattering can be obtained from the literature or computer programs, for example, the ELSEPA software [67]. Note that $F(s)$ is exactly the atomic form factor for X-ray scattering, and the term $Z - F(s)$ in Eq. (8.5) represents the screening of the nucleus' Coulomb potential by the electrons. The atomic form factor $f(s)$ is real under the first-order Born approximation [3, 68]. Schomaker and Glauber first investigated the validity of the first-order Born approximation in the 1950s [68]. They found that the structural refinement of many compounds can be improved by introducing a scattering phase (second-order Born approximation), making $f(s)$ a complex function. This scattering phase increases with $s$ and atom weight, and decreases with the kinetic energy of the scattering electrons.

For a molecule, the elastic scattering amplitude under the IAM can be written as

$$\tilde{f}(\vec{s}) = \sum_m f_i(s) e^{i\vec{s}\cdot\vec{r_m}} \tag{8.7}$$

where the sum is over all the atoms in the molecule, $f_i(s)$ is the atomic form factor of the $i$th atom, and $\vec{r_i}$ is the vector pointing from a randomly chosen origin to the $i$th atom. To clarify the nomenclature, in this chapter we use $\tilde{f}(s)$ to represent the molecular scattering amplitude and $f(s)$ to represent the atomic scattering amplitude. Therefore, the elastic scattering intensity for a molecule is given by

$$I_{\text{elastic}} = |f(\vec{s})|^2 = \sum_m |f_m(s)|^2 + \sum_{m \neq n} f_i^*(s) f_j(s) e^{i\vec{s}\cdot\vec{r_{mn}}} = I_{at} + I_{mol} \tag{8.8}$$

where $\vec{r_{mn}} = \vec{r_m} - \vec{r_n}$ is the vector pointing from the $m$th atom to the $n$th atom. The first term $I_{at}$ is scattering from individual atoms and the second term $I_{\text{mol}}$ is the interference term between atoms that contains structural information.

For a molecular ensemble in the gas phase, the total scattering is an incoherent sum of all the molecules, because there's no fixed phase relation between the scattered waves from different molecules. In addition, molecules are randomly oriented in such an ensemble and, thus, we can integrate over angular coordinates,

$$\begin{aligned} I_{\text{mol}}^{\text{ran}} &= \frac{1}{4\pi} \int_0^{2\pi} d\varphi \int_0^{\pi} \sum_{m \neq n} f_m^*(s) f_n(s) e^{isr_{mn}\cos\theta} \sin\theta d\theta \\ &= \sum_{m \neq n} f_m^*(s) f_n(s) \frac{\sin(sr_{mn})}{sr_{mn}}. \end{aligned} \tag{8.9}$$

here the polar angle $\theta$ and azimuthal angle $\varphi$ are defined with respect to the vector $\vec{s}$. After the integral $I_{\text{mol}}^{\text{ran}}$ is only a function of $s$, the magnitude of vector $\vec{s}$, meaning that the diffraction pattern is isotropic. It has been found that for static electron diffraction from gas molecules, molecular vibration is also an important factor to account for, especially for high momentum transfer [4]. If we assume that each atom pair distance $mn$ follows a Gaussian probability density with a mean square amplitude of vibration $l_{mn}^2$, the scattering pattern takes the following form:

$$I_{\text{mol}}^{\text{ran}} = \sum_{m \neq n} f_m^*(s) f_n(s) \frac{\sin(sr_{mn})}{sr_{mn}} \exp\left(-\frac{1}{2} l_{mn}^2 s^2\right). \tag{8.10}$$

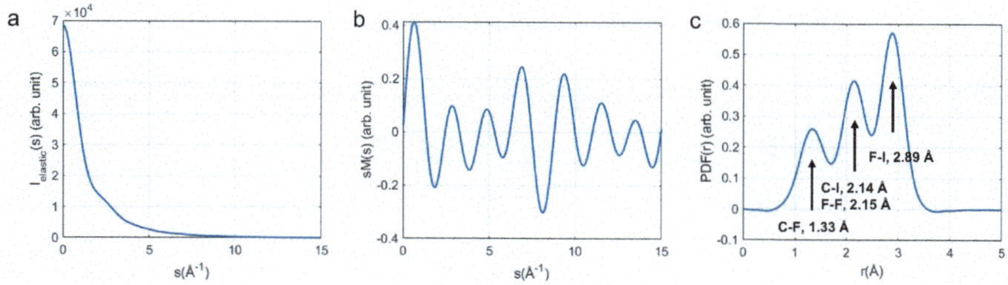


**Fig. 8.1** Simulated static diffraction from gas phase $CF_3I$ molecules. **a** $I_{elastic}(s)$, **b** sM(s), **c** PDF(r). The corresponding atom pairs in PDF(r) are marked

This term has a similar form (and same physical origin) as the famous Debye–Waller factor in condensed matter physics. This term effectively adds a "fuzziness" to the real-space distribution. For photo-excited molecules, it might be used to describe a hot molecule if the excited state population is not too far from a Gaussian distribution, but cannot describe large-scale dynamics such as photodissociation, or complicated dynamics such as wavepacket branching.

In electron scattering, the form factors $f(s)$ have a $s^{-2}$ dependence (Eq. 8.5). Therefore $I_{\mathrm{mol}}^{\mathrm{ran}}$ has a ~ $s^{-5}$ dependence[1]. The scattering intensity drops so fast that it is usually challenging to read information from a plot of $I_{\mathrm{mol}}^{\mathrm{ran}}$ on a linear scale over a typical $s$ range (see Fig. 8.1a). A convenient convention is to use the modified diffraction intensity sM(s) (see Fig. 8.1b), defined as

$$sM(s) = \frac{I_{\mathrm{mol}}^{\mathrm{ran}}}{I_{at}} s = \frac{\sum_{m \neq n} f_m{}^*(s) f_n(s) \frac{\sin(sr_{mn})}{r_{mn}}}{\sum_m f_m{}^*(s) f_m(s)}. \tag{8.11}$$

For homoatomic molecules, the form factors cancel out, and each atom pair $mn$ gives rise to a sinusoidal term $\sin(sr_{mn})$ with an amplitude of $1/r_{mn}$. For heteroatomic molecules, since the "shape" of atomic form factors $f_m(s)$ are typically similar, the form factors can still mostly cancel out. Assuming there exists a common shape $f_o(s)$ for all atoms, we express the $f_i(s)$ with the relation $f_i(s) \approx c_i f_o(s)$, where $c_i$ is a coefficient for atom $i$ that is roughly proportional to the square root of its atomic elastic scattering cross section. Inserting this into (8.11) yields

$$sM(s) \approx \frac{\sum_{m \neq n} c_m c_n \frac{\sin(sr_{mn})}{r_{mn}}}{\sum_m c_m^2}. \tag{8.12}$$

[1] $I_{mol}^{ran}$ Actually drops slower than $s^{-5}$, because the term $Z\text{-}F(s)$ in Eq. (8.5) increases with $s$.

#### 8.2.1.2 Structural Retrieval from Static Electron Diffraction Patterns

In this section, we will discuss the method to retrieve real-space information from static diffraction patterns.

The most straightforward way to retrieve real-space information is by calculating the pair distribution function (PDF) from the modified diffraction intensity sM(s), using the following formula:

$$PDF(r) = \int_0^\infty sM(s)\sin(sr)ds. \tag{8.13}$$

Combining Eqs. 8.12 and 8.13 and assuming the sM is measured over all s, one can immediately get

$$PDF(r) \approx \frac{\sum_{m\neq n} c_m c_n \frac{1}{r_{mn}} \delta(r - r_{mn})}{\sum_m c_m^2} \tag{8.14}$$

where $\delta(r)$ is the Dirac delta function. Therefore, the PDF consists of individual peaks at distances equal to the distance between each atom pair. The amplitude of each peak is proportional to the product of the scattering amplitudes of the two atoms, and inversely proportional to their distance (see Fig. 8.1c).

Calculating a PDF using Eq. (8.12) requires access to sM(s) in the full range ($0 < s < \infty$), which is not feasible experimentally. For static GED experiments, the maximum s range is approximately 30 $Å^{-1}$ [4]. For GUED, most experiments reach a maximum s of ~10 $Å^{-1}$ [8, 49, 50]. Thus, the integral in Eq. 8.13 is performed over a limited s-range, which determines the broadening of the peaks corresponding to interatomic distances. Furthermore, a damping factor is added to the sine transform to minimize artifacts due to edge effects:

$$PDF(r) = \int_0^{s_{\max}} sM(s)\sin(sr)e^{-ks^2}ds \tag{8.15}$$

where k is a damping factor. The damping factor is usually chosen so that $e^{-ks_{max}^2}$ is somewhere between 0.01 and 0.1. Another issue is that the low s ($s <$ ~2 $Å^{-1}$) signal is often not available experimentally, or deviates from the IAM model [4]. When calculating the PDF, a common workaround is to add the low s signal with modeled values in the IAM framework [4].

A real-space PDF can provide a direct and intuitive view of molecular structure. However, it is inherently one-dimensional information, therefore it usually cannot be directly translated to 3-D molecular structure. To extract the 3-D molecular structure, a $\chi^2$ refinement on sM(s) is often used. The method usually compares the simulated and experimental sM(s) while keeping scanning structural parameters until the best match is achieved. Using $\chi^2$ refinement, static GED can typically reach a structural precision of better than 0.005 Å in bond length and 0.5° in bond angle [4]. In addition, direct structural retrieval in sM(s) uses only the experimentally available $s$ range, therefore avoids the issue of the finite experimental $s$ range. Once the sM(s) is optimized, the missing s range can be filled in with simulated values for calculating PDF(r).

### *8.2.2 Ultrafast Electron Diffraction from a Gas Ensemble*

In GUED, the initial structure of the molecule is usually known, and the goal is to find out how the structure changes after excitation. Diffraction patterns are recorded as a function of pump-probe delay time, and the change in the diffraction pattern (named difference pattern hereafter) is the key observable, which can be written as

$$\Delta I(s; t) = I(s; t) - I(s; t < 0), \tag{8.16}$$

where $t$ is the pump-probe delay time, and $I(s; t < 0)$ is a reference pattern that is taken before the arrival of the pump laser. The advantage of using difference patterns is that all the contribution that does not change with pump-probe delay, such as experimental background and the diffraction signal from unexcited molecules, cancels out by the subtraction. In other words, only those molecules that undergo structural changes contribute to the difference signal.

When the difference pattern is isotropic, one can use Eq. (8.15) to calculate the difference PDF, as in the static case. The difference PDF then reflects the gain and loss of atom pair distances and can be used to determine the transient structure of the photoexcited molecule [48, 50].

When the difference pattern is not isotropic, the structural retrieval gets slightly more complicated. For example, when a linearly polarized laser pulse is used as the pump, the selection rule of the photoexcitation process preferentially creates an angularly anisotropic ensemble in the excited state (and a corresponding "hole" in the ground state). This leads to an anisotropic diffraction pattern that encodes the 3-D structural information of the excited state [35, 36]. In addition, one can also actively align the molecules with an intense IR pulse, and the corresponding diffraction pattern can be used to retrieve the 3-D structure of the molecule on the ground state [31, 33]. Here we introduce only the simplest case, where anisotropy is introduced by photoselection through a single photon excitation.

Assuming $\gamma$ is the angle between the laser polarization and the transition dipole moment (TDM) of the molecule, the angular shape of the excited state molecular ensemble will follow a $\cos^2\gamma$ distribution in a single-photon excitation process [35]. This anisotropy is transient and will dephase and rephase as the rotational wavepacket evolves. For asymmetric-top molecules, the rephase is typically only partial [69]. The typical dephasing timescale is a few hundred femtoseconds to a few picoseconds, depending on the moment of inertia of the system.

For each atom pair, the angular distribution also depends on its relative orientation to the TDM of the molecule. For example, assuming $\Theta$ is the angle between the atom pair and the laser polarization, for an atom pair that is parallel to the TDM, the angular distribution is $\cos^2\Theta$. For an atom pair that is perpendicular to the TDM, the angular distribution will become $\sin^2\Theta$. In general, for a one photon transition and a specific atom pair $ij$ pointing at a specific direction $\hat{r}_{ij}(t)$ at time $t$, the angular distribution of this pair was derived by Baskin and Zewail [35, 36]:

$$A_{ij}(t;\Theta) = P_0[\cos\Theta] + 2P_2\left[\hat{r}_{ij}(t)\cdot\hat{\mu}(0)\right]P_2[\cos\Theta], \tag{8.17}$$

where $P_0$ and $P_2$ are the zeroth and the second order Legendre polynomial, and $\hat{\mu}(0)$ is the TDM orientation at time zero. The caret symbol means a unit vector. The angular distribution is always uniform along the azimuthal angle in the lab frame because of cylindrical symmetry with respect to the laser polarization.

For an atom pair with an angular distribution given by Eq. (8.17), the diffraction pattern under the IAM can be written as [35, 36]

$$\begin{aligned} I_{ij}(t;s,\phi) = {} & Re\left(f_i^* f_j\right) j_0\left(sr_{ij}\right) P_0[\cos\phi] \\ & - 2Re\left(f_i^* f_j\right) P_2\left[\hat{r}_{ij}(t)\cdot\hat{\mu}(0)\right] j_2\left(sr_{ij}\right) P_2[\cos\phi] \end{aligned} \tag{8.18}$$

where $j_0$ and $j_2$ are the zeroth and the second-order spherical Bessel functions of the first kind, $\phi$ is the azimuthal angle on the diffraction pattern, and $f_x$ is the atomic scattering form factor of atom $x$. Equation (8.18) assumes the polar diffraction angle $\theta$ is small enough that $\cos\left(\frac{\theta}{2}\right) = 1$ is a good approximation, which holds reasonably well for both keV and MeV electrons.

Since the angular component of the diffraction pattern can be represented by the zeroth and the second-order Legendre polynomial, a useful way to remove background noise is to apply a Legendre polynomial decomposition on the angular coordinate and keep only these two terms [49, 70].

To retrieve the angular-resolved *PDF* (ARPDF) from the anisotropic scattering signal described by Eq. (8.18), one can use an inverse 2-D Fourier transform followed by an inverse Abel transform [49]. Since the angular dependence of the real space signal contains only the zeroth and the second-order Legendre polynomials, it naturally fits to the pBasex Abel inversion algorithm [71]. An example is given in Fig. 8.2. Figure 8.2a shows a simulated diffraction pattern for an ensemble of isolated $CF_3I$ molecules with a $\cos^2\Theta$ angular distribution. Its 2-D Fourier transform is given in Fig. 8.2b, and the ARPDF obtained by Abel inversion is shown in Fig. 8.2c. Figure 8.2d shows the ARPDF in polar coordinates, where the C-I, F-I appear preferentially along the parallel direction, and C-F and F-F appear preferentially along the perpendicular direction. Since both Fourier and Abel transforms are "passive" mathematical transforms, this example shows that ARPDF can retrieve multidimensional molecular structural information without any a priori knowledge about the molecule.

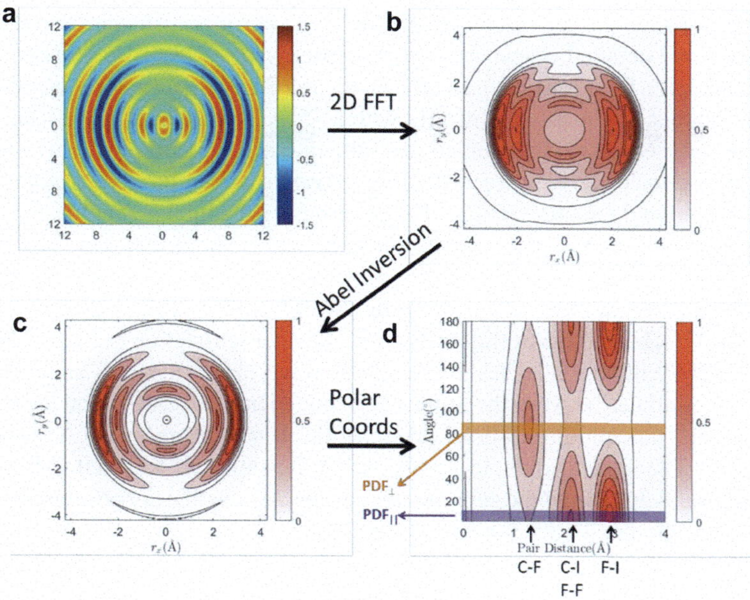


**Fig. 8.2** An illustration of the angular-dependent Fourier analysis. **a** Simulated diffraction pattern of gas phase $CF_3I$ with a $\cos^2 \Theta$ angular distribution. **b** The 2D FFT of part (**a**). **c** ARPDF, obtained by a pBasex Abel inversion of part (**b**). **d** ARPDF represented in polar coordinates, where the parallel (purple stripe) and perpendicular (brown stripe) directions show very different structures. From Ref. [49]. Reprinted with permission from AAAS

## 8.3 Case Study

In this chapter, we will show two GUED experiments on excited state molecular dynamics. The first proof-of-principle experiment studied coherent vibrational dynamics on the B ($^3\Pi_{0u}$) excited state of iodine [48]. The vibration of a diatomic molecule is one of the simplest molecular motion, and the iodine atom is a very strong scatterer. Nevertheless, resolving the vibrational motion of a diatomic molecule is a direct realization of making a so-called "molecular movie". The second experiment shown is on a polyatomic molecule, $CF_3I$ [49]. In this case, complicated dynamics such as the recoil motion during photodissociation and wavepacket branching through a conical intersection is resolved with rich details.

### *8.3.1 The Vibrational Wavepacket in the B State Iodine*

We now briefly describe the setup. In the iodine experiment, a 530 nm femtosecond pump laser pulse, generated from a commercial optical parametric amplifier (OPA) pumped by an 800 nm laser pulse, is used to excite iodine molecules onto the B ($^3\Pi_{0u}$)

state. The experimental setup at the SLAC MeV UED facility is shown schematically in Fig. 8.3a and detailed elsewhere [38, 39]. Briefly, the $I_2$ gas is introduced to the interaction region by a pulsed nozzle, the incident 3.7 meV electron beam and pump laser beam propagate nearly collinearly with a ~5° angle between them. The pump laser is introduced to and outcoupled from the interaction region by holey mirrors, and the diffraction patterns are recorded on a phosphor screen-based detector at ~3 m distance from the interaction region. The electron beam contains roughly $1.2 \times 10^4$ electrons per pulse and is collimated to a spot size of about 200 μm. The nozzle orifice has an opening of 100 μm, the gas jet has a diameter of roughly 300 um FWHM, and the laser focal size is also roughly 300 μm FWHM. The whole system operates at 120 Hz, and the overall instrumental response function is 230 fs FWHM.

The photoexcitation launches a vibrational wavepacket on the B state surface, which then oscillates between the ground state bond length of 2.66 Å and the turning point at 4 Å. The equilibrium bond length of the B state is around 3 Å, as shown in Fig. 8.3b. The simulated nuclear wavepacket dynamics on the excited B state is shown in Fig. 8.3c, where the vibrational wavepacket oscillates between the inner and outer turning point with a period of ~400 fs.

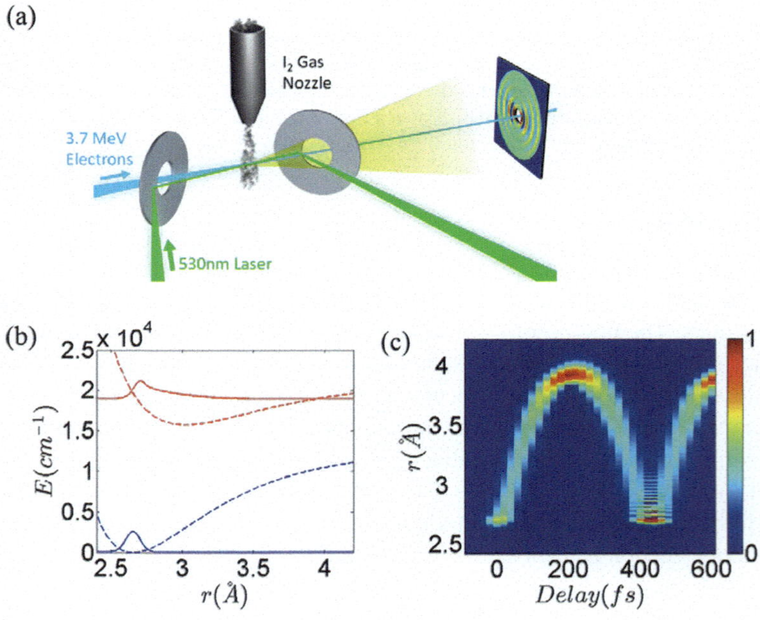


**Fig. 8.3** Experimental setup and excited state dynamics of gas phase iodine. **a** A schematic drawing of the experimental setup. **b** The potential energy surface (dashed lines) and the nuclear wavepacket in the Franck–Condon region (solid lines). The ground state is shown in blue and the excited B state is shown in red. **c** A simulation of the probability density of the coherent nuclear wavepacket in false color representation in the first 600 fs after photoexcitation. The figure is reproduced from Ref. [48]

The experimental diffraction-difference pattern ΔI, averaged over time delays between 50 and 550 fs, is shown in Fig. 8.4a, and the corresponding simulated diffraction pattern is shown in Fig. 8.4b. The simulation is performed by the combination of the quantum wavepacket in Fig. 8.3c and the IAM simulation given in Eq. (8.8), assuming a $\cos^2\gamma$ angular distribution. The experimental ΔI is available in an s range of 1.6 to 10 $Å^{-1}$, and the simulation is cut to match the s range of the experiment. Both the radial and the angular distribution of the two patterns match relatively well. The ΔI at individual pump-probe delays of −184, 17, 217, and 417 are shown in parts (c) to (f) of Fig. 8.4. The anisotropy reflects the selection rule of the photoexcitation, and can, in principle, be used to extract the 3-D structural information. However, since the diatomic $I_2$ is a one-dimensional molecule, this extra information is not necessary in this specific case. Therefore, we only apply the isotropic analysis given by Eqs. (8.10) to (8.15) to this dataset. The anisotropic analysis will be utilized in the next case study, $CF_3I$, to show the power of retrieving 3-D structural dynamics.

To process this signal using the isotropic 1-D analysis algorithm, we first radially average the 2-D ΔI(s,φ) signal to generate ΔI(s), then generate ΔsM(s) using Eq. (8.11). The experimental and simulated ΔsM(s) is then shown in Fig. 8.5a–b. The appearance of stripes at around time zero represents the beginning of a bond length change. To better determine the bond length at every delay time, we first extract the sM(s). The contribution of sM(s) from the ground state is added back to the ΔsM(s) signal to get the excited state sM(s). An excitation ratio is needed to properly account for the fraction of excitation. This ratio is determined to be ~15% by comparing the experimental and simulated ΔsM(s). The resulting excited state sM(s) is shown in

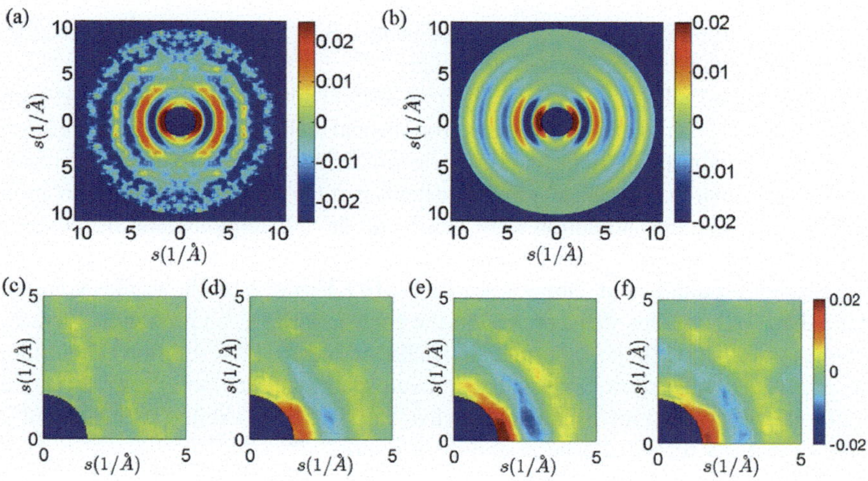


**Fig. 8.4** Difference diffraction pattern ΔI. **a** Experimental ΔI averaged between 50 and 550 fs. **b** Simulated ΔI averaged between 50 and 550 fs. **c**–**f** Experimental ΔI at **c** −184 fs, **d** 17 fs, **e** 217 fs, and **f** 417 fs, respectively. The figure is reproduced from Ref. [48]

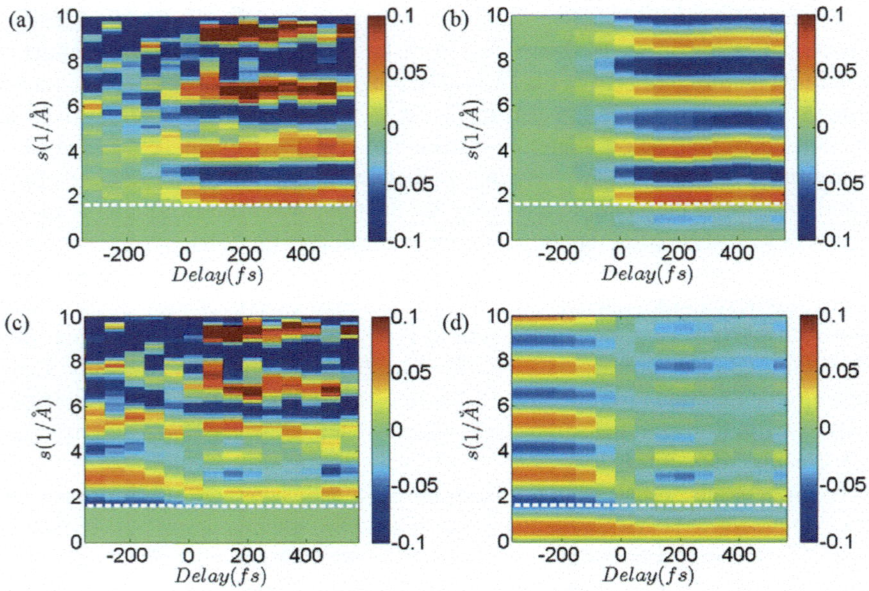


**Fig. 8.5** Dynamics of $I_2$ in B state in reciprocal s space. **a–b** Experimental (**a**) and simulated (**b**) ΔsM(s), obtained by a radial average of difference pattern ΔsM(s,φ). **c–d**, Experimental (**c**) and simulated (**d**) sM(s), obtained by adding a known contribution from the ground state sM(s) to ΔsM(s). The white dashed line at $s = 1.6$ Å$^{-1}$ shows the lower limit of the experimentally measured s range. The simulation is blurred by a 230 fs FWHM Gaussian running average to account for the experimental IRF. Figure is reproduced from Ref. [48]

Fig. 8.5c–d. Around time zero, the positive peak at 3 Å$^{-1}$ moves to a smaller value, indicating that the $I_2$ molecule has elongated.

Excited state PDFs can be calculated from sM(s) using Eq. (8.15). It can be seen from Eq. (8.14) that the PDF is inversely proportional to the pair distance r, therefore a real-space probability distribution P(r) is constructed by $P(r) = r \cdot PDF(r)$. The resultant P(r) is shown in Fig. 8.6. Figure 8.6a shows that the experimental peak position of P(r) reproduces the simulation well. This result directly demonstrates the concept of a "molecular movie", where atoms moving with respect to each other, can be resolved both spatially and temporally with GUED. Figure 8.6b shows the evolution of the shape of P(r). This shape represents the nuclear wavepacket probability density $|\psi(r)|^2$, convolved with the experimental spatial and temporal resolution. Therefore, this experiment also shows that with improved spatial and temporal resolution, the amplitude of the quantum mechanical nuclear wavepacket of excited state molecules could be directly retrieved experimentally in future.

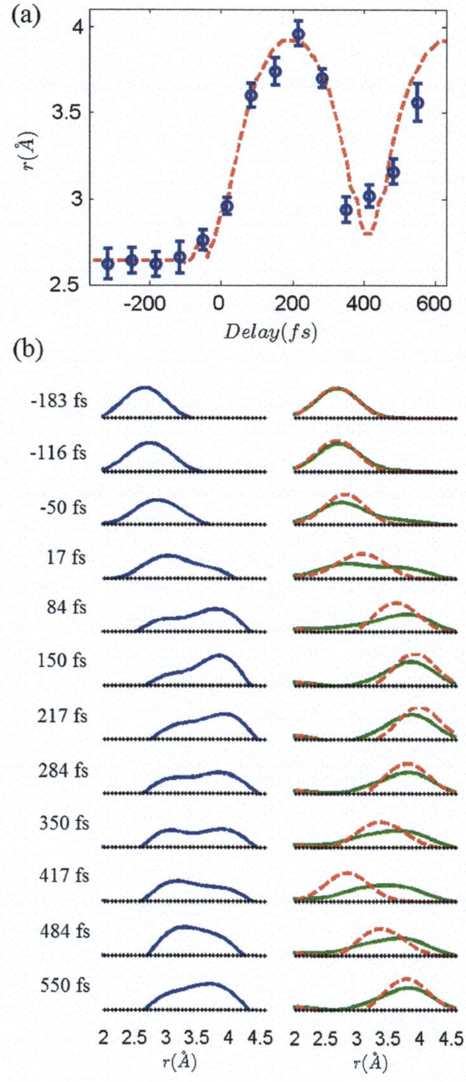


**Fig. 8.6** Dynamics of $I_2$ in the B state in the real space. **a** Retrieved peak position of the real-space probability density P(r) from experiment (circles with error bars) and the quantum wavepacket simulation (dashed line) as a function of delay time. **b** The evolution of the shape of P(r) from experiment (left), simulation (right red dashed line), and simulation with running average (right green solid line). The figure is reproduced from Ref. [48]

## 8.3.2 *The Multidimensional Imaging of $CF_3I$ Photodissociation and Rydberg State Dynamics*

$CF_3I$ is a simple alkyl halide with $C_{3v}$ symmetry. It is a prototypical target for studying photodissociation and intersystem crossing. A single photon of 266 nm promotes the molecule to its $^3Q_0$ state. Due to the strong spin–orbit coupling promoted by the heavy iodine atom, dissociation occurs either to $CF_3I + I^*(^2P_{1/2})$ or after an intersystem crossing to the singlet $^1Q_1$ state into $CF_3I + I(^2P_{3/2})$. This reaction channel has been extensively studied both theoretically and experimentally [72–77]. A two-photon excitation channel, which directly promotes the molecule onto the $[5p\pi^3, {}^2\Pi_{1/2}]$ (7 s) Rydberg state (referred to as Rydberg 7 s state), has been observed in several studies, mostly through resonance-enhanced multiphoton ionization [78–81], but the details of this reaction pathway have not been retrieved. Here, we show that through the combination of multi-dimensional imaging of GUED and *ab-initio* nonadiabatic dynamics simulations, we are able to follow both the one-photon and the two-photon pathways in great detail.

The one-photon and the two-photon excitations have different TDM—the one photon excitation is preferentially along the symmetry axis of the molecule, while two-photon excitation is preferentially perpendicular to the symmetry axis. Since the iodine atom is much heavier (therefore a stronger scatterer) than any other atoms in $CF_3I$, the C-I and F-I distances will be the dominating observables, both along the TDM direction (Fig. 8.7). Therefore, the dominating one-photon signal will be along the direction parallel to the laser polarization while the dominating two-photon signal will be along the perpendicular direction.

The angularly-resolved ΔPDF is shown in Fig. 8.8, each panel displays an average over a 10° angle, with 0° and 90° being parallel and perpendicular to the laser polarization, respectively. It is obvious that the parallel and perpendicular directions show very different structural dynamics, with features in the parallel slices significantly stronger and simpler than those on the perpendicular side.

The $\Delta PDF_{||}$, taken by averaging the ΔARPDF between 0° and 10°, is shown in Fig. 8.9a, where the two strong bleaching bands at ~2.1 Å and ~2.9 Å reflect the loss of C-I (2.14 Å) and F-I (2.89 Å) atom pair distances in $CF_3I$ upon dissociation. These two features are reproduced by ab-initio multiple spawning (AIMS) molecular

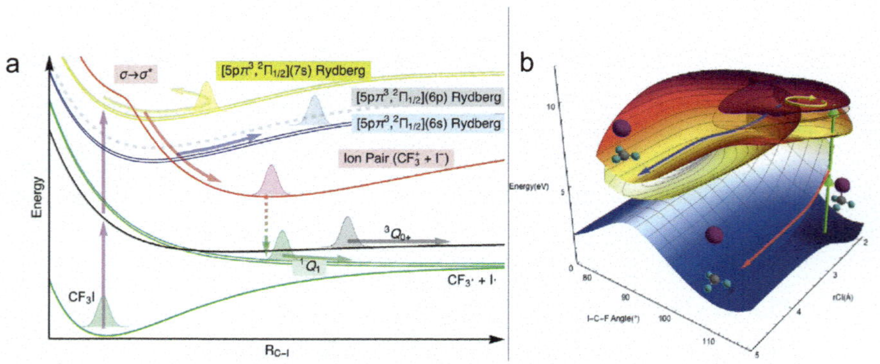


**Fig. 8.7** Potential energy surface and reaction pathway for $CF_3I$. **a** As a function of the C-I bond length, **b** additionally as a function of the I-C-F bond angle. Part (**a**) is color coded with electronic state character: the one-photon excitation creates a dissociative wavepacket on the $^3Q_{0+}$ state (grey), and the two-photon excitation creates vibrational wavepacket on the Rydberg 7 s surface (yellow). The two-photon reaction pathway involves an ion-pair state (red), the Rydberg 6 s state (blue), ground state and the valence open-shell states (green). Part **a** was reproduced from Ref. [49]. Reprinted with permission from AAAS

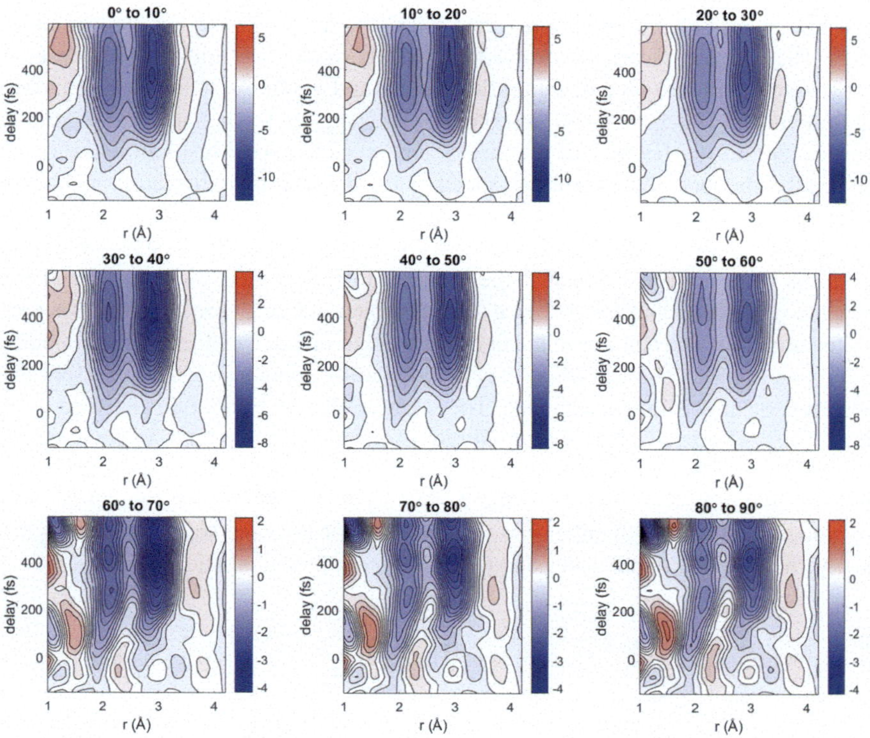


**Fig. 8.8** Experimental ΔARPDF as a function of delay time. Each panel represents angular average over a 10° cone. Note the color scale in each row is different from Ref. [49]. Reprinted with permission from AAAS

dynamics simulation shown in Fig. 8.9c. A more careful analysis of the two bands reveals that the C-I pair is lost slightly earlier (~30 fs) than the F-I pair, as shown in Fig. 8.9b. This subtle difference can be attributed to the fact that the C-I bond breaking starts with a strong recoil of the carbon atom, because the iodine atom is a factor of ~10 heavier than carbon. This recoil of the carbon atom immediately alters the original C-I distance, while the change of the F-I distance is delayed. The AIMS simulation reveals a ~16 fs delay between the two bleaching signals, which confirms this interpretation. In addition, this strong recoil causes an immediate and strong umbrella opening of the $CF_3$ fragment. This signal appears mostly in the $\Delta PDF_{\perp}$ (averaging the ΔARPDF between 80° and 90°) as the bond direction is close to the direction perpendicular to the transition dipole. A $\chi^2$ fitting on the $\Delta PDF_{\perp}$ shows that the umbrella angle opens immediately by roughly 4° (after taking into account the 150 fs experimental IRF), with a slower expansion of C-F bond length of roughly 0.03 Å. These two motions are also confirmed by AIMS simulation. The simulation seems to underestimate the C-F bond lengthening amplitude, which might be caused by various approximations adopted in the simulation. The simulation shows that both the umbrella bending vibration and the C-F stretching have a period of under 50 fs. They are too fast to be resolved in our experiment and thus we only observe an averaged motion with a 4° umbrella opening and a 0.03 Å C-F bond elongation. The previously observed ~10% intersystem crossing to $^1Q_1$ is not resolved in this experiment, mostly because of the insufficient spatiotemporal resolution to discern the small difference of the travelling speed of the dissociating iodine atom in the two channels.

The experimental $\Delta PDF_{\perp}$ is shown in Fig. 8.10a. Figure 8.10b shows the time evolution of three pair distances—the initial C-I distance of 2.14 Å, the initial F-I distance of 2.90 Å, and the distance 2.52 Å which exhibits zero amplitude before photoexcitation. The features in the first 200 fs are dominated by the umbrella motion of the one-photon channel, while the rich oscillating signals after 200 fs reveal the existence of an additional channel. The counter-phased oscillation between 2.14 and 2.52 Å reflects a C-I stretching vibration with a ~200 fs period, in good agreement with the expected vibrational period of C-I stretching mode on Rydberg states [82]. Additionally, a recurrence of the 2.90 Å signal at ~500 fs does not agree with any anticipated feature in either dissociative states or Rydberg states, indicating the involvement of additional states and the existence of nonadiabatic dynamics.

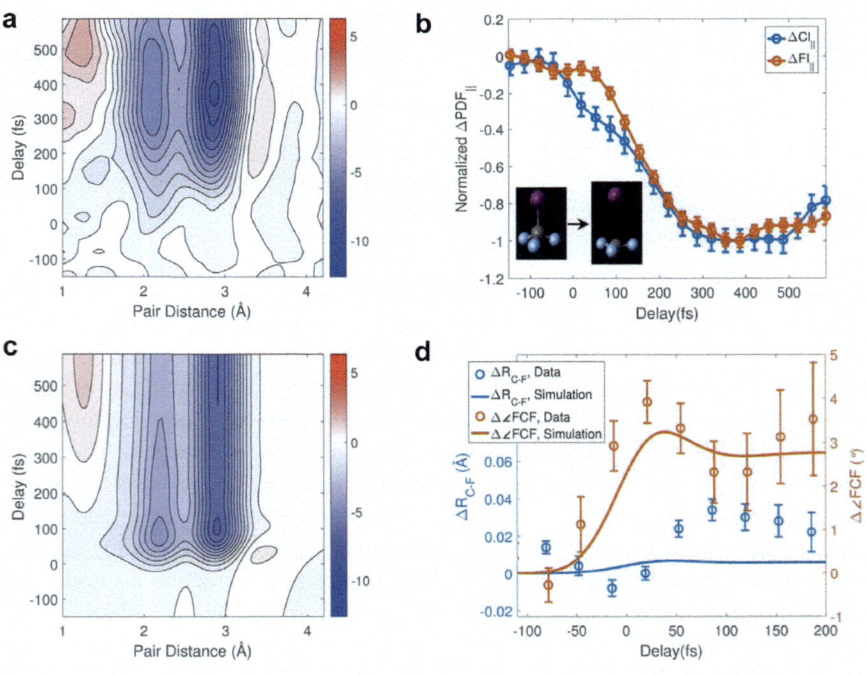


**Fig. 8.9** One-photon dissociation pathway. **a** Experimental $\Delta PDF_{\parallel}$. **b** The loss of the C-I and F-I atom pair, each normalized to the minimum value. **c** Simulated $\Delta PDF_{\parallel}$. **d** A $\chi^2$ fitting on $\Delta PDF_{\perp}$ shows an umbrella opening and a C-F bond lengthening. From Ref. [49]. Reprinted with permission from AAAS

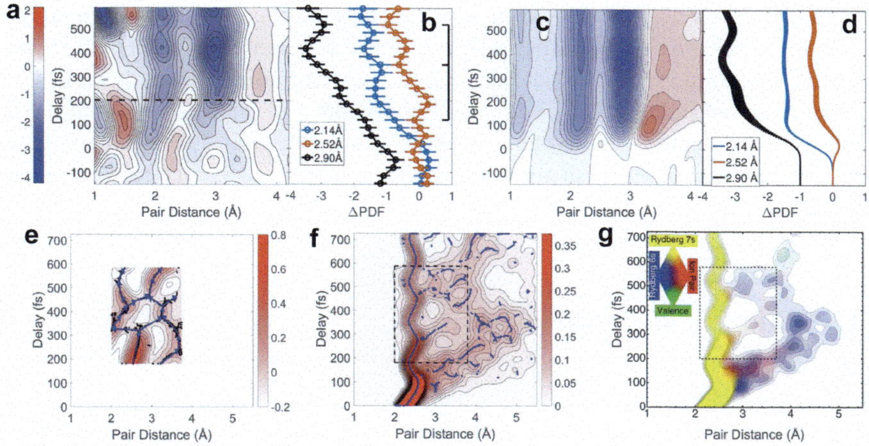


**Fig. 8.10** Two-photon Rydberg pathway. **a** Experimental $\Delta PDF_{\perp}$. **b** The time evolution at 2.14, 2.52, and 2.90 Å from experimental data. **c** Simulated $\Delta PDF_{\perp}$. **d** The time evolution at 2.14 Å, 2.52 Å, and 2.90 Å from simulation. **e** The experimental $PDF_{\perp}$ obtained by removing a common decaying signal from $\Delta PDF_{\perp}$. Black dots are obtained by a ridge finding algorithm, and blue arrows guide the reaction pathway by connecting the dots. **f** The simulated nuclear wavepacket along the C-I coordinate, the blue dots are obtained by a ridge finding algorithm. The dashed box corresponds to the window of observation in experiment. **g** Similar with (**f**), but the color coding indicates the electronic state character from Ref. [49]. Reprinted with permission from AAAS

The AIMS simulation reveals strong nonadiabtic events after Rydberg 7 s state excitation, as shown in Fig. 8.10g. The wavepacket remaining on the Rydberg 7 s state (yellow) continues to vibrate with a ~200 fs period. This motion corresponds to the counter-phased oscillation between 2.14 and 2.52 Å in the $\Delta PDF_{\perp}$. A conical intersection between an ion-pair state and the Rydberg 7 s state gives rise to strong internal conversion to the ion pair (red) and subsequently Rydberg 6 s (blue) states. Since the Rydberg 6 s state is still a bound state, the first outgoing wave is trapped on the 6 s surface and eventually returns close to the Franck–Condon point at around 500 fs, leading to a strong recurrence at 2.9 Å. This recurrence is reproduced in the simulated $\Delta PDF_{\perp}$, shown in Fig. 8.10c–d.

The experimental $\Delta PDF_{\perp}$ has captured the signature of both the Rydberg state C-I vibration and the nonadiabatic wavepacket recurrence. To further extract the dynamics from the $PDF_{\perp}$, we remove a common decaying signal, mostly coming from the rotational dephasing in the one-photon channel, from the $\Delta PDF_{\perp}$. The result is given in Fig. 8.10e. A ridge-finding algorithm is applied for visual guidance, with the same algorithm applied to the simulation shown in Fig. 8.10f. In both figures, one can identify two wavepacket bifurcation events occurring at conical intersections: ~2.7 Å/300 fs and ~2.4 Å/420 fs in the experiment, and at 2.6 Å/270 fs and 2.5 Å/460 fs in the simulation. In addition, a two-branch crossover can be found at ~3.3 Å/400 fs in experiment and ~3.6 Å/420 fs in simulation. This comparison shows that GUED can directly capture wavepacket motion through conical intersections.

In summary, in this section we showed two photoexcited molecular dynamical processes captured by GUED experiments: a simple coherent oscillation in the bond length of a diatomic molecule, and the dissociation and nonadiabatic dynamics of a polyatomic molecule $CF_3I$. The anisotropy introduced by photoselection provides a 3-D view of the structural dynamics, which is not only useful in visualizing the overall structural evolution, but also powerful in separating different photoexcitation channels.

## 8.4 Summary and Outlook

Through the technical development of the past 30 years, GUED has emerged as a tool to meet its promise—resolving the motion of atoms during excited state photochemical reactions. In comparison to the existing large variety of ultrafast laser spectroscopy methods, GUED serves as a complementary tool to observe changes in molecular structures. In addition, the data treatment is typically more straightforward in GUED in comparison to most spectroscopic techniques, because the key step to convert diffraction patterns to PDF and ARPDF is simply a Fourier transformation, and no a priori information about the target system is needed.

GUED has so far been limited to small molecules (up to ~20 atoms) in the gas phase. For larger and more complicated molecules, in principle the same experiment could be performed, provided that the vapor pressure is sufficiently high for reaching the desired target number density ($\sim 10^{16}/\text{cm}^3$ for most existing experiments). In addition, the data interpretation would be more challenging, because many degrees of freedom can potentially lead to similar-looking signals. Combining GUED with both time-resolved spectroscopic methods and MD simulations would be a potential direction to gain additional information.

The current state-of-the-art temporal resolution, 150 fs FWHM [38], is only fast enough to resolve the vibration of heavy atoms. To resolve motion of light atoms such as carbon and oxygen, ideally a time resolution of 10–20 fs is needed. This would require the bunch length of electron beam, pump laser, and the time-of-arrival jitter between the pump and probe beams to be all on the order of 10 fs or less. Maxon et al. have demonstrated experimentally that relativistic electron beams can be compressed down to sub-10 fs with an RF cavity [43]. Using a half-cycle THz field, Zhao et al. [83] and Li et al. [84] have shown that the time-of-arrival jitter can be measured with few- and sub-femtosecond precision on a shot-to-shot basis. A direct electron detector that is capable of recording MeV diffraction pattern with single-electron detection capability and fast frame rate is also being actively developed [85], which could be used for time-stamping similar to the one that is implemented in X-ray free-electron-lasers [86]. With the combination of the technologies above, it can be expected that the next breakthrough in UED temporal resolution can be reached within a decade.

Another important future perspective is the information encoded within diffraction pattern that is beyond the IAM framework. The IAM ignores the contribution from the formation of chemical bonds [61], which contains information from both electron binding and electron correlation [59, 62, 87]. For time-resolved X-ray diffraction (TRXD), a number of theoretical works have started to explore the information content beyond IAM [88–90]. The binding effect was recently observed in a TRXD experiment in a Rydberg state [57]. Within the IAM, the information content of UED and TRXD is exactly the same—both methods reveal nuclear motion in space and time. It would be very interesting to study and compare the information content for UED and TRXD beyond IAM.

## References


1. C. Davisson, L.H. Germer, Diffraction of electrons by a crystal of nickel. Phys. Rev. **30**, 705–740 (1927)
2. H. Mark, R. Wierl, Über Elektronenbeugung am einzelnen Molekül. Naturwissenschaften **18**, 205 (1930)
3. L.O. Brockway, Electron diffraction by gas molecules. Rev. Mod. Phys. **8**, 0231–0266 (1936)
4. L. Schafer, Electron-diffraction as a tool of structural chemistry. Appl. Spectrosc. **30**, 123–149 (1976)
5. A.H. Zewail, Femtochemistry: atomic-scale dynamics of the chemical bond using ultrafast lasers—(Nobel lecture). Angew. Chem. Int. Ed. **39**, 2587–2631 (2000)
6. A.A. Ischenko et al., A stroboscopical gas-electron diffraction method for the investigation of short-lived molecular-species. Appl. Phys. B **32**, 161–163 (1983)
7. G. Mourou, S. Williamson, Picosecond electron-diffraction. Appl. Phys. Lett. **41**, 44–45 (1982)
8. R. Srinivasan, V.A. Lobastov, C.Y. Ruan, A.H. Zewail, Ultrafast electron diffraction (UED)—a new development for the 4D determination of transient molecular structures. Helv. Chim. Acta **86**, 1763–1838 (2003)
9. H. Ihee, B.M. Goodson, R. Srinivasan, V.A. Lobastov, A.H. Zewail, Ultrafast electron diffraction and structural dynamics: transient intermediates in the elimination reaction of C2F4I2. J. Phys. Chem. A **106**, 4087–4103 (2002)
10. H. Ihee, J.S. Feenstra, J.M. Cao, A.H. Zewail, Ultrafast electron diffraction of transient cyclopentadienyl radical: a dynamic pseudorotary structure. Chem. Phys. Lett. **353**, 325–334 (2002)
11. J. Cao, H. Ihee, A.H. Zewail, Ultrafast electron diffraction: determination of radical structure with picosecond time resolution. Chem. Phys. Lett. **290**, 1–8 (1998)
12. H. Ihee, J. Cao, A.H. Zewail, Ultrafast electron diffraction: structures in dissociation dynamics of Fe(CO)(5). Chem. Phys. Lett. **281**, 10–19 (1997)
13. J.C. Williamson, A.H. Zewail, Ultrafast electron-diffraction. 4. Molecular-structures and coherent dynamics. J. Phys. Chem. **98**, 2766–2781 (1994)
14. M. Dantus, S.B. Kim, J.C. Williamson, A.H. Zewail, Ultrafast electron-diffraction. 5. Experimental time resolution and applications. J. Phys. Chem. **98**, 2782–2796 (1994)
15. J.C. Williamson, M. Dantus, S.B. Kim, A.H. Zewail, Ultrafast diffraction and molecular structure. Chem. Phys. Lett. **196**, 529–534 (1992)
16. J.C. Williamson, J. Cao, H. Ihee, H. Frey, A.H. Zewail, Clocking transient chemical changes by ultrafast electron diffraction. Nature **386**, 159–162 (1997)
17. H. Ihee, J.M. Cao, A.H. Zewail, Ultrafast electron diffraction of transient [Fe(CO)(4)]: determination of molecular structure and reaction pathway. Angew. Chem. Int. Ed. **40**, 1532-+ (2001)
18. J.M. Cao, H. Ihee, A.H. Zewail, Ultrafast electron diffraction and direct observation of transient structures in a chemical reaction. Proc. Natl. Acad. Sci. U.S.A. **96**, 338–342 (1999)

19. H. Ihee et al., Direct imaging of transient molecular structures with ultrafast diffraction. Science **291**, 458 (2001)
20. R.C. Dudek, P.M. Weber, Ultrafast diffraction imaging of the electrocyclic ring-opening reaction of 1,3-cyclohexadiene. J. Phys. Chem. A **105**, 4167–4171 (2001)
21. C.-Y. Ruan et al., Ultrafast diffraction and structural dynamics: The nature of complex molecules far from equilibrium. Proc. Natl. Acad. Sci. U.S.A. **98**, 7117 (2001)
22. V.A. Lobastov et al., Ultrafast diffraction of transient molecular structures in radiationless transitions. J. Phys. Chem. A **105**, 11159–11164 (2001)
23. R. Srinivasan, J.S. Feenstra, S.T. Park, S. Xu, A.H. Zewail, Dark structures in molecular radiationless transitions determined by ultrafast diffraction. Science **307**, 558–563 (2005)
24. B.J. Siwick, J.R. Dwyer, R.E. Jordan, R.J.D. Miller, An atomic-level view of melting using femtosecond electron diffraction. Science **302**, 1382–1385 (2003)
25. H. Jean-Ruel et al., Ring-closing reaction in diarylethene captured by femtosecond electron crystallography. J. Phys. Chem. B **117**, 15894–15902 (2013)
26. R.J. Miller, Mapping atomic motions with ultrabright electrons: the chemists' gedanken experiment enters the lab frame. Annu. Rev. Phys. Chem. **65**, 583–604 (2014)
27. G. Sciaini et al., Electronic acceleration of atomic motions and disordering in bismuth. Nature **458**, 56–59 (2009)
28. G. Sciaini, R.J.D. Miller, Femtosecond electron diffraction: heralding the era of atomically resolved dynamics. Rep. Prog. Phys. **74**, 096101 (2011)
29. A.A. Ischenko, P.M. Weber, R.J.D. Miller, Capturing chemistry in action with electrons: realization of atomically resolved reaction dynamics. Chem. Rev. **117**, 11066–11124 (2017)
30. P. Reckenthaeler et al., Time-resolved electron diffraction from selectively aligned molecules. Phys. Rev. Lett. **102**, 213001 (2009)
31. C.J. Hensley, J. Yang, M. Centurion, Imaging of isolated molecules with ultrafast electron pulses. Phys. Rev. Lett. **109**, 133202 (2012)
32. J. Yang, J. Beck, C.J. Uiterwaal, M. Centurion, Imaging of alignment and structural changes of carbon disulfide molecules using ultrafast electron diffraction. Nat. Commun. **6**, 8172 (2015)
33. J. Yang, V. Makhija, V. Kumarappan, M. Centurion, Reconstruction of three-dimensional molecular structure from diffraction of laser-aligned molecules. Struct. Dyn. **1**, 044101 (2014)
34. J.P.F. Nunes, M. Centurion, in *Advances in Atomic, Molecular, and Optical Physics*, vol. 68, ed. by L.F. Dimauro, H. Perrin, S.F. Yelin (Academic Press, 2019), pp. 39–73
35. J.S. Baskin, A.H. Zewail, Ultrafast electron diffraction: oriented molecular structures in space and time. Chem. Phys. Chem. **6**, 2261–2276 (2005)
36. J.S. Baskin, A.H. Zewail, Oriented ensembles in ultrafast electron diffraction. Chem. Phys. Chem. **7**, 1562–1574 (2006)
37. K. Hoshina, K. Yamanouchi, T. Ohshima, Y. Ose, H. Todokoro, Alignment of CS2 in intense nanosecond laser fields probed by pulsed gas electron diffraction. J. Chem. Phys. **118**, 6211–6221 (2003)
38. X. Shen et al., Femtosecond gas-phase mega-electron-volt ultrafast electron diffraction. Struct. Dyn. **6**, 054305 (2019)
39. S.P. Weathersby et al., Mega-electron-volt ultrafast electron diffraction at SLAC national accelerator laboratory. Rev. Sci. Instrum. **86**, 073702 (2015)
40. P. Musumeci, J.T. Moody, C.M. Scoby, Relativistic electron diffraction at the UCLA Pegasus photoinjector laboratory. Ultramicroscopy **108**, 1450–1453 (2008)
41. T. van Oudheusden et al., Compression of subrelativistic space-charge-dominated electron bunches for single-shot femtosecond electron diffraction. Phys. Rev. Lett. **105**, 264801 (2010)
42. R.P. Chatelain, V.R. Morrison, B.L. Klarenaar, B.J. Siwick, Coherent and incoherent electron-phonon coupling in graphite observed with radio-frequency compressed ultrafast electron diffraction. Phys. Rev. Lett. **113**, 235502 (2014)
43. J. Maxson et al., Direct measurement of sub-10 fs relativistic electron beams with ultralow emittance. Phys. Rev. Lett. **118**, 154802 (2017)
44. O. Zandi, K.J. Wilkin, Y.W. Xiong, M. Centurion, High current table-top setup for femtosecond gas electron diffraction. Struct. Dyn. **4** (2017)

45. L. Waldecker, R. Bertoni, R. Ernstorfer, Compact femtosecond electron diffractometer with 100 keV electron bunches approaching the single-electron pulse duration limit. J. Appl. Phys. **117**, 044903 (2015)
46. C. Gerbig, A. Senftleben, S. Morgenstern, C. Sarpe, T. Baumert, Spatio-temporal resolution studies on a highly compact ultrafast electron diffractometer. New J. Phys. **17**, 043050 (2015)
47. J. Yang et al., Diffractive imaging of a rotational wavepacket in nitrogen molecules with femtosecond megaelectronvolt electron pulses. Nat. Commun. **7**, 11232 (2016)
48. J. Yang et al., Diffractive imaging of coherent nuclear motion in isolated molecules. Phys. Rev. Lett. **117**, 153002 (2016)
49. J. Yang et al., Imaging CF3I conical intersection and photodissociation dynamics with ultrafast electron diffraction. Science **361**, 64–67 (2018)
50. T.J.A. Wolf et al., The photochemical ring-opening of 1,3-cyclohexadiene imaged by ultrafast electron diffraction. Nat. Chem. **11**, 504–509 (2019)
51. K.J. Wilkin et al., Diffractive imaging of dissociation and ground-state dynamics in a complex molecule. Phys. Rev. A **100**, 023402 (2019)
52. P. Emma et al., First lasing and operation of an ångstrom-wavelength free-electron laser. Nat. Photonics **4**, 641–647 (2010)
53. J. Küpper et al., X-ray diffraction from isolated and strongly aligned gas-phase molecules with a free-electron laser. Phys. Rev. Lett. **112**, 083002 (2014)
54. M.P. Minitti et al., Imaging molecular motion: femtosecond X-ray scattering of an electrocyclic chemical reaction. Phys. Rev. Lett. **114**, 255501 (2015)
55. H. Yong et al., Determining orientations of optical transition dipole moments using ultrafast X-ray scattering. J. Phys. Chem. Lett. **9**, 6556–6562 (2018)
56. J.M. Ruddock et al., Simplicity beneath complexity: counting molecular electrons reveals transients and kinetics of photodissociation reactions. Angew. Chem. Int. Ed. **58**, 6371–6375 (2019)
57. B. Stankus et al., Ultrafast X-ray scattering reveals vibrational coherence following Rydberg excitation. Nat. Chem. **11**, 716–721 (2019)
58. J.M. Ruddock, et al., A deep UV trigger for ground-state ring-opening dynamics of 1,3-cyclohexadiene. Sci. Adv. **5**, eaax6625 (2019)
59. L.S. Bartell, R.M. Gavin, Effects of electron correlation in X-ray and electron diffraction. J. Am. Chem. Soc. **86**, 3493–3498 (1964)
60. M. Breitenstein, R.J. Mawhorter, H. Meyer, A. Schweig, Theoretical-study of potential-energy differences from high-energy electron-scattering cross-sections of $CO_2$. Phys. Rev. Lett. **53**, 2398–2401 (1984)
61. S. Shibata, H. Sekiyama, K. Tachikawa, M. Moribe, Chemical bonding effect in electron scattering by gaseous molecules. J. Mol. Struct. **641**, 1–6 (2002)
62. J.J. Mcclelland, M. Fink, Electron correlation and binding effects in measured electron-scattering cross-sections of $CO_2$. Phys. Rev. Lett. **54**, 2218–2221 (1985)
63. P. Pulay, R. Mawhorter, D.A. Kohl, M. Fink, Abinitio Hartree-Fock calculation of the elastic electron-scattering cross-section of sulfur-hexafluoride. J. Chem. Phys. **79**, 185–191 (1983)
64. D.A. Kohl, L.S. Bartell, Electron densities from gas-phase electron diffraction intensities. 2. Molecular Hartree-Fock cross sections. J. Chem. Phys. **51**, 2896 (1969)
65. Y. Sasaki, H. Takeuchi, S. Konaka, M. Kimura, Small-angle electron-scattering by formaldehyde and ketene—effects of electron correlation and chemical-binding. Int. J. Quantum Chem. **43**, 701–712 (1992)
66. S. Shibata, F. Hirota, N. Kakuta, T. Muramatsu, Electron-distribution in water by high-energy electron-scattering. Int. J. Quantum Chem. **18**, 281–285 (1980)
67. F. Salvat, A. Jablonski, C.J. Powell, ELSEPA—dDirac partial-wave calculation of elastic scattering of electrons and positrons by atoms, positive ions and molecules. Comput. Phys. Commun. **165**, 157–190 (2005)
68. V. Schomaker, R.O.Y. Glauber, The born approximation in electron diffraction. Nature **170**, 290–291 (1952)

69. L. Holmegaard et al., Control of rotational wave-packet dynamics in asymmetric top molecules. Phys. Rev. A **75**, 051403 (2007)
70. J.M. Glownia et al., Self-referenced coherent diffraction X-ray movie of angstrom- and femtosecond-scale atomic motion. Phys. Rev. Lett. **117**, 153003 (2016)
71. G.A. Garcia, L. Nahon, I. Powis, Two-dimensional charged particle image inversion using a polar basis function expansion. Rev. Sci. Instrum. **75**, 4989–4996 (2004)
72. F. Aguirre, S.T. Pratt, Photoionization of vibrationally hot $CH_3$ and CF3. J. Chem. Phys. **122** (2005)
73. F. Aguirre, S.T. Pratt, Velocity map imaging of the photo dissociation of CF3I: vibrational energy dependence of the recoil anisotropy. J. Chem. Phys. **225**, U475–U475 (2003)
74. M. Zahedi, J.A. Harrison, J.W. Nibler, 266-Nm $CH_3$i photodissociation—$CH_3$ spectra and population-distributions by coherent raman-spectroscopy. J. Chem. Phys. **100**, 4043–4055 (1994)
75. M.D. Person, P.W. Kash, L.J. Butler, The influence of parent bending motion on branching at a conical intersection in the photodissociation of $CH_3I$, $CD_3I$, $CF_3I$. J. Chem. Phys. **94**, 2557–2563 (1991)
76. A. Gedanken, The magnetic circular-dichroism of the a-band in $CF_3I$, $C_2H_5I$ and t-BuI. Chem. Phys. Lett. **137**, 462–466 (1987)
77. A.B. Alekseyev, H.P. Liebermann, R.J. Buenker, Potential energy surfaces for ground and excited electronic states of the $CF_3I$ molecule and their relevance to its A-band photodissociation. Phys. Chem. Chem. Phys. **15**, 6660–6666 (2013)
78. W.G. Roeterdink, M.H.M. Janssen, Femtosecond velocity map imaging of dissociative ionization dynamics in $CF_3I$. Phys. Chem. Chem. Phys. **4**, 601–612 (2002)
79. W.G. Roeterdink, M.H.M. Janssen, Velocity map imaging of femtosecond photodynamics in $CF_3I$. Chem. Phys. Lett. **345**, 72–80 (2001)
80. H.P. Liu, Z.G. Sun, S.D. Hogan, N.Q. Lou, Photodissociation dynamics of $CF_3I$ investigated by two-color femtosecond laser pulses. Eur. Phys. J. D **40**, 357–362 (2006)
81. S.H. Yin et al., Femtosecond pump-probe mass spectra on the dissociative photoionization of $CF_3I$. Chem. Phys. Lett. **372**, 904–910 (2003)
82. S. Eden, P. Limao-Vieira, S.V. Hoffmann, N.J. Mason, VUV photoabsorption in $CF_3X$ (X = Cl, Br, I) fluoro-alkanes. Chem. Phys. **323**, 313–333 (2006)
83. L. Zhao et al., Terahertz streaking of few-femtosecond relativistic electron beams. Phys. Rev. X **8**, 021061 (2018)
84. R.K. Li et al., Terahertz-based subfemtosecond metrology of relativistic electron beams. Phys. Rev. Accel. Beams **22**, 012803 (2019)
85. T. Vecchione, et al., A direct electron detector for time-resolved MeV electron microscopy. Rev. Sci. Instrum. **88** (2017)
86. M. Harmand et al., Achieving few-femtosecond time-sorting at hard X-ray free-electron lasers. Nat. Photonics **7**, 215–218 (2013)
87. J.H. Wang, A.N. Tripathi, V.H. Smith, Chemical-binding and electron correlation-effects in X-ray and high-energy electron-scattering. J. Chem. Phys. **101**, 4842–4854 (1994)
88. K. Bennett, M. Kowalewski, J.R. Rouxel, S. Mukamel, Monitoring molecular nonadiabatic dynamics with femtosecond X-ray diffraction. Proc. Natl. Acad. Sci. U.S.A. **115**, 6538–6547 (2018)
89. M. Kowalewski, K. Bennett, S. Mukamel, Monitoring nonadiabatic avoided crossing dynamics in molecules by ultrafast X-ray diffraction. Struct. Dyn. **4**, 054101 (2017)
90. A.M. Carrascosa, H.W. Yong, D.L. Crittenden, P.M. Weber, A. Kirrander, Ab initio calculation of total X-ray scattering from molecules. J. Chem. Theory Comput. **15**, 2836–2846 (2019)

# Chapter 9
# Ultrafast X-ray Spectroscopy for Probing a Nuclear Wavepacket in Photoexcited Molecular Complexes

**Tetsuo Katayama, Thomas J. Penfold, and Christian Bressler**

**Abstract** Revealing details of coherent nuclear motion during photochemical reactions on the femtosecond timescale is indispensable for understanding the reaction mechanism in the excited state manifold. Owing to the improvement in time resolution and data quality associated with the advent of X-ray free-electron lasers (XFELs), this can now be achieved with atomic structural sensitivity using time-resolved X-ray absorption near edge structure (TR-XANES) spectroscopy. In this chapter, we describe how vibrational motions are observed and interpreted with this X-ray technique. Femtosecond vibrational modes are studied on two prototypical transition metal complexes, $[Fe(bpy)_3]^{2+}$ (bpy = 2,2′-bipyridine) and $[Cu(dmphen)_2]^+$ (dmphen = 2,9-dimethyl-1,10-phenanthroline), and future perspectives exploiting ultrafast X-ray spectroscopies with high repetition rate XFEL machines are presented.

**Keywords** Nuclear wavepacket · Nonadiabatic chemical reaction · Transition metal complex · Ultrafast X-ray spectroscopy

T. Katayama (✉)
Japan Synchrotron Radiation Research Institute, Kouto 1-1-1, Sayo 679-5198, Hyogo, Japan
e-mail: tetsuo@spring8.or.jp

RIKEN, SPring-8 Center, 1-1-1 Kouto, Sayo 679-5148, Hyogo, Japan

T. J. Penfold
Chemistry-School of National and Environmental Sciences, Newcastle University, Newcastle Upon-Tyne NE1 7RU, UK

C. Bressler
European XFEL, 22869 Schenefeld, Germany

The Hamburg Centre of Ultrafast Imaging, 22761 Hamburg, Germany

Department of Physics, Universität Hamburg, 22607 Hamburg, Germany

 
K. Ueda (ed.), *Ultrafast Electronic and Structural Dynamics*,
https://doi.org/10.1007/978-981-97-2914-2_9

## 9.1 A Nuclear Wavepacket in Ultrafast Nonadiabatic Chemical Reactions

During nonadiabatic chemical reactions, the electronic and nuclear degrees of freedom cannot be described independently as fast nuclear motion leads to a breakdown of the Born–Oppenheimer approximation and coupling between close-lying photoexcited states. This means that in contrast to the simple Jablonski-type diagram depicting the relative energies of the involved excited states at a fixed nuclear geometry, one needs to consider the ladder of vibronic energy levels formed by the mixing of the electronic and vibrational wavefunctions [1]. To fully understand the relaxation processes after photoexcitation, it is crucial to track the coupled electronic and vibrational energy flow over complex multi-dimensional potential energy surfaces in order to identify the dominant or relevant pathways of a photoexcited wavepacket. A detailed understanding of this energy flow into, within and also out of the molecular system is required for any attempt to modulate and enhance functional properties of molecules and materials, e.g., in solar energy conversion and magnetic storage applications, in photocatalytic processes, and for light-emitting devices.

Capturing the motion of nuclear wavepackets in real time has been matured using femtosecond (fs) pump-probe optical spectroscopies over the past few decades, following the pioneering work of Ahmed Zewail [2], the 1999 Novel Prize laureate in Chemistry. However, although these tools with ultrashort optical laser pulses are sensitive to valence electronic states and provide time-resolved spectral information revealing specific vibrational motions, it is often difficult to gain the structural details, i.e., the precise intramolecular bond lengths or bond angles of reacting polyatomic molecules, which involve multi-dimensional nuclear coordinates. In addition, optical tools lack direct atomic structural resolution.

## 9.2 Time-Resolved Hard X-ray Absorption Spectroscopy

One promising method to achieve direct structural insight is time-resolved X-ray absorption near edge structure (TR-XANES) that is sensitive not only to the local electronic structure changes but also to the structural dynamics in the vicinity of the selected absorbing atom. XANES is applicable to all matter and no long-range order information is required. It is therefore highly suited to investigate ultrafast structural dynamics in disordered systems without any crystalline order, e.g., of solvated transition metal complexes in solution phase.

In the early 2000s, chemical dynamics research [3–8] with structural-sensitive X-ray tools at 3rd generation synchrotron radiation (SR) sources on transition metal complexes revealed the potential of these tools to complement our knowledge from optical spectroscopy. However, the X-ray pulse duration with SR is limited to tens of picoseconds (ps) due to the equilibrated electron bunch length inside storage rings, which prohibit the observation of nuclear wavepackets with their typically femtosecond vibrational periods. Right before the advent of femtosecond X-ray free-electron laser (XFEL) sources, the so-called laser-electron slicing scheme [9] inside a synchrotron storage ring delivered femtosecond flashes of X-ray pulses, but at the cost of a largely reduced X-ray flux, down to about ten (for hard X-rays) or a few hundred (for soft X-rays) photons per pulse at typically kHz repetition rates. This limited feasible experiments to those not requiring a large integrated X-ray intensity, e.g., the measurements of coherent optical phonons excited in solid-state crystalline materials [10, 11]. For dilute and disordered systems, structural dynamics were observed with 250–300 fs time resolution (as in the solid-state experiments), but this demanded a careful design of the setup and the chosen X-ray tool next to long data collection times [12]. The direct observation of nuclear wavepacket signals was not achieved with the slicing scheme due to the reduced signal quality and the time resolution on the order of 250–300 fs.

The advent of femtosecond X-ray pulses generated by XFELs [13–17] thus marked a striking breakthrough in this situation, since single pulse intensities were increased by 10–12 orders of magnitude compared to the slicing scheme. These extremely bright X-ray sources with an ultrashort pulse duration on the order of 10 fs have meanwhile paved the way for TR-XANES to achieve large signal-to-noise ratios with the required high time resolution. These ingredients allow scientists to track the position and dispersion of the nuclear wavepacket in the excited state manifold. In the following, we present recent milestone studies illustrating the state of the art in TR-XANES to gain new insight into ultrafast nonadiabatic chemical reactions.

## 9.3 Application of TR-XANES to Probe Nuclear Wavepackets

### 9.3.1 *The Case of [Fe(bpy)$_3$]$^{2+}$ (Bpy = 2,2′-Bipyridine)*

Several Fe(II) compounds have stable low-spin (LS) and high-spin (HS) potential energy surfaces with a rather small LS-HS energy difference [18]. This gives rise to the phenomenon called light-induced excited spin state trapping (LIESST), which allows to shuttle the system back and forth between the LS (S = 0) $^1A_{1g}$ ground state and the HS (S = 2) $^5T_{2g}$ state by applying pressure or temperature changes, or via absorption of light. Usually, this requires cryogenic temperatures to keep the system in the selected spin state configuration. The system [Fe(bpy)$_3$]$^{2+}$ (bpy = 2,2′-bipyridine) cannot be stabilized in this way, even at the lowest temperatures the HS state relaxes within microseconds into the LS ground state. This is due to the large energy difference between both states (0.6 eV), which allows an efficient crossing into the LS ground state potential. However, on the fs timescale of the wavepacket dynamics the light-driven HS formation is stable, even at room temperature the HS lifetime is still on the order of 0.6 ns, and therefore this system can be used to study

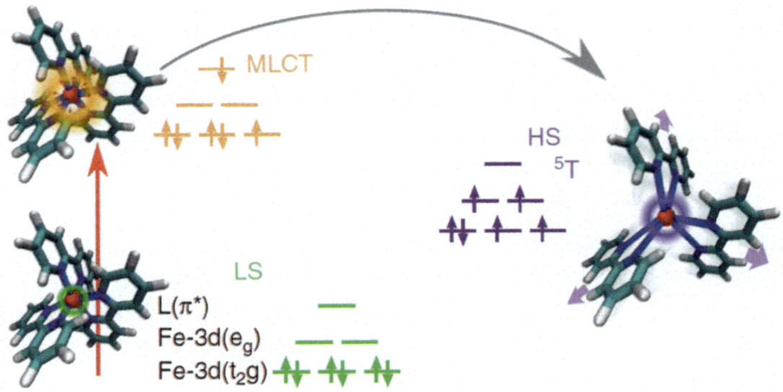


**Fig. 9.1** Photoexcitation and subsequent steps to drive $[Fe(bpy)_3]^{2+}$ into its HS state: After photoexcitation from the LS ($t_{2g}{}^6e_g{}^0L^0$) ground state into the MLCT ($t_{2g}{}^5e_g{}^0L^1$) manifold, the system decays within 120 fs into the HS ($t_{2g}{}^4e_g{}^2L^0$) state, with an elongated molecular structure

the nuclear wavepacket formation and decay, when driving it optically from the LS ground state into the HS state.

The dynamics of this light-induced spin-transition process has been characterized by ultrafast optical and X-ray spectroscopies, revealing that after photoexcitation into the metal-to-ligand-charge-transfer (MLCT) state, $[Fe(bpy)_3]^{2+}$ transfers into the nonemissive HS electronic state (see Fig. 9.1) within less than 130 fs, and the energy relaxation is accompanied by coherent vibrations within the first picoseconds. This LS-HS transition involves two rather prompt spin-forbidden transitions, which meanwhile have been observed for several Fe(II) compounds [19]. However, the relevant vibrational modes involved in such structural modifications and reaction intermediates in the spin transition dynamics are still under debate [12, 20–21].

Recently Lemke et al. exploited the high peak power of the LCLS XFEL beam and probed the wavepacket dynamics of $[Fe(bpy)_3]^{2+}$ dissolved in water with TR-XANES [22]. This experiment demonstrated the power of ultrafast X-ray spectroscopy to capture nuclear motion with the fs time resolution. The time-dependent transient Fe K-edge XANES signals (defined as the normalized difference XANES after (laser on) and before (laser off) laser excitation, or $\Delta I/I_{off}$) were measured at selected probe energies ranging from 7113.5 to 7164.0 eV around the Fe K edge (Fig. 9.2). The observed damped oscillations (due to coherent molecular vibration) were observed throughout this energy range, except for the weak pre-edge region (7113.5 eV), which may have suffered from poor signal quality. Lemke et al. confirmed the optically measured 120 fs MLCT lifetime [23]. Following the departure from the MLCT state into the HS state, the average Fe–N bond length ($r$) is elongated by ~0.2 Å [24, 25] and coherent molecular vibrations within the HS state were observed. The frequency of the oscillatory signal was 126 $cm^{-1}$ (oscillation period of 265 fs), which was attributed to the breathing mode of all Fe–N bonds, and based on the symmetry considerations of the vibrational modes that the XANES signal can pick up as a function of $r$. They also extracted information about the size of the wavepacket, i.e., the degree of delocalization of the molecular ensemble along $r$ in the HS potential energy manifold (Fig. 9.3). Although TR-XANES did not disentangle the triplet intermediate state ($^3$T) that may contribute to the relaxation dynamics, Lemke et al. discussed that the transient spectra were consistent with a very short-lived $^3$T state (<50 fs) which is fed by the 120 fs MLCT decay time in agreement with earlier estimates using time-sliced SR [12]. Similar spin dynamic estimates were reported by Zhang et al. when analyzing fs time-resolved X-ray emission spectroscopy [26].

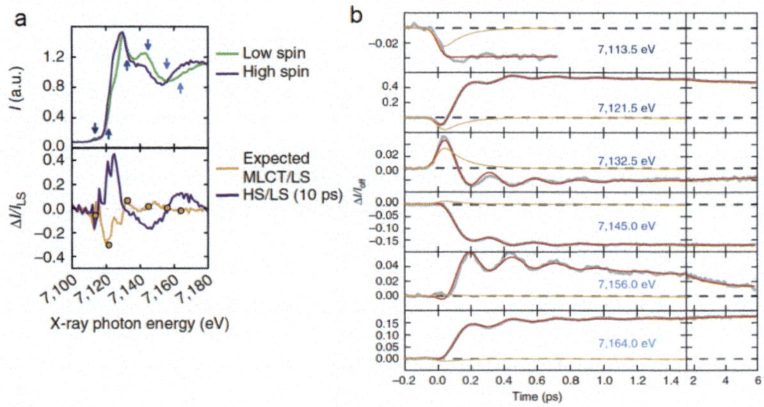


**Fig. 9.2** The fs transient Fe K-edge XANES. **a** The top shows the measured spectra for the ground LS state (green) and the excited HS state (purple). The arrows indicate the energy positions where the fs temporal traces were recorded. The bottom corresponds to the transient change ratio of the HS state (purple) and the calculated MLCT (orange) state with respect to the ground LS state. **b** Time scan data (solid gray lines) measured at selected photon energies. Red lines correspond to the global fitting with an empirical model. Orange lines represent the MLCT contributions

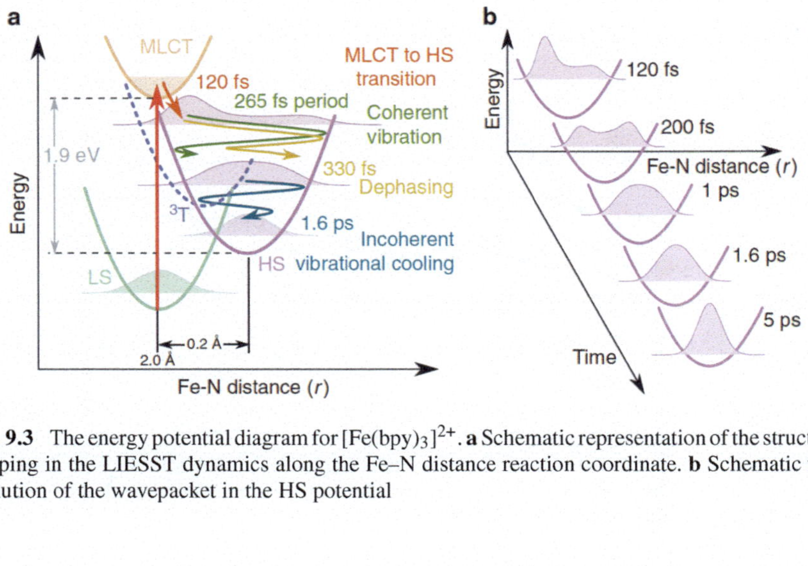


**Fig. 9.3** The energy potential diagram for $[Fe(bpy)_3]^{2+}$. **a** Schematic representation of the structural trapping in the LIESST dynamics along the Fe–N distance reaction coordinate. **b** Schematic time evolution of the wavepacket in the HS potential

### 9.3.2 The Case of $[Cu(dmphen)_2]^+$ (Dmphen = 2,9-Dimethyl-1,10-Phenanthroline)

Cu(I) diimine coordination complexes have been studied as a promising alternative to traditional polypyridyl Ru(II) complexes and are considered as useful candidates for photosensitizer materials based on earth-abundant or first-row transition metals. Extensive photophysical and photochemical studies of different Cu(I) diimine complexes show that—similar to Ru complexes—(i) the molecule is excited from its electronic ground ($S_0$) state to the MLCT manifold by absorption of a visible photon, which oxidizes the central Cu ion from its Cu(I)($d^{10}$) to a Cu(II)($d^9$) electronic configuration. In contrast to Ru complexes (ii) the initially accessed MLCT state is accompanied by a significant structural change for this new Cu(II) electronic configuration, which flattens the dihedral angle between both ligands [27, 28]. This structural reorganization is known as a pseudo Jahn–Teller (PJT) distortion (Fig. 9.4) and plays an important role in solute–solvent effects and also alters its photophysical properties.

Among the Cu(I) diimine complexes, the $[Cu(dmphen)_2]^+$ (dmphen = 2,9-dimethyl-1,10-phenanthroline) complex was used as a model system. Previous ultrafast optical absorption and emission spectroscopies by Iwamura et al. and Chen et al. reported that the MLCT state of $[Cu(dmphen)_2]^+$ relaxes within ~10 ps into the lowest triplet ($T_1$) state with its flattened ligand structure (via PJT distortion) [8, 29]. Furthermore, Iwamura et al. observed coherent oscillations with a damping time of ~1 ps in their transient optical absorption spectra [30]. The oscillations themselves were composed of the superposition of multiple normal modes, dominated by two 125 and 290 $cm^{-1}$ vibrational frequencies, and assigned to the Metal–Ligand breathing and Ligand-twisting modes, respectively.

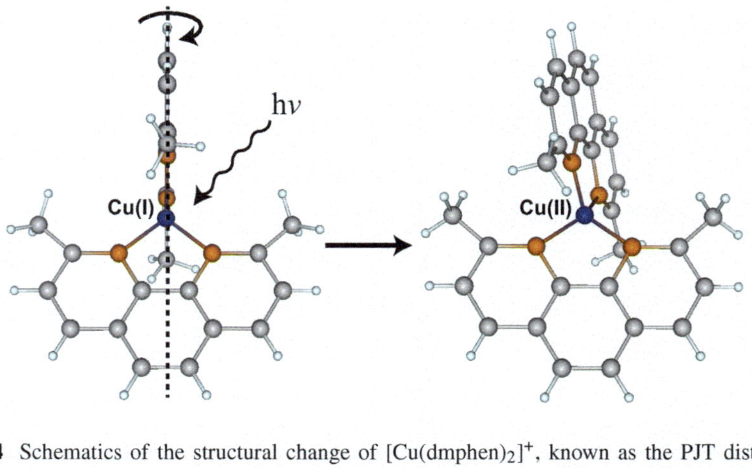


**Fig. 9.4** Schematics of the structural change of $[Cu(dmphen)_2]^+$, known as the PJT distortion. The symmetry is reduced from $D_{2d}$ to $D_2$ by the flattening of the dihedral angle between two dimethyl-phenanthroline ligands

Katayama et al. performed a TR-XANES study of $[Cu(dmphen)_2]^+$ in acetonitrile solution using the SACLA XFEL beam [31]. The Cu K-edge XANES spectra with and without laser pulses exciting the molecular system, their constructed difference spectra, and the temporal evolution of the XANES (measured at three selected probe energies) are shown in Fig. 9.5. While oscillatory signals were indeed observed at the pre-edge (8979.5 eV) and at the rise of the absorption edge (8985.0 eV), no oscillatory features were observed in the temporal trace at the largest transient XANES signal at 8986.5 eV. This is different to the case [22] of $[Fe(bpy)_3]^{2+}$ and indicates that the correct choice of the incident X-ray energy is crucial for observing the wavepacket dynamics. Indeed, the transient XANES difference spectra are dominated by the dynamic blue shift of the absorption edge associated following the initial MLCT excitation [32]: the nuclear wavepacket oscillations are then superimposed on this large background signal, and the energy positions of the prominent wavepacket signals are either too weak in comparison or do not necessarily correspond to those observed for the strongest electronic changes [33].

The small oscillatory signals in the time traces were extracted from the residuals in the global fitting analysis (Fig. 9.6). At the pre-edge (8979.5 eV), three oscillations were found with frequencies in the 83–122, 165–195, and 269–287 $cm^{-1}$ regions, which are assigned to the normal modes of $v_8$ (the in-phase breathing mode), $v_{21}$ (the symmetry-breaking ligand mode associated with the PJT flattening), and $v_{25}$ (another symmetry-breaking mode), respectively [33, 34]. At the rising absorption edge (8985.0 eV), only one single frequency was found around 100–122 $cm^{-1}$ (the corresponding breathing mode). The amplitude of the breathing mode corresponds to a $\pm$ 0.1 Å amplitude motion around the Cu–N equilibrium distance of (~2 Å) within the first 0.5 ps and decays exponentially with a time constant of 0.54 ps, while the symmetry-breaking modes decay within ~0.2 ps, which is faster than the PJT flattening dynamics (0.34 ps [29]). The observed energy-dependent sensitivity of the transient XANES signal to the different nuclear motions is rationalized as follows: At the pre-edge, *3d-4p* mixing adds dipole character to this transition (otherwise 1s-3d is dipole-forbidden), which is enhanced by the symmetry-breaking of the molecule (out of its quasi-tetrahedral symmetry) and/or modulated by the degree of metal–ligand covalency. Therefore, both the symmetry-breaking and the breathing modes become observable at the pre-edge. On the other hand, the rising-edge intensity derives from the shift of the absorption edge reflecting the effective charge state of the Cu ion, which changes as a function of the average Cu–N bond length. This explains why only (or mainly) the breathing mode is observed at the absorption edge itself. Katayama et al. also discussed the connection between these nuclear motions and the PJT ligand-flattening, based on the difference of the observed vibrational decay times: the symmetry-breaking modes are strongly coupled to the flattening motion, while the breathing mode ($v_8$) does not change the molecular symmetry but still dominates the wavepacket dynamics. The schematic potential energy surface landscape in Fig. 9.7 summarizes these observations.

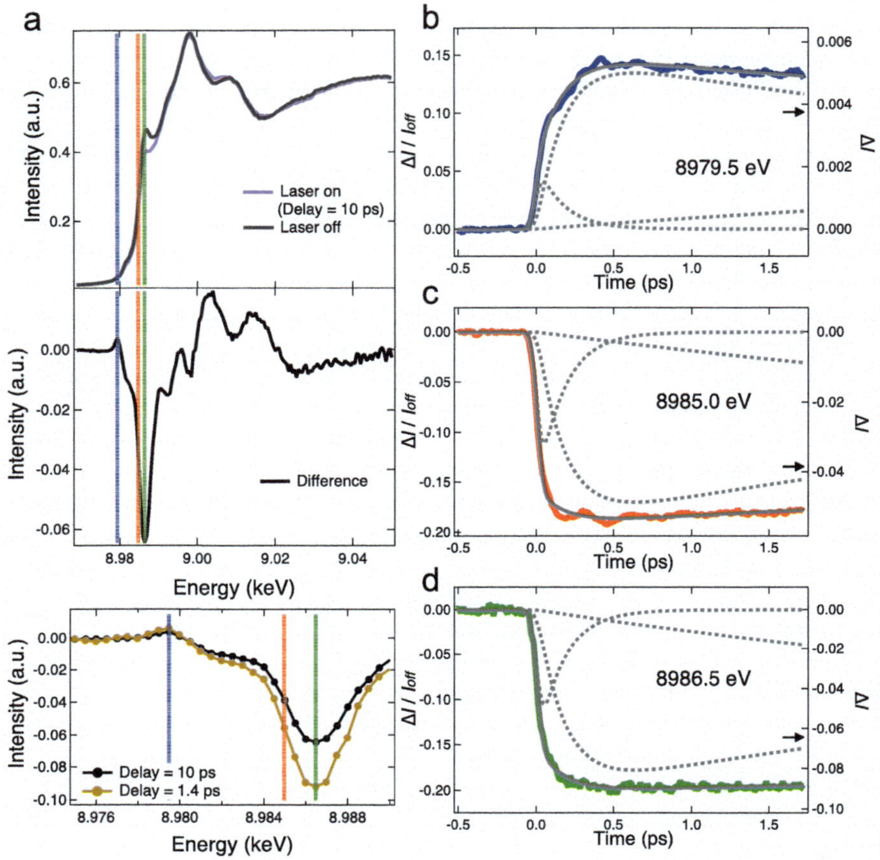


**Fig. 9.5** The fs transient Cu K-edge XANES. **a** Top: Cu K-edge XANES spectra of the $[Cu(dmphen)_2]^+$ ground state (a black line) and the $T_1$ state (a purple line). Middle: Difference spectrum between these two spectra. Bottom: The zoomed view of the difference spectra, measured at 1.4 ps and 10 ps after the optical laser irradiation. The blue, red, and green dotted lines indicate the photon energy positions where the fs temporal traces were measured. **b–d** Time dependences of the transient XANES signals measured at **b** 8979.5 eV, **c** 8985.0 eV, and **d** 8986.5 eV, respectively. The gray solid and dotted lines are the results of the global fitting analysis and the individual components, respectively. The arrows correspond to the signal intensities at 10 ps after photoexcitation

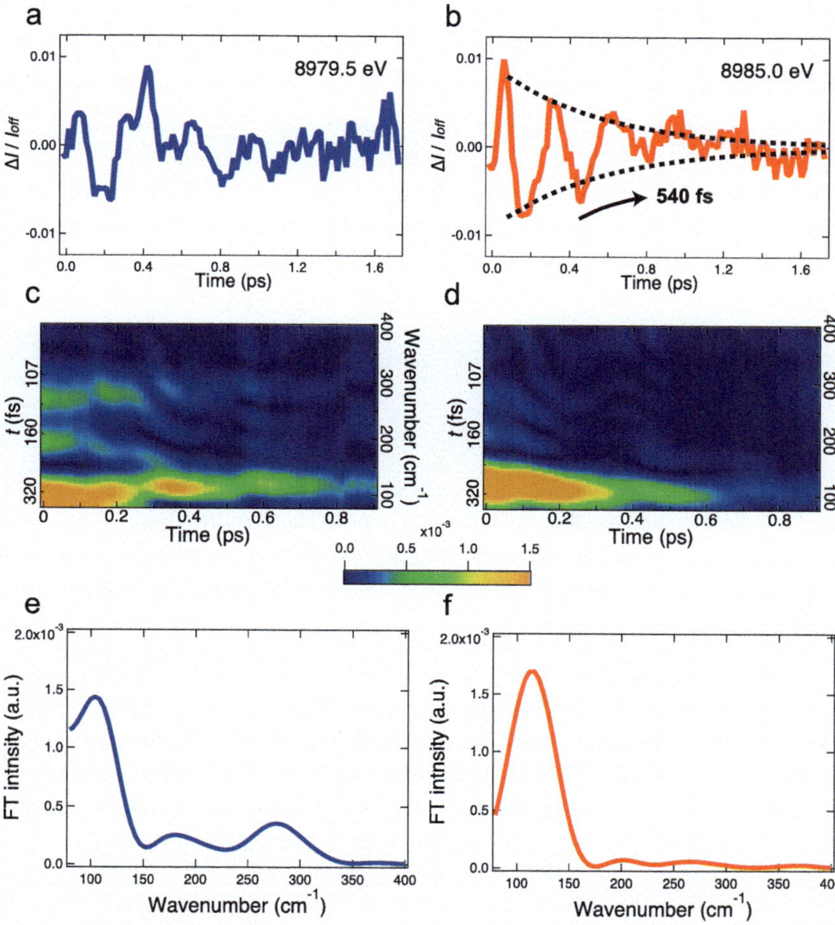


**Fig. 9.6** Extracted coherent nuclear wavepacket. **a**, **b** The residual profiles after the global fitting analysis. **c**, **d** The time-dependent Fourier transforms of (**a**, **b**). While three bands were observed at 83–122, 165–195, and 269–287 $cm^{-1}$ in (**c**), only single band was observed at 100–122 $cm^{-1}$ in (**d**). **e**, **f** The vertical projections of **c**, **d** with a time window of 0–0.4 ps

## 9.4 Potential of a High Repetition XFEL Machine

The observed nuclear wavepacket dynamics were possible due to the superior time resolution of 100 fs (or better) and the superior signal quality, extracted from intense XFELs. With the application of these XFELs over the past ten years in chemical dynamics experiments and the upcoming extension of available photon energies spanning the soft to hard X-ray ranges (0.1–20 keV) [13–17, 35–38], entirely new experiments in ultrafast photochemistry are now feasible and thus underway. This has been successfully demonstrated by a multitude of studies [22, 26, 31, 39–52], focusing on tracking correlated electronic and nuclear motion in a strongly nonadiabatic regime

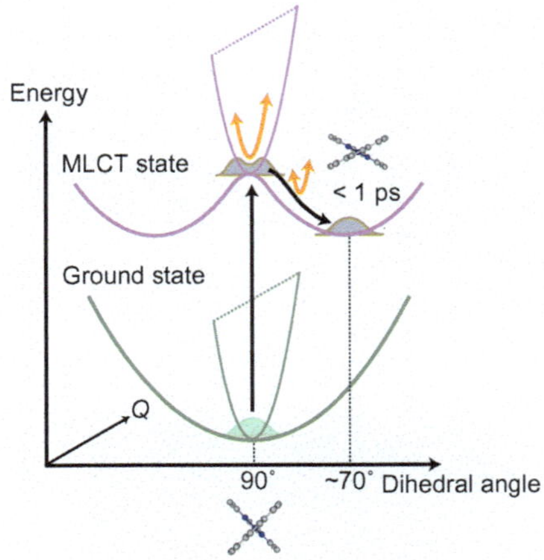


**Fig. 9.7** Potential energy surface landscape of $[Cu(dmphen)_2]^+$

during the formation and breaking of chemical bonds. These studies looked at atomic movements during chemical reactions with an atomic spatio-temporal resolution while tracking the excited state dynamics of reacting molecules, as illustrated in the above Sect. 9.3.

Since 2017, the European X-ray Free-Electron Laser (European XFEL) user facility [9] offers X-ray pulses with the world's highest average brilliance due to its superconducting accelerator that allows for a remarkably high 4.5 MHz repetition rate pulse pattern—and its delivery of up to 2700 intense, ultrashort, transversely coherent X-ray pulses per second in 10 Hz bursts, or up to 27,000 pulses per second. This is more than a 100-fold of what other FEL facilities can offer to date. In combination with advanced scientific instrumentation and sample preparation techniques, this unmatched average X-ray flux opens unprecedented opportunities for the investigation of time-dependent phenomena, such as solution-phase photochemistry and material science studies. One example requiring a large incident intensity is hard X-ray Raman scattering (XRS): one can seek to use every single X-ray pulse at 4.5 MHz to collect the required statistics, thus obtaining the element-specific information of low-Z elements (such as N, C, and O), usually extracted from soft X-ray spectra with their low penetration depths, while maintaining the experimental benefits of hard X-ray techniques [53, 54]. Moreover, the high flux enables time-resolved studies on very dilute solutions, a typical challenge for biologically relevant systems. The scientific instrument FXE [55, 56] is designed to advance into this uncharted territory, utilizing hard X-rays to make fs-resolved "molecular movies" of ensuing nuclear and electronic dynamics while exploiting the uniquely high average flux of European XFEL. The scientific instrument FXE (Fig. 9.8) seeks to make use of the large flux available at European XFEL. The pulse sequence (Fig. 9.9) can be

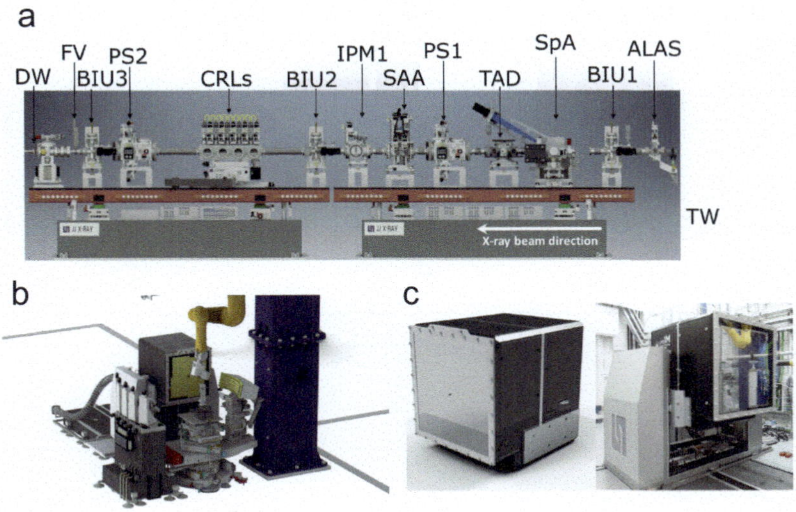


**Fig. 9.8** The scientific instrument FXE. **a** The X-ray optics of the FXE beamline. Legend: DW = diamond window, FV = fast valve, BIU = beam imaging unit, PS = power slit, CRL = beryllium lenses, IPM = intensity/position monitor, SAA = solid attenuator assembly, TAD = time arrival detector, SpA = single-shot spectrum analyzer, ALAS = alignment laser, TW = tunnel wall. **b** Illustration of the sample interaction area at FXE with a setup for the liquid sample. Two X-ray spectrometers, von Hamos (right) and Johann (left), and the Large Pixel Detector (LPD) for forward scattering/diffraction experiments surround the sample. **c** The design and picture of LPD having 1 million pixels, each with $0.5 \times 0.5$ mm$^2$ square size. The detector can store a maximum of 510 images sampled with a frequency of 4.5 MHz. Each single-shot image has a dynamic range of about $1 \times 10^5$ 12 keV photons

tailored to the experimental needs, which consist of (i) sample refreshing rate, (ii) detector readout times, and—most importantly—(iii) the intrinsic lifetimes of the processes to record. At FXE, the sample can be refreshed for each X-ray shot within one microsecond (and in some cases even below) due to the use of high-speed liquid jets. Femtosecond timing information between the exciting laser and XFEL pulses can be extracted utilizing new schemes of femtosecond timing tools, which operate reliably and precisely at MHz repetition rates [57–60].

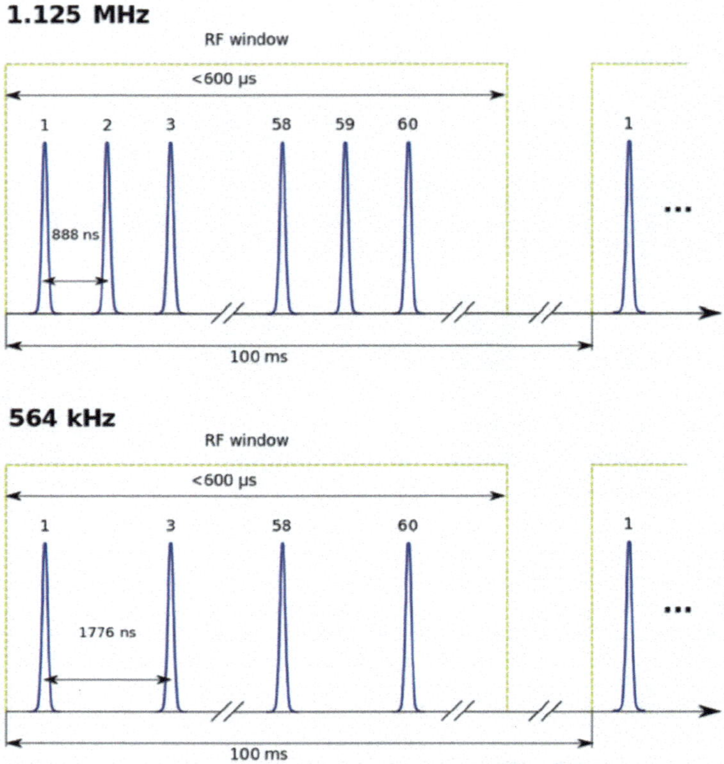


**Fig. 9.9** Pulse train filling patterns with two different intra-train repetition rates at European XFEL. Currently, the repetition rate is 1.125 MHz, or one pulse every ~888 ns, with a maximum of 120 bunches per pulse train. Lower intra-train repetition rates are obtained by removing electron bunches from a train, thus reducing the total number of pulses to obtain the lower repetition rate in the experiment

## References

1. T.J. Penfold et al., Spin-vibronic mechanism for intersystem crossing. Chem. Rev. **118**, 6975–7025 (2018)
2. A.H. Zewail, Femtochemistry: atomic-scale dynamics of the chemical bond. J. Phys. Chem. A **104**, 5660–5694 (2000)
3. M. Chergui, Picosecond and femtosecond X-ray absorption spectroscopy of molecular systems. Acta Cryst. **A66**, 229–239 (2010)
4. L.X. Chen et al., Excited-state molecular structures captured by X-ray transient absorption spectroscopy: a decade and beyond. Acta Cryst. **A66**, 240–251 (2010)
5. L.X. Chen et al., Capturing a photoexcited molecular structure through time-domain X-ray absorption fine structure. Science **292**, 262–264 (2001)
6. C. Bressler et al., Toward structural dynamics in condensed chemical systems exploiting ultrafast time-resolved X-ray absorption spectroscopy. J. Chem. Phys. **116**, 2955–2966 (2002)
7. M. Saes et al., Observing photochemical transients by ultrafast X-ray absorption spectroscopy. Phys. Rev. Lett. **90**, 047403 (2003)

8. L.X. Chen et al., MLCT state structure and dynamics of a copper(I) diimine complex characterized by pump-probe X-ray and laser spectroscopies and DFT calculations. J. Am. Chem. Soc. **125**, 7022–7034 (2003)
9. R.W. Schoenlein et al., Generation of femtosecond pulses of synchrotron radiation. Science **287**, 2237–2240 (2000)
10. P. Beaud et al., Spatiotemporal stability of a femtosecond hard-X-ray undulator source studied by control of coherent optical phonons. Phys. Rev. Lett. **99**, 174801 (2007)
11. S.L. Johnson et al., Direct observation of non-fully-symmetric coherent optical phonons by femtosecond X-ray diffraction. Phys. Rev. B **87**, 054301 (2013)
12. C. Bressler et al., Femtosecond XANES study of the light-induced spin crossover dynamics in an iron(II) complex. Science **323**, 489–492 (2009)
13. P. Emma et al., First lasing and operation of an angstrom wavelength free-electron laser. Nat. Photon. **4**, 641–647 (2010)
14. T. Ishikawa et al., A compact X-ray free-electron laser emitting in the sub-angstrom region. Nat. Photon. **6**, 540–544 (2012)
15. H. Kang et al., Hard X-ray free-electron laser with femtosecond-scale timing jitter. Nat. Photon. **11**, 708–713 (2017)
16. M. Altarelli, The European X-ray free-electron laser facility in Hamburg. Nucl. Instrum. Methods Phys. Res. Sect. B **269**, 2845–2849 (2011)
17. C. Milne et al., SwissFEL: the Swiss X-ray free electron laser. Appl. Sci. **7**, 720 (2017)
18. J.K. McCusker et al., Sub-picosecond $\Delta S=2$ intersystem crossing in low-spin ferrous complexes. J. Am. Chem. Soc. **114**, 6919–6920 (1992)
19. M.C. Carey, S.L. Adelman, J.K. McCusker, Insights into the excited state dynamics of Fe(II) polypyridyl complexes from variable-temperature ultrafast spectroscopy. Chem. Sci. **10**, 134–144 (2019)
20. G. Auböck, M. Chergui, Sub-50-fs photoinduced spin crossover in $[Fe(bpy)_3]^{2+}$. Nat. Chem. **7**, 629–633 (2015)
21. K.S. Kjær et al., Finding intersections between electronic excited state potential energy surfaces with simultaneous ultrafast X-ray scattering and spectroscopy. Chem. Sci. **10**, 5749–5760 (2019)
22. H.T. Lemke et al., Coherent structural trapping through wave packet dispersion during photoinduced spin state switching. Nat. Commun. **8**, 15342 (2017)
23. W. Gawelda et al., Ultrafast nonadiabatic dynamics of $[Fe(bpy)_3]^{2+}$ in solution. J. Am. Chem. Soc. **129**, 8199–8206 (2007)
24. W. Gawelda et al., Structural determination of a short-lived excited iron(II) complex by picosecond X-ray absorption spectroscopy. Phys. Rev. Lett. **98**, 057401 (2007)
25. W. Gawelda et al., Structural analysis of ultrafast extended x-ray absorption fine structure with subpicometer spatial resolution: application to spin crossover complexes. J. Chem. Phys. **130**, 124520 (2009)
26. W. Zhang et al., Tracking excited-state charge and spin dynamics in iron coordination complexes. Nature **509**, 345–348 (2014)
27. M.W. Mara, K.A. Fransted, L.X. Chen, Interplays of excited state structures and dynamics in copper(I) diimine complexes: Implications and perspectives. Coord. Chem. Rev. **282–283**, 2–18 (2015)
28. M. Iwamura, S. Takeuchi, T. Tahara, Ultrafast excited-state dynamics of copper(I) complexes. Acc. Chem. Res. **48**, 782–791 (2015)
29. M. Iwamura, S. Takeuchi, T. Tahara, Real-time observation of the photoinduced structural change of Bis(2,9-dimethyl-1,10-phenanthroline)copper(I) by femtosecond fluorescence spectroscopy: aA realistic potential curve of the Jahn-Teller distortion. J. Am. Chem. Soc. **129**, 5248–5256 (2007)
30. M. Iwamura et al., Coherent nuclear dynamics in ultrafast photoinduced structural change of Bis(diimine)copper(I) complex. J. Am. Chem. Soc. **133**, 7728–7736 (2011)
31. T. Katayama et al., Tracking multiple components of a nuclear wavepacket in photoinduced Cu(I)-phenanthroline complex using ultrafast X-ray spectroscopy. Nat. Commun. **10**, 3606 (2019)

32. T.J. Penfold et al., Solvent-induced luminescence quenching: static and time-resolved X-ray absorption spectroscopy of a copper(I) phenanthroline complex. J. Phys. Chem. A **117**, 4591–4601 (2013)
33. G. Capano et al., Probing wavepacket dynamics using ultrafast X-ray spectroscopy. J. Phys. B At. Mol. Opt. Phys. **48**, 214001 (2015)
34. G. Capano et al., A quantum dynamics study of the ultrafast relaxation in a prototypical Cu(I)-phenanthroline. J. Phys. Chem. A **118**, 9861–9869 (2014)
35. J. Ullrich, A. Rudenko, R. Moshammer, Free-electron lasers: new avenues in molecular physics and photochemistry. Annu. Rev. Phys. Chem. **63**, 635–660 (2012)
36. L. Gallmann, C. Cirelli, U. Keller, Attosecond science: recent highlights and future trends. Annu. Rev. Phys. Chem. **63**, 447–469 (2012)
37. W. Ackermann et al., Operation of a free-electron laser from the extreme violet to the water window. Nat. Photon. **1**, 336–342 (2007)
38. T. Shintake et al., A compact free-electron laser for generating coherent radiation in the extreme ultraviolet region. Nat. Photon. **2**, 555–559 (2008)
39. M.W. Mara et al., Metalloprotein entatic control of ligand-metal bonds quantified by ultrafast X-ray spectroscopy. Science **356**, 1276–1280 (2017)
40. N.A. Miller et al., Polarized XANES monitors femtosecond structural evolution of photoexcited vitamin $B_{12}$. J. Am. Chem. Soc. **139**, 1894–1899 (2017)
41. W. Zhang et al., Manipulating charge transfer excited state relaxation and spin crossover in iron coordination complexes with ligand substitution. Chem. Sci. **8**, 515–523 (2017)
42. T.B. van Driel et al., Atomistic characterization of the active-site solvation dynamics of a model photocatalyst. Nat. Commun. **7**, 13678 (2016)
43. M.L. Shelby et al., Ultrafast excited state relaxation of a metalloporphyrin revealed by femtosecond X-ray absorption spectroscopy. J. Am. Chem. Soc. **138**, 8752–8764 (2016)
44. E. Biasin et al., Femtosecond X-ray scattering study of ultrafast photoinduced structural dynamics in solvated $[Co(terpy)_2]^{2+}$. Phys. Rev. Lett. **117**, 013002 (2016)
45. K. Pande et al., Femtosecond structural dynamics drives the trans/cis isomerization in photoactive yellow protein. Science **352**, 725–729 (2016)
46. T.R.M. Barends et al., Direct observation of ultrafast collective motions in CO myoglobin upon ligand dissociation. Science **350**, 445–450 (2015)
47. J. Kern, V.K. Yachandra, J. Yano, Metalloprotein structures at ambient conditions and in real-time: biological crystallography and spectroscopy using X-ray free electron lasers. Curr. Opin. Struct. Biol. **34**, 87–98 (2015)
48. W. Zhang, K.J. Gaffney, Mechanistic studies of photoinduced spin crossover and electron transfer in inorganic complexes. Acc. Chem. Res. **48**, 1140–1148 (2015)
49. K.H. Kim et al., Direct observation of bond formation in solution with femtosecond X-ray scattering. Nature **518**, 385–389 (2015)
50. S.E. Canton et al., Visualizing the non-equilibrium dynamics of photoinduced intramolecular electron transfer with femtosecond X-ray pulses. Nat. Commun. **6**, 6359 (2015)
51. M. Levantino et al., Ultrafast myoglobin structural dynamics observed with an X-ray free-electron laser. Nat. Commun. **6**, 6772 (2015)
52. D. Arnlund et al., Visualizing a protein quake with time-resolved X-ray scattering at a free-electron laser. Nat. Methods **11**, 923–926 (2014)
53. U. Bergmann et al., X-ray Raman spectroscopy at the oxygen K edge of water and ice: implications on local structure models. Phys. Rev. B **66**, 092107 (2002)
54. J. Szlachetko et al., A dispersive inelastic X-ray scattering spectrometer for use at X-ray free electron lasers. Appl. Sci. **7**, 899 (2017)
55. A. Galler et al., Scientific instrument femtosecond X-ray experiments (FXE): instrumentation and experimental capabilities. J. Synchrotron Rad. **26**, 1432–1447 (2019)
56. D. Khakhulin et al., Ultrafast X-ray photochemistry at European XFEL: capabilities of the femtosecond X-ray experiments (FXE) instrument. Appl. Sci. **10**, 995 (2020)
57. M. Diez et al., A self-referenced in-situ arrival time monitor for X-ray free-electron lasers. Sci. Rep. **11**, 3562 (2021)

58. M. Diez et al., A sensitive high repetition rate arrival time monitor for X-ray free electron lasers. Nat. Commun. **14**, 2495 (2023)
59. A. Burnett et al., UK XFEL Science Case. Science and Technology Facility Council
60. M. Dunne, LCLS-II commissioning, first light, and future prospects (Conference Presentation), in *Proceedings of the SPIE PC12581, X-Ray Free-Electron Lasers: Advances in Source Development and Instrumentation VI, PC1258101* (9 June 2023)

# Chapter 10
# Ultrafast X-Ray Probes of Dynamics in Solids

**Mariano Trigo, Mark P. M. Dean, and David A. Reis**

Advances in our ability to understand and utilize the world around us have always relied on the development of advanced tools for probing and manipulating material properties. X-ray matter interactions played a critical role in the development of the modern theory of solid-state materials that has continued over more than a century. The development of ever-brighter X-ray sources has facilitated ever more sensitive X-ray scattering and spectroscopy measurements that are able to probe not just the lattice structure, but also the spectrum of elementary excitations in complex materials. The interactions underlying the electronic, magnetic, and thermal properties of solids tend to be associated with lengthscales comparable to the atomic separation and timescales ranging from femtosecond to picoseconds. The short wavelength, femtosecond pulses from X-ray free-electron lasers (XFEL) therefore offer unprecedented opportunities to probe and understand material properties on their natural lengthscales and timescales of these processes. This chapter reviews a number of exemplary XFEL-based experiments on lattice and electronic dynamics, and their microscopic interactions both near and out of equilibrium. We conclude with a brief discussion of new forms of spectroscopy enabled by the combination of high flux and short pulse duration and give an outlook for how the field will develop in the future.

---

M. Trigo · D. A. Reis (✉)
Stanford PULSE Institute and Stanford Institute for Materials and Energy Sciences (SIMES), SLAC National Accelerator Laboratory, Menlo Park, CA 94025, USA
e-mail: dreis@stanford.edu

M. Trigo
e-mail: mtrigo@slac.stanford.edu

M. P. M. Dean
Condensed Matter Physics and Materials Science Department, Brookhaven National Laboratory, Upton, NY 11973, USA
e-mail: mdean@bnl.gov

 
K. Ueda (ed.), *Ultrafast Electronic and Structural Dynamics*,
https://doi.org/10.1007/978-981-97-2914-2_10

## 10.1 Introduction

X-ray radiation has proven invaluable for studying the structure and dynamics of matter on the atomic scale. A variety of ever-more-sophisticated techniques have been developed over the past 100 and 25 years concomitant with improvements in X-ray sources, optics, and detectors. These methods make use of the short wavelength (comparable to the size of atoms), atomic and chemical specificity, as well as the penetrating power of X-rays. Each new advance has led to important discoveries across a wide range of disciplines. In the case of condensed matter physics, the high brightness and tunability of storage-ring-based synchrotron radiation sources enabled the development of ultra-sensitive X-ray methods including non-resonant inelastic X-ray scattering (IXS) [1, 2] and resonant inelastic X-ray scattering (RIXS) [3–5]. These techniques have been transformational for the study of collective modes of quantum materials. Here dedicated instruments have been developed that push the resolution in energy and momentum that allow for the study of small crystals and heterostructures, in a variety of environments. The use of such sources for studying the nonequilibrium dynamics of quantum materials in the time domain is limited by the pulse duration of the sources which are typically tens to a hundred picoseconds per burst. However, dynamics at the inter-atomic scale, including the highest frequency vibrations and the making and breaking of chemical bonds, occurs on a few femtosecond timescale, while the dynamics associated with electron correlation and exchange interaction can be even faster.

Extremely bright femtosecond XFEL sources are transforming our ability to follow the dynamics of solid-state materials on the relevant timescales of their low-lying elementary excitations, as well as to study their nonequilibrium behavior on the atomic scale [6–10]. The first hard X-ray XFEL, the Linac Coherent Light Source (LCLS) became operational over a decade ago, initially lasing at a wavelength of 1.5 Å, immediately exceeding the peak brightness of the most intense X-ray light sources in operation by 9–10 orders of magnitude [11]. This dramatic increase in peak brightness was due to a combination of the ultrashort pulse duration of $\sim$100 fs, large pulse energy $\sim$mJ, and near-transform-limited transverse spatial coherence characteristic of the high-gain self-amplified spontaneous emission (SASE) process [12–14]. Since this time a number of hard X-ray free-electron laser (FEL) facilities have been built around the world [15–18]. The capabilities and performance of the various XFELs have also grown including self-seeding [19, 20], shorter pulses of a few tens of femtoseconds down to sub-femtosecond (attosecond) regime [21, 22], tunability of X-ray energy up to $\sim$20 keV [23], as well as two-color operation [24]. The next-generation superconducting LINAC-based XFELs will operate at much higher average power and repetition rates [25, 26].

We will not discuss the physics of the FEL process here, as there are a number of excellent references for the interested reader [27–31]. Instead we will focus on describing a sub-set of experiments that make use of the unique properties of the XFEL for studying the ultrafast dynamics of quantum materials. After a brief discussion of the general experimental considerations, we focus on discussing different classes of dynamics in optical-laser-pumped materials probed by both non-resonant and resonant scattering techniques. We begin by considering examples where non-resonant time-resolved diffraction probes coherent excitation of lattice modes and electron-phonon coupling in the linear-response regime, both at the zone center and, in the case of diffuse X-ray scattering, for two-phonon excitations spanning the Brillouin zone. We then go on to discuss examples where these methods have been used to uncover the nonlinear couplings between multiple degrees of freedom, as well as cases where photoexcitation induces excitations well beyond linear response using non-resonant diffuse scattering to look at metal–insulator transitions and resonant diffraction to study changes in charge and orbital ordering. Finally, we describe experiments where time-resolved resonant inelastic scattering provides combined momentum and energy-resolved information on magnetic, orbital, and charge excitations. We have not attempted to be comprehensive and notably have left out a number of important applications and techniques including lensless imaging [32], X-ray photon correlation spectroscopy [33], and (for the most part) X-ray nonlinear optics in solids [34–39].

Finally, we note that even before the XFEL, femtosecond duration X-ray pulses were produced in laser–plasma interactions [40, 41], laser slicing on storage rings [42], and via electron compression in linear-accelerator-based spontaneous synchrotron sources [43–45]. The relatively low flux and availability of these sources have limited their utility. Nonetheless these sources were critical in developing the nascent field of ultrafast X-ray science [46–49] including the experiments described here.

## 10.2 X-Ray Free-Electron Laser-Based Pump–Probe Methods

The different experiments discussed in this chapter on solid-state dynamics all make use of the combination of the short pulse duration and high flux available on the XFEL. In particular, each derives its temporal resolution through stroboscopic, pump–probe, methods, whereby a femtosecond long-wavelength pump (often a near-IR laser) excites the material which is subsequently probed by the X-rays captured on a relatively slow detector. In this manner the dynamics are built up statistically as a function of the relative timing of the pump and probe, and the resolution is given by the combination of the pump duration, the probe duration, and our ability to control (or measure) the relative delay. The earlier XFEL experiments were limited by both the timing jitter and drift between the optical and X-ray laser, although subsequently nearly pulse-length-limited resolution has been achieved using a single-shot X-ray arrival time monitor [50, 51]. In addition, because the SASE process starts up from noise all the various properties of the X-ray pulse fluctuate from shot-to-shot, such that in addition to the timing, a typical experiment needs to monitor the pulse properties and reads the detectors on a shot-by-shot basis, at a few tens to 120 Hz repetition rate. Here care needs to be taken to minimize systematic errors over the course of an experiment, as well as to handle the processing and analysis of the data (which can easily exceed several TB/hour).

## 10.3 Coherent Phonon Spectroscopy in Linear Response

We begin our discussion by considering time-domain measurements of coherently excited collective lattice modes (phonons). Coherent phonon spectroscopy [52, 53] has been an invaluable tool in the optical domain for probing the vibrational properties of solids. An attractive feature of time-domain, coherent phonon spectroscopy is that, similar to nuclear magnetic resonance (NMR), the time-domain sampling of the oscillations produced by a coherent vibration allows for frequency resolution comparable or surpassing the natural oscillator linewidth. X-ray scattering has several advantages over optical methods for coherent phonon spectroscopy, including direct measurements of lattice displacements and momentum resolution that is inaccessible with long-wavelength probing. Much of the early work on time-resolved X-ray scattering focused on scattering from coherent acoustic phonons [46, 48, 49]. X-ray FELs have the advantage of much better time resolution and flux than any previous source, so it is no surprise that they are powerful tools for measuring coherent phonons with much better sensitivity and from a wider range of materials than previously possible. For example, Singer et al. used ultrafast X-ray diffraction at the LCLS to probe the dynamics of the charge density wave (CDW) in a 28-nm-thick film of elemental chromium (Cr) [54]. This material exhibits an incommensurate spin density wave (SDW) below $T_N = 290$ K accompanied by a CDW at $2q$, where $q$ is the magnitude of the SDW wavevector. In the sample studied in [54], the CDW wavevector, $(0\,0\,2q)$, was perpendicular to the film. The finite thickness of the film gives rise to finite-size side bands of the crystal Bragg peak as shown in Fig. 10.1a. These finite-size satellites align and are commensurate with the CDW wavevector due to it being pinned to the film surface. Importantly, these finite-size side bands at $-2q$ and $+2q$ interfere constructively (destructively) with the X-ray field scattered by the CDW, resulting in an increase (decrease) of the CDW peak intensity, respectively. This interference, akin to a heterodyne measurement, makes the intensity depend linearly on the atomic displacement associated with the CDW, $\delta y$ [55] and the measurement is sensitive to the sign of the atomic motion with $\Delta I(t) = \pm 1$ corresponding to $\delta y = \pm 1$ in normalized units, respectively. The dynamics of the normalized intensity change for the $\mathbf{Q} = (0\,0\,k)$ peak with $k = 2 \pm 2q$ are shown in Fig. 10.1b for several incident pump fluences. The observed oscillations with period ~0.45 ps (frequency 2.2 THz) originate from the amplitude mode of the CDW.

The CDW dynamics in Fig. 10.1b follow the displacive excitation of coherent phonons (DECP) [53, 56] common among many charge-ordered systems under ultrafast excitation. In DECP, the coherent motion of the mode is initiated by a sudden displacement of the parabolic potential as shown in Fig. 10.1c. The general solution for the atomic displacement $\delta y(t > 0)$ under a displaced harmonic potential is [56]

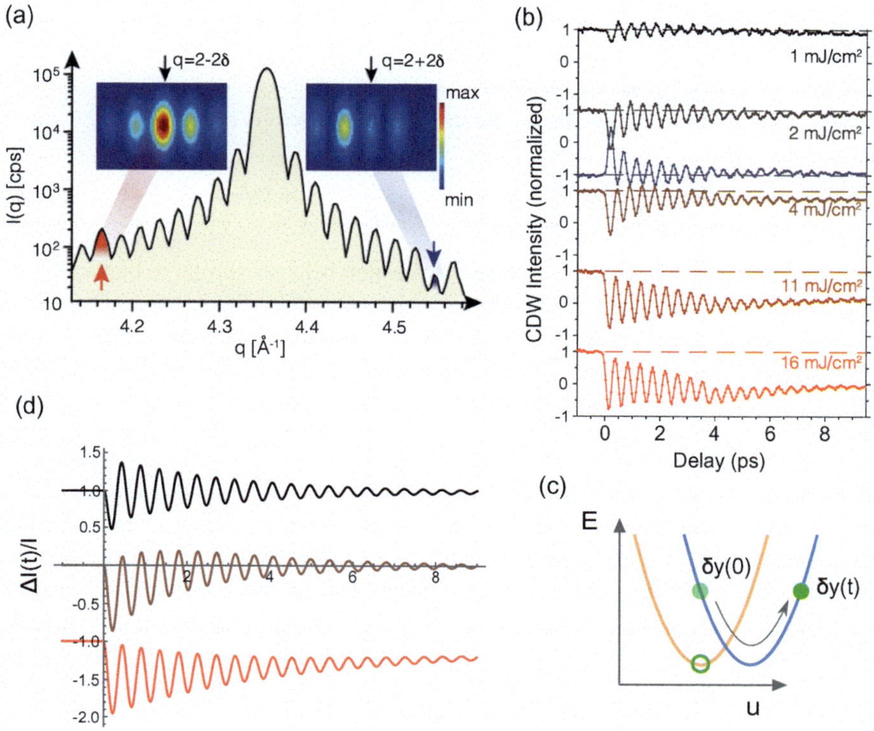


**Fig. 10.1** Example of coherent phonons probed by ultrafast X-ray diffraction. **a** The diffraction pattern around $Q = (0\ 0\ 2q)$ of a 28-nm-thick Cr film. The small oscillations are sidebands due to finite size. The inset shows a detector image at the CDW wavevector. **b** Dynamics of the normalized CDW intensity for several incident fluences. **c** Schematic of the displacive excitation of a coherent phonon. **d** Calculation of the atomic displacement from Eq. 10.1 using parameters from [54]. The curves are displaced vertically for clarity. (**a, b**) Reprinted figure elements with permission from Ref. [54] Copyright 1996 by the American Physical Society

$$\delta y(t) = \frac{A}{2}\frac{\Omega^2}{\Omega^2+\beta^2-2\gamma\beta}\left\{e^{-\gamma t}\left(\cos\Omega t - \frac{\beta-\gamma}{\Omega}\sin\Omega t\right) - e^{-\beta t}\right\}, \tag{10.1}$$

where $\Omega$, $A$, and $\gamma$ are the frequency, amplitude, and damping constant of the oscillator, and $\beta^{-1}$ is the timescale for return to equilibrium of the displaced harmonic potential. If $\beta - \gamma \ll \Omega$, i.e., a harmonic system with a slow recovery of the potential compared to the frequency, Eq. 10.1 can be simplified to

$$\delta y(t) = \frac{A}{2}e^{-\gamma t}\left(\cos\Omega t - e^{-(\beta-\gamma)t}\right) \tag{10.2}$$

which is an oscillatory (cosine) motion with a slowly recovering equilibrium offset. In the limit of slow relaxation of the potential $\beta \approx 0$ and $\delta y(t) = \frac{A}{2}(e^{-\gamma t}\cos\Omega t - 1)$ corresponding to a cosine-like phase.

The opposite limit where the displaced potential recovers quickly is $\beta \gg \gamma$ and we get

$$\delta y(t) = \frac{A}{2}\frac{\Omega\beta}{\beta^2 + \Omega^2 - 2\gamma\beta}e^{-\gamma t}\sin\Omega t. \tag{10.3}$$

In this case the oscillatory motion is sine-like which behaves similar to that achieved by an impulsive excitation. This is the behavior of the low-fluence trace from [54] shown in Fig. 10.1. In Fig. 10.1d we show the normalized dynamics (i.e., setting the amplitude $A = 1$ for all traces for clarity) of $\delta y(t)$ predicted by Eq. 10.1 with parameters corresponding to fluences of 1, 4, and 11 mJ/cm$^2$ (see Table S1 in the Supplementary Material of [54]) [same color code as Fig. 10.1b]. The crossover between sine-like and cosine-like motion is clear as the fluence increases and $\beta$ decreases.

Generally, under the influence of a coherent phonon the electronic bands shift in energy due to their coupling with the lattice, quantified by the deformation potential. Gerber et al. [57] used time- and angle-resolved photoemission spectroscopy (ARPES) to measure $\Delta E_{xz/yz}$ and $\Delta E_{z^2}$, the shift in the $d_{xz/yz}$ and $d_{z^2}$ bands of FeSe due to a coherent lattice vibration. They used femtosecond hard X-ray pulses from the LCLS to measure precisely the displacement of the Se atom $\Delta z_{\text{Se}}$ associated with the coherent $A_{1g}$ phonon mode induced by the pump pulse which modulates the electronic bands. In their diffraction experiment, the phonon dynamics were excited by an IR laser with 1.55 eV photons and 40 fs in duration and the diffraction of the (0 0 4) Bragg peak was probed by a delayed 8.7 keV X-ray pulse of 25 fs duration from the LCLS. Figure 10.2a shows the dynamics of the integrated intensity normalized to the laser-off intensity when pumped with a fluence of 1.83 mJ/cm$^2$. The oscillations are due to a 5.3 THz $A_{1g}$ phonon coherently modulating the $z$-position of the selenium atoms, depicted in the inset of the top panel. The spectrum in the inset shows the Fourier transform of the background-subtracted traces in Fig. 10.2a. Using information on the equilibrium atomic coordinates and the $A_{1g}$ mode eigenvector, they relate the measured intensity to the absolute atomic motion involving the Se atom away from the Fe, which at this fluence is $\Delta z_{\text{Se}}$ ~2 pm near $\Delta t = 0$ ps [see top curve in Fig. 10.2b]. In their time-resolved ARPES measurement, Gerber et al. observed that the $d_{z^2}$ and $d_{xz/yz}$ bands shifted downward in energy as the Se atom moved away from the Fe atom. In Fig. 10.2b, we show $\Delta z_{\text{Se}}$ (top curve, left y-axis) and $\Delta E_{xz/yz}$, $\Delta E_{z^2}$ (middle and lowest curve, right y-axis), as well as the schematic of the corresponding atomic and band motion. A careful check of the linear fluence dependence ensures that the measurements are representative of the equilibrium properties. Thus by fitting the linear fluence dependence of both measurements they were able to obtain the deformation potentials of the two bands $\Delta E_{xz/yz}/\Delta z_{Se} = -13 \pm 2.5$ and $\Delta E_{z^2}/\Delta z_{Se} = -16.5 \pm 3.2$ meV/pm. These values are significantly higher than those predicted by DFT calculations [57]. Their interpretation is that electronic correlations enhance the electron–phonon coupling

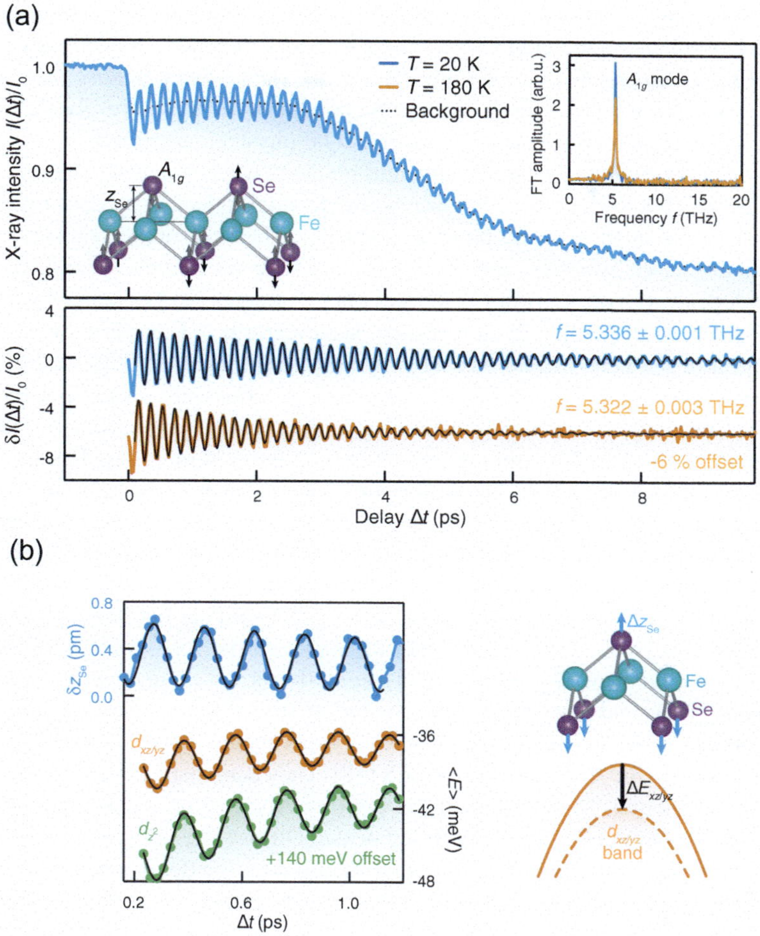


**Fig. 10.2** Ultrafast X-ray diffraction from coherent phonons in FeSe. **a** The top panel shows the integrated intensity of the (0 0 4) Bragg peak after ultrafast excitation with 1.55 eV photons. The lower panel shows the oscillatory component with a slowly varying background subtracted. The inset in the top panel shows the Fourier transform of the oscillatory component in the lower panel. **b** The calibrated $\Delta z_{\mathrm{Se}}$ (top curve, left y-axis), and corresponding $\Delta E_{xz/yz}$ and $\Delta E_{z^2}$ extracted from time-resolved ARPES (middle and lower curves, right y-axis). Also shown is a schematic of the motion and the electronic band displacement. Adapted from [57]. Reprinted with permission from AAAS

in FeSe. They argue that such concerted electron–electron and electron–phonon interactions may enhance superconductivity in FeSe, which depends exponentially on the value of the electron–phonon coupling. This work shows that a precise measure of atomic motion in the linear-response regime combined with time-resolved ARPES enables model-free measurements of fundamental physical quantities.

The discussion above considers coherent phonons where the macroscopic mode amplitude $\langle u_{\mathbf{q}} \rangle(t)$, with $u_{\mathbf{q}}$ the microscopic mode amplitude at wavevector $\mathbf{q}$, oscillates coherently in time at the phonon frequency. Here the angle brackets correspond to ensemble averages, appropriate for macroscopic samples measured with many independent X-ray exposures. These coherent modes can be excited by a long-wavelength near-IR pump only at near-zero wavevector $\mathbf{q} \approx 0$ because of the small momentum of the near-IR photons. This is the case in materials with well-defined translational symmetry (continuous or discrete). Alternatively, in the presence of weak disorder with Fourier component at $\mathbf{q} \neq 0$, coherent modes, $\langle u_{\mathbf{q}} \rangle(t)$, can be generated as crystal momentum only needs to be conserved up to $\hbar\mathbf{q}$. This can happen, for example, when a static CDW develops, which can provide the momentum to "fold" a large wavevector $\mathbf{q}$ back onto the Brillouin zone center, as discussed above.

Optical excitation of a material also produces a different kind of phonon dynamics with the average $\langle u_{\mathbf{q}} \rangle = 0$, but where the variance $\langle u_{-\mathbf{q}} u_{\mathbf{q}} \rangle(t)$ oscillates in time at *twice* the phonon frequency. Wavevector conservation still applies here, but because this motion involves coherence between $u_{\mathbf{q}}$ and $u_{-\mathbf{q}}$, the total wavevector is still $\mathbf{q} - \mathbf{q} = 0$. In this case, the coherence in the mode variance can span the Brillouin zone. Trigo, Reis, and coworkers demonstrated in 2013 that femtosecond X-ray diffuse scattering is sensitive to these dynamics in experiments on near-infrared pumped germanium using the XPP station on LCLS [58]. They demonstrated temporal coherences due to phonons across an extended region of the Brillouin zone by recording the diffuse X-ray intensity between Bragg peaks. Diffuse intensity, often associated with an unwanted background, contains information on the deviations from perfect crystallinity, as the measured intensity is related to the projection of the mean-square phonon displacements on the momentum transfer; in the case of non-thermal vibrations, it originates from temporal correlations in $\langle u_{-\mathbf{q}} u_{\mathbf{q}} \rangle(t)$ induced by the pump. On the other hand, when the correlations are thermal there is no well-defined phase relationship between the modes, and the diffuse X-ray intensity is time independent [59]. Instead, in a pump–probe experiment the correlations are time dependent because the sudden excitation of the crystal induces a well-defined phase between $u_{\mathbf{q}}$ and $u_{-\mathbf{q}}$.

In Fig. 10.3a, we show the X-ray diffuse intensity of a single crystal of germanium with 10 keV X-ray photons at LCLS [58]. The orientation was such that multiple Brillouin zones were intercepted by the Ewald sphere and covered by the detector. The two brightest spots correspond to regions near the (0 2 2) and (1 1 3) reciprocal lattice points (conventional unit cell), with the brightest pixels closer to the corresponding zone center. Importantly, the geometry was chosen such that the Bragg condition was not satisfied, and the reciprocal lattice points are offset in the direction perpendicular to the image [58]. In Fig. 10.3b, we show the dynamics of the intensity at the two locations in the image labeled $u$, near, and $v$, away, from the zone center, respectively.

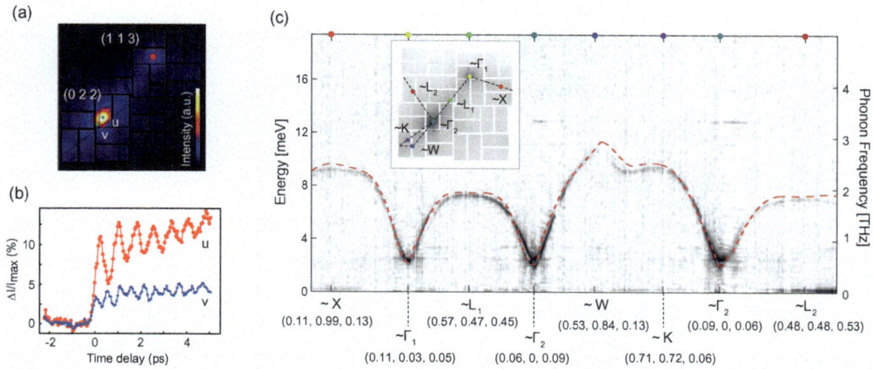


**Fig. 10.3** **a** Static thermal diffuse intensity of germanium at room temperature. The two closest reciprocal lattice points are labeled by their Miller indices in the conventional cubic basis. **b** Oscillatory dynamics of germanium after photoexcitation for representative pixels labeled u and v in (**a**) [adapted by permission from Springer Nature [58]]. **c** transverse acoustic phonon dispersion along the wavevectors marked on the image on the inset. The closest high-symmetry points are labeled and the corresponding coordinates are shown in reciprocal lattice units. The dashed line shows a calculation of the lowest frequency TA branch in Ge. Reprinted figure with permission from [60] Copyright 2015 by the American Physical Society

In Fig. 10.3c, we show the dispersion obtained from a Fourier transform of the time-domain dynamics along wavevectors indicated in the inset, with labels indicating the closest high-symmetry point in the Brillouin zone of Ge [60]. The frequency in Fig. 10.3c is obtained from a Fourier transform of the oscillatory signal and the color intensity is related to the amplitude of oscillation. The number of branches visible in this type of experiment is a function of the strength of the coupling to the pump and the polarization sensitivity of the X-ray scattering to the corresponding motion. The time resolution also determines the highest frequency that can be sampled. For the same reason, the frequency resolution is governed by the longest delay scanned. In this way, sub-meV resolution for the excited phonons was achieved by measuring oscillations up to 10 ps. The amplitude of the oscillations as a function of $\mathbf{q}$ is related to how strongly the various phonon modes couple to the photoexcited charge density, and thus provide a way to access $\mathbf{q}$-dependent electron–phonon matrix elements. It also allows for nonequlibrium lattice dynamics measurements, and the extraction of transient photoexcited forces[61, 62].

## 10.4 Couplings Among Multiple Degrees of Freedom

In complex materials, the strong interaction among coupled charge, spin, orbital, and lattice degrees of freedom can lead to nearly degenerate phases with different broken symmetries [63]. Optical excitation has been used to alter this delicate balance and produce properties not accessible in equilibrium. In this section, we review how

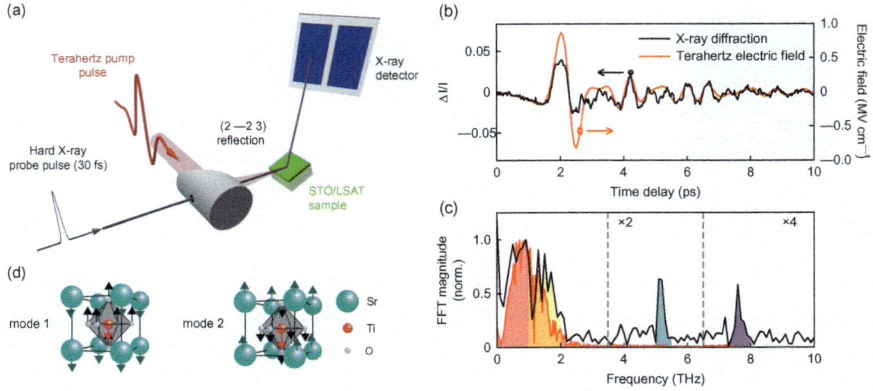


**Fig. 10.4** **a** Schematic of the single-cycle THz-pump X-ray-probe experiment on $SrTiO_3$ in Ref. [66]. **b** Dynamics of the intensity of the $(2\,\bar{2}\,3)$ Bragg peak of $SrTiO_3$ (black curve). Also shown is the electric field of the THz pulse (red curve). **c** Fourier transform of the $(2\,\bar{2}\,3)$ Bragg peak response. The red and yellow indicate the spectral regions of the THz field and the soft phonon, respectively. Green and purple indicate the peaks due to anharmonically coupled modes. **d** Illustration of the soft mode (mode 1) and the ~5 THz mode (mode 2). Adapted by permission from Springer Nature [66], Copyright (2019)

ultrafast X-ray scattering is used to probe this coupling in complex materials. We focus first on the coupling among phonon degrees of freedom, before moving onto coupled charge and spin collective modes. While near-IR and optical excitation can transiently modify the electronic distribution of a material, which in turn can change structural and magnetic properties, recent development in mid-infrared lasers enables tuning the central frequency of the incident laser pulse to be in resonance with an optically active phonon in the THz range [64–66]. This provides a potentially attractive avenue to tune material properties by interacting directly with the lattice modes.

We consider anharmonic couplings between phonon coordinates $Q_1$ and $Q_2$. We write a generic expansion of the anharmonic potential in powers of quasi-harmonic eigenmodes as

$$V(Q_1, Q_2) = \sum_{n,m} g_{n,m} Q_1^n Q_2^m \tag{10.4}$$

where some terms may be forbidden by the crystal symmetry. In the following we discuss experimental manifestations due to anharmonic terms in this expansion achieving mode couplings and structural control.

Kozina et al. used such a source of intense THz radiation to drive a particularly anharmonic soft mode in the incipient ferroelectric $SrTiO_3$ (STO) into a regime of large amplitude [66]. The large motion induces upconversion into a different phonon mode at higher frequency at the same wavevector [67]. The displacements internal to the STO unit cell were observed by ultrafast X-ray diffraction at the LCLS [66]. Figure 10.4a shows a schematic of their pump–probe experiment. A single-cycle THz pulse with duration <1 ps and spectral content between 0.2 and 2.5 THz is produced by optical rectification of a femtosecond Ti:sapphire pulse in a single crystal of $LiNbO_3$ (not shown), and focused on the sample by an off-axis parabolic mirror. The sample, a 50 nm film of STO on an LSAT substrate, is oriented such that the $(2\,\bar{2}\,3)$ Bragg peak (referred to a cubic Bravais lattice) diffracts the delayed X-ray pulse onto the detector. Because of its strong anharmonicity the frequency of the STO soft mode is strongly temperature dependent. Kozina et al. used this feature to tune the frequency of the mode to the peak of their pump spectrum near ~1 THz. Figure 10.4b shows the dynamics of the integrated intensity of the $(2\,\bar{2}\,3)$ peak (black curve) together on the measured electric field of the THz field (red curve). The Fourier transform of these traces is shown in Fig. 10.4c. The spectrum of both X-ray and THz field has a strong peak at low frequencies <1 THz, indicating that the atomic motion follows the incident field in phase [66]. More notably, unlike the THz field, the spectral content of the XRD response has strong features in the range $1-2$ THz associated with a strongly damped soft mode and peaks at 5.15 and 7.6 THz corresponding to two additional phonon modes of STO. Since the THz field has no spectral content overlapping these modes, the excitation of these must be due to anharmonic coupling with the soft mode at low frequency. Also, the 7.6 THz mode is not IR active, at least in the bulk. The mechanism was also checked by tuning the mode frequency away from resonance with the THz field by changing temperature. At 250 K the mode frequency is at 2.5 THz and the field does not induce strong anharmonic motion, none of the features are present in the Fourier transform of the X-ray diffraction at this temperature (not shown) [66].

While a full treatment of the potential contains multiple terms, in Ref. [66], they also consider a simplified coupling between the low-frequency soft ferroelectric mode $Q_1$ and the 5.15 THz mode, $Q_2$ that explains the observed dependence on the field strength. Here, $Q_1$ couples directly to the strong THz field as $\sim Z_1 E_{\mathrm{THz}}(t)$, with $Z_1$ a mode effective charge, while $Q_2$ couples to $Q_1$ through anharmonic terms in the Hamiltonian $V(Q_1, Q_2) \sim 1/4 g_{4,0} Q_1^4 + g_{3,1} Q_1^3 Q_2$. The equations of motion for this model are

$$\ddot{Q}_1 + \gamma_1 \dot{Q}_1 + (\Omega_1^2 + g_{4,0} Q_1^2) Q_1 = Z_1 E_{\mathrm{THz}}(t) \tag{10.5}$$

$$\ddot{Q}_2 + \gamma_2 \dot{Q}_2 + \Omega_2^2 Q_2 = -g_{3,1} Q_1^3, \tag{10.6}$$

where any other nonlinear terms in the equation for $Q_2$ are neglected. Since for stability reasons $g_{4,0} > 0$, the corresponding cubic term in the equation for $Q_1$ makes the frequency higher for large amplitude motion, making it detune from resonance as temperature increases, as observed experimentally. The term $\sim -g_{3,1} Q_1^3$ in the equation for $Q_2$ produces harmonics of $\Omega_1$, which can overlap with $\Omega_2$ and resonantly drive $Q_2$.

An especially intriguing example of nonlinear THz excitation relates to work on the high-temperature cuprates superconductors, whose crystal structures feature stacks of copper-oxide square-net planes. Prior work has studied the transient state of these systems after photoexcitation with radiation centered at 20 THz (~ 15 μm

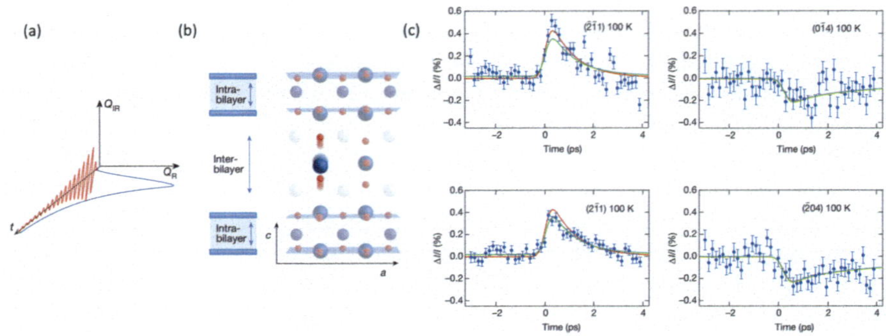


**Fig. 10.5** Ultrafast modification of the $YBa_2Cu_3O_{6.5}$ crystal structure after multi-cycle terahertz excitation. **a** Illustration of coupling between infrared and Raman phonon modes through nonlinear phononics. **b** Depiction of the intra-bilayer and inter-bilayer distortions in $YBa_2Cu_3O_{6.5}$. **c** Diffraction intensity of various Bragg peaks. The solid lines represent fits to the data based on the coupling of the $B_{1u}$ phonon to $A_{1g}$ phonons. The green line considers all the $A_{1g}$ phonons, whereas the red line only includes the four most strongly coupled modes. Adapted by permission from Springer Nature [65], Copyright (2014)

wavelength) detecting increased interlayer transport reminiscent of enhanced superconductivity [68]. Mankowskii et al. have studied $YBa_2Cu_3O_{6.5}$ after photoexcitation with $c$-axis-polarized THz radiation [65]. This radiation was chosen to maximize the interaction with a $B_{1u}$ infrared-active phonon associated with $c$-axis atomic displacements and driven with high fluence in order to maximize nonlinear phonon interactions [64]. Figure 10.5 illustrates the concepts and data behind this work. The photoexcited $B_{1u}$ mode interacts with $A_g$ phonons through nonlinear effects and induces changes in the crystal structure, which were probed by ultrafast diffraction at LCLS through changes in Bragg intensity. By fitting different models that consider these phonon excitations to their data, they conclude that the intra-$CuO_2$-plane distance is increased within the transient state at the expense of reducing the inter-$CuO_2$-plane distance. Such a change would increase the contribution of the Cu $d_{xy}$ orbital to low-energy electronic states, which provides a possible explanation for the proposed enhancement of superconductivity [65].

In general the anharmonic interaction of phonons is not restricted to the coupling between modes at the same wavevector. A classic example is the anharmonic decay of phonons [69, 70] which is the dominant mechanism determining the thermal conductivity of insulators and often plays a central role in structural phase transitions. In the context of anharmonic decay, first principles density-functional perturbation theory (DFPT) calculations have been used to calculate individual mode couplings for a couple of decades [71–73]. While more traditional inelastic neutron and X-ray scattering methods can give information about the lifetime of modes at a given wavevector, measurements of their underlying microscopic decay channels have not been possible before the advent of the XFELs.

In an experiment by Teitelbaum et al., researchers directly measured the anharmonic decay of a coherent zone-center phonon in photoexcited bismuth [74]. Bismuth is the prototypical Peierls distorted material (which can be viewed as a one-dimensional commensurate charge density wave) and one of the first materials to show displacive-like excitation of coherent phonons [56, 75] the dynamics of which at zone center has been well studied by both ultrafast optical [76–80] and ultrafast X-ray diffraction using plasma sources [81], the SPPS linac-based femtosecond spontaneous synchrotron source [82] and laser slicing in synchrotrons [83]. Here the phonon coordinate for the zone-center, fully symmetric mode $A_{1g}$ serves as a proxy for the Peierls distortion. Photoexcitation tends to reduce the bandgap as it lessens the energetic advantage for a structural distortion leading to the displacive excitation of coherent phonons. The decay of the coherent phonon proceeds largely through anharmonic coupling with pairs of acoustic phonons of equal and opposite momenta through a parametric resonance process akin to parametric downconversion [84].

The microscopic decay channels and their coupling constants can be measured directly in the time domain through the build-up of squeezed phonon oscillations in femtosecond time-resolved X-ray diffuse scattering measurements when compared to time-resolved diffraction. As described in Sect. 10.3, the diffuse X-ray intensity measures the mode variance $\langle u_{-\mathbf{q}} u_{\mathbf{q}} \rangle(t)$, which is driven parametrically by the large amplitude $A_{1g}$ mode. In the X-ray experiments of Teitelbaum et al., the degenerate decay of the $A_{1g}$ oscillations are clearly resolved across a wide range of the Brillouin zone. One particular decay channel into two acoustic modes at a specific $\mathbf{q}$ is shown schematically in Fig. 10.6. The coupling constant is determined from the ratio of amplitudes of the squeezed phonon oscillations in $\langle u_{-\mathbf{q}} u_{\mathbf{q}} \rangle(t)$ at a particular wavevector and the amplitude of the $A_{1g}$ coherent phonon oscillation as measured simultaneously in time-resolved diffraction at $\mathbf{q} = 0$. In the linear-response regime, the decay rate of a coherent phonon is expected to be the same as the anharmonic decay rates obtained by perturbation theory, which in the case of bismuth originates from the third-order force constants [84]. The experimental results give a value for the coupling constant that is within an order of magnitude of DFPT-based calculations, but systematically lower across all the measured channels. While the reason for the discrepancy is not yet understood the results mark the first momentum-resolved measurement of anharmonic coupling channels and open the door to similar studies in a broad class of complex materials such as described above, where couplings off-zone center can be critically important.

## 10.5 Beyond Linear Response

In this section, we describe experiments where the perturbation is pushed beyond linear response. In these experiments, the pump intensity is sufficient to induce non-adiabatic dynamics not accessible thermally, exemplified by lattice symmetry changes. We begin with systems that can be described by a single or few time-dependent order parameter(s). Then we show experiments where fluctuations from this single coordinate along multiple degrees of freedom can be central to the physics of the material, such as the case of $VO_2$.

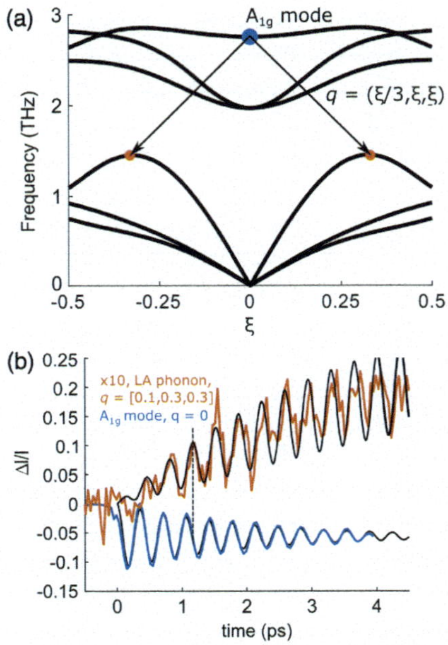


**Fig. 10.6** Downconversion of $A_{1g}$ mode in bismuth. **a** Phonon dispersion relation in bismuth along the $\boldsymbol{q} = (\frac{1}{3}\xi\ \xi\ \xi)$ direction showing one possible decay channel of the $A_{1g}$ mode into a degenerate pair of LA phonons at $\boldsymbol{q} = \pm(0.1\ 0.3\ 0.3)$ and **b** the experimental signature corresponding to this particular channel. The blue curve shows the relative intensity change of the (2 3 2) Bragg peak from which the $A_{1g}$ mode amplitude is measured. The orange curve shows the relative intensity change of the relevant region of diffuse scattering in the (0 1 1) zone. The black lines are simulations as described in the paper. Note the dashed line indicates a $\pi/2$ phase shift between the $A_{1g}$ mode and the target mode which is expected on parametric resonance. Reprinted figure with permission from [74] Copyright 2018 by the American Physical Society

As one of the simplest examples of dynamics far from harmonic behavior we consider the dynamics of the charge density wave order in $SmTe_3$. In this section, we first describe the experiment in $SmTe_3$ [85] and introduce the simplest Ginzburg–Landau formalism that results in anharmonic dynamics. We later review experiments where this formalism is extended to multiple degrees of freedom in $Pr_{0.5}Ca_{0.5}MnO_3$, and finally discuss $VO_2$ where the dynamics cannot be described by a few degrees of freedom but involve a continuum of modes.

$SmTe_3$ exhibits an incommensurate CDW below $T = 416$ K [86] corresponding to a modulation of the Te–Te planes along the $c$-axis [86]. The dynamics of the integrated intensity of the (1 7 $q_{\mathrm{cdw}}$) CDW Bragg peak are shown in Fig. 10.7a for several incident fluences [85]. Here $q_{\mathrm{cdw}} \approx 2/7$ is the CDW wavevector. At low fluence, the response shows oscillations at 1.6 THz, related to the DECP of the amplitude mode, Eq. 10.1, representing a small deviation from the equilibrium amplitude of the CDW, $\delta y(t)$. On the other hand, for strong photoexcitation at fluences $\geq$0.5 mJ/cm$^2$, the peak is strongly suppressed and the dynamics become overdamped and the recover

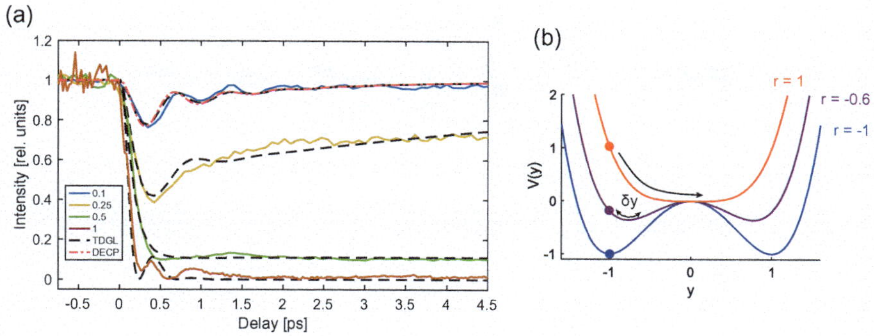


**Fig. 10.7** **a** Solid curves: dynamics of the integrated intensity of the (1 7 $q_{\mathrm{cdw}}$) of $SmTe_3$ at room temperature for several incident pump fluences (in $mJ/cm^2$). Dashed lines are the dynamics computed from the Ginzburg–Landau model described in the text for the corresponding values of $\eta$ (see text). **b** Sketch of the effective potential $V(y)$ for several excitation levels, represented by $r$ in Eq. 10.7. Reprinted figure with permission from [85] Copyright 2019 by the American Physical Society

timescale becomes longer. This is a quite common feature in the photoexcited dynamics of broken symmetry states.

At a phenomenological level the dynamics can be described by an extension of the Ginzburg–Landau theory for second-order phase transitions [87] to the time domain [85, 88–90]. The simplest description considers a scalar order parameter $y$, representing both the amplitude of the CDW lattice distortion and the charge density (or the value of the gap), where in normalized units $|y| = 1$ and $y = 0$ represent the distorted low-symmetry phase or undistorted high-symmetry phase, respectively. The potential energy is taken of the form

$$V(y) = V_0 \left(2ry^2 + y^4\right). \tag{10.7}$$

In equilibrium, the parameter $r = T/T_c - 1$, with $T_c$ the transition temperature, controls the relative stability of the two phases. The low-symmetry phase with $y = \pm 1$ is the stable minimum of $V(y)$ for $r = -1$ $(T = 0)$ (Fig. 10.7b (blue curve)). When $r \geq 0$ the potential has a single minimum at $y = 0$, and the symmetric phase is stable (Fig. 10.7b). Importantly, to lowest order the intensity of the CDW diffraction peaks is $I_{\mathrm{cdw}} \propto y^2$. As expected, diffraction alone cannot distinguish between $y = -1$ and $y = +1$, and also $I_{\mathrm{cdw}} = 0$ for the high-symmetry phase with $y = 0$.

At ultrafast timescales the temperature is not a well-defined quantity, however a reasonable assumption is that the parameter $r = r(t)$ is related to the photoexcited charge density with a good approximation being $r(t) = \eta\Theta(t)e^{-t/\tau} - 1$ [89], where $\Theta(t)$ is a unit step function, $\tau$ is the decay constant of the force, and $\eta$ is controlled by the incident pump fluence [85]. At room temperature $SmTe_3$ is in the low-symmetry (CDW-distorted) phase and we take $y(t < 0) = -1$ [$y(t < 0) = +1$ is equivalent], representing the static distorted lattice. After photoexcitation, $r(t >$

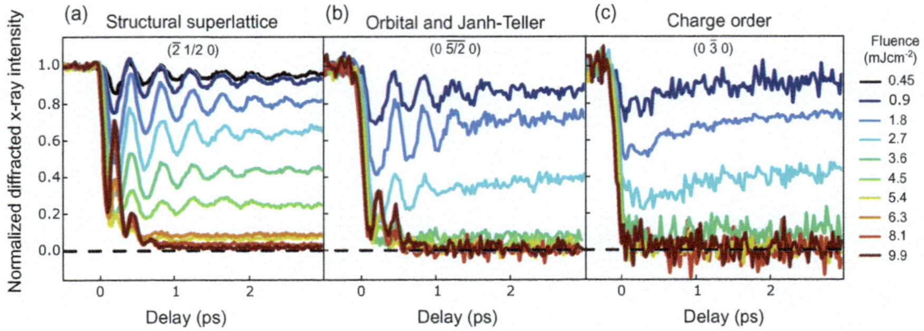


**Fig. 10.8** Dynamics of multiple orders in $Pr_{0.5}Ca_{0.5}MnO_3$ at different incident fluences. **a** Bragg peak from a crystal superstructure. **b** Orbital order peak also associated with a Jahn–Teller distortion of the $MnO_6$ octahedra. **c** charge order peak. Adapted by permission from Springer Nature [90], Copyright (2014)

$0) = \eta e^{-t/\tau} - 1$, if the strength of excitation $\eta$ is small we can consider a small perturbation of the potential $V(y)$, which can be a DECP dynamics for the CDW mode [85], with a DECP displacement as in Eq. 10.1. At higher fluences (Fig. 10.7a green and red traces), the anharmonic behavior is most severe when $\eta \geq 1$, $(r \geq 0)$. Here the potential has a single minimum at $y = 0$ initially, and the system can reach the symmetric configuration where the distortion associated to $y$ and the diffraction intensity is suppressed completely. This is the essence of the high-fluence behavior in Fig. 10.7 a as well as more complex materials like $Pr_{0.5}Ca_{0.5}MnO_3$ to be discussed next (shown in Fig. 10.8a, b). More sophisticated models build-up from this simple model [88, 91], but the common feature is some form of time-dependent coefficient $r(t)$, representing a phenomenological perturbation to the electronic properties. In all cases, the anharmonic lattice dynamics results from nonlinear (usually quartic) terms in the potential, Eq. 10.7.

The Ginzburg–Landau description can be extended to systems with multiple order parameters, for example, representing coupled spin, lattice and charge degrees of freedom [92]. Beaud and colleagues used the LCLS to probe the dynamics of the phase transformation of $Pr_{0.5}Ca_{0.5}MnO_3$ whose complex phase diagram is linked to a coupling between lattice, spin, charge and orbital degrees of freedom [90]. Resonant diffraction where the photon energy is tuned near a core-hole absorption edge (of Mn in this case) can enhance the contribution from valence electrons. This is the basis for novel spectroscopic tools (see more in the resonant inelastic X-ray scattering Sect. 10.6) making diffraction sensitive to lattice (Fig. 10.8a), orbital (Fig. 10.8b) and charge order (Fig. 10.8c). Below 5 mJ/cm$^2$ the charge order is partially suppressed and the orbital and structural superlattice peaks show strong oscillations from a 2.45 THz phonon mode oscillating coherently. At fluence > 5 mJ/cm$^2$ the three types of order are fully suppressed. Notably, at this fluence the structural and orbital orders exhibit an approximate doubling of the 2.45 THz frequency when the intensity reaches zero, indicative of the $\sim y^2$ scaling of the Bragg intensity when the relevant order parameter approaches the symmetric potential minimum near $y = 0$ [89, 90]. They model the structural and Jahn–Teller dynamics using a phenomenological Ginzburg–Landau potential similar to the one described above but with four effective lattice degrees of freedom [90] representing combinations of phonon modes of $Pr_{0.5}Ca_{0.5}MnO_3$. Using resonance diffraction to enhance sensitivity to charge order, this experiment showed that although the dynamics are highly out of equilibrium, the system has a well-defined order parameter across the symmetry breaking phase transition [90].

In another experiment exploiting resonant scattering to enhance the sensitivity of valence electrons, Lee et al. used resonant diffraction to probe the dynamics of charge ordering in $La_{1.75}Sr_{0.25}NiO_4$ (LSNO) [93]. Unlike the PCMO experiment, where the coherent dynamics can be accounted for by the amplitude of the order parameter alone, in LSNO the pump suppresses the diffraction intensity of the charge-ordered peaks without affecting the peak width. This contrasts with the thermodynamic transition where the peak width (the inverse CDW correlation length) broadens with increasing temperature through $T_c$. This indicates that, for the fluence range studied, the photoexcited transient of the CDW does not contain topological defects [93] in stark contrast to the expected thermal critical behavior. These results in LSNO suggest that photoexcitation populates phase modes at wavevectors other than the nominal ordering wavevector, and these suppress the CDW diffraction peak through a Debye–Waller effect.

Just as with any other type of disorder, phase fluctuations should manifest as an increase in the diffuse intensity at Fourier components away from $\mathbf{q}_{\mathrm{cdw}}$. Wall et al. used diffuse X-ray scattering at LCLS to probe the photoinduced phase transition of $VO_2$ [94]. This material undergoes an insulator–metal transition accompanied by a structural change between monoclinic ($M_1$) and tetragonal structure (Rutile, R) [95, 96] at $T_c = 340$ K (Fig. 10.9a). The $M_1$ has approximately twice the volume of the Rutile unit cell, and additional commensurate Bragg peaks indicate cell-doubling and long-range ordering of V–V dimers [94, 96]. The V–V distances break into long ($d_L$) and short ($d_S$) bonds, which are $d_L = d_S = d_R$ in the R structure, as indicated in Fig. 10.9a. The top panel of Fig. 10.9b shows representative snapshots of the differential diffuse intensity $\Delta I(\mathbf{Q}, t)$ covering multiple Brillouin zones of $VO_2$. Also indicated are sharp, localized features from the $(\bar{1}\ 0\ \bar{2})$, $(\bar{1}\ 2\ 2)$ and $(\bar{1}\ 2\ 0)$ peaks of the $M_1$ phase (indices of the monoclinic structure), which become suppressed in $\sim$140 fs as shown in the lower panel. The diffuse intensity integrated between the Bragg peaks, representative of dynamic disorder of the V–V pairs associated with the $M_1$ superstructure and peaks, increases with the same timescale (Fig. 10.9b, purple symbols). The distribution of diffuse intensity in $\mathbf{Q}$-space resembles the distinct R-phase pattern [94], which originates from transverse acoustic modes that are highly anharmonic, akin to dynamic disorder, and account for large fraction of the lattice entropy [97]. The transient Rutile diffuse pattern develops extremely fast, indicating that the lattice reaches the disorder in the lattice expected for the Rutile phase in 140 fs. These results were confirmed by ab-initio molecular dynamics (AIMD), which showed that the local potential of the V–V pairs (purple double-well trace in Fig. 10.9c) is switched quickly to a nearly flat, quartic potential (pink trace in Fig. 10.9c) that allows the V atoms to quickly fill it by randomizing their positions.

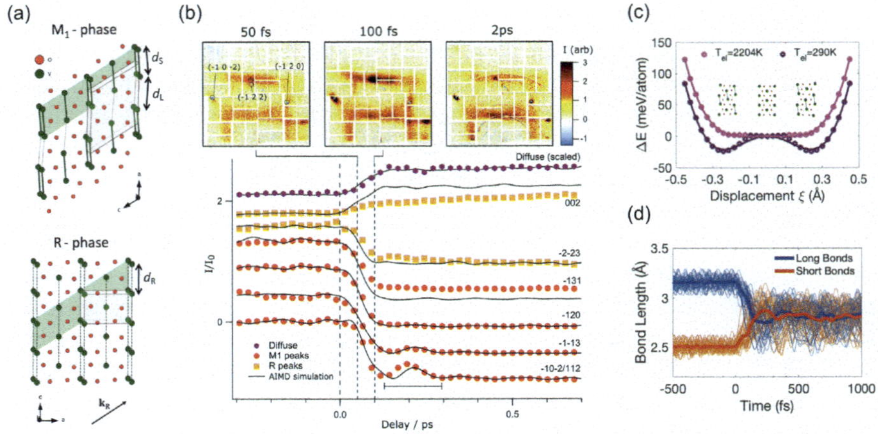

**Fig. 10.9** **a** Structure of the $M_1$ and $R$ phases of $VO_2$. The unit cell is indicated with the gray shaded area. **b** top panel: snapshots of the pump-induced change in diffuse intensity $\Delta I(\mathbf{Q}, t)$ for representative delays. Bragg peaks corresponding to the $M_1$ cell doubling are indicated ($M_1$ Miller indices). (**b**) lower panel: dynamics of the integrated intensity of $M_1$ peaks (red symbols), $R$ peaks (orange symbols) and the integrated diffuse intensity (purple symbols). **c** the computed V–V potential for the initial $M_1$ phase ($T_{el} = 290$ K) and the computed potential under an electronic temperature of $T_{el} = 2204$ K. **d** AIMD trajectories of the loss of V–V dimerization. From [94]. Reprinted with permission from AAAS

The absorption of the near-IR pump is modeled by introducing a fictitious electronic temperature $T_{el}$. Figure 10.9d shows individual trajectories of this process from the AIMD simulation [94]. The implications of these results extend beyond $VO_2$: Many ultrafast experiments focus on the behavior of Bragg peaks, while ignoring the role of disorder and fluctuations. This work shows that disorder plays an important role in the lattice of $VO_2$ and ultrafast diffuse scattering not only provides a visualization of the transformation pathway, but it can also yield an alternative view of the thermodynamic transition.

## 10.6 Resonant Inelastic X-ray Scattering

An important sub-set of modern X-ray experiments takes advantage of X-ray core-hole resonances in order to obtain more information than is otherwise possible. Far from resonance X-rays interact with all the electrons in the sample and tend to provide information about structure and structural dynamics such as phonons. Tuning the X-ray energy to a core-hole resonance opens routes to couple to other degrees of freedom. RIXS, as illustrated in Fig. 10.10, involves measuring the X-ray energy loss as a function of momentum in order to characterize collective electronic excitations and gain insights into short-range correlations and interactions. This can be thought of as the inelastic version of the resonant diffraction experiments described

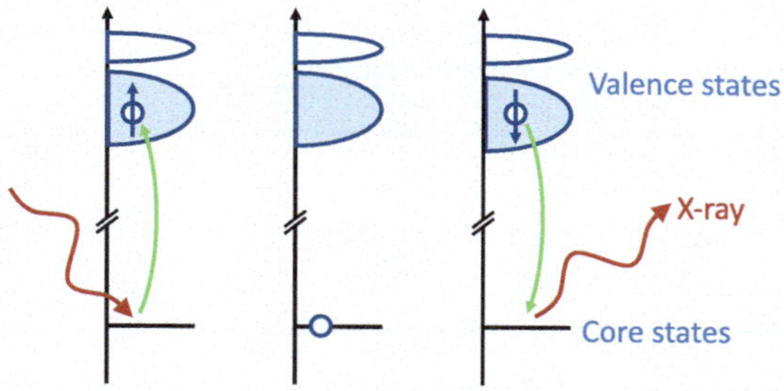


**Fig. 10.10** A schematic of a direct RIXS process showing the initial state, the intermediate state in which a core electron has been excited, and final states in which an elementary quasiparticle has been created from left to right. The energy and momentum change of the X-ray photon encodes the dispersion of the quasiparticle

previously (Fig. 10.8 [90] and Ref. [93]), where the change in X-ray energy is directly measured in order to infer the properties of excitations beyond simple measurements of the order parameter. Alternatively, it can be conceptualized as an IXS experiment where an X-ray resonance is used to probe spin, charge, or orbital excitations as well as phonons.

The RIXS process is codified in the Kramers–Heisenberg equation, which describes the second-order perturbation theory for the interaction between the X-ray photon and a material [98]. The coupling induced in the scattering process can be expressed in terms of electric and magnetic dipole and quadrupole operators (and in special cases to higher order still) [3, 5, 99]. These magnetic interactions occur due to the fact that core-hole spin-orbit coupling can, if present, exchange the orbital angular momentum of the photon with the spin in the valence band of the material. For this reason, standard magnetic dipole resonant scattering involves a rotation of the X-ray polarization—something that can be exploited in order to identify and isolate magnetic scattering in experiments. Other types of coupling, such as orbital scattering, also occur in well-defined channels in the X-ray polarization.

Syncrotron RIXS has grown dramatically in the past decades driven primarily by improvements in energy resolution that have helped access the low-energy degrees of freedom of correlated materials [3–5]. Performing RIXS at XFELs requires the same complicated spectrometer setup to be integrated with an XFEL often together with schemes for photoexciting the sample. We are, nonetheless, seeing successful examples of time-resolved RIXS, which have targeted charge, orbital, and spin degrees of freedom [10, 100–104].

Understanding transient magnetism is an especially important area for ultrafast research [105, 106]. Not only is magnetism central to many phenomena including unconventional superconductivity, charge-stripe correlations, and quantum spin liquids, but it is also one of the main means for information storage in computers. Recent work by Dean and collaborators performed the first time-resolved RIXS measurement of magnons [100]. In this study, antiferromagnetic $Sr_2IrO_4$ was studied after photoexcitation via a 100 fs, 2 $\mu$m pulse tuned to excite carriers across the bandgap.

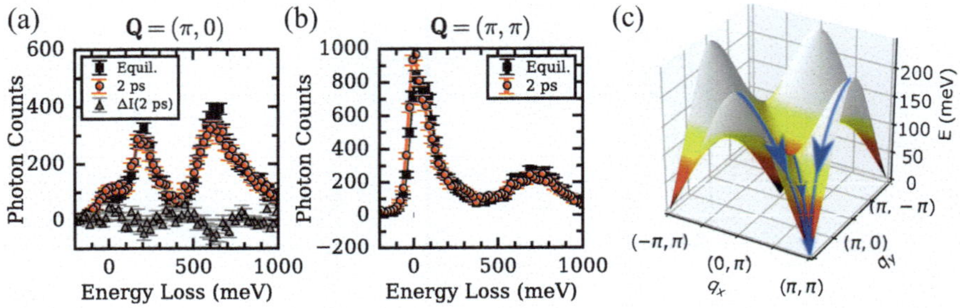


**Fig. 10.11** Example of ultrafast magnetism probed by RIXS. **a**, **b** show RIXS spectra of $Sr_2IrO_4$ in equilibrium and 2 ps after photoexcitation. Magnetic modes are seen in the 0–200 meV energy window and orbital excitations are seen around 600 meV. The specific $Q$ points measured were (**a**) $(\pi, 0)$ and **b** $(\pi, \pi)$ [100]. **c** The magnetic dispersion of $Sr_2IrO_4$ showing magnons at around 200 meV at $(\pi, 0)$ and almost gapless modes at $(\pi, \pi)$. The blue arrow illustrates the decay of magnons toward the energy minimum at $(\pi, 0)$ [101]. Adapted by permission from Springer Nature [100], Copyright (2016)

It was found that a fluence of 6 mJ/cm$^2$ was sufficient to almost completely suppress long-range magnetic order. The authors proceeded to study the form of the collective magnons in this transient state 2 ps after photoexcitation. As can be seen in Fig. 10.11, a very similar magnon is observed at the $(\pi, 0)$ reciprocal space point before and after photoexcitation, which suggests that short-range magnetic correlations can continue to exist even while long-range order is strongly suppressed. Since the fluence of the initial excitation process excites a very large fraction of the total magnetic sites, it was hypothesized that the $(\pi, 0)$ magnon might have been initially suppressed, but that it recovers much faster than 2 ps. The low-energy magnon at $(\pi, \pi)$ is modified with respect to the equilibrium configuration at 2 ps (see Fig. 10.11b) and recovers on a few ps timescale. Based on these results, together with resonant elastic scattering data, a picture of a two-stage magnetic recovery was developed in which magnetic correlations are first recovered within the 2D $IrO_2$ planes of $Sr_2IrO_4$ and then the registry between planes in achieved in order to return to the ground state. More recently, experiments have been preformed on bilayer antiferromagnetic iridate $Sr_3Ir_2O_7$. This system features a very large spin gap and a much more gentle dispersion. In this case, magnons at both $(\pi, 0)$ and $(\pi, \pi)$ reciprocal space points were disturbed in the transient state, which was interpreted in terms of a "spin-bottleneck" effect in which the large spin gap trapped the transient magnons over all reciprocal space [101].

RIXS is also known to be highly sensitive to diffuse charge density wave order [3–5, 107]. Mitrano et al. exploited this effect to analyze the CDW dynamics in cuprate superconductor $La_{2-x}Ba_xCuO_4$ $x = 0.125$ as shown in Fig. 10.12. They observed that the CDW intensity returned to equilibrium without exhibiting any oscillatory behavior. This is consistent with overdamped, diffusive CDW dynamics. The authors outlined a scaling model for the momentum and time dependence of the dynamics [103]. A small, but non-negligible change in the CDW $q$-vector was also observed,

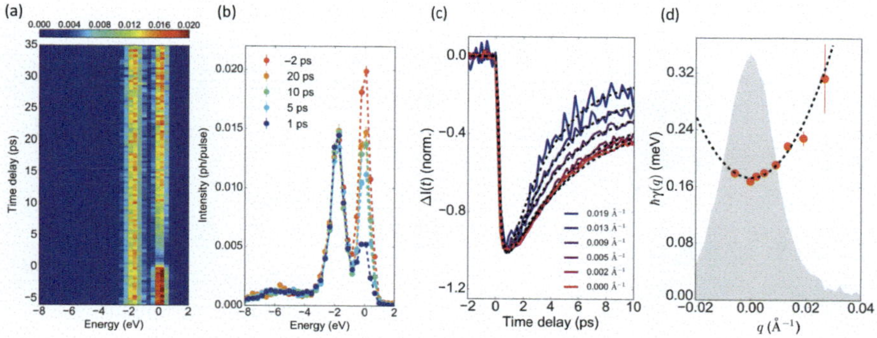


**Fig. 10.12** RIXS measurements of charge density wave (CDW) dynamics in $La_{2-x}Ba_xCuO_4$ $x = 0.125$ [103]. **a**, **b** show RIXS spectra at (0.23, 0, 1.5) as a function of time delay after a 0.1 mJ/cm$^2$ optical pump pulse represented as either (**a**) a colormap or **b** over-plotted spectra. The quasi-elastic CDW scattering is seen to first be suppressed and then to recover. **c** time traces of the recovery as a function of Q relative to the peak in the ordering wavevector. **d** The exponential recovery parameter $\hbar\gamma(q)$ as a function of the momentum relative to the peak in the ordering wavevector. The gray shaded area represents the intensity of the CDW in equilibrium. From [103]. Reprinted with permission from AAAS

which was interpreted in terms of a "Doppler effect", in which the pump imparts momentum to the CDW condensate.

To date, the majority of RIXS experiments has been analyzed assuming that the X-rays represent a small perturbation to the material. In this approximation, the spectrum can be interpreted in terms of a single photon interacting with the sample, since the incident photon density is small enough that there is a negligible probability of one photon encountering the effects of other photons. Pioneering work has demonstrated that FELs now deliver peak photon intensity sufficient to stimulate processes [108–110]. Such work has progressed from demonstrations in gases to work on solids, which are the subject of this chapter and the energy employed in the demonstration has extended from extreme ultraviolet energies, through soft X-ray into work with hard X-rays. Figure 10.13 reproduces work by Yoneda and collaborators, which realizes lasing with the $K_\alpha$ emission line of Cu films [110]. The authors used a two-stage system to focus the beam from the SACLA XFEL to a 120 nm spot on the sample and monitored the emitted X-rays with a flat silicon crystal spectrometer. Amplified spontaneous emission of X-ray was shown by demonstrating a super-linear dependence of the emitted X-ray intensity as a function of the incident intensity, as plotted in Fig. 10.13c. A notable broadening of the spectrum was reported at high fluences which was attributed to the presence of $3d$ core holes states as well as the target $1s$ core holes excited. Further demonstration of the nonlinear interactions was provided by demonstrating seeded emission in which two different color pulses are used, the setup for which is illustrated in Fig. 10.13b. The results (as shown in Ref. [110]) demonstrate that such a seeded approach can amplify a specific energy emission line, as chosen by providing a narrow-bandwidth seed pulse [110]. As well as being of intrinsic interest in terms of photonics, such approaches provide new

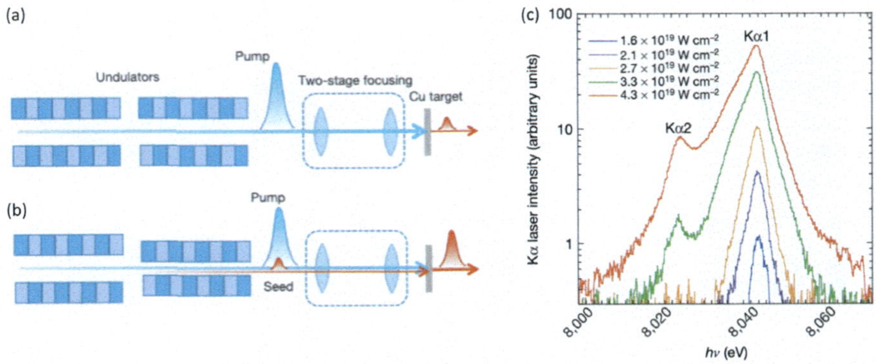


**Fig. 10.13** Nonlinear processes in the X-ray regime. **a** The experimental setup for demonstrating amplified spontaneous emission, by population inversion for Cu $K_\alpha$ emission. X-rays are produced by an undulator and focused to a 120 nm spot on a copper foil. **b** The setup showing amplified emission, in which the pump pulse is accompanied by a seed pulse. **c** Measured Cu $K_\alpha$ emission spectra at different pump intensities. The emitted intensity increases exponentially with incident pump intensity. Adapted by permission from Springer Nature [110], Copyright (2015)

possibilities to perform more efficient measurement that mitigate beam damage or use the enhanced efficiency to access weak multipolar order parameters [111].

## 10.7 Outlook

In quantum materials, the complex microscopic interactions result in strong competition between multiple nearly degenerate ordered phases. Ultrafast pulses provide a strategy for selectively perturbing microscopic interaction parameters to transiently coerce materials into a desired phase [112]. XFEL sources will soon offer high repetition rate and high average power [25, 26]. The development of cavity-based FEL [113] would allow for transform-limited pulses with narrow bandwidth ~1 meV in the case of an XFEL oscillator (XFELO) [114, 115] and higher peak power with moderate ~100 meV bandwidth [113, 116]in the case of an XFEL regenerative amplifier (RAFEL). These developments would complement the broadband ~10 eV pulses that allow for near-transform-limited attosecond pulses [21, 22]. These developments will provide higher-stability and control over the X-ray properties with applications that require either high spectral or temporal resolution, as well as provide multi-color, multi-pulse capabilities for X-ray pump and X-ray probe measurements of material dynamics.

XFELs will enable comprehensive microscopic characterization of nonequilibrium dynamics in complex materials, by providing unprecedented sensitivity to the collective modes. For example, high repetition rate sources will facilitate visualization of nanoscale heat conduction with potential prospects for microscopic control of energy flow. Novel approaches have emerged for control of sound and heat exploiting (coherent) interferences of waves within structures comparable to the wavelength [117]. The impact of new XFEL sources for probing nanoscale heat transport will be in probing the dynamics of high wavevector phonons with nm wavelengths at femtosecond timescales, yielding a microscopic view of wave propagation and heat conduction.

In addition, the combination of high repetition rate and tunable photon energy is well matched to studying novel light-driven quantum states through resonant scattering, while multi-color, multi-pulse, and attosecond operation will enable a much deeper understanding of nonlinear light–matter interactions in quantum materials. Time-resolved inelastic scattering measurements will be possible in a variety of sample environments such as inside a diamond anvil cell. This will enable detailed measurements of quantum critical fluctuations of charge and spin degrees of freedom, their associated collective modes and their couplings, while tuning continuously through a quantum critical point. Such unique capabilities will improve our understanding of the connection between superconductivity and the fluctuations in other order parameters [118, 119]. These are just some examples of how the revolutionary properties of X-ray free-electron lasers are likely to make a lasting impact on important problems in materials science and condensed matter physics.

This work was supported by the U.S. Department of Energy, Office of Science, Office of Basic Energy Sciences through the Division of Materials Sciences and Engineering under Contract Nos. DE-AC02-76SF00515 (MT and DAR) and DE-desc0012704 (MPMD).

## References

1. M. Krisch, F. Sette, in *Light Scattering in Solid IX* (Springer, 2007), pp. 317–370
2. A.Q.R. Baron, in *Synchrotron Light Sources and Free-Electron Lasers: Accelerator Physics, Instrumentation and Science Applications*, ed. by E. Jaeschke, S. Khan, J.R. Schneider, J.B. Hastings (Springer International Publishing, 2015), pp. 1–68
3. L.J.P. Ament, M. van Veenendaal, T.P. Devereaux, J.P. Hill, J. van den Brink, Rev. Mod. Phys. **83**, 705 (2011)
4. M. Dean, J. Magn. Magnet. Mater. **376**, 3 (2015)
5. M. Mitrano, S. Johnston, Y.J. Kim, M.P.M. Dean, Exploring quantum materials with resonant inelastic x-ray scattering (2024)
6. S. Wall, M. Trigo, Synchrotron Radiation News **29**(5), 13 (2016)
7. A.M. Lindenberg, S.L. Johnson, D.A. Reis, Ann. Rev. Mater. Res. **47**(1), 425 (2017)
8. M. Buzzi, M. Först, R. Mankowsky, A. Cavalleri, Nat. Rev. Mater. **3**(9), 299 (2018)
9. M. Dunne, Nat. Rev. Mater. **3**(9), 290 (2018)
10. Y. Cao, D.G. Mazzone, D. Meyers, J.P. Hill, X. Liu, S. Wall, M.P.M. Dean, Philosophical transactions of the royal society a: mathematical. Phys. Eng. Sci. **377**(2145), 20170480 (2019)
11. P. Emma, R. Akre, J. Arthur, R. Bionta, C. Bostedt, J. Bozek, A. Brachmann, P. Bucksbaum, R. Coffee, F.J. Decker, et al., Nature Photonics **4**(9), 641 (2010)
12. A. Kondratenko, E. Saldin, Part. Accel. **10**, 207 (1980)
13. R. Bonifacio, C. Pellegrini, L. Narducci, in *AIP Conference Proceedings*, vol. 118 (American Institute of Physics, 1984), vol. 118, pp. 236–259

14. R. Bonifacio, C. Pellegrini, L.M. Narducci, Opt. Commun. **50**(6), 373 (1984)
15. T. Ishikawa, H. Aoyagi, T. Asaka, Y. Asano, N. Azumi, T. Bizen, H. Ego, K. Fukami, T. Fukui, Y. Furukawa et al., Nat. Photonics **6**(8), 540 (2012)
16. S. Abeghyan, M. Bagha-Shanjani, G. Chen, U. Englisch, S. Karabekyan, Y. Li, F. Preisskorn, F. Wolff-Fabris, M. Wuenschel, M. Yakopov, J. Pflueger, J. Synchrotron Radiation **26**(2), 302 (2019)
17. H.S. Kang, C.K. Min, H. Heo, C. Kim, H. Yang, G. Kim, I. Nam, S.Y. Baek, H.J. Choi, G. Mun et al., Nat. Photonics **11**(11), 708 (2017)
18. C.J. Milne, T. Schietinger, M. Aiba, A. Alarcon, J. Alex, A. Anghel, V. Arsov, C. Beard, P. Beaud, S. Bettoni et al., Appl. Sci. **7**(7), 720 (2017)
19. J. Amann, W. Berg, V. Blank, F.J. Decker, Y. Ding, P. Emma, Y. Feng, J. Frisch, D. Fritz, J. Hastings et al., Nat. Photonics **6**(10), 693 (2012)
20. I. Inoue, T. Osaka, T. Hara, T. Tanaka, T. Inagaki, T. Fukui, S. Goto, Y. Inubushi, H. Kimura, R. Kinjo, H. Ohashi, K. Togawa, K. Tono, M. Yamaga, H. Tanaka, T. Ishikawa, M. Yabashi, Nat. Photonics **13**(5), 319 (2019)
21. S. Huang, Y. Ding, Y. Feng, E. Hemsing, Z. Huang, J. Krzywinski, A.A. Lutman, A. Marinelli, T.J. Maxwell, D. Zhu, Phys. Rev. Lett. **119**, 154801 (2017)
22. J. Duris, S. Li, T. Driver, E.G. Champenois, J.P. MacArthur, A.A. Lutman, Z. Zhang, P. Rosenberger, J.W. Aldrich, R. Coffee, G. Coslovich, F.J. Decker, J.M. Glownia, G. Hartmann, W. Helml, A. Kamalov, J. Knurr, J. Krzywinski, M.F. Lin, J.P. Marangos, M. Nantel, A. Natan, J.T. O'Neal, N. Shivaram, P. Walter, A.L. Wang, J.J. Welch, T.J.A. Wolf, J.Z. Xu, M.F. Kling, P.H. Bucksbaum, A. Zholents, Z. Huang, J.P. Cryan, A. Marinelli, Nat. Photonics **14**(1), 30 (2020)
23. K. Tono, T. Togashi, Y. Inubushi, T. Sato, T. Katayama, K. Ogawa, H. Ohashi, H. Kimura, S. Takahashi, K. Takeshita, H. Tomizawa, S. Goto, T. Ishikawa, M. Yabashi, New J. Phys. **15**(8), 083035 (2013)
24. T. Hara, Y. Inubushi, T. Katayama, T. Sato, H. Tanaka, T. Tanaka, T. Togashi, K. Togawa, K. Tono, M. Yabashi, T. Ishikawa, Nat. Commun. **4**(1), 2919 (2013)
25. J. Galayda, et al., in *9th International Particle Accelerator Conference (IPAC'18), Vancouver, BC, Canada, April 29-May 4, 2018* (2018), pp. 18–23
26. Z. Zhu, Z. Zhao, D. Wang, Z. Liu, R. Li, L. Yin, Z. Yang, *Proceedings of the FEL2017*, Santa Fe, NM, USA pp. 20–25 (2017)
27. Z. Huang, K.J. Kim, Phys. Rev. ST Accel. Beams **10**, 034801 (2007)
28. E.L. Saldin, E.A. Schneidmiller, M. Yurkov, New J. Phys. **12**(3), 035010 (2010)
29. B.W.J. McNeil, N.R. Thompson, Nat. Photonics **4**(12), 814 (2010)
30. C. Pellegrini, A. Marinelli, S. Reiche, Rev. Mod. Phys. **88**, 015006 (2016)
31. E. Seddon, J. Clarke, D. Dunning, C. Masciovecchio, C. Milne, F. Parmigiani, D. Rugg, J. Spence, N. Thompson, K. Ueda et al., Reports Progress Phys. **80**(11), 115901 (2017)
32. K.A. Nugent, Adv. Phys. **59**(1), 1 (2010)
33. O.G. Shpyrko, J. Synchrotron Radiation **21**(5), 1057 (2014)
34. T.E. Glover, D.M. Fritz, M. Cammarata, T.K. Allison, S. Coh, J.M. Feldkamp, H. Lemke, D. Zhu, Y. Feng, R.N. Coffee, M. Fuchs, S. Ghimire, J. Chen, S. Shwartz, D.A. Reis, S.E. Harris, J.B. Hastings, Nature **488**(7413), 603 (2012)
35. S. Shwartz, M. Fuchs, J.B. Hastings, Y. Inubushi, T. Ishikawa, T. Katayama, D.A. Reis, T. Sato, K. Tono, M. Yabashi, S. Yudovich, S.E. Harris, Phys. Rev. Lett. **112**, 163901 (2014)
36. K. Tamasaku, E. Shigemasa, Y. Inubushi, T. Katayama, K. Sawada, H. Yumoto, H. Ohashi, H. Mimura, M. Yabashi, K. Yamauchi, T. Ishikawa, Nat. Photon **8**(4), 313 (2014)
37. M. Fuchs, M. Trigo, J. Chen, S. Ghimire, S. Shwartz, M. Kozina, M. Jiang, T. Henighan, C. Bray, G. Ndabashimiye, P.H. Bucksbaum, Y. Feng, S. Herrmann, G.A. Carini, J. Pines, P. Hart, C. Kenney, S. Guillet, S. Boutet, G.J. Williams, M. Messerschmidt, M.M. Seibert, S. Moeller, J.B. Hastings, D.A. Reis, Nat. Phys. **11**(11), 964 (2015)
38. S. Ghimire, M. Fuchs, J. Hastings, S.C. Herrmann, Y. Inubushi, J. Pines, S. Shwartz, M. Yabashi, D.A. Reis, Phys. Rev. A **94**, 043418 (2016)

39. M. Fuchs, D.A. Reis, in *Young, L. et al., Roadmap of ultrafast X-ray atomic and molecular physics*, vol. 51 (Journal of Physics B: Atomic, Molecular and Optical Physics, 2018), p. 032003
40. M.M. Murnane, H.C. Kapteyn, M.D. Rosen, R.W. Falcone, Science **251**(4993), 531 (1991)
41. K. Sokolowski-Tinten, D. von der Linde, J. Phys. Condensed Matter **16**(49), R1517 (2004)
42. R. Schoenlein, S. Chattopadhyay, H. Chong, T. Glover, P. Heimann, C. Shank, A. Zholents, M. Zolotorev, Science **287**(5461), 2237 (2000)
43. L. Bentson, P. Bolton, E. Bong, P. Emma, J. Galayda, J. Hastings, P. Krejcik, C. Rago, J. Rifkin, C. Spencer, Nuclear instruments and methods in physics research section a: accelerators. Spectrometers Detectors Assoc. Equip. **507**(1–2), 205 (2003)
44. A.M. Lindenberg, J. Larsson, K. Sokolowski-Tinten, K.J. Gaffney, C. Blome, O. Synnergren, J. Sheppard, C. Caleman, A.G. MacPhee, D. Weinstein, D.P. Lowney, T.K. Allison, T. Matthews, R.W. Falcone, A.L. Cavalieri, D.M. Fritz, S.H. Lee, P.H. Bucksbaum, D.A. Reis, J. Rudati, P.H. Fuoss, C.C. Kao, D.P. Siddons, R. Pahl, J. Als-Nielson, S. Duerster, I. R., H. Schlarb, H. Schulte-Schrepping, T. Tschentscher, J. Schneider, D. von der Linde, O. Hignette, F. Sette, H.N. Chapman, R.W. Lee, T.N. Hansen, S. Techert, J.S. Wark, M. Bergh, G. Huldt, D. van der Spoel, T. N., J. Hajdu, R.A. Akre, E. Bong, P. Krejcik, J. Arthur, S. Brennan, L. K., J.B. Hastings, Science **308**, 392 (2005)
45. A.L. Cavalieri, D.M. Fritz, S.H. Lee, P.H. Bucksbaum, D.A. Reis, J. Rudati, D.M. Mills, P.H. Fuoss, G.B. Stephenson, C.C. Kao, D.P. Siddons, D.P. Lowney, A.G. MacPhee, D. Weinstein, R.W. Falcone, R. Pahl, J. Als-Nielsen, C. Blome, S. Dürsterer, R. Ischebeck, H. Schlarb, H. Schulte-Schrepping, T. Tschentscher, J. Schneider, O. Hignette, F. Sette, k. Sokolowski-Tinten, H.N. Chapman, R.W. Lee, T.N. Hansen, O. Synnergren, J. Larsson, S. Techert, J. Sheppard, J.S. Wark, M. Bergh, C. Caleman, G. Huldt, D. van der Spoel, N. Timneanu, J. Hajdu, R.A. Akre, E. Bong, P. Emma, P. Krejcik, J. Arthur, S. Brennan, K.J. Gaffney, A.M. Lindenberg, K. Luening, J.B. Hastings, Phys. Rev. Lett. **94**, 114801 (2005)
46. A. Rousse, C. Rischel, J.C. Gauthier, Rev. Mod. Phys. **73**, 17 (2001)
47. A. Cavalleri, C. Blome, P. Forget, J. Kieffer, S. Magnan, C. Siders, K. Sokolowski-Tinten, J. Squier, C. Tóth, D.V. Linde, Phase Transitions **75**(7–8), 769 (2002)
48. D.A. Reis, A.M. Lindenberg, in *Light Scattering in Solid IX* (Springer, 2006), pp. 371–422
49. M. Bargheer, N. Zhavoronkov, M. Woerner, T. Elsaesser, Chemphyschem: A Europ. J. Chem. Phys. Phys. Chem. **7**(4), 783 (2006)
50. M. Bionta, H. Lemke, J. Cryan, J. Glownia, C. Bostedt, M. Cammarata, J. Castagna, Y. Ding, D. Fritz, A. Fry, J. Krzywinski, M. Messerschmidt, S. Schorb, M. Swiggers, R. Coffee, Opt. Express **19**(22), 21855 (2011)
51. M. Chollet, R. Alonso-Mori, M. Cammarata, D. Damiani, J. Defever, J.T. Delor, Y. Feng, J.M. Glownia, J.B. Langton, S. Nelson, K. Ramsey, A. Robert, M. Sikorski, S. Song, D. Stefanescu, V. Srinivasan, D. Zhu, H.T. Lemke, D.M. Fritz, J. Synchrotron Radiation **22**(3), 503 (2015)
52. L. Dhar, J.A. Rogers, K.A. Nelson, Chem. Rev. **94**(1), 157 (1994)
53. R. Merlin, Solid State Commun. **102**(2–3), 207 (1997)
54. A. Singer, S.K.K. Patel, R. Kukreja, V. Uhlíř, J. Wingert, S. Festersen, D. Zhu, J.M. Glownia, H.T. Lemke, S. Nelson, M. Kozina, K. Rossnagel, M. Bauer, B.M. Murphy, O.M. Magnussen, E.E. Fullerton, O.G. Shpyrko, Phys. Rev. Lett. **117**, 056401 (2016)
55. M. Kozina, M. Trigo, M. Chollet, J.N. Clark, J.M. Glownia, A.C. Gossard, T. Henighan, M.P. Jiang, H. Lu, A. Majumdar, D. Zhu, D.A. Reis, Struct. Dyn. **4**(5), 054305 (2017)
56. H.J. Zeiger, J. Vidal, T.K. Cheng, E.P. Ippen, G. Dresselhaus, M.S. Dresselhaus, Phys. Rev. B **45**, 768 (1992)
57. S. Gerber, S.L. Yang, D. Zhu, H. Soifer, J.A. Sobota, S. Rebec, J.J. Lee, T. Jia, B. Moritz, C. Jia, A. Gauthier, Y. Li, D. Leuenberger, Y. Zhang, L. Chaix, W. Li, H. Jang, J.S. Lee, M. Yi, G.L. Dakovski, S. Song, J.M. Glownia, S. Nelson, K.W. Kim, Y.D. Chuang, Z. Hussain, R.G. Moore, T.P. Devereaux, W.S. Lee, P.S. Kirchmann, Z.X. Shen, Science **357**(6346), 71 (2017)
58. M. Trigo, M. Fuchs, J. Chen, M.P. Jiang, M. Cammarata, S. Fahy, D.M. Fritz, K. Gaffney, S. Ghimire, A. Higginbotham, S.L. Johnson, M.E. Kozina, J. Larsson, H. Lemke, A.M. Lindenberg, G. Ndabashimiye, F. Quirin, K. Sokolowski-Tinten, C. Uher, G. Wang, J.S. Wark, D. Zhu, D.A. Reis, Nat. Phys. **9**(12), 790 (2013)

59. B.E. Warren, *X-Ray Diffraction* (Dover, New York, 1969)
60. D. Zhu, A. Robert, T. Henighan, H.T. Lemke, M. Chollet, J.M. Glownia, D.A. Reis, M. Trigo, Phys. Rev. B **92** (2015)
61. S.W. Teitelbaum, T.C. Henighan, H. Liu, M.P. Jiang, D. Zhu, M. Chollet, T. Sato, E.D. Murray, S. Fahy, S. O'Mahony, T.P. Bailey, C. Uher, M. Trigo, D.A. Reis, Phys. Rev. B **103**, L180101 (2021)
62. Y. Huang, S. Teitelbaum, S. Yang, G. De la Peña, T. Sato, M. Chollet, D. Zhu, J.L. Niedziela, D. Bansal, A.F. May, A.M. Lindenberg, O. Delaire, M. Trigo, D.A. Reis, Phys. Rev. Lett. **131**, 156902 (2023)
63. M. Imada, A. Fujimori, Y. Tokura, Rev. Mod. Phys. **70**, 1039 (1998)
64. M. Först, C. Manzoni, S. Kaiser, Y. Tomioka, Y. Tokura, R. Merlin, A. Cavalleri, Nat. Phys. **7**(11), 854 (2011)
65. R. Mankowsky, A. Subedi, M. Först, S.O. Mariager, M. Chollet, H.T. Lemke, J.S. Robinson, J.M. Glownia, M.P. Minitti, A. Frano, M. Fechner, N.A. Spaldin, T. Loew, B. Keimer, A. Georges, A. Cavalleri, Nature **516**(7529), 71 (2014)
66. M. Kozina, M. Fechner, P. Marsik, T. van Driel, J.M. Glownia, C. Bernhard, M. Radovic, D. Zhu, S. Bonetti, U. Staub, M.C. Hoffmann, Nat. Phys. **15**(4), 387 (2019)
67. D.M. Juraschek, S.F. Maehrlein, Phys. Rev. B **97**, 174302 (2018)
68. D. Fausti, R. Tobey, N. Dean, S. Kaiser, A. Dienst, M.C. Hoffmann, S. Pyon, T. Takayama, H. Takagi, A. Cavalleri, Science **331**(6014), 189 (2011)
69. R. Orbach, Phys. Rev. Lett. **16**, 15 (1966)
70. P.G. Klemens, Phys. Rev. **148**, 845 (1966)
71. A. Debernardi, S. Baroni, E. Molinari, Phys. Rev. Lett. **75**, 1819 (1995)
72. A. Debernardi, Phys. Rev. B **57**, 12847 (1998)
73. A. Togo, L. Chaput, I. Tanaka, Phys. Rev. B **91**, 094306 (2015)
74. S.W. Teitelbaum, T. Henighan, Y. Huang, H. Liu, M.P. Jiang, D. Zhu, M. Chollet, T. Sato, E.D. Murray, S. Fahy, S. O'Mahony, T.P. Bailey, C. Uher, M. Trigo, D.A. Reis, Phys. Rev. Lett. **121**, 125901 (2018)
75. T.K. Cheng, S.D. Brorson, A.S. Kazeroonian, J.S. Moodera, G. Dresselhaus, M.S. Dresselhaus, E.P. Ippen, Appl. Phys. Lett. **57**, 1004 (1990)
76. M.F. DeCamp, D.A. Reis, P.H. Bucksbaum, R. Merlin, Phys. Rev. B. **64**, 092301 (2001)
77. M. Hase, K. Mizoguchi, H. Harima, S. Nakashima, M. Tani, K. Sakai, M. Hangyo, Appl. Phys. Lett. **69**(17), 2474 (1996)
78. M. Hase, M. Kitajima, S. Nakashima, K. Mizoguchi, Phys. Rev. Lett. **88**(6), 067401 (2002)
79. E.D. Murray, D.M. Fritz, J.K. Wahlstrand, S. Fahy, D.A. Reis, Phys. Rev. B **72**(6), 060301 (2005)
80. S.W. Teitelbaum, T. Shin, J.W. Wolfson, Y.H. Cheng, I.J. Porter, M. Kandyla, K.A. Nelson, Phys. Rev. X **8**(3), 031081 (2018)
81. K. Sokolowski-Tinten, C. Blome, J. Blums, A. Cavalleri, C. Dietrich, A. Tarasevitch, I. Uschmann, E. Förster, M. Kamller, M. Horn von Hoegen, D. von der Linde, Nature **422**, 287 (2003)
82. D.M. Fritz, D.A. Reis, B. Adams, R.A. Akre, J. Arthur, C. Blome, P.H. Bucksbaum, A.L. Cavalieri, S. Engemann, S. Fahy, R.W. Falcone, P.H. Fuoss, K.J. Gaffney, M.J. George, J. Hajdu, M.P. Hertlein, P.B. Hillyard, M. Horn-von Hoegen, M. Kammler, J. Kaspar, R. Kienberger, P. Krejcik, S.H. Lee, A.M. Lindenberg, B. McFarland, D. Meyer, T. Montagne, É.D. Murray, A.J. Nelson, M. Nicoul, R. Pahl, J. Rudati, H. Schlarb, D.P. Siddons, K. Sokolowski-Tinten, T. Tschentscher, D. von der Linde, J.B. Hastings, Science **315**(5812), 633 (2007)
83. S.L. Johnson, P. Beaud, C.J. Milne, F.S. Krasniqi, E.S. Zijlstra, M.E. Garcia, M. Kaiser, D. Grolimund, R. Abela, G. Ingold, Phys. Rev. Lett. **100**, 155501 (2008)
84. S. Fahy, E.D. Murray, D.A. Reis, Phys. Rev. B **93**, 134308 (2016)
85. M. Trigo, P. Giraldo-Gallo, M.E. Kozina, T. Henighan, M.P. Jiang, H. Liu, J.N. Clark, M. Chollet, J.M. Glownia, D. Zhu, T. Katayama, D. Leuenberger, P.S. Kirchmann, I.R. Fisher, Z.X. Shen, D.A. Reis, Phys. Rev. B **99**, 104111 (2019)

86. N. Ru, C.L. Condron, G.Y. Margulis, K.Y. Shin, J. Laverock, S.B. Dugdale, M.F. Toney, I.R. Fisher, Phys. Rev. B **77**, 035114 (2008)
87. N. Goldenfeld, *Lectures on phase transitions and the renormalization group*. Frontiers in physics (Addison-Wesley, Advanced Book Program, 1992)
88. H. Schaefer, V.V. Kabanov, J. Demsar, Phys. Rev. B **89**, 045106 (2014)
89. T. Huber, S.O. Mariager, A. Ferrer, H. Schäfer, J.A. Johnson, S. Grübel, A. Lübcke, L. Huber, T. Kubacka, C. Dornes, C. Laulhe, S. Ravy, G. Ingold, P. Beaud, J. Demsar, S.L. Johnson, Phys. Rev. Lett. **113**, 026401 (2014)
90. P. Beaud, A. Caviezel, S. Mariager, L. Rettig, G. Ingold, C. Dornes, S. Huang, J. Johnson, M. Radovic, T. Huber et al., Nat. Mater. **13**(10), 923 (2014)
91. Y.D. Chuang, W.S. Lee, Y.F. Kung, A.P. Sorini, B. Moritz, R.G. Moore, L. Patthey, M. Trigo, D.H. Lu, P.S. Kirchmann, M. Yi, O. Krupin, M. Langner, Y. Zhu, S.Y. Zhou, D.A. Reis, N. Huse, J.S. Robinson, R.A. Kaindl, R.W. Schoenlein, S.L. Johnson, M. Först, D. Doering, P. Denes, W.F. Schlotter, J.J. Turner, T. Sasagawa, Z. Hussain, Z.X. Shen, T.P. Devereaux, Phys. Rev. Lett. **110**, 127404 (2013)
92. Y.D. Chuang, W.S. Lee, Y.F. Kung, A.P. Sorini, B. Moritz, R.G. Moore, L. Patthey, M. Trigo, D.H. Lu, P.S. Kirchmann, M. Yi, O. Krupin, M. Langner, Y. Zhu, S.Y. Zhou, D.A. Reis, N. Huse, J.S. Robinson, R.A. Kaindl, R.W. Schoenlein, S.L. Johnson, M. Först, D. Doering, P. Denes, W.F. Schlotter, J.J. Turner, T. Sasagawa, Z. Hussain, Z.X. Shen, T.P. Devereaux, Phys. Rev. Lett. **110**, 127404 (2013)
93. W.S. Lee, Y.D. Chuang, R.G. Moore, Y. Zhu, L. Patthey, M. Trigo, D.H. Lu, P.S. Kirchmann, O. Krupin, M. Yi, M. Langner, N. Huse, J.S. Robinson, Y. Chen, S.Y. Zhou, G. Coslovich, B. Huber, D.A. Reis, R.A. Kaindl, R.W. Schoenlein, D. Doering, P. Denes, W.F. Schlotter, J.J. Turner, S.L. Johnson, M. Först, T. Sasagawa, Y.F. Kung, A.P. Sorini, A.F. Kemper, B. Moritz, T.P. Devereaux, D.H. Lee, Z.X. Shen, Z. Hussain, Nat. Commun. **3**(1), 838 (2012)
94. S. Wall, S. Yang, L. Vidas, M. Chollet, J.M. Glownia, M. Kozina, T. Katayama, T. Henighan, M. Jiang, T.A. Miller, D.A. Reis, L.A. Boatner, O. Delaire, M. Trigo, Science **362**(6414), 572 (2018)
95. F.J. Morin, Phys. Rev. Lett. **3**, 34 (1959)
96. A. Cavalleri, C. Tóth, C.W. Siders, J.A. Squier, F. Ráksi, P. Forget, J.C. Kieffer, Phys. Rev. Lett. **87**, 237401 (2001)
97. J.D. Budai, J. Hong, M.E. Manley, E.D. Specht, C.W. Li, J.Z. Tischler, D.L. Abernathy, A.H. Said, B.M. Leu, L.A. Boatner et al., Nature **515**(7528), 535 (2014)
98. J. Sakurai, *Advanced Quantum Mechanics* (Incorporated, Always learning (Pearson Education, 1967)
99. Y. Joly, S.D. Matteo, O. Bunău, Europ. Phys. J. Special Topics **208**(1), 21 (2012)
100. M.P.M. Dean, Y. Cao, X. Liu, S. Wall, D. Zhu, R. Mankowsky, V. Thampy, X.M. Chen, J.G. Vale, D. Casa, J. Kim, A.H. Said, P. Juhas, R. Alonso-Mori, J.M. Glownia, A. Robert, J. Robinson, M. Sikorski, S. Song, M. Kozina, H. Lemke, L. Patthey, S. Owada, T. Katayama, M. Yabashi, Y. Tanaka, T. Togashi, J. Liu, C. Rayan Serrao, B.J. Kim, L. Huber, C.L. Chang, D.F. McMorrow, M. Först, J.P. Hill, Nat. Mater. **15**, 601 (2016)
101. D.G. Mazzone, D. Meyers, Y. Cao, J.G. Vale, C.D. Dashwood, Y. Shi, A.J.A. James, N.J. Robinson, J. Lin, V. Thampy, Y. Tanaka, A.S. Johnson, H. Miao, R. Wang, T.A. Assefa, J. Kim, D. Casa, R. Mankowsky, D. Zhu, R. Alonso-Mori, S. Song, H. Yavas, T. Katayama, M. Yabashi, Y. Kubota, S. Owada, J. Liu, J. Yang, R.M. Konik, I.K. Robinson, J.P. Hill, D.F. McMorrow, M. Först, S. Wall, X. Liu, M.P.M. Dean, Proc. Nat. Acad. Sci. **118**(22), e2103696118 (2021)
102. M. Dell'Angela, F. Hieke, M. Malvestuto, L. Sturari, S. Bajt, I. Kozhevnikov, J. Ratanapreechachai, A. Caretta, B. Casarin, F. Glerean et al., Scientific Reports **6**, 38796 (2016)
103. M. Mitrano, S. Lee, A.A. Husain, L. Delacretaz, M. Zhu, G. de la Peña Munoz, S.X.L. Sun, Y.I. Joe, A.H. Reid, S.F. Wandel, G. Coslovich, W. Schlotter, T. van Driel, J. Schneeloch, G.D. Gu, S. Hartnoll, N. Goldenfeld, P. Abbamonte, Sci. Adv. **5**(8) (2019)

104. S. Parchenko, E. Paris, D. McNally, E. Abreu, M. Dantz, E.M. Bothschafter, A.H. Reid, W.F. Schlotter, M.F. Lin, S.F. Wandel, G. Coslovich, S. Zohar, G.L. Dakovski, J.J. Turner, S. Moeller, Y. Tseng, M. Radovic, C. Saathe, M. Agaaker, J.E. Nordgren, S.L. Johnson, T. Schmitt, U. Staub, Phys. Rev. Res. **2**, 023110 (2020)
105. A. Kirilyuk, A.V. Kimel, T. Rasing, Rev. Mod. Phys. **82**, 2731 (2010)
106. M. Gandolfi, G.L. Celardo, F. Borgonovi, G. Ferrini, A. Avella, F. Banfi, C. Giannetti, Physica Scripta **92**(3), 034004 (2017)
107. H. Miao, J. Lorenzana, G. Seibold, Y. Peng, A. Amorese, F. Yakhou-Harris, K. Kummer, N. Brookes, R. Konik, V. Thampy et al., Proc. Nat. Acad. Sci. **114**(47), 12430 (2017)
108. N. Rohringer, D. Ryan, R.A. London, M. Purvis, F. Albert, J. Dunn, J.D. Bozek, C. Bostedt, A. Graf, R. Hill et al., Nature **481**(7382), 488 (2012)
109. M. Beye, S. Schreck, F. Sorgenfrei, C. Trabant, N. Pontius, C. Schüßler-Langeheine, W. Wurth, A. Föhlisch, Nature **501**(7466), 191 (2013)
110. H. Yoneda, Y. Inubushi, K. Nagamine, Y. Michine, H. Ohashi, H. Yumoto, K. Yamauchi, H. Mimura, H. Kitamura, T. Katayama et al., Nature **524**(7566), 446 (2015)
111. Y.L. Wang, G. Fabbris, D. Meyers, N.H. Sung, R.E. Baumbach, E.D. Bauer, P.J. Ryan, J.W. Kim, X. Liu, M.P.M. Dean, G. Kotliar, X. Dai, Phys. Rev. B **96**, 085146 (2017)
112. D.N. Basov, R.D. Averitt, D. Hsieh, Nat. Mater. **16**, 1077 EP (2017)
113. G. Marcus, J. Anton, L. Assoufid, F.J. Decker, G. Gassner, K. Goetze, A. Halavanau, J. Hastings, Z. Huang, W. Jansma, et al., in *39th Free Electron Laser Conf.(FEL'19), Hamburg, Germany, 26-30 August 2019* (JACOW Publishing, Geneva, Switzerland, 2019), pp. 282–287
114. K.J. Kim, Y. Shvyd'ko, S. Reiche, Phys. Rev. Lett. **100**, 244802 (2008)
115. K.J. Kim, Y.V. Shvyd'ko, Phys. Rev. Special Topics-Accelerators Beams **12**(3), 030703 (2009)
116. Z. Huang, R.D. Ruth, Phys. Rev. Lett. **96**(14), 144801 (2006)
117. M. Maldovan, Nature **503**(7475), 209 (2013)
118. Y.I. Joe, X.M. Chen, P. Ghaemi, K.D. Finkelstein, G.A. de la Peña, Y. Gan, J.C.T. Lee, S. Yuan, J. Geck, G.J. MacDougall, T.C. Chiang, S.L. Cooper, E. Fradkin, P. Abbamonte, Nat. Phys. **10**(6), 421 (2014)
119. Y. Feng, J. van Wezel, J. Wang, F. Flicker, D.M. Silevitch, P.B. Littlewood, T.F. Rosenbaum, Nat. Phys. **11**(10), 865 (2015)

# Chapter 11
# Ultrafast Transient Absorption Spectroscopy for Probing Primary Photochemical Reaction of Proteins

**Atsushi Yabushita**

**Abstract** Photochemical reactions of proteins are playing important roles for all life on earth; retinal proteins in retina work as light sensor to provide vision, phytochromes in plants regulate the germination of seeds, photoprotein in fireflies and jelly fishes generates bioluminescence, and chlorophyll triggers carbonic acid assimilation by photosynthesis reaction. Exploration of materials which have better efficiency for those photochemical reactions cannot be performed efficiently by haphazard approach. Elucidation of the primary reaction process is expected to provide key information for the development of future materials. The primary reaction could be studied in detail by performing ultrafast spectroscopy which visualizes spectral change during ultrafast transition between electronic states after photoexcitation. Using ultrashort laser pulse whose duration is shorter than the molecular vibration period, the observed transient absorption signal shows modulation reflecting the real-time motion of the molecular vibration in time domain. Time-gated Fourier analysis of the signal elucidates the time development of the molecular vibration frequency which elucidates molecular structure change during the photo-reaction. Ultrashort pulse lasers in visible and ultraviolet spectral region were developed to be applied for ultrafast spectroscopy of the protein samples to elucidate their primary reactions. For the study of molecular vibrational dynamics, signal fluctuation of the transient absorption should be suppressed to observe fine signal modulation caused by the molecular vibration. However, irradiation of the ultrashort laser pulse required for the ultrafast spectroscopy gives serious damage accumulation especially in the protein samples, which degrades signal quality of the transient absorption. Developing fastscan ultrafast spectroscopy system, we have succeeded to improve the signal-to-noise ratio of the transient absorption signal able to study molecular vibrational dynamics. The ultrashort laser pulse and fastscan ultrafast spectroscopy system to study the primary reaction dynamics of the protein samples. Here we describe some of those works for samples of a light-driven proton pump (bacteriorhodopsin) and a heme protein (nitric oxide synthase).

A. Yabushita (✉)
Department of Electrophysics, National Yang Ming Chiao Tung University, Hsinchu, Taiwan
e-mail: yabushita@nycu.edu.tw

Research Institute for Engineering, Kanagawa University, Yokohama, Japan



K. Ueda (ed.), *Ultrafast Electronic and Structural Dynamics*,
https://doi.org/10.1007/978-981-97-2914-2_11

**Keywords** Ultrafast spectroscopy · Non-collinear optical parametric amplifier · Vibrational dynamics · Rhodopsin · Hemoglobin

## 11.1 Introduction

Proteins play important roles for all life on earth; rhodopsins in retina work as light sensor to provide vision [1–3], bacteriorhodopsins in membrane of bacterium pumps out proton through the membrane [4–6], phytochromes in plants regulate the germination of seeds [7–10], photoprotein in jelly fishes generates bioluminescence [11–14], heme proteins delivers molecules as seen in oxygen carrier of haemoglobin [15, 16] and nitric oxide synthesized in nitric oxide synthase [17–20], and chlorophyll triggers carbonic acid assimilation by photosynthesis reaction [21–23].

Development of artificial materials which can proceed the same reaction of protein is one of the important studies. Another important topic is development of chemical materials or mutant proteins which can control the reaction. However, those developments cannot be performed efficiently by haphazard approach. Elucidation of the primary reaction process is expected to provide key information for the development of future materials.

Light pulse was firstly applied for visualization of chemical reaction by Norrish and Porter in microsecond time-resolved flash photolysis [24]. The time-resolution of the observed phenomena is limited by the pulse duration of the light pulse and the measurement scheme. When the transient signal is time-resolved by electronic detector, the time-resolution is typically in the order of sub-nanosecond. When optical pump-probe method is used to time-resolve the transient signal, the time-resolution is limited by the pulse duration of the light pulse used the pump-probe measurement. Thus, for the measurement of ultrafast phenomena to observe primary reaction, it is necessary to perform the pump-probe measurement utilizing ultrashort laser pulse.

Development of laser [25] enabled us to produce intense laser light with enough energy to perform the pump-probe measurement. The pulse duration of the laser pulse could be shortened by inventing mode-locking technique [26, 27]. The shortest limit of the laser pulse duration is inversely proportional to the spectral bandwidth of the laser. Usage of dye solution with colliding pulse mode-locking (CPM) method has generated laser pulse as short as 100 fs duration [28]. Kerr lens mode-locking (KLM) method in solid laser gain medium can generate the ultrashort pulse with much better stability than the CPM laser. Titanium-doped sapphire (Ti:sapphire) laser was found as a broadband gain medium suitable to generate near infrared (NIR) laser pulse [29] and 60 fs pulse was generated by the KLM Ti:sapphire laser [30]. The KLM Ti:sapphire laser is now most commonly used as a stable commercial light source of femtosecond laser pulse.

The primary reaction could be studied in detail by performing ultrafast transient absorption (TA) spectroscopy which visualizes spectral change during ultrafast transition between electronic states after photoexcitation (ultrafast electronic dynamics). Using ultrashort laser pulse whose duration is shorter than the molecular vibration period, the observed transient absorption signal shows modulation reflecting the real-time motion of the molecular vibration in time domain. Time-gated Fourier analysis of the signal elucidates the time development of the molecular vibration frequency which elucidates molecular structure change during the photo-reaction (ultrafast vibrational dynamics). Thus, using ultrashort pulse laser, TA spectroscopy can study both of ultrafast electronic dynamics and ultrafast vibrational dynamics simultaneously from the same scan of measurement data, i.e., under the same measurement condition.

In the present work, we have developed ultrashort pulse lasers in visible and ultraviolet spectral region, then applied them for ultrafast spectroscopy of the protein samples to elucidate their primary reactions. For the study of molecular vibrational dynamics, signal fluctuation of the transient absorption should be suppressed to observe fine signal modulation caused by the molecular vibration. However, irradiation of the ultrashort laser pulse required for the ultrafast spectroscopy gives serious damage accumulation especially in the protein samples, which degrades signal quality of the transient absorption. Developing fastscan ultrafast spectroscopy system, we have succeeded to improve the signal-to-noise ratio of the transient absorption signal able to study molecular vibrational dynamics.

## 11.2 Ultrafast Transient Absorption (TA) Spectroscopy

Dynamics of the photo-excited state in the photo-reaction can be studied by observing transient absorption (TA) signal using pump-probe method, which was firstly demonstrated by Porter et al. in flash photolysis [24]. In the pump-probe measurement, transmission change of the probe light induced by the pump light irradiation is measured as a function of probe wavelength λ and delay of the probe pulse from the pump pulse τ as $\Delta T(\lambda, \tau)$. Using the measured transmission change, the TA signal can be calculated as $\Delta A(\lambda, \tau) = -\log_{10}\left(1 + \frac{\Delta T(\lambda,\tau)}{T(\lambda)}\right)$ where $T(\lambda)$ is transmission spectrum of the probe pulse through the sample. Reflecting the population excited in the electronic excited state, the TA signal includes three signal components of induced absorption (IA), stimulated emission (SE), and photo-bleaching (PB). The dynamics of IA, SE, and PB to be observed in the TA signal are schematically shown in Fig. 11.1.

The TA signal of IA has positive sign which reflects appearance of an absorption band corresponding to absorption spectrum of the electronic excited state. Meanwhile, SE and PB result in negative TA signal by increase of transmission. Figure 11.1g shows schematic TA trace for each of these three components of IA, SE, and PB; however, general TA trace is a linear combination of these three terms and hard to differentiate each contribution from the others.

Spectral resolution of the TA signal by observation of TA spectroscopy can help the differentiation of each signal contribution as follows. The PB signal is caused

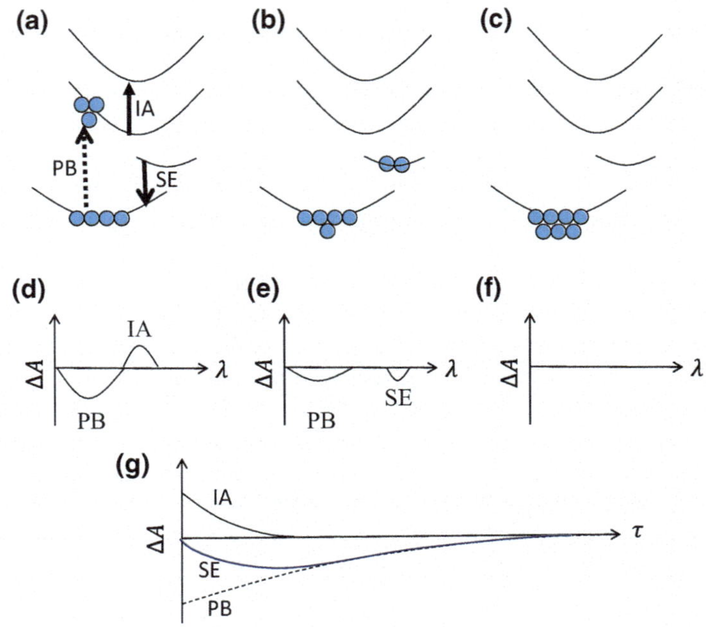


**Fig. 11.1** Schematic figure to show dynamics of pump-probe signal observed in photobleaching (PB), induced absorption (IA), and stimulated emission (SE). Electronic population **a** just after pump irradiation, **b** after transiting to intermediate state, and **c** finally relaxing back to the electronic ground state. **d–f** Transient absorption spectrum observed for the delay time corresponding to (**a–c**), respectively. **g** Transient absorption trace as a function of delay time

by ground state depletion after photoexcitation; therefore, the TA component corresponding to PB resembles the spectrum of stationary absorption spectrum of the electronic ground state. Meanwhile, the SE signal is generally emitted with lower photon energy than the photon energy of the stationary absorption band (see Fig. 11.1e). Thus, even though both of PB and SE show same negative TA signal, spectral difference helps to differentiate each contribution in the TA signal. IA spectrum is also generally shifted from the stationary absorption band; therefore, IA spectral component can also be easily differentiated from other contributions not only by (positive) sign of TA signal but also by spectral information.

Typical signal intensity of TA signal is $10^{-3}$ or smaller. Considering typical fluctuation of probe light intensity is $10^{-2}$ or more, TA signal measurement requires extremely high signal-to-noise (S/N) ratio in the measurement result. One of the method to improve S/N ratio is lock-in amplification, which can extract a signal from a noise environment [31–33]. The lock-in detection of the TA signal could be performed modulating pump pulse by utilizing a mechanical chopper with a rotating blade or an electro-optic modulator (EOM). The mechanical chopper has an advantage that the modulator does not add any material chirp on the pump pulse and a disadvantage that the modulation frequency cannot be higher than 10kHz. Generally, higher frequency region has lower environmental noise and EOM is preferred for the lock-in detection modulating signal at frequency of MHz range. However, when ultrashort femtosecond laser pulse is used for the pump-probe measurement, lock-in detection can be performed only by using the mechanical chopper. The TA spectroscopy utilizing lock-in detection is discussed in Sect. 11.5.

Another method to improve S/N ratio is high frequency measurement of transmission spectrum. The probe light intensity is fluctuated by environmental condition change like temperature, air flow, and humidity. Those environmental condition changes in relatively slow time scale in the order of seconds. Therefore, high frequency measurement of probe transmission spectrum can study TA signal caused by the pump pulse irradiation not by environmental change. Section 11.6 describes the detail of the single-shot based TA spectroscopy utilizing fast framerate line-sensor.

When the pump-probe measurement is performed using ultrashort laser pulse, molecular vibration mode whose vibration period is slower than the pump pulse duration can be excited coherently to produce wavepacket on the energy potential surface, which was firstly demonstrated for malachite green using 40-fs laser pulse by Tang et al. [34] followed by Shank el al. using 6-fs laser pulse [35]. As shown in schematic figure of Fig. 11.2a, the wavepacket oscillates on the potential energy surface with the period of molecular vibration. Thus, the TA signal shows oscillation with this period (see Fig. 11.2b).

Simulated data of the TA signal measured by the ultrashort laser pulse was calculated using the following equation.

$\Delta \mathrm{A}(\mathrm{t}) = \mathrm{A}_0 \exp\left(-\frac{t}{\tau_1}\right)\{1 + A_1 \cos 2\pi t f(t)\}(\mathrm{t} \geq 0)$ and $\Delta \mathrm{A}(\mathrm{t}) = 0(\mathrm{t} < 0)$, where $\mathrm{f}(t) = \mathrm{f}_0 + df \exp\left(-\frac{t}{\tau_2}\right)$. Parameters were set as $(\mathrm{A}_0, \mathrm{A}_1, \tau_1, \tau_2, f_0, df) = (0.01, 0.2, 500, 500, 0.01, 0.002)$ adding simulated noise with amplitude of 0.003 to $\Delta \mathrm{A}(\mathrm{t})$. The simulated data is plotted as black curves in Fig. 11.3a.

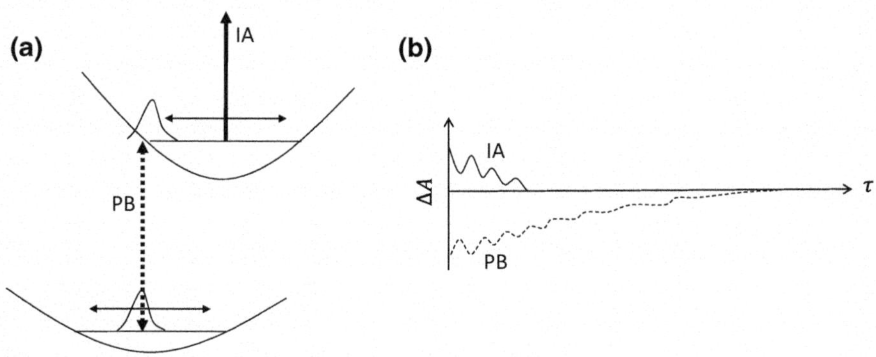


**Fig. 11.2** **a** Schematic figure of the wavepacket produced by the ultrashort pump laser pulse in the electronic ground state and in the electronic excited state. **b** TA signal of induced absorption (IA) and photo-bleaching (PB) to be observed under excitation by the ultrashort pump laser pulse, which is modulated by the wavepacket motion oscillating at the period of the molecular vibration

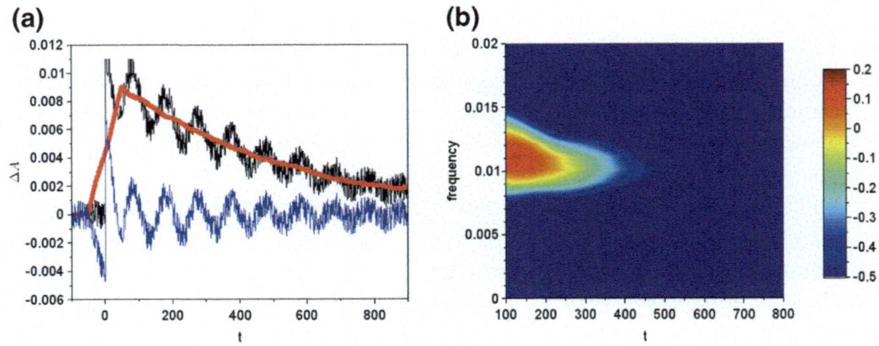


**Fig. 11.3** **a** Simulated data of the TA signal measured by the ultrashort laser pulse (black) and its neighboring average for 101 data points around each delay point (red). Their difference is oscillating component (blue). **b** Spectrogram trace calculated from the oscillating component after t = 0, where color scale is plotted in logscale

The TA signal (black curves of Fig. 11.3a) was subtracted by its neighboring average (red curves of Fig. 11.3a) to extract the oscillating component of the TA signal (blue curves of Fig. 11.3a). The oscillating component of the TA signal can be analyzed by time–frequency analysis [36, 37] to calculate instantaneous frequency of molecular vibration, which elucidates molecular vibration dynamics after photoexcitation.

In the time–frequency analysis of the present work, we have applied a method of spectrogram analysis, which was originally developed for sound analysis [38, 39] and is now widely used as a common tool of the time–frequency analysis. The spectrogram, which is the magnitude-square of the short-time Fourier transform of the signal, can be mathematically calculated as $S(\tau, \omega) = \left|\int x(t)h(t+\tau)e^{-i\omega t}dt\right|^2$, , where x(t) and h(t) are the signal and a window function, respectively. Dynamic range and frequency resolution depend on the type and width of the window function. In the present work, we use a Blackman window function given by $h(t) = 0.42 - 0.5\cos\frac{2\pi t}{w} + 0.08\cos\frac{4\pi t}{w}$, , where w is width of the window function and full width of half maximum (FWHM) of the window is $\sim 0.4w$. . Using the oscillating component after $t = 0$ (blue curve in Fig. 11.3a, spectrogram trace was calculated as shown in Fig. 11.3b using the Blackman window with FWHM of 160. The calculated spectrogram trace shows that the peak frequency is decreasing with *t* as set in f(t) of the simulation function.

The neighboring average of the TA signal (red curves of Fig. 11.3a) represents slow relaxation component reflecting the electronic dynamics. Meanwhile, the oscillating component of the TA signal (blue curves of Fig. 11.3a) can elucidate molecular vibration dynamics by the spectrogram analysis (see Fig. 11.3b). Thus, the pump-probe measurement of TA signal by using ultrashort laser pulse enables us to study both of electronic dynamics and molecular vibration dynamics simultaneously. Typical molecular vibration period is about 20 fs or longer, thus for the study of ultrafast vibrational dynamics, pump-probe spectroscopy should be performed using ultrashort laser pulse whose pulse duration is as short as ~10 fs.

The 10-fs pulse laser is commercially available in NIR spectral region using KLM Ti:sapphire laser; however, this NIR 10-fs pulse laser is not suitable to study most of photoreactions because the transition energy of those photoreactions is in visible spectral region unable to be excited by the NIR light.

Therefore, we have developed 10-fs visible pulse laser as described in the next section.

## 11.3 10-Fs Visible Pulse Laser

One of our targets of the present work is the study of a membrane protein for light-driven proton pump, bacteriorhodopsin (BR); however, it has absorption band only in visible spectral region not in NIR spectral region. Therefore, as follows, we have developed 10-fs visible pulse laser suitable for the TA visible spectroscopy. Light source of the 10-fs visible pulse laser used in the present work is a Ti:sapphire regenerative amplifier (Coherent Inc., Legend-USP-HP). The NIR laser pulse generated by the Ti:sapphire regenerative amplifier has center wavelength of 800 nm, bandwidth of 30 nm, pulse duration of 35 fs, repetition rate of 5 kHz, and pulse energy of 0.5 mJ.

The femtosecond NIR laser pulse was focused into a sapphire plate with thickness of 2mm to cause self-phase modulation (SPM) [40] by the high peak intensity of the femtosecond pulse. SPM has broadened the NIR spectrum covering all of the visible spectrum region (see Fig. 11.4a).

Efficiency of the SPM to convert into visible spectrum is less than $10^{-2}$ still including enormous remainder of the NIR component. Passing the SPM pulse through a short pass filter (SPF) which suppress the NIR component down to $10^{-6}$, visible laser spectrum not including the NIR component was obtained with average power of $2\mu$W.

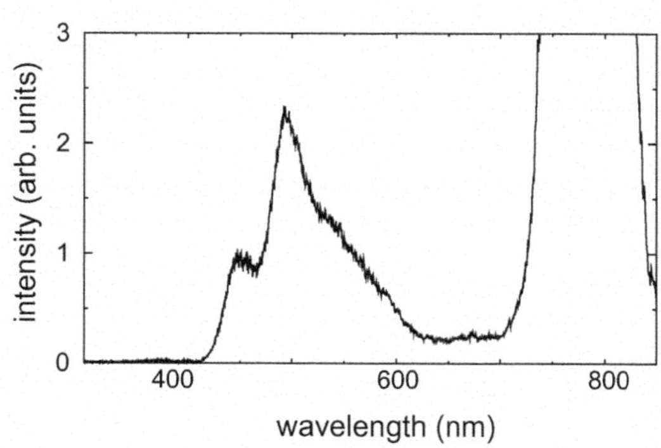


**Fig. 11.4** Spectrum of the femtosecond NIR pulse broadened by SPM

The pulse energy of this broadband visible laser is not enough for the TA spectroscopy and needs to be amplified. Optical parametric amplifier (OPA) is a common method to amplify the spectrum which is not overlapped with the spectrum of the light source. OPA for the visible spectrum requires second harmonic (SH) of the femtosecond NIR pulse to be used as a pump pulse. The NIR source laser pulse was focused into a 0.1 mm-thick $\beta - BaB_2O_4$ (BBO) crystal to generate the SH (pump) pulse with average power of 100 mW with center wavelength of 400 nm. The broadband visible pulse to be amplified by the OPA process is called seed pulse.

The seed pulse can be amplified in the OPA process, which converts one pump photon into a seed photon and another (idler) photon. When both of the pump pulse and the seed pulse are focused into a BBO crystal, which can cause OPA process under the condition satisfying "temporal and spatial overlap of the two pulses (pump pulse and seed pulse) in the crystal" and "conservation of photon energy $\left(\frac{\mathrm{hc}}{\lambda_\mathrm{p}} = \frac{hc}{\lambda_s} + \frac{hc}{\lambda_i}\right)$ and photon momentum $\left(\Delta\vec{k} = \vec{\mathrm{k}}_p - \vec{\mathrm{k}}_s - \vec{\mathrm{k}}_i = 0\right)$, where $\lambda_{\mathrm{p,s,i}}$ is wavelength and $\vec{\mathrm{k}}_{p,s,i}$ is angular wave number vector. Suffix of p, s, and, I, represents pump, seed, and idler, respectively.

When the momentum conservation is not satisfied $\left(\Delta\vec{\mathrm{k}} \neq 0\right)$, OPA efficiency, which is proportional to $\mathrm{L}^2\mathrm{sinc}^2\frac{\Delta k \cdot L}{2}$, decreases, where L is thickness of the BBO crystal to be used for the OPA process. The component along the propagation direction of the pump pulse for the photon momentum conservation can be written as $\frac{\mathrm{n_p}(\theta_\mathrm{p})}{\lambda_\mathrm{p}} = \frac{\mathrm{n_s}(\theta_\mathrm{s})}{\lambda_\mathrm{s}}\cos\phi_s + \frac{\mathrm{n_i}(\theta_\mathrm{i})}{\lambda_\mathrm{i}}\cos\phi_i$, where $\mathrm{n_{p,s,i}}$ is refractive index, $\phi_{\mathrm{s,i}}$ is the angle between the propagation direction of the corresponding pulse and that of the pump pulse, and $\theta_{\mathrm{p,s,i}}$ is angle between the light and the optical axis of the crystal. Generally, this equation cannot be satisfied because of monotonous dependence on wavelength of the refractive index which is called normal dispersion. However, in birefringent crystal, this equation can be satisfied utilizing combination of different polarizations of lights.

In the present work, we have used configuration of type-I phase-matching condition in which pump pulse has extraordinary polarization and other two (seed and idler) pulses have ordinary polarization, where ordinary (extraordinary) polarization is the one in (orthogonal to) the plane containing the light propagation axis and the optical axis of the crystal. Refractive index of the BBO crystal for extraordinary polarization depends on light wavelength, $\lambda$, and angle between the light polarization and the optical axis of the BBO crystal, $\theta$, which can be represented as $\mathrm{n_e}(\lambda, \theta)$. Meanwhile, that for ordinary polarization only depends on light wavelength as $\mathrm{n_o}(\lambda)$ because the light polarization is always orthogonal to the optical axis of the BBO crystal. Those refractive index of the BBO crystal could be calculated using Sellmeier's equations with parameters reported by Kato [41].

At first, we have considered for the case when all of the three beams (pump, seed, and idler) runs collinearly into the BBO crystal, i.e., $\theta_\mathrm{p} = \theta_s = \theta_i$. Using the refractive index, $\Delta\mathrm{k} \cdot \mathrm{L}$ was calculated as a function of $\lambda_\mathrm{s}$ scanning the value of $\theta_\mathrm{p} = \theta_s = \theta_i$ (see Fig. 11.5a). Here, we assume all of three pulse of pump, seed, and

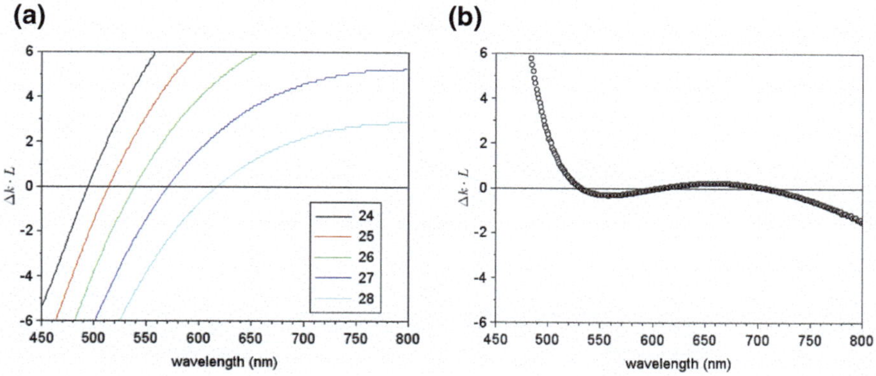


**Fig. 11.5** Phase mismatch of $\Delta k \cdot L$ as a function of seed wavelength when the pump and seed pulses run **a** collinear $\left(\theta_p = 24^\circ \sim 28^\circ, \alpha = 0^\circ\right)$ and **b** non-collinear $\left(\theta_p = 31^\circ, \alpha = 6^\circ\right)$

idler run collinearly, i.e., $\phi_s = \phi_i = 0$. Ordinary polarization of seed and idler pulse gives $\theta_s = \theta_i = 90^\circ$.

As you can see in Fig. 11.5a, the condition of $\Delta k \cdot L \approx 0$ can be satisfied in narrow spectral region indicating that the OPA process gives narrow band visible spectrum. Theoretically calculated gain spectrum of the OPA process is shown in Fig. 11.6.

Smaller L gives smaller $\Delta k \cdot L$ however decreases OPA efficiency $\left(\propto L^2 \text{sinc}^2 \frac{\Delta k \cdot L}{2}\right)$. Thus, conventional OPA process cannot be used to generate intense broadband visible pulse useful for the TA spectroscopy.

Spectrum of the light is Fourier transform of the electric field, which lets duration of the shortest pulse (transform limited pulse) be inversely proportional to the spectral bandwidth of the laser pulse. Thus, the development of the 10-fs visible pulse requires to generate broadband visible laser spectrum whose bandwidth is as broad as 100 nm and the TA spectroscopy requires pulse energy to be higher than 10 nJ to obtain measurable TA signal.

In case of non-collinear OPA (NOPA), the pump beam and the seed beam incident on the BBO crystal has finite angle. The angle between the two beams measured outside of the BBO crystal is called non-collinear angle $\alpha$. When $\theta_p = 31^\circ$ and $\alpha = 6^\circ$, we found the phase-matching condition $(\Delta k \cdot L \approx 0)$ can be satisfied in broadband region (see Fig. 11.6b).

This method called non-collinear OPA (NOPA) developed by several groups almost at the same time independently [42–44]. The gain spectrum corresponding to (collinear) OPA, NOPA with broadest band, and NOPA scanning the external angle are plotted in Fig. 11.6. With the broadest band condition, NOPA could produce visible broadband spectrum with pulse energy of 800 nJ repeating the NOPA process twice, whose optical setup is shown in Fig. 11.7.

The output pulse of the NOPA was separated into two pulses using a 1mm-thick glass plate with power ratio of 10:1. The pulse with higher and lower power was used as pump and probe pulse in the pump-probe measurement for the TA spectroscopy. These pulses have large positive chirp caused by transmission through

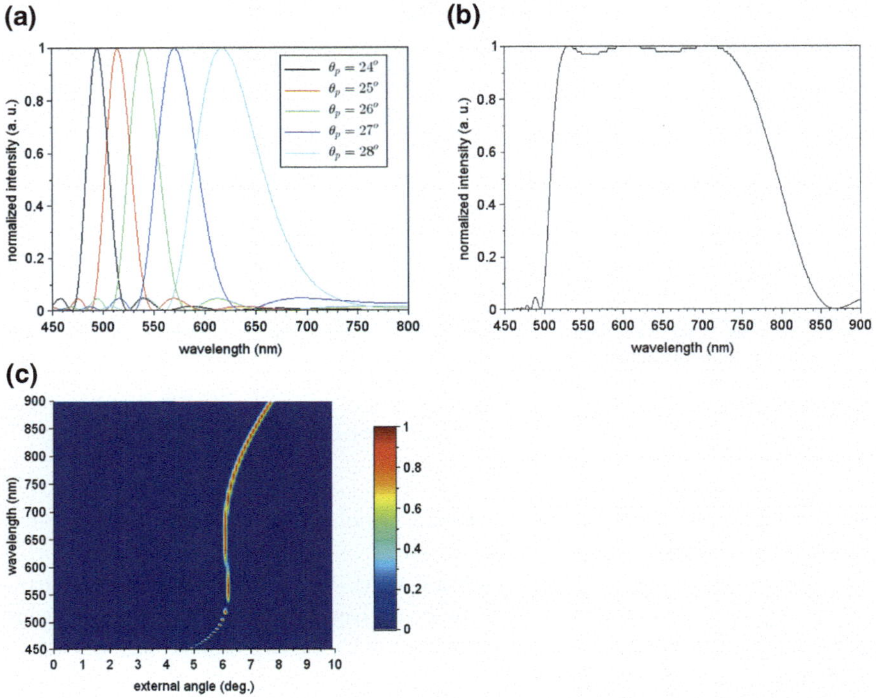


**Fig. 11.6** Gain spectrum of the OPA process when the pump and seed pulses run **a** collinearly $(\theta_p = 24° \sim 28°, \alpha = 0°)$ and **b** non-collinearly $(\theta_p = 31°, \alpha = 6°)$. **c** Gain spectrum calculated scanning $\alpha = 0° \sim 10°$ fixing $\theta_p = 31°$

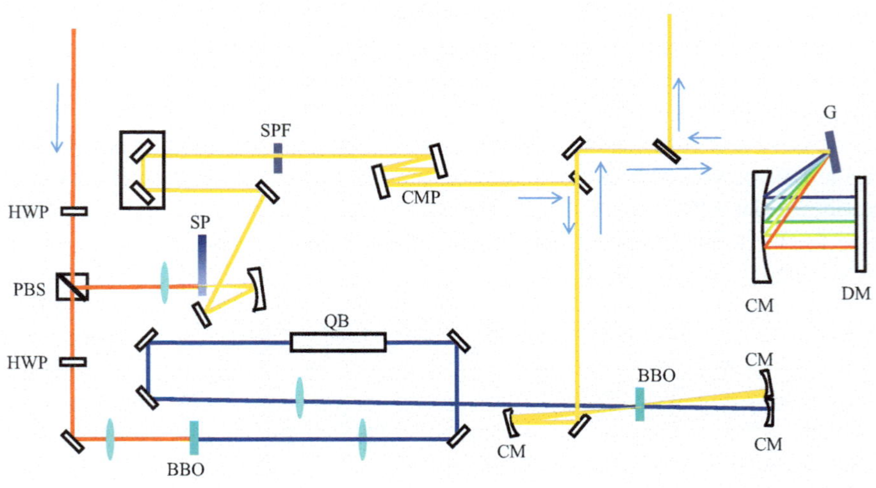


**Fig. 11.7** Optical setup of NOPA. HWP: half-wave plate, PBS: polarization beam splitter, BBO: a BBO crystal, SP: sapphire plate, SPF: short-pass filter $(\lambda < 750\text{nm})$, QB: quartz block, CM: curved mirror, CMP: chirped-mirror pair, G: grating, DM: deformable mirror

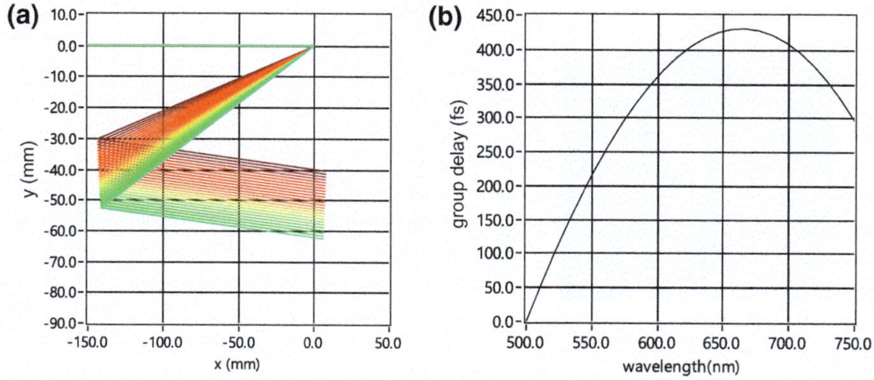


**Fig. 11.8** **a** Result of raytracing for the pulse compressor and **c** calculated chirp to be added by the pulse compressor

sapphire plate, SPF, BBO crystal for NOPA, and a glass plate used to produce the pump and probe pulses. Therefore, we have developed a pulse compressor which consists of a diffraction grating and a deformable mirror. It is not put in 4-f configuration and gives negative chirp to the pulse. The result of raytracing and calculated chirp are shown in Fig. 11.8.

Thus, we could compress the pulse duration of the intense visible broadband laser pulse as short as 10 fs estimated by SH frequency resolved optical gating (FROG) method. The measured FROG trace, the calculated pulse train, and spectrum are shown in Fig. 11.9.

## 11.4 10-Fs Ultraviolet Pulse Laser

For the study of the heme protein which has absorption band in near ultraviolet (NUV) region, we have developed a 10-fs ultraviolet pulse laser using Ti:sapphire regenerative amplifier as a light source. The regenerative amplifier generates NIR pulse with center wavelength of 800nm, pulse duration of 35 fs, average power of 5 W, and repetition rate of 1 kHz. SH of the NIR pulse was generated by a 0.2mm-thick BBO crystal with average power of 110 mW and bandwidth of 10 nm centered at 400 nm (see Fig. 11.10).

The bandwidth of this SH pulse is not enough to be compressed shorter than 10 fs; therefore, it is necessary to broaden the bandwidth. We have guided the SH pulse into a hollow-core fiber filled with argon gas with pressure of $8.0 \times 10^4$ Pa, which has caused SPM in the core of the hollow-core fiber with internal dimeter of 0.14 mm and length of 60 cm. The generated broadened spectrum is shown in Fig. 11.10. The pulse duration of the broad NUV spectrum was compressed using a prism pair, a diffraction grating, and a deformable mirror. The schematic figure of the optical setup is shown in Fig. 11.11.

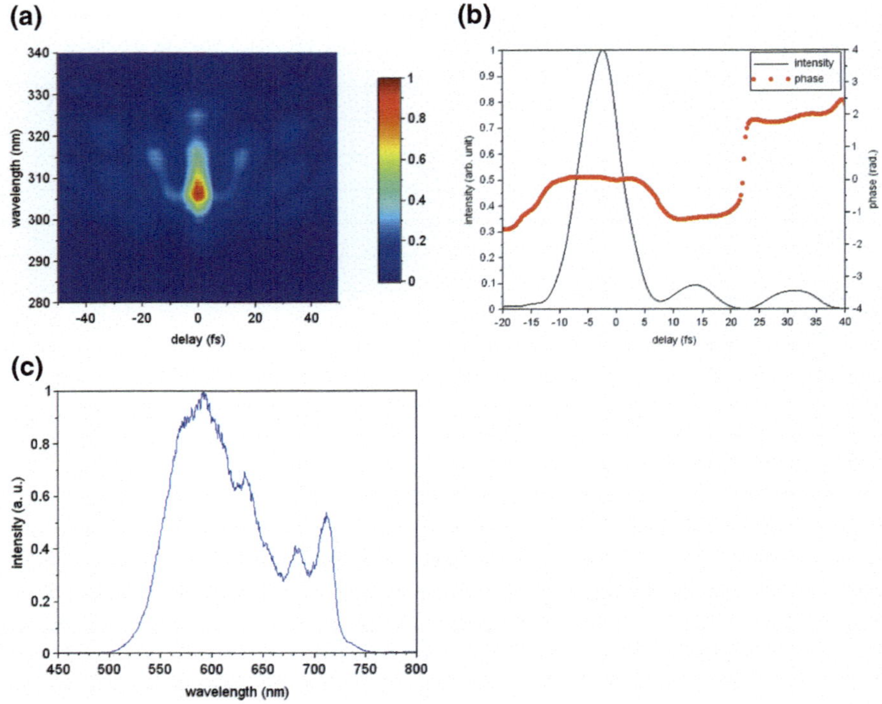


**Fig. 11.9** **a** Measured SH-FROG trace, **b** pulse train retrieved from the FROG trace, and **c** spectrum of the visible broadband 10-fs laser pulse

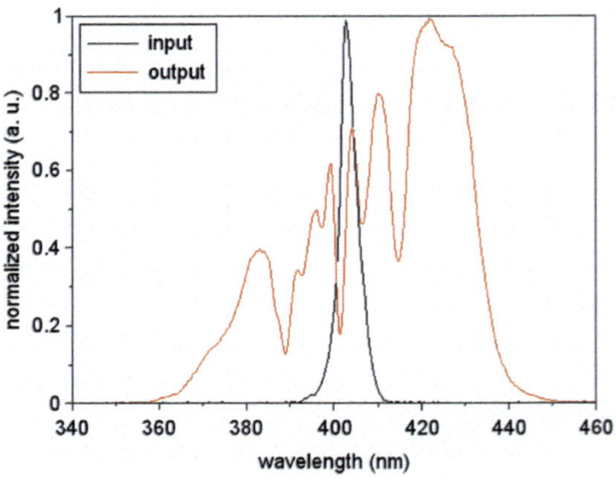


**Fig. 11.10** SH spectrum generated by a 0.2mm-thick BBO crystal has a bandwidth of about 7 nm which was broadened in a hollow-core fiber filled with argon gas to have broadband width of 90 nm for generation of ultrashort NUV laser pulse

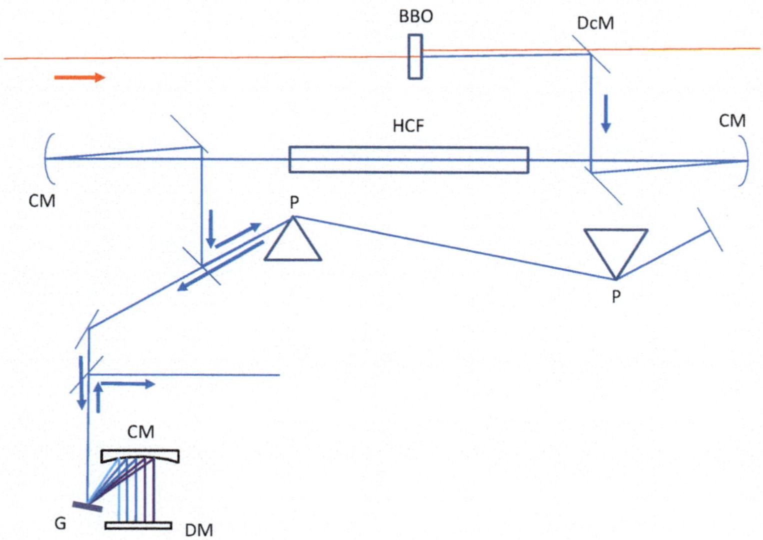


**Fig. 11.11** Schematic figure of NUV 10-fs pulse laser. BBO: a BBO crystal, DcM: dichroic mirror (high reflection for SH and high transmission for NIR fundamental light), CM: curved mirror, P: prism pair, G: diffraction grating, DM: deformable mirror

The pulse duration of the NUV pulse was estimated by self-diffraction FROG (SD-FROG) whose measured trace and retrieved pulse train are shown in Fig. 11.12.

The retrieved result shows that the broadband NUV pulse was compressed shorter than 10 fs being suitable to temporally resolve molecular vibration signal of heme protein samples.

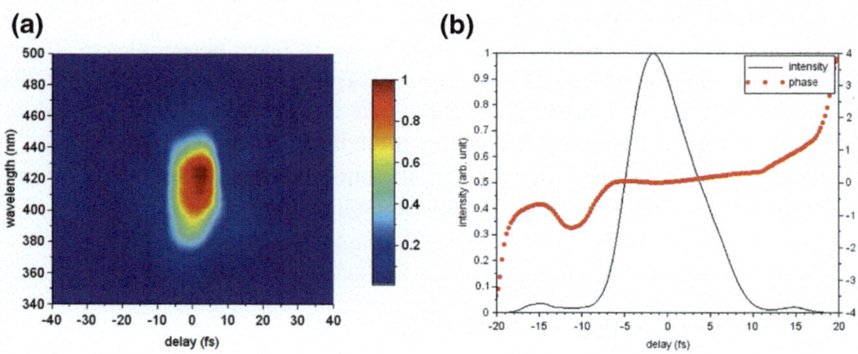


**Fig. 11.12** **a** Measured SD-FROG trace and **b** pulse train retrieved from the SD-FROG trace

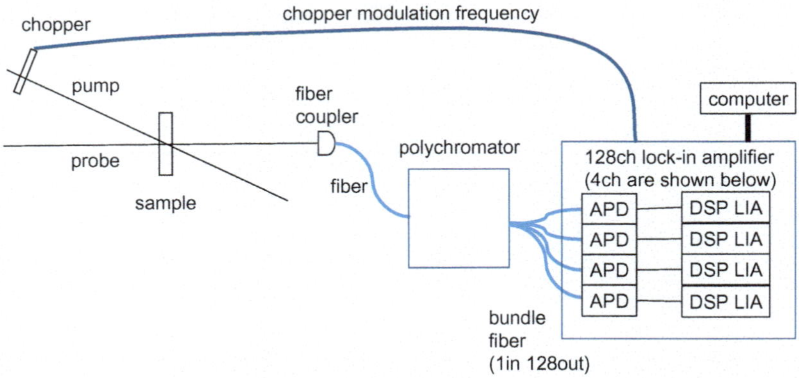


**Fig. 11.13** Multi-channel lock-in amplifier system which measures TA trace at 128 probe wavelengths simultaneously. APD = avalanche photodiode, DSP LIA = digital signal processing lock-in amplifier

## 11.5 Multichannel Lock-In Spectroscopy

The broadband visible 10-fs laser pulse and the broadband NUV 10-fs laser pulse were used for the TA spectroscopy. For the detection of fine change of absorption, it is necessary to detect the signal by lock-in amplification. Conventional lock-in TA measurement could be performed selecting probe wavelength by using a monochromator; therefore, TA spectroscopy requires to repeat the lock-in measurement for each probe wavelength.

In the simultaneous measurement of electronic dynamics and vibrational dynamics by the ultrafast TA spectroscopy, the delay should be scanned as long as a few ps and the delay step should be as small as a few fs. Thus, the delay scanning stage should have high accuracy as good as 10 nm which can be stabilized in 500–600 ms typically, and the scan up to a few ps takes about 30 min. If TA measurement should be repeated for each probe wavelength, 100 probe wavelengths take several days which does not provide reliable data because of accumulation of irradiation damage and intensity fluctuation of the light source.

We have developed a multi-channel lock-in amplifier system which can perform TA measurement for all probe wavelengths simultaneously (see Fig. 11.13).

## 11.6 TA Spectroscopy by Fast Line-Sensor

The MLA system has enabled us to study electronic dynamics and vibrational dynamics simultaneously from a single scan of TA spectroscopy. The measurement of the all probe wavelengths could be accomplished in about 30 min of typical measurement time. Utilizing the lock-in amplification method, TA signal as small as $10^{-4}$ could be detected. However, the MLA system has three major disadvantages. First disadvantage is the expensive cost as high as ~0.1 M USD (united states dollars). Second disadvantage is that the MLA system is home-made not commonly available. Third disadvantage is the bulky size of the system as large as $0.7\text{m}^3 (= 1.5 \times 0.8 \times 0.6m^3$. To overcome these three advantages, we have developed a TA spectroscopy system using a fast scan rate line-sensor whose detail is described below and schematic figure of the system is shown in Fig. 11.14.

The spectrum of the probe pulse was spectrally resolved by using a polychromator (HORIBA, CP 140–104) which has average dispersion of 24.3 nm/mm and wavelength range of 250–850 nm. The spatially dispersed probe pulse was directly coupled on the surface of the fast scan rate line-sensor (Entwicklungsburo Stresing, S2000). The line-sensor has a sensor of a fast line-sensor (Hamamatsu Photonics, S8380-256Q) whose pixel width, pixel height, number of pixels, maximum pixel scan rate, and maximum line scan rate are $50\mu$m, 2.5 mm, 256, 2 MHz, and 8 kHz, respectively.

The timing of the laser irradiation at *f* kHz was sent from the laser controller unit to the camera to be used as a trigger signal, which let the camera record probe pulse spectrum on irradiation of each probe pulse on the line-sensor. When the probe spectrum was recorded on the camera, the camera also records a chopper state signal (running at *f/2* kHz), which shows whether the chopper blade was open or closed at the timing of the data collection. Thus, we could know whether the recorded signal spectrum is for the case when the sample was excited by the pump pulse or not. The mechanical chopper in the pump light path switches whether the chopper blade blocks or not for each pump pulse. Thus, successive two probe light spectra recorded by the camera includes one from the sample excited by the pump pulse, $\text{T}(\lambda, \tau) + \Delta\text{T}(\lambda, \tau)$, and one from the sample which was not excited by the pump pulse, $\text{T}(\lambda, \tau)$. From the two successive two probe light spectra, i.e., at *f/2* kHz, transient absorption spectrum can be calculated as $\Delta\text{A}(\lambda, \tau) = -\log_{10}\left(1 + \frac{\Delta T(\lambda,\tau)}{T(\lambda)}\right)$. Thus, each of the calculated TA spectra is not affected by intensity fluctuation of $\text{T}(\lambda, \tau)$ and $\Delta\text{T}(\lambda, \tau)$ whose typical modulation time order (~one minute) is much slower than *f/2* kHz.

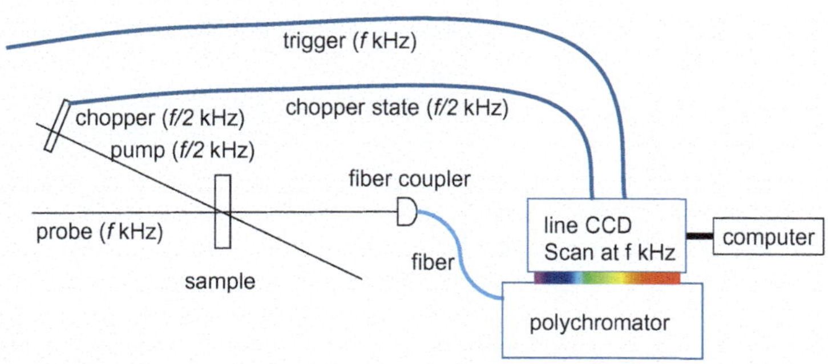


**Fig. 11.14** Schematic figure of the TA spectroscopy system using a fast scan rate line-sensor

Accumulating the TA signal for half second at each delay point, we could measure TA spectrum as small as $10^{-3}$ using the single-shot line-sensor based TA spectroscopy system. Its total cost is about 10,000 USD which is about ten times less than that of the MLA based TA spectroscopy, and it only includes commercially available instruments. Total spatial volume of this system is about $8 \times 10^{-3} \text{m}^3 (= 0.2 \times 0.2 \times 0.2\text{m}^3)$ which is about 100 times smaller than the MLA based TA spectroscopy system with volume of $7 \times 10^{-1} \text{m}^3 (= 0.8 \times 0.6 \times 1.5m^3)$.

## 11.7 Fastscan Ultrafast TA Spectroscopy System

Using the ultrashort pulse lasers, both of the MLA based TA spectroscopy system (Sect. 11.5) and the single-shot line-sensor based TA spectroscopy system (Sect. 11.6) could study ultrafast electronic dynamics and ultrafast vibrational dynamics simultaneously by scanning the data with high delay accuracy resolving the molecular vibration in time domain. To scan up to a few ps using high accuracy delay stage, typically it takes about 30 min for single scan of measurement.

The relatively long measurement time can cause two several troubles. First point is the irradiation damage on the sample irradiated by the ultrashort pulse laser. Because of the ultrashort duration of the laser pulse, its high peak intensity can cause serious irradiation damage especially for polymer film samples and protein samples. Second point is that the measurement data can be fluctuated by instability of the light source. The ultrashort laser pulses are generated by using several step of non-linear optical processes each of which can be affected by environmental change. Any change of air for temperature, humidity, flow can cause modulation of optical path length and light path which can affect efficiency of those non-linear optical processes, which results in instability appearing in time scale of minutes.

To overcome the difficulty caused by the long measurement time, we have developed fastscan ultrafast TA spectroscopy system for the MLA based TA spectroscopy system as follows. The schematic figure of the optical system is shown in Fig. 11.15.

The stage to scan the optical delay is a fast-scan stage (ScanDelay-15, APE-Berlin) whose position can be set by analog voltage sent from computer. The analog voltage was sent from a digital-to-analog converter (DAC) (LPC-361316, Interface Inc.) to scan the whole range of 1.5 ps in five seconds for 500 delay points, i.e., 10 ms for each delay point. To obtain the data in MLA synchronized with the delay scanning motion, the D/A converter also sends a TTL pulse to MLA which triggers data collection in MLA at the timing when it sends each analog voltage to the fast-scan stage. The data collection in MLA could be accomplished in 10 ms developing built-in memory in MLA which can store 4000 data points in maximum.

This scan speed of five seconds is 360 times faster than the typical measurement time of 30 min; however, the data obtained in the single scan of fastscan TA spectroscopy do not have enough quality for analysis and have taken average of 24 scans of the fastscan TA spectroscopy result.

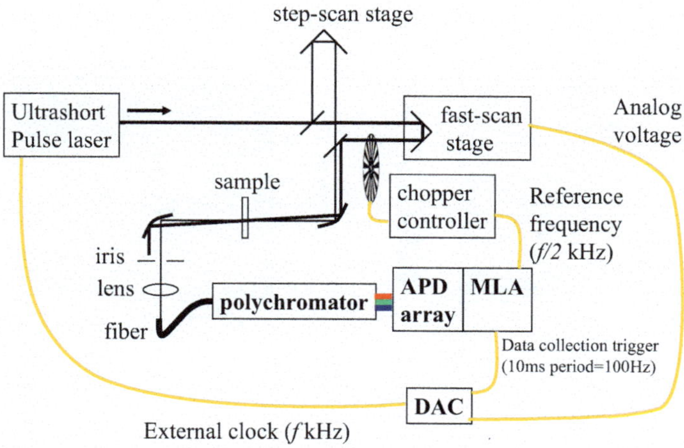


**Fig. 11.15** Schematic figure of the fastscan MLA-based TA spectroscopy system. Orange lines represent path way of electric signal. DAC: dignal-to-analog converter

A motorized stage (PFS-1020, Sigma-Tech Inc.) which is put in the probe light path of the measurement system with name of stepscan stage has positioning accuracy of 10 nm (1/15 fs). It can temporally resolve the signal caused by molecular vibration in time domain. The stepscan stage was used for calibration of the positioning of the fastscan stage and also for measurement of conventional TA spectroscopy to be compared with the fastscan TA spectroscopy result.

Figure 11.16 shows the result of TA spectroscopy for a popular electroluminescent material of a poly[2-methoxy-5-(20-ethyl-hexyloxy)-p-phenylene vinylene] (MEH-PPV) measured by using the 10-fs visible pulse laser and the conventional TA spectroscopy.

The sample film was prepared by spin coating a chloroform solution of MEH-PPV to form a film on a 1mm-thick quartz plate with a film thickness of 0.5–1.0 $\mu$m. Because of poor stability of light source whose intensity modulates in the time scale of minutes, each of the TA traces is seriously fluctuated during the 30min-scan of TA spectroscopy. The result of forward scan, which scans delay in increasing order, and that of backward scan decreasing delay do not show reproducible result. The backward scan was performed after the forward scan which results in less than half of TA signal amplitude in the backward scan compared with that in the forward scan because of the irradiation damage accumulation in the forward scan. Note that the TA traces of both scans show periodical modulation caused by C=C stretching vibration mode. Fourier power spectrum of the TA traces show same peak at the C=C stretching frequency of 1588 cm$^{-1}$.

Using the same sample of the MEH-PPV film, we have performed fastscan TA spectroscopy whose result is shown in Fig. 11.17.

Each of the scan was measured in five seconds starting from the backward scan followed by the forward scan. The results shown in Fig. 11.17 are the average of

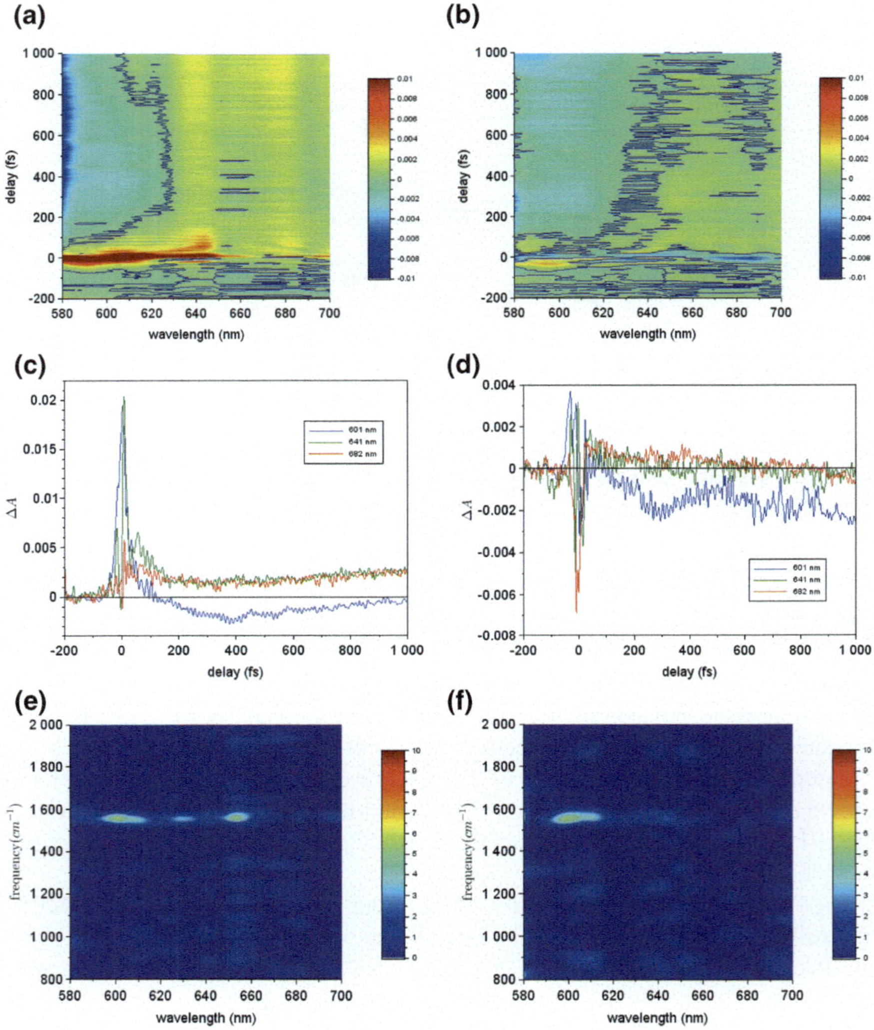


**Fig. 11.16** TA spectroscopy signal measured by the conventional TA spectroscopy system, which has scanned the delay time in **a** increasing order and **b** decreasing order. **c**, **d** are the TA trace at three probe wavelengths corresponding to the data of (**a**) and (**b**), respectively. **e**, **f** are Fourier power spectrum at every probe wavelengths calculated from the data of (**a**) and (**b**), respectively

24 scans for each of the forward and backward scan. The signal amplitudes of TA trace are reproducible between both of the scans with error smaller than 10%, which indicates that the damage accumulation during each of five second scan is negligible. TA spectra agreement in the whole delay region (see Fig. 11.17a, b) reflects that the stability of the light source in each of five second scan of fastscan TA spectroscopy is much better than that in 30 min scan of traditional TA spectroscopy. Fourier power

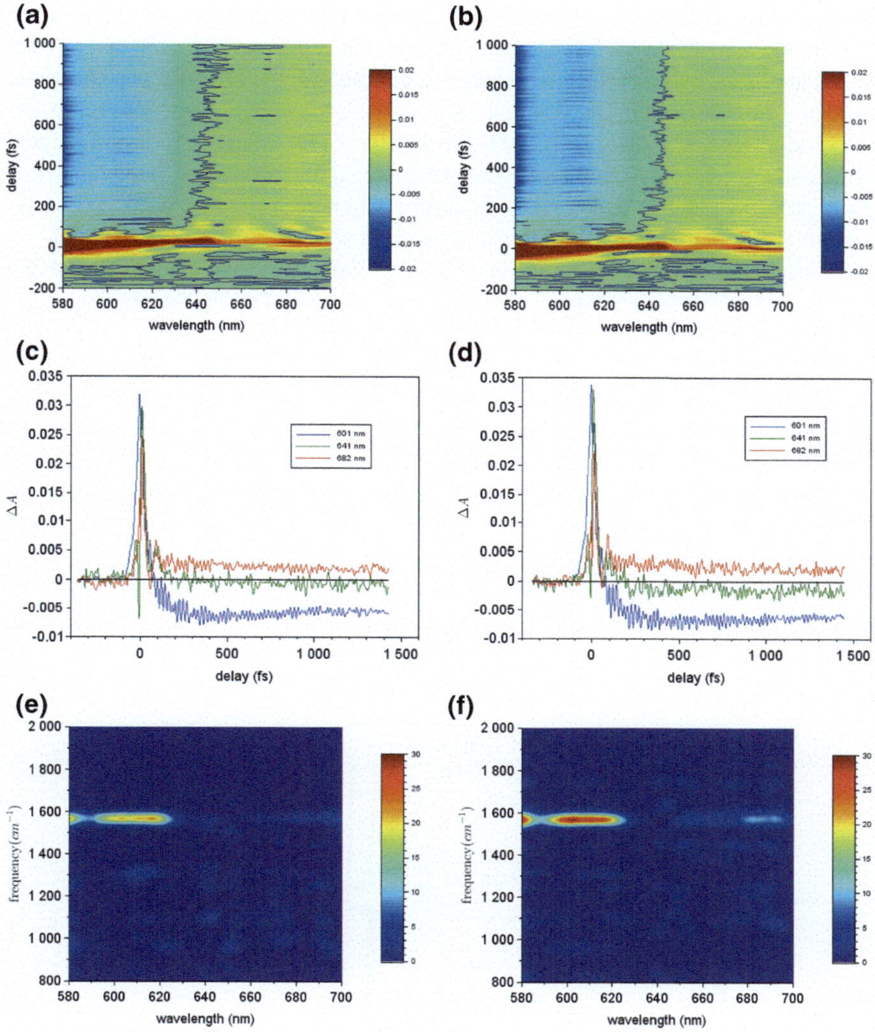


**Fig. 11.17** TA spectroscopy signal measured by the fastscan TA spectroscopy system, which has scanned the delay time in **a** increasing order and **b** decreasing order. **c**, **d** are the TA trace at three probe wavelengths corresponding to the data of (**a**) and (**b**), respectively. **e**, **f** are Fourier power spectrum at every probe wavelengths calculated from the data of (**a**) and (**b**), respectively

spectrum for each probe wavelength of the TA traces also shows reproducible result between both of the scan directions. Detail can be found in Ref. [45].

We also have developed the line-sensor-based fastscan TA spectroscopy system which uses single-shot line-sensor. In case of the MLA-based fastscan TA spectroscopy system, data collection trigger to MLA was sent at 10 ms period. In the line-sensor-based fastscan system, data collection trigger (DCT) should be sent to the line-sensor synchronized with the timing of the laser pulse. Generation of the DCT should start when each scan of the fastscan starts and end with the end of each scan. Repetition timing of the laser cannot be directly used as DCT because it cannot be set on or off depending on whether the fastscan is running or not. Therefore, we have used a logic AND IC (74HCT08) for sending an incoming pulse timing signal to line camera as DCT signal only when the gate signal is on. The gate signal was set on and off by the DAC board at the timing when each of scan starts and stops, respectively. Schematic figure of the experimental setup is shown in Fig. 11.18.

We also have performed the measurement of the MEH-PPV sample film using this line-sensor-based fastscan TA spectroscopy system. The measurement results shows reproducibility between the forward scan result and the backward scan result as seen in Fig. 11.19.

Each of the scan was measured in five seconds starting from the backward scan followed by the forward scan. The results shown in Fig. 11.19 are also average of 24 scans for each of the forward and backward scan. The signal amplitudes of TA trace and Fourier power spectrum for each probe wavelength of the TA traces shows reproducible result between the forward scan result and the backward scan result. Detail can be found in Ref. [46].

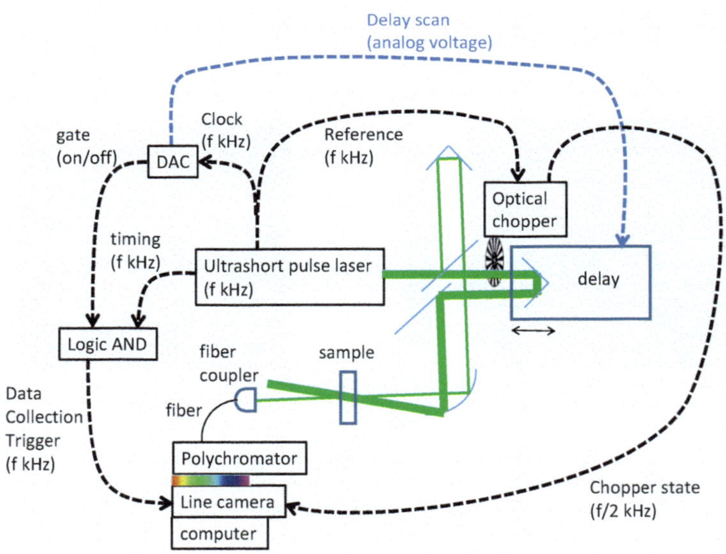


**Fig. 11.18** Schematic figure of the line-sensor-based fastscan TA spectroscopy system using single-shot line-sensor

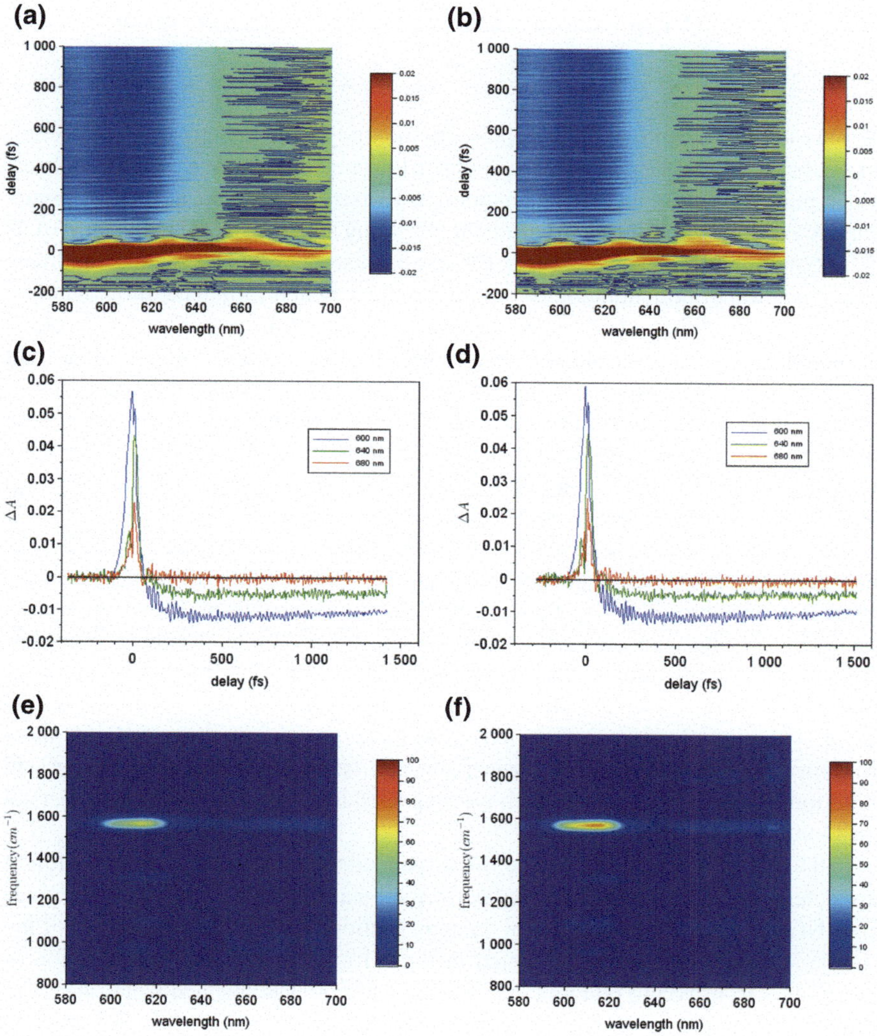


**Fig. 11.19** TA spectroscopy signal measured by the line-sensor-based fastscan TA spectroscopy system, which has scanned the delay time in **a** increasing order and **b** decreasing order. **c**, **d** are the TA trace at three probe wavelengths corresponding to the data of (**a**) and (**b**), respectively. **e**, **f** are Fourier power spectrum at every probe wavelengths calculated from the data of (**a**) and (**b**) respectively

## 11.8 Ultrafast Photoisomerization in Bacteriorhodopsin (BR)

We have applied the ultrashort laser pulse for the study of TA spectroscopy of bacteriorhodopsin (BR) [4, 47, 48]. BR is a protein in the purple membrane used by Archaea, which pumps a proton through the membrane to extracellular side excited by pump. The proton transfer produces pH gradient triggering synthesis of ATP. Because of its stability and ultrafast photoreaction, BR has been studied as a promising candidate for application of optical memory [49] and optical switch [50]. The initial reaction just after photoexcitation of BR is ultrafast trans–cis photoisomerization of retinal chromophore in femtosecond region (see Fig. 11.20a), which is similar to a light sensor protein, rhodopsin (RH), whose initial reaction is ultrafast cis–trans photoisomerization of the retinal chromophore. BR could be studied as a model protein of RH because ultrafast dynamics of RH is hard to be studied easily damaged by irradiation of the ultrashort laser pulse. Thus, initial reaction mechanism of BR has been studied widely in theory [6, 51] and experiment [49, 52–57].

Intermediates appear after photoexcitation of BR is named such as H, I, J, K intermediates. They were reported to appear 0 fs, 200 fs, 500 fs, 3 ps, respectively, after photoexcitation showing their specific absorption spectrum. However, assignment of each intermediate is still controversial. For example, Atkinson el al. has claimed the isomerization is not taking place even in the J intermediate [58] meanwhile the J intermediates are often assigned to a ground-state species whose retinal chromophore is already photoisomerized to have cis-configuration [59]. TA spectroscopy of BR using ultrashort laser pulse can visualize the spectral change and vibrational frequency change, which was expected to clarify the above controversial questions.

In the present work, we have used BR suspended in water buffered at pH 7 without any detergent. TA spectroscopy was performed using ultrashort visible laser pulse from NOPA and MLA spectroscopy system. Stationary absorption spectrum of the sample and laser spectrum are plotted in Fig. 11.20b. The measurement region of spectrum is from 505 to 664 nm, and the delay region was scanned from -400 fs to 800 fs. Two dimensional view of the measured TA spectra is shown in Fig. 11.20c.

TA spectra show difference in temporal and spectral behavior in four spectral regions; 505–530 nm (S), 540–600 nm (M), 610–630 nm (L1), and 635–664 nm (L2). The blue colored area observed among M, L1, L2 regions has negative signal ($\Delta A < 0$) reflecting two signal contributions. A main contribution is from bleaching by depletion of all-trans ground state whose absorption peak is around 562 nm. Another minor contribution is from stimulated emission of HI state, which is the state between H and I state. Positive signal of TA observed among S, L1, L2, regions represents induced absorption of probe pulse in transition from 1st electronic excited state to higher level of electronic excited state.

Taking average of TA spectra for each 100 fs of delay region, delay time dependence of TA spectra was plotted as shown in Fig. 11.20d which clearly shows the spectral change observed in femtosecond region. In S region, the TA signal partially

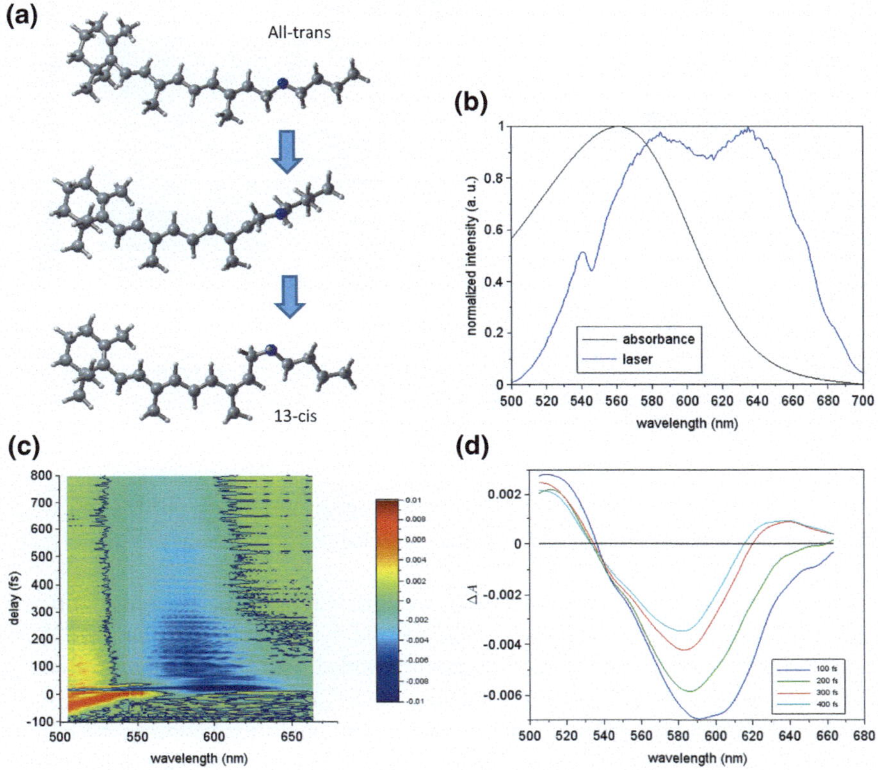


**Fig. 11.20** **a** Schematic figure of photo-isomerization of retinal chromophore. Gray: carbon atom, Light gray: hydrogen atom, Blue: nitrogen atom in Schiff base. Broken line is an eye guide showing the location where the photo-isomerization takes plate. **b** Stationary absorption spectrum of the sample (black) and laser spectrum (blue). **c** Two dimensional view of the measured TA spectra where black curves represent the point where $\Delta A = 0$. **d** Delay time dependence of TA in femtosecond region

decays in about 200 fs corresponding to the change of induced absorption spectrum in transition from H state to I state. Negative TA spectrum observed in M region reflects ground state depletion which does not decay in femtosecond region and stimulated emission of HI state which decays in femtosecond region. TA component in L1 and L2 regions changes from negative to positive which is caused by the appearance of positive TA signal (induced absorption of I state) and the disappearance of negative TA signal (simulated emission of HI state).

Time trace of the measured TA signal at three probe wavelegnths of 515, 585, and 640 nm are shown in Fig. 11.21a. Slow relaxation on these TA traces reflect signal decay or growth on electronic transition. Fast modulation on these TA traces are caused by wavepacket motion at the period of molecular vibration, thus time-gated Fourier transform of the TA traces can visualize time dependent change of molecular vibration reflecting molecular structure change during the reaction. The time-gated

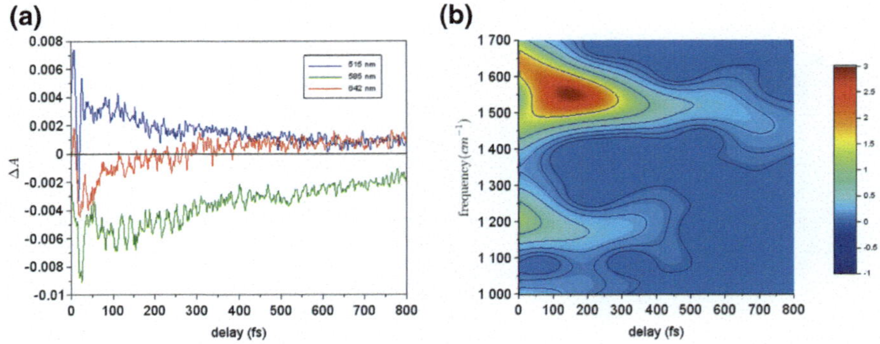


**Fig. 11.21** **a** Measured TA traces at three probe wavelegnths of 515, 585, and 640 nm, which were plotted taking cross-section of the data shown in Fig. 11.20a. **b** Time-gated Fourier transform of the TA trace probed at 577 nm

Fourier power spectrum of the TA traces, called spectrogram traces [37, 39], were calculated applying Blackman gate function with full width half maximum of 120 fs (see Fig. 11.21b).

The spectrogram trace of Fig. 11.21b shows signals at frequencies of around 1000, 1250, 1520, and 1640 $cm^{-1}$ corresponding to “the hydrogen out-of-plane (HOOP) mode”, “the in-plane C = C-H bending mode coupled with C–C stretching mode”, “C=C stretching mode”, and “C=N stretching mode”, respectively.

The C=N stretching mode was observed just after photo-excitation with lifetime of about 30 fs, then the C=C stretching mode has appeared. It implies that the first configuration change occurs at protonated C=N bond followed by the torsion around the C=C bond resulting in photo-isomerization.

The HOOP mode and the in-plane bending mode were found to mix to each other at around 200 fs, which separates to each other later. These two modes become indistinguishable when the retinal chromophore has non-planer structure by torsion around the C=C bond under photo-isomerization, then they become distinguishable when the photo-isomerization has completed transferring to J state. Thus, this result implies that the torsion occurs in 200 fs for the photo-isomerization. When the torsion around the C=C bond occurs, the bond order of the C=C bond is thought to decrease from double bond like to single bond like resulting in decrease of vibrational frequency. We have found the C=C stretching frequency has decreased at the timing of mixture of the HOOP mode and the in-plane bending mode, then the C=C stretching frequency has recovered later, which supports the assignment that the ultrafast vibrational dynamics of 200 fs represents the torsion around the C=C bond.

Detail data and discussion of this work can be found in [60].

## 11.9 BR Mutants to Study Effect of Residues Around Chromophore

The proton transfer channel in BR consists of two sub-channels; an extracellular (EC) channel and a cytoplasmic (CP) channel. The EC channel connects the Schiff base of the retinal chromophore with the EC medium. The CP channel connects the Schiff base with the cytoplasm. In BR from an archaeon of Haloquadratum walsbyi (HwBR), aspartic acids 93 and 104 (D93 and D104) play a central role in the proton transfer for the EC channel and the CP channel, respectively. D93 is deprotonated to accept a proton from the Schiff base and D104 works as a proton donor for the Schiff base. We have performed ultrafast TA spectroscopy of three kinds of HwBR for its wild-type, D93N mutant, and D104N mutant to elucidate the effect of D93 and D104 on the ultrafast dynamics of the retinal chromophore.

TA spectroscopy of the three samples of HwBR for its wild-type, D93N mutant, and D104N mutant were performed for femtosecond region (from -350 fs to 1450 fs) and picosecond region (from 1 ps to 14.5 ps) using ultrashort visible laser pulse and the fastscan MLA-based TA spectroscopy system (see Fig. 11.22).

In each of the three samples, TA spectra shows similar spectral characters as follows. Negative TA signal was found in middle spectral region where the stationary absorption spectrum of the samples overlaps, which reflects the photo-bleaching single caused by ground state depletion. In both of the edge spectral region, positive TA signal was observed reflecting the induced absorption in transition from the first electronic excited state to higher electronic excited states. The decay lifetime of the TA signal reflects the lifetime of each intermediate states during the photo-reaction, which were found to be different between the three samples. The TA spectra was fitted by the following tri-exponential function using global fitting method:

$$\begin{aligned}\Delta A(\lambda, t) =& \Delta A_0(\lambda) + \Delta A_1(\lambda)\exp\left(-\frac{t}{\tau_1}\right)\\ &+ \Delta A_2(\lambda)\exp\left(-\frac{t}{\tau_2}\right) + \Delta A_3(\lambda)\exp\left(-\frac{t}{\tau_3}\right)\end{aligned}$$

where $\tau_1 < \tau_2 < \tau_3$. The estimated lifetimes $\tau_i (i = 1, 2, 3)$ are plotted in Fig. 11.23.

The estimated three lifetimes of $\tau_1$, $\tau_2$, $\tau_3$ reflect the transition rate between intermediate states for Hstate → Istate, Istate → Jstate, Jstate → Kstate, respectively. These H, I, J, K states are assigned to the Franck–Condon state, the state at the conical intersection between the electronic ground state and first excited state, the hot vibrational state formed after photoisomerization of retinal chromophores, and a thermalized state, respectively [61].

Compared with wild-type, mutation on D104 did not exhibit difference for the lifetimes of $\tau_1$ and $\tau_2$. It means that the elimination of negative charge on D104 does not affect the dynamics of ultrafast photoisomerization.

The lifetime of $\tau_2$ of the D93N mutant was found to be ~60% longer than that of the wild-type and D104N mutant, implying that the photoisomerization proceeds slower

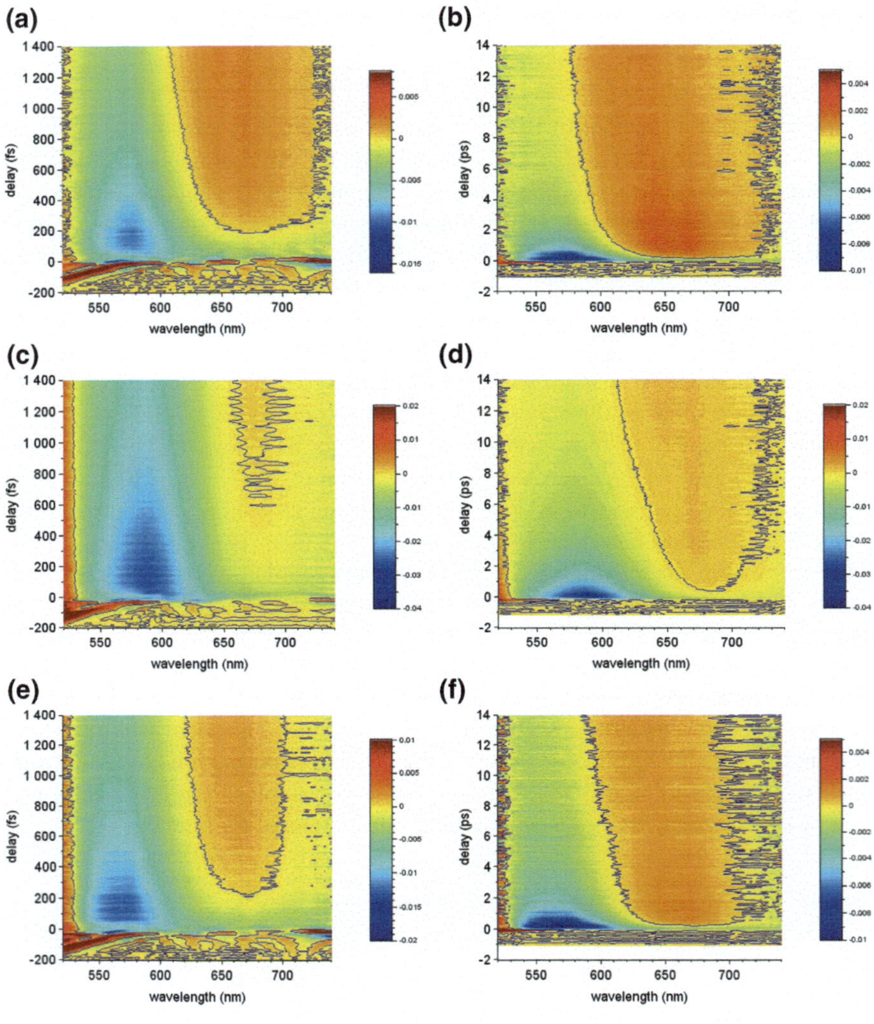


**Fig. 11.22** Two dimensional view of the measured TA spectra scanned in **a**, **c**, **e** femtosecond region and **b**, **d**, **f** picosecond region for the HwBR samples of (**a**, **b**) wild-type, (**c**, **d**) D93N mutant, and (**e**, **f**) D104N mutant. Black curves represent the point where $\Delta A = 0$

by inactivation of D93 (the Schiff base proton acceptor). During the photoisomerization of the retinal chromophore, a spatial electrostatic potential (ESP) is thought to show ultrafast and substantial change resulting in destabilization of hydrogen bonding network (HBN) [62]. The difference found on the D93N indicates that the negative charge from the D93 affects the substantial ultrafast change in ESP associated with photoisomerization process.

Following the ultrafast ESP change, the chromophore cavity HBN is reconstructed in picosecond time scale. The picosecond lifetime of $\tau_3$ was found to be ~50% longer

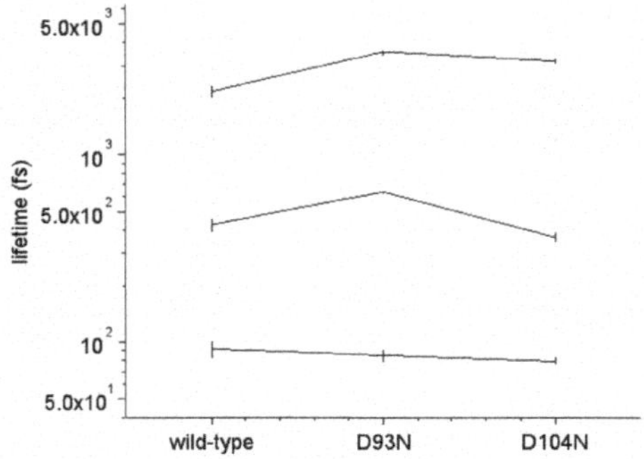


**Fig. 11.23** Three lifetimes of the HwBR samples (wild-type, D94N mutant, D104 mutant) estimated by tri-exponential fit of the TA traces

in the mutants of D93N and D104N indicating that the Shiff base proton donor (D104) assists to reconstruct HBN when the vibrational hot state (J state) is cooled down to be the thermalized state (K state).

These TA spectroscopy of HwBR samples using 10-fs visible laser pulse show periodical modulation on the TA traces reflecting time-dependent molecular vibration frequency (see Fig. 11.24).

The measured TA traces were analyzed by time-gated Fourier transform using the Blackman window function with FWHM of 260 fs. The result is plotted as spectrogram trace in Fig. 11.24. The result shows that the frequency of C=C stretching mode was recovered at around 300 fs in wild-type sample but the recovery takes about 700 fs in D93N mutant. It supports that the conclusion from the electronic dynamics saying that the negative charge from the D93 affects the substantial ultrafast change in ESP associated with photoisomerization process. Note that in D104N visibility of the C=C stretching mode was not enough to discuss its dynamics. Detail discussion can be found in our published work [63].

## 11.10 Ultrafast Photodissociation in Nitric Oxide Synthase

Concentration of nitric oxide (NO) in living organisms controls various important functions such as vasodilation to control blood pressure, neurotransmission to transfer information between synapse, and immune mechanism of macrophage [64, 65]. Nitric oxide synthases (NOS) in mammals biosynthesizes the NO from L-arginine [66]. There are known to be three isoforms of NOS [67] each of which produces NO for blood pressure control in endothelial NOS (eNOS) [18, 68], for neurotransimission in neuronal NOS (nNOS) [69], and for trigger of immune defense in inducible NOS (iNOS) [70], respectively. Each of NOS is homodimeric enzyme which contains heme, Flavin adenine nucleotide (FAD), Flavin mono-nucleotide (FMN), and tetrahydrobiopterin ($BH_4$) [66, 68, 71–73]. In C-terminal half of NOS works as electron donor called reductase domain containing nicotinamide adenine dinucleotide phosphate, FAD, and FMN recognition sites. In N-terminal half of NOS accepts electron

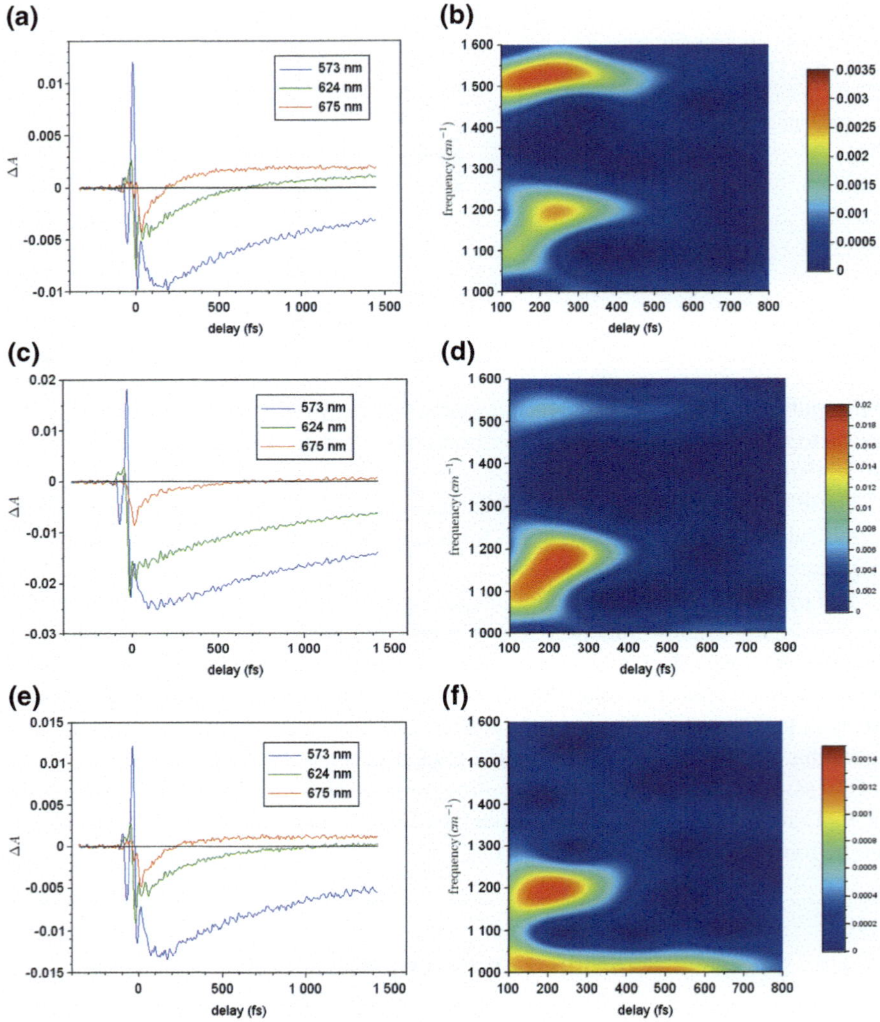


**Fig. 11.24** **a**, **c**, **e** TA traces and **b**, **d**, **f** their spectrogram traces analyzed by time-gated Fourier transform for the HwBR samples of (**a**, **b**) wild-type, (**c**, **d**) D93N mutant, and (**e**, **f**) D104N mutant

called oxygenase domain which contains heme, $BH_4$, and L-argine binding site [74]. The electron transfer from the reductase domain to the oxygenase domain proceeds via a $Ca^{2+}$/calmodulin binding site connecting both of the two domains.

The production of NO in NOS occurs in heme active site of the NOS oxygenase domain (NOS-oxy). Thiolate ligand is in proximal side of heme and the distal heme pocket can bind ligand or substrate. Addition of imidazole and L-arginine to NOS was reported to change spin state of heme as six-coordinate low-spin state and five-coordinate high-spin state, respectively [19]. Without those additive, water molecule is thought to be weakly bound in the distal heme pocket.

In the present work, we have performed ultrafast TA spectroscopy of eNOS-oxy for original one (eNOS-oxy), one with imidazole (eNOS-oxy/Im), and one with L-arginine (eNOS-oxy/L-Arg). Light source is an ultrashort NIR pulse (pulse duration of 35 fs, center wavelength of 800 nm, repetition rate of 5 kHz) generated by a Ti:sapphire regenerative amplifier (Coherent Inc., Legend-USP-HP). The NIR pulse was focused into a 0.5mm-thick BBO crystal to generate second harmonic (SH) pulse at wavelength of 400 nm. Remainder of the NIR pulse was removed passing the laser pulse through a short pass filter transmitting wavelength shorter than 750 nm.

The SH pulse was separated by a beam sampler with power ratio of 9:1 whose higher intensity pulse and lower intensity pulse were used as pump pulse and probe pulse, respectively, in the TA spectroscopy. The measured TA traces for the three samples of eNOS-oxy, eNOS-oxy/Im, and eNOS-oxy/L-Arg are plotted in Fig. 11.25 with fitting curves.

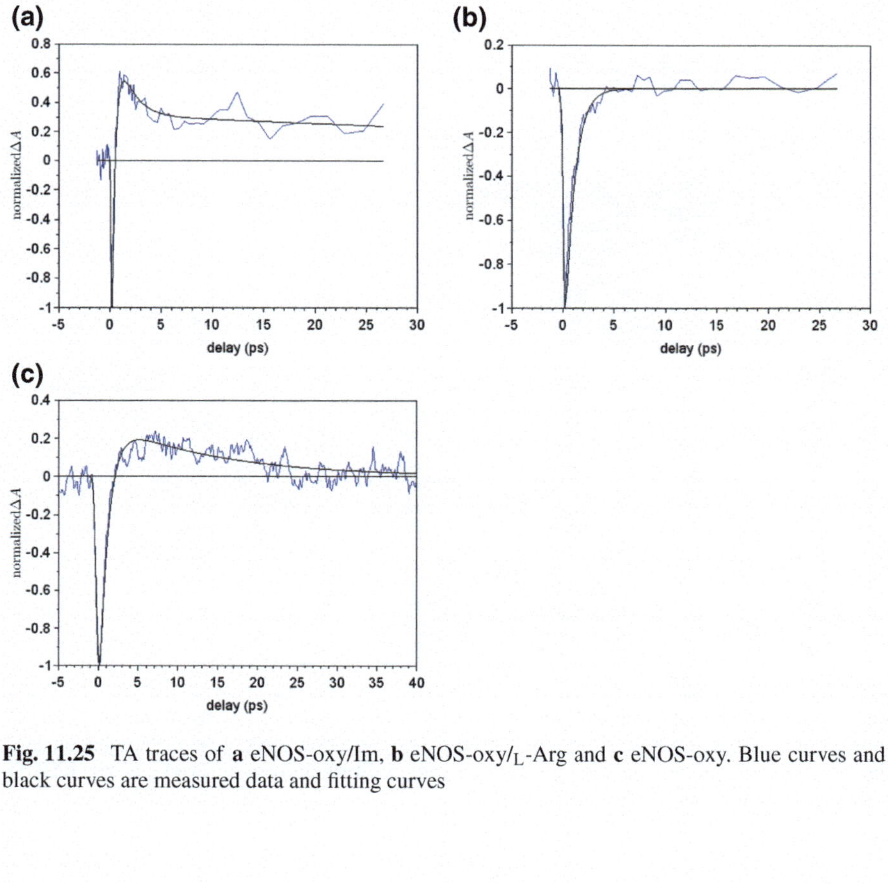


**Fig. 11.25** TA traces of **a** eNOS-oxy/Im, **b** eNOS-oxy/L-Arg and **c** eNOS-oxy. Blue curves and black curves are measured data and fitting curves

The TA trace of eNOS-oxy/Im shows negative signal just after photoexcitation followed by positive signal (see Fig. 11.25a). Adding Imidazole, the heme has six-coordinate structure binding imidazole as a ligand on the distal heme pocket. Thus, the photoexcitation is thought to cause photo-dissociation of imidazole from heme. The ultrafast lifetime of the initial negative signal was estimated to be 0.36 ps which can be assigned to heme-ligand dissociation changing heme structure from six-coordinate to five-coordinate, and the protein conformation change causes protein quake [75–77]. The positive signal is thought to be reflecting induced absorption of the product of five-coordinate heme. The estimated time constant for the heme-ligand dissociation is comparable with the ones previously reported for a heme protein of cytochrome(Cyt)-c in theoretical work [78] and experiments [79–81]. The TA trace was fitted by a tri-exponential function, which shows that the rapid decay in 0.36 ps is followed by relaxation with lifetimes of 1.2 ps and 144 ps. The former one of 1.2 ps reflects vibrational cooling and the latter one of 144 ps was assigned to heme-imidazole recombination and relaxation of the protein quake. Figure 11.26 shows the schematic figure of heme group binding imidazole and its relaxation dynamics estimated here.

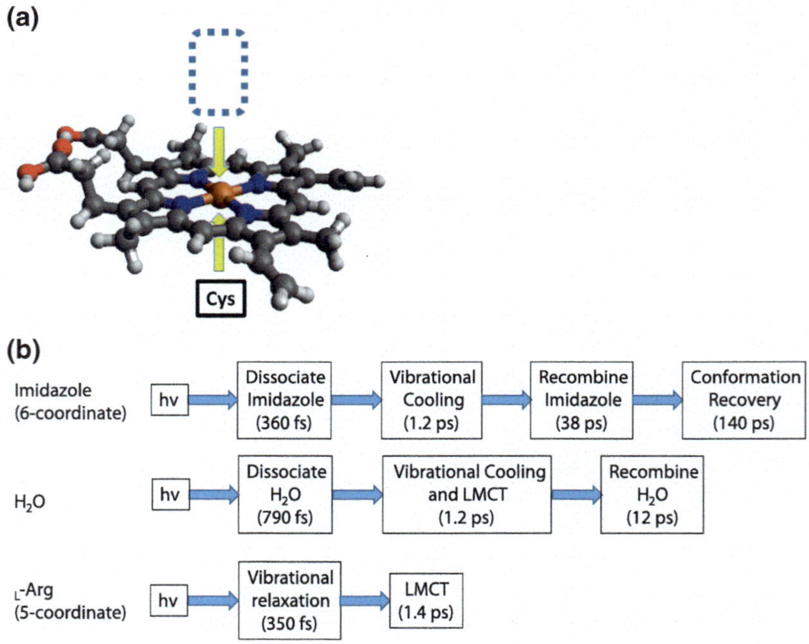


**Fig. 11.26** **a** Schematic figure of heme group in eNOS-oxy. Blue round square indicates the distal binding site of heme where ligand can be bound. Cys represents cysteine bound to the proximal binding site of heme. Dark gray: carbon, light gray: hydrogen, red: oxygen, blue: nitrogen, orange: iron atom. **b** Relaxation dynamics dependent on the additive (Imidazole, $H_2O$, L-Arg). LMCT: ligand (porphyrin ring) to metal charge transfer

In the TA trace of eNOS-oxy/L-Arg (see Fig. 11.25b), negative signal was observed for whole delay region. Fitting the TA trace by a bi-exponential function, decay lifetimes were estimated to be 0.35 ps and 1.4 ps fitting the TA trace by bi-exponential function. The additive of L-Arg does not allow ligand to be bound at the distal heme pocket, thus these two decay lifetimes reflect the relaxation dynamics of five-coordinate heme. The lifetime of 0.35 ps reflects the vibrational relaxation in the Soret band after photo-excitation, which is followed by metal-ring charge transfer proceeding in 1.4 ps [82].

The TA trace of eNOS-oxy shown in Fig. 11.25c shows initial negative signal which gradually changes to positive signal. Tri-exponential function was used to fit the TA trace estimating the lifetimes as 0.79, 4.8, and 12 ps. In the eNOS-oxy, water molecule is weakly bound at the distal heme pocket and its dynamics consists that of six-coordinate heme (seen in eNOS-oxy/Im) and that of five-coordinate heme (seen in eNOS-oxy/L-Arg). Thus, the initial time constant of 0.79 ps reflects the photo-dissociation of water molecule to produce five-coordinate heme. The second time constant of 4.8 ps can be assigned to the relaxation of five-coordinate via vibrational cooling with metal-ring charge transfer. It is comparable with the lifetime reported to be 3 ps in Cyt-c [79] and 6 ps in myoglobin [83]. The last time constant of 12 ps corresponds to rebinding of water molecule, which is comparable with the value of 20 ps reported for heme-based sensor proteins [20, 84, 85].

Detail of this our work (NUV pump-NUV probe TA spectroscopy of eNOS-oxy samples with and without additives) can be found in [86].

After the heme protein is excited into the Soret band of NUV spectral region [20, 79, 82, 87, 88], it decays into Q band and CT band in visible spectral region [69, 89–91]. Stationary absorption spectrum studied in the present work is shown in Fig. 11.27 showing peaks at 405 nm (Soret band), 550 nm (Q band), and 625–675 nm (CT band). The peak around 405 nm was found to be broader than that of five-coordinate heme or six-coordinate heme, which indicates the coexistence of both of two differently coordinated heme in the system [19, 92, 93].

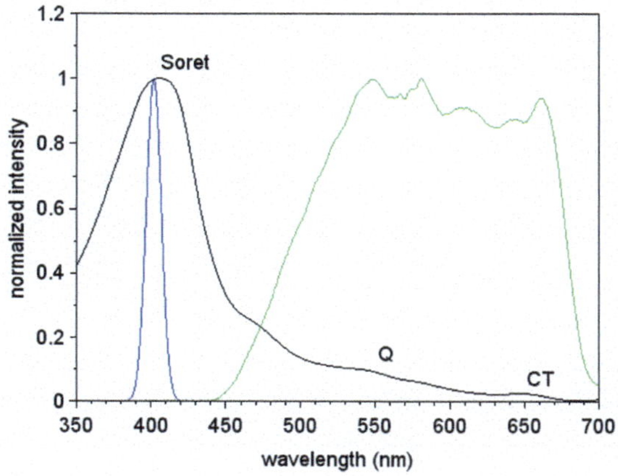


**Fig. 11.27** Stationary absorption spectrum of the eNOS-oxy (black curves), NUV pump laser spectrum (blue curves), and visible broadband probe laser spectrum (green curves). Soret: Soret band, Q: Q band, CT: charge-transfer band

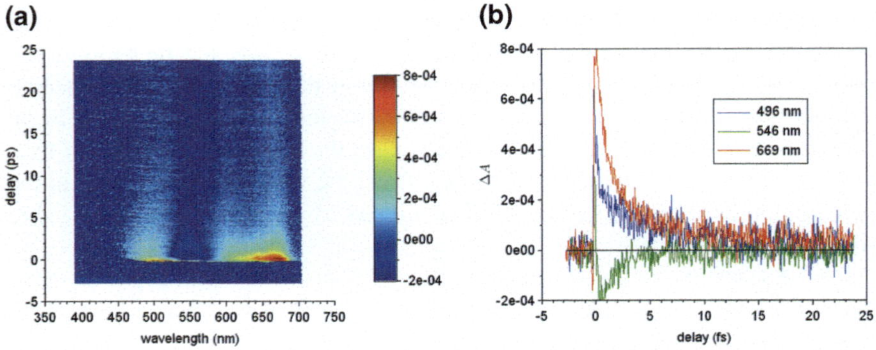


**Fig. 11.28** **a** Two-dimensional view of the TA spectroscopy of eNOS-oxy measured by NUV pump-visible probe measurement. **b** TA trace probed at 496 nm, 546 nm, and 669 nm

To observe the ultrafast dynamics in transition from the Soret band to Q band and CT band, we have performed NUV pump-visible probe TA spectroscopy of eNOS-oxy as follows.

The NUV pump pulse is same as the one used in the NUV pump-NUV probe TA spectroscopy of eNOS-oxy samples. The visible broadband pulse was generated as same as the seed pulse used for NOPA (in which the ultrashort NIR laser pulse from the Ti:sapphire regenerative amplifier was focused into a sapphire plate with thickness of 2 mm). The visible broadband pulse generated by the sapphire plate was used as a probe pulse without being amplified. Note that visible pump-visible probe TA spectroscopy required the visible broadband pulse to be amplified by NOPA because the visible pump pulse requires high intensity to excite the sample.

The measurement result of the TA spectroscopy of eNOS-oxy is shown in Fig. 11.28a.

The measured TA spectra show negative signal in the spectral region of 530–570 nm (M) and positive signal at both sides of 450–520 nm (S) and 580–700 nm (L). The TA time traces for the three spectral regions probed at 496 nm, 546 nm, and 669 nm are plotted in Fig. 11.28b.

The observed TA traces were showing femtosecond and picosecond relaxations. To estimate those relaxation lifetimes and their spectral components, we have analyzed the measured TA traces for all probe wavelengths by global fitting method utilizing a software called Glotaran [94]. Before performing the global fitting, we have applied singular value decomposition (SVD) filter to the measured data to improve its signal to noise ratio. The left singular vector for the fifth singular value only shows contribution from coherent artifact around zero delay. The left singular vector for sixth and larger singular values were randomly modulating reflecting noise. Thus, the signal components corresponding to fifth or larger singular values were removed from the observed TA traces applying the SVD filter to suppress the coherent artifact and the noise.

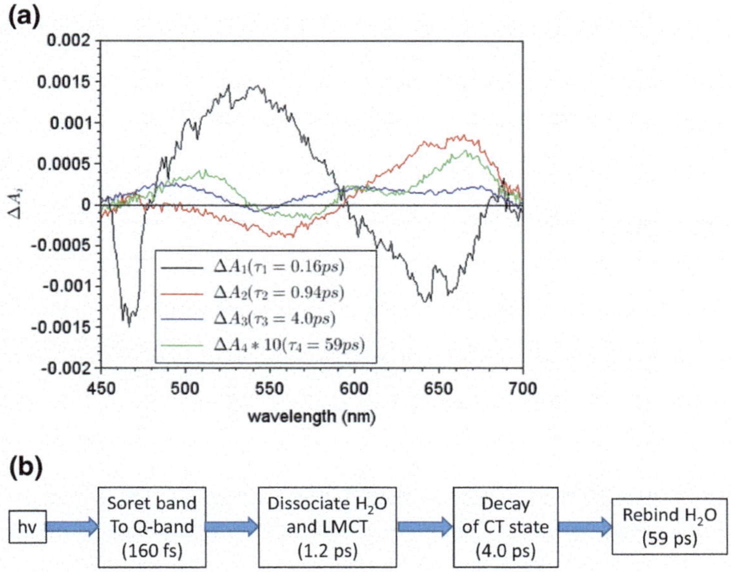


**Fig. 11.29** **a** Decay associated spectra obtained by global fitting analysis of the measured TA traces for the whole probe wavelength region shown in Fig. 11.28a. Amplitude of $\Delta A_4$ was multiplied with ten to be plotted in the figure. **b** Decay dynamics assigned for the estimated lifetimes

The global fitting have estimated four lifetimes of $(\tau_1, \tau_2, \tau_3, \tau_4) = (0.16, 0.94, 4.0, 59ps)$. A spectral component of TA signal corresponding to each lifetime, $\Delta A_i(\lambda)(i = 1, 2, 3, 4)$, is called decay associated spectra (DAS) simultaneously estimated in the global fitting analysis. The calculated DAS are plotted in Fig. 11.29.

The DAS corresponding to $\tau_2$, $\Delta A_2$, has negative signal around 550 nm coinciding with the peak of Q-band. The negative TA signal around 550 nm was also found in $\Delta A_3$ and $\Delta A_4$ reflecting the bleaching of the Q band. Opposite (positive) sign of signal around 550 nm was found in $\Delta A_1$ indicating that the negative TA signal grows in $\tau_1$ and decays in $\tau_2$. It shows that the electronic population excited in Soret band transfers to Q band in $\tau_1$ causing the decrease of absorption (the growth of the negative TA signal) around the absorption peak of Q band.

Considering the result of NUV pump-NUV probe TA spectroscopy of eNOS-oxy samples, the lifetime of $\tau_2$ could be assigned to the photo-dissociation of water molecule at the distal heme pocket. Thus, above assignment of $\tau_1$ (to the transition from Soret band to Q band) implies that the ligand dissociation proceeds in Q band.

Metal-ring charge transfer (CT) in heme protein produces CT state. The lifetime of the CT state was reported to be about 3 ps and 5 ps in oxy-hemoglobin [82] and oxy-myoglobin [16], respectively. The estimated time constant of $\tau_3$ is thought to be corresponding to the lifetime of the CT state. It means that the ligand dissociation occurring in $\tau_2$ proceeds with the metal-ring charge transfer producing the CT state.

The recombination of ligand to the distal heme pocket was reported to take ~40 ps [95]. The estimated lifetime of $\tau_4$ is thought to be corresponding to the ligand rebinding process.

The detail description about this work can be found in [96].

## 11.11 Summary

We have developed ultrashort pulse lasers in visible and ultraviolet spectral region with broadband spectral bandwidth. Performing ultrafast TA spectroscopy in the whole broadband probe spectrum, primary reactions of the protein samples were visualized by observing the TA spectral dynamics reflecting the ultrafast electronic transitions in the reaction.

Using ultrashort laser pulse whose duration is shorter than the molecular vibration period, the observed TA signal shows modulation reflecting the real-time motion of the molecular vibration in time domain. Time-gated Fourier analysis of the signal calculated the instantaneous frequency for the molecular vibration modes to elucidate molecular structure change during the photo-reaction.

To observe fine signal modulation caused by the molecular vibration, it was necessary to suppress signal fluctuation affected by environment around the light source. Irradiation of the ultrashort laser pulse required for the ultrafast spectroscopy gives serious damage accumulation especially in the protein samples, which degrades signal quality of the transient absorption. Both of the problems (signal fluctuation by environment and damage accumulation on sample) were solved by developing fastscan ultrafast spectroscopy system.

In the study of the light-driven proton pump of BR and its mutants, we could elucidate how the ultrafast photo-isomerization dynamics of the chromophore is affected by the amino acid residues around the chromophore. The ultrafast TA spectroscopy of NOS with additives have clarified the reaction dynamics dependent on coordination condition of heme.

We acknowledge that this work was supported by JPSP KAKENHI under Grant No. 23K03349 and by National Science and Technology Council (NSTC), R.O.C. under Grant No. 111–2811-M-A49 -511.

## References

1. L. Salem, P. Bruckmann, Conversion of a photon to an electrical signal by sudden polarisation in the N-retinylidene visual chromophore. Nature **258**, 526–528 (1975). https://doi.org/10.1038/258526a0
2. A. Warshel, Bicycle-pedal model for the first step in the vision process. Nature **260**, 679–683 (1976). https://doi.org/10.1038/260679a0

3. H. Chosrowjan, N. Mataga, Y. Shibata, S. Tachibanaki, H. Kandori, Y. Shichida, T. Okada, T. Kouyama, Rhodopsin emission in real time: a new aspect of the primary event in vision. J. Am. Chem. Soc. **120**, 9706–9707 (1998). https://doi.org/10.1021/JA981659W
4. W.T. Pollard, C.H.B. Cruz, C.V. Shank, R.A. Mathies, Direct observation of the excited-state cis–trans photoisomerization of bacteriorhodopsin: multilevel line shape theory for femtosecond dynamic hole burning and its application. J. Chem. Phys. **90**, 199 (1989). https://doi.org/10.1063/1.456658
5. T. Kobayashi, M. Kim, M. Taiji, T. Iwasa, M. Nakagawa, M. Tsuda, Femtosecond spectroscopy of halorhodopsin and rhodopsin in a broad spectral range of 400–1000 nm. J. Phys. Chem. B. 102 (1998)
6. R. González-Luque, M. Garavelli, F. Bernardi, M. Merchán, M. a Robb, M. Olivucci, Computational evidence in favor of a two-state, two-mode model of the retinal chromophore photoisomerization. Proc. Natl. Acad. Sci. USA **97**, 9379–9384 (2000). https://doi.org/10.1073/pnas.97.17.9379
7. M.G. Müller, I. Lindner, I. Martin, W. Gärtner, A.R. Holzwarth, Femtosecond kinetics of photoconversion of the higher plant photoreceptor phytochrome carrying native and modified chromophores. Biophys. J. **94**, 4370–4382 (2008). https://doi.org/10.1529/biophysj.106.091652
8. S.B. Hendricks, H.A. Borthwick, The function of phytochrome in regulation of plant growth. Proc. Natl. Acad. Sci. USA **58**, 2125–2130 (1967). https://doi.org/10.1073/pnas.58.5.2125
9. M. Furuya, Phytochromes: their molecular species, gene families, and functions. Annu. Rev. Plant. Physiol. Plant. Mol. Biol. **44**, 617–645 (1993). https://doi.org/10.1146/annurev.pp.44.060193.003153
10. J. Chory, M. Chatterjee, R.K. Cook, T. Elich, C. Fankhauser, J. Li, P. Nagpal, M. Neff, A. Pepper, D. Poole, J. Reed, V. Vitart, From seed germination to flowering, light controls plant development via the pigment phytochrome. Proc. Natl. Acad. Sci. USA **93**, 12066–12071 (1996). https://doi.org/10.1073/pnas.93.22.12066
11. M. Chalfie, Y. Tu, G. Euskirchen, W.W. Ward, D.C. Prasher, Green fluorescent protein as a marker for gene expression. Science (80), 263, 802–805 (1994). https://doi.org/10.1126/science.8303295
12. H. Morise, O. Shimomura, F.H. Johnson, J. Winant, Intermolecular energy transfer in the bioluminescent system of aequorea. Biochemistry **13**, 2656–2662 (1974). https://doi.org/10.1021/bi00709a028
13. O. Shimomura, Bioluminescence in the sea: photoprotein systems (1985). http://www.ncbi.nlm.nih.gov/pubmed/2871634. Accessed October 8, 2019
14. A. Miyawaki, J. Llopis, R. Heim, J. Michael McCaffery, J.A. Adams, M. Ikura, R.Y. Tsien, Fluorescent indicators for Ca2+ based on green fluorescent proteins and calmodulin. Nature **388**, 882–887 (1997). https://doi.org/10.1038/42264
15. J.W. Petrich, C. Poyart, J.L. Martin, Photophysics and reactivity of heme proteins: a femtosecond absorption study of hemoglobin, myoglobin, and protoheme. Biochemistry **27**, 4049–4060 (1988). https://doi.org/10.1021/bi00411a022
16. S. Ishizaka, T. Wada, N. Kitamura, Femtosecond transient absorption study on relaxation intermediates in oxymyoglobin. Photochem. Photobiol. Sci. **4**, 10–16 (1999). https://doi.org/10.1016/j.chemphys.2011.04.003
17. O. Kajimoto, K. Honma, T. Kobayashi, Formation of electronically excited NO and energy partitioning in the 193-nm photolysis of the NO dimer. J. Phys. Chem. **89** (1985)
18. P.F. Chen, A.L. Tsai, V. Berka, K.K. Wu, Endothelial nitric-oxide synthase: evidence for bidomain structure and successful reconstitution of catalytic activity from two separate domains generated by a baculovirus expression system. J. Biol. Chem. **271**, 14631–14635 (1996). https://doi.org/10.1074/jbc.271.24.14631
19. P.F. Chen, A.L. Tsai, V. Berka, K.K. Wu, Mutation of Glu-361 in human endothelial nitric-oxide synthase selectively abolishes L-arginine binding without perturbing the behavior of heme and other redox centers. J. Biol. Chem. **272**, 6114–6118 (1997). https://doi.org/10.1074/jbc.272.10.6114

20. S.M. Kapetanaki, S.J. Field, R.J.L. Hughes, N.J. Watmough, U. Liebl, M.H. Vos, Ultrafast ligand binding dynamics in the active site of native bacterial nitric oxide reductase. Biochim. Biophys. Acta Bioenerg. **1777**, 919–924 (2008). https://doi.org/10.1016/j.bbabio.2008.03.012
21. F. Pellegrino, Ultrafast energy transfer processes in photosynthetic systems probed by picosecond fluorescence spectroscopy. Opt. Eng. **22**, 225508 (1983). https://doi.org/10.1117/12.7973190
22. Y. Inoue, T. Kobayashi, T. Ogawa, K. Shibata, A short lived intermediate in the photoconversion of protochlorophyllide to chlorophyllide a, Plant Cell Physiol. **22** (1981)
23. B. Panneton, S. Guillaume, G. Samson, J.-M. Roger, Discrimination of corn from monocotyledonous weeds with ultraviolet (UV) induced fluorescence. Appl. Spectrosc. **65**, 10–19 (2011). https://doi.org/10.1366/10-06100
24. R.G.W. Norrish, G. Porter, Chemical reactions produced by very high light intensities. Nature. **164**, 658 (1949). https://doi.org/10.1038/164658a0
25. T.H. Maiman, Stimulated optical radiation in ruby. Nature **187**, 493–494 (1960). https://doi.org/10.1038/187493a0
26. W.E. Lamb, Theory of an optical maser. Phys. Rev. **134** (1964). https://doi.org/10.1103/PhysRev.134.A1429
27. L.E. Hargrove, R.L. Fork, M.A. Pollack, Locking of he–ne laser modes induced by synchronous intracavity modulation. Appl. Phys. Lett. **5**, 4–5 (1964). https://doi.org/10.1063/1.1754025
28. R.L. Fork, B.I. Greene, C.V. Shank, Generation of optical pulses shorter than 0.1 psec by colliding pulse mode locking. Appl. Phys. Lett. **38**, 671–672 (1981). https://doi.org/10.1063/1.92500
29. P. Moulton, Ti-doped sapphire: tunable solid-state laser. Opt. News. **8**, 9 (1982). https://doi.org/10.1364/ON.8.6.000009
30. D.E. Spence, P.N. Kean, W. Sibbett, 60-fsec pulse generation from a self-mode-locked Ti:sapphire laser. Opt. Lett. **16**, 42 (1991). https://doi.org/10.1364/OL.16.000042
31. W.C. Michels, N.L. Curtis, A pentode lock-in amplifier of high frequency selectivity. Rev. Sci. Instrum. **12**, 444–447 (1941). https://doi.org/10.1063/1.1769919
32. C.R. Cosens, A balance-detector for alternating-current bridges. Proc. Phys. Soc. **46**, 818–823 (1934). https://doi.org/10.1088/0959-5309/46/6/310
33. C.A. Stutt, Low-frequency spectrum of lock-in amplifiers research laboratory of electronics massachusetts institute of technology (1949)
34. M.J. Rosker, F.W. Wise, C.L. Tang, Femtosecond relaxation dynamics of large molecules. Phys. Rev. Lett. **57**, 321–324 (1986). https://doi.org/10.1103/PhysRevLett.57.321
35. H.L. Fragnito, J.Y. Bigot, P.C. Becker, C.V. Shank, Evolution of the vibronic absorption spectrum in a molecule following impulsive excitation with a 6 fs optical pulse. Chem. Phys. Lett. **160**, 101–104 (1989). https://doi.org/10.1016/0009-2614(89)87564-5
36. I. Shafi, J. Ahmad, S.I. Shah, F.M. Kashif, Techniques to obtain good resolution and concentrated time-frequency distributions: a review. EURASIP J. Adv. Signal Process. **2009**, 673539 (2009). https://doi.org/10.1155/2009/673539
37. L. Cohen, *Time-Frequency Analysis*. Prentice Hall PTR (1995)
38. A.M. Dziewonski, S. Bloch, M. Landisman, A technique for the analysis of transient seismic signals (1969)
39. A.V. Oppenheim, Speech spectrograms using the fast Fourier transform: Increased flexibility and the capability for on-line analysis are the two primary reasons for utilizing a digital computer for the generation and display of speech spectrograms. IEEE Spectr. **7**, 57–62 (1970). https://doi.org/10.1109/MSPEC.1970.5213512
40. Y.R. Shen, G.-Z. Yang, Theory of self-phase modulation and spectral broadening, in *The Supercontinuum Laser Source*, Springer New York, pp. 1–32 (1989). https://doi.org/10.1007/978-1-4757-2070-9_1
41. K. Kato, Second-Harmonic generation to 2048 Å in β-BaB<inf>2</inf>O<inf>4</inf>. IEEE J. Quant. Electron. **22**, 1013–1014 (1986). https://doi.org/10.1109/JQE.1986.1073097
42. T. Wilhelm, J. Piel, E. Riedle, Sub-20-fs pulses tunable across the visible from a blue-pumped single-pass noncollinear parametric converter. Opt. Lett. **22**, 1494–1496 (1997). https://doi.org/10.1364/OL.22.001494

43. G. Cerullo, M. Nisoli, S. Stagira, S. De Silvestri, Sub-8-fs pulses from an ultrabroadband optical parametric amplifier in the visible. Opt. Lett. **23**, 1283–1285 (1998). https://doi.org/10.1364/OL.23.001283
44. T. Kobayashi, A. Shirakawa, Tunable visible and near-infrared pulse generator in a 5 fs regime. Appl. Phys. B Lasers Opt. 70 (2000). https://doi.org/10.1007/s003400000325
45. A. Yabushita, Y.-H. Lee, T. Kobayashi, Development of a multiplex fast-scan system for ultrafast time-resolved spectroscopy. Rev. Sci. Instrum. **81**, 063110 (2010). https://doi.org/10.1063/1.3455809
46. A. Yabushita, C.-H.C.-H. Kao, Y.-H. Lee, T. Kobayashi, Development and demonstration of table-top synchronized fast-scan femtosecond time-resolved spectroscopy system by single-shot scan photo detector array. Jpn. J. Appl. Phys. **54**, 072401 (2015). https://doi.org/10.7567/JJAP.54.072401
47. R.H. Lozier, R.A. Bogomolni, W. Stoeckenius, Bacteriorhodopsin: a light-driven proton pump in Halobacterium Halobium. Biophys. J. **15**, 955–962 (1975). https://doi.org/10.1016/S0006-3495(75)85875-9
48. D. Oesterhelt, W. Stoeckenius, Isolation of the cell membrane of Halobacterium halobium and its fractionation into red and purple membrane. Methods Enzymol. **31**, 667–678 (1974). https://doi.org/10.1016/0076-6879(74)31072-5
49. J.A. Stuart, D.L. Marcy, K.J. Wise, R.R. Birge, Volumetric optical memory based on bacteriorhodopsin, in: Synth. Met., pp. 3–15 (2002). https://doi.org/10.1016/S0379-6779(01)00586-0
50. Y. Huang, S.-T. Wu, Y. Zhao, All-optical switching characteristics in bacteriorhodopsin and its applications in integrated optics. Opt. Express. **12**, 895–906 (2004). http://www.ncbi.nlm.nih.gov/pubmed/19474900
51. W. Humphrey, H. Lu, I. Logunov, H.J. Werner, K. Schulten, Three electronic state model of the primary phototransformation of bacteriorhodopsin. Biophys. J. **75**, 1689–1699 (1998). https://doi.org/10.1016/S0006-3495(98)77611-8
52. S. Schenkl, F. van Mourik, G. van der Zwan, S. Haacke, M. Chergui, Probing the ultrafast charge translocation of Photoexcited retinal in bacteriorhodopsin. Science (80), 309, 917–920 (2005). https://doi.org/10.1126/science.1111482
53. J. Herbst, K. Heyne, R. Diller, Femtosecond infrared spectroscopy of bacteriorhodopsin chromophore isomerization. Science **297**, 822–825 (2002). https://doi.org/10.1126/science.1072144
54. G. Haran, K. Wynne, A. Xie, Q. He, M. Chance, R.M. Hochstrasser, Excited state dynamics of bacteriorhodopsin revealed by transient stimulated emission spectra. Chem. Phys. Lett. **261**, 389–395 (1996). https://doi.org/10.1016/0009-2614(96)01017-2
55. F. Gai, K.C. Hasson, J.C. McDonald, P. a Anfinrud, Chemical dynamics in proteins: the photoisomerization of retinal in bacteriorhodopsin. Science. **279**, 1886–1891 (1998). https://doi.org/10.1126/science.279.5358.1886
56. Li Song, M.A. El-Sayed, Primary step in bacteriorhodopsin photosynthesis: bond stretch rather than angle twist of its retinal excited-state structure. J. Am. Chem. Soc. **120**, 8889–8890 (1998). https://doi.org/10.1021/JA980390D
57. M. Du, G.R. Fleming, Femtosecond time-resolved fluorescence spectroscopy of bacteriorhodopsin: direct observation of excited state dynamics in the primary step of the proton pump cycle. Biophys. Chem. **48**, 101–111 (1993). https://doi.org/10.1016/0301-4622(93)85002-Y
58. A.C. Terentis, Y. Zhou, G.H. Atkinson, L. Ujj, Picosecond time-resolved coherent anti-stokes raman spectroscopy of the artificial bacteriorhodopsin pigment. BR6.11 †, J. Phys. Chem. A. **107**, 10787–10797 (2003). https://doi.org/10.1021/jp030612g
59. A. Kahan, O. Nahmias, N. Friedman, M. Sheves, S. Ruhman, Following photoinduced dynamics in bacteriorhodopsin with 7-fs impulsive vibrational spectroscopy. J. Am. Chem. Soc. **129**, 537–546 (2007). https://doi.org/10.1021/ja064910d
60. A. Yabushita, T. Kobayashi, Primary conformation change in bacteriorhodopsin on photoexcitation. Biophys. J. **96**, 1447–1461 (2009). https://doi.org/10.1016/j.bpj.2008.10.050

61. H. Abramczyk, Femtosecond primary events in bacteriorhodopsin and its retinal modified analogs: Revision of commonly accepted interpretation of electronic spectra of transient intermediates in the bacteriorhodopsin photocycle. J. Chem. Phys. **120**, 11120–11132 (2004). https://doi.org/10.1063/1.1737731
62. F. Melaccio, N. Calimet, I. Schapiro, A. Valentini, M. Cecchini, M. Olivucci, Space and time evolution of the electrostatic potential during the activation of a visual pigment. J. Phys. Chem. Lett., 2563–2567 (2016). https://doi.org/10.1021/acs.jpclett.6b00977
63. C.-C. Hung, X.-R. Chen, Y.-K. Ko, T. Kobayashi, C.-S. Yang, A. Yabushita, Schiff Base proton acceptor assists Photoisomerization of retinal chromophores in bacteriorhodopsin. Biophys. J. 112 (2017). https://doi.org/10.1016/j.bpj.2017.05.015
64. S. Moncada, E.A. Higgs, The discovery of nitric oxide and its role in vascular biology. Br. J. Pharmacol. **147**, S193–S201 (2009). https://doi.org/10.1038/sj.bjp.0706458
65. H.H.H.W. Schmidt, U. Walter, NO at work. Cell **78**, 919–925 (1994). https://doi.org/10.1016/0092-8674(94)90267-4
66. D.J. Stuehr, N.S. Kwon, C.F. Nathan, O.W. Griffith, P.L. Feldman, J. Wiseman, N(ω)-hydroxy-L-arginine is an intermediate in the biosynthesis of nitric oxide from L-arginine. J. Biol. Chem. **266**, 6259–6263 (1991)
67. L.J. Ignarro, Nitric oxide : biology and pathobiology. Academic (2010)
68. J.S. Pollock, U. Förstermann, J.A. Mitchell, T.D. Warner, H.H.H.W. Schmidt, M. Nakane, F. Murad, Purification and characterization of particulate endothelium-derived relaxing factor synthase from cultured and native bovine aortic endothelial cells. Proc. Natl. Acad. Sci. USA **88**, 10480–10484 (1991). https://doi.org/10.1073/pnas.88.23.10480
69. I. Mikula, S. Durocher, P. Martasek, B. Mutus, A. Slama-Schwok, Isoform-specific differences in the nitrite reductase activity of nitric oxide synthases under hypoxia. Biochem. J. **418**, 673–682 (2009). https://doi.org/10.1042/BJ20080987
70. F. Terenzi, M.J.M. Diaz-Guerra, M. Casado, S. Hortelano, S. Leoni, L. Bosca, Bacterial lipopeptides induce nitric oxide synthase and promote apoptosis through nitric oxide-independent pathways in rat macrophages. J. Biol. Chem. **270**, 6017–6021 (1995). https://doi.org/10.1074/jbc.270.11.6017
71. J.M. Hevel, K.A. White, M.A. Marletta, Purification of the inducible murine macrophage nitric oxide synthase: identification as a flavoprotein. J. Biol. Chem. **266**, 22789–22791 (1991)
72. K. Mcmillan, D.S. Bredt, D.J. Hirsch, S.H. Snyder, J.E. Clark, B.S.S. Masters, Cloned, expressed rat cerebellar nitric oxide synthase contains stoichiometric amounts of heme, which binds carbon monoxide. Proc. Natl. Acad. Sci. USA **89**, 11141–11145 (1992). https://doi.org/10.1073/pnas.89.23.11141
73. D.J. Stuehr, M. Ikeda-Saito, Spectral characterization of brain and macrophage nitric oxide synthases. Cytochrome P-450-like hemeproteins that contain a flavin semiquinone radical. J. Biol. Chem. **267**, 20547–20550 (1992)
74. J.S. Nishimura, P. Martasek, K. McMillan, J.C. Salerno, Q. Liu, S.S. Gross, B.S.S. Masters, Modular structure of neuronal nitric oxide synthase: Localization of the arginine binding site and modulation by pterin. Biochem. Biophys. Res. Commun. **210**, 288–294 (1995). https://doi.org/10.1006/bbrc.1995.1659
75. A. Ansari, J. Berendzen, S.F. Bowne, H. Frauenfelder, I.E. Iben, T.B. Sauke, E. Shyamsunder, R.D. Young, Protein states and proteinquakes. Proc. Natl. Acad. Sci. USA **82**, 5000–5004 (1985). https://doi.org/10.1073/pnas.82.15.5000
76. G. Dadusc, J.P. Ogilvie, P. Schulenberg, U. Marvet, R.J.D. Miller, Diffractive optics-based heterodyne-detected four-wave mixing signals of protein motion: From "protein quakes" to ligand escape for myoglobin. Proc. Natl. Acad. Sci. USA **98**, 6110–6115 (2001). https://doi.org/10.1073/pnas.101130298
77. M.H. Vos, Ultrafast dynamics of ligands within heme proteins. Biochim. Biophys. Acta Bioenerg. **1777**, 15–31 (2008). https://doi.org/10.1016/j.bbabio.2007.10.004
78. H.M. Marques, K.L. Brown, Molecular mechanics and molecular dynamics simulations of porphyrins, metalloporphyrins, heme proteins and cobalt corrinoids. Coord. Chem. Rev. **225**, 123–158 (2002). https://doi.org/10.1016/S0010-8545(01)00411-8

79. M. Negrerie, S. Cianetti, M.H. Vos, M. Jean-Louis, S.G. Kruglik, Ultrafast heme dynamics in ferrous versus ferric cytochrome c studied by time-resolved resonance Raman and transient absorption spectroscopy. J. Phys. Chem. B **110**, 12766–12781 (2006). https://doi.org/10.1021/jp0559377
80. W. Wang, X. Ye, A.A. Demidov, F. Rosca, T. Sjodin, W. Cao, M. Sheeran, P.M. Champion, Femtosecond multicolor pump–probe spectroscopy of ferrous cytochrome c. J. Phys. Chem. B **104**, 10789–10801 (2000). https://doi.org/10.1021/jp0008602
81. C. Zang, J.A. Stevens, J.J. Link, L. Guo, L. Wang, D. Zhong, Ultrafast Proteinquake dynamics in cytochrome *c*. J. Am. Chem. Soc. **131**, 2846–2852 (2009). https://doi.org/10.1021/ja8057293
82. S. Franzen, L. Kiger, C. Poyart, J.L. Martin, Heme photolysis occurs by ultrafast excited state metal-to-ring charge transfer. Biophys. J. **80**, 2372–2385 (2001). https://doi.org/10.1016/S0006-3495(01)76207-8
83. M. Lim, T.A. Jackson, P.A. Anfinrud, Femtosecond near-IR absorbance study of photoexcited myoglobin: dynamics of electronic and thermal relaxation. J. Phys. Chem. **100**, 12043–12051 (1996). https://doi.org/10.1021/jp9536458
84. H.J. Park, C. Suquet, J.D. Satterlee, C.H. Kang, Insights into signal transduction involving PAS domain oxygen-sensing Heme Proteins from the X-ray crystal structure of Escherichia coli dos Heme Domain (Ec DosH). Biochemistry **43**, 2738–2746 (2004). https://doi.org/10.1021/bi035980p
85. H. Kurokawa, D.S. Lee, M. Watanabe, I. Sagami, B. Mikami, C.S. Raman, T. Shimizu, A redox-controlled molecular switch revealed by the crystal structure of a bacterial Heme PAS sensor. J. Biol. Chem. **279**, 20186–20193 (2004). https://doi.org/10.1074/jbc.M314199200
86. C.C.C.-C. Hung, A. Yabushita, T. Kobayashi, P.-F.P.F. Chen, K.S.K.S. Liang, Ultrafast dynamics of ligand and substrate interaction in endothelial nitric oxide synthase under Soret excitation. Biophys. Chem. **214–215**, 11–16 (2016). https://doi.org/10.1016/j.bpc.2016.05.001
87. H.S. Eom, S.C. Jeoung, D. Kim, J.H. Ha, Y.R. Kim, Ultrafast vibrational relaxation and ligand photodissociation/photoassociation processes of nickel(II) porphyrins in the condensed phase. J. Phys. Chem. A **101**, 3661–3669 (1997). https://doi.org/10.1021/jp962374d
88. M.H. Vos, A. Battistoni, C. Lechauve, M.C. Marden, L. Kiger, A. Desbois, E. Pilet, E. De Rosny, U. Liebl, Ultrafast heme-residue bond formation in six-coordinate heme proteins: implications for functional ligand exchange. Biochemistry **47**, 5718–5723 (2008). https://doi.org/10.1021/bi800288z
89. H.S. Cho, N.W. Song, Y.H. Kim, S.C. Jeoung, S. Hahn, D. Kim, S.K. Kim, N. Yoshida, A. Osuka, Ultrafast energy relaxation dynamics of directly linked porphyrin arrays. J. Phys. Chem. A **104**, 3287–3298 (2000). https://doi.org/10.1021/jp9942623
90. A. Yabushita, T. Kobayashi, Ultrafast spectroscopy of Oxyhemoglobin during Photodissociation. J. Phys. Chem. B **114**, 11654–11658 (2010). https://doi.org/10.1021/jp103593q
91. C. Consani, G. Auböck, O. Bräm, F. Van Mourik, M. Chergui, A cascade through spin states in the ultrafast haem relaxation of met-myoglobin. J. Chem. Phys. 140 (2014). https://doi.org/10.1063/1.4861467
92. F.J.M. Chartier, M. Couture, Stability of the heme environment of the nitric oxide synthase from Staphylococcus aureus in the absence of pterin cofactor. Biophys. J. **87**, 1939–1950 (2004). https://doi.org/10.1529/biophysj.104.042119
93. R. Huszánk, G. Lendvay, O. Horváth, Air-stable, heme-like water-soluble iron(II) porphyrin: In situ preparation and characterization. J. Biol. Inorg. Chem. **12**, 681–690 (2007). https://doi.org/10.1007/s00775-007-0217-y
94. J. Snellenburg, S. Laptenok, R. Seger, K. Mullen, I. Van Stokkum, Glotaran: A Java-based graphical user interface for the R package TIMP. J. Stat. Softw. 49 (2012). https://doi.org/10.18637/jss.v049.i03
95. M. Négrerie, V. Berka, M.H. Vos, U. Liebl, J.C. Lambry, A.L. Tsai, J.L. Martin, Geminate recombination of nitric oxide to endothelial nitric-oxide synthase and mechanistic implications. J. Biol. Chem. **274**, 24694–24702 (1999). https://doi.org/10.1074/jbc.274.35.24694
96. C.-C. Hung, A. Yabushita, T. Kobayashi, P.-F. Chen, K.S. Liang, Ultrafast relaxation dynamics of nitric oxide synthase studied by visible broadband transient absorption spectroscopy. Chem. Phys. Lett. 683 (2017). https://doi.org/10.1016/j.cplett.2017.04.081

# Chapter 12
# Time-Resolved Raman Mapping of Energy Flow in Proteins

**Yasuhisa Mizutani, Satoshi Yamashita, and Misao Mizuno**

**Abstract** We have summarized our work on time-resolved Raman mapping of protein energy flow. It is not yet clear how energy migrates through proteins. Anti-Stokes ultraviolet resonance Raman (UVRR) spectroscopy has been used to develop techniques to address the characteristics of energy flow. One of the key advantages of UVRR spectroscopy is its high sensitivity to aromatic side chains due to resonance Raman enhancement. This enhancement allows you to observe specific sites on large protein molecules at the level of a single amino acid residue. In addition, anti-Stokes intensity is a selective probe of vibrationally excited populations. These advantages make time-resolved anti-Stokes UVRR spectroscopy ideal for studying the vibrational energy flow of proteins. Our studies on heme proteins demonstrated that the major channel of the excess energy transfer is not through covalent bonds of the main chain but through van der Waals atomic contacts between heme and the probe residue. It was also shown that anti-Stokes spectra of tryptophan residues serve an excellent spectroscopic "thermometer" in terms of high sensitivity and straightforward interpretation.

**Keywords** Resonance Raman spectroscopy · Anti-Stokes intensity · Vibrational energy transfer · Heme proteins

## 12.1 Introduction

Energy is the origin of many fascinating dynamics in nature. The energy flow in a condensed phase is fundamentally crucial for chemical dynamics [1–6]. It is very challenging to provide a microscopic picture of energy dissipation on the spatial scale of atoms. At the atomic scale, approximations of continuum media under Fourier's law describing macroscopic heat dissipation [7] are not applicable.

---

Y. Mizutani (✉) · S. Yamashita · M. Mizuno
Department of Chemistry, Graduate School of Science, Osaka University, 1-1 Machikaneyama, Toyonaka 560-0043, Osaka, Japan
e-mail: mztn@chem.sci.osaka-u.ac.jp



K. Ueda (ed.), *Ultrafast Electronic and Structural Dynamics*,
https://doi.org/10.1007/978-981-97-2914-2_12

Temporally resolved mapping of energy flow on the atomic scale is a prerequisite for understanding the microscopic mechanism of energy flow. To elucidate the energy flow pathways in a condensed phase on an atomic space scale, we need an energy source molecule (*heater* molecule) and an energy probe molecule (*probe* molecule), with a well-defined intermolecular distance and relative orientation. However, in solution, it is difficult to control the distance and relative orientation between a pair of molecules. We have solved this problem by using heme proteins as the observation system.

Heme proteins are ideal molecular systems to study energy flow because a photoexcited heme group is converted to an electronically ground state via ultrafast internal conversion [8], and thus excess vibrational energy as high as 25,000 $cm^{-1}$ can be deposited locally at the heme site in proteins immediately after the photoexcitation via the Soret transition. In consequence, heme acts as a very efficient energy convertor. In fact, we showed that the excess energy is released in 1–2 ps from the heme site [9–12], demonstrating that heme acts as a heater molecule.

To observe subsequent energy relaxation processes, we adopted anti-Stokes ultraviolet resonance Raman (UVRR) spectroscopy [13–16]. Because of the resonance effect, UVRR spectroscopy probes Raman bands of aromatic amino acid residues with high selectivity [17–21], allowing site-selective detection of energy at the level of a single amino acid residue in a large protein molecule. Besides, time-resolved anti-Stokes Raman spectroscopy is selective for vibrationally excited populations and is suitable for studying vibrational energy.

In heme proteins, the distance and relative orientation between heme and amino acid residues are well defined in their stable structures based on X-ray crystallographic data. Thus, the protein moiety of a heme protein is a "quasi-solvent," with a well-characterized three-dimensional structure, where the distance between heme and the amino acid residues can be as long as 20 Å. In addition, the structure can be modified by site-directed mutagenesis. Therefore, studies using our methodology based on heme proteins can provide new insights for understanding the mechanism of vibrational energy transfer in condensed phases.

In this chapter, we first described our time-resolved UVRR apparatus with time resolution of a picosecond and wide tunabilities both of pump and probe pulses. Usefulness of the apparatus to time-resolved anti-Stokes UVRR spectroscopy was demonstrated by our proof-of-principle experiments. Then, the enegy flow in heme proteins were discussed in terms of anisotropy and transfer mechanism. We also showed that anti-Stokes spectra of tryptophan residues serve an excellent spectroscopic "thermometer" in terms of high sensitivity and straightforward interpretation.

## 12.2 Time-Resolved UVRR Apparatus

After successful developments of the time-resolved visible resonance Raman spectrometer [22, 23], we extended the wavelength region of the probe pulse to the UV and far UV regions by constructing an apparatus consisting of a widely tunable light source in the UV and far UV regions using a 1-kHz picosecond Ti:sapphire laser/regenerative amplifier system [24–27]. Details of the time-resolved UVRR apparatus was described elsewhere [26]. Figure 12.1a shows a schematic of the time-resolved UVRR measurement apparatus in our laboratory. A Ti:sapphire oscillator (Tsunami pumped by Millennia-Vs, Spectra-Physics) and amplifier (Spitfire pumped by Evolution-15, Spectra-Physics) system operating at 1 kHz provided 778–820 nm pulses, each with an energy of about 0.8 mJ and duration of 2.5 ps under operation at 1 kHz. In the probe arm, the second harmonic of the laser output was focused into a cell filled with methane or hydrogen gas to generate the first-order Stokes stimulated scattering in 439–494 nm. For example, a UV probe pulse at 230 nm used in the time-resolved UVRR measurements described in this chapter was generated with a BBO crystal as the second harmonic of the 460-nm output. In this way, a UV pulse was generated in the wavelength range of 220–247 nm. Sum frequency generation between the second harmonic and the stimulated Raman scattering was performed to produce a UV probe pulse in the range of 206–218 nm. This wide tunability is quite important for site-specific observation of the protein. By properly selecting the wavelength of the probe pulse, we can observe resonantly-enhanced Raman bands of aromatic side chains. The tunability of the probe pulse is shown in the right panel of Fig. 12.1b. In the pump arm, a pump pulse of 530–600, 439–494, and 389–410 nm was generated by optical parametric generation and amplification, stimulated Raman scattering in compressed methane or hydrogen gas, and the frequency doubling of the amplified laser output, respectively. The tunability of the pump pulse is shown in the right panel of Fig. 12.1c.

After the pump and probe beams were made co-propagating using a dichroic mirror, they were focused with a spherical lens onto a flowing thin film of the sample solution formed by a wire-guided jet nozzle [28]. Sample solution was continuously circulated to avoid possible damage due to laser illumination during the measurements. Light components other than the probe pulse were eliminated spatially with a Pellin-Broca prism and spectrally with dichroic mirrors. At the sample, the energies of the probe and pump pulses were attenuated to 0.5 and 5 μJ, respectively, using Cr-coated quartz ND filters. The two beams were configured for 135° backscattering illumination and collection. A cross-correlation trace of the pump and probe pulses measured by difference frequency generation with a thin BBO crystal indicated a pulse width of 3.0–3.7 ps. Raman scattered light was collected by two quartz achromatic doublet lenses onto the entrance slit of a Czerny-Turner configured Littrow prism prefilter [29] coupled to a 50-cm single spectrograph (500 M, SPEX). The spectrograph was equipped with a 1200-grooves/mm, 500-nm blazed grating operating in the second order, or a 2400-grooves/mm, 250-nm blazed grating operating in the first order. Dispersed light in the spectrograph was detected with a liquid-nitrogen-cooled

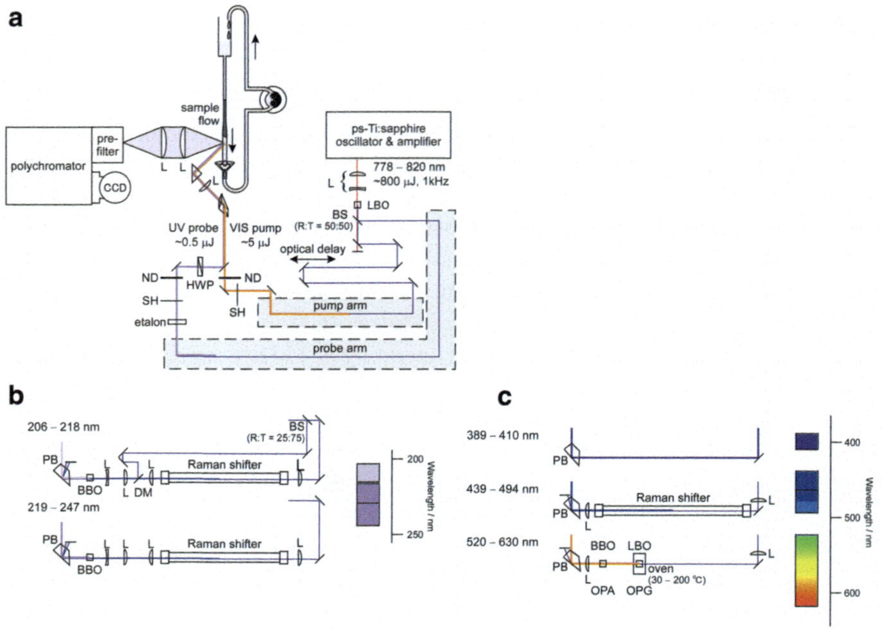


**Fig. 12.1** Time-resolved UVRR spectrometer. **a** Optical setup of the spectrometer. **b** Optical configuration in the probe arm. **c** Optical configuration in the pump arm. BBO = β-barium borate, BS = beam splitter, DM = dichroic mirror, HWP = half wave plate, L = lens, LBO = lithium triborate, ND = neutral-density filter, PB = Pellin-Broca prism, SH = mechanical shutter. (Adapted with permission from M. Mizuno, Y. Mizutani, In Recent Progress in Colloid and Surface Chemistry with Biological Applications, American Chemical Society, 2015, Vol. 1215, Chap. 16, p. 329; Copyright 2015 American Chemical Society.)

CCD detector (SPEC-10:400B/LN, Roper Scientific) with Unichrome UV-enhancing coating.

## 12.3 Energy Flow in Heme Proteins

### *12.3.1 Proof-of-Principle Experiments*

Anti-Stokes UVRR spectroscopy on amino acid residues has the potential ability to provide information on the excess energy in the residues. However, no measurements were reported to probe the excess energy diffusion in the protein moiety by selectively observing the anti-Stokes resonance Raman spectra of Trp residues in the protein. In a "proof-of-principle" report, we accomplished direct observation of vibrational energy flow in cytochrome *c* and demonstrated that time-resolved anti-Stokes UVRR Raman spectroscopy is a powerful tool for monitoring vibrational energy flow in proteins [14]. Cytochrome *c* was chosen because it has a single tryptophan residue (Trp59 in bovine cytochrome *c*) near the heme group, as shown in Fig. 12.2a. Temporal evolutions of anti-Stokes UVRR intensities of Trp59 revealed the kinetics of energy flow from the heme group and energy release of the residue in cytochrome *c*.

Figure 12.2b shows the time-resolved anti-Stokes UVRR difference spectra obtained with pump and probe pulses with wavelengths of 405 and 230 nm, respectively. The top trace is a probe-without-photolysis spectrum, which represents the anti-Stokes UVRR spectrum for ferric cytochrome *c*, and contains the UVRR bands for Trp at 759 (W18), 876 (W17), and 1015 $cm^{-1}$ (W16), and a band for Tyr at 1181 $cm^{-1}$ (Y9a). The band at 1599 $cm^{-1}$ was not assigned to any modes of aromatic amino acid residues. The band at 983 $cm^{-1}$ indicated by the asterisk was due to the sulfate ion added as an internal standard for determining Raman intensity. The anti-Stokes bands in the probe-without-photolysis spectrum arouse from thermal populations of the vibrationally excited states at room temperature. Pump-induced difference spectra were obtained for cytochrome *c* (Fig. 12.2b). The 405-nm pump pulse excited the heme into an electronically excited state. The spectrum at 5 ps shows pump-induced positive difference bands observed for the Trp and Tyr anti-Stokes Raman bands, which disappeared within 30 ps in the anti-Stokes UVRR difference spectra. The temporal evolutions of the integrated intensity of the W18 and W16 bands in the time-resolved difference spectra relative to those in the probe-without-photolysis spectrum are shown in Fig. 12.2b and c, respectively. Intensity changes in the W16 and W18 bands were fitted by a convolution of the instrument response with an exponential rise and an exponential decay, $I\left[\exp\left(-t/\tau_{decay}\right) - \exp(-t/\tau_{rise})\right]$. Time constants of rise and decay for the anti-Stokes W16 band were 5.67 ± 3.0 and 5.68 ± 3.0 ps, respectively. For the anti-Stokes W18 band, the time constants of 5.5 ± 2.5 and 5.6 ± 2.5 ps were obtained for the rise and decay, respectively. Upon the photoexcitation, Stokes W16 and W18 bands also changed their intensities, presumably due to a hydrogen bond between heme and Trp59, indicating that the Raman cross-sections were temporally changed. By considering the of intensity changes of the Stokes W16 and W18 bands rates for inflow and outflow of the vibrational energy for Trp59 to be estimated to be 1–3 and ~8 ps, respectively, were obtained. The data demonstrated that our technique is powerful for studying vibrational energy flow in proteins.

Energy dissipation from the protein to the solvent water has been studied using femtosecond time-resolved infrared [30] and transient phase grating techniques [31, 32] that monitored the heating of the solvent water caused by photoexcitation of protein. These studies showed that the excess energy was transferred to the water interface through the protein matrix in less than 20 ps. The result that the time constant of energy release from Trp59 was smaller than that of water heating is consistent with the supposition that the energy released from Trp59 is transferred to the solvent water through the remaining protein matrix. Direct energy transfer from Trp59 to the solvent water was very unlikely because Trp59 has no exposed contact to the solvent.

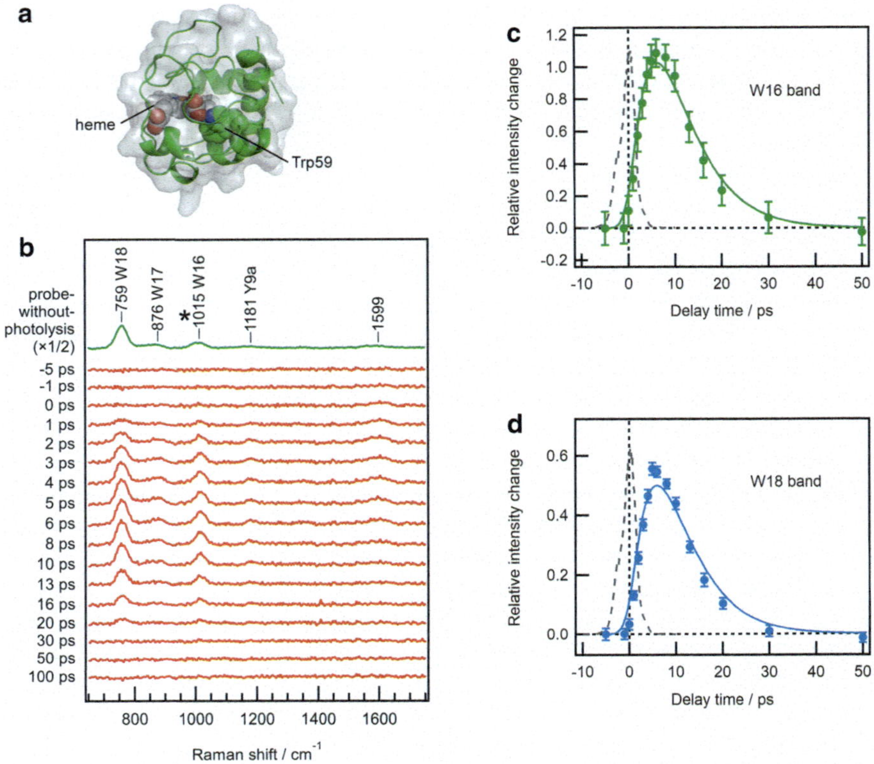


**Fig. 12.2** Energy dissipation in ferric cytochrome *c*. **a** Crystallographic structure of bovine heart ferric cytochrome *c*. Heme and Trp59 are shown as space-filling spheres, and the protein is shown as a green ribbon with a grey surface representation superimposed. **b** Time-resolved anti-Stokes UVRR spectra of ferric cytochrome *c* for time delays from −5 ps–100 ps. Probe and pump wavelengths were 230 and 405 nm, respectively. Top trace is the probe-without-photolysis spectrum corresponding to the anti-Stokes UVRR spectrum of ferric cytochrome *c* divided by a factor of 2. Other spectra are time-resolved difference spectra generated by subtracting the probe-without-photolysis spectrum from the pump-probe spectrum at each delay time. The asterisk represents the sulfate band at 983 $cm^{-1}$ as an intensity standard. **c** and **d** Temporal intensity changes in anti-Stokes **c** W16 and **d** W18 bands in the range of −5–50 ps. Circles indicate band intensity measured at each delay time relative to the band intensity in the probe-without-photolysis spectrum. Solid lines were fit to a double exponential function convoluted with the instrument response function. Lines shown in the panels were obtained using parameters of $\tau_{rise} = 5.5 \pm 2.5$ ps and $\tau_{decay} = 5.6 \pm 2.5$ ps for the W18 band, and $\tau_{rise} = 5.6_7 \pm 3.0$ ps and $\tau_{decay} = 5.6_8 \pm 3.0$ ps for the W16 band. (Adapted with permission from N. Fujii, M. Mizuno, Y. Mizutani, *J. Phys. Chem. B* 2011, 115, 13,057; copyright 2011 American Chemical Society; Y. Mizutani, *Bull. Chem. Soc. Jpn.* 2017, 90, 1344; copyright 2017 The Chemical Society of Japan.)

### 12.3.2 Anisotropy in Energy Flow

We demonstrated that time-resolved anti-Stokes UVRR spectroscopy is a powerful tool for monitoring vibrational energy flow in a protein. This technique was further improved by combining it with site-directed mutagenesis [13]. The position of a Trp residue in a protein can be changed by amino acid substitution. Thus, by comparing data from mutants with different positions of the Trp residue with respect to the heme, mapping energy flow in a protein is possible by moving the position of the probe residue. Thus, heme proteins provide significant advantages for studying energy flow in the condensed phase because the distance and relative orientation of the heater (heme) and probe groups (Trp residue) can be fixed in proteins. This is opposed to the solution phase where the two molecules would diffuse freely. Wild-type sperm whale myoglobin (Mb) has two Trp residues, Trp7 and Trp14. First, mutants devoid of Trp residues were prepared by replacing Trp7 and Trp14 with Tyr and Phe, respectively [33]. Then, we prepared three mutants with a single Trp at a specific position. Figure 12.3a shows the Mb mutants prepared. One mutant had a Trp residue at position 68, which is in the vicinity of the heme. The center-to-center distance between the heme and Trp68 is 6.8 Å (Protein Data Band entry, 2OH9). Another mutant has a Trp residue at position 28, with a distance to the heme of 12.4 Å (Protein Data Band entry, 2OH8). In the third mutant, Trp7 was replaced with Tyr while Trp14 remained unaltered; the distance from the heme to Trp14 is 15.0 Å. The Trp residues in the three mutants were located in a similar direction from the heme, but their distances from the heme differed significantly. Crystallographic data of these mutants [34] showed that the three-dimensional structures of the mutants are very close to that of wild-type Mb. Accordingly, a comparison of the data using these three mutants allowed examination of the distance dependence of energy flow from the heme to the Trp residues.

Figure 12.3b and c depict the temporal evolution of the anti-Stokes intensity of the W18 and W16 bands in the time-resolved difference spectra relative to that in the probe-without-photolysis spectrum. The anti-Stokes intensities of the mutants reflected the amount of energy delivered to the Trp residues at different positions in the protein. Data in Fig. 12.3b show that anti-Stokes intensities decreased as heme–Trp distance increased. This behavior can be qualitatively explained by the classical thermal diffusion model showing that excess energy becomes spatially less dense as the energy diffuses in the protein. It is interesting to compare time constants of the rise at the different positions. For the W18band, time constants of $3.0 \pm 0.4$ and $4.0 \pm 0.6$ ps for the rise were obtained for the Trp68 and Trp28 mutant, respectively. For the W16 band, time constants of the rise were $3.0 \pm 0.7$ and $4.9 \pm 1.1$ ps for the Trp68 and Trp28 mutant, respectively. The rise time for Trp28 was consistently longer than that for Trp68, indicating that excess energy takes longer to arrive at position 28 (Trp28, 12.4 Å) than at position 68 (Trp68, 6.8 Å). Thus, the observed distance dependence is qualitatively consistent with classical thermal diffusion. This study is the *first* example of the direct space-resolved observation of energy dissipation in a protein moiety with the spatial resolution of an amino acid residue.

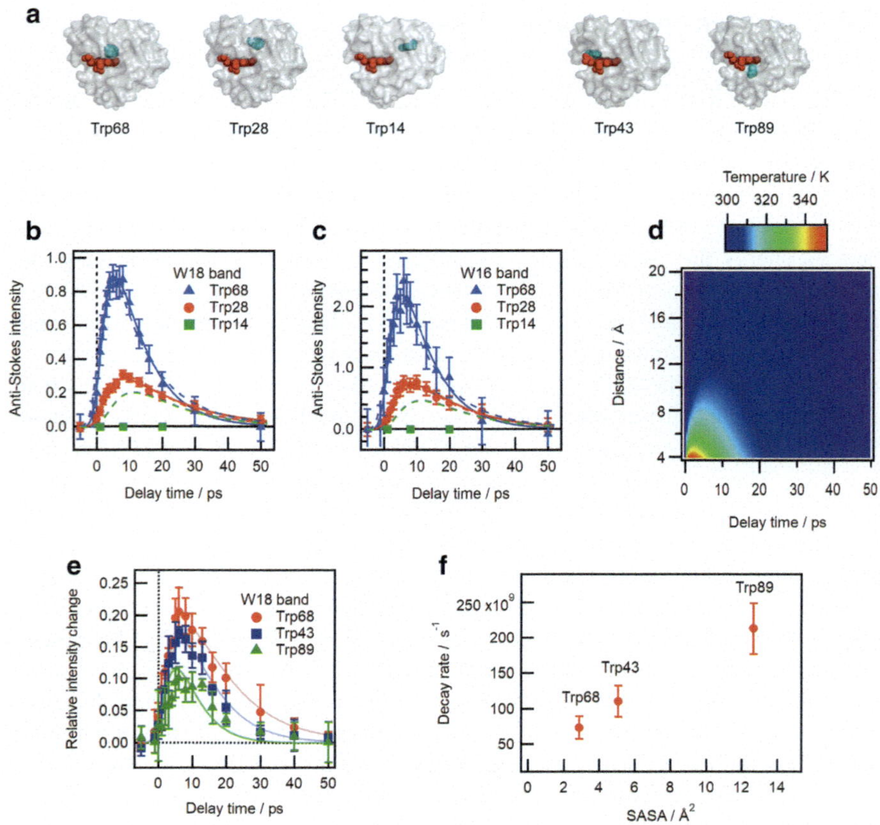


**Fig. 12.3** Energy dissipation in Mb having the probe Trp residue at different positions. **a** Crystallographic structure of sperm whale Mb. Heme and probe Trp residues are represented by space-filling spheres in red and cyan, respectively. **b** and **c** Temporal changes in anti-Stokes **b** W18 and **c** W16 band intensities in the range of −5–50 ps upon excitation at 405 nm. Closed blue triangles, red circles, and green squares indicate band intensity of the Trp68, Trp28, and Trp14 mutants, respectively, measured at each delay time relative to that in the probe-without-photolysis spectrum. Solid lines were fit to a double exponential function convoluted with the instrument response function. Broken lines indicate the Boltzmann factor based on temperature calculated in the two-boundary classical heat transport model. Time constants of the rise and decay of the Trp68 mutant were 3.0 ± 0.4 and 9.6 ± 1.0 ps, respectively. For the Trp28 mutant, time constants of 4.0 ± 0.6 and 19.2 ± 2.7 ps were obtained for the rise and decay, respectively. Time constants of the rise and decay of the Trp68 mutant were 3.0 ± 0.7 and 9.2 ± 1.8 ps, respectively. For the Trp28 mutant, time constants of 4.9 ± 1.1 and 14.9 ± 3.4 ps were obtained for the rise and decay, respectively. **d** Temperature as a function of position and time calculated from the two-boundary classical heat transport model for a solvated hemeprotein. **e** Temporal changes of the anti-Stokes W18 band intensities of the Trp43, Trp 68, and Trp89 residues in the range from −5 to 50 ps upon photoexcitation at 405 nm. Solid lines were fit to a double-exponential function convoluted using the instrument response function. **f** Decay rate of the photoinduced anti-Stokes W18 band intensity change of the Trp43, Trp68, and Trp89 residues plotted against the SASA value at each Trp position. (Adapted with permission from N. Fujii, M. Mizuno, H. Ishikawa, Y. Mizutani, *J. Phys. Chem. Lett.* 2014, 5, 3269, copyright 2014 American Chemical Society; M. Kondoh, M. Mizuno, Y. Mizutani, *J. Phys. Chem. Lett.* 2016, 7, 1950, copyright 2016 American Chemical Society; Y. Mizutani, *Bull. Chem. Soc. Jpn.* 2017, 90, 1344; copyright 2017 The Chemical Society of Japan.)

Next, these observational data were compared with quantitative data calculated based on a classical two-boundary thermal transport model by Li and Champion [35] to simulate the thermal dynamics of transient cooling in chromophoric biomolecules. We calculated the temperature distribution in the protein based on the two-boundary heat transport model. Figure 12.3e depicts the calculated temperature as a function of time and the heme–Trp distance. The temperature dependence of the anti-Stokes Raman intensity was expressed by the Boltzmann factor. Based on the calculated temperature, the Boltzmann factor for the W18 and W16 modes was determined to investigate the temporal evolution of the anti-Stokes intensities. The broken lines in Fig. 12.3b and c represent temporal profiles of the Boltzmann factor based on the model. The model reproduced the temporal evolution of the anti-Stokes intensities of the W16 and W18 bands for both the Trp68 and Trp28 mutants, but did not reproduce temporal evolution of the Trp14 mutant. Thus, the classical heat transport model is unable to reproduce the entire data set, suggesting that protein matrices are never uniform as thermal conductor. These experimental data validate a conclusion by theoretical work showing that energy flow is intrinsically anisotropic due to the geometry of proteins [36].

### *12.3.3 Atomic Contacts in Energy Transfer*

Energy transfer to amino acid residues contacting the heme group was investigated to examine the pathway of the heme–polypeptide energy transfer. We prepared three Mb mutants, F43W, V68W, and L89W. The Trp residue at position 43, 68, or 89 in each mutant is located close to the heme, represented by space-filling purple spheres in Fig. 12.3a. The distance from the heme to the Trp43, Trp68, or Trp89 residue in each mutant was 6–7 Å. In these Mb mutants, the original Trp residues (Trp7 and Trp14) were not replaced because these residues made almost no contribution to the pump-induced anti-Stokes intensity changes and the replacements of the residues significantly reduced the efficiency of protein expression. The anti-Stokes W18 band intensities of the Trp43, Trp 68, and Trp89 residues in the time-resolved difference spectra of each mutant were plotted against delay time (Fig. 12.3e), to compare their temporal evolutions. The vertical axis represents a pump-induced change in the W18 band intensity relative to that in the probe-without-photolysis spectrum. Temporal changes were similar for all mutants; band intensity increased up to 6–8 ps, and then decayed to almost zero within 50 ps. This implies that vibrational energy was transferred from the heme to the three positions (43, 68, and 89) with similar rates.

These data give us an insight about the mechanism of energy flow in proteins. The heme group is linked to polypeptide chain through a histidine residue (His93). Thus, the distances between heme and Trp43, Trp68, and Trp89 along the main chain of the protein are very different. If the excess energy is transferred to Trp residues through the main chain, the energy transfer rate would be completely different among the three positions, but that was not observed. Nonbonded contacts of Trp43, Trp68, and Trp89 with the heme group were evident from the X-ray crystallographic data

(Fig. 12.3a) [34, 37]. The similar rates of heme–Trp energy transfer suggest that the energy is transferred not through the heme–His93 covalent linkage and the protein main chain but through atomic contacts between the heme group and the residues.

The three Trp residues at the different positions showed the distinctly different amplitudes of band intensity changes (Fig. 12.3e): amplitude was most significant in the order Trp68 > Trp43 > Trp89. This observed difference suggests that energy flow at the three positions is also different. Two possibilities exist for the origin of this difference: the rate of energy flow into or out of each Trp residue is different. A careful examination of the X-ray crystallographic structures of the mutants revealed that amplitude increased as Trp residue exposure to the solvent water decreased: Trp68 and Trp43 were buried inside the protein, while Trp89 was near the protein surface, facing the solvent [34, 37]. Amplitudes of the anti-Stokes band intensity changes (Fig. 12.3e) were inversely correlated with values for solvent accessible surface area (SASA): the larger the SASA value, the smaller the amplitude. This correlation was explained by assuming that the solvent water efficiently accepts the excess energy of the Trp residues. This assumption is consistent with our results on the role of propionate groups in vibrational energy relaxation [9, 10]. Computational studies indicated that the propionate groups of heme function as an efficient channel for energy transfer from heme to solvent water [38, 39]. The two propionate groups protrude on the protein surface and have contacts with the solvent water. Inspired by these studies, the vibrational energy relaxation of heme lacking propionate group(s) were investigated, with results demonstrating that the rate of the heme energy relaxation is decelerated upon removal of the propionate groups [9, 10]. Therefore, solvent water is an efficient acceptor of vibrational energy.

The study using the mutants of Trp43, Trp68, and Trp89 proposed the importance of atomic contacts in vibrational energy flow in proteins. The dominant channel for energy flow from the heme group to the protein moiety is not through the covalent linkage of heme–His93 and the protein main chain, but instead through atomic contacts between the heme and residues. To directly examine the contribution of the linkage between the heme group and the polypeptide chain to the energy transfer pathway, we investigated the vibrational energy flow in a Mb mutant lacking the covalent linkage [16]. In the Mb mutant constructed in the study, H93G(Im), His93 is replaced by glycine. This replacement abolishes the sole covalent bond between the polypeptide chain and heme. The resultant cavity can be occupied by exogenous ligands such as imidazole (Im). H93G(Im) lacks the covalent bond between the five-membered ring of the His93 and the polypeptide. To probe the energy flow from heme to the protein moiety, we introduced a Trp reside at position 68 of the protein. We examined the effect of the covalent linkage between heme and the polypeptide chain on the energy flow by comparing the data obtained for the V68W (Fig. 12.4a) and V68W/H93G(Im) mutants (Fig. 12.4b).

Temporal evolutions of the anti-Stokes W18 band intensities of the V68W and V68W/H93G(Im) mutants in the time-resolved difference spectra were indicated in Fig. 12.4c. The ordinate represents the ratio of the pump-induced intensity change of the W18 band to the band intensity in the steady-state spectrum. Following the photoexcitation, the anti-Stokes band intensity increased up to 8 ps and then decayed

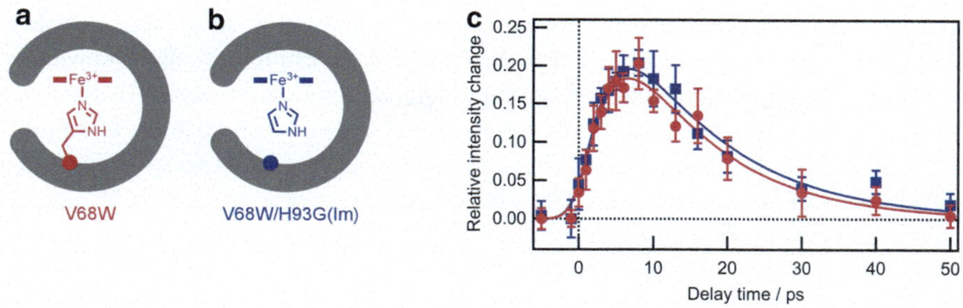


**Fig. 12.15.4** Energy dissipation in Mb lacking His93-polypeptide covalent bond. **a** and **b** Schematic views around heme in the Mb mutants. In the V68W/H93G(Im) mutant (**c**), the proximal histidine is replaced by glycine. This replacement abolishes the sole covalent connection between the polypeptide chain and heme and creates a cavity that can be occupied by exogenous ligands such as imidazole. In the V68W mutant (**a**), this covalent bond is intact. **c** Temporal profiles of the pump-induced anti-Stokes W18 band intensity in the range from −5 to 50 ps following the photoexcitation. Solid lines show the best fits to a double-exponential function convoluted using the instrument response function. (Adapted with permission from S. Yamashita, M. Mizuno, D. P. Tran, H. Dokainish, A. Kitao, Y. Mizutani *J. Phys. Chem. B* 2018, 122, 5877, copyright 2018 American Chemical Society.)

to nearly zero at 50 ps for both mutants. These results showed that the V68W and V68W/H93G(Im) mutants exhibited identical temporal profiles of the anti-Stokes intensity changes within experimental uncertainty, indicating that the covalent bond between the five-membered ring of His and the polypeptide chain plays negligible role on the energy flow from the heme group to Trp68 and the van der Waals contacts are the dominant pathway for energy migration from heme to the Trp residue.

Kitao and his coworkers collaborarted with us and conducted molecular dynamics simulations to compare the three-dimensional structures of the V68W and V68W/H93G(Im) mutants because the X-ray crystallographic data were not available for the V68W/H93G(Im) mutant. The calculated structures indicate that the backbone structures and the heme–Trp distances are very similar between the two mutants. Therefore, it is highly likely that the heme group in the V68W/H93G(Im) mutant forms van der Waals contacts to the surrounding, which is similar to those in the V68W mutant. Cleavage of the covalent bond between the proximal His and the polypeptide chain did not affect the rate of the heme relaxation. This result means that the van der Waals contacts between heme and its surroundings are the dominant pathway for the vibrational energy relaxation of heme. This situation is similar to that of solutions, where the excess energy in solute molecules is exchanged dynamically through collisions with the first solvation shell, then exchanged dynamically through collisions of the first solvation shell with the second layer of solvation molecules, and so forth [40, 41].

The packing density of the interior of a native protein is high much like dense liquid [42]. For a given protein, the polypeptide chain is folded into a characteristic compact structure in aqueous solution. The packing density of the protein is nearly maximized, as indicated by the isothermal compressibilities of proteins being one order of magnitude smaller than those of organic liquids [43]. In the heme pocket of Mb, there are many atomic contacts between the protein moiety and the heme–His group. The high packing density prompted the hypothesis that not only covalent and hydrogen bonds, but also non-bonded contacts, play essential roles in energy flow in proteins. The absence of internal void spaces is crucial for successive structural changes to propagate to spatially distinct sites, i.e., proteins are molecules of *functional compactness* [27].

It should also be pointed out that there are both regions of low density as well as regions that are highly compact in the protein structure. The packing density inside proteins was quantitatively analyzed and appeared to be inhomogeneous [44, 45]. The inhomogeneity of the packing density in the protein structure gives rise to anisotropy of structural change [46] and that of energy flow [13]. The role of packing and non-bonded contacts in energy transfer identified in our study would further indicate how variation in packing within a protein regulates anisotropic energy flow. Energy flow in biomolecules has been studied in numerous systems using various computational approaches to discern the contributions from through-bond and through-space transport mechanisms [47–52]. We experimentally demonstrated that atomic contacts between heme and nearby amino acid residues are the dominant channel for energy flow from the heme group to the protein moiety. The results of our studies will stimulate theoretical investigation of the mechanism of the vibrational energy flow in proteins. For example, Reid et al. discussed ingenious use of time-resolved vibrational spectroscopy to determine change in conformational entropy with change in protein functional state and provided an expression quantifying the connection [53].

## 12.4 UVRR Spectroscopic Thermometer

Anti-Stokes intensity tells us the extent of vibrationally excited populations. So far, we discussed anti-Stokes intensities of particular modes. One may wonder if the extent of vibrational excitation in particular modes reflect the extent of excitation in the whole molecule. Strictly speaking, the temperature of a molecular system is defined only when the entire system is in thermodynamic equilibrium. However, the transiently excited molecules discussed in this chapter are not in thermodynamic equilibrium. It is nontrivial to answer to what extent we are allowed to talk of an internal temperature after vibrational excitation of the molecules. To discuss these questions, we compare occupation probabilities in equilibrium and those in nonequilibrium.

In equilibrium, the occupation probability $N(\nu_i)$ of a vibrational state decreases exponentially with increasing frequency $\nu_i$ according to the Boltzmann factor.

$$N(\nu_\mathrm{i}) \propto \exp\left(-\frac{h\nu_\mathrm{i}}{k_\mathrm{B}T}\right) \tag{12.1}$$

For nonequilibrium state, we calculated the occupation probability in nonequilibrium under a set of assumptions. After excitation of a molecule, the vibrational energy is rapidly redistributed over a huge number of fundamental and isoenergetic combination states. Each state is assumed to have equal weight. The redistribution is faster than the energy exchange among the amino acid residues.

In nonequilibrium states, the probability of finding a molecule in the range of a total energy between $E$ and $E + dE$ in the mode $i$ where the $v = 1$ level is occupied is given by,

$$N_E(\nu_{\rm i})dE = \frac{\rho_i(h\nu_{\rm i})\rho_{\rm r}(E-h\nu_{\rm i})}{\rho(E)}dE. \tag{12.2}$$

$\rho(E)$ is the density of states at the energy $E$. $\rho_i(h\nu_{\rm i})$ and $\rho_{\rm r}(E - h\nu_{\rm i})$ are the density of states of the mode $i$ and that of the rest of the modes, respectively. The occupation probability of the vibrational state of the mode $i$ is given by integration of $N_{\rm E}(\nu_{\rm i})$, weighted by $P(E)$ over $E$.

$$N(\nu_{\rm i}) = \int \frac{\rho_i(h\nu_{\rm i})\rho_{\rm r}(E-h\nu_{\rm i})}{\rho(E)} P(E)dE \tag{12.3}$$

The energy distribution $P(E)$ is a probability of finding the molecules with a total energy $E$.

We calculated the total occupation probability for 3-methyl indole, which is a model compound of the Trp side chain. Figure 12.5a shows its total energy distributions. Frequencies of fundamental modes of 3-methyl indole were obtained by density functional theoretical calculations. The geometry optimization and normal mode analysis were performed at the B3LYP/6-311G(d,p) level. All the calculations were conducted with Gaussian16W program package. Densities of the vibrational sates below 6000 m$^{-1}$ were calculated based on the normal mode frequencies. The blue line shows the total energy distribution in equilibrium at 300 K. The distribution was calculated as a product of the density of states and the Boltzmann distribution at 300 K. The distribution shown in the green line is similar to the blue one but shifted to a higher energy side by 1000 cm$^{-1}$. Red line shows the total energy distribution in equilibrium at 387 K. We calculated the distribution at 387 K because the introduction of the 1000-cm$^{-1}$ energy raises temperature of 3-methyl indole from 300 to 387 K in equilibrium.

Figure 12.5b shows the total occupation probability in the logarithmic scale as a function of the vibrational frequency. The total occupation probabilities for the equilibrium states are identical to the Boltzmann factor at a given temperature. The blue and red lines show the total occupation probabilities at 300 and 387 K, respectively. The green markers in Fig. 12.5b show the total occupation probabilities for the total energy distribution shown by the green line in Fig. 12.5a. Although this energy distribution is obviously in nonequilibrium, the total occupation probability below 2000 cm$^{-1}$ is nearly identical to that in the Boltzmann distribution at 387 K. This means that the vibrationally excited Trp residue behaves as if it had a high temperature in a canonical ensemble, which is called the internal temperature of the excited molecules. In a large polyatomic molecule at high excitation, the partial

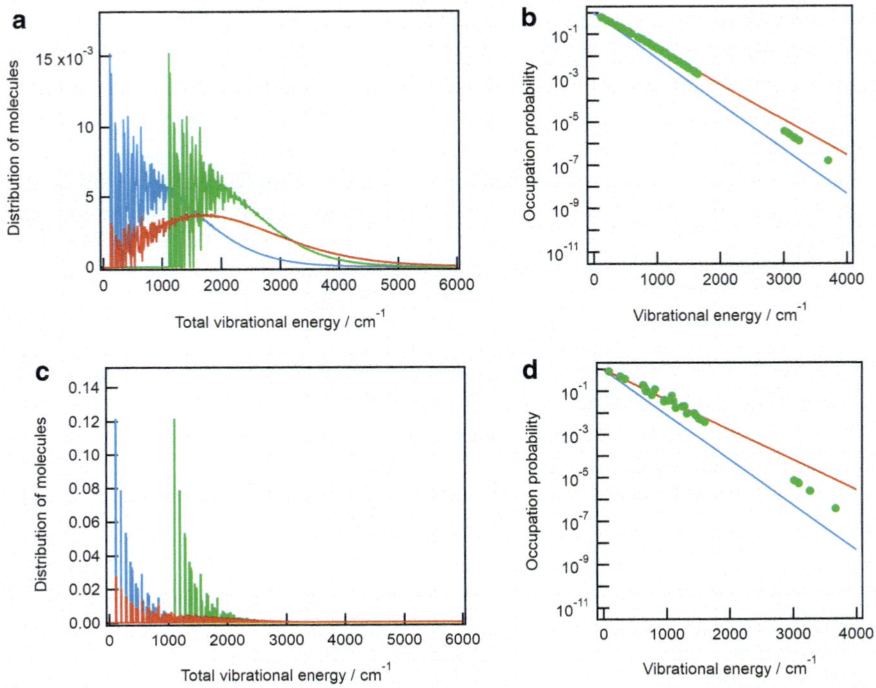


**Fig. 12.15.5** Vibrational energy distributions of side chains of Trp and His. **a** and **c** Total energy distribution of 3-methyl indole and 4-methyl imidazole, respectively. In panel (**a**), the blue and red lines show the total energy distribution in equilibrium at 300 and 387 K, respectively. In panel (**b**), the blue and red lines show the total energy distribution in equilibrium at 300 and 387 K, respectively. The green line shows the total energy distribution in nonequilibrium, where the distribution has a same form as one shown in the blue line and is shifted to higher energy side by 1000 $cm^{-1}$. **b** and **d** Total occupation probability of 3-methyl indole and 4-methyl imidazole, respectively. The blue and red lines show the total occupation probabilities at 300 and 449 K, respectively. The green markers show the total occupation probabilities for the total energy distribution shown by the green line in panel (**a**) and (**c**). (Adapted with permission from S. Yamashita, M. Mizuno, Y. Mizutani *J. Chem. Phys.* 2022, 156, 075,101, copyright 2022 American Institute of Physics.)

energy distributions of a selected oscillator with small excitations per oscillator agree for microcanonical and canonical total distributions, provided that the average total energies are identical. Therefore, an anti-Stokes intensity is a measure of the energy locally deposited to Trp residues in proteins. This conclusion results from that Trp residues have the largest side chain and the highest density of states among the 20 amino acids composing proteins. The situation depends on the number of fundamental modes or the density of states. For example, for 4-methyl imidazole, which is a model compound of the His side chain, the total occupation probability for the nonequilibrium states deviated from that in the Boltzmann distribution (Fig. 12.5c and d). The deviation is due to that the density of states of the His side chain is not high enough.

We utilized the anti-Stokes intensities of the Trp bands to study the energy flow in proteins. At the initial stage of the research project, we have chosen the Trp bands as a probe of the vibarationally excited populations because they are most intense in the UVRR spectra. Recently, we found that the Trp bands are useful as the local temperature probe in proteins because the Trp side chain has the highest density of states [54, 55]. Accordingly, anti-Stokes spectra of Trp residues serve an excellent spectroscopic "thermometer" in terms of high sensitivity and straightforward interpretation.

## 12.5 Summary

In this chapter, we described our recent work on the energy flow in the proteins using time-resolved UVRR spectroscopy. The technique of energy mapping based on the anti-Stokes UVRR intensity of Trp residues fully utilizes two unique advantages of Raman spectroscopy. One is site-specific observation of the resonance Raman effect and the other is selective observation of the vibrationally excited populations through anti-Stokes scattering. We further developed this technique by combining it with site-directed mutagenesis, which allows us to map energy flow in a protein with a spatial resolution of a single amino acid residue. The Trp side chain has so high density of vibrational states that its anti-Stokes intensities provide a measure of the local temperature. Systematic application of the time-resolved UVRR mapping to proteins with different structural motifs will provide a greatly increased understanding of energy flow in proteins.

Our studies provided direct experimental evidence showing that the covalent bond plays a negligible role and that the van der Waals contacts are the dominant pathway for energy migration from heme to the polypeptide. The heme group in the heme pocket approximates a solute in solution because it is relatively isolated from the rest of the protein and it has van der Waals contacts with amino acid residues. Vibrational energy exchange between heme and adjacent residues occurs predominantly through van der Waals contacts, in analogy to solute–solvent exchange processes in solutions. Heme proteins have an advantage that the distance between the heater and the probe molecules can be maintained and accurately characterized in the stable protein structure. Consequently, insights obtained from studies on heme proteins provide a general view of vibrational energy transfer in condensed phases.

**Acknowledgements** We are grateful to Dr. Haruto Ishikawa, Dr. Masato Kondoh, Mr. Naoki Fujii, Prof. Akio Kitao, Dr. Duy Phuoc Tran, and Dr. Hisham Dokainish for their contributions to the earlier works of the energy flow in heme proteins. This work was supported by the Grants-in-Aid for Scientific Research on Innovative Areas "Soft Molecular Systems" from the Ministry of Education, Culture, Sports, Science and Technology (MEXT) of Japan to Y. M. (No. JP25104006) and a Grant-in-Aid for Scientific Research (B) from the Japan Society for the Promotion of Science to Y. M. (No. JP26288008). Additional support was provided from the Japan Society for the Promotion of Science through a Grant-in-Aid for JSPS Fellows (No. JP19J20486) to S. Y.

## References

1. M. Abe, T. Kitagawa, K. Kyogoku, Resonance Raman spectra of octaethylporphyrinato-Ni(II) and meso-deuterated and $^{15}$N substituted derivatives. II. A normal coordinate analysis. J. Chem. Phys. **69**(10), 4526–4534 (1978)
2. T. Elsaesser, W. Kaiser, Vibrational and vibronic relaxation of large polyatomic molecules in liquids. Annu. Rev. Phys. Chem. **42**(1), 83–107 (1991)
3. D.M. Leitner, J.E. Straub, *Proteins: Energy, Heat and Signal Flow. Computation in Chemistry*. CRC Press, Boca Raton (2009)
4. J.C. Owrutsky, D. Raftery, R.M. Hochstrasser, Vibrational relaxation dynamics in solutions. Annu. Rev. Phys. Chem. **45**(1), 519–555 (1994)
5. A. Seilmeier, W. Kaiser, in *Ultrashort Laser Pulses*. ed. by W. Kaiser (Springer, New York, 1993), p.279
6. W. Zinth, W. Kaiser, in *Ultrashort Laser Pulses*. ed. by W. Kaiser (Springer, New York, 1993), p.235
7. H.S. Carslaw, J.C. Jaeger, *Conduction of Heat in Solids* (Oxford University Press, Oxford, 1959)
8. P.M. Champion, R. Lange, On the quantitation of light emission from cytochrome *c* in the low quantum yield limit. J. Chem. Phys. **73**(12), 5947–5957 (1980)
9. Y. Gao, M. Koyama, S.F. El-Mashtoly, T. Hayashi, K. Harada, Y. Mizutani, T. Kitagawa, Time-resolved raman evidence for energy 'Funneling' through propionate side chains in heme 'Cooling' upon photolysis of carbonmonoxy myoglobin. Chem. Phys. Lett. **429**(1–3), 239–243 (2006)
10. M. Koyama, S. Neya, Y. Mizutani, Role of heme propionates of myoglobin in vibrational energy relaxation. Chem. Phys. Lett. **430**(4–6), 404–408 (2006)
11. Y. Mizutani, T. Kitagawa, Direct observation of cooling of heme upon photodissociation of carbonmonoxy myoglobin. Science **278**(5337), 443–446 (1997)
12. Y. Mizutani, T. Kitagawa, Ultrafast dynamics of myoglobin probed by time-resolved resonance raman spectroscopy. Chem. Rec. **1**(3), 258–275 (2001)
13. N. Fujii, M. Mizuno, H. Ishikawa, Y. Mizutani, Observing vibrational energy flow in a protein with the spatial resolution of a single amino acid residue. J. Phys. Chem. Lett. **5**(18), 3269–3273 (2014)
14. N. Fujii, M. Mizuno, Y. Mizutani, Direct observation of vibrational energy flow in cytochrome c. J. Phys. Chem. B **115**(44), 13057–13064 (2011)
15. M. Kondoh, M. Mizuno, Y. Mizutani, Importance of atomic contacts in vibrational energy flow in proteins. J. Phys. Chem. Lett. **7**(11), 1950–1954 (2016)
16. S. Yamashita, M. Mizuno, D.P. Tran, H. Dokainish, A. Kitao, Y. Mizutani, Vibrational energy transfer from heme through atomic contacts in proteins. J. Phys. Chem. B **122**(22), 5877–5884 (2018)
17. S.A. Asher, UV resonance raman studies of molecular structure and dynamics: applications in physical and biophysical chemistry. Annu. Rev. Phys. Chem. **39**(1), 537–588 (1988)
18. I. Harada, H. Takeuchi, Raman and ultraviolet resonance raman spectra of proteins and related compounds, in *Spectroscopy of Biological Systems*. ed. by R.J.H. Clark, R.E. Hester (John Wiley & Sons, Chichester, 1986), pp.113–175
19. T. Kitagawa, Investigation of higher order structures of proteins by ultraviolet resonance raman spectroscopy. Prog. Biophys. Mol. Biol. **58**(1), 1–18 (1992)
20. H. Takeuchi, Raman structural markers of tryptophan and histidine side chains in proteins. Biopolymers **72**(5), 305–317 (2003)
21. H. Takeuchi, UV Raman markers for structural analysis of aromatic side chains in proteins. Anal. Sci. **27**(11), 1077–1077 (2011)
22. Y. Uesugi, Y. Mizutani, T. Kitagawa, Developments of widely tunable light sources for picosecond time-resolved resonance raman spectroscopy. Rev. Sci. Instrum. **68**(11), 4001–4008 (1997)

23. Y. Uesugi, Y. Mizutani, S.G. Kruglik, A.G. Shvedko, V.A. Orlovich, T. Kitagawa, Characterization of stimulated raman scattering of hydrogen and methane gases as a light source for picosecond time-resolved raman spectroscopy. J. Raman Spectrosc. **31**(4), 339–348 (2000)
24. A. Sato, Y. Gao, T. Kitagawa, Y. Mizutani, Primary protein response after ligand photodissociation in carbonmonoxy myoglobin. Proc. Natl. Acad. Sci. U.S.A. **104**(23), 9627–9632 (2007)
25. A. Sato, Y. Mizutani, Picosecond structural dynamics of myoglobin following photodissociation of carbon monoxide as revealed by ultraviolet time-resolved resonance raman spectroscopy. Biochemistry **44**(45), 14709–14714 (2005)
26. M. Mizuno, Y. Mizutani, Protein response to chromophore isomerization in microbial rhodopsins revealed by picosecond time-resolved ultraviolet resonance raman spectroscopy: a review, in *Recent Progress in Colloid and Surface Chemistry with Biological Applications*, vol. 1215, (American Chemical Society, ACS Symposium Series, 2015), pp.329–353
27. Y. Mizutani, Time-resolved resonance raman spectroscopy and application to studies on ultrafast protein dynamics. Bull. Chem. Soc. Jpn **90**(12), 1344–1371 (2017)
28. M.J. Tauber, R.A. Mathies, X. Chen, S.E. Bradforth, Flowing liquid sample jet for resonance raman and ultrafast optical spectroscopy. Rev. Sci. Instrum. **74**(11), 4958–4960 (2003)
29. S. Kaminaka, R.A. Mathies, High-throughput large-aperture prism prefilter for ultraviolet resonance raman spectroscopy. Appl. Spectrosc. **52**(3), 469–473 (1998)
30. T. Lian, B. Locke, Y. Kholodenko, R.M. Hochstrasser, Energy flow from solute to solvent probed by femtosecond IR spectroscopy: malachite green and heme protein solutions. J. Phys. Chem. **98**(45), 11648–11656 (1994)
31. L. Genberg, Q. Bao, S. Gracewski, R.J.D. Miller, Picosecond transient thermal phase grating spectroscopy: a new approach to the study of vibrational energy relaxation processes in proteins. Chem. Phys. **131**(1), 81–97 (1989)
32. L. Genberg, F. Heisel, G. McLendon, R.J.D. Miller, Vibrational energy relaxation processes in heme proteins: model systems of vibrational energy dispersion in disordered systems. J. Phys. Chem. **91**(22), 5521–5524 (1987)
33. I. Sirangelo, S. Tavassi, P.L. Martelli, R. Casadio, G. Irace, The effect of tryptophanyl substitution on folding and structure of myoglobin. Eur. J. Biochem. **267**(13), 3937–3945 (2000)
34. J.S. Olson, J. Soman, G.N. Phillips, Ligand pathways in myoglobin: a review of trp cavity mutations. IUBMB Life **59**(8–9), 552–562 (2007)
35. P. Li, P.M. Champion, Investigations of the thermal response of laser-excited biomolecules. Biophys. J. **66**(2), 430–436 (1994)
36. D.M. Leitner, Energy flow in proteins. Annu. Rev. Phys. Chem. **59**(1), 233–259 (2008)
37. Y. Watanabe, H. Nakajima, T. Ueno, Reactivities of oxo and peroxo intermediates studied by hemoprotein mutants. Acc. Chem. Res. **40**(7), 554–562 (2007)
38. I. Okazaki, Y. Hara, M. Nagaoka, On vibrational cooling upon photodissociation of carbonmonoxymyoglobin and its microscopic mechanism from the viewpoint of vibrational modes of heme. Chem. Phys. Lett. **337**, 151–157 (2001)
39. D.E. Sagnella, J.E. Straub, Directed energy "Funneling" mechanism for heme cooling following ligand photolysis or direct excitation in solvated carbonmonoxy myoglobin. J. Phys. Chem. B **105**, 7057–7063 (2001)
40. K. Iwata, H.-O. Hamaguchi, Microscopic mechanism of solute−solvent energy dissipation probed by picosecond time-resolved raman spectroscopy. J. Phys. Chem. A **101**(4), 632–637 (1997)
41. Y. Mizutani, T. Kitagawa, A role of solvent in vibrational energy relaxation of metalloporphyrins. J. Mol. Liq. **90**(1), 233–242 (2001)
42. J. Tsai, R. Taylor, C. Chothia, M. Gerstein, The packing density in proteins: standard radii and volumes. J. Mol. Biol. **290**(1), 253–266 (1999)
43. K. Gekko, Y. Hasegawa, Compressibility-structure relationship of globular proteins. Biochemistry **25**(21), 6563–6571 (1986)

44. M.B. Enright, D.M. Leitner, Mass fractal dimension and the compactness of proteins. Phys. Rev. E **71**(1), 011912 (2005)
45. J. Liang, K.A. Dill, Are proteins well-packed? Biophys. J. **81**(2), 751–766 (2001)
46. L.U.L. Brinkmann, J.S. Hub, Ultrafast anisotropic protein quake propagation after CO photodissociation in myoglobin. Proc. Natl. Acad. Sci. U.S.A. **113**(38), 10565–10570 (2016)
47. S. Buchenberg, D.M. Leitner, G. Stock, Scaling rules for vibrational energy transport in globular proteins. J. Phys. Chem. Lett. **7**(1), 25–30 (2016)
48. D.M. Leitner, S. Buchenberg, P. Brettel, G. Stock, Vibrational energy flow in the villin headpiece subdomain: master equation simulations. J. Chem. Phys. **142**(7), 075101 (2015)
49. P.H. Nguyen, P. Derreumaux, G. Stock, Energy flow and long-range correlations in guanine-binding riboswitch: a nonequilibrium molecular dynamics study. J. Phys. Chem. B **113**(27), 9340–9347 (2009)
50. L. Martínez, A.C.M. Figueira, P. Webb, I. Polikarpov, M.S. Skaf, Mapping the intramolecular vibrational energy flow in proteins reveals functionally important residues. J. Phys. Chem. Lett. **2**(16), 2073–2078 (2011)
51. N. Ota, D.A. Agard, Intramolecular signaling pathways revealed by modeling anisotropic thermal diffusion. J. Mol. Biol. **351**(2), 345–354 (2005)
52. A.A.S.T. Ribeiro, V. Ortiz, Energy propagation and network energetic coupling in proteins. J. Phys. Chem. B **119**(5), 1835–1846 (2015)
53. K.M. Reid, T. Yamato, D.M. Leitner, Scaling of rates of vibrational energy transfer in proteins with equilibrium dynamics and entropy. J. Phys. Chem. B **122**(40), 9331–9339 (2018)
54. S. Yamashita, M. Mizuno, Y. Mizutani, High suitability of tryptophan residues as a spectroscopic thermometer for local temperature in proteins under nonequilibrium conditions. J. Chem. Phys. **156**(7), 075101 (2022)
55. Y. Mizutani, M. Mizuno, Time-resolved spectroscopic mapping of vibrational energy flow in proteins: understanding thermal diffusion at the nanoscale. J. Chem. Phys. **157**(24), 240901 (2022)

# Chapter 13
# Ultrafast Two-Dimensional Spectroscopy of Photosynthetic Systems

**Donatas Zigmantas and Tomáš Mančal**

**Abstract** This chapter starts with a review of the theoretical foundation for understanding two-dimensional electronic spectroscopy (2DES) signals in molecular systems. We derive and motivate key properties of 2DES, demonstrating its ability to yield complex information about energy transfer processes and couplings within molecular assemblies, such as photosynthetic antennae. We continue with a discussion encompassing crucial aspects of experimental implementations of 2DES, with particular attention to polarization-controlled experiments. Through illustrative applications of 2DES in studies of light-harvesting functions in isolated complexes and an intact photosyntetic unit in green sulfur bacteria the chapter reveals the extensive insights into the photophysical functions of photosynthetic machinery that can be obtained. Emphasis is placed on identifying exciton coupling, delineating energy transfer pathways, and quantifying rates and efficiencies. The chapter concludes by anticipating future 2DES developments, e.g., the studies on intact photosynthetic units at physiological temperatures, that will contribute to a more holistic comprehension of primary photosynthetic functions, particularly in light harvesting and charge separation.

## 13.1 Introduction

### *13.1.1 Photosynthesis*

Photosynthesis is a biological process which directly or indirectly provides energy to almost all living organisms and makes our planet steaming with life. It can be therefore considered the most important photophysical/biochemical process on Earth.

D. Zigmantas (✉)
Chemical Physics, Lund University, Lund, Sweden
e-mail: donatas.zigmantas@chemphys.lu.se

T. Mančal (✉)
Faculty of Mathematics and Physics, Charles University, Prague, Czech Republic
e-mail: mancal@karlov.mff.cuni.cz

 
K. Ueda (ed.), *Ultrafast Electronic and Structural Dynamics*,
https://doi.org/10.1007/978-981-97-2914-2_13

Photosynthesis starts with the absorption of sunlight and involves multiple photophysical and chemical steps, which in the end lead to the conversion of sunlight into chemical energy, often stored in the energy-rich molecules, such as sugars. First photosynthetic organisms appeared on Earth around 3.5 billion years ago, and during their subsequent evolution nature has diversified and optimized the ways photosynthesis is carried out. Today it can be found in a variety of plants, algae and bacteria, and it can be assumed with a reasonable certainty that during the course of natural selection it has been highly optimized for the needs of photosynthetic organisms. This optimization includes light-conversion efficiency, robustness and ability to regulate the energy flow, when absorbed energy exceeds the capacity of the photosynthetic system to utilize it.

The photosynthetic process can be divided into two parts: (i) primary events in photosynthesis which involve absorption of sunlight, transfer of the absorbed energy to the reaction center (RC) and charge separation inside RC, (ii) the secondary events involve a cascade of biochemical reactions, which use the electrochemical potential, created in primary events, resulting in synthesis of energy-rich molecules. Subsequently these molecules are used to fuel all the processes in photosynthetic organisms. In this chapter we focus on the studies that explore the primary processes in photosynthesis, and special attention is paid to the light-harvesting function and energy transfer phenomenon.

### *13.1.2 Structure and Functions of Photosynthetic Complexes and Their Components*

Generally photosynthetic unit [1], where the primary functions of photosynthesis take place, consists of light-harvesting (LH) complexes and RC(s), which are located in the membrane of photosynthetic cells. Both are protein complexes, with multiple embedded pigment molecules. Light is usually absorbed by the LH complexes and its energy is then transferred to the RCs. These carry out charge separation across the photosynthetic membrane. Interestingly, in terms of architecture and pigment composition, photosynthetic organisms have evolved a large variety of LH complexes, but the basic structure of reaction centers has been preserved.

Three classes of pigments can be found in photosynthetic organisms, namely chlorophyllides, carotenoids and phycobilins. These pigments are bound in pigment-protein complexes (PPC) in specific ways that result in a composite structure, which is often highly optimized for energy and electron transfer. The proteins do not simply serve as a passive scaffold for the pigments—their additional function, together with the embedded water molecules is to tune the photophysical and photochemical properties of the pigments. This tuning includes creating polar environments, making hydrogen bonds with the pigments, as well as providing large number of vibrational degrees of freedom to facilitate energy dissipation whenever needed.

The main roles of the photosynthetic pigments are light absorption, energy and electron transfer, quenching of singlet oxygen and providing structural stability to PPCs. The latter two functions are mostly attributable to carotenoids. In this chapter we focus on photosynthetic complexes containing chlorophyllides, which are the most abundant photosynthetic pigments, and are responsible for energy and electron transfer. Interestingly, these pigments do not "work" alone—in many PPCs they are located close to each other and form molecular dimers or larger aggregates. The coupling of transition dipoles between the pigments in aggregates results in modification of their photophysical properties. In such a situation excitations cannot be assigned to individual pigments, instead they are delocalized among the coupled molecules. Such a relation between two excitations is often referred to as coherence between the individual molecular excitations [2]. Together the delocalized excitations form quasiparticles called *excitons*. A molecular exciton comprises a rather closely bound electron–hole pair, and its delocalization is relatively small. Usually it does not extend over more than a few molecules, in contrast to electron delocalization in semiconductor crystals. Molecular aggregates support the so-called Frenkel type of excitons [3].

### *13.1.3 Photosynthetic Complexes as Open Quantum Systems*

To understand primary photosynthetic functions, namely the energy and electron transfer processes, theoretical modeling is indispensable, and a proper framework, in which theoretical models are set, is required. The PPC systems are formed of a large number of atoms and an immense number of coupled degrees of freedom, and it is therefore not feasible to create a full closed quantum system model based on the first principles. Thus the modeling using quantum theory has to be done on phenomenological level, where molecules are described as multilevel quantum systems and multiple degrees of freedom of the environment are treated as a collective *bath*. Since we are interested in light-triggered processes, the light-matter interaction has to be included in the modeling as well. This is usually done by a semiclassical approach (quantum molecules interacting with classical light) and perturbation theory.

As we have seen, the electronic excitations in photosynthetic complexes have to be considered as Frenkel excitons. These represent, to a good approximation, the electronic eigenstates of the photosynthetic complexes, which interact resonantly with light, and which are thus revealed in optical spectra. The other essential part of the picture is the coupling of the excitons to their molecular environment (bath), leading to the phenomena of dephasing and energy dissipation, which greatly influence and condition the processes of interest to us, namely the energy and electron transfers. The terms bath and environment comprise all the degrees of freedom of the photosynthetic system, which do not directly interact with light. For instance, the pigment molecules possess many vibrational degrees of freedom, which contribute to the environment/bath of the electronic excitations in these molecules.

In a photosynthetic system containing multiple interacting pigments, the Frenkel excitons form a characteristic band structure, with individual bands corresponding to a given multiplicity of excitons (single-, double-, triple-exciton bands) separated roughly by the single pigment transition energy. Energy transfer (accompanied by a partial energy dissipation) between excitons occurs almost exclusively between the states within the same band. For the sake of simplicity, in our considerations we will ignore the effects of exciton-exciton annihilation, and related multi-exciton effects.

## 13.2 Theory of the Nonlinear Spectroscopy Signals

### *13.2.1 Nonlinearity and Perturbation Theory*

Traditional spectroscopic techniques of time-resolved spectroscopy in optical domain are intimately linked with the validity of perturbation theory with respect to the strength of the light-matter interaction. Low order perturbation theory in the light electric field supports a simplified theoretical picture, which enables researchers to intuitively predict the outcome of experiments. Interestingly, some of these approximations do not represent limitations of the theoretical and conceptual description of the time-resolved experiments, rather, their role is to *define* the corresponding experimental techniques. When experiments using these techniques are performed, the experimentalists must ensure that the conditions required for the validity of the defining approximations are fulfilled. For example, when a theorist calculates linear absorption coefficient, she can be absolutely sure that by applying first order (i.e. linear) perturbation theory, she provides the complete description of the experiment with respect to its dependence on the electric field of the light.

The two-dimensional electronic spectroscopy (2DES) discussed in this chapter, belongs to the class of the *third order* nonlinear optical experiments. Just like the (linear) absorption, this class of experiments is defined by its specific order of dependence on the light field. In the third order experiments, the interesting spectroscopic signal is of the third order in the electric field of light, and its intensity therefore depends on the third power of the intensity of the exciting light-field. The experiments are performed with short laser pulses (typically 10s of femtoseconds in duration), which can be represented with a high degree of precision by their classical time-dependent electric fields. The light-matter interaction Hamiltonian operator of a molecular system at the position $\boldsymbol{r}$ in a diluted sample of molecules is then assumed to be of the form

$$H_I(\boldsymbol{r}, t) = -\boldsymbol{\mu} \cdot \boldsymbol{E}(\boldsymbol{r}, t), \tag{13.1}$$

where $\boldsymbol{E}(\boldsymbol{r}, t)$ is the classical electric field of the exciting light (expectation value of the electric field), and $\boldsymbol{\mu}$ is the transition dipole moment operator of the molecular system found at the position $\boldsymbol{r}$. Most importantly, this operator comprises the transition dipole moment for the optical transitions of the studied molecular system. The incident electric field $\boldsymbol{E}_0(\boldsymbol{r}, t)$ will be, in general, assumed to be composed of several laser pulses with different envelops $\mathrm{E}_n(t)$, polarization vectors $\boldsymbol{e}_n$, carrier frequencies $\omega_n$ and directions of propagation $\boldsymbol{s}_n$ such that the pulse wave vectors read $\boldsymbol{k}_n = \frac{\omega_n n(\omega_n)}{c}\boldsymbol{s}_n$ with the refraction index $n(\omega)$:

$$\boldsymbol{E}_0(\boldsymbol{r}, t) = \sum_{n=1}^{N_p} \boldsymbol{e}_n \mathrm{E}_n(t - \tau_n)\left(e^{-i\omega_n(t-\tau_n)+i\boldsymbol{k}_n\cdot\boldsymbol{r}} + e^{i\omega_n(t-\tau_n)-i\boldsymbol{k}_n\cdot\boldsymbol{r}}\right). \quad (13.2)$$

The semiclassical theory of non-liner response of the material containing molecules of interest is formulated using classical Maxwell equations. These link self-consistently the macroscopic electric field $\boldsymbol{E}(\boldsymbol{r}, t)$ and the macroscopic polarization $\boldsymbol{P}(\boldsymbol{r}, t)$. The latter is induced by the incident fields. Nonlinear signals of certain order are linked to a nonlinear polarization of the same order in the exciting field.

### *13.2.2 Linear Absorption of an Ultrashort Pulse*

To set up the stage, it is worth discussing the well-known linear absorption experiment. To measure the absorption coefficient, we may apply monochromatic light of varying wavelength to a sample of molecules and record absorption at one wavelength at a time. Because of the linearity of the problem (we measure linear absorption coefficient), we can just as well apply the field, which contains all wavelength in a single ultrashort shot, and decompose afterwards the signal into individual components in a spectrometer. The idealized field of an ultrashort pulse containing all wavelengths of light in phase has the Dirac delta-function as its time-dependent envelop, $\boldsymbol{E}_0(\boldsymbol{r}, t) = \boldsymbol{e}E_0\delta(t - t_0)e^{i\boldsymbol{k}\cdot\boldsymbol{r}-i\omega_0(t-t_0)}$. As is usual in the theory of linear optical phenomena, we work with the complex fields, assuming that the results are obtained by taking their real parts. The Fourier decomposition of the ultrashort pulse into frequency components reads as

$$\boldsymbol{E}_0(r, \omega) = \int_{-\infty}^{\infty} dt \boldsymbol{E}_0(r, t)e^{i\omega t} = \boldsymbol{e}E_0 e^{i\boldsymbol{k}\cdot\boldsymbol{r}+i\omega_0 t_0}, \quad (13.3)$$

and, as a function of the frequency $\omega$, the field $\boldsymbol{E}_0(r, \omega)$ is constant. We will assume a field propagating along the $y$-axis of the Cartesian coordinate. For simplicity we ignore the vectorial character of the field, assuming that its polarization remains intact during the propagation. We can therefore discuss propagation of a scalar field $E(y, t) = \boldsymbol{e}\cdot\boldsymbol{E}(t)$, and link it through the Maxwell equations to the scalar polarization $P(y, t) = \boldsymbol{e}\cdot\boldsymbol{P}(y, t)$ of the material. The polarization is induced by the external electric field, and it feeds back into the propagation of this field through the material. It is this polarization that provides for us the doorway, through which we will introduce microscopic (quantum) theory of light-matter interaction into our description.

We can formally decompose the total electric field $E(y, t)$ into the exciting field, which is assumed to propagate in the medium without loss (only influenced by the refraction index $n(\omega)$), and the field generated by the polarization (which is itself generated in the material by the exciting field). By definition, the total field thus reads $E(y, t) = E_0(y, t) + E^{(1)}(y, t)$. The boundary condition of the problem is such that at $y = 0$, i.e. when the external field enters the sample (contained e.g. in an optical cell), $E(y = 0, t) = E_0(y = 0, t)$, so that $E^{(1)}(y = 0, t) = 0$. The field $E^{(1)}(y, t)$ is generated by the polarization, and interferes with the exciting field. In an absorption experiment, when the intensity of the total field is detected in a frequency resolved manner (by a square-law detector) after light passes a layer of material with the thickness $h$, we get

$$I(h, \omega) \approx |E_0(h, \omega)|^2 + 2\mathrm{ReE}_0^*(h, \omega)E^{(1)}(h, \omega) + \cdots . \tag{13.4}$$

We neglected the term containing the square of the weak first order field $E^{(1)}(y, t)$. The detected change in the light intensity is therefore a result of interference between the exciting incident field and the newly generated field. Equation (13.4) forms the basis of signal detection in the 2DES we discuss in this chapter.

To discuss the absorption, we switch to frequency domain by applying the Fourier transform (Eq. 13.3) to the wave equation for the electric field (see e.g. [4]). We will assume that the polarization has a leading term, which is a linear function of the electric field. In the frequency domain, the relation is extremely simple $P(\boldsymbol{r}, \omega) = \epsilon_0 \chi(\omega) E(\boldsymbol{r}, \omega) + P^{(nl)}(\boldsymbol{r}, \omega)$, where $P^{(nl)}(\boldsymbol{r}, \omega)$ represents nonlinear polarization (neglected in the linear experiments). For our particular problem of absorption of a field travelling along the $y$-axis, we will assume

$$P(y, \omega) = \epsilon_0 \chi(\omega) E(y, \omega) + P^{(nl)}(\boldsymbol{r}, \omega)|_y, \tag{13.5}$$

where $P^{(nl)}(\boldsymbol{r}, \omega)$ is allowed to have a wavevector pointing in direction different from the $y$-axis and $P^{(nl)}(\boldsymbol{r}, \omega)|_y$ represents the component of this polarization traveling into the direction of the $y$-axis. We assume that linear susceptibility $\chi(\omega)$ is a complex quantity, and it can be written using its real and imaginary parts $\chi(\omega) = \chi'(\omega) + i\chi''(\omega)$. For later use, we can define the index of refraction $n(\omega) = \sqrt{1 + \chi'(\omega)}$. Because the absorption is assumed to be weak, the amplitude of the field $E^{(1)}(y, \omega)$ changes only very slowly along the coordinate $y$, and correspondingly, when

$$E^{(1)}(y, \omega) = \tilde{E}(y, \omega) e^{iky}, \tag{13.6}$$

the quantity $\tilde{E}(y, \omega)$ has a negligible second derivative according to the coordinate $y$. The Maxwell equations then yield an equation for the slowly changing amplitude $\tilde{E}(y, \omega)$ in the form of

$$\frac{\partial \tilde{E}(y,\omega)}{\partial y} = -\gamma(\omega)\big(E_0 + \tilde{E}(y,\omega)\big) + \frac{i\omega}{2\epsilon_0 c n(\omega)} P^{(nl)}(\mathbf{r},\omega)|_y e^{-iky}, \tag{13.7}$$

where $\gamma(\omega) = \frac{\omega}{2cn(\omega)}\chi''(\omega)$. According to Eq. (13.7), the total field $E_0(y,\omega) + E^{(1)}(y,\omega)$ decays with the rate $\gamma(\omega)$, and the amplitude of the new field $E^{(1)}(y,\omega)$ changes due to the nonlinear polarization term $P^{(nl)}(\mathbf{r},\omega)$. In the case of linear experiments, we assume the latter to be negligible. The solution of Eq. (13.7) without the nonlinear term can be easily found in a form of $\tilde{E}(y,\omega) = E_0(e^{-\gamma(\omega)y} - 1)$, which yields $E^{(1)}(y,\omega) = E_0(e^{-\gamma(\omega)y} - 1)e^{iky}$ according to Eq. (13.6). The total field propagating through the material reads then as

$$E(y,t) = E_0 e^{-\gamma(\omega)y - iky}. \tag{13.8}$$

Assuming that the initial intensity of the light is $I_0 \approx |E_0|^2$, after passing a slab of the thickness $h$, we have the intensity $I(h,\omega) = I_0 e^{-2\gamma(\omega)h}$. For the absorption coefficient defined by Lambert–Beer law as $\alpha(\omega) = -\frac{1}{h}\ln\frac{I}{I_0}$ we finally get:

$$\alpha(\omega) = \frac{\omega}{cn(\omega)}\chi''(\omega). \tag{13.9}$$

Equation (13.9) links the absorption coefficient to the imaginary part of the linear susceptibility.

### 13.2.3 Reaction of a Molecular System to an Ultrashort Pulse

So far, we avoided almost all details of the molecules excited by light. In this section, we will look at a very much simplified picture of a molecular system, and we refine the notion of resonance between the exciting field and the molecular transitions. Nonlinear spectroscopy is a technique, which measures the response of molecular systems to manipulation by light. With the nonlinear spectroscopy in mind, we will look at a change induced by an ultrashort pulse to a system, which starts from an arbitrary state, not necessarily a ground state. Photosynthetic systems exhibit a series of excitonic bands (see Fig. 13.1a), of which the single- and two-exciton bands are relevant for the spectroscopic signals discussed here.

Let us consider a three-state system (see Fig. 13.1b) with energy gaps nearly resonant with carrier frequency of the excitation field $E_{ex}(t) = \mathrm{E}(t)\big(e^{-i\omega t + i\mathbf{k}\cdot\mathbf{r}} + e^{i\omega t - i\mathbf{k}\cdot\mathbf{r}}\big)$. We are preparing to enter the nonlinear realm, where we need to consider both complex conjugated components of the real fields. The function $\mathrm{E}(t)$ represents the field envelop, which does not need to be very short. The state of the system is described by the coefficients $c_g(t)$, $c_e(t)$ and $c_f(t)$ of the state vector $|\psi(t)\rangle = c_g(t)|g\rangle + c_e(t)|e\rangle + c_f(t)|f\rangle$, which evolves according to the Schrödinger equation (see Fig. 13.1c). For simplicity, we ignore the polarization of light and the fact that

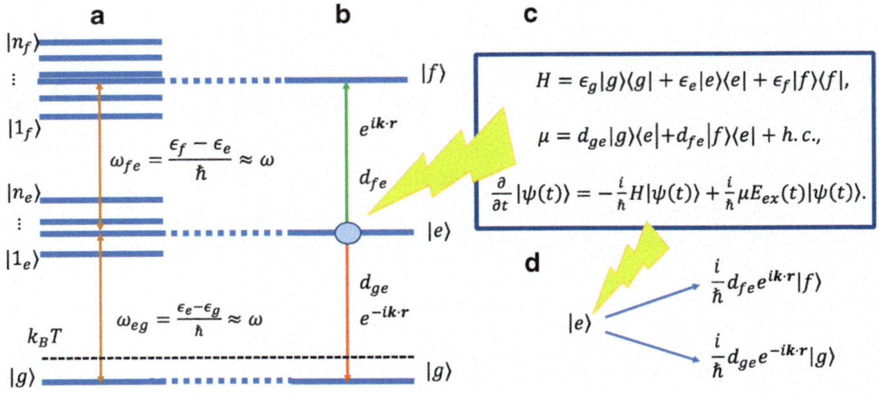


**Fig. 13.1** **a** Photosynthetic systems exhibit a number of bands of states separated by a characteristic frequency corresponding to the transition frequency of the individual pigments. **b** Three state model contains the ground state $|g\rangle$ and representative states $|e\rangle$ and $|f\rangle$ of the single- and two-exciton bands, respectively. Section **c** summarizes the Hamiltonian, dipole moment operator and the governing Schrödinger equation. **d** Starting from the state $|e\rangle$, the perturbation leads to transitions into states $|g\rangle$ and $|f\rangle$ characterized by specific phase factors $e^{ik\cdot r}$ and $e^{-ik\cdot r}$ (also depicted in part **b** of the figure)

the transition dipole moment operator $\mu$ is a vector. With the electric field absent, the coefficients evolve as

$$c_g(t) = c_g^0,\ c_e(t) = c_e^0 e^{-i\omega_{eg}t} \text{ and } c_f(t) = c_f^0 e^{-i\omega_{fg}t}. \tag{13.10}$$

Let us assume that the system starts from the excited state $|\psi(0)\rangle = |e\rangle$, i.e. $c_g(0) = 0$, $c_e(0) = 1$ and $c_f(0) = 0$. After external field was switch on, we expect the general oscillating character of the time evolution to be preserved, with the coefficients $c_g^0$, $c_e^0$ and $c_f^0$ of Eq. (13.10) now becoming time dependent. At the short time $t$, when almost no changes of population of the three states have occurred, the Schrödinger equation leads to the following approximate expressions for the coefficients $c_i^0(t)$:

$$\frac{\partial}{\partial t}c_g^0(t) = \frac{i}{\hbar}d_{ge}E_{ex}(t)c_e^0(t)e^{-i\omega_{eg}t}, \tag{13.11}$$

$$\frac{\partial}{\partial t}c_e^0(t) = 0, \tag{13.12}$$

$$\frac{\partial}{\partial t}c_f^0(t) = \frac{i}{\hbar}d_{fe}E_{ex}(t)e^{i\omega_{fg}t}c_e^0(t)e^{-i\omega_{eg}t} = \frac{i}{\hbar}d_{fe}E_{ex}(t)e^{i\omega_{fe}t}c_e^0(t). \tag{13.13}$$

In all three equations we neglected terms containing $c_g^0(t)$ and $c_f^0(t)$ on the right-hand-side as they are nearly zero. The solutions of Eqs. (13.11–13.13) can be obtained simply by integrating them, leading e.g. to $c_g^0(t) = \frac{i}{\hbar}d_{ge}\int_0^t d\tau E_{ex}(\tau)e^{-i\omega_{eg}\tau}c_e^0(0)$, where we assumed that $c_e^0$ does not evolve in time (see Eq. (13.12)). The electric field contains two oscillating terms. Assuming that all the frequencies involved are approximately equal $\omega \approx \omega_{eg} \approx \omega_{fe}$ we get

$$\int_0^t d\tau E_{ex}(\tau) e^{-i\omega_{eg}\tau} \approx \int_0^t d\tau \ \mathrm{E}(t) \left( e^{-i2\omega t + i\boldsymbol{k}\cdot\boldsymbol{r}} + e^{-i\boldsymbol{k}\cdot\boldsymbol{r}} \right) \approx e^{-i\boldsymbol{k}\cdot\boldsymbol{r}} \int_0^t d\tau \, \mathrm{E}(t). \tag{13.14}$$

The integral thus occurs over a function oscillating with twice the optical frequency, and over the field envelop alone. The former term can be neglected, because the integral is itself oscillating (it does not grow), while the latter integrates into a finite value, depending on the shape of the envelop. Interestingly, in Eq. (13.13), we have the opposite frequency under the integral and correspondingly we get a complex conjugated phase factor for $c_f^0(t)$. After the pulse is over, the coefficients read (assuming the overall change is only small)

$$c_g^0(t) = e^{-i\boldsymbol{k}\cdot\boldsymbol{r}} \frac{i}{\hbar} d_{ge} \int_0^t d\tau \ \mathrm{E}(t), \, c_e^0(t) = c_e^0, \, c_f^0(t) = e^{i\boldsymbol{k}\cdot\boldsymbol{r}} \frac{i}{\hbar} d_{fe} \int_0^t d\tau \ \mathrm{E}(t), \tag{13.15}$$

We can conclude that both the ground state $|g\rangle$ and the second excited state $|f\rangle$ received an equal share of population (up to a different value of the transition dipole moment), and the "deexcitation" and "excitation" processes carry different phase factors, namely $e^{i\boldsymbol{k}\cdot\boldsymbol{r}}$ for "excitation" and $e^{-i\boldsymbol{k}\cdot\boldsymbol{r}}$ for "deexcitation". In the rest of this Chapter, we will assume that the pulses are so short that no dynamics occurs during their duration, and they are spectrally broad so that all the states in the excited state manifold are covered by their spectrum. The most important conclusion we make here is that a short light pulse imprints its spatial phase factors $e^{i\boldsymbol{k}\cdot\boldsymbol{r}}$ and $e^{-i\boldsymbol{k}\cdot\boldsymbol{r}}$ onto the state of the system.

### *13.2.4 Multi-level Systems Embedded in an Environment*

Generalization of the above rules for a photosynthetic system, i.e. a molecular system with many levels in each band, separated roughly by the optical frequency, is straightforward. To have our description suitable for real-world spectroscopic experiments, we need to include the fact that electronic states of the photosynthetic systems interact with many molecular degrees of freedom, both in the molecules and in their protein environment. For the sake of clarity, we will adopt a phenomenological description of the protein influence on the signals detected in nonlinear spectroscopy. As a consequence, only the electronic degrees of freedom, which are directly addressed by light in a spectroscopic experiment, will be included in the Hamiltonian and the Schrödinger equation. The quantum expressions for the polarization of the material, foundational to our microscopic theory of spectroscopic signals, will be corrected phenomenologically to reflect the effects of dephasing and decoherence. These effects originate from the interaction of the pigment molecules with their protein environment, as well as their vibrational degrees of freedom. Both processes contribute to the decay of polarization.

The initial state of the system can be characterized by a single electronic state $|g\rangle$, because thermal excitation is highly unlikely for optical energy gaps (see Fig. 13.1a for comparison with the thermal energy $k_B T$). The light-induced transitions between electronic states are mediated by the transition dipole vectors $\boldsymbol{d}_{ng} = \langle n|\boldsymbol{\mu}|g\rangle$ and $\boldsymbol{d}_{Mn} = \langle M|\boldsymbol{\mu}|n\rangle$ for the transition between the ground state $|g\rangle$ and the excited state $|n\rangle$, and between $|n\rangle$ and the states $|M\rangle$ of the second band of states, respectively. When the transition dipole moment operator acts on the ground state, it projects on an excited state (with the weight given by the transition dipole moment), and when it acts on an excited state, it projects both back on the ground state and on the second excited state, analogously.

The action of a pulse with the wave-vector $\boldsymbol{k}$, arriving at the time $t_0$ results in a perturbed state with a small probability of being excited, and with an appropriate phase factor

$$|\psi(t < t_0)\rangle = |\phi_0\rangle|g\rangle \rightarrow |\psi(t_0)\rangle = |\phi_0\rangle|g\rangle + \frac{i}{\hbar}\sum_{n=1}^{N} \boldsymbol{d}_{ng} \cdot \boldsymbol{E}_0 e^{i\boldsymbol{k}\cdot\boldsymbol{r}}|\phi_0\rangle|n\rangle + \cdots . \tag{13.16}$$

Here $|\phi_0\rangle$ is the initial environment state and N is the number of the singly excited states, the dots represent higher order contributions. The system state then evolves $(t > t_0)$ according to the Schrödinger equation into

$$|\psi(t)\rangle = U_g(t - t_0)|\phi_0\rangle|g\rangle + \sum_{n=1}^{N} a_n(t - t_0)U_n(t - t_0)\frac{i}{\hbar}\boldsymbol{d}_{ng} \cdot \boldsymbol{E}_0 e^{i\boldsymbol{k}\cdot\boldsymbol{r}}|\phi_0\rangle|n\rangle + \cdots , \tag{13.17}$$

where $U_n(t)$ represents the free evolution operator of the state $|n\rangle$. It contains the phase factor $e^{-i\omega_{ng}(t-t_0)}$ and the propagation of the initial state of the protein environment $|\phi_0\rangle$. This propagation leads to the effects of dephasing and decoherence, which we will account for phenomenologically. The evolution operators will only function here as labels to keep track of the information about the environment effect that we need to account for later. The coefficients $a_n(t)$ keep track of the redistribution of the excitation among the excited states should it occur, and their squares give relative probability of finding the system in a state $|n\rangle$, after it was excited. We have $a_n(0) = 1$, while the probability $p_n(t_0)$ of initially populating the state $|n\rangle$ reads as $p_n(t_0) = \frac{|\boldsymbol{d}_{ng} \cdot \boldsymbol{E}_0|^2}{\hbar^2}$.

### 13.2.5 *Microscopic Theory of a Linear Absorption Coefficient*

In order to construct a theory of the spectroscopic signals, we need to be able to calculate the polarization created by the pulse at the position **r**. The polarization is given by the expectation value of the polarization (dipole moment) operator of a molecule residing at the position $r$, which reads as

$$\boldsymbol{P}(\mathbf{r}, t > t_0) = \langle \psi(t) | \boldsymbol{\mu}_{\mathbf{r}} | \psi(t) \rangle. \tag{13.18}$$

If we write Eq. (13.17) as an expansion in orders of the field, i.e. $|\psi(t)\rangle = |\psi^{(0)}(t)\rangle + |\psi^{(1)}(t)\rangle + |\psi^{(2)}(t)\rangle + \cdots$, we get two complex conjugated terms for the linear polarization, namely$\boldsymbol{P}^{(1)}(\mathbf{r}, t) = \langle \psi^{(0)} | \boldsymbol{\mu}_{\mathbf{r}} | \psi^{(1)} \rangle + \langle \psi^{(1)} | \boldsymbol{\mu}_{\mathbf{r}} | \psi^{(0)} \rangle$, where the first term reads as

$$\left\langle \psi^{(0)} \middle| \boldsymbol{\mu}_{\mathbf{r}} \middle| \psi^{(1)} \right\rangle = \frac{i}{\hbar} \boldsymbol{d}_{gn} \left( \boldsymbol{d}_{ng} \cdot \boldsymbol{E}_0 \right) e^{i\boldsymbol{k}\cdot\boldsymbol{r}} \langle \phi_0 | U_g^{+}(t) U_n(t) | \phi_0 \rangle. \tag{13.19}$$

The scalar product $\langle \phi_0 | U_g^{+}(t) U_n(t) | \phi_0 \rangle$ decays rather rapidly, and it determines the absorption lineshape. The decay process can be in general captured by a formula involving the so-called *line shape function* $g_n(t)$ ($n$ refers to the electronic state):

$$\langle \phi_0 | U_g^{+}(t) U_n(t) | \phi_0 \rangle = e^{-g_n(t) - i\omega_{ng} t}; \; t > 0. \tag{13.20}$$

Taking a Fourier transform of the polarization we obtain

$$\begin{aligned} \boldsymbol{P}(\boldsymbol{r}, \omega) &= \frac{i}{\hbar} \sum_n \boldsymbol{d}_{gn} \left( \boldsymbol{d}_{ng} \cdot \boldsymbol{E}_0 \right) e^{i\boldsymbol{k}\cdot\boldsymbol{r}} \int\limits_0^{\infty} d\tau e^{-g_n(\tau) + i(\omega - \omega_{ng})\tau} + c.c. \\ &= \epsilon_0 \overleftrightarrow{\chi}(\omega) \boldsymbol{E}_0 e^{i\boldsymbol{k}\cdot\boldsymbol{r}} + c.c.. \end{aligned} \tag{13.21}$$

Here, $c.c.$ represents complex conjugated term. The quantity $\overleftrightarrow{\chi}(\omega)$ is the tensor of linear susceptibility

$$\overleftrightarrow{\chi}(\omega) = \frac{i}{\hbar \epsilon_0} \sum_n \boldsymbol{d}_{gn} \int\limits_0^{\infty} d\tau e^{-g_n(\tau) + i(\omega - \omega_{ng})\tau} \boldsymbol{d}_{ng} \cdot, \tag{13.22}$$

where the dot at the end of Eq. (13.22) represents the scalar product, which is taken with the vector on which the tensor $\overleftrightarrow{\chi}(\omega)$ acts. The imaginary part of the susceptibility enters the absorption coefficient $\alpha(\omega)$, Eq. (13.9), which then reads as

$$\alpha(\omega) = \frac{\omega}{\hbar \epsilon_0 c n(\omega)} \sum\nolimits_n \langle \boldsymbol{d}_{gn} \boldsymbol{d}_{ng} \cdot \rangle_{\Omega} \mathrm{Re} \int_0^{\infty} d\tau e^{-g_n(\tau) + i(\omega - \omega_{ng})\tau}. \tag{13.23}$$

Here, $\langle \boldsymbol{d}_{gn}\boldsymbol{d}_{ng}\cdot\rangle_{\Omega}$ is the projection $\boldsymbol{d}_{gn}\boldsymbol{d}_{ng}\cdot\boldsymbol{e}$, where $\boldsymbol{e}$ is the polarization vector of the exciting field, back on $\boldsymbol{e}$, averaged over isotropic random distribution of the orientations of the vector $\boldsymbol{d}_{gn}$. Such an averaging yields $\langle \boldsymbol{d}_{gn}\boldsymbol{d}_{ng}\cdot\rangle_{\Omega} = \frac{1}{3}\left|\boldsymbol{d}_{gn}\right|^2$. To complete the microscopic derivation of absorption spectrum, we would need to provide the theory of the line-shape functions $g_n(t)$. The line-shape function is related to the so-called energy gap correlation function $C_n(t)$, which describes the fluctuations of the energy gap between the ground state and the state $|n\rangle$. We will, however, avoid discussion of microscopic models of the bath by pointing to specialized literature [5]. For later use, we will denote the integral in Eq. (13.23) as $G_n(\omega - \omega_{ng})$, i.e.

$$G_n(\omega) = \int_0^{\infty} d\tau e^{-g_n(\tau)-i\omega\tau}. \tag{13.24}$$

The absorption coefficient now reads as

$$\alpha(\omega) = \frac{\omega}{3\hbar\epsilon_0 cn(\omega)} \sum_n \left|\boldsymbol{d}_{gn}\right|^2 \mathrm{Re}G_n(\omega - \omega_n), \tag{13.25}$$

where the real part of the function $G_n(\omega)$ represents the absorption line shape.

### *13.2.6 Nonlinear Generalization*

Time resolution in spectroscopy can only be achieved by applying more than one pulse. The time axis on which we perform the measurement is defined by the delay between the pulses. An intuitive nonlinear spectroscopy extension of the linear absorption measurement, is the so-called *transient absorption* (TA). The idea is that the first pulse induces certain changes in the studied molecular system—by exciting it—and a second pulse measures the absorption of the system with varying delay after the first pulse. Subtracting from this signal the absorption measured without the first exciting pulse, we isolate the change in absorption due to the presence of the first pulse. Despite the intuitive character of this line of thought, we will not follow it directly. We will rather continue thinking in terms of manipulations performed on the state of a photosynthetic system by weak short pulses.

We know now that if we act with a short pulse on a molecular system, the phase factor $e^{ik\cdot r}$ will be imprinted on its state. This factor will transfer to the polarization (see Eq. (13.21)) which acts as a source of a new electric field, which then interferes with the one exciting the material. The nonlinear polarization in Eq. (13.7) acts as a source of nonlinear fields. The setting in which an experimenter has the most control over the nonlinear signal would be the one in which each pulse contributed a single order of nonlinearity. The nonlinear polarization $\boldsymbol{P}^{(nl)}$ would therefore ideally be a result of several linear contributions originating from different pulses, traveling in different directions. Let us assume a sequence of three pulses traveling in directions given by wave-vectors $\boldsymbol{k}_1$, $\boldsymbol{k}_2$ and $\boldsymbol{k}_3$. For simplicity, let us assume that the pulses have the same carrier frequencies $\omega_p = \omega_1 = \omega_2 = \omega_3$. The nonlinear polarization generated by these three pulses has all possible orders $n = 2, 3, \ldots$, but let us concentrate on the third order, and let us require that the polarization oscillates on the same optical frequency $\omega$ as the three pulses. As each spatial phase factor comes with a corresponding time-dependent phase factor—there is always $e^{i\boldsymbol{k}\cdot\boldsymbol{r}-i\omega_p t}$ and $e^{-i\boldsymbol{k}\cdot\boldsymbol{r}+i\omega_p t}$—we can assume that the polarization with a phase factor $\boldsymbol{k}_s = \pm\boldsymbol{k}_1 \pm \boldsymbol{k}_2 \pm \boldsymbol{k}_3$ will be generating a field with a frequency $\omega_s = \pm\omega_1 \pm \omega_2 \pm \omega_3$. To obtain a signal frequency equal to $\omega$ we need to combine the directions so that there is one minus contribution, e.g.

$$\boldsymbol{k}_s = -\boldsymbol{k}_1 + \boldsymbol{k}_2 + \boldsymbol{k}_3. \tag{13.26}$$

This direction is generally different from any of the directions of the exciting field, and we can therefore arrange Eq. (13.7) for the generation of a new field $\tilde{E}^{(3)}(y, \omega; t_2, t_1)$ in a form

$$\frac{\partial \tilde{E}^{(3)}(y, \omega; t_2, t_1)}{\partial y} + \gamma(\omega)\tilde{E}^{(3)}(y, \omega; t_2, t_1) = \frac{i\omega}{2\epsilon_0 c n(\omega)}\tilde{P}^{(3)}(\omega; t_2, t_1)e^{i(k_s-k)y}. \tag{13.27}$$

Here, $k_s = \boldsymbol{k}_s \cdot \hat{y}$ (with $\hat{y}$ representing a unit vector in the direction of the axis $y$) and $\tilde{P}^{(3)}(\omega; t_2, t_1)$ is the amplitude of the third order polarization $P^{(3)}(\boldsymbol{r}, \omega; t_2, t_1) = \tilde{P}^{(3)}(\omega; t_2, t_1)e^{i\boldsymbol{k}\cdot\boldsymbol{r}}$. In all the expressions above we explicitly denoted the parametric dependence of the polarization on the delays between exciting pulses. It can be shown [4] that the generated field $\tilde{E}^{(3)}(y, \omega; t_2, t_1)$ peaks sharply when the direction of the $y$-axis is chosen along the wave-vector $\boldsymbol{k}_s$. Very little or no field is generated in directions other than the given phase matching condition, Eq. (13.26). The class of experiments in which three pulses are used to generate a fourth field in a new direction is often denoted as *four-wave-mixing experiments*. If we orient the $y$-axis along $\boldsymbol{k}_s$ and neglect the absorption (by setting $\gamma(\omega) = 0$), Eq. (13.27) yields

$$\tilde{E}^{(3)}(y, \omega; t_2, t_1) = \frac{i\omega}{2\epsilon_0 c n(\omega)}\tilde{P}^{(3)}(\omega; t_2, t_1)y. \tag{13.28}$$

We assume that $\tilde{P}^{(3)}(\omega; t_2, t_1)$ does not depend on the location (phase factor has already been accounted for), because we neglect the absorption of the exciting fields, too. The third order field $\tilde{E}^{(3)}(y, \omega; t_2, t_1)$ thus grows with the distance travelled through the sample (note the factor $y$ in Eq. (13.28)). The frequency dependence of the pre-factor can usually be neglected for the relatively narrow frequency interval defined by the laser pulse width, and we can assume that we measure the third order field$\tilde{E}^{(3)}(y, \omega; t_2, t_1)$, which is a function of detection frequency $\omega$, and the two delays $t_1$ and $t_2$ between the three exciting pulses, and which is directly proportional to the third order polarization as

$$\tilde{E}^{(3)}(h, \omega; t_2, t_1) \approx i\tilde{P}^{(3)}(\omega; t_2, t_1)h. \tag{13.29}$$

The signal field defined in Eq. (13.29) can indeed be experimentally determined and detected as a function of $\omega$, which is conventionally denoted as $\omega_3$. It travels in a background free direction given by Eq. (13.26), where it is detected as a field with its phase and amplitude resolved in a scheme similar to Eq. (13.4) with the field $\boldsymbol{E}_0$ replaced by a controlled field $\boldsymbol{E}_{LO}$, the so-called local oscillator. The remaining question of the spectroscopic method outlined here is therefore: what kind of information is carried by the third order polarization $\tilde{P}^{(3)}(\omega; t_2, t_1)$, and how to best represent this three-dimensional quantity, to make its content more obvious and useful. To answer these questions, let us first study its microscopic origin.

### 13.2.7 Third-Order Response

We have chosen to investigate the third-order polarization generating a field in the direction given by the combination of wave-vectors, Eq. (13.26). We will assume that the pulses are characterized by delta-like envelops centered at times $\tau_1 = -t_2 - t_1$, $\tau_2 = -t_2$ and $\tau_3 = 0$. Because the signal is generated only after the third pulse arrives, setting time zero to the arrival of the third pulse ensures that the signal starts at $t = 0$. For simplicity, we assume equal amplitudes $E_1 = E_2 = E_3 = E_0$, equal frequencies $\omega_p = \omega_1 = \omega_2 = \omega_3$ and different wave-vectors $\boldsymbol{k}_1$, $\boldsymbol{k}_2$ and $\boldsymbol{k}_3$ for the three pulses. The delays $t_1$ and $t_2$ between the pulses are positive, i.e. the pulses arrived at the order $1 - 2 - 3$. Let us now generate the state of the system up to the third order in electric field.

The zero order state is the initial state of the system before any exciting field arrives, and it reads $|\psi_0(t)\rangle = U_g(t)|\phi_0\rangle|g\rangle$. It is denoted in the upper left corner of Fig. 13.2 by its electronic part $|g\rangle$ only. After the first pulse arrives, we obtain a state vector in the form of Eq. (13.17), composed of the zero order and the first order in the pulse with the wave-vector $\boldsymbol{k}_1$. The time $t_0$ equals to $t_0 = -t_2 - t_1$. The first pulse can only excite the system from the ground-state and correspondingly, only the factor $e^{i\boldsymbol{k}_1 \cdot \boldsymbol{r}}$ can appear. The state of the system after the action of the first pulse is depicted by the first grey layer (labelled 1) of the Fig. 13.2. When the second pulse arrives at the molecular system, it will perturb the zero-order term of the layer 1, and it will create a first order term with the same structure as in Eq. (13.17) in the layer labelled 2. We need to keep track of the fact that between the two pulses, the corresponding state evolved on the electronic ground state. This fact can be expressed by two consecutive evolution operators, in this case $U_n(t + t_2)U_g(t_1)|\phi_0\rangle$, each representing the system evolving in a different electronic state. Evolution operators are also denoted in Fig. 13.2. We will continue building the theory of spectroscopic signals *as if no spontaneous transitions between eigenstates of the molecular system occurred* and we will fix the lack of such transitions phenomenologically, once the theory is complete. From now on, we will therefore use evolution operators $U_g$, $U_n$ and $U_M$ to describe evolution of the protein environment, when the system is in electronic states $|g\rangle$, $|n\rangle$ and $|M\rangle$, respectively, assuming e.g. that $U(t)|e\rangle = U_e(t)|e\rangle$.

In the perturbed wavefunction, every occurrence of the wavevector with the positive sign means an excitation to the higher state, every occurrence of the wavevector with the negative sign means a deexcitation to a lower lying state. We will keep track of time as an argument of the wavefunction, and we will separate the time intervals between instances, in which the system was excited/deexcited to different electronic states, by a semicolon. After two interactions with light, we get several contributions, including the two second order ones $|\psi_2^{k_1+k_2}(t+t_2;t_1)\rangle$ and $|\psi_2^{k_1-k_2}(t+t_2;t_1)\rangle$ corresponding to excitation into the second excited manifold (phase factor $e^{i(\boldsymbol{k}_1+\boldsymbol{k}_2)\cdot\boldsymbol{r}}$), and deexcitation back into the ground state (phase factor $e^{i(\boldsymbol{k}_1-\boldsymbol{k}_2)\cdot\boldsymbol{r}}$), respectively.

The last pulse with the wave-vector $\boldsymbol{k}_3$ arrives at $t = 0$, and it will produce the third grey layer of Fig. 13.2, which contains 12 numbered contributions. There are seven new contributions created by the third pulse, one first order term $|\psi_1^{k_3}(t;t_1+t_2)\rangle$ (contribution 2 in Fig. 13.2), four second order contributions $|\psi_2^{k_1+k_3}(t;t_1+t_2)\rangle$, $|\psi_2^{k_1-k_3}(t;t_1+t_2)\rangle$, $|\psi_2^{k_2+k_3}(t;t_2;t_1)\rangle$ and $|\psi_2^{k_2-k_3}(t;t_2;t_1)\rangle$ (4, 5, 7 and 8 in Fig. 13.2), and also two relevant third order contributions, both finishing in the first exciting band state $|e\rangle$ and carrying the total phase factors $e^{i(\boldsymbol{k}_1-\boldsymbol{k}_2+\boldsymbol{k}_3)\cdot\boldsymbol{r}}$ and $e^{i(\boldsymbol{k}_1+\boldsymbol{k}_2-\boldsymbol{k}_3)\cdot\boldsymbol{r}}$ (contributions 10 and 12). The state of the system after interaction with three pulses up to the third order reads

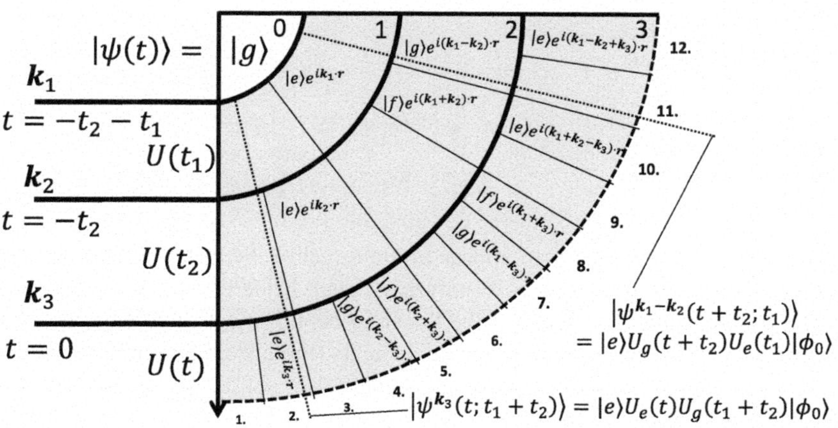


**Fig. 13.2** Construction of a three-times perturbed wavefunction of a molecular system. Three interactions with the electric field occur at times $t = -t_2 - t_1$, $-t_2$ and 0. Initial state corresponds the electronic ground state and the corresponding initial state of the bath, i.e. it reads as $|\phi_0\rangle|g\rangle$. Each interaction adds a perturbed term to the original state, where the electronic state is shifted and a corresponding phase factor is added. In a three-level system of Fig. 13.1 there are twelve different contributions. Between the interactions, the state evolves by the evolution operator $U(t)$. The response function of Eq. (13.33) is constructed by the contributions 2 and 11, which are stated explicitly including the evolution operators, but without spatial phase factors

$$
\begin{aligned}
|\psi_3(t)\rangle = & |\psi_0(t+t_1+t_2)\rangle + \left|\psi_1^{\boldsymbol{k}_1}(t+t_1+t_2)\right\rangle \\
& + \left|\psi_1^{\boldsymbol{k}_2}(t+t_2;t_1)\right\rangle + \left|\psi_1^{\boldsymbol{k}_3}(t;t_1+t_2)\right\rangle \\
& + \left|\psi_2^{\boldsymbol{k}_1+\boldsymbol{k}_2}(t+t_2;t_1)\right\rangle + \left|\psi_2^{\boldsymbol{k}_1-\boldsymbol{k}_2}(t+t_2;t_1)\right\rangle \\
& + \left|\psi_2^{\boldsymbol{k}_1+\boldsymbol{k}_3}(t;t_1+t_2)\right\rangle + \left|\psi_2^{\boldsymbol{k}_2+\boldsymbol{k}_3}(t;t_2;t_1)\right\rangle \\
& + \left|\psi_2^{\boldsymbol{k}_1-\boldsymbol{k}_3}(t;t_1+t_2)\right\rangle + \left|\psi_2^{\boldsymbol{k}_2-\boldsymbol{k}_3}(t;t_2;t_1)\right\rangle \\
& + \left|\psi_3^{\boldsymbol{k}_1-\boldsymbol{k}_2+\boldsymbol{k}_3}(t;t_1;t_2)\right\rangle + \left|\psi_3^{\boldsymbol{k}_1+\boldsymbol{k}_2+\boldsymbol{k}_3}(t;t_1;t_2)\right\rangle + \cdots, \qquad (13.30)
\end{aligned}
$$

where the dots represent third order terms, where any of the pulses contributes twice or three times. Our task now is to calculate the component of the third order polarization

$$
\tilde{P}^{(3)}(t;t_2,t_1) = \langle\psi_3(t)|\boldsymbol{\mu}|\psi_3(t)\rangle, \qquad (13.31)
$$

which has the correct phase factor given by Eq. (13.26). Because the first pulse must occur as $-\boldsymbol{k}_1$, and all other wave-vectors occur with a plus sign, we have only three possible terms of Eq. (13.30), which could be placed on the left of the operator $\boldsymbol{\mu}$ in Eq. (13.31), namely $\langle\psi_1^{\boldsymbol{k}_1}(t+t_1+t_2)|$, $\langle\psi_2^{\boldsymbol{k}_1-\boldsymbol{k}_3}(t;t_1+t_2)|$ and $\langle\psi_2^{\boldsymbol{k}_1-\boldsymbol{k}_2}(t+t_2;t_1)|$ (contributions 6, 7 and 11). These terms have to be complemented by other terms to form the required phase factor. The required component of the third order polarization therefore reads:

$$
\begin{aligned}
P^{-\boldsymbol{k}_1+\boldsymbol{k}_2+\boldsymbol{k}_2}(t;t_2,t_1) = & \left\langle\psi_1^{\boldsymbol{k}_1}(t+t_1+t_2)\middle|\boldsymbol{\mu}\middle|\psi_2^{\boldsymbol{k}_2+\boldsymbol{k}_3}(t;t_2)\right\rangle \\
& + \left\langle\psi_2^{\boldsymbol{k}_1-\boldsymbol{k}_3}(t;t_1+t_2)\middle|\boldsymbol{\mu}\middle|\psi_1^{\boldsymbol{k}_2}(t+t_2)\right\rangle \\
& + \left\langle\psi_2^{\boldsymbol{k}_1-\boldsymbol{k}_2}(t+t_2;t_1)\middle|\boldsymbol{\mu}\middle|\psi_1^{\boldsymbol{k}_3}(t)\right\rangle. \qquad (13.32)
\end{aligned}
$$

The signal that we expect to appear in the direction $\boldsymbol{k}_S$ (Eq. (13.26)) has three distinct components, which we will analyze in detail. All terms contain the factor $\boldsymbol{E}_0^3$ and the phase factor $e^{i\boldsymbol{k}_s\cdot\boldsymbol{r}}$. Just like the linear susceptibility, its nonlinear version is a tensor. It contains four dipole moment elements (three from the interactions with the fields and one because we take an expectation value of the dipole moment operator). The third order susceptibility (and the response function) is a tensor, which is multiplied by three vectors (three electric fields) to produce a single resulting polarization vector. If we assume that the molecules in each macroscopic point are randomly oriented, we can average over their random orientations. This produces a scalar factor, which depends on the polarizations of the incident pulses, and the polarization at which the signal is detected. The dependence of this factor on polarization of the pulses can be utilized to selectively enhance or suppress contributions to the nonlinear response (see *Polarization control in 2DES* section below). For now, we will just assume that such factor $\Omega(\boldsymbol{e}_1,\boldsymbol{e}_2,\boldsymbol{e}_3,\boldsymbol{e}_d)$, or simply $\Omega$, exists. The last factor, which needs to be accounted for, is the imaginary factor $\frac{i}{\hbar}$, which comes with a plus sign when it belongs to the ket vector of the first order (like in $|\psi_1^{\boldsymbol{k}_3}(t)\rangle\rangle$) and with a minus sign, when it belongs to a bra vector (such as $\langle\psi_1^{\boldsymbol{k}_1}(t+t_1+t_2)|$). Second order vectors always come with the factors $-\frac{1}{\hbar^2}$. For the first analysis, we will limit the number $N$ of states in the first excited state band to one ($n = e$), and we will start with the response function:

$$R_2(t, t_2, t_1) = \left\langle\left\langle\psi_2^{\boldsymbol{k}_1-\boldsymbol{k}_2}(t+t_2; t_1)\middle|\boldsymbol{\mu}\cdot\boldsymbol{e}_d\middle|\psi_1^{\boldsymbol{k}_3}(t)\right\rangle\right\rangle_\Omega e^{-i\boldsymbol{k}_s\cdot\boldsymbol{r}}/E_0^3, \tag{13.33}$$

where $\langle\cdots\rangle_\Omega$ denotes averaging over the random distribution of molecular orientations. In Eq. (13.33), we divided the polarization by electric field, which yields the so-called *response function*, related by Fourier transform to the nonlinear susceptibility. One can show that there are four types of response functions, conventionally denoted as $R_1$, $R_2$, $R_3$ and $R_4$. Equation (13.32) comprises three of four types of response.

Before we look in detail to the other two contributions to the third order field in the direction $\boldsymbol{k}_s$, we will try to understand and interpret the response function $R_2$.

The wave-vectors $\boldsymbol{k}_n$ in Eq. (13.33) reflect the history of interactions between the molecular system and light. First pulse, represented by the vector $\boldsymbol{k}_1$, excites the system from the electronic ground state $|g\rangle$, second pulse deexcites the system back to the ground state, third pulse excites to the excited state and the polarization is generated by the transition dipole moment $\boldsymbol{\mu}_{ge} = d_{ge}\boldsymbol{n}_{ge}$ between the ground- and excited states $|g\rangle$ and $|e\rangle$, respectively. Here, $\boldsymbol{n}_{ge}$ is a unity vector in the direction of the transition dipole and $d_{ge}$ is its length. Different field interactions, however, enter the expression for the expectation value of the polarization from different "sides". In order to understand the sequence of interactions, we will construct a diagram (Fig. 13.3a) visualizing the response function $R_2$ of Eq. (13.33). In this diagram, the time runs from bottom to top. Each interaction on the left- and the right-hand sides of the diagram is marked by an arrow and a corresponding final state. The state is propagated by the corresponding evolution operator $U(t_n)$, which is also denoted in the diagram. Each arrow representing excitation, or deexcitation is labeled with the corresponding phase factor. At the top of the diagram, we form the expectation value of the dipole moment operator. The whole diagram, which represents the polarization, gets one more imaginary unity to produce the electric field (see Eq. 13.29). Collecting all these factors we can calculate the overall sign of the contribution, which turns out to be $+1$ for the response function $R_{2g}$. We added an index $g$ to distinguish pathways involving only state $|g\rangle$ and $|e\rangle$ from those involving the state $|f\rangle$ in addition to them. The latter pathway would be denoted as $R_{2f}$.

Adding the factors $a(t)$, which track the relative probabilities of finding the system in a particular electronic state, the response function of Eq. (13.33) reads as

$$R_{2g}(t, t_2, t_1) = -\frac{i}{\hbar^3}|d_{ge}|^4 a_e^*(t_1)a_e(t)\Omega\langle\phi_0|U_e^+(t_1)U_g^+(t_2)$$
$$U_g^+(t)U_e(t)U_g(t_2+t_1)|\phi_0\rangle, \tag{13.34}$$

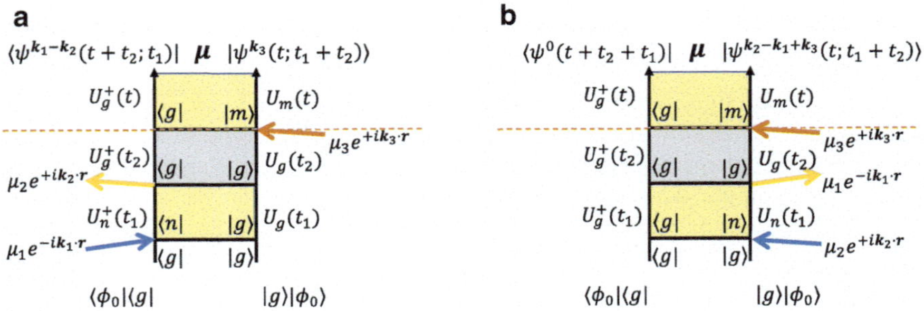


**Fig. 13.3** Diagrams corresponding to the rephasing response function $R_2$ **a** and its non-rephasing counterpart, the response function $R_4$ (introduced in the *Rephasing and Non-rephasing Signals* section below) **b**. The diagrams share the same spatial phase factor $k_s = -k_1 + k_2 + k_3$, in both the middle interval $t_2$ (grey) corresponds to the evolution of the system on the electronic ground state, they differ however, in the order of the first two pulses

where we defined $\Omega = \langle(\boldsymbol{n}_{ge}\cdot\boldsymbol{e}_1)(\boldsymbol{n}_{eg}\cdot\boldsymbol{e}_2)(\boldsymbol{n}_{ge}\cdot\boldsymbol{e}_d)(\boldsymbol{n}_{eg}\cdot\boldsymbol{e}_3)\rangle_\Omega$. The influence of the environment can be evaluated exactly with a specific model of the environment [4]. Here, we will only try to understand the overall time dependence of $R_2$ in order to provide an interpretation of nonlinear spectra. During the interval $t_1$ the first pulse is absorbed. The dependence of the function $R_{2g}$ on $t_1$ resembles a complex conjugated version of Eq. (13.20), from which we eventually calculated the absorption coefficient. The Fourier transform of $G_n(t_1)$ yields the absorption line shape $G_n(\omega_1)$. The dependence of the factor $R_{2g}$ on $t_2$ corresponds to the change of probability of finding the system in the excited state. We assumed preliminarily that no transitions between the states occur, and consequently we will assume that $R_{2g}$ does not depend on $t_2$. Finally, the dependence of $R_{2g}$ on $t$ resembles again Eq. (13.20), which enters the absorption coefficient, this time with the time dependent function $G_e(t)$. This leads to a simplified expression for the response function $R_{2g}$

$$R_{2g}(t,t_2,t_1) \approx -\frac{i}{\hbar^3}|d_{eg}|^4 a_e^*(t_1)a_e(t)G_e(t)G_e^*(t_1)e^{-i\omega_{eg}(t-t_1)}. \tag{13.35}$$

For the rest of our discussion, we will stay at this level of approximation.

The other two response functions, which form the signal travelling into the direction $\boldsymbol{k}_S$ are usually denoted as $R_{3g}(t,t_3,t_1)$ and $R_{1f}^*(t,t_3,t_1)$, where the star is a part of the standard notation (see e.g. [6]), and it denotes complex conjugation. At our level of approximation, these response functions yield

$$R_{3g}(t,t_2,t_1) \approx -\frac{i}{\hbar^3}|d_{eg}|^4 a_e^*(t_1)|a_e(t_2)|^2 a_e(t)G_e(t)G_e^*(t_1)e^{-i\omega_{eg}(t-t_1)}, \tag{13.36}$$

and

$$R_{1f}^*(t,t_2,t_1) \approx \frac{i}{\hbar^3}|d_{eg}|^2|d_{fe}|^2 a_f^*(t_1)|a_e(t_2)|^2 a_e(t)G_{fe}(t)G_e^*(t_1)e^{-i\omega_{fe}t+i\omega_{eg}t_1}, \tag{13.37}$$

where $G_{fe}(t)$ is connected by the Fourier transform with the line-shape $G_{fe}(\omega)$ of the absorption from the excited state $|e\rangle$ into a higher excited state $|f\rangle$.

The three contributions we have identified can be given an interpretation by identifying their overall signs and the excitation and deexcitation processes they describe. The responses $R_{3g}$ and $R^*_{1f}$ originate from the excited state, the former adding to the signal and the latter subtracting from it, in other words, the former is in phase and the latter is out of phase with the generating field. Given the states, into which the system is promoted after the interval $t_2$ by the action of the third pulse, (the ground state in the case of $R_{3g}$ and the higher excited state in the case of $R^*_{1f}$), we can interpret the contribution $R_{3g}$ as a *stimulated emission* (SE). This yields a field which adds to the generating field. The response $R^*_{1f}$ has a negative sign and therefore removes signal, i.e. it yields more absorption. It is correspondingly interpreted as an absorption into a higher excited state, the so-called *excited state absorption* (ESA). The remaining contribution, the response $R_{2g}$, also adds to the exciting field, just like the stimulated emission, but it involves the ground state. It turns out that it can be interpreted as the so-called *ground state bleach* (GSB), i.e. the diminished absorption due to depletion of the ground state. SE, GSB and ESA are three basic types of signals known in third order nonlinear spectroscopy.

We said that the response functions $R_{3g}$ and $R^*_{1f}$ depend on time $t_2$ only weakly, but this was conditioned by the requirement that there are no transfers between the states. If such transfers occur, the coefficients $a_n(t)$ do change in time. If a state $|n\rangle$ loses its population by a rate $K$, the probability of $P_n(t) = |a_n(t)|^2$ decreases exponentially as $P_n(t) = P_n(0)e^{-Kt}$. This means that $a_n(t)$ must decay as $a_n(t) = e^{-Kt/2}$(we defined $a_n(t)$ so that $a_n(0) = 1$). We assume that such a process leads to a dephasing between different electronic states and correspondingly, if it occurs during the time interval $t_1$ or $t$, it leads to an irreversible loss of the signal. When the transfer occurs during the interval characterized by time $t_2$, it is assumed to occur simultaneously on both bra and ket sides of the response. The signal is lost from the state $|n\rangle$, but it rises at a state $|m\rangle$, if this is the state to which the system transferred. Assuming that the populations $P_k(t)$ of excited states follow some set of rate equations with $K_{kl}$, the rate of energy transfer from $|l\rangle$ to state $|k\rangle$, we can define conditional populations $G_{kl}(t_2) = P_k(t_2; P_l(0) = 1)$ that a state $|k\rangle$ is populated at time $t_2$ when the state $|l\rangle$ was populated at $t_2 = 0$ with probability 1. Using this quantity, we can account for the change of the excited state, from which the signal originates during the time interval $t_2$. For the total probability to remain normalized, the rate $K_{kk}$ (the case when $k = l$) must correspond to the negative sum of the rates from $|k\rangle$ to all other states. We will use it to express coefficients $a_n(t)$, as discussed above. To simplify our formulae, we absorb the corresponding decay factor into the function $G$. This results in an absorption lineshape, broadened by energy transfer processes, defined as $\tilde{G}_{ng}(t) = G_{ng}(t)a_n(t) = G_{ng}(t)e^{-K_{nn}t/2}$.

Now we will have a set of response functions for each pair of the excited states $|n\rangle$ and $|m\rangle$, e.g.

$$R_3^{nm}(t, t_2, t_1) \approx -\frac{i}{\hbar^3}\left|d_{mg}\right|^2\left|d_{ng}\right|^2\tilde{G}_m(t)G_{mn}(t_2)\tilde{G}^*_n(t_1)e^{-i\omega_{mg}t+i\omega_{ng}t_1}. \quad (13.38)$$

Similarly, for the excited state absorption response we get

$$R_{1f}^{*nm}(t, t_2, t_1) \approx \frac{i}{\hbar^3} |d_{ng}|^2 |d_{fm}|^2 \tilde{G}_{fm}(t) G_{mn}(t_2) \tilde{G}_n^*(t_1) e^{-i\omega_{fm}t + i\omega_{ng}t_1}, \quad (13.39)$$

where $f$ represents any of the two-exciton states $|M\rangle$. As we can see, the excited state absorption signal is also influenced by energy transfer in the excited state manifold. Equations (13.35), (13.38) and (13.39) give the polarization divided by the exciting field. According to Eq. (13.29), when this is multiplied by imaginary unit it yields the third order electric field, divided by the exciting field. This is how we define our time dependent signal:

$$S(t, t_2, t_1) = \tilde{E}^{(3)}(t, t_2, t_1) \frac{e^{-i\boldsymbol{k}_s \cdot \boldsymbol{r}}}{E_0^3} = i\Big(R_{2g}(t, t_2, t_1) + R_{3g}(t, t_2, t_1) + R_{1f}^*(t, t_2, t_1)\Big), \quad (13.40)$$

where the $\tilde{E}^{(3)}(t, t_2, t_1)$ is the time domain version of the field defined in Eq. (13.29) and the $R$ functions represent the responses summed over all available excited states $|n\rangle$ and $|M\rangle$. To single out individual contributions to the signal $S$, we will denote its parts using the same notation as for the responses, e.g. $S_{2g}^{nm} = iR_{2g}^{nm}$. In order to best represent the signal, Eq. (13.40), we consider its frequency-resolved detection

$$S(\omega_3, t_2, t_1) = 2\mathrm{Re} \int_0^\infty dt S(t, t_2, t_3) e^{i\omega_3 t}, \quad (13.41)$$

in the so-called heterodyne detection scheme (see *Description of the 2DES experiment* section). This Fourier transform yields absorption line-shapes $G_n(\omega_3)$. Interestingly, Fourier transform according to time $t_1$ would yield a complex conjugated line-shape $G_n^*(\omega_1)$, if we define

$$S(\omega_3, t_2, \omega_1) = 2\mathrm{Re} \int_0^\infty dt \int_0^\infty dt_1 S(t, t_2, t_3) e^{i\omega_3 t - i\omega_1 t_1}. \quad (13.42)$$

The signal component $S_3^{nm}$ now reads as

$$S_{3g}^{nm}(\omega_3, t_2, \omega_1) = \frac{1}{\hbar^3} |d_{mg}|^2 |d_{ng}|^2 2\mathrm{Re} G_m(\omega_3) G_{nm}(t_2) G_n^*(\omega_1). \quad (13.43)$$

This is very close to be interpreted as a product of two absorption lineshapes (see for example Eq. 13.25). However, the fact that $G_n$ appears in Eq. (13.43) in a complex conjugated form means that in the real part of the expression for $S_{3g}^{nm}$, we have contributions from imaginary parts of both complex line-shapes $G_n$ and $G_m$. The components of the signal $S_{3g}(\omega_3, t_2, \omega_1)$ do not look like products of two absorption lineshapes, and without further knowledge, the two lineshapes derived from $G_n$ and $G_m$ cannot be disentangled from Eq. (13.43). The problem would be solved, however, if we could find a signal contribution, which would look exactly the same as Eq. (13.43), except for the complex conjugation in $G_n$. Remarkably, such signal exists among the fields generated in the very same experiment, except for the reversed order of the first and the second pulses. Without going into details, we notice that reversing the order of the first and second pulses in the diagram of Fig. 13.3a leads to diagrams of Fig. 13.3b. The two diagrams differ in the complex conjugation on interval $t_1$, only. Every signal contribution we discussed so far has its counterpart with the complex conjugated first $t_1$ interval.

We interpreted $S_{3g}$ signal corresponding to the response $R_{3g}$ as the stimulated emission (SE). Among all the available signal contributions, we can find a signal $S_{1g}$ (originating from the response function conventionally denoted $R_{1g}$) which has the same stimulated emission interpretation, but carries a complex conjugated lineshape in the variable $t_1$, so that we obtain a quantity such as

$$
\begin{aligned}
S_{3,SE}(\omega_3, t_2, \omega_1) &= \sum_{nm} \left( S_{3g}^{nm}(\omega_3, t_2, \omega_1) + S_{1g}^{nm}(\omega_3, t_2, \omega_1) \right) \\
&= \sum_{nm} \frac{1}{\hbar^3} \left|d_{mg}\right|^2 \left|d_{ng}\right|^2 2\mathrm{Re}G_m(\omega_3) G_{nm}(t_2) 2\mathrm{Re}G_n(\omega_1) \\
&\approx \sum_{nm} \alpha_m(\omega_3) G_{nm}(t_2) \alpha_n(\omega_1).
\end{aligned}
\tag{13.44}
$$

This newly formed signal can be interpreted as a product of two absorption lineshapes, although it represents a process of absorption on frequency $\omega_1$, followed by a SE on frequency $\omega_3$. Similarly, we get a signal for the GSB and ESA, by combining the signals from the directions $\boldsymbol{k}_S$ with the order of pulses 1–2–3 and 2–1–3. The line-shapes $\alpha_m(\omega_3)$ and $\alpha_n(\omega_1)$ are connected by the conditional probability $G_{nm}(t_2)$ for the transfer of population from the state $|m\rangle$ to $|n\rangle$. The signal $S_{3,SE}(\omega_3, t_2, \omega_1)$ thus correlates the absorption and emission frequencies, and allows us to observe excitation transfer as a function of the delay time $t_2$. We often refer to the spectrum defined by Eq. (13.44) as the absorptive spectrum. If in Eq. (13.42), instead of the real part, we take the imaginary part of the response functions, we obtain a spectrum corresponding to the product of the absorptive lineshape $\alpha_n(\omega_1)$ and a refractive (related to the refraction index $n(\omega)$ instead of absorption) lineshape $\mathrm{Im}G_m(\omega_3)$. This spectrum is often denoted as the refractive spectrum.

### *13.2.8 Rephasing and Non-rephasing Signals*

We have noticed in the previous section that only a spectrum constructed as a special combination of the third-order signals can be interpreted as an absorption–emission correlation plot. Only when each of the GSB, SE and ESA contributions to the signal in the direction of $\boldsymbol{k}_S$ where combined with conjugated contributions travelling in the same directions except for the reverse order of the first two pulses, we obtained the signal as a product of two absorption spectra. The signal corresponding to the reversed order of pulses is obtained by substituting the negative of the time delay $t_1$ into Eqs. (13.36), (13.39) and (13.40). The signal in the direction $\boldsymbol{k}_S$, corresponding to the pulse $\boldsymbol{k}_1$ arriving first, is characterized by an overall time-dependent phase factor$e^{-i\omega_m t+i\omega_n t_1}$, in which the two phases evolve in the opposite direction. This type of signal is usually termed *rephasing signal.* It is responsible for the well-known optical *photon echo,* because signals from molecules with different transition frequencies behave at the time $t = t_1$ as if they all had the phase equal to zero. The signal measured in the direction $\boldsymbol{k}_S$, with the pulse $\boldsymbol{k}_2$ arriving first, is characterized by an overall time-dependent phase factor$e^{-i\omega_m t-i\omega_n t_1}$, which does not have the rephasing character, and it is correspondingly termed *non-rephasing signal.* We now have at least two possible classifications of nonlinear signals: a) three physical types of signals (GSB, SE and ESA), and b) the rephasing and non-rephasing signals. The latter two can be measured separately as they correspond to different ordering of pulses, but they have to be added in order to yield a spectrum with a suitable physical interpretation. The real part of the so-called *total 2D spectrum* (rephasing and non-rephasing parts added) can be interpreted as an absorption–emission correlation plot. The word emission comprises also the effects of ground state bleach and excited state absorption, because they also correspond to a generated nonlinear signal with a specific phase. Furthermore, total 2D spectrum can be directly compared to more familiar TA spectrum (more about it in the *Two-dimensional Electronic Spectroscopy* section).

### 13.2.9 Additivity of 2D Spectrum

Linear signals, such as linear absorption have an important property that signals produced by many non-interacting systems are additive. This means that a spectrum of a mixture of two molecular species will correspond to a spectrum obtained by adding (with appropriate weights) the spectra of individual species. In a nonlinear signal, this cannot be in general expected. Nevertheless, it can be shown that 2D spectrum as conceived here is perfectly additive for non-interacting systems. The non-additivity can in fact be interpreted as a sign of interaction. One important consequence of additivity and an advantage gained by the 2D character of the spectra is the ability of 2D spectrum to reveal and to some extent overcome the inhomogeneity of the molecular ensemble on which we perform the measurement. Let us assume, for simplicity, a molecule with one transition $|g\rangle \rightarrow |e\rangle$ with frequency $\omega_{eg}$. In a macroscopic molecular sample it is likely that individual molecules will differ slightly by their transition frequencies (following some distribution around the characteristic frequency $\omega_{eg}$). According to the theory presented here, the molecule which absorbs with frequency $\omega_{eg}+\delta\omega$ will also emit on the same frequency. In fact, if we set $t_2 = 0$ this will be true also for molecules, which have more than one excited state, and could in principle undergo a transition to a different state for $t_2 > 0$. Correspondingly, if the distribution of frequencies has the width comparable or larger than the lineshape of individual molecules, the 2D spectrum of an ensemble will become elongated along its diagonal line $\omega_1 = \omega_3$. Unlike in absorption spectrum, in the 2D spectrum one can distinguish the so-called *homogeneous line-width* originating from individual molecules, from the so-called *inhomogeneous line-width* which is produced by the disorder in an ensemble of molecules. The homogeneous lineshape can be read from the spectrum by looking at its anti-diagonal width, while the inhomogeneous lineshape is characterized by the diagonal width. This applies to both the positive (GSB and SE) and the negative (ESA) signals.

### 13.2.10 Cross-Peaks for Monitoring Excitonic Energy Transfer

Let us finally consider a molecular system, such as a photosynthetic antenna, composed of interacting pigment molecules. By interaction we mean mutual influence of molecular electronic transitions. We will roughly divide these interactions into two classes: first, the molecules interact weakly, so that their optical transitions are only weakly affected (we actually assume that they are not affected at all) but the molecules can exchange energy by some relatively slow energy transfer process. A typical example of this type of interaction is Förster resonance energy transfer. As a consequence of the weakness of interaction, the 2D spectrum at $t_2 = 0$ will look like a spectrum of non-interacting molecules, but at $t_2 > 0$, we will observe effects of energy transfer between the molecules. The second class of interactions will comprise all situations, in which the molecules interact strongly, such that the electronic transitions on the individual molecules are significantly affected by the interaction. We will discuss the situation frequently occurring in photosynthetic systems, namely that the individual molecular transitions combine to form delocalized excitons, i.e., states shared across several molecules.

We will first discuss the weak coupling situation in a Donor–Acceptor dimer (see Fig. 13.4).

At $t_2 = 0$ the spectra of two molecular species, which form the dimer, are additive, and we see only two diagonal peaks. To understand properly the origin of 2D spectra features marking the energy transfer in both the weak and the strong coupling limits, we need to explain in more detail, how the spectrum in a dimer case comes about. Figure 13.4 presents the dimer of molecules by two equivalent state diagrams. The diagram Fig. 13.4a represents the two molecules by their molecular states, while Fig. 13.4b shows the *collective states* of the dimer. By generalization of the problem to $N$ molecules, we arrive at the band structure shown in Fig. 13.1a. The collective states can be always defined, no matter whether we work with states of interacting or non-interacting entities. A collective ground state is a state of the dimer, in which both molecules are in their ground states. The two collective *singly excited states*

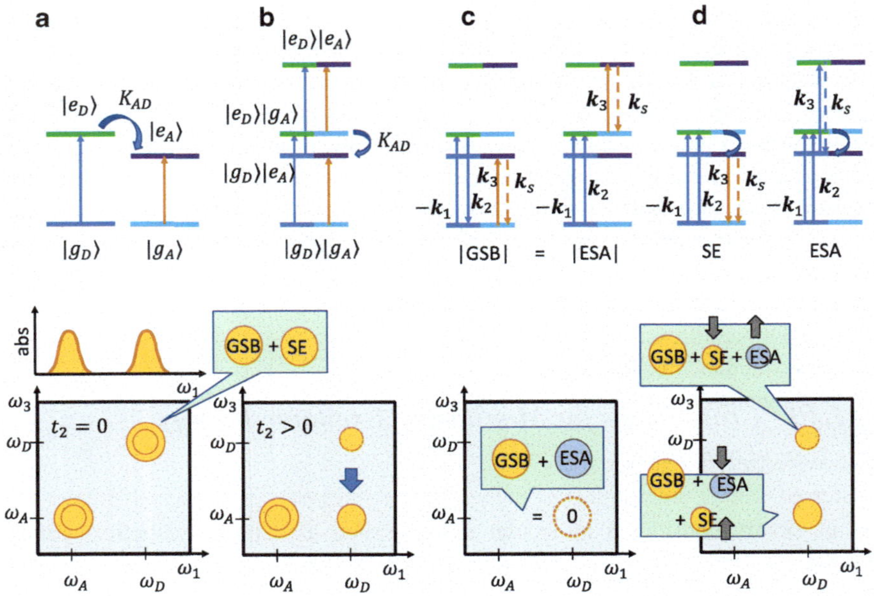


**Fig. 13.4** Weakly coupled Donor–Acceptor dimer in the molecular states diagram **a** and the collective states diagram **b**. Throughout the figures we denote the Donor transition by blue arrow and the Acceptor transition by orange arrow. Different molecular states in both molecular and collective state diagrams are represented by different colors. At the bottom part of parts **a** and **b**: absorption spectrum and $t_2 = 0$ 2D spectrum of the dimer, and the $t_2 > 0$ 2D spectrum with a relaxation cross-peak. In part **c** we demonstrate that the ground state bleach (GSB) and the excited state absorption (ESA) signals are composed from exactly the same transitions, leading to two signals, which differ only by the overall sign, and correspondingly they cancel each other—bottom section of part **c**. In Part **d**, energy transfer occurs with the energy transfer rate $K_{AD}$ (see Parts **a** and **b**), which leads to a new SE signal (left-hand-side of the Part **d** and a new ESA signal (right-hand-side of the Part **d**). Bottom section of Part **d** shows the composition of the Donor diagonal peak, and the transfer cross-peak. The grey arrows in the 2D spectrum denotes the tendencies of the signals to rise (arrow pointing up) or decay (arrow pointing down) with time

are states, in which one (or the other) molecule is excited, and there is also a *doubly excited state*, in which both molecules are excited. Such states can be formally formed by any number of molecules, however distant, interacting or non-interacting. Using the collective and the molecular state diagrams is completely equivalent in the non-interacting case, however, the advantages of the collective picture become apparent when applied to strongly interacting systems, where the molecular state diagrams lose their validity entirely.

Let us concentrate on the time $t_2 = 0$ and the rephasing signal only. We have three contributions to the signal (GSB, ESA and SE), and we are interested in the contributions, which combine the two molecules (signal originating from one of the two molecules exclusively are already additive, and of no concern here). The GSB signal, for instance, can be formed from excitation and deexcitation on frequency $\omega_D$ (pulses $\boldsymbol{k}_1$ and $\boldsymbol{k}_2$) and an excitation at the frequency $\omega_A$ (pulse $k_3$). This is denoted in Fig. 13.4c. The transition, which produces the signal, originates on the Acceptor molecule. Similarly, there is an excited state absorption contribution, which starts with two excitations on the Donor molecule at frequency $\omega_D$ and ends with the Acceptor transition at frequency $\omega_A$. This contribution involves precisely the same transitions, and it yields the signal with the opposite sign to the one of the GSB. As a consequence, the two signals cancel. No cross-peak therefore appears in the 2D spectrum as presented at the bottom section of Fig. 13.4c. This cancellation of signals is exact and occurs for all combined diagrams involving any pair of non-interacting molecules. One might be inclined to conclude that these signals are therefore only a product of our theoretical apparatus, and have no reality on their own. However, recent experiments with fluorescence-detected 2DES experiment strongly suggests that we indeed deal with true cancelling signals [7, 8].

When energy transfer takes place in a system with weak coupling, the signal contributions involving the excited state (ESA and SE) decay. Simultaneously, new pathways emerge, generating SE and ESA from the state to which the system is transferred. The 2D spectrum with energy transfer at $t_2 > 0$ is depicted in Fig. 13.4b, bottom section. Figure 13.4d presents the corresponding signals in a collective state diagram, from which we can see that the new signals occur on different detection frequency than the original ones (before the energy transfer). They therefore contribute to different $\omega_3$ frequencies. Bottom section of Fig. 13.4d demonstrates the composition of the energy transfer cross-peak, where there was originally a cancellation between GSB and ESA signals. With the rise of the cross-peak, we can see a decay of the Donor diagonal peak. If the energy transfer is such that the Donor gets completely depleted at very long times, $t_2 \to \infty$, the Donor diagonal peak will be composed of the GSB and the completely transferred ESA. These signals will perfectly cancel, and the diagonal peak disappears, being effectively completely transfers into the cross-peak.

### *13.2.11 Cross-Peaks for Monitoring Excitonic Coupling*

As we have seen, energy transfer in the weak coupling regime results in a cross-peak at $t_2 > 0$. The absence of a signal combining the two molecules was a result of cancellation of certain GSB and ESA contributions. The cross-peak can also appear, if this cancellation is perturbed by the interaction between the molecules. Figure 13.5 depicts just such a situation. Due to the strong resonance interaction, the transition dipole moments between the collective transitions are redistributed, and the ESA and the GSB on the cross-peaks both above and below diagonal do not cancel any more at $t_2 = 0$ From now on, we will refer to the two states of the system as Upper and Lower states as the collective states cannot be strictly identified with the molecular states of the Donor nor the Acceptor and both correspond to their superpositions. The amplitude of the cross-peaks depends on the strengths and the mutual orientation of the transition dipole moments involved in the transitions. In general, however, one can conclude that the stronger the cross-peak, the stronger the coupling. 2D electronic spectroscopy observes the eigenstates of the system, and correspondingly the cross-peaks do not directly reflect the coupling between the observed states, but rather their mutual correlation due to the fact that they are composed of the coupled localized electronic transitions.

As the time $t_2$ progresses, the peaks redistribute in a similar way as in the case of the weakly coupled molecules. In the case of weakly coupled molecules, all diagonal peaks corresponding to molecules, which lose their excitation either due to deexcitation or due to excitation transfer, will be absent at $t_2 \to \infty$. For a strongly coupled system, however, cross-peaks due to the exciton coupling remain present also at very long $t_2$ times.

In Fig. 13.6 we present all rephasing diagrams that lead to four peaks, assuming that there was a complete transfer of energy to the lower lying state. To gauge the relative strengths of the cross-peaks at long times, let us assume that the Upper state transition, which is strengthened in our case, has twice stronger absorption than the Lower state, i.e. it has a transition dipole moment by the factor of square root of two larger. The diagonal peak at the Lower state spectral region has the GSB and SE contributions, both produced by the dipole which was weakened by the coupling between the molecules. Let us assume this peak has a weight of one. The diagonal peak at the Upper state region has two contributions: ESA, composed of one strengthened and one weakened transition and a GSB contribution, produced by two strengthened transitions. The ESA will be twice as strong in absolute value as the Lower state peak on the diagonal, and the GSB will be four times as strong as the Lower state diagonal peak. The total positive amplitude of the Upper state peak

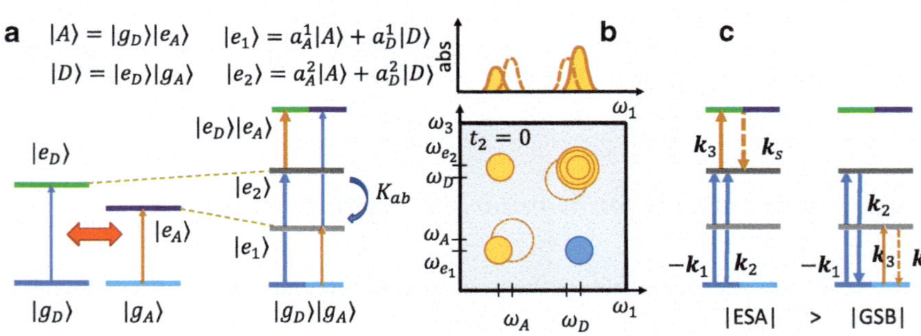


**Fig. 13.5** Strongly coupled dimer of molecules. **a** The resonance coupling between the molecular transitions (red double-sided arrow) leads to a level repulsion and formation of new collective eigenstates $|e_1\rangle$ and $|e_2\rangle$ of the dimer with new shifted transition frequencies $\omega_{e_1}$ and $\omega_{e_2}$. Also the transition dipole moments of the transitions are redistributed depending on the strength of the coupling and the mutual orientation of the transition dipole moments of the involved molecules. We have chosen the situation in which the transition into the upper state is enhanced. In this case the ESA from the upper state is also enhanced. All enhancements are depicted by thicker arrows. The positions of the absorption and 2D diagonal peaks are shifted with respect to the uncoupled situation (depicted in Part **b** by dashed lines), and their amplitudes are changed. Part **c** shows the transitions corresponding to the lower-left cross-peak ($\omega_1$ roughly corresponding to Donor frequency). Clearly the ESA contribution is stronger than the GSB contribution to the cross-peak therefore it is negative (the cross-peak below the diagonal in **b**)

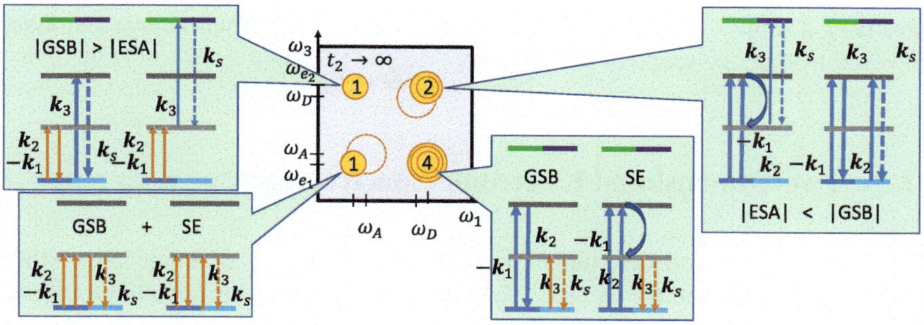


**Fig. 13.6** Cross-peak amplitudes at very long $t_2$ times for strong resonance coupling case. Assuming that the Upper state transition dipole moment was strengthened by the resonance interaction between Donor and Acceptor by a factor of $\sqrt{2}$ we obtain four positive peaks at very long $t_2$ times. The peaks in the upper row (Donor detection frequency $\omega_3 \approx \omega_D$) are of excitonic origin, similarly to the cross-peaks appearing at $t_2 = 0$

will be (4–2 = 2) twice as large as the one of the diagonal Lower state peak. The cross-peak excited at the Upper state and emitted at the Lower state frequency has two positive contributions (GSB and SE), both twice as large as the Lower state peak. The total relative amplitude of this cross-peak is therefore four. Finally, the cross-peak excited at the Lower state and emitted at the Upper state frequency consists of ESA with a relative weight of one, and GSB with a relative weight of two. This cross-peak will have a total relative amplitude of one. In contrast to the weak resonant coupling case, we have non-zero cross-peaks at the Upper state detection ($\omega_3$) frequency, while assuming that the Upper state is entirely depopulated. The effect is clearly originating in the excitonic coupling between the molecules. The dependence of the cross-peak magnitude on the coupling between the pigments in the photosynthetic complex is highly non-trivial and we can only hope to determine the couplings indirectly by fitting the spectra to specific structural models.

We have discussed in detail how the 2D spectrum reflects the energy transfer between individual electronic states in a molecular system. We have seen that not only the states and their time-dependent populations are revealed by the 2D spectra, but also certain correlations between the states, which originate in the delocalization of the observed states, due to the interaction between spatially distinct molecules. There are further details which can be learned from the spectrum. We can, for instance, follow the time evolution of the photo-induced time-dependent electronic coherence between the delocalized excited states of the photosynthetic system. The influence of these coherences on the 2D spectrum can be incorporated into our formalism by considering special types of nonlinear response, which involves a superposition of a pair of states during the delay $t_2$ (unlike the responses considered here, which always correspond to a single state populated during the time interval $t_2$). An important distinction is that the time evolution of a coherence between the two states leads to oscillations in the 2D spectra as a function the delay $t_2$. Such nonlinear signals have been a subject of intense study in recent years [9, 10]. Their interpretation is, however, beyond the scope of this chapter.

## 13.3 Two-Dimensional Electronic Spectroscopy

2DES allows for a full determination of the third-order polarization induced by the excitation pulses in the sample. Thus both absorptive and refractive properties of the sample can be studied simultaneously. In the applications of the technique to investigate photosynthetic systems, only absorptive part of the signal is usually analyzed (the imaginary part of dielectric permittivity), because it directly relates to the molecular properties of the system under investigation, e.g. reveals populations of the electronic states.

Even though 2DES is a third-order nonlinear optical spectroscopy technique, just like the more-standard TA, what distinguishes 2DES from other nonlinear spectroscopy techniques is that it provides an access to unprecedented amount of detail about the system under investigation [11]. 2DES has been inspired by the tremendous success of the multidimensional nuclear magnetic resonance spectroscopy [12], and similarly represents the measured signals on the 2D dimensional maps with excitation and detection frequency axes. Most importantly, this representation enables facile determination of correlations between different transitions (or resonances) detected in the system under investigation. These transitions can be excited both from the ground and higher states of the molecules. Another important aspect of the technique is that the correlation maps are obtained in a single experiment, and thus 2D spectra provide an overview of all the transitions present in the frequency range accessible in the experiment. In comparison, TA can usually provide the same information, but it requires measuring multiple spectra with tuned narrow-band excitation pulses. We are largely interested in observing dynamics of the system, following excitation by light, and this is revealed in 2DES by obtaining a sequence of the 2D spectra with changing population time, in analogy to TA measurements. Thus, the evolution of each spectral feature on the 2D map, related to the population of states and correlations between the states, can be followed. This is arguably the most useful feature when studying PPCs.

Besides studying population of different electronic states, another major application of 2DES in photosynthesis research, as well as in molecular studies in general, is recording coherence signals. Coherences, which report on the degree of superposition of quantum states, appear in 2DES (and any other third-order nonlinear spectroscopy) measurements as spectral signatures oscillating with increasing population time. Because signals are spectrally resolved on excitation and detection frequency axes in 2DES, the spectral shapes and patterns of the coherence signals can be readily analyzed. This type of information is not accessible in TA or time-resolved fluorescence measurements. Analysis of these patterns allows, for example, for determining the physical origin of coherences. A reader interested in learning more about the potential role of coherences (especially electronic coherences) are referred to [10].

Another advantage of the 2DES spectroscopy is that, in some implementations, polarization of each of the three excitation pulses, as well as of the detected signal can be controlled. In TA, in contrast, polarization of only two pulses, pump and probe, can be separately set. Thus intricate polarization schemes can be arranged in 2DES permitting a range of different anisotropy measurements. These, in turn, enable reducing the number of signals contributing to 2D spectra, which otherwise can be highly congested. Applications of polarization-controlled 2DES include filtering out only certain coherence signals, or enhancing correlation signals (see *Polarization control in 2DES* section below).

### 13.3.1 Implementations of 2DES

There are several implementations of 2DES, some of which use direct detection of the optical field. Others rely on measuring the so-called action signal of one kind or other (for an overview see [13]). In the most common implementation of the action-detected 2DES, fluorescence signal is measured [14], however, detection of photocurrent [15], photo emitted electrons [16] and ions [17] have been realized as well. In terms of generating necessary time delays between the pulses, some 2DES techniques use mechanical delay stages, others use pulse-shapers. There are also different implementations with respect to the geometrical arrangement of the exciting beams: collinear, pump-probe geometry (involving two beam directions), and fully non-collinear. In this chapter we will focus on the 2DES implementation, which detects coherent optical signal and exploits fully non-collinear geometry of a four-wave-mixing experiment. This is the type of 2DES technique, which is employed very broadly for studying photosynthetic and other molecular systems.

### 13.3.2 Description of a 2DES Experiment

In standard 2DES experiments, three pulses are used for the excitation of the sample and provide three interactions with the system (Fig. 13.7, also see *Third order response* section). The fourth pulse, the so-called local oscillator (LO), is used for implementing heterodyne detection of the emitted signal. First, two pulses are generated by making a pulse replica with the help of a plate beamsplitter, and then a transmission grating (or alternatively another plate beamsplitter) is employed to split each of these pulses into further two, thus generating two pulse pairs $k_1$&$k_2$ and $k_3$&$k_{LO}$. The direction of the generated signal in the sample is dictated by the phase matching condition $\boldsymbol{k}_s = -\boldsymbol{k}_1 + \boldsymbol{k}_2 + \boldsymbol{k}_3$. The LO pulse is send collinearly (usually through the sample) with the signal (Fig. 13.7).

The LO and the signal are then spectrally dispersed in a spectrometer and their interference is recorded by a multichannel detector, such as a CCD. Spectral interferometry is used to extract full information about the signal, i.e. its amplitude and

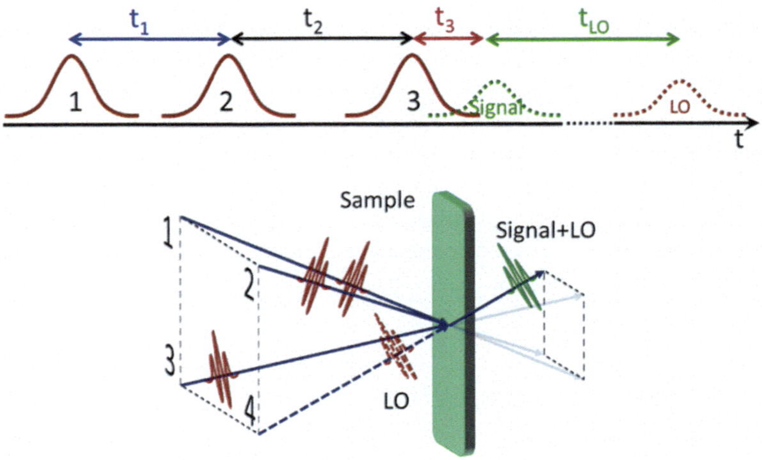


**Fig. 13.7** Arrangement of the pulses in time and in space in the standard non-collinear 2DES experiment

phase. Two time delays, denoted as *coherence* time, $t_1$, and *population* time, $t_2$, are varied during the experiment (Fig. 13.7). In spectral interferometry procedure, back and forward Fourier transforms are carried out with an appropriate windowing in the time domain to extract signal dependence on the detection frequency $\omega_3$ . The excitation frequency, $\omega_1$, dependence is obtained by scanning coherence time $t_1$ and performing Fourier transform along it. In any 2DES experiment there is a stringent requirement regarding the accuracy of the time delays $t_1$ and $t_{LO}$. The former is often scanned by gradually inserting a glass wedge into the beam path, which can provide attosecond resolution. The phase stability between pulses in pairs $k_1$&$k_2$ and $k_3$&$k_{LO}$ is assured by the geometry of the setup and the use of a transmission grating beamsplitter [6].

To obtain the total 2D spectrum, which is the sum of the rephasing and non-rephasing parts [4] (see also *Rephasing and non-rephasing signals* section) the coherence time, $t_1$, has to be scanned to both negative and positive ranges. It is important to note that only the total absorptive 2D spectra provide intuitive presentation of the signals, and allows for straight forward comparison to TA signals. For 2DES studies of photosynthetic systems the coherence time $t_1$ is typically scanned in the range of $\pm 50$ (ambient temperature) to $\pm 500$ fs (cryogenic temperature). Population time is usually scanned with the help of a mechanical translation stage, which can easily achieve accuracy of <2 fs and a scanning range of 1 ns. The time resolution of the 2DES experiment depends on the length of the pulses employed and typically is <20 fs. The frequency range covered in 2DES depends on the bandwidth of the pulses and often spans the wavenumber range of >1000 $cm^{-1}$. In most experiments, replicas of a pulse with the same spectrum are used, but two-color 2DES has also been implemented [18], where $k_1$&$k_2$ pulses have one spectrum and $k_3$&$k_{LO}$ another. In this way carrying out several single-color and two-color 2DES experiments enables for correlating optical transitions in any spectral region of interest.

The spectral resolution on the excitation frequency axis, $\omega_1$, in the 2DES experiment is given by the scanning range of the coherence time, whereas resolution of the detection frequency axis, $\omega_3$ is usually limited by the windowing function applied in the time domain to reject noise and to extract the signal from spectral interferograms. Typical resolution on the two frequency axes is in the range of 20 – 80 $cm^{-1}$. The common practice in the 2DES community is to use Fourier zero padding interpolation to upscale the true resolution, which produces smooth 2D spectra, which are visually more appealing.

Since 2DES is a phase-sensitive technique and the phases of laser pulses are imprinted on to the signal (see the *Reaction of a molecular system to an ultrashort pulse* section), the absolute phase of the signal has to be known to extract the real (absorptive) and imaginary (refractive) parts of the signal. To obtain the absolute phase of the signal the time delays $t_1$ and $t_{LO}$ have to be known with the absolute accuracy, which is highly challenging. Therefore, 2D spectra are generally "phased" by comparing to a signal which is intrinsically phased. The projection-slice theorem of multidimensional Fourier analysis is used [19] and appropriate projection of the 2D spectrum is compared to the TA signal measured under the same conditions. TA is a self-heterodyne technique, where, in the response functions formalism language, the first two interactions come from the pump pulse and therefore have the same phase, and the third interaction together with the LO are provided by the probe pulse, again having the same phase. Thus TA reports absorptive part of the signal in a manner similar to linear absorption (see Eq. (13.4)).

2DES is not self-normalizing technique and therefore each of the three interactions with laser pulses "imprints" the laser spectrum on the signal. In principle, this can be corrected for, but a common practice is to report the 2D spectra without normalization and in the electric field units. Thus care has to be taken when comparing 2D spectrum projections to TA signals, which are independent of the probe spectrum and are reported as unitless differential absorption or transmission.

### *13.3.3 Polarization Control in 2DES*

As already mentioned, 2DES allows for individual control of polarizations of all four interactions with the field. Setting different linear polarization can help to suppress and enhance different signals that correspond to certain interaction pathways. The signal selection principle is based on the orientational averaging over all the molecules or molecular systems present in the sample, which is usually isotropic. This averaging results in different orientational factors for different response functions, some of which can be enhanced, suppressed or even eliminated. If polarization of four pulses (fields) are **a**, **b**, **c**, **d**, and orientation of transition dipole moments that each fields interact with for any response signal are $\alpha, \beta, \gamma, \delta$ , the orientational factor $\Omega$ (in *Third-order response* section, Eq. (13.33) we introduced this factor for a special case of $\mathbf{a} = \mathbf{e}_1$, $\mathbf{b} = \mathbf{e}_2$, $\mathbf{c} = \mathbf{e}_3$, $\mathbf{d} = \mathbf{e}_\mathbf{d}$ and $\alpha = \beta = \gamma = \delta = \mathbf{n}_{eg}$) of a particular signal will be $\Omega = \langle a_\alpha b_\beta c_\gamma d_\delta \rangle$ [20]. It can be expressed in terms of projections between relative polarization orientations of laser pulses and relative polarization between transition dipole moments as

$$\begin{aligned}\langle a_\alpha b_\beta c_\gamma d_\delta \rangle = \frac{1}{30}\{&\langle cos\theta_{\alpha\beta} cos\theta_{\gamma\delta}\rangle(4cos\theta_{ab}cos\theta_{cd} - cos\theta_{ac}cos\theta_{bd} - cos\theta_{ad}cos\theta_{bc})\\ &+ \langle cos\theta_{\alpha\gamma} cos\theta_{\beta\delta}\rangle(4cos\theta_{ac}cos\theta_{bd} - cos\theta_{ab}cos\theta_{cd} - cos\theta_{ad}cos\theta_{bc})\\ &+ \langle cos\theta_{\alpha\delta} cos\theta_{\beta\gamma}\rangle(4cos\theta_{ad}cos\theta_{bc} - cos\theta_{ab}cos\theta_{cd} - cos\theta_{ac}cos\theta_{bd})\}\end{aligned}$$

Here $\theta_{ij}$ are the angles between any of the corresponding vector directions and $\langle\rangle$ denotes the orientational averaging. This equation allows for determining the orientational factor for any signal. The most useful observation here is that depending on the polarization sequence used in the experiment, some of the signals will have the orientational factor equal to zero, and thus can be completely suppressed.

Let us assume two weakly coupled transition dipoles (1 and 2). In this situation we excite localized molecular energy states. Let us consider only a few sequences of transition dipoles interrogated by pulses, namely: (1111), (1122), (1212) and (1221), and also the ones that can be obtained by exchanging all 1s to 2s and 2s to 1s. Considering the order of pulse-matter interactions in the nonlinear responses (see e.g. Figure 13.4 in the *Cross-peaks for monitoring excitonic energy transfer* section), it is easy to see that (1111) corresponds to the diagonal peaks, including both GSB and SE contributions, as well as purely vibrational coherences [21]. (1122) corresponds to energy transfer and correlation cross-peaks, whereas (1212) and (1221) correspond to coherence signals, which require first two interactions to be of different frequency and with different transition dipole moment orientations.

Using considerations regarding symmetries of the signals, some laser polarization sequences can be identified, which are the most useful for obtaining targeted information. The most commonly used polarization sequence is (0° 0° 0° 0°). Since in the standard experiment polarizations of all laser pulses are anyway parallel to each other, it is easy to set and usually results in the strongest signals in 2D spectra. However, this polarization sequence does not provide any signal selectivity and the measured signal is sensitive to depolarization of the signal, for example due to the rotation of the molecules in solution. Better polarization sequence to study population dynamics is the magic angle set (54.7° 54.7° 0 °0°), which does not provide any signal selectivity either, but does not suffer from the sensitivity to depolarization effects.

We list a couple of polarization sequences, which provide selectivity for particular signals: the cross-peak specific $S_{CPS}$ = (60° −60° 0° 0°) and the double crossed polarization $S_{DC}$ = (45° −45° 90° 0°). $S_{CPS}$ removes all (1111) type contributions, and therefore only cross-peaks should preferentially remain in the spectra [22]. It should be noted that the same signal can be reconstructed from the two *standard* measurements (0° 0° 0° 0°) and (90° 90° 0° 0°) [23]. Unlike $S_{CPS}$, which does not have a clear phasing procedure (see the *Description of the 2DES experiment section*), the latter two are straightforward to phase. However, the challenge when using two or even more different polarization sequences, which are then used in a linear combination, is to keep the relative amplitude of each measured sequence the same.

The $S_{DC}$ polarization sequence is even more selective, as it additionally removes the (1122) contributions and only the (1212) and (1221) type signals remain [22, 24]. Closer inspection reveals that only electronic coherences or coherences excited via vibronically coupled transitions contribute to the signal in these experiments. Any monotonously decaying signals, corresponding to population dynamics, or purely vibrational coherences are strongly suppressed. This polarization sequence is especially well suited for studies of inter-exciton coherences and vibronic coupling in the molecules and molecular systems.

### 13.3.4 Potential Challenges for 2DES

It should come as no surprise that such a complex spectroscopy technique, as 2DES, comes with a whole array of potential artefacts, and only the most common will be mention here. First, even though 2DES is a background free method, scattering from the exciting pulses poses a serious problem, since measured signals are usually orders of magnitude weaker. One method which has been successfully employed to enable measurements of scattering media, such as whole cells, is double-frequency lock-in detection technique [25].

Another challenge is the presence of contributions from the signals coming from "incorrect" ordering of pulses during the pulse overlap. These signals can be much stronger than the "proper" ones from the correct pulse ordering, especially when detecting very week signals in polarization-selected sequences (see the section above) [26]. Contribution from the incorrect pulse ordering artefacts dominates in the pulse overlap region. However, because of often unfavorable relative signal amplitudes, as well as pulses having wings in their temporal profile, they can be detected well outside the pulse overlap window. Prominent presence of such signals in the negative population time range (usually within the optical dephasing time of the system), also means that there is no clean onset of the signals of interest when crossing the population time zero, whereas such signal onset is clearly seen in TA or time-resolved fluorescence measurements. The possible way to deal with incorrect pulse sequence signals is to explicitly take them into account when modeling 2DES signals.

The third challenge is the presence of the so-called non-resonant nonlinear signals in the pulse overlap range [27]. These are dominated by the cross-phase modulation signals from the solvent and glass walls of the optical cell. These artefacts are third-order nonlinear signals and therefore scale in the same way as 2DES signals. Thus reducing intensity of laser pulses does not result in their suppression. Fortunately, because of their non-resonant nature, these signals are not present at all in the action-detected 2DES.

The common way of dealing with incorrect pulse ordering and non-resonant signals in coherent 2DES is by removing the pulse overlap region from consideration. This, of course, then excludes studying dynamic processes on the time scale comparable with the pulse length.

## 13.4 2DES Application Examples to Photosynthetic Aparatus of Green Sulfur Bacteria

We have described theoretical background and experimental details of carrying out 2DES studies of molecular systems. In the following we provide a couple of illustrative applications of 2DES to the studies of photosynthetic systems. Here we will focus on photosynthetic apparatus of green sulfur bacteria (GrSB) *Chlorobaculum tepidum*, which is an anoxygenic photoautotroph found in hot springs. Representatives of GrSB thrive at the see bottom near the back smokers, where light levels are extremely low, and therefore it can be assumed that these organisms evolved to capture and utilize scarcely available light very efficiently. Photosynthetic apparatus of these organisms, schematically depicted in Fig. 13.8 comprise light-harvesting antenna chlorosome, energy-transfer complex Fenna-Matthews-Olson (FMO) and RC. It has been estimated that about 30 RCs and 60 FMO complexes are connected to one chlorosome [28].

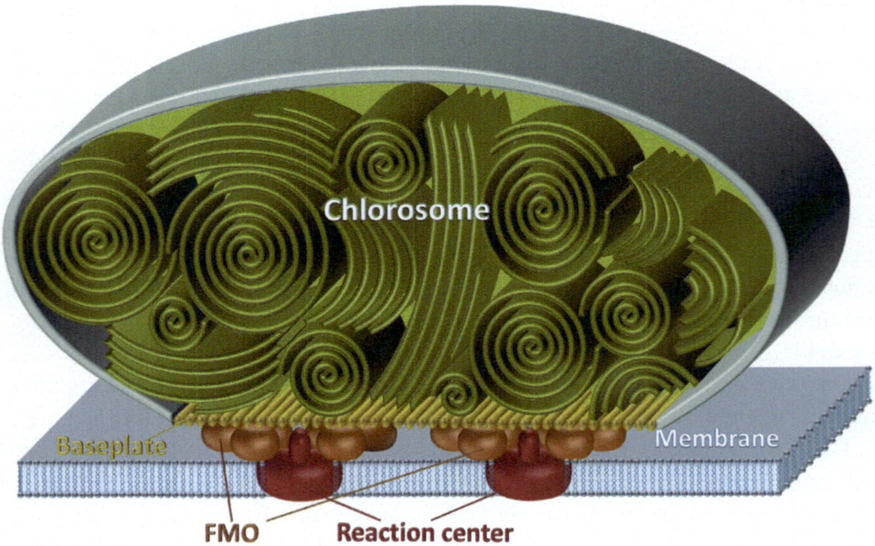


**Fig. 13.8** Schematic depiction of the photosynthetic apparatus found in green sulfur bacteria. The chlorosome is located on the cytoplasmic side of the membrane. The electron shuttles that complement the electron transfer chain, ferredoxin and cytochrome c, are not shown. Adapted with permission from Zigmantas et al. 2022. Copyright (2022) AIP Publishing

Chlorosome is the biggest light-harvesting antenna found in nature as it contains hundreds of thousands of bacteriochlorophyll (Bchl) molecules, which are self-aggregated in lamella and rod-like structures. Since the choromophores are tightly packed, excitations in chlorosome form excitons delocalized over multiple Bchl molecules. Because of the high density of molecules, chlorosome features a very high absorption cross section and serves as extremely efficient antenna. At the side of the chlorosome facing the FMO complexes, a semi-two-dimensional structure, called baseplate, is found. The structure of baseplate is comprised of the monomeric protein units binding Bchl *a* molecules. Since the states found in the baseplate are lower in energy than those found in the chlorosome aggregate, light absorbed anywhere in the chlorosome aggregate is eventually transferred to the baseplate.

Electron microscopy studies indicate that the FMO complexes are found just above the membrane in between the baseplate and the RCs [28, 29]. From FMO the excitation energy is transferred to the RC, where the charge separation takes place. Whereas the primary electron donor in the reaction center is the "special pair" formed by two Bchl *a* molecules, the final electron acceptors are iron-sulfur clusters. Electron donor and acceptor are located at the opposite sides of the photosynthetic membrane.

One of the goals of the primary photosynthetic functions studies is to track the energy transfer through the whole photosynthetic apparatus. This entails resolving the energy transfer rates and efficiencies both inside each of the constituting parts and in between them. During the past few decades each subunit of the photosynthetic machinery of GrSB has been extensively studied using various spectroscopic techniques. Here we highlight 2DES studies of the FMO complex and of the intact photosynthetic unit at cryogenic temperatures.

### *13.4.1 Excitonic Structure and Energy Transfer in the FMO Complex*

Owning to the fact that the FMO complex structure has been determined in seventies, and due to its relatively small size amendable to modeling, it has been one of the favorite objects for time-resolved spectroscopy studies in photosynthesis. Perhaps not surprisingly the first 2DES measurements on the biological systems were carried out on FMO. This study demonstrated the ability of 2DES to clearly show correlations between excitonic transitions [30].

The FMO complex comprises three identical protein units, each binding eight Bchl *a* molecules (Fig. 13.9a). The distances between the pigments in FMO are 11 Å and larger, which are short enough for Bchls to form a manifold of exciton states with transition energies found in the rage of 780 to 835 nm (11,980–12,820 $cm^{-1}$) (Fig. 13.9b).

In a more recent FMO 2DES study at 77 K, polarization sequences and global analysis of the 2D data were employed with the aim to gain more accurate electronic

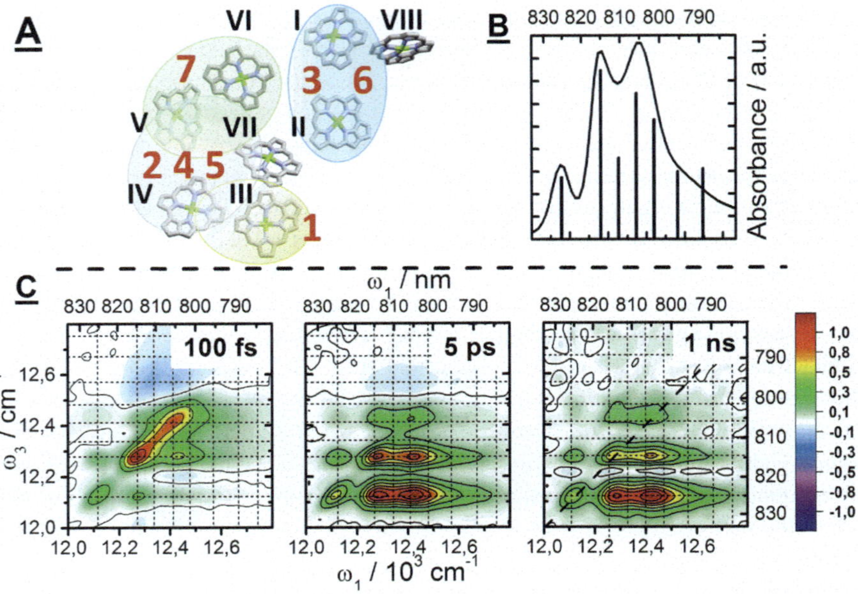


**Fig. 13.9** **a** arrangement of the Bchl *a* molecules in an FMO monomer. Ellipses indicate the delocalization of the exciton states. **b**: absorption spectrum of FMO at 77 K **c**: Absorptive magic angle 2D spectra measured at 77 K at indicated population delays. Dashed lines indicate excitonic energies. Adapted with permission from Thyrhaug et al. 2016. Copyright (2016) American Chemical Society

structure and the full network of energy transfer pathways [23]. Representative 2D spectra measured at 100 fs, 5 ps, and 1 ns are shown in Fig. 13.9c. Some information about the FMO complex can be gained directly by inspecting these 2D spectra. We recall here that the signals, which are seen in the 2D spectra, correspond to the electronic Hamiltonian eigenstates, which are exciton states in the case of FMO (see the *Photosynthetic complexes as open quantum systems* section). First, the abundance of the cross-peaks shows ubiquitous correlations between different exciton states, which implies presence of the coupling between the Bchl molecules (see the *Cross-peaks for monitoring excitonic coupling* section). In particular, occurrence of the positive cross-peaks above the diagonal directly reports on exciton correlation. In any of the shown 2D spectra, these cross-peaks are clearly visible between exciton states 1–2, 1–4 and 2–4. Additionally, excitonic coupling can be explored by inspecting the late population time, when all the energy relaxation within the exciton manifold is finished (~3 ps in the case of FMO). If there was no excitonic coupling, the only positive signals present in the 1 ns 2D spectrum would be GSB and SE contributions on the lowest row, corresponding to the detection frequency, $\omega_3$ of the 1$^{st}$ exciton. The clearly visible multiple peaks both on and off the diagonal again point to the appreciable excitonic coupling between the molecules. We would like to note here that presently a direct quantitative evaluation of the excitonic coupling from the amplitudes in 2D spectra is not attainable. However, coupling strength could be in principle estimated by carrying out full modelling of the system and fitting 2D spectra (see *Cross-peaks for monitoring excitonic coupling* section).

The second question regarding the FMO complex addresses electronic structure. Using excitonic calculations of FMO, absorption spectrum has been modelled and exciton state energies extracted [31]. This can be refined using 2DES data, because the presence of multiple correlation peaks allows for reading out of the energy states directly. For this purpose, cross-peak specific polarization sequence $S_{CPS}$ has been employed (see the *Polarization Control in 2DES* section), which removes strong diagonal peak contributions and allows for more clear "triangulation" of the cross-peaks. Analysis of both the diagonal and off-diagonal peaks in the 2D spectra obtained with different polarization sequences enabled determination of all the exciton states. This includes the eighth exciton (corresponding mostly to the 8th Bchl *a*), which was elusive for a long time. Via the correlation of the 8th exciton transition to the exciton 3, it was found that its energy is the highest of all the transitions in FMO, and is found at 12,750 $cm^{-1}$ (784 nm).

Finally, with the exciton energies and their correlations in place, energy transfer network through the FMO complex could be determined. This information is encoded in the evolution of the 2D spectrum with the population time $t_2$ from which one could in principle determine the complete information on the excited state dynamics. Even after a quick glance at the three 2D spectra presented in Fig. 13.9c, the downhill energy transfer can be recognized as weakening of the diagonal peaks and appearance of the cross-peaks below the diagonal. The latter peaks originate for the SE signals from the lower exciton states being populated by the energy transfer from the initially excited higher energy states. Tracing evolution of various peaks in the 2D spectra provides energy transfer rates and pathways. The best way to analyze the energy transfer network is to apply global analysis to the complete sequence of the measured 2D spectra. In [23] it was done by carrying out the global fitting of kinetic rate equations connecting each exciton state with all others. This procedure requires data with good signal-to-noise ratio, and, if successful, provides unique solution to the kinetic problem. For the FMO complex full energy transfer network was thus determined using global fitting. It should be mentioned that most of the energy transfer rates have been obtained earlier by TA spectroscopy. However, in [23] a full energy transfer network was obtained for the first time for any light-harvesting complex. The pairwise energy transfer rates between exciton pairs were found to proceed largely on sub-picosecond timescales, and ranged from 16 to 0.7 $ps^{-1}$. "Multi-branching", interconnected network of energy transfer pathways, where each state may transfer its population to a number of other states, was revealed. The energy flow branching is governed by the spatial location of excitons in the complex and their overlap. Knowing full energy transfer network, energy flow through space in the FMO complex can be visualized [10].

Yet another piece of information regarding the energy transfer was gained by observing the change of the lineshape of the lowest exciton peak. FMO complexes form trimers, where the distance between pigments in different monomers are larger than closest distances inside each monomer. Still, excitation energy can equilibrate between the different monomers in the trimer. By closely inspecting 2D spectra in Fig. 13.9c it can be seen that the lowest exciton diagonal peak in the 1 ns 2D spectrum is much more round than in the 100 fs or 5 ps spectra. The loss of diagonal elongation indicates the loss of excitation frequency memory, and this is exactly what is expected when energy is exchanged between the FMO monomers in the trimer, because each can have slightly different lowest exciton energy. From the lineshape dynamics it was estimated that the internal energy equilibration within the trimer was ~25 ps.

### *13.4.2 Energy Flow Through the Intact Photosynthetic Unit of Green Sulfur Bacteria*

The ultimate goal of the spectroscopic studies of photosynthetic light-harvesting is to obtain the full information about the energy transfer pathways, rates and efficiencies in the intact photosynthetic apparatus. This is a daunting task for any spectroscopic method, because it entails measuring the whole photosynthetic organism or intact and fully-functioning parts of it. Because cells are opaque and contain many organelles, which have dimensions comparable to the wavelength of light, they are infamous for their scattering properties. Time-resolved fluorescence spectroscopy is tolerant to scattering and was therefore previously employed for measuring either the intact photosynthetic cells or photosynthetic membranes containing most of the photosynthetic apparatus [32, 33]. However, in this method the excitation frequency resolution is poor, only emitting species are detected, and time resolution is limited. These shortcomings restrict the information that can be obtained on energy transfer processes.

2DES spectroscopy has the ability to correlate all the (allowed) transitions covered by the laser spectrum, thus each energy transfer pathway can be resolved inside the photosynthetic complexes and between them, if they have spectrally separable features. The scattering problem of the intact cells can be overcome by implementing scatter-resistant 2DES techniques (see for example [25]). Here we present an example of the 2DES study at 77 K of intact cells of GrSB *Chlorobaculum tepidum* [34], the photosynthetic unit of which is depicted in Fig. 13.8. The 2D spectra measured at selected population times from this study are shown in Fig. 13.10.

First, the 2D spectrum at early population time (30 fs) can be inspected. Here mostly diagonal features are observed, which correspond to absorption bands of the intact cells. The transitions that can be clearly identified correspond to a broad band of the chlorosome aggregate, three transitions of the FMO complexes and one peak of RC. Thus in photosynthetic unit of GrSB (Fig. 13.8) all major complexes connected in the energy transfer chain are spectrally separable. We note that the chlorosome baseplate signal "hides" below the FMO transitions, and thus cannot be probed separately. Additionally, besides the lowest energy transition seen at 11,950 $cm^{-1}$ in Fig. 13.10, RC features several more transitions from the RC antenna part, which overlap with stronger transitions of the FMO complex.

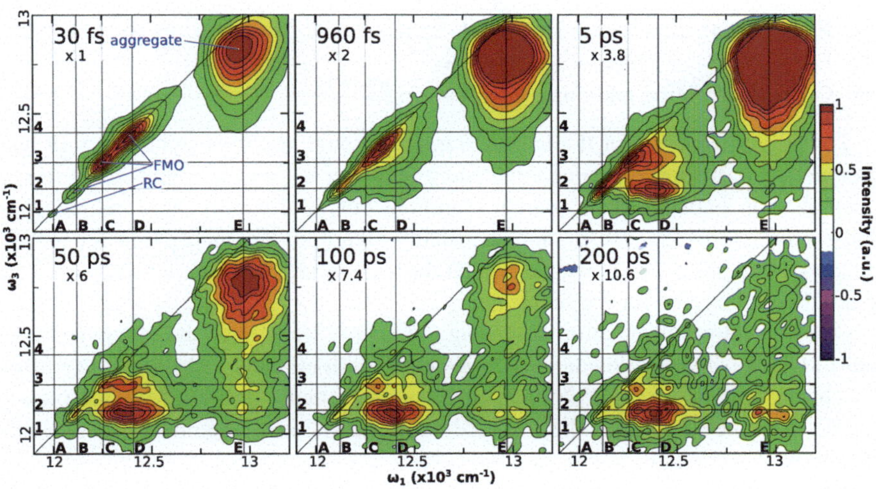


**Fig. 13.10** Normalized absorptive 2D spectra of the intact GrBS cells at indicated population times measured at 77 K. The horizontal and vertical lines indicate the distinct peaks, corresponding to transitions in the reaction center (**A**), FMO complex (**B**–**D**) and chlorosome (**E**). The excitation energy transfer is seen with increasing population times as a decay of the diagonal peaks and rise of the cross-peaks below the diagonal. Adapted with permission from Dostál et al. 2016. Copyright (2016) Macmillan Publishers Limited

By inspecting 2D spectra at different population times the energy transfer can be mapped through the whole photosynthetic unit. Note that the sequence of 2DES allows for simultaneous tracking of energy transfer after excitation to any of the subunits. For example, if the FMO excitation range is explored (excitation frequencies B, C, D in Fig. 13.10) similar exciton relaxation dynamics can be observed as presented in the 2DES study of isolated FMO complexes above. However, in addition to that, energy transfer to the RC can be observed as the appearance of weak cross-peaks at B1, C1 and D1. To follow the energy flow through the whole system, the 2D spectral range where the chlorosome is excited (excitation frequency E) has to be explored. Here, with increasing population time, it can be seen how the chlorosome diagonal peak disappears and the cross-peaks at FMO detection frequencies (2,3,4) appear. The final energy transfer step from FMO to RC is obscured here.

To get the clearer picture of all the energy transfer steps and rates, global analysis has to be employed. Here decay associate spectra (DAS) analysis was carried out [35]. This type of analysis globally fits all points in the 2D spectra by a sum of exponential functions. The outcome is that for each exponential decay or rise component an amplitude is assigned for each point in the 2D spectra, which together form a DAS spectrum. Thus one gets several 2D DAS, with corresponding exponential time constants. By applying DAS analysis to separate 2D spectral regions of the intact cells data the following energy transfer steps have been identified and characterized:

(1) Energy relaxation within the chlorosome aggregates: 100 fs – 1 ps.
(2) Chlorosome aggregate → baseplate → FMO energy transfer: ~**70 ps**.

(3) Energy relaxation within FMO: 0.1–25 ps.
(4) Energy transfer FMO → RC: **~17 ps**.
(5) Charge separation and other processes in RC: <1 ps.

Thus the full picture of the energy transfer through the photosynthetic unit has been revealed. It was ascertained that energy transfer within the complexes is much faster than in between them, which identified potential energy flow bottlenecks. Substantial part of the information, especially regarding the energy transfer within the complexes has been known from previous studies, only the transfer steps marked in bold were revealed for the first time.

One interesting observation regarding the energy transfer in the photosynthetic unit of GrSB is that while energy is transferred with efficiency close to unity from the chlorosome to FMO, 25% of excitations get "stuck" on the FMO complexes and do not transfer to RC. This is puzzling, because in the organism that evolved to survive in extremely low-light conditions it is expected that the energy transfer processes are highly optimized. Interestingly, in all the other studies of energy transfer between FMO and RC, mostly in FMO-RC complexes, estimated energy transfer efficiency has been similar or even lower, see for example [36]. This raises an intriguing question – how important was for photosynthetic organisms to develop highly-efficient energy harvesting machinery during evolutionary process. Perhaps other evolutionary drives ensuring survival of species have been more important.

## 13.5 Summary and Outlook

In this chapter we provided theoretical background for the description and understanding of 2DES signals from molecular systems. We also discussed aspects of experimental implementation of 2DES. By providing a couple of examples of application of 2DES to study photosynthetic systems we demonstrated the wealth of information that can be learned about the functions of photosynthetic machinery. This includes identifying exciton coupling between pigments, energy transfer pathways, rates and efficiencies. We focused on the studies of population dynamics directly related to the understanding of the energy harvesting function of photosynthesis. Even though coherence studies in photosynthesis garnered a lot of attention recently, we left them out of the present discussion, because their direct role for photosynthetic functions is highly uncertain [10].

Theory, experimental implementations and analysis tools of 2DES are constantly developing. This spectroscopic method will certainly keep revealing new information about photosynthetic systems in the future. Novel developments of 2DES include microscopy applications, which enable studies of inhomogeneous properties of photosynthetic complexes. Polarization control schemes in 2DES will be likely more widely employed to reduce the congestion and reveal weaker, but more informative signals. We also think that detection and analysis of coherence signals will be used more as a tool for learning subtle properties of biological systems, such as identifying [37] and eventually quantifying mixing of electronic and nuclear degrees of freedom (vibronic mixing) and understanding its role for energy and electron transfer processes. Future 2DES studies of intact photosynthetic units, preferably at physiological temperatures, will help us to obtain more holistic picture of primary photosynthetic functions of light harvesting and charge separation.

## References


1. R. Emerson, W. Arnold, J. Gen. Physiol. **16**, 191–205 (1932)
2. H. Sumi, J. Phys. Chem. B **103**, 252–260 (1999)
3. H. van Amerongen, R. van Grondelle, L. Valkunas, *Photosynthetic Excitons* (World Scientific, 2000)
4. S. Mukamel, *Principles of Nonlinear Optical Spectroscopy* (Oxford University Press, New York, 1999)
5. L. Valkunas, D. Abramavicius, T. Mancal, *Molecular Excitation Dynamics and Relaxation: Quantum Theory and Spectroscopy* (Wiley-VCH, Weinheim, 1 edition., 2013).
6. T. Brixner, T. Mančal, I.V. Stiopkin, G.R. Fleming, J. Chem. Phys. **121**, 4221–4236 (2004)
7. P. Malý, T. Mančal, J. Phys. Chem. Lett. **9**, 5654–5659 (2018)
8. O. Kühn, T. Mančal, T. Pullerits, J. Phys. Chem. Lett. **11**, 838–842 (2020)
9. G.D. Scholes et al., Nature **543**, 647–656 (2017)
10. J. Cao et al., Sci. Adv. **6**, eaaz4888 (2020)
11. D. Zigmantas, T. Polívka, P. Persson, V. Sundström, Chem. Phys. Rev. **3**, 041303 (2022)
12. R.R. Ernst, *Principles of Nuclear Magnetic Resonance in One and Two Dimensions* (Oxford University Press, Oxford, 1990)
13. F.D. Fuller, J.P. Ogilvie, Annu. Rev. Phys. Chem. **66**, 667–690 (2015)
14. P.F. Tekavec, G.A. Lott, A.H. Marcus, J. Chem. Phys. **127**, 214307 (2007)
15. G. Nardin, T.M. Autry, K.L. Silverman, S.T. Cundiff, Opt. Express **21**, 28617–28627 (2013)
16. M. Aeschlimann et al., Science **333**, 1723 (2011)
17. S. Roeding, T. Brixner, Nat. Commun. **9**, 2519 (2018)
18. J.A. Myers, K.L. Lewis, P.F. Tekavec, J.P. Ogilvie, Opt. Express **16**, 17420–17428 (2008)
19. D.M. Jonas, Annu. Rev. Phys. Chem. **54**, 425–463 (2003)
20. R.M. Hochstrasser, Chem. Phys. **266**, 273–284 (2001)
21. V. Butkus, D. Zigmantas, L. Valkunas, D. Abramavicius, Chem. Phys. Lett. **545**, 40–43 (2012)
22. M.T. Zanni, N.-H. Ge, Y.S. Kim, R.M. Hochstrasser, Proc. Natl. Acad. Sci. **98**, 11265 (2001)
23. E. Thyrhaug, K. Žídek, J. Dostál, D. Bína, D. Zigmantas, J. Phys. Chem. Lett. **7**, 1653–1660 (2016)
24. S. Westenhoff, D. Paleček, P. Edlund, P. Smith, D. Zigmantas, J. Am. Chem. Soc. **134**, 16484–16487 (2012)
25. R. Augulis, D. Zigmantas, Opt. Express **19**, 13126–13133 (2011)
26. D. Paleček, P. Edlund, E. Gustavsson, S. Westenhoff, D. Zigmantas, J. Chem. Phys. **151**, 024201 (2019)
27. J. Dostál, Spectrochim. Acta. A. Mol. Biomol. Spectrosc. **267**, 120441 (2022)
28. D. Bína, Z. Gardian, F. Vácha, R. Litvín, Photosynth. Res. **128**, 93–102 (2016)
29. H. Xie et al., Proc. Natl. Acad. Sci. **120**, e2216734120 (2023)
30. T. Brixner et al., Nature **434**, 625–628 (2005)
31. J. Adolphs, T. Renger, Biophys. J. **91**, 2778–2797 (2006)
32. M. Mimuro et al., J. Phys. Chem., 7503–7509 (1989)
33. M.G. Müller, K. Griebenow, A.R. Holzwarth, Biochim. Biophys. Acta **1144**, 161–169 (1993)
34. J. Dostál, J. Pšenčík, D. Zigmantas, Nat. Chem. **8**, 705–710 (2016)
35. J.A. Myers et al., J. Phys. Chem. Lett. **1**, 2774–2780 (2010)

36. N.C.M. Magdaong, D.M. Niedzwiedzki, R.G. Saer, C. Goodson, R.E. Blankenship, Biochim. Biophys. Acta Bioenerg. **1859**, 1180–1190 (2018)
37. E. Bukarte, A. Haufe, D. Palecek, C. Buechel, D. Zigmantas, Chem. Phys. **530**, 110643 (2020)

# Chapter 14
# Vibrational Coherence and Tunneling in Proteins

**Abdelkrim Benabbas and Paul M. Champion**

**Abstract** This chapter discusses the use of vibrational coherence and ultrafast wide-dynamic-range population kinetics to probe biological molecules. We show how impulsive stimulated Raman scattering can be used to develop the method of vibrational coherence spectroscopy, which reveals both the structural and functional aspects of the difficult to detect low-frequency modes ($h\nu \lesssim k_B T$) in proteins. Studies of electron tunneling in cytochrome c as well as the kinetics of the methionine-heme binding reaction are emphasized. Several ultrafast kinetic studies of heme proteins are used to infer the adiabaticity of ligand-heme binding reactions as well as the potential role of heavy atom tunneling (at temperatures below ~60 K). We also examine vibrational coherence and its potential participation in the excited state proton transfer of green fluorescent protein (GFP). We compare three independent observations of vibrational coherence in GFP and conclude that coherent motion does not affect the excited state proton transfer rate that occurs on the ps timescale. For the ground state proton back-transfer reaction, we find that (incoherent) vibrationally assisted proton tunneling is the dominant transport channel and that the tunneling rate is ~400 ps at room temperature. These studies suggest how serine and/or threonine residues may play an important role in controlling biological proton transport along water-based proton wires.

**Keywords** Ibrational coherence · Impulsive Raman · Electron tunneling · Proton tunneling · Heavy atom tunneling · Heme protein kinetics · GFP kinetics · Cytochrome c

---

A. Benabbas · P. M. Champion (✉)
Department of Physics and Center for Interdisciplinary Research on Complex Systems, Northeastern University, Boston, MA 02115, USA
e-mail: p.champion@northeastern.edu

 
K. Ueda (ed.), *Ultrafast Electronic and Structural Dynamics*,
https://doi.org/10.1007/978-981-97-2914-2_14

## 14.1 Introduction

Protein dynamics extend over a broad time window (fs to ms) [1] with many of the associated motions playing essential roles in function [2] and folding [3]. An important experimental and theoretical challenge is to unravel the structural reorganizations and time scales that are relevant to the biological function of a given protein, i.e., "functionally important dynamics" [4]. Sub-Ångstrom motions that take place on sub-picosecond timescales at a protein active site can be crucial in facilitating a wide range of chemical reactions. Common examples include proton tunneling [5–10], photoisomerization [11, 12] electron transfer [13–15], and bond cleavage or formation [16–20]. Within the protein interior, these rapid and small length-scale fluctuations are often underdamped and are best described as low-frequency molecular vibrations. Thus, the analysis of the thermally accessible ($\lesssim$200 cm$^{-1}$) vibrational modes at the active site, and their interaction with the surrounding amino acids, is key to determining the mechanisms of chemical reaction in proteins and to understanding how dynamic structures relate to the protein's function. Infrared and resonance Raman spectroscopy cannot reliably detect protein modes below ~150 cm$^{-1}$ in the aqueous phase, owing to the strong absorption of water and its quasi-elastic light scattering as well as to intense resonance Rayleigh scattering from the resonant chromophore [21]. In contrast, impulsive stimulated Raman scattering [12, 22–24] or vibrational coherence spectroscopy (VCS) [17, 20, 25–29] makes it possible to extract low-frequency vibrational modulations of the third-order polarization of the target molecule or resonant chromophore. This technique, which is briefly discussed in Sect. 14.2, provides access to the relatively unexplored region involving vibrational modes below 200 cm$^{-1}$ that undergo significant thermal excitation and, therefore, potentially play important functional roles. An ancillary goal of this chapter is to examine how quantum mechanical tunneling in proteins (including electrons, protons, and heavy atoms) can be modulated by these low-frequency vibrations and associated structural distortions.

It is worth noting at the outset that, although the excitation of coherent vibrational motions is utilized to probe the spectral content at low frequency, this by no means implies that vibrational coherence should necessarily be associated with functionality. Although there are conceivable exceptions for certain photoreactions with timescales shorter than the decoherence time (which is typically ~ps in the condensed phase), most low-frequency functional vibrations (and certainly all those associated with "dark" reactions) are excited non-coherently by stochastic interactions with the thermal bath.

Proton transfer (PT) is widespread in biology and it underpins many fundamental life processes [30–39] such as oxidative phosphorylation, photosynthesis, catalysis, and acid–base reactions. Some proton transfer processes in proteins or protein complexes occur by means of "proton wires", but such wires contain fluctuating or transient water molecules, and their composition can thus be difficult to resolve using traditional structural methods. These wires are often modeled using classical methods so that ionization-resistant amino acid residues, such as serine and threonine, with their large barriers for proton dissociation, are rarely (if ever) included as active transport elements. On the other hand, if these residues have sufficiently fast vibrationally assisted proton tunneling rates, they can be gainfully included in proton wire motifs as active transport elements. This suggests novel functional possibilities for serine and threonine going beyond their more traditional role as structural components that simply help to stabilize water molecules.

Surprisingly fast (sub-ns) proton tunneling rates at room temperature, even in the presence of the high potential energy barrier presented by serine, have recently been predicted [8] and observed [5] for the ground state proton back-reaction of the green fluorescent protein (GFP). Thermally excited low-frequency donor–acceptor (D-A) vibrational motions play a dynamical role that leads to time-dependent variations in the D-A hydrogen bond length, transiently reducing the proton tunneling distance and enhancing the tunneling rate by orders of magnitude [5–8, 40, 41]. Because of this, the GFP system offers an important model for the investigation of proton transfer within or through proteins via proton wires. One of GFP's most important properties arises from its internal tri-peptide (Thr-Tyr-Gly) chromophore that can be used to optically trigger PT on a structurally well-characterized proton wire [5, 42, 43]. This allows both the forward, non-equilibrium, excited state PT reaction, as well as the thermally equilibrated ground state PT "reset" back-reaction, to be monitored over a wide range of time and temperature [5].

The proton wire in GFP includes the chromophore Tyr residue along with water and glutamate, which are also common to other biological proton wires [44]. Additionally, the wire in GFP includes a high-p$K_a$ serine residue (Ser205) which nature has clearly included as an active proton transport element. The detailed experiments and theoretical investigations presented in Sect. 14.3 provide insight into the structure and function of such wires and they highlight the central role of low-frequency D-A vibrational motion in boosting the proton tunneling rate into the sub-nanosecond regime at room temperature, even though the equilibrium structure involves typical OH···O D-A distances (~2.7–2.8 Å).

Sections 14.4 and 14.5 of this chapter are devoted to the investigation of low-frequency vibrations and the mechanisms of electron transport and ligand binding in heme proteins. The active site of this class of proteins consists of a heme chromophore (iron protoporphyrin IX, or FePPIX). For the "b-type" heme (e.g., as found in myoglobin and hemoglobin), the iron atom is covalently linked to the protein only through the axial ligand(s) to the heme iron. For the "c-type" heme (found in cytochrome c), there are two additional thioether bonds from the heme periphery to the protein amino acid backbone.

One of the major outstanding issues in our understanding of heme proteins concerns how they structurally tune the properties of a given type of heme to efficiently carry out their broad range of biological functions. Although heme ligation and the protein structure surrounding the heme are undoubtedly important, protein-induced distortions of the heme plane along the low-frequency out-of-plane vibrational normal modes characterized as "doming", "ruffling" and "saddling" can also play an important role in tuning reduction potentials and various reaction rates (including electron tunneling), thus helping to optimize heme functionality [18, 25, 45, 46].

Systematic analysis of X-ray crystal structures of heme proteins has shown that the proteins belonging to the same functional class share similar out-of-plane (OOP) heme distortions [45, 47, 48]. These protein-induced OOP distortions are energetically unfavorable for the heme, and their evolutionary conservation indicates that they have biological significance. For example, doming and ruffling have been reasonably well characterized and correlated with protein functions [18, 45, 46, 49]. In Sect. 14.4 we will demonstrate how heme ruffling modulates electron tunneling in the photoreduction of ferric c-type cytochromes [46]. Doming is typically observed in ligand storage or transport proteins such as hemoglobin [50, 51] and myoglobin [52] and we will discuss in Sect. 14.5 how the distribution of heme doming geometries leads to kinetic inhomogeneity, strongly affecting the rebinding of carbon monoxide (CO) in heme systems [18, 49, 53–55]. We will also present evidence that ligand binding in heme proteins is adiabatic [53, 56], rather than non-adiabatic, and that reaction control by spin-selection rules, which is often assumed [57–59], does not appear to play an essential or particularly important role.

More specifically, we examine in detail the case of CO rebinding to bare protoheme as well as to the CO oxidation activator heme protein (CooA), which is the sensor responsible for inducing transcription of the proteins that certain bacteria utilize to oxidize CO as an energy source. We also investigate the photolysis of Met80, the endogenous methionine-heme ligand in ferrous cytochrome c (cyt c), and analyze its ultrafast rebinding kinetics. In addition, we report on the geminate rebinding kinetics of several heme ligands at very low temperature, which do not seem to follow the expected over-barrier classical predictions. The analysis suggests that quantum mechanical tunneling involving the heme doming coordinate may be an active channel for ligand binding at very low temperature [53, 56].

## 14.2 Vibrational Coherence Spectroscopy

Vibrational coherence spectroscopy (VCS) [20, 27–29, 60, 61] is an ultrafast pump–probe technique that is equivalent to impulsive stimulated Raman scattering [22–24, 62] whereby coherent vibrational states of molecules or resonant chromophores can be excited and studied. Due to the large spectral bandwidth of the ultrafast optical pump pulse, electric fields of different frequencies within the pulse interact with the molecular system to generate non-stationary superpositions of vibrational eigenstates. The subsequent coherent nuclear dynamics modulates the third-order polarization of the target molecule, which induces damped oscillatory changes in the temporal evolution of the optical response, as detected by the self-heterodyned probe pulse [60, 63, 64].

A diagram of a degenerate pump-probe VCS setup used in some of the work presented here is shown in Fig. 14.1. The laser system consists of a tunable (750–960 nm) Ti–Sapphire oscillator (MIRA 900; Coherent, Santa Clara, CA) pumped by a diode laser (Verdi 10; Coherent). Laser pulses of 50–100 fs duration and 76 MHz repetition rate, with an energy of ~10 nJ/pulse are generated by the oscillator. The pulses are usually frequency-doubled in a 250 μm β-barium borate crystal and then chirp-compensated by a pair of SF10 prisms. Subsequently, the laser light is split with a ratio of 2:1 for the pump and probe beams respectively. An acousto-optic modulator (NEOS Technologies, Melbourne, FL) is used to modulate the pump beam at 1.5 MHz. Before entering the sample, the pump and probe beam polarizations are adjusted to be perpendicular to one another. The time delay between the pump and probe pulse is controlled by a translation stage. After the sample, the beams are re-collimated and the pump light is spatially blocked (using a pinhole) and extinguished by a polarization analyzer so that only the probe light is detected.

Two common methods are used for detection: the "open band" and the "detuned" scheme [27, 29]. In the open-band experimental scheme, the detected signal is the temporal evolution of the differential transmission of the full bandwidth of the probe light passing through the sample; whereas in the detuned measurement, we evaluate the temporal evolution of a spectral slice within the probe bandwidth by using a monochromator just prior to the photon detector. The open-band measurements are more effective for the identification of very low-frequency modes but they are also affected by the presence of low-frequency quasi-elastic interactions and ultrafast non-radiative population decay, which sometimes make the signal analysis more challenging. The higher frequency coherences are significantly attenuated relative to the lower frequencies within the open-band detected signal. On the other hand,

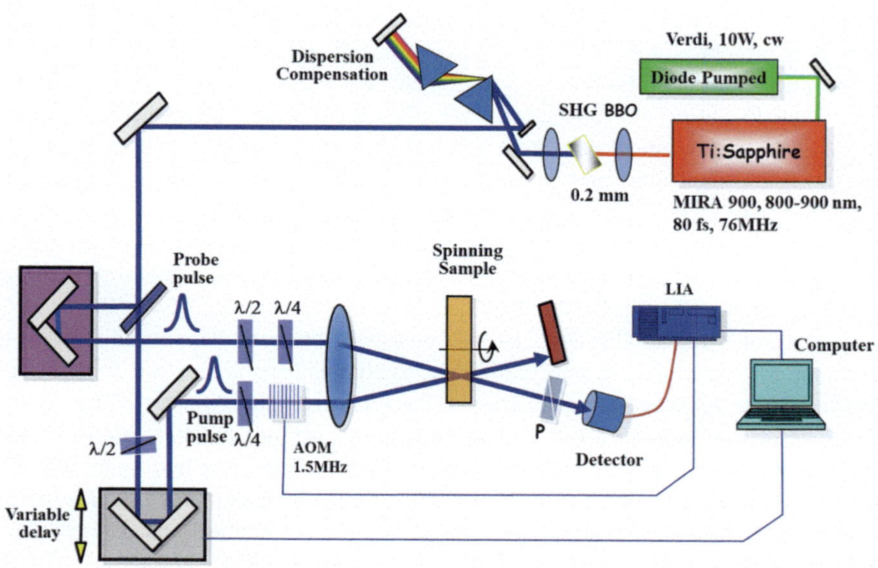


**Fig. 14.1** Schematic diagram of the vibrational coherence spectroscopy setup. The spinning sample can also be replaced by a stationary sample and the combined beams can be rapidly scanned across the sample using a set of x–y galvo mirrors

in the detuned scheme, the lower frequencies in the measured signal are attenuated relative to the somewhat higher frequencies that correspond to the monochromator detuning from the optical carrier frequency of the laser pulse. Thus, the two detection schemes are complementary and help resolve different ranges of frequencies below 200 $cm^{-1}$.

When the detuned measurement is Fourier transformed or analyzed with linear predictive singular value decomposition (LPSVD) [61], the highest frequencies in the detuned VCS measurement can be correlated with the lowest frequencies in the traditional frequency domain resonance Raman spectrum. Usually good correspondence is obtained, and this gives confidence in the time domain data analysis protocol. During a normal set of measurements, the detuned VCS data act as a bridge between the very low-frequency open-band VCS data and the higher frequency resonance Raman results. There can be significant spectral overlap with the various types of experiment, as can be seen in Fig. 14.2. In general, there is a very good correlation between the results of frequency domain techniques and the time domain measurements (when using both the open-band and detuned configurations) [17, 25, 46, 65].

## 14.3 Vibrational Coherence Measurements and Proton Tunneling in GFP

In this section, studies of GFP are presented as a prototypical example of proton transfer in proteins. GFP is an excellent experimental system for study because near-uv pulsed excitation of its chromophore initiates a photocycle (Fig. 14.3a) that starts with excited state proton transfer (ESPT). This is followed by green fluorescence decay to the ground electronic state and a thermally equilibrated "reset" reaction that involves ground state proton transfer (GSPT) along a well-characterized proton wire [5, 42, 43]. The GSPT is particularly interesting because it takes place under the "dark" thermal equilibrium conditions that are common to most biological PT reactions. As a result, we focus primarily on GSPT in this section, although we also offer some perspectives related to ESPT.

The proton wire in GFP (Fig. 14.3b) consists of phenolic oxygen on the chromophore, a water molecule, plus the serine and glutamate residues [66]. To describe the GFP photocycle, we start in the ground state ($A$) where photon absorption leads to the excited electronic state ($A^*$). This photoexcitation is immediately (within ~10 ps) followed by an excited state proton transfer reaction through a hydrogen-bonded proton wire to the glutamate (Fig. 14.3c), forming the $I^*$-state. Green fluorescence from $I^*$ to $I$ takes place with a typical radiative time constant of 3 ns. The protons (with the system now in the ground electronic state) then reset to their initial $A$-state from the $I$-state. Near room temperature (~300 K) the proton reset reaction takes place on a time scale that is faster than the fluorescence lifetime, although the deuteron reset reaction is much slower. It turns out that this back-transfer reset reaction involves

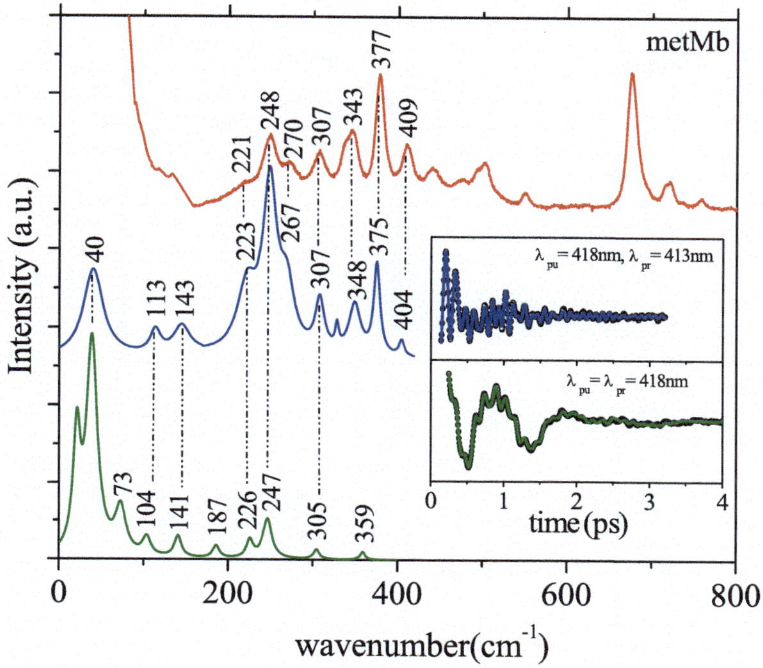


**Fig. 14.2** Correlation between frequency domain (resonance Raman) and time domain (VCS) spectra for ferric myoglobin (metMb). Resonance Raman, open band, and detuned VCS spectra are represented by red, green, and blue curves, respectively. The inset displays the corresponding real-time VCS oscillatory signals (open circles) and the LPSVD fits (solid lines). The wavelength for the resonance Raman spectrum is 413 nm and the carrier wavelength for the VCS measurements is 418 nm and the monochromator is set to 413 nm for the detuned VCS measurement. Reprinted from reference [65] with permission from Elsevier

tunneling that is driven by thermally equilibrated vibrational excitations. At lower temperatures, the vibrational excited state populations are reduced and the rate of proton tunneling from the $I$-state to the $A$-state slows so that it can be differentiated from the temperature-independent fluorescent population decay [5, 42].

### 14.3.1 Low-Frequency Vibrational Coherence in GFP

Recently there has been interest in the low-frequency modes of GFP because of their potential role in distorting the chromophore and optimizing the system for ESPT [67]. Low-frequency modes are usually anharmonic, and they can couple to the reaction coordinates of biological processes. In the case of GFP, Fang et al. [68], used femtosecond stimulated Raman spectroscopy (FSRS) to study the time-resolved Raman spectra of the excited state, $A^*$, and they proposed that a 120 $\pm$

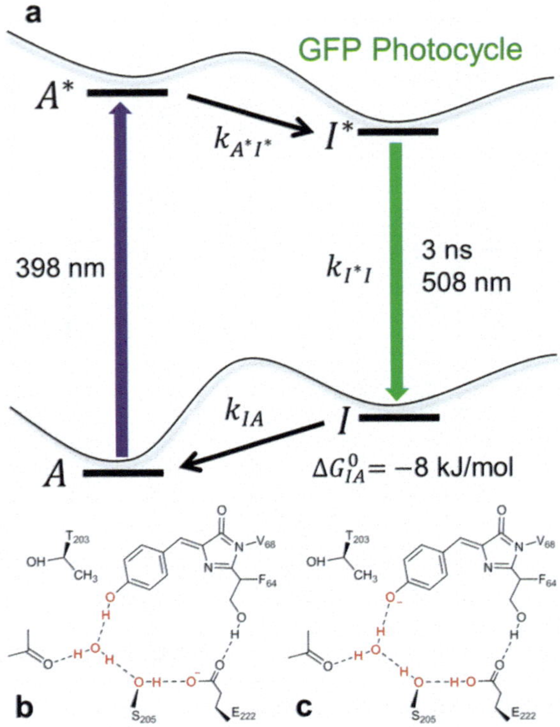


**Fig. 14.3** The photocycle and chemical structure of GFP. **a** Upon excitation of the absorption band associated with the $A - A^*$ transition using 398 nm light, the chromophore undergoes picosecond ESPT to form an intermediate state, $I^*$, with an anionic chromophore (this is a bi-exponential ultrafast process denoted here simply as $k_{A^*I^*}$). After fluorescence emission to the anionic ground state, $I$, the chromophore is reprotonated via ground state proton transfer with a rate denoted as $k_{IA}$. **b** Chemical structure of the $A$ state with a protonated chromophore. **c** Chemical structure of the $I$ state with an anionic chromophore. Reprinted from reference [5] with permission

20 cm$^{-1}$ mode is impulsively excited and optimizes the chromophore geometry for ESPT. It was suggested that this mode is associated with the wagging motion of the phenolic chromophore ring, which modulates the hydrogen-bonding geometry along the proton wire, making proton transfer possible. In their measurement [68] the 120 cm$^{-1}$ mode was not measured directly, rather it was inferred from the frequency modulation of multiple high-frequency bands. In contrast, VCS measurements can be used to directly probe low-frequency vibrational modes and when applied to GFP, such modes are clearly observed.

VCS spectra of GFP at 300 K are shown in Fig. 14.4 with data that was collected in the open-band configuration. Spectra are presented using carrier frequency excitation at 405, 420, and 435 nm, located on the red side of the 395 nm $A$-state absorption band, and multiple modes in the region below 300 cm$^{-1}$ are observed. The strong feature seen at ~120 cm$^{-1}$ is consistent with the anharmonic modulation of high-frequency modes observed previously [68]. Interestingly, another study [69] using somewhat different excitation conditions found this mode at a lower frequency (104 cm$^{-1}$) and showed that it modulated the intensity of the 1142 cm$^{-1}$ phenolic CH bending mode in the $A^*$ state. The spectra in Fig. 14.4 display a subset of VCS data obtained at various excitation wavelengths within the $A$-state absorption band. In contrast to methods that monitor modulations of high-frequency modes, the VCS data yield a

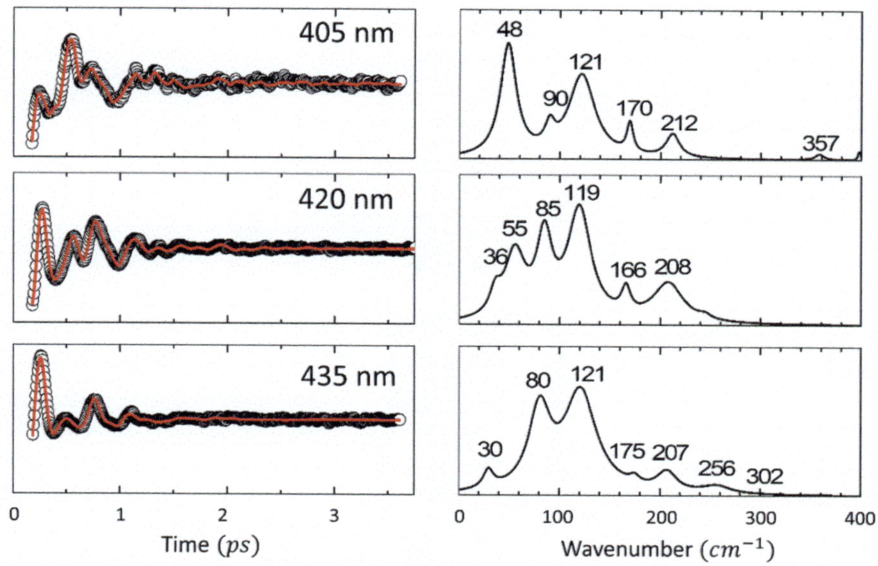


**Fig. 14.4** Coherent response (left) and spectral content (right) of GFP at 295 K with pump-probe excitation at 405 nm, 420 nm, and 435 nm, respectively. The solid red lines are the linear predictive singular value decomposition (LPSVD) that fits the oscillatory content. The frequencies and damping of these oscillations are plotted in the right panels. Reprinted from reference [67] with permission

*direct* readout of all Raman active, thermally accessible, low-frequency modes that can potentially aid ESPT in GFP [67].

Density functional theory (DFT) can be used to better understand the normal modes of the GFP chromophore, hydroxybenzylidene imidazolinone (HBDI). The normally planar chromophore is modified in the protein by conformational forces and the ensuing distortions can be analyzed by normal coordinate structural decomposition (NSD) [48]. NSD is based on the structural change between the distorted and undistorted chromophore that can be found in the X-ray structure. This difference can be expressed as a linear combination of displacements along the orthogonal low-frequency normal modes. The low-frequency normal mode activity in VCS is expected to scale roughly quadratically with the magnitude of its displacement in the NSD analysis [25].

The NSD results for the HBDI chromophore in GFP are shown in Fig. 14.5 where distortions extracted from the protein crystal structure [70] have been projected along the six lowest frequency DFT normal modes of HBDI. Diagrams showing the displacement vectors for some of the modes are shown on the right side of Fig. 14.5. Comparing Figs. 14.4 and 14.5, we find a reasonable agreement between the calculated mode frequencies (given in parentheses) and the experimental observations. For example, the mode at 106 $cm^{-1}$ involves the wagging motion of the phenol ring that was previously suggested to have functional significance [68].

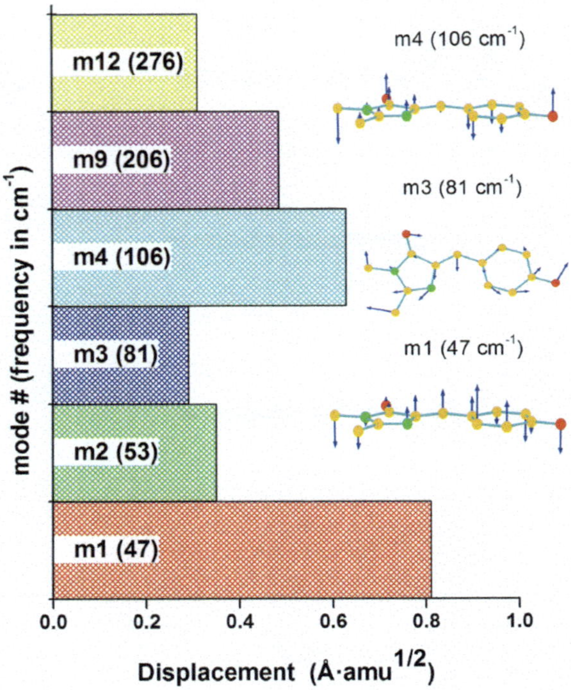


**Fig. 14.5** NSD analysis of GFP (left) and the HBDI normal mode diagrams (right). The frequencies of modes m1-m12 are given in parentheses

Additional comparisons between the DFT calculations and the various low-frequency spectra using open band, detuned, and traditional Raman measurements reveal a reasonable set of matching frequencies. However, the fact that the VCS and Raman measurements are carried out under optical resonance conditions means that only a subset of the DFT derived mode frequencies is expected to show correlations with the experimental spectra (i.e., those modes that are "coupled" to the resonant electronic transition). Resonance excitation profile effects [60, 63] can be clearly observed in Fig. 14.4, where coherent activity near ~50 cm$^{-1}$ at 405 nm diminishes as the excitation moves to 435 nm and the mode near 80 cm$^{-1}$ increases. In the Raman spectrum taken at 356 nm (not shown), a weak vibrational mode is observed at 110 cm$^{-1}$, which might correlate (given the resonance excitation difference) with the coherences that have been observed at 104 cm$^{-1}$ and 120 cm$^{-1}$ in prior work [68, 69] and in Fig. 14.4. It is also noteworthy that there is a clearly defined vibrational band near 145 cm$^{-1}$, in the room temperature Raman spectrum taken at 356 nm, that may be related to features at ~135–150 cm$^{-1}$ that appear in the VCS spectrum of GFP at low temperature (120 K).

Importantly, the VCS measurements are consistent with the existence of a low-frequency mode near ~120 cm$^{-1}$, confirming the earlier report, where this mode was extracted based on modulation of high-frequency modes [68]. As mentioned above, a more recent study [69] on GFP using time-resolved impulsive stimulated Raman spectroscopy (TR-ISRS) has also probed the dynamic evolution of the $A^*$ and $I^*$ states and found anharmonic vibrational coupling between the high-frequency phenolic CH bending mode of the chromophore and a low-frequency mode at ~104 $cm^{-1}$ (which was suggested to be analogous to the mode at 120 $cm^{-1}$). Going beyond spectral characterization, this investigation [69] directly addresses the issue of whether such a low-frequency mode might actually have an effect on the efficiency and rate of ESPT. It was shown that the intensity of the high-frequency phenolate CO stretching band, which is a measure of $I^*$ population development, is not modulated by any low-frequency mode. This experiment demonstrates that the coherence of the 104 $cm^{-1}$ mode (or the 120 $cm^{-1}$ mode) does not significantly affect ESPT efficiency during the formation of the $I^*$-state. An alternative hypothesis suggested that the ESPT is driven by incoherent thermally excited motions, including those of the proton wire components, rather than by a single low-frequency mode of the chromophore [69]. In addition, we must also acknowledge that the incoherent thermal excitations will likely also involve non-equilibrium thermal transients arising from the redistribution of excess photon energy that is deposited in the vibrational manifold of the $A^*$ state.

Finally, we note that, at 295 K, the coherence of the ~120 $cm^{-1}$ mode shown in Fig. 14.4 has almost completely dissipated within ~1–2 ps. This coherence decay timescale is faster than the ~3 ps and ~10 ps ESPT transition from $A^* \rightarrow I^*$. Thus, the suggestion [68] that coherent motion of the ~120 $cm^{-1}$ mode has an important effect on the ESPT reaction appears unlikely. The differences in the reported frequency (120 $cm^{-1}$ vs 104 $cm^{-1}$) obtained in these three separate experiments (i.e., the results shown in Fig. 14.4 and references [68, 69]) remain unresolved.

In related earlier work, Leiderman et al. [71]. proposed a vibrationally assisted tunneling model for ESPT in GFP by invoking a ~200 $cm^{-1}$ promoting mode. In principle, the promoting "mode" may represent a combination of coherently excited, as well as incoherently excited, motions of the chromophore and the surrounding hydrogen-bonded residues. These vibrational motions lead to compression of the distance between the donor and acceptor atoms that helps to efficiently drive the proton transport. Although the temperature dependence of the kinetics of the ESPT process has been measured [5], a complete kinetic analysis has not yet been reported. Assuming over-barrier transport, we note that at $T = 295$ K there is an enthalpic barrier of ~2.4 kJ/mol associated with the slower ~10 ps kinetic phase, while the faster ~3 ps ($T = 295$ K) kinetic phase would have a significantly smaller barrier (~0.5 kJ/mol). The presence of the two kinetic phases is not fully understood, but the relative amplitudes are nearly temperature independent, suggesting parallel processes. The possibility of non-equilibrium thermal excitation by excess photon energy complicates the excited state analysis, but the possibility of excited state tunneling cannot be excluded [67]. In the following section, we present a proton tunneling analysis of the ground state GFP kinetics that describes the thermally equilibrated $I \rightarrow A$ proton reset reaction.

### 14.3.2 Kinetics of the GFP Photocycle

The kinetics of the GFP photocycle was measured using a unique ultrafast two-color pump-probe setup [5, 72]. This instrument is based on two synchronized ultrafast lasers, and it has the capability to monitor the temperature-dependent kinetics of biomolecules over a broad time range extending from ~100 fs to hundreds of microseconds [5, 72]. The kinetics of the GFP photocycle are displayed in Fig. 14.6, which presents the absorption changes as a function of temperature on a logarithmic time scale. At lower temperatures, the thermally driven ground state proton transfer from $I$ to $A$ can be clearly observed at the longer timescales. The absorbance in Fig. 14.6 is probed at the 420 nm $A^*$–$A$ isosbestic point so that the cycle $A^* \rightarrow I^* \rightarrow I \rightarrow A$ is observed. The absorbance changes were also followed by using the $I^*$–$A$ isosbestic point near 405 nm, confirming the observed population dynamics [5].

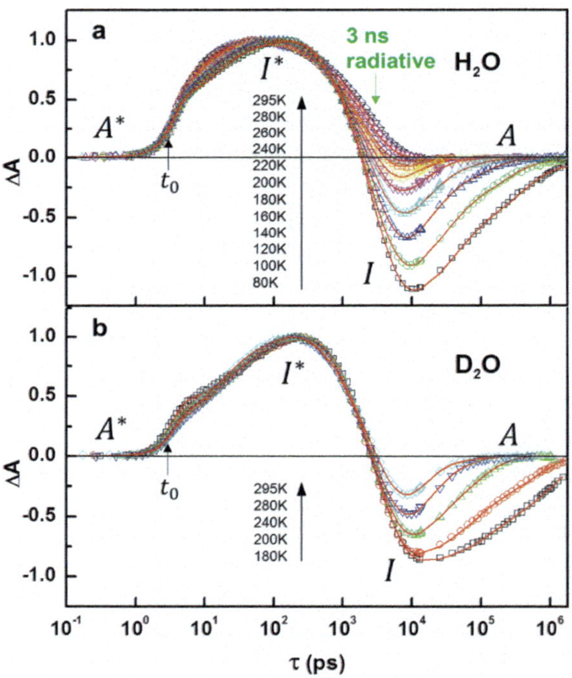


**Fig. 14.6** Transient absorption measurements as a function of temperature for GFP in **a** $H_2O$ and **b** $D_2O$. A 150 fs pump pulse at 405 nm was used to excite the sample and a 3 ps pulse tuned to 420 nm, the isosbestic point[10] between $A$ and $A^*$, was used to probe the sample. The red lines fit to the data over the full photocycle using exponentials to account for the two phases of decay from $A^* \rightarrow I^*$ the radiative decay from $I^* \rightarrow I$, and the slightly non-exponential proton transport from $I \rightarrow A$. A Gaussian instrument function (~3 ps) was also included in the fits and the system time-zero is denoted as $t_0$ (the data have been shifted by 3 ps to allow visualization of probe-before-pump times on the logarithmic time plot). Reproduced from reference [5] with permission

Exited state proton transfer $A^* \rightarrow I^*$ is fit with a bi-exponential, which is most evident at lower temperatures. Similar amplitudes of the two exponentials are maintained at all temperatures, which suggests (as noted above) that two parallel channels of excited state transport are involved. The extracted time constants at room temperature are ~3 ps and ~10 ps, consistent with other studies [5, 42, 43]. At lower temperatures, the slower decay rate is reduced by an order of magnitude leading to a decay time of ~100 ps at ~120 K, while the ~3 ps decay remains nearly independent of temperature. The $I^*$ state decay to the $I$-state is governed by temperature-independent spontaneous green fluorescence with a time constant of ~3 ns [5, 42]. At 295 K, the $I^* \rightarrow I$ process overlaps with the oppositely signed $I \rightarrow A$ process making detection of the ground state proton transfer rate quite difficult. However, by inducing stimulated emission from $I^* \rightarrow I$, it is possible to move the population from $I^*$ to the $I$-state more rapidly, so that the $I \rightarrow A$ process can be exposed at room temperature using a "pump-dump-probe" protocol [42]. Such an experiment leads to the rate, $k_{IA}$, that is shown in Fig. 14.7 as an open diamond [42]. Because $k_{IA}$ slows at lower temperatures, it no-longer overlaps with the temperature-independent decay of $I^*$ by spontaneous emission. Thus, it can be measured at lower temperatures. Similarly, when deuterons are exchanged for protons, the heavier mass dramatically slows $k_{IA}$ over the whole temperature range, demonstrating that this kinetic step is due to a tunneling reaction [5].

### *14.3.3 Ground State Proton Tunneling in GFP*

An Arrhenius plot of $k_{IA}$ is shown in Fig. 14.7 with black symbols for the proton (triangles) and deuteron (squares) rates. The green dots depict the temperature-independent $k_{I^*I}$ resulting from the fluorescent decay. The full proton kinetic cycle is extracted by using a global fit to the time-dependent absorbance change at each temperature (global fit shown as red lines). A slight decrease in $k_{I^*I}$ at higher temperatures ($T$>200 K) is observed for what should be a temperature-independent spontaneous emission rate (green dots). This apparent decrease is due to the inability of the fitting program to resolve the oppositely signed $I \rightarrow A$ process that temporally overlaps $I^* \rightarrow I$ at these higher temperatures. At lower temperatures, the kinetic cycle and the $I \rightarrow A$ process is more clearly resolved. Between 295 K (where the timescale for $I \rightarrow A$ proton transfer was found to be ~400 ps [42]) and 160 K we find a large kinetic isotope effect (KIE) for $k_{IA}$, which varies from ~40 to 73.

Prior proton tunneling studies on enzymes have demonstrated temperature-dependent KIEs that involve proton-coupled electron transfer (PCET) [73–76]. The much smaller temperature range near room temperature that has been accessed in these pioneering proton tunneling experiments is shaded gray in Fig. 14.7. Although classical D-A distributions can often be used near room temperature to account for the quasi-linear temperature dependence of the KIE within this narrow range [40, 74, 77], at lower temperatures a quantum description is required [8]. The enzyme studies also generally involve much slower rates [74] and often involve a non-adiabatic

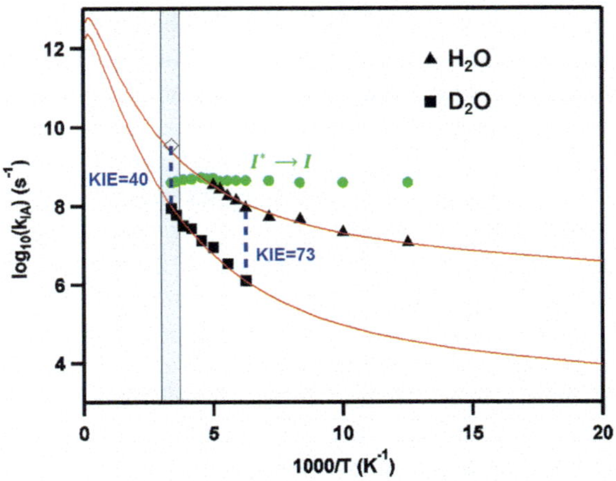


**Fig. 14.7** Arrhenius plot of the ground state ($I \rightarrow A$) proton transfer rates in GFP. The associated error bars are smaller than the data points and are not shown. The typical temperature range for studies of biological proton transport is shaded in gray. The rate for the $H_2O$ samples (in 50% glycerol buffer) increases at higher temperature and becomes inseparable from the 3 ns $I^* \rightarrow I$ population transfer due to fluorescence decay (green points). The apparent decrease in the $I^* \rightarrow I$ rates at higher temperature in $H_2O$ is due to the underlying oppositely signed $I \rightarrow A$ process. A measurement [42] of the rate at room temperature in pure buffer (0% glycerol) using the pump-dump-probe protocol is shown with the open diamond and has been scaled by a small factor to correct for the measured solvent differences. The experimental KIEs of 73 (160 K) and 40 (295 K), are depicted by the blue dashed lines. The global fits to the kinetic data are shown as the red lines and the fitting parameters are listed in Table 6.4. Reproduced from reference [5] by permission

PCET mechanism [41]. The rate expression for enzymes is further complicated by the presence of low probability conformational interconversions that are needed to find "tunnel ready" states with a donor–acceptor hydrogen-bonding distance that is small enough to allow tunneling [77, 78]. Nevertheless, when this is considered, the underlying tunneling process of the enzyme system at room temperature turns out to have a timescale (~1 ns) that is similar to that found for GFP (~400 ps). In contrast to the enzyme system, the GFP beta barrel structure is rigidly fixed, and the D-A hydrogen bonding is structurally stabilized. Thus, the $I \rightarrow A$ process in GFP directly monitors the tunneling reaction rate, without the need to consider the conformational interconversions found in an enzyme system.

Due to the wide experimental temperature range utilized in the GFP experiments (Fig. 14.7), the D-A vibrational motion must be treated using a quantized approach [5–8]. The effective D-A vibrational frequency involves all protein normal modes [5] and relies on an Einstein oscillator approximation, $\omega_{DA} \cong \overline{\omega}_E$. Moreover, the rate limiting event for the GSPT reaction in GFP involves tunneling between only one of the three pairs of OH--O atoms in the GFP "wire" [5, 78]. It appears that the remaining two protons immediately follow the tunneling step via an over-barrier hopping mechanism. This can be characterized as a concerted asynchronous reaction (i.e., tunneling with asynchronous hopping) [5, 78]. This scenario is inferred, based on the longer effective tunnel distances and the lower frequencies associated with the four body correlated oxygen motions that would be required to account for a simultaneous (i.e., synchronous) three-proton tunneling event [5, 78]. Attempts to fit the GFP data set using an increased effective tunnel distance, associated with a synchronous three-proton tunneling model, were not able to account for both the absolute magnitude and temperature dependence of the experimental tunneling rates [5].

An electronically adiabatic tunneling expression was used to globally fit the H and D tunneling data in Fig. 14.7 (red lines), which is appropriate for a single isolated double well electronic potential. The proton level tunnel splitting was calculated in closed form using the WKB approach and was found to be much smaller than $k_B T$. This allows the use of an electronically adiabatic and proton level non-adiabatic tunneling rate expression [5, 8, 78]. For a quantized D-A coordinate this is given by the following Golden Rule expression:

$$k_{IA}(\overline{\ell}_0, \omega_{DA}, \lambda_s) = \sum_n \sum_m P_n \frac{2\pi}{\hbar} \left| \frac{\Delta_0^{nm}}{2} \right|^2 \frac{1}{\sqrt{4\pi\lambda_s k_B T}} \exp\left[ -\frac{\Delta G_{nm}^{\dagger}}{k_B T} \right] \tag{14.1}$$

with

$$\Delta G_{nm}^{\dagger} = \frac{\left(\Delta G^0 + \hbar\omega_{DA}(m-n) + \lambda_s\right)^2}{4\lambda_s} \text{ and } P_n = e^{-\frac{n\hbar\omega_{DA}}{k_B T}} \left[1 - e^{-\frac{\hbar\omega_{DA}}{k_B T}}\right]. \tag{14.2}$$

The quantity $\Delta_0^{nm}$ is the tunnel splitting for a degenerate pair of reactant and product D-A vibrational states $(n, m)$. The free energy of the reaction including the surrounding environmental coordinate is given by $\Delta G^0$ and the barrier to reach degeneracy for a given pair of states is denoted by $\Delta G_{nm}^{\dagger}$. The overall reorganization energy of the reaction is denoted by $\lambda_s$.

Many of the parameters that are needed for fitting the GFP data using Eqs. 14.1 and 14.2 can be extracted from other independent measurements. For example, the free energy of the GFP proton reset reaction is determined by monitoring the absorption bands to find the temperature-dependent equilibrium populations of reactant and product states. This leads to $\Delta G_{IA}^0 = -665\ \text{cm}^{-1}$ [5]. The O–H mode frequency, $\omega_1$, which is used to find the tunnel splitting, is found from an experimentally determined correlation between $\omega_1$ and the OH--O oxygen separation [79]. Given that the OH covalent bond length is 0.99 Å, this directly links $\omega_1$ to the equilibrium tunneling distance, $\overline{\ell}_0$.

A quasi-harmonic potential with a cusp was employed for the WKB calculation, but the barrier height dictated by the cusp plays only a minor role because the barrier width near the degenerate zero-point states is the primary factor in determining the tunnel splitting. The global least squares fitting results for GFP are given in Table 14.1 where only three independent parameters ($\omega_{DA}$, $\overline{\ell}_0$, and $\lambda_s$) are needed to provide an excellent simultaneous fit to all (H and D) tunneling data. Additionally, it should be noted that both $\overline{\ell}_0$ and $\omega_{DA}$ are tightly constrained by other independent experiments.

**Table 14.1** Parameters for ground state proton transfer kinetics in GFP

| Unconstrained | Value | Description |
|---|---|---|
| $\lambda_s$ | $1165 \pm 20$ cm$^{-1}$ | Reorganization energy |
| Constrained[a] | | |
| $\omega_{DA}/2\pi c$ | $273 \pm 7$ cm$^{-1}$ | D-A mode frequency (fit) |
| $\overline{\ell}_0$ | 0.78Å ($\sigma_{\ell_0}$= 0.02Å) | Equilibrium tunnel distance |
| Fixed[b] | | |
| $2b_0/d_0$ | <0.2 | Relative width of parabolic cusp |
| $\Delta G^0$ | $-665$ cm$^{-1}$ | Free energy of $I \rightarrow A$ reaction |
| $\omega_1(H)/2\pi c$ | 3330 cm$^{-1}$ | OH mode frequency |
| $\mu_{DA}$ | $8.5 \pm 0.5$ amu | Reduced mass of D-A mode |
| $\theta_H$ | $19.2° \pm 0.5°$ | Duschinsky rotation for H |
| $\theta_D$ | $26.5° \pm 0.7°$ | Duschinsky rotation for D |
| $\chi^2_{red}$ | $8.0 \times 10^{-3}$ | Reduced chi-squared |

[a] The value of $\overline{\ell}_0$ is constrained by X-ray data and I-state ONIOM minimization and $\sigma_{\ell_0}$ is the width of $\overline{\ell}_0$ distribution. The value of $\omega_{DA} \cong \overline{\omega}_E$ is constrained to 264–289 cm$^{-1}$ based on normal mode analysis [5]

[b] The O–H frequency of the proton, $\omega_1(H)$, was correlated [79] to the O--O distance and fixed by the value of $\overline{\ell}_0$. The parameter $b_0$ locates the attachment of a continuous inverted parabola at the top of the barrier. The effect is negligible on the other fitting parameters if $2b_0/d_0 < 0.25$ ($d_0 = \sqrt{\mu_L}\ell_0$). The effective mass for the D-A oxygen motion is $\mu_{DA} \cong 8.5 \pm 0.5$ amu. The Duschinsky rotation angles between the reactant and product potential energy surfaces are given by $\theta_L \cong \sqrt{\mu_L/\mu_{DA}}$. The uncertainty in $\mu_{DA}$ generates uncertainty in the other parameters. The tunnel particle reduced mass is $\mu_L$ ($L$=$D$ or $H$). Table 14.1 is reprinted from reference [5] by permission

For example, the value determined for $\overline{\ell}_0 = 0.78$ Å dictates an equilibrium D-A distance of $\overline{R}_0$=2.76Å, which is very close to both the X-ray [70] and the DFT energy minimized $I$-state [80] values ($\overline{R}_0 \sim 2.7$Å). The magnitude of $\omega_{DA}$ is constrained by an independent analysis of the $I$-state normal modes in GFP obtained using a Gaussian hybrid ONIOM(QM/MM) approach [80]. This theoretical study generates an Einstein oscillator mode with frequency $\overline{\omega}_E/2\pi c \cong 285$ cm$^{-1}$ and reduced mass $\mu_{DA} = 8.5 \pm 0.5$ amu for the D-A coordinate involving the serine and water oxygen atoms [5]. This protein-driven D-A atomic motion controls proton tunneling in GFP and the calculated Einstein frequency compares well with the experimental result $\omega_{DA} = 273$ cm$^{-1}$ found from the kinetic data. Within this analytical model, the reorganization energy, $\lambda_s = 1165$ cm$^{-1}$, is the only completely free fitting parameter.

Proton tunneling in GFP near room temperature is surprisingly fast (~400 ps), particularly upon realizing that typical values for the H-bonding distance and the OH--O vibrational frequency are involved. In the GFP system, we have assigned [5] the rate limiting $I \rightarrow A$ kinetic step to proton tunneling from the oxygen on the serine residue (considered to be un-ionizable with a $pK_a$ ~ 16) to the adjoining water molecule (see Fig. 14.3c). The other protons in the GFP wire immediately follow ("asynchronously") via an over-barrier concerted hopping process.

Serine and threonine residues are often found in close proximity to biological proton "water wires" in proteins [38, 39, 81–84]. Based on the GFP tunneling results, this has led us to suggest [5, 78] that these residues can play a direct and more general role in biological proton transfer processes. Serine and threonine residues are normally not included classical molecular dynamics (MD) simulations of proton water wires [82, 83, 85–87] because their classical ionization potential is so energetically costly. However, tunneling can overcome this problem and the rapid serine-to-water tunneling timescale (~400 ps) at room temperature for the GFP proton wire [5] suggests that quantum tunneling makes these residues particularly relevant for PT in biology.

Although the proton tunneling rate is much slower than the ps timescale of proton hopping in water, ionization-resistant residues like serine and threonine can be used as control elements that utilize tunneling to bias the direction of proton flow [5, 78]. Proton wire tunneling involving serine has been experimentally observed in the GFP system. This is based on both the X-ray analysis of the wire structure [70] and the very large temperature-dependent KIE that is observed for GSPT [5]. It was previously suggested [87] that alternative water wires may be present in GFP that do not include the serine, and would therefore transfer protons via a traditional hopping mechanism. However, such a wire is not consistent with the large KIE that is experimentally observed [5] or with the room temperature Laue crystal structures that do not show additional water molecules in the vicinity of the wire (J. van Thor, private communication).

Thus, we suggest that serine and threonine residues can play a role in protein water wires that go beyond simply "stabilizing" the water molecules. They can also act as control elements within the "wire" by kinetically trapping each proton after it has tunneled to the "downstream" water molecule adjacent to the serine in the wire. Proton pumping directionality can be maintained if the tunneling rate from the serine oxygen in the preferred "downstream" direction is significantly larger (via a smaller $\overline{\ell}_0$ and $\omega_{DA}$) than the tunneling rate back to the preceding water molecule on the "upstream" side of the serine oxygen. The differential (upstream vs downstream) conditions on the relative values of $\overline{\ell}_0$ and $\omega_{DA}$ only need to hold during a timescale that is longer than the ps hopping times of the non-serine protons that do not engage in tunneling. The kinetic trapping can become even more robust if the upstream tunneling reaction timescale becomes longer than the hydrogen bond reorientation reaction [78]. These ideas are depicted in Fig. 14.8.

The tunneling-based kinetic trapping is realized because the water molecules, located on either side of the serine residue, interact with different $sp^3$ orbitals of the serine oxygen. In the forward or downstream direction, the water molecule acts as a proton acceptor from the protonated serine oxygen orbital, while the water molecule on the upstream side is initially a proton donor (via concerted hopping) to a different (initially unprotonated) serine oxygen orbital (i.e., the proton transferring from water hops to an unprotonated lone pair oxygen orbital on the serine just as the proton on the serine orbital pointing downstream tunnels forward to the adjacent

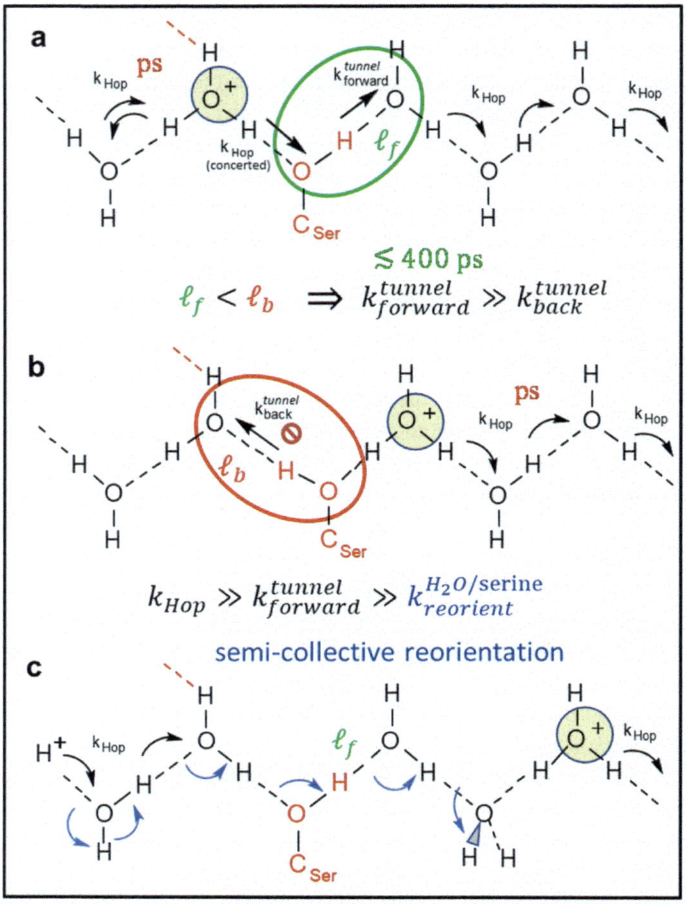


**Fig. 14.8** Model for kinetic trapping in protein-based water wires. Concerted asynchronous proton tunneling and hopping can bias transport in the forward direction (green) vs the backward direction (red). The differential tunnel lengths $\ell_f$ and $\ell_b$ are shown explicitly, but the relative D-A vibrational frequencies in the forward and backward direction can also play a major role in biasing the tunneling rates. Parts (a) and (b) depict the forward and backward steps, respectively. When hopping rates are much faster than the backward tunneling rate, the proton is kinetically trapped so that the proton (yellow shading) moves forward. Part (c) depicts the semi-collective H-bond reorientation step, which is relatively slow and is the overall rate limiting step for continuous proton pumping. This "masks" the KIE associated with the ~400 ps tunneling step, making it difficult to detect experimentally

hydrogen-bonded water molecule [78]). This transport process can be characterized as "concerted asynchronous tunneling and hopping".

For the reverse process, the upstream water oxygen becomes the proton acceptor for the serine proton, which (after the initial step) now points in the upstream direction. However, the upstream tunneling rate can be orders of magnitude slower, even if there is only a small (~tenths of Å) increase in the average upstream tunneling distance, or if the effective D-A frequency of the upstream oxygen pair is higher. The latter condition holds because a higher $\omega_{DA}$ frequency serves to narrow the width of the D-A distance distribution $\left(\sigma_{DA}^2 \cong k_B T/\mu_{DA}\omega_{DA}^2\right)$, which also slows tunneling. If a significant disparity in the forward and reverse tunneling rates persists for time scales that are longer than the downstream ps proton hopping time, the serine tunneling step acts as a kinetic trap that helps to bias the proton movement in the preferred (downstream) direction.

Finally, it is also noteworthy that a longer H-bond pointing in the upstream direction also indicates a less stable serine orientation, which would presumably facilitate H-bonding reorientation and the formation of the stronger H-bond that points in the forward downstream direction. It is known that collective water reorientation and reorganization can be hindered by reduced mobility within the protein interior [88]. This reorientation reaction is usually considered to be the overall rate limiting step for continuous proton pumping because the H-bonds of the serine and the water molecules in the wire must return to their preferred forward proton pumping orientations so that repetitive proton pumping can take place.

Thus, concerted asynchronous proton tunneling and hopping that involves serine (or threonine) would extend the traditional "hop-and-turn" proton pumping mechanism [88, 89] to a "tunnel-hop-and-turn" scenario. Importantly, when the "turn" (i.e., H-bond reorientation) becomes the rate limiting step (as is very likely), the KIE of the overall continuous proton pumping process will be "masked" because the ~400 ps tunneling process is not the rate limiting step. This means that the crucial underlying role of tunneling-based kinetic trapping, which can act to bias the proton transport direction in protein water wires [78], would not be detected by isotope-dependent rate measurements.

## 14.4 Heme Ruffling and Electron Transfer in Cytochrome c

Cytochrome c (cyt c) is primarily known as an electron-carrying mitochondrial protein. The transition of cytochrome c between the ferrous and ferric states within the cell makes it an efficient biological electron-transporter and it plays a vital role in cellular oxidation/reduction reactions in both plants and animals. Its main function in cellular respiration is to transport electrons from cytochrome c reductase (Complex III) to cytochrome oxidase (Complex IV). The c-type heme group is the functional center of cyt c and, in the solution state of the native protein, the heme iron is axially coordinated to His18 and Met80. The periphery of the c-type heme is also covalently anchored to the cyt c protein material by two thioether linkages (Cys-14 and Cys-17) found within a Cys-X-X-Cys-His (CXXCH) pentapeptide motif that is a distinguishing feature shared by nearly all c-type heme proteins.

### 14.4.1 Heme Ruffling Mode

This CXXCH motif, in conjunction with the protein fold, leads to the heme in cyt c having a geometry that is dominated by a large ruffling distortion where the pyrrole-ring is twisted about the Fe–N bond [46, 90, 91]. This tilts the $p_z$ orbitals of the porphyrin nitrogen atoms away from the heme normal, increasing the overlap of the porphyrin $a_{2u}$ and iron $d_{xy}$ orbitals. Both NMR experiments and DFT calculations [92] have demonstrated that heme ruffling increases the electron density in the $d_\pi$ ($d_{xz}, d_{yz}$) orbitals of the heme iron. Ruffling also reduces porphyrin meso-carbon electron donation to the iron $d_{xy}$ orbital [92]. All three occupied Fe 3d-based molecular orbitals in the $t_{2g}$ subset are destabilized by ruffling and this reduces the respective positive and negative spin density on the β-pyrrole and meso-carbons [92]. The resulting ruffling-induced electron localization at the iron atom suggests that both the electron tunneling rate from an external reductant to the heme periphery and its reduction potential will be anti-correlated with the magnitude of the ruffling distortion [46, 92]. This suggestion is consistent with observations showing that increased ruffling reduces the reduction potential of ferric cyt c [93–95].

To quantitatively investigate the correlation between heme distortion and electron transfer in this class of proteins, *Pseudomonas aeruginosa (Pa)* cyt $c_{551}$ and its F7A mutant were studied using absorption spectroscopy, resonance Raman spectroscopy, and VCS [46]. *Pa* cyt $c_{551}$ and its F7A mutant have very similar crystal structures but a significant structural difference is found in the heme out-of-plane (OOP) distortion along the ruffling coordinate. This difference is revealed by using the normal coordinate structural decomposition (NSD) method [45, 47, 48], as discussed above and applied to GFP. For application to heme systems, the central iron atom must be included along with a reference structure that is based on the optimized planar structure of the $D_{4h}$ iron-based porphine [25].

As can be seen from Fig. 14.9, the native wild type (WT) cyt $c_{551}$ shows a strong ruffling distortion of 1.6 amu$^{1/2}$ Å; whereas, its distortions along the other normal coordinates are less than 0.5 amu$^{1/2}$ Å. The F7A mutation generates a 1.4 amu$^{1/2}$ Å increase in the ruffling distortion compared to the WT. The difference along the ruffling coordinate accounts for ~90% of the overall difference between WT and the F7A mutant, and this offers an excellent way to quantitatively assess the effect of ruffling on electron transfer in c-type hemes. In Fig. 14.9 we also display the NSD analysis of the hh cyt c distortions where we observe even more ruffling than found in the F7A mutant of cyt $c_{551}$.

Figure 14.10 displays a comparison between the low-frequency open-band VCS spectra of WT and the F7A mutant of *Pa* cyt $c_{551}$ as well as the VCS spectrum of hh cyt c [46]. We previously showed that the LPSVD spectra derived from the time domain data agree well with frequency domain resonance Raman spectra in the region of overlap between 200–400 cm$^{-1}$. As can be seen in the figure, the modes $\gamma_b$, $\gamma_c$, 155 cm$^{-1}$, 183 cm$^{-1}$, 240 cm$^{-1}$ ($\gamma_{24}$), and 350 cm$^{-1}$ ($\upsilon_8$), are in excellent agreement among the three proteins [46]. It has also been suggested that the lowest frequency mode, labeled $\gamma_a$ in Fig. 14.10, has significant ruffling content [90] and

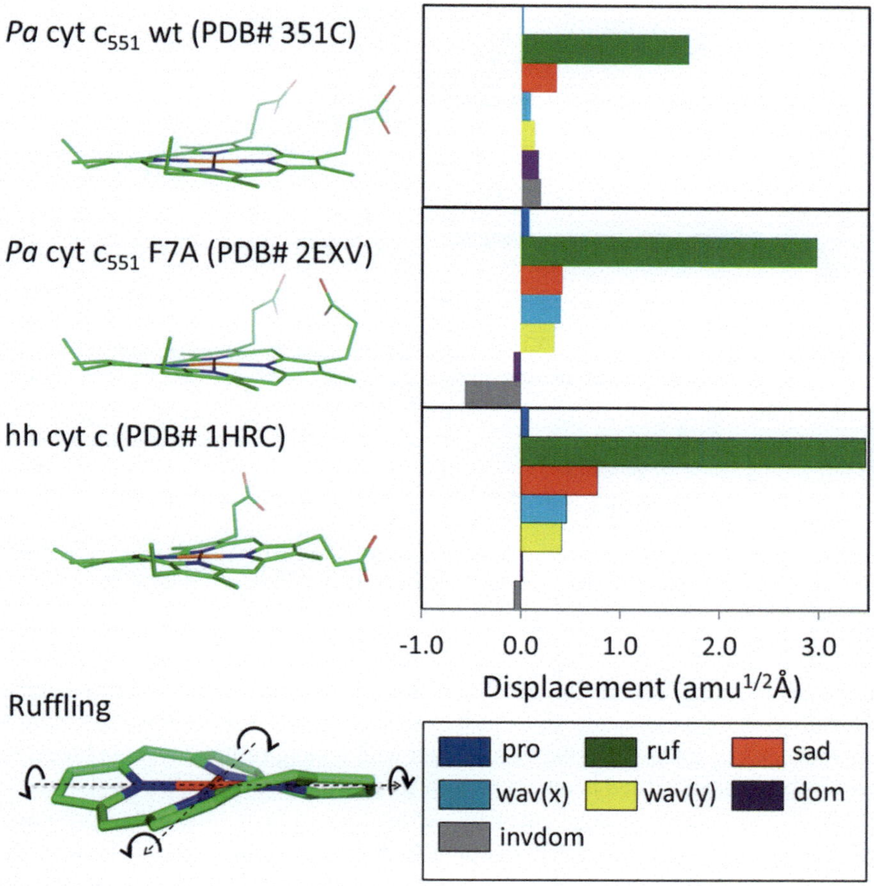


**Fig. 14.9** Crystal structure and NSD analysis of hemes in ferric Pa cyt $c_{551}$ and its F7A mutant are compared with hh cyt c. The minus sign of displacement is defined only for doming and inverse doming to indicate the direction of Fe displacement (+, proximal; −, distal). The ruffling mode is shown at the lower left part of the figure and the arrows indicate the rotation of pyrrole rings with respect to the Fe–N axis (dotted black lines). Reproduced from reference [46] with permission

the amplitude ratios $\gamma_a(WT)/\gamma_a(F7A)$ and $\gamma_b(WT)/\gamma_b(F7A)$ were found to be 0.2 and 0.9, respectively, under the same experimental conditions [90]. The amplitude ratio of $\gamma_b$ near unity is consistent with other studies where $\gamma_b$ was found to be insensitive to heme structural changes along the ruffling coordinate as the protein unfolds [90]. Thus, it is assumed that $\gamma_b$ has very little or no ruffling content. The ratio of $\gamma_a(WT)/\gamma_a(F7A) = 0.2$ is close to what is obtained by squaring the relative ruffling distortion of WT and F7A ($\sim 0.5^2 = 0.25$). This supports the assignment of $\gamma_a$ as primarily a ruffling mode and is also consistent with VCS Raman activity of "soft" low-frequency modes [25] that is proportional to the square of the protein-induced distortion from the planar reference state. As can be seen from Fig. 14.10 the

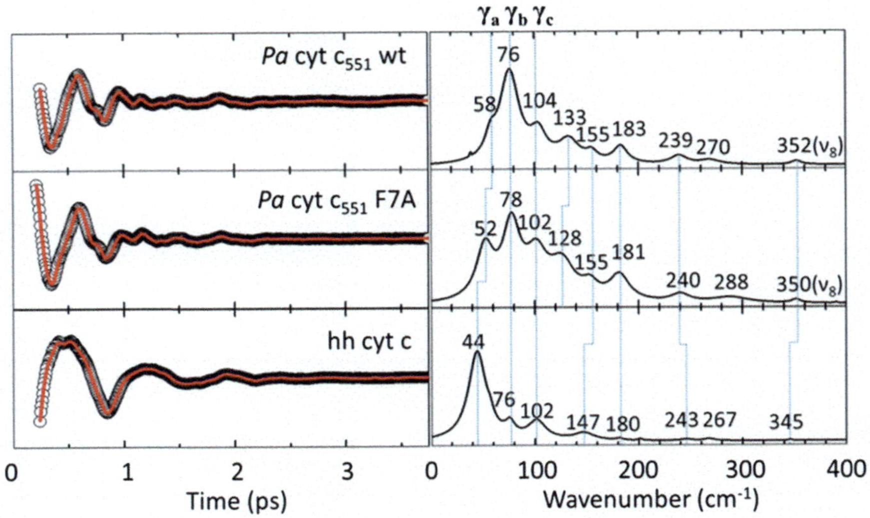


**Fig. 14.10** Open-band VCS spectra of ferric Pa cyt $c_{551}$ WT and its F7A mutant at 412 nm. LPSVD fits (Left) and LPSVD power spectra (Right) are shown. Reproduced from reference [46] with permission

frequency of $\gamma_a$ and the magnitude of the ruffling distortion are anti-correlated. For the less ruffled WT cyt $c_{551}$ heme, $\gamma_a$ is located at 58 $cm^{-1}$ and it shifts down to 52 $cm^{-1}$ and increases its intensity in the F7A mutant with a more ruffled heme. In hh cyt c, with its even more ruffled heme, $\gamma_a$ is located at a still lower frequency (44 $cm^{-1}$) and it displays much stronger intensity. A detailed analysis [46] of the proposed anharmonic heme potential, and its coupling to external protein forces, leads naturally to an inverse correlation between the magnitude of the ruffling distortion and its observed frequency.

### *14.4.2 Heme Ruffling Modulates Photoreduction Tunneling Distance*

In Fig. 14.11a we plot kinetic measurements [46] of photoreduction in cyt $c_{551}$ and cyt c. The photoreduction cross section as a function of the ruffling distortion, denoted as$q_0$, is shown in Fig. 14.11b. This figure demonstrates that the photoreduction cross section of *Pa* cyt $c_{551}$ is nearly two orders of magnitude larger than that of hh cyt c. The heme ruffling distortion in hh cyt c is over two times larger than that of WT *Pa* cyt $c_{551}$, but it is only ~17% larger than the F7A mutant [46]. This suggests an exponential dependence on the electron transport rate and the magnitude of the protein-induced ruffling. The photoreduction cross sections, obtained from the kinetic data, were therefore fit using an exponential function that has the form, $\sigma_r = Ce^{-\gamma q_0}$, where $\gamma$

includes the factors that scale the exponential dependence of the rate as a function of the ruffling distortion. The electron transfer rate,$k_{et}$, can be written as:

$$k_{et} = \frac{2\pi}{\hbar} T_{DA}^2 \frac{1}{\sqrt{4\lambda_s k_B T}} e^{-\left(\Delta G_{red}^0 + \lambda_s\right)^2 / 4\lambda_s k_B T} \tag{14.3}$$

where $T_{DA}$ is the electron tunneling overlap matrix element and $\Delta G_{red}^0$ and $\lambda_s$ are the free energy of reduction and reorganization energy, respectively. This expression contains several factors that might be affected by a change in the ruffling distortion.

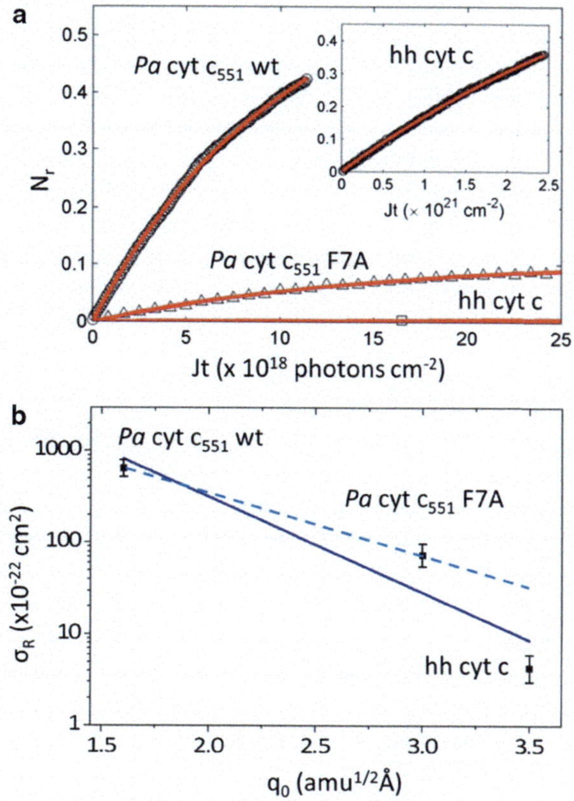


**Fig. 14.11** **a** Kinetic measurements of photoreduction. The reduced population, $N_r(t)$, increases with irradiation time and is fitted with $N_r(t) = N_r(\infty)\left[1 - e^{-\sigma Jt}\right]$ and $N_r(\infty) = \sigma_r/\sigma$ with $\sigma = \sigma_r + \sigma_o$. The quantities $\sigma_r$ and $\sigma_o$ are the photoreduction and photooxidation cross sections, respectively. The data for Pa cyt $c_{551}$ WT, F7A, and cyt c are shown as circles, triangles, and squares, respectively. The fitted curves are shown in red. **b** The photoreduction cross-sections are plotted as a function of ruffling distortion and fit using an exponential function (solid line) for all three samples (including hh cyt c) and when hh cyt c is excluded (dashed line). Reproduced from reference [46] with permission

The most important factor is the tunneling matrix element [96]:

$$T_{DA} \sim (n_D n_A)^{-1/2} T_0 e^{-\beta(r_0+\alpha q_0)/2}, \tag{14.4}$$

where the effect of the tunneling distortion, $\alpha q_0$, on the total tunneling distance, $r_0+\alpha q_0$, is explicitly shown and $\alpha$, which maps the mass-weighted ruffling distortion into an effective cartesian distance, is explicitly introduced.

In the ferric heme, the three $t_{2g}$ orbitals of the iron atom are occupied by 5 d-electrons and the single $t_{2g}$ orbital that remains unoccupied is the final (acceptor) state of the tunneling electron during the heme iron reduction process. This orbital vacancy is sometimes denoted as a "hole" state. Thus, when the heme is planar, the "hole" acceptor state is delocalized to the heme periphery via the iron $d_{xz}$ and $d_{yz}$ (i.e., $d_\pi$) orbitals and their interaction with the π-orbitals of the porphyrin. This spatially extends the acceptor state to the edge of the heme and leads to a tunneling distance, from the donor orbital state to the heme edge, which we denote as $r_0$. The exponent in Eq. 14.4 is written to expose the possibility that an extra "effective" tunneling distance $\alpha q_0$, beyond the distance ($r_0$) between the electron donor orbital and the heme edge, will result from hole state and electron density localization onto the central iron atom as the heme becomes more ruffled and the Fe ($d_\pi$)-porphyrin($\pi$) overlap is disrupted. It is noteworthy that, because the ruffling mode can effectively modulate the donor–acceptor tunneling distance, its vibrational excitation by the thermal bath can also affect the temperature dependence of the electron tunneling rate.

The normalization factor, $(n_D n_A)^{-1/2}$, in Eq. 14.4 takes into account [96] the delocalization of the electron wavefunction over the number of donor atoms ($n_D$) and acceptor atoms ($n_A$) that can be affected by ruffling. However, this factor does not have an exponential dependence. Another factor that can depend upon ruffling, but which might have a weak exponential effect, is the environmental reorganization energy, $\lambda_s$. This parameter could increase to some degree in response to increased charge localization on the iron atom when the heme is ruffled. For a fixed reduction free energy, $\Delta G^0_{red}$, the increased $\lambda_s$ would act to increase the semiclassical Marcus barrier, $\Delta G^\ddagger_s = \left(\Delta G^0_{red} + \lambda_s\right)^2/4\lambda_s$, assuming the electron transfer process occurs in the "normal" region [97]. The quantity, $\Delta G^0_{red}$ may also become less negative as the heme ruffles, consistent with the cyt $c_{551}$ reduction potentials [93]. Both of these effects could lead to a weak exponential decrease in the electron transport rate as the ruffling distortion is increased.

Because there are additional factors in Eqs. 14.3 and 14.4, beyond the total tunneling distance, $r_0 + \alpha q_0$, that might reduce the rate as the ruffling distortion ($q_0$) is increased, we proceed to estimate an upper limit for the maximum allowed tunnel distance increase due to ruffling. We denote this maximum distance change as $\alpha_{max} q_0$, and proceed to determine if it falls within the range of the heme radial dimension ($\lesssim$5 Å). To do this, we assume that the full exponential decrease in the photoreduction rate arises from the increased tunneling distance that occurs due to hole localization from the ruffling distortion. We choose a value for $\beta$ ~2.0 Å$^{-1}$ in

Eq. 14.4 that is typical for through space aromatic donor-acceptor systems [98–100] although certain ionic donors and covalent bonding can lead to smaller values for $\beta \sim 1.0$–$1.5$ Å$^{-1}$ [101]. If we choose a value of $\beta{\sim}1.5$ Å$^{-1}$, it increases the upper limit, $\alpha_{max}q_0$, by a factor of 1.33 and therefore we consider a range for $\beta \sim 1.5$–$2.0$ Å$^{-1}$ in our estimate. The photoreduction rate is found from the cross section, $\sigma_r$, and the laser flux (J) so that a fit of the data [46] using Eqs. 14.3 and 14.4 (Fig. 14.11b) depends only on the choice of $\alpha_{max}$. This leads to $\alpha_{max}$ = [1.6–1.2] $amu^{-1/2}$ for $\beta$ = [1.5–2.0] Å$^{-1}$. When the β-dependent range of $\alpha_{max}$ is used to evaluate $\alpha_{max}q_0$, we can extract the maximum possible ruffling-induced tunneling distance changes for each sample. This leads to the following ranges: [1.9–2.5] Å for *Pa* cyt $c_{551}$, [3.6–4.8] Å for F7A *Pa* cyt $c_{551}$, and [4.2–5.6] Å hh cyt c,. Remarkably, these upper limit values for the change in the tunnel distance are on the order of the ~5 Å distance between the Fe atom and the β carbons on the edge of the heme. Thus, the experimentally extracted distance changes are consistent with the hypothesis that the d-orbital hole associated with the ferric iron (which is the acceptor for the tunneling electron) becomes more localized on the iron atom as the ruffling distortion increases. Because of the complexity of the tunneling expression, these can only be considered as order-of-magnitude estimates. However, they demonstrate that the observed changes in the photoreduction rate are consistent with hole delocalization on a length scale that does not significantly exceed and, in fact, seems to be surprisingly similar to the heme dimension. Moreover, if (as expected) ruffling also increases the environmental barrier, $\Delta G_s^{\ddagger}$, to some degree, the estimates will be reduced so that they lie fully within the heme dimension.

## 14.5 Adiabatic Ligand Binding and Heavy Atom Tunneling in Heme Proteins

Ligand binding to heme systems belongs to a large and important class of chemical reactions called "spin-forbidden reactions" [58, 59, 102–115] where the total spin-state of the system changes ($\Delta S \neq 0$) in going from reactant to product. Spin-forbidden reactions are quite common among the many reactions that involve open shell transition metals and are found widely in catalysis and bioorganic chemistry. However, the current computational methods that deal with these reactions are still not as well established as for the ($\Delta S = 0$) spin-conserving reactions. One difficulty in the theoretical understanding of spin-changing reactions arises from the need for complicated electronic structure calculations in which spin–orbit coupling and electron correlations must be accurately included. The calculation of the spin–orbit coupling matrix elements $V_{soc}$, using ab-initio methods, is still quite challenging. Thus, most current theoretical approaches simply treat these reactions [58, 59, 102–111] as non-adiabatic, particularly those that involve ligand binding to heme proteins [58, 59, 102, 109, 116].

However, there is no strong experimental evidence that justifies this assumption for heme proteins except for the unusually small Arrhenius prefactor that has been observed in the reaction where CO binds to the heme of myoglobin (Mb) [49, 117–119]. Non-adiabatic calculations usually focus on locating the geometry of the lowest energy crossing point between the diabatic spin-state potential energy surfaces (PESs) of the reactant and the product. As depicted in Fig. 14.12, this is called the minimum energy crossing point (MECP) [58, 59, 102–111]. These treatments, generally minimize the role of spin-orbit coupling, which can lead to mixed and/or intermediate spin ground states having a lowered energy barrier that allows the reaction to proceed adiabatically as it passes through the transition state region. The saddle point on a mixed-spin adiabatic ground state with a large $V_{soc}$ can be considerably lower than the MECP that is found at the intersection of the two diabatic spin PESs (Fig. 14.12). This can introduce errors in the estimation of the activation energy barrier and, consequently, a significant underestimation of the reaction rate. Spin-changing reactions are mainly governed by: (i) the amount of energy needed for the system to reach a configuration where the potential energies of two spin states along the reaction path are close: i.e., the activation energy barrier, and (ii) the magnitude of spin-orbit coupling between the spin states, $V_{soc}$. If $V_{soc}$ is weak compared to $k_B T$ the reaction will continue on a non-adiabatic trajectory; whereas, if it is sufficiently strong, the reaction becomes adiabatic. In this later case, the Arrhenius prefactor of the reaction no-longer includes $V_{soc}$ and the reaction becomes independent of spin-selection rules. Theoretical models [120, 121] for "spin-forbidden" reactions that include spin–orbit coupling and electronic correlation effects have only been developed very recently and applied to small molecules. Hopefully, these promising techniques will mature in the near future so they can be applied to larger systems such as ligand binding to heme proteins.

We have recently investigated the temperature dependence of ultrafast CO binding to heme proteins [53] and heme model compounds [18], as well as the ultrafast rebinding of Met80 in cyt c [56], over broad temperature ranges. Analysis of the experimental data has demonstrated that ligand binding to heme, in contrast to the commonly held view [57–59, 102, 109, 116, 122], is an adiabatic reaction with a spin-independent prefactor. Instead of resulting from spin selection, the much smaller prefactor found for CO binding to Mb is attributed to the effect of the histidine in the distal binding pocket of the protein. This "distal histidine" increases the enthalpic barrier for CO rebinding to the heme iron and slows the reaction to the point that entropy development, associated with the orientational disorder of the CO ligand and its position in the distal pocket, becomes a major factor. Entropy production on these longer timescales generates an additional entropic barrier to CO rebinding that reduces the prefactor by about two orders of magnitude compared to heme proteins without histidine in the distal pocket [18, 54]. Our studies of heme systems with ultrafast room temperature CO rebinding kinetics clearly demonstrate that the often assumed "spin-forbidden" nature of the CO-heme binding reaction does not affect the rate. During these same studies, we also found several systems where the temperature dependence of the rebinding rate becomes anomalous at very low temperature ($\lesssim$ 60 K). One possible explanation for the anomalous temperature dependence is that

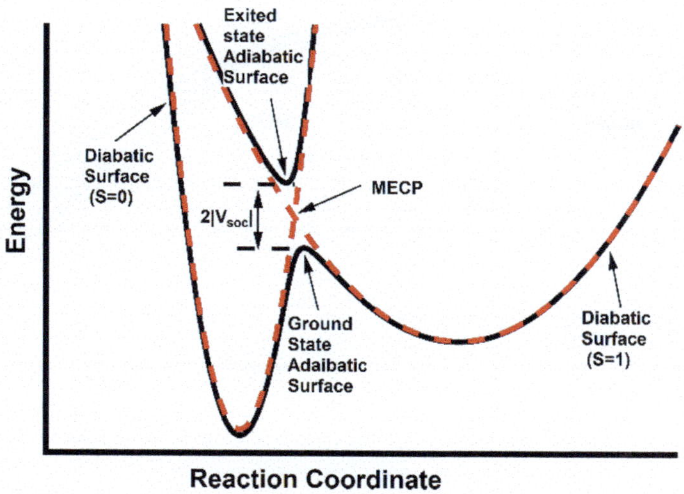


**Fig. 14.12** Typical PES for a spin-forbidden reaction with a spin change from S = 1 to S = 0. The diabatic PESs are red dashed lines and the corresponding adiabatic PESs are shown as black solid lines. Reproduced from reference [53] with permission of the American Chemical Society

heavy atom quantum mechanical tunneling becomes an important reaction channel at these temperatures [53, 56].

Figure 14.13 displays CO geminate rebinding kinetics to the carbon monoxide oxidation activator heme protein from the thermophilic anaerobic bacterium Carboxydothermus hydrogenoformans (ChCooA) at temperatures above the glass transition, $T_g \sim 180$ K. The kinetic response at each temperature was fit (solid lines) using a model based on the relaxation of an inhomogeneous ensemble of heme doming geometries, which has been described in detail elsewhere [18, 49, 53]. In this temperature range, CO rebinding kinetics slows down as the temperature increases. This anti-Arrhenius behavior is indicative of an increase in the average rebinding barrier that arises from protein relaxation processes that begin to occur following photolysis when $T \geq T_g$. The relaxation leads to an increase in the equilibrium position of the heme doming coordinate at higher temperatures. The non-exponential kinetic response observed for CO rebinding to ChCooA at temperatures above $T_g$ is attributed to rebinding of the photolyzed CO molecules to an ensemble of heme doming geometries that is interconverting more slowly than the rebinding timescales. There is an increase in the average heme doming displacement as the temperature is raised above $T_g$, but the ensemble still remains heterogeneous on the ultrafast time scale of the rebinding kinetics [18, 49].

Similar non-exponential CO rebinding on ultrafast timescales has been observed for the bare heme chromophore (FePPIX) [18] where temperature-dependent anharmonic effects also lead to a more domed heme geometry at elevated temperatures [18]. The evolution of the heme photoproduct ensemble leads to the iron atom moving out-of-plane to a more domed position with an average value denoted as, $a_0(T)$.

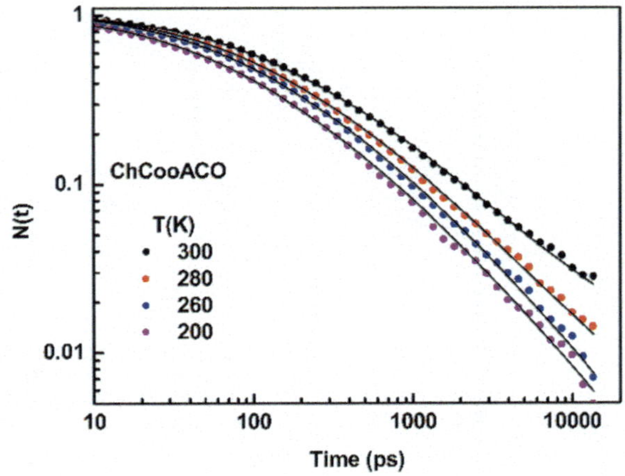


**Fig. 14.13** Geminate rebinding kinetics of CO in ChCooA above $T_g$. Reproduced from reference [53] with permission of the American Chemical Society

As the temperature is raised, the evolution of $a_0(T)$ has also been correlated with the position of the structure-sensitive Band III (~760 nm) of the heme group [18]. Myoglobin has much slower CO rebinding kinetics than FePPIX due to the presence of the distal histidine, and it also shows anti-Arrhenius behavior at temperatures above the glass transition. However, the kinetic inhomogeneities within the MbCO ensemble, arising from differing protein conformers, are thermodynamically averaged because of the slower rebinding timescale so that a single exponential rebinding response is observed above $T_g$ [117, 123].

In ferrous cyt c, high quantum yield photolysis of the endogenous Met80 ligand is observed, followed by ultrafast geminate rebinding [124]. The kinetics of this process above $T_g$ are displayed in Fig. 14.14. At these higher temperatures, the Met80 geminate recombination is only weakly temperature dependent and the kinetics can be fit [56] using a single exponential having a time constant of ~7 ps, in very good agreement with other investigations [124–127]. Spectroscopic [124, 126, 127] and ultrafast X-ray absorption [125] have shown that methionine dissociation and rebinding involve a $\Delta S = 2$ spin transition, which is the same as CO binding to heme except for cyt c the sulfur atom of Met80 binds in the sixth ligand position.

However, the analysis used in the X-ray work [125] contains serious flaws [56]. These arise from the erroneous assumption that the ligand-bound and unbound populations are in full thermodynamic equilibrium on the ps timescale. Non-equilibrium population changes due to the ligand rebinding that follows photolysis are not considered. Instead, a time-dependent effective temperature [125] is used to attribute <u>all</u> the observed ps population changes to transient cooling of the system temperature, which alters "equilibrium" populations of the ligand-bound and unbound states. These populations, along with estimates of the time-dependent system temperature are then used to create a Van't Hoff plot that reflects the free energy difference between the methionine bound and unbound states within the protein structure [125]. This analysis is incorrect because non-equilibrium unidirectional ligand binding is clearly taking place on the 1–10 ps timescale. It is the free energy <u>barrier</u> to the

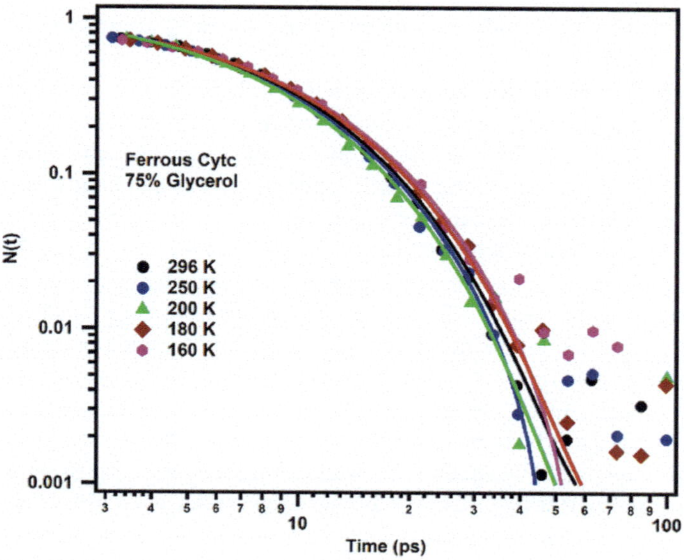


**Fig. 14.14** Geminate rebinding kinetics of Met80 in ferrous cyt c above $T_g$. Reproduced from reference [56] with permission of the American Chemical Society

ligand binding reaction that is crucially important in determining the relative bound and unbound populations on this timescale, not the overall reaction free energy.

In order to explain the nearly temperature-independent Met80 rebinding kinetics of cyt c, we have suggested that the endogenous nature of the Met80 ligand could affect the reaction coordinate and reduce the proximal enthalpic barrier, $H_P$, by tightly constraining the iron-ligand distance [56]. The enthalpic barrier is related to the energy needed to vacate the two ($e_g$) d-orbitals as the heme iron is driven from a domed high-spin $(t_{2g})^4(e_g)^2 = (d_{xy})^2(d_{xz})^1(d_{yz})^1(d_{z^2})^1(d_{x^2-y^2})^1$ quintet configuration to the planar ligand-bound low-spin singlet $(t_{2g})^6(e_g)^0 = (d_{xy})^2(d_{xz})^2(d_{yz})^2(d_{z^2})^0(d_{x^2-y^2})^0$ where the d-orbital $t_{2g}$ subset is filled. During the binding reaction, there are at least two possible intermediate-state triplet configurations: $(t_{2g})^5(d_{z^2})^1(d_{x^2-y^2})^0$ and $(t_{2g})^5(d_{z^2})^0(d_{x^2-y^2})^1$, which we abbreviate as $(d_{x^2-y^2})^0$ and $(d_{z^2})^0$, respectively. The intervention of an intermediate triplet state configuration means that the state mixing can be governed by a first-order, rather than by a second order, spin–orbit coupling (the latter is needed for a direct $\Delta S = 2$ transition). The first-order spin–orbit matrix element is much larger and can be strong enough to generate an adiabatic reaction surface, leading to spin-independent selection rules, as recently observed experimentally [18, 53, 56]. Moreover, because there can be a strong dependence of the transition metal spin-state on relatively small structural perturbations under biologically relevant circumstances, it is likely that spin-state entanglement occurs [128–130]. In this treatment, we invoke only simple crystal field concepts, but we note that the interplay of Coulomb and Hund's rule interactions must be carefully considered when attempting to develop an accurate theoretical picture [129].

Many studies have shown that CO rebinding to heme proteins and heme model compounds, at both room- and low temperature, is consistent with a distribution of transition state enthalpic barriers that are quasi-static on the CO rebinding timescale [49, 117, 119, 131]. The resulting kinetic heterogeneity is quantitatively consistent with a distribution of out-of-plane heme doming conformations, $P(a)$, where $a$ is the heme doming equilibrium position [18, 49, 53, 54]. On the other hand, the endogenous Met80 rebinding of cyt c has an unusually low, and nearly constant, rebinding barrier at higher temperatures [56]. Because of this, we have suggested that the covalent connection of the c-type heme and the Met80 to the folded protein leaves the endogenous Met80 in tight proximity to the heme iron following photo-dissociation. The on-axis Met80 sulfur electron density so close to the iron atom may lead to a relatively high $d_{z^2}$ orbital energy, even in the transient 5C domed photoproduct. This would allow for the formation of a more "reactant-like" transition state [132, 133], where the $(d_{z^2})^0$ triplet is able to mix with the singlet and quintet states. Such a scenario could lead to a substantial reduction of the enthalpic energy barrier relative to an exogenous ligand like CO, which is free to move into the distal pocket away from the iron atom. The ultrafast X-ray absorption measurements [125] find the Fe-$S_{Met80}$ distance in the high-spin photoproduct state of cyt c to be only 3.04 Å. Based on the 1.8 Å van der Waals radius of sulfur and the 1.4 Å atomic radius (or the 2.05 Å van der Waals radius [134]) of iron, this indicates very close contact without bond formation.

Below $T_g$, the solvent is frozen and protein interconversions that might affect the structural conformation of the embedded heme are suppressed. In this temperature range, the kinetics of CO rebinding to ChCooA, as well as Met80 ligand rebinding in cyt c, follows the expected thermodynamic behavior where the rebinding slows down as the temperature decreases (Fig. 14.15). However, the kinetics of the rebinding process are distinctly non-exponential, reflecting a quenched distribution of rebinding barriers.

As can be seen from Fig. 14.16, the values of both the first ($a_0$) and second ($\sigma_a$) moments of the heme doming distribution, $P(a)$, which are used to describe the ensemble of heme conformations following CO dissociation, are only weakly dependent on temperature from ~180 K to ~80 K. This behavior is observed for CO binding to both ChCooA and to the bare heme (FePPIX) [18] as shown in Fig. 14.16. As noted above, the CO binding kinetic response of these heme systems is analogous to what is observed for CO binding to Mb, but on a much faster time scale [49, 119]. The Arrhenius prefactors for CO rebinding to ChCooA and for Met80 to cyt c are $k_0 = 1.2 \times 10^{11}$ s$^{-1}$ [53] and $1.6 \times 10^{11}$ s$^{-1}$ [56], respectively. Such values of $k_0$ are very similar to that found for CO binding to the bare protoheme ($1.5 \times 10^{11}$ s$^{-1}$) with either 2-methyl imidazole (2MeIm) or water as the heme proximal ligand [18]. Significantly, these $k_0$ values are over two orders of magnitude larger than the prefactor observed for CO binding to Mb ($k_0 \sim 10^9$ s$^{-1}$) [49, 117–119].

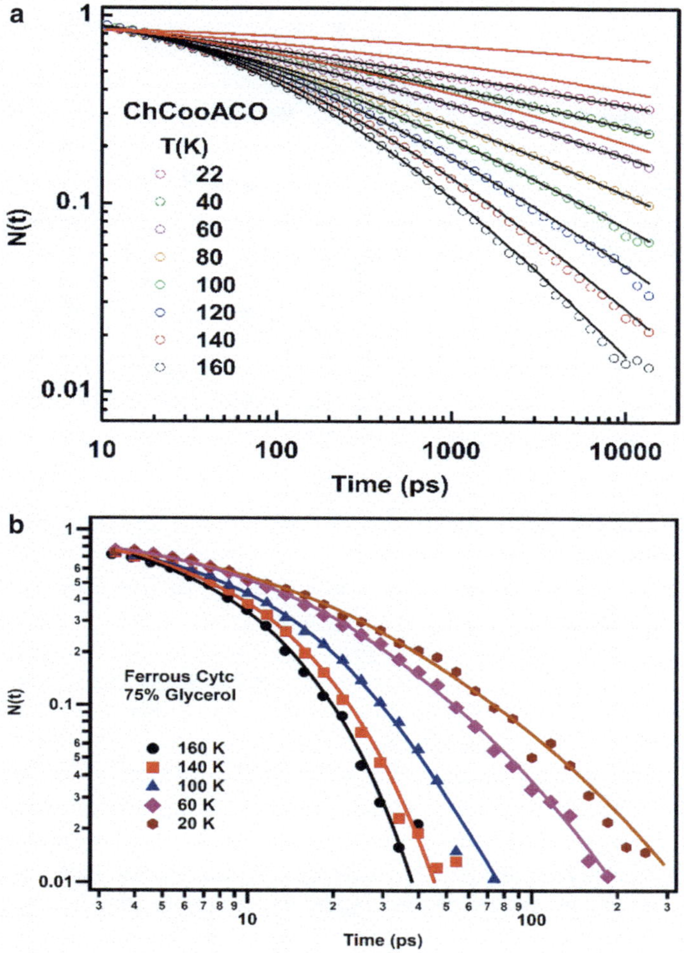


**Fig. 14.15** Geminate rebinding kinetics of **a** CO to ChCooA and **b** Met80 to ferrous cyt c below $T_g$. Reproduced from references [53, 56] with permission of the American Chemical Society

Within an adiabatic model for heme-CO binding, the Arrhenius prefactor includes an entropic barrier, $\Delta S^\dagger$, and can be written as: $k_0 \propto \nu_0 exp\left(\frac{\Delta S^\dagger}{k_B}\right) = \nu_0 \frac{\Omega^\dagger}{\Omega_0}$ where $\nu_0$ is an attempt frequency and $\Omega_0$ and $\Omega^\dagger$ are the number of protein–ligand micro-states in the dissociated system and in the rebinding transition state, respectively. Because the overall electronic spin-state transition ($\Delta S = 2$) is the same for the CO binding reaction in ChCooA, protoheme, and Mb, the large difference in their prefactors cannot be due to spin-selection rules as is often suggested [57–59, 102, 116, 122]. Instead, the differences are explained by changes in either $\Omega_0$ and $\Omega^\dagger$, or both. In the case of ChCooA and bare protoheme, the entropic barriers for CO rebinding must be much lower in comparison to that of Mb. This means that the activation entropy

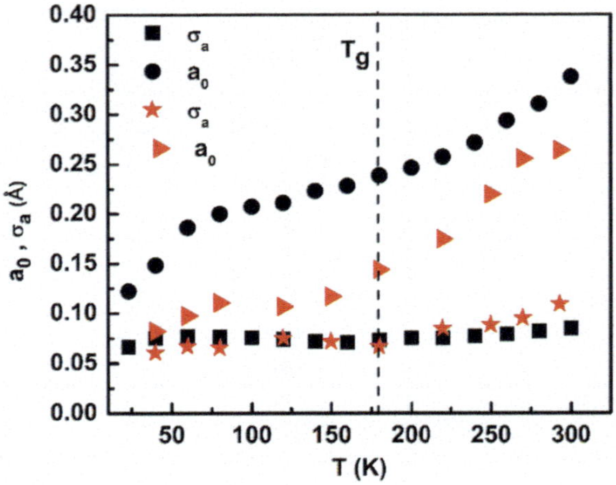


**Fig. 14.16** Temperature dependence of the average iron out-of-plane displacement $a_0$ and its variance $\sigma_a$ as a function of temperature for CO rebinding to ChCooA (black) and $H_2O$-FePPIX (red). Reproduced from reference [53] with permission of the American Chemical Society

can strongly affect the CO rebinding rate of heme proteins, and this will depend on the architecture of the heme distal pocket.

Moreover, the large reduction in the activation entropy for CO binding to ChCooA and protoheme compared to Mb is also dependent on the relative timescales of the reactions. The entropic barrier is synergistically dependent on the enthalpic barrier for ligand binding. This occurs because, as the enthalpic barrier is increased, the reaction slows and this allows more time for the entropic barrier to develop [18]. Due to the absence of the distal histidine, which must be moved for CO binding to occur in Mb, the average enthalpic barrier for ChCooA (found [56] to be $\langle H \rangle =$ 7.5 kJ/mol at 300 K and $\langle H \rangle = 3$ kJ/mol below $T_g$) is much smaller than for Mb (found [56] to be $\langle H \rangle = 18$ kJ/mol at room $T$ and $\langle H \rangle = 12$ kJ/mol below $T_g$). The absence of the distal histidine greatly speeds up CO geminate rebinding in ChCooA, and there is not enough time for entropy production to significantly increase $\Omega_0(t)$ in the ChCooA distal pocket. This significantly reduces the size of the entropic barrier, which is a non-equilibrium, time-dependent, variable on these ultrafast timescales. Basically, because CO rebinds so fast following the photolysis of ChCooA-CO, the system is unable to fully sample the entire range of possible dissociated states. Thus, in situations where the rebinding time scale is faster than the dissociated state entropy production time, the entropic barrier will be significantly reduced and the observed Arrhenius prefactor will be substantially increased [18].

### *14.5.1 Spin–Orbit Coupling and the Non-adiabatic Reaction Model*

When a non-adiabatic model is used to explain the heme-ligand binding reaction, the Arrhenius prefactor will depend strongly on the spin-selection rules that govern the reaction [116]. Such a model can be applied when friction is low [118] and the matrix elements, involving spin–orbit coupling between different spin states of the reacting system, are small (i.e., $V_{soc} \lesssim k_B T$). Several quantum calculations that focus on CO binding to heme have been carried out in order to better understand how changes in the heme doming structure and the ligand position will lead to intersections of the various electronic spin-state surfaces in these systems [58, 59, 102]. The outcome depends to some degree on the DFT functionals and basis sets that are employed [58, 102]. A "direct" two-state ($\Delta S = 2$) non-adiabatic heme-ligand binding model is often taken as the scenario for rebinding; however, given the entropy-based explanation for the small prefactor observed for the MbCO rebinding reaction, it is by no means obvious that a direct $\Delta S = 2$ reaction is the appropriate reaction pathway.

If the non-adiabatic reaction mechanism is to be taken as a starting point for CO binding in heme proteins, the calculations must also somehow account for the ultrafast reaction channel that has been observed for ChCooA [53, 54] and the bare heme (FePPIX) [18], as well as in a variety of other heme systems [135, 136]. One possible ultrafast non-adiabatic mechanism involves sequential ($\Delta S = 1$) transitions where the iron triplet ($S = 1$) spin-state is a weakly coupled intermediate (see Fig. 14.17). As suggested by others [57–59, 102, 116, 122], the explanation for the slower CO rebinding observed for Mb and Hb can be ascribed to a "direct" non-adiabatic $\Delta S = 2$ transition. The action of a second-order spin–orbit coupling matrix element [116] at the diabatic crossing point between the $S = 2$ and the $S = 0$ states significantly reduces the prefactor relative to the *first-order* matrix elements that are depicted in Fig. 14.17 (small circles). However, to properly explain the ultrafast heme-CO rebinding reactions, the barrier between the $S = 1$ and the $S = 0$ states must be small, as shown in Fig. 14.17a. This is necessary in order to account for the absence of any detectible optical intermediate other than the quintet and singlet states in the ultrafast CO rebinding reaction, where a clean isosbestic point in the rebinding reaction is observed [18]. Within the model presented in Fig. 14.17a, a vanishing barrier between the $S = 1$ and the $S = 0$ states will eliminate any transient triplet state population so that the rate limiting step involves only the initial first-order ($\Delta S = 1$) iron transition between the $S = 2$ and the $S = 1$ states.

However, DFT-based calculations that evaluate the minimum energies for the quintet-triplet, triplet-singlet, and quintet-singlet crossing points have found [58, 102] that the direct ($\Delta S = 2$) quintet-singlet crossing is the lowest energy surface crossing. Thus, the $\Delta S = 2$ reaction has been assigned as the primary route for heme-CO rebinding using the non-adiabatic reaction model [58, 102]. Such a choice conveniently explains the small prefactors that are observed for Mb and Hb [49, 117–119], but it does not address the much larger prefactors that are observed in most other heme-CO reactions [53, 55, 56].

Importantly, the DFT results also find that the minimum energy crossing point between the $S = 1$ and the $S = 0$ states lies at a *higher* energy than the minimum energy crossing point of the $S = 2$ and the $S = 1$ states [109]. This situation is depicted schematically in Fig. 14.17b and it presents a serious problem for the sequential non-adiabatic model that has been established based on DFT calculations. Under the condition shown in Fig. 14.17b, some population in the triplet state should

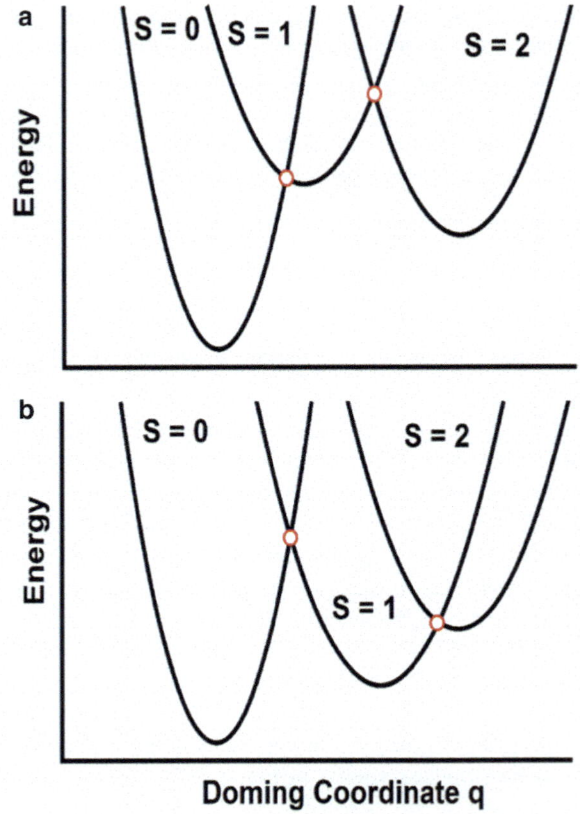


**Fig. 14.17** Schematic representation of the diabatic electronic spin states for the heme-CO system along the doming coordinate. **a** The barrier for the transition 2 → 1 is greater than that for 1 → 0 so that the first transition becomes rate limiting and no intermediate S = 1 population can be detected; **b** The barrier for the transition 2 → 1 is smaller than that for 1 → 0 so that the population in the S = 1 state should be detectable. Reproduced from reference [53] with the permission of the American Chemical Society

be observable as an intermediate during the ultrafast heme-CO rebinding reaction. However, optical experiments reveal no evidence of a potential $S = 1$ intermediate when the ultrafast rebinding is followed using a broad spectral bandwidth probe [18, 135, 136]. Instead, a clear isosbestic point is observed [18, 135, 136] between the reactant ($S = 2$ quintet state) and the product ($S = 0$ singlet state). Thus, the non-adiabatic sequential model, based on currently available DFT calculations, is not consistent with observed measurements of the ultrafast heme-CO rebinding reaction in multiple systems.

Alternatively, an adiabatic model can describe the ultrafast CO rebinding of these heme systems as well as the slower CO rebinding in Mb and Hb. The adiabatic model introduces an entropic barrier to CO binding, $\Delta S^{\dagger}$, which develops on longer timescales and is synergistic with the size of the enthalpic barrier and the rebinding timescale. The adiabatic model, in contrast to non-adiabatic models based on spin-selection rules, self-consistently accounts for both the reduced prefactors of Mb and Hb [49, 117–119, 123] and the much larger prefactors observed in many other heme-CO reactions [18, 53, 55, 56].

### 14.5.2 Ligand Binding Models for Heme Proteins

As suggested above, CO binding to the heme iron appears to be an adiabatic process [53], which shifts the commonly held paradigm that CO rebinding is non-adiabatic due to spin-forbidden ($\Delta S = 2$) selection rules [57, 59, 102, 109, 116, 122]. In addition to CO, the binding of endogenous heme ligands also involves a $\Delta S = 2$ reaction and, for the case of the endogenous Met80 ligand in cyt c, it has been demonstrated [56] that the Arrhenius prefactor is $k_0$ ~$1.5 \times 10^{11}$ $s^{-1}$. This is very close to the value found for CO [18, 53] and NO [18, 132] binding in a variety of other heme proteins. The notable exceptions are MbCO and HbCO, where the prefactor is reduced [49, 117–119] due to the suggested development of an entropic barrier [18] that has time to develop because of the added enthalpic barrier from the distal histidine, which slows the rebinding significantly. The rebinding of NO, which has a spin of $S = \frac{1}{2}$ that combines with the $S = 2$ spin of the iron atom ($S_{tot} = \frac{3}{2}, \frac{5}{2}$), can take place through an obvious a first-order spin–orbit coupling channel ($\frac{3}{2} \rightarrow \frac{1}{2}$) where$\Delta S = 1$. This fast channel for the NO reaction [132] has an Arrhenius prefactor ($\sim 10^{11}$ $s^{-1}$) that is very similar to what is found for the ($\Delta S = 2$) CO binding reaction of heme systems lacking the distal histidine. This comparison provides additional evidence that the heme-ligand binding reaction is independent of spin-selection rules. Other heme rebinding reactions with endogenous ligands also involve a $\Delta S = 2$ transition and they also display a dominant and universal ultrafast time constant of ~5–7 ps at room temperature [54, 56, 126, 137]. It is likely that all endogenous ligand rebinding reactions are analogous to what has been found for Met80 rebinding in cyt c and their generic Arrhenius prefactors are ~$10^{11}$ $s^{-1}$. These various kinetic observations appear to be completely consistent with the proposition that ligand binding in ferrous heme proteins is an adiabatic process with essentially no dependence on spin-selection rules [53, 56].

For the case of endogenous ligand binding, the proximity of the endogenous ligand to the iron tends to increase the spin–orbit coupling matrix elements [100, 138] and, for first-order intersections (between $S = 2$ and $S = 1$ or $S = 0$ and $S = 1$), the reacting states are mixed more strongly, leading to an adiabatic reaction surface. In this situation, the electronic state is nearly instantaneously following the nuclear motion during the reaction. Finally, it is noteworthy that a variety of other effects can also serve to move the ligand binding reaction of heme proteins toward the adiabatic limit [118].

As an example, a schematic adiabatic surface [56] is shown in Fig. 14.18 where a "reactant-like" transition state is shown at the intersection of the quintet and triplet state. Relatively strong spin-state mixing, which is first-order in spin–orbit coupling, leads to the adiabatic ground state reaction surface. In the absence of the triplet state intersection, much weaker non-adiabatic coupling is expected for the direct $\Delta S = 2$ reaction because only second-order spin–orbit coupling is available to mix the reacting states. Because there are two possible triplet states, $(d_{z^2})^0$ and $\left(d_{x^2-y^2}\right)^0$ as noted above, Fig. 14.18 shows how different intersections might occur, depending on the relative energies of these states. This suggests how the CO (exogenous) and

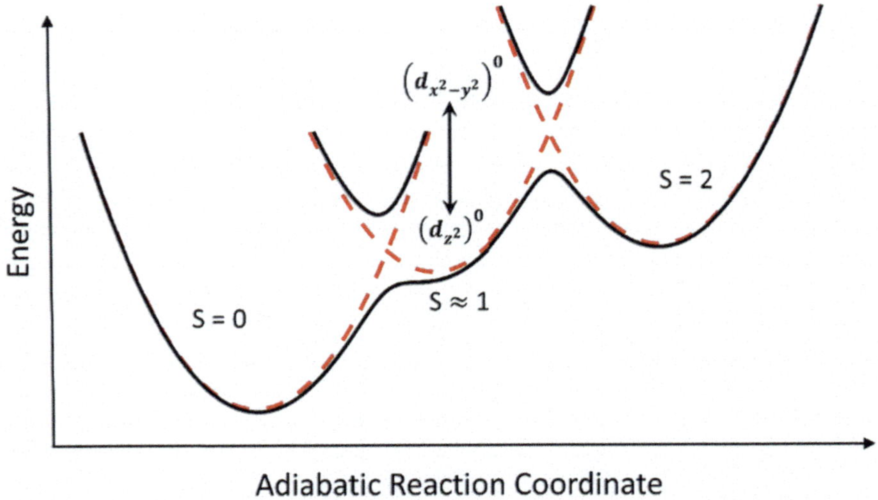


**Fig. 14.18** A proposed potential energy surface where the iron triplet state intersects along the reaction coordinate so that first-order spin–orbit coupling strongly mixes the spin states and leads to an ultrafast binding reaction that proceeds exclusively along the ground state adiabatic potential energy surface shown as the lower black solid line (not to scale). Reproduced from reference [56] with the permission of the American Chemical Society

methionine (endogenous) ligand rebinding reactions can both remain adiabatic but have different reaction barriers and electronic state compositions along the reaction path. At a minimum, the reaction path involves both heme doming and the iron-ligand distance, which means that the binding of an endogenous ligand that is structurally positioned very near to the heme iron may lead to a different energetic ordering of the iron triplet states relative to the situation where a diatomic ligand resides at, and rebinds from, a more distant location.

For Met80 binding in cyt c, the sulfur ligand will be held in close proximity to the iron atom by the protein structure and its thioether bonds to the heme periphery. This could enhance the depopulation of the $d_{z^2}$ orbital so that the $(d_{z^2})^0$ triplet intersects with the quintet surface at lower energy compared to the $\left(d_{x^2-y^2}\right)^0$ triplet. A lower reaction barrier would likely result for the endogenous methionine ligand because the heme can remain more domed in the transition state. Only a small fluctuation along the doming coordinate is needed to reduce the distance between the iron and the endogenous Met80 ligand, allowing covalent bond formation to take place. This scenario is analogous to the temperature-independent ultrafast rebinding reaction for NO that has been described elsewhere [132].

On the other hand, the $\left(d_{x^2-y^2}\right)^0$ triplet may be more relevant to the adiabatic CO rebinding reaction. In this case, the photolyzed ligand can move to a more distant docking site allowing the $d_{z^2}$ orbital to drop in energy so that the $\left(d_{x^2-y^2}\right)^0$ triplet configuration intersects the quintet surface. In this situation, more in-plane heme motion along the doming coordinate is required to reach the transition state. Because it requires more energy to reach the planar transition state, a larger enthalpic barrier would be observed for the exogenous ligand compared to the endogenous ligand. Here we want to re-emphasize that ultrafast optical studies have so far not revealed any evidence for the transient population of an independent triplet state [18, 124, 127, 135, 136]. However, in the adiabatic picture, the triplet state should not be considered as a separate, or independent, state. Rather, it is positioned and strongly admixed in the transition state region, as indicated by Fig. 14.18, so that, in effect, only two-state behavior is observed.

### 14.5.3 *Quantum Tunneling at Low Temperature*

When the sample temperature decreases below ~60 K, the non-exponential kinetic responses of CO rebinding to ChCooA [53] and Met80 rebinding to cyt c [56] are much faster than what is predicted using the classical heme doming model with a quenched barrier distribution [49, 53, 56]. It is possible that the faster than expected low-temperature kinetics could be related to incomplete heme photoproduct relaxation at very low temperature or to the existence of a quantum mechanical tunneling channel that competes with the classical over-barrier process [53, 56]. Previous studies have also invoked tunneling effects [116] to explain CO [139, 140] and water [141] binding to Mb and heme at temperatures below 25 K.

The tunneling process in heme proteins that involves ligand binding to the heme iron (with an overall $\Delta S = 2$ spin-state change) can be considered as a combination of ligand tunneling through the distal barrier of height, $H_D$, and tunneling of the iron through the proximal barrier, $H_P$, from the domed out-of-plane position to the final planar position. These tunneling processes can be either concerted or sequential. When a significant barrier exists due to the distal histidine, as observed for CO binding to myoglobin, CO tunneling through the distal barrier becomes the dominant low-temperature tunneling process [140]. On the other hand, for heme proteins where the distal reaction barrier is small or non-existent ($H_D \sim 0$), such as CO binding to ChCooA [53] or Met80 binding to cyt c [56], iron tunneling from an out-of-plane to an in-plane heme position can also potentially take place at low temperature [53, 116].

When an adiabatic reaction surface is used to describe ligand binding to heme, the Bell tunneling model [142] can be used to crudely estimate the temperature, $T_t$, where the quantum mechanical tunneling rate starts to compete with the classical over-barrier process. The high-temperature heme-CO binding rate for a classical transition,$k_c$, is assumed to be given by an Arrhenius expression $k_c = k_0 e^{-H/k_B T}$, where $H$ is the enthalpic barrier height. At very low temperatures, the rate is assumed to be dominated by tunneling and is approximated [142] as: $k_t = k_0 e^{(-\pi a \sqrt{2\mu H}/2\hbar)}$, where $a$ and $\mu$ are the tunneling distance and the reduced mass of the tunneling particle, respectively. In the absence of a distal barrier ($H_D$), as found for Met80 binding to cyt c and for CO binding to ChCooA and PPIX, only the possibility of tunneling along the doming coordinate needs to be considered. In such a situation, the tunneling distance ($a$) can be taken as the heme out-of-plane displacement and $\mu$

is the reduced mass of the nuclei involved in the doming motion (this may involve the proximal histidine ligand, the iron atom, the porphyrin, and potentially even some of the surrounding protein residues). Although the rate can be calculated numerically [142], we simply assume the total rate at any temperature is the sum of the tunneling and classical channels,$k \approx k_c + k_t$. The crossover temperature,$T_t$, where the rates for the two channels are equal is then given as

$$T_t = \frac{\hbar}{\pi a k_B}\sqrt{\frac{2H}{\mu}} \tag{14.5}$$

and, if we take [53] $H = H_p \cong \frac{1}{2}Ka^2$, we find

$$T_t = \frac{\hbar}{\pi k_B}\sqrt{\frac{K}{\mu}} = \frac{\hbar\omega}{\pi k_B} \tag{14.6}$$

where the effective force constant for the doming mode is $K$ and $\omega$ is its frequency. It is important to note that, along with $\mu$, the parameter, $a$, the heme out-of-plane equilibrium position, drops out of the expression for $T_t$. This means that even in a heterogeneous ensemble, where $a$ is a distributed parameter, the entire distribution will have roughly the same crossover temperature. This may explain why there is such a rapid onset of the anomalously fast rebinding kinetics as temperatures fall below ~60 K.

We note that the doming frequency of photolyzed hemes at low temperature is potentially somewhat higher than the value found at room temperature because the heme photoproduct is not able to fully relax. As observed in VCS studies of ligand photolysis in Mb [28], there is a strong coherence amplitude at both ~80 $cm^{-1}$ and ~40 $cm^{-1}$ that is attributed to heme doming motions. Only the latter frequency is observed for the fully relaxed deoxy Mb heme. The higher frequency component has been suggested to arise from the photoproduct, Mb*, where photolysis has taken place but the protein is not able to fully relax on the ultrafast timescale [28]. At temperatures below $T_g$, photolysis will also generate an unrelaxed species because the frozen protein cannot rearrange its conformation to accommodate the photolyzed heme. Thus, an increased heme doming frequency, relative to the equilibrium value at room temperature, may apply when the heme is photolyzed at low temperature. Analogous to what is observed for Mb, VCS measurements of ChCooA-CO [17] reveal a mode at ~100 $cm^{-1}$ following CO photolysis in addition to the ~50 $cm^{-1}$ mode that has been assigned to the relaxed heme doming frequency in the VCS spectrum of un-photolyzed ChCooA. Thus, we take the frequency for the ChCooA doming mode following photolysis at low temperature to be on the order of ~100 $cm^{-1}$. Using this value for the doming frequency, $\omega$, in Eq. 14.6 leads to a crossover temperature $T_t \sim$ 45 K where the tunneling rate is roughly the same as the adiabatic barrier crossing rate. This very simplified analysis suggests that iron tunneling along the doming coordinate of the heme may begin to play a significant role in the ligand binding process as the temperatures begin to approach ~45 K. In addition, tunneling can be enhanced by thermally driven oscillations of the doming mode around its out-of-plane equilibrium position, $a$. These vibrationally enhanced tunneling effects have been ignored here in order to obtain a simple estimate of the crossover temperature using the Bell model. Their inclusion would serve to increase the effect of tunneling on the ligand binding rate and increase the crossover temperature. We conclude that heavy atom tunneling could potentially help to account for deviations from the classical rebinding model, and the more rapid than expected rates observed at low temperature ($\lesssim$ 60 K) for both CO binding in ChCooA [53] and Met 80 binding in cyt c [56]. Thus, it is possible that heavy atom tunneling at low temperature is a more widespread phenomenon in biomolecules than is currently assumed.

**Acknowledgements** This work was supported by the NSF CHE-1764221 and MCB-2231080. The authors thank numerous co-authors, especially Yuhan Sun, Bridget Salna, Tim Sage, and Jasper van Thor, for their contributions to this work.

## References

1. V.L. Schramm, Accounts Chem. Res. **48**, 1032 (2015)
2. T.H. Kim et al., Science **355** (2017)
3. A. Sali, E. Shakhnovich, M. Karplus, Nature **369**, 248 (1994)
4. H. Frauenfelder, G. Chen, J. Berendzen, P.W. Fenimore, H. Jansson, B.H. McMahon, I.R. Stroe, J. Swenson, R.D. Young, P Natl. Acad. Sci. USA **106**, 5129 (2009)
5. B. Salna, A. Benabbas, J.T. Sage, J. van Thor, P.M. Champion, Nat. Chem. **8**, 874 (2016)
6. P.M. Kiefer, J.T. Hynes, J. Phys. Org. Chem. **23**, 632 (2010)
7. P.M. Kiefer, J.T. Hynes, J. Phys. Chem. A **108**, 11793 (2004)
8. A. Benabbas, B. Salna, J.T. Sage, P.M. Champion, J Chem Phys **142** (2015)
9. A.V. Soudackov, S. Hammes-Schiffer, Faraday Discuss. **195**, 171 (2016)
10. J.P. Klinman, A.R. Offenbacher, Acc. Chem. Res. **51**, 1966 (2018)
11. T. Kobayashi, T. Saito, H. Ohtani, Nature **414**, 531 (2001)
12. Q. Wang, R.W. Schoenlein, L.A. Peteanu, R.A. Mathies, C.V. Shank, Science **266**, 422 (1994)
13. M.H. Vos, C. Rischel, M.R. Jones, J.L. Martin, Biochemistry **39**, 8353 (2000)
14. S. Nakashima, Y. Nagasawa, K. Seike, T. Okada, M. Sato, T. Kohzuma, Chem. Phys. Lett. **331**, 396 (2000)
15. M.H. Vos, J.C. Lambry, S.J. Robles, D.C. Youvan, J. Breton, J.L. Martin, P Natl. Acad. Sci. USA **88**, 8885 (1991)
16. Y.H. Sun, A. Benabbas, W.Q. Zeng, S. Muralidharan, E.M. Boon, P.M. Champion, J. Phys. Chem. B **120**, 5351 (2016)
17. V. Karunakaran, A. Benabbas, H. Youn, P.M. Champion, J. Am. Chem. Soc. **133**, 18816 (2011)
18. X. Ye, D. Ionascu, F. Gruia, A. Yu, A. Benabbas, P.M. Champion, Proc. Natl. Acad. Sci. USA **104**, 14682 (2007)
19. P.M. Champion, Science **310**, 980 (2005)
20. L.Y. Zhu, J.T. Sage, P.M. Champion, Science **266**, 629 (1994)
21. L. Reinisch, K.T. Schomacker, P.M. Champion, J. Chem. Phys. **87**, 150 (1987)
22. L. Dhar, J.A. Rogers, K.A. Nelson, Chem. Rev. **91**, 157 (1994)
23. A. Kahan, O. Nahmias, N. Friedman, M. Sheves, S. Ruhman, J. Am. Chem. Soc. **129**, 537 (2007)

24. M.J. Rosker, F.W. Wise, C.L. Tang, Phys. Rev. Lett. **57**, 321 (1986)
25. M. Kubo, F. Gruia, A. Benabbas, A. Barabanschikov, W.R. Montfort, E.M. Maes, P.M. Champion, J. Am. Chem. Soc. **130**, 9800 (2008)
26. F. Gruia, M. Kubo, X. Ye, D. Ionascu, C. Lu, R.K. Poole, S.R. Yeh, P.M. Champion, J. Am. Chem. Soc. **130**, 5231 (2008)
27. F. Rosca, A.T.N. Kumar, D. Ionascu, T. Sjodin, A.A. Demidov, P.M. Champion, J. Chem. Phys. **114**, 10884 (2001)
28. F. Rosca et al., J. Phys. Chem. A **106**, 3540 (2002)
29. F. Rosca, A.T.N. Kumar, X. Ye, T. Sjodin, A.A. Demidov, P.M. Champion, J. Phys. Chem. A **104**, 4280 (2000)
30. I. Belevich, M.I. Verkhovsky, M. Wikstrom, Nature **440**, 829 (2006)
31. M. Saraste, Science **283**, 1488 (1999)
32. H. Luecke, H.T. Richter, J.K. Lanyi, Science **280**, 1934 (1998)
33. C.T. Wolke et al., Science **354**, 1131 (2016)
34. F. Dahms, B.P. Fingerhut, E.T.J. Nibbering, E. Pines, T. Elsaesser, Science **357**, 491 (2017)
35. M. Baroncini, S. Silvi, A. Credi, Chem. Rev. **120**, 200 (2020)
36. A.N. Bondar, M.J. Lemieux, Chem. Rev. **119**, 6162 (2019)
37. M. Wikstrom, K. Krab, V. Sharma, Chem. Rev. **118**, 2469 (2018)
38. M.Y. Okamura, M.L. Paddock, M.S. Graige, G. Feher, Bba-Bioenergetics **1458**, 148 (2000)
39. M.Y. Okamura, G. Feher, Annu. Rev. Biochem. **61**, 861 (1992)
40. A.M. Kuznetsov, J. Ulstrup, Can. J. Chem. **77**, 1085 (1999)
41. S. Hammes-Schiffer, Acc. Chem. Res. **39**, 93 (2006)
42. J.T.M. Kennis, D.S. Larsen, I.H.M. van Stokkum, M. Vengris, J.J. van Thor, R. van Grondelle, P Natl. Acad. Sci USA **101**, 17988 (2004)
43. M. Chattoraj, B.A. King, G.U. Bublitz, S.G. Boxer, P Natl. Acad. Sci. USA **93**, 8362 (1996)
44. C.A. Wraight, Bba-Bioenergetics **1757**, 886 (2006)
45. J.A. Shelnutt, X.Z. Song, J.G. Ma, S.L. Jia, W. Jentzen, C.J. Medforth, Chem. Soc. Rev. **27**, 31 (1998)
46. Y.H. Sun, A. Benabbas, W.Q. Zeng, J.G. Kleingardner, K.L. Bren, P.M. Champion, P Natl. Acad. Sci. USA **111**, 6570 (2014)
47. W. Jentzen, J.G. Ma, J.A. Shelnutt, Biophys. J. **74**, 753 (1998)
48. W. Jentzen, X.Z. Song, J.A. Shelnutt, J. Phys. Chem. B **101**, 1684 (1997)
49. V. Srajer, L. Reinisch, P.M. Champion, J. Am. Chem. Soc. **110**, 6656 (1988)
50. M.F. Perutz, A.J. Wilkinson, M. Paoli, G.G. Dodson, Annu. Rev. Bioph. Biom. **27**, 1 (1998)
51. M.F. Perutz, Annu. Rev. Biochem. **48**, 327 (1979)
52. S.V. Evans, G.D. Brayer, J. Mol. Biol. **213**, 885 (1990)
53. A. Benabbas, Y.H. Sun, T.L. Poulos, P.M. Champion, J. Am. Chem. Soc. **139**, 15738 (2017)
54. A. Benabbas, V. Karunakaran, H. Youn, T.L. Poulos, P.M. Champion, J. Biol. Chem. **287**, 21729 (2012)
55. A. Benabbas, X. Ye, M. Kubo, Z.Y. Zhang, E.M. Maes, W.R. Montfort, P.M. Champion, J. Am. Chem. Soc. **132**, 2811 (2010)
56. A. Benabbas, P.M. Champion, J. Phys. Chem. B **122**, 11431 (2018)
57. S. Franzen, Proc. Natl Acad Sci U S A **99**, 16754 (2002)
58. N. Strickland, J.N. Harvey, J. Phys. Chem. B **111**, 841 (2007)
59. J.N. Harvey, Faraday Discuss. **127**, 165 (2004)
60. A.T.N. Kumar, F. Rosca, A. Widom, P.M. Champion, J. Chem. Phys. **114**, 701 (2001)
61. L. Zhu, P. Li, M. Huang, J.T. Sage, P.M. Champion, Phys. Rev. Lett. **72**, 301 (1994)
62. S.L. Dexheimer, Q. Wang, L.A. Peteanu, W.T. Pollard, R.A. Mathies, C.V. Shank, Chem. Phys. Lett. **188**, 61 (1992)
63. A.T.N. Kumar, F. Rosca, A. Widom, P.M. Champion, J. Chem. Phys. **114**, 6795 (2001)
64. S. Mukamel, *Principles of Nonlinear Optical Spectroscopy* (Oxford University Press, 1999)
65. F. Gruia, M. Kubo, X. Ye, P.M. Champion, Biophys. J. **94**, 2252 (2008)
66. F. Yang, L.G. Moss, G.N. Phillips, Nat. Biotechnol. **14**, 1246 (1996)
67. J.J. van Thor, P.M. Champion, Annu. Rev. Phys. Chem. **74**, 123 (2023)

68. C. Fang, R.R. Frontiera, R. Tran, R.A. Mathies, Nature **462**, 200 (2009)
69. T. Fujisawa, H. Kuramochi, H. Hosoi, S. Takeuchi, T. Tahara, J. Am. Chem. Soc. **138**, 3942 (2016)
70. J.J. van Thor, G.Y. Georgiev, M. Towrie, J.T. Sage, J. Biol. Chem. **280**, 33652 (2005)
71. P. Leiderman, R. Gepshtein, I. Tsimberov, D. Huppert, J. Phys. Chem. B **112**, 1232 (2008)
72. A.C. Yu, X. Ye, D. Ionascu, W.X. Cao, P.M. Champion, Rev. Sci. Instrum. **76** (2005)
73. D. Roston, Z. Islam, A. Kohen, Molecules **18**, 5543 (2013)
74. M.J. Knapp, K. Rickert, J.P. Klinman, J. Am. Chem. Soc. **124**, 3865 (2002)
75. J.C. Nesheim, J.D. Lipscomb, Biochemistry **35**, 10240 (1996)
76. T. Jonsson, D.E. Edmondson, J.P. Klinman, Biochemistry **33**, 14871 (1994)
77. B. Salna, A. Benabbas, D. Russo, P.M. Champion, J. Phys. Chem. B **121**, 6869 (2017)
78. P.M. Champion, A. Benabbas, in *Tunneling in Molecules*, edited by J. Kaestner, and S. Kozuch (Royal Society of Chemistry, Cambridge, **2020**)
79. A. Novak, Struct. Bond. **18**, 177 (1974)
80. L.M. Thompson, A. Lasoroski, P.M. Champion, J.T. Sage, M.J. Frisch, J.J. van Thor, M.J. Bearpark, J. Chem. Theory Comput. **10**, 751 (2014)
81. M. Wikstrom, V. Sharma, V.R.I. Kaila, J.P. Hosler, G. Hummer, Chem. Rev. **115**, 2196 (2015)
82. P. Goyal, S. Yang, Q. Cui, Chem. Sci. **6**, 826 (2015)
83. B. Ginovska-Pangovska, M.H. Ho, J.C. Linehan, Y.H. Cheng, M. Dupuis, S. Raugei, W.J. Shaw, Bba-Bioenergetics **1837**, 131 (2014)
84. K. Saito, A.W. Rutherford, H. Ishikita, P Natl. Acad. Sci. USA **110**, 954 (2013)
85. A.V. Pisliakov, P.K. Sharma, Z.T. Chu, M. Haranczyk, A. Warshel, Proc. Natl. Acad. Sci. USA **105**, 7726 (2008)
86. J. Xu, G.A. Voth, Proc. Natl. Acad. Sci. USA **102**, 6795 (2005)
87. A. Shinobu, N. Agmon, J. Chem. Theory Comput. **13**, 353 (2017)
88. R. Pomes, B. Roux, Biophys. J. **71**, 19 (1996)
89. J.F. Nagle, H.J. Morowitz, Proc. Natl. Acad. Sci. USA **75**, 298 (1978)
90. Y.H. Sun, V. Karunakaran, P.M. Champion, J. Phys. Chem. B **117**, 9615 (2013)
91. J.G. Ma, J.M. Vanderkooi, J. Zhang, S.L. Jia, J.A. Shelnutt, Biochemistry **38**, 2787 (1999)
92. M.D. Liptak, X. Wen, K.L. Bren, J. Am. Chem. Soc. **132**, 9753 (2010)
93. S.J. Takayama, K. Irie, H.L. Tai, T. Kawahara, S. Hirota, T. Takabe, L.A. Alcaraz, A. Donaire, Y. Yamamoto, J. Biol. Inorg. Chem. **14**, 821 (2009)
94. J.G. Ma, J. Zhang, R. Franco, S.L. Jia, I. Moura, J.J.G. Moura, P.M.H. Kroneck, J.A. Shelnutt, Biochemistry **37**, 12431 (1998)
95. Y. Matsuura, T. Takano, R.E. Dickerson, J. Mol. Biol. **156**, 389 (1982)
96. J.J. Hopfield, Proc. Natl. Acad. Sci. USA **71**, 3640 (1974)
97. R.A. Marcus, N. Sutin, Biochim. Biophys. Acta **811**, 265 (1985)
98. D.M.A. Smith, K.M. Rosso, M. Dupuis, M. Valiev, T.P. Straatsma, J. Phys. Chem. B **110**, 15582 (2006)
99. F.D. Lewis, J.Q. Liu, W. Weigel, W. Rettig, I.V. Kurnikov, D.N. Beratan, P Natl. Acad. Sci. USA **99**, 12536 (2002)
100. R.J. Cave, P. Siders, R.A. Marcus, J. Phys. Chem-Us **90**, 1436 (1986)
101. H.B. Gray, J.R. Winkler, Chem. Phys. Lett. **483**, 1 (2009)
102. A. Labas, D.K. Menyhard, J.N. Harvey, J. Olah, Chem-Eur. J. **24**, 5350 (2018)
103. D.A. Fedorov, S.R. Pruitt, K. Keipert, M.S. Gordon, S.A. Varganov, J. Phys. Chem. A **120**, 2911 (2016)
104. G.E. dePolo, D.S. Kaliakin, S.A. Varganov, J Phys Chem A **120**, 8691 (2016)
105. J.N. Harvey, Wires. Comput. Mol. Sci. **4**, 1 (2014)
106. J.N. Harvey, D.P. Tew, Int. J. Mass Spectrom. **354**, 263 (2013)
107. V.W. Manner, A.D. Lindsay, E.A. Mader, J.N. Harvey, J.M. Mayer, Chem. Sci. **3**, 230 (2012)
108. M. Besora, J.N. Harvey, J Chem Phys **129** (2008)
109. J.N. Harvey, Phys. Chem. Chem. Phys. **9**, 541 (2007)
110. J.N. Harvey, R. Poli, Dalton T, 4100 (2003)
111. J.N. Harvey, M. Aschi, Faraday Discuss. **124**, 129 (2003)

112. C. Rue, P.B. Armentrout, I. Kretzschmar, D. Schroder, J.N. Harvey, H. Schwarz, J. Chem. Phys. **110**, 7858 (1999)
113. J.N. Harvey, M. Aschi, Phys. Chem. Chem. Phys. **1**, 5555 (1999)
114. M. Aschi, J.N. Harvey, C.A. Schalley, D. Schroder, H. Schwarz, Chem Commun, 531 (1998)
115. G.P. Smith, R.N. Zare, J. Am. Chem. Soc. **97**, 1985 (1975)
116. J. Jortner, J. Ulstrup, J. Am. Chem. Soc. **101**, 3744 (1979)
117. W.D. Tian, J.T. Sage, V. Srajer, P.M. Champion, Phys. Rev. Lett. **68**, 408 (1992)
118. H. Frauenfelder, P.G. Wolynes, Science **229**, 337 (1985)
119. R.H. Austin, K.W. Beeson, L. Eisenstein, H. Frauenfelder, I.C. Gunsalus, Biochemistry **14**, 5355 (1975)
120. B. Yang, L. Gagliardi, D.G. Truhlar, Phys. Chem. Chem. Phys. **20**, 4129 (2018)
121. C.A. Gaggioli, L. Belpassi, F. Tarantelli, J.N. Harvey, P. Belanzoni, Chem. Eur. J. **24**, 5006 (2018)
122. M.H. Redi, B.S. Gerstman, J.J. Hopfield, Biophys. J. **35**, 471 (1981)
123. A. Ansari, C.M. Jones, E.R. Henry, J. Hofrichter, W.A. Eaton, Science **256**, 1796 (1992)
124. W. Wang, X. Ye, A.A. Demidov, F. Rosca, T. Sjodin, W.X. Cao, M. Sheeran, P.M. Champion, J. Phys. Chem. B **104**, 10789 (2000)
125. M.W. Mara et al., Science **356**, 1276 (2017)
126. M.H. Vos, A. Battistoni, C. Lechauve, M.C. Marden, L. Kiger, A. Desbois, E. Pilet, E. de Rosny, U. Liebl, Biochemistry **47**, 5718 (2008)
127. M. Negrerie, S. Cianetti, M.H. Vos, J.L. Martin, S.G. Kruglik, J. Phys. Chem. B **110**, 12766 (2006)
128. S. Sharma, K. Sivalingam, F. Neese, G.K.-L. Chan, Nat. Chem. **6**, 927 (2014)
129. C. Weber, D.D. O'Regan, N.D. Hine, P.B. Littlewood, G. Kotliar, M.C. Payne, Phys. Rev. Lett. **110**, 106402 (2013)
130. Y. Kurashige, G.K.-L. Chan, T. Yanai, Nat. Chem. **5**, 660 (2013)
131. W.D. Tian, J.T. Sage, P.M. Champion, E. Chien, S.G. Sligar, Biochemistry **35**, 3487 (1996)
132. D. Ionascu, F. Gruia, X. Ye, A.C. Yu, F. Rosca, C. Beck, A. Demidov, J.S. Olson, P.M. Champion, J. Am. Chem. Soc. **127**, 16921 (2005)
133. A. Szabo, Proc Natl Acad Sci U S A **75**, 2108 (1978)
134. S.S. Batsanov, Inorg Mater+ **37**, 871 (2001)
135. L. Bouzhir-Sima, R. Motterlini, J. Gross, M.H. Vos, U. Liebl, J. Phys. Chem. B **120**, 10686 (2016)
136. L. Lobato, L. Bouzhir-Sima, T. Yamashita, M.T. Wilson, M.H. Vos, U. Liebl, J. Biol. Chem. **289**, 26514 (2014)
137. S. Cianetti, M. Negrerie, M.H. Vos, J.L. Martin, S.G. Kruglik, J. Am. Chem. Soc. **126**, 13932 (2004)
138. R.J. Cave, M.D. Newton, J. Chem. Phys. **106**, 9213 (1997)
139. J.O. Alben et al., Phys. Rev. Lett. **44**, 1157 (1980)
140. N. Alberding, R.H. Austin, K.W. Beeson, S.S. Chan, L. Eisenstein, H. Frauenfelder, T.M. Nordlund, Science **192**, 1002 (1976)
141. D.C. Lamb, J. Kriegl, K. Kastens, G.U. Nienhaus, J. Phys. Org. Chem. **13**, 659 (2000)
142. H.H. Limbach, J.M. Lopez, A. Kohen, Philos. T R Soc. B **361**, 1399 (2006)

# Chapter 15
# Time-Resolved Studies of Protein Structural Dynamics

**Allen M. Orville, Eriko Nango, So Iwata, Sandra Mous, Joerg Standfuss, Przemyslaw Nogly, Michihiro Suga, Jian-Ren Shen, and Minoru Kubo**

**Keywords** X-ray free electron lasers · X-ray crystallography · Time-resolved serial femtosecond crystallography · Protein dynamics · Rhodopsin · Photosystem II · Cytochrome P450nor

---

A. M. Orville
DiamondLight Source Ltd., Didcot, UK

Harwell Science and Innovation Campus, Didcot, UK

E. Nango (✉) · S. Iwata
RIKEN SPring-8 Center, 1-1-1 Kouto, Sayo-cho, Sayo-gun, Hyogo 679-5148, Japan
e-mail: eriko.nango.c4@tohoku.ac.jp

E. Nango
Institute of Multidisciplinary Research for Advanced Materials, Tohoku University, 2-1-1 Katahira, Aoba-ku, Sendai 980-8577, Miyagi, Japan

S. Iwata
Department of Cell Biology, Graduate School of Medicine, Kyoto University, Yoshidakonoe-cho, Sakyo-ku, Kyoto 606-8501, Japan

S. Mous
Linac Coherent Light Source, SLAC National Accelerator Laboratory, Menlo Park, USA

J. Standfuss
Laboratory of Biomolecular Research, Biology and Chemistry Division, Paul Scherrer Institute, Villigen PSI, Switzerland

P. Nogly
Dioscuri Centre for Structural Dynamics of Receptors, Faculty of Biochemistry, Biophysics and Biotechnology, Jagiellonian University, Gronostajowa 7, 30-387, Kraków, Poland

M. Suga · J.-R. Shen
Graduate School of Environmental, Life, Natural Science and Technology, Research Institute for Interdisciplinary Science, Okayama University, Okayama, Japan

M. Kubo
Graduate School of Science, University of Hyogo, Ako 678-1297, Hyogo, Japan

 
K. Ueda (ed.), *Ultrafast Electronic and Structural Dynamics*,
https://doi.org/10.1007/978-981-97-2914-2_15

Proteins change their conformations in a sophisticated manner when they perform their functions. Time-resolved serial femtosecond crystallography is a potent tool for determining protein dynamic structures. In this chapter, we introduce the principles of time-resolved serial femtosecond crystallography and actual analyses using the technique. Section 15.1 is the introduction by Allen M. Orville. In Sect. 15.2, Eriko Nango and So Iwata describe time-resolved studies of bacteriorhodopsin. In Sect. 15.3, Sandra Mous, Joerg Standfuss, and Przemyslaw Nogly introduce time-resolved studies on the ultrafast dynamics of bacteriorhodopsin. Michihiro Suga and Jian-Ren Shen explain time-resolved studies of photosystem II in Sect. 15.4. Finally, Minoru Kubo describes time-resolved studies of cytochrome P450nor with caged substrate.

## 15.1 Introduction

Crystallographers frequently observe microcrystal showers measuring only a few microns on a side that arise from sparse matrix screens used early in most projects. These conditions are "optimized" to yield large single crystals typically measuring ~25–100 s μm or more on at least two sides. During typical data collection at synchrotrons with samples held at 100 K, the X-ray dose is distributed throughout the entire crystal volume by combinations of rotation and translation [1–8]. And so, although microcrystals are ubiquitous, they are frequently overlooked because they can be difficult to use and/or perceived to be inappropriate for structural analysis.

However, the characteristics of XFELs change the sample requirements for macromolecular crystallography (MX). Indeed, showers of microcrystals are ideal for XFELs that deliver hard X-rays with high flux density in a very well-focused fs-long pulse. Consequently, XFELs offer new opportunities in structural biology, especially for samples at near-physiological conditions and for time-resolved studies that link together the analysis of structure and function within the same samples [9–12]. The unparalleled XFEL intensity reduces the crystal size requirements such that even submicron crystals can yield high-quality diffraction data [13, 14]. The fs pulse duration also provides sharp temporal resolution, without sufficient time for radiation-induced alterations in most cases.

Serial femtosecond crystallography (SFX) is a new technique developed to exploit the fs pulses from XFELs and to use thousands of micron-size crystals or smaller [9–12]. It is the dominant method in life sciences at all five hard X-ray energy XFEL sources since it was first reported in 2011 by Chapman and colleagues at the LCLS [15, 16]. The impact of SFX is summarized in Table 15.1, wherein atomic models released by the protein databank (PDB) are a measure of impact [17]. These atomic models are almost always accompanied by a primary citation. It is noteworthy that ~60% of the primary citations since 2011 were in high profile publications such as *Science*, *Nature* or *Cell*, etc. These results often highlight the unique characteristics of SFX methods afforded by XFEL facilities. This includes an almost complete lack of radiation-induced alterations in the atomic models, exploitation of micron to submicron-size crystals, and exquisitely sharp temporal resolution in time-resolved studies. However, under some conditions, samples are clearly perturbed by the intense X-ray photon beam, especially in metalloenzymes and at heavy atoms sites within macromolecules [18–22]. Together, these demonstrate impressive advances in only a few years of operations; not surprisingly, the biology community is very eager for more XFEL access.

SFX studies are very often conducted at room temperature, from which one still diffraction pattern is recorded from each microcrystal in a random orientation. XFEL beams are typically well-focused to deliver submicron or 1–3 μm spot size at the sample and are about nine orders of magnitude brighter than synchrotrons. Because so much energy is deposited into the sample, it explodes [23]. Consequently, SFX methods require a unique sample for each diffraction pattern and the whole dataset is merged from thousands of still images. To these ends, research and development in sample delivery is a very active effort in the field. The aim is to rapidly and efficiently

**Table 15.1** Macromolecular atomic models linked to data collected at an XFEL released by the PDB[a]

| |
|---|
| **LCLS** (USA, 2011 to date; CXI, XPP, MFX beamlines) |
| • 322 atomic models released |
| • 122 primary citations |
| • 77 high profile citations[b] |
| **SACLA** (Japan, 2014 to date; BL2, BL3 beamlines) |
| • 256 atomic models released |
| • 73 primary citations |
| • 27 high profile citations[b] |
| **European XFEL** (Germany, 2018 to date; SPB/SFX or FXE instruments) |
| • 35 atomic models released |
| • 11 primary citations |
| • 8 high profile citations[b] |
| **PAL-XFEL** (South Korea, 2019 to date; NCI—SFX beamline) |
| • 31 atomic models released |
| • 15 primary citations |
| • 2 high profile citations[b] |
| **SwissFEL** (Switzerland, 2019 to date; Aramis, Bernina beamlines) |
| • 126 atomic model released |
| • 13 primary citations |
| • 10 high profile citations[b] |

[a] Compiled on 20 May 2024

[b] Primary citation appeared in either *Science, Nature, Cell, Proc Nat Acad Sci USA, Nature Methods, Nature Communications, Nature Chemistry,* or *Structure*

deliver sample, without wasting precious material, at a rate that matches the XFEL pulse frequency and/or the detector characteristics.

A variety of methods have been developed to deliver a slurry of microcrystals into the XFEL interaction region, which includes flow-focusing gas dynamic virtual nozzles (ff-GDVN) and liquid jets [15, 24–27], viscous media extruders [28–31], a concentric-flow electrokinetic injector [14, 32, 33], on-demand micro-droplets that may be coupled to a conveyor belt transport system [34–43], fixed targets that raster a sample array through the X-ray beam [39, 44–53], and goniometer-based methods [54–58]. Many of these methods are readily adaptable to time-resolve studies in which the reaction is triggered by either light or mixing as discussed below.

Time-resolved serial MX naturally blends functional and structural analyses into the same sample and experiment. This is sometimes described as *Dynamic Structural Biology*. It exploits slurries of microcrystals at room temperature and is increasingly prevalent at XFELs and synchrotrons. Although relatively rare in biology, light-activated systems are easy to initiate and many have the potential to probe isomerization rates from fs and slower. Many scientists differentiate ultrafast applications from more general time-resolved SFX experiments by the observational time scales and the fact that decoherence of the experiment/sample is typically completed within a few picoseconds.

Ultrafast time-resolved spectroscopic experiments performed on photoactive proteins in solution have provided many important mechanistic insights into the very early events after absorbing a photon. Because XFELs are still relatively new, analogous time-resolved SFX experiments are still emerging. It is common for authors to compare time-resolved SFX and spectroscopic results. However, although the two methods may have used similar pump-probe delay times, the comparisons frequently do not yield one-to-one correlations. One possible explanation is that the crystal versus solution conditions are sufficiently different that ultrafast spectroscopic experiments have not been performed at the same pH, viscosity, ionic strength, among many other potential variables that were present in the SFX experiments. Furthermore, the desire of structural biologists to observe and maximize illumination-dependent differences in electron density maps has often pushed experimental conditions into the multiphoton regime [59]. Recently it has become clear that visible light power dependence studies are critically important to nearly all light-activated time-resolved SFX experiments. Unfortunately, in part because XFEL beamtime is so rare, scientists may underprioritize these “control” experiments, and as a result these types of data may be sacrificed under time-pressure situations.

Examples of XFEL experiments involving light-activated systems include photosystem I [16, 60–64], photosystem II [32, 34, 36, 37, 43, 56, 65–76], photoactive yellow protein [77–79], human rhodopsin [80, 81], bacteriorhodopsins [82–88], light-activated ion channels [89], fluorescent proteins [90–93], several phytochromes [36, 94, 95], and photo-dissociation studies of myoglobin-CO [96] or cytochrome c oxidase-CO [97, 98]. Of these systems, photosystems I and II have evolved mechanisms to diffuse the excess energy absorbed in multiphoton events through the network of internal chromophores. Therefore, they are less susceptible to the impact of single versus multiphoton time-resolved SFX studies. The photo-dissociation of

CO from metalloproteins can be considered outside their normal function and less constrained by illumination conditions. The remaining systems evolved to react to visible photons, and single photon methods are likely to be the most physiological. Moreover, many of these systems experience photo-driven conformational changes that are often linked to photoisomerization of the chromophore. Such events are often very fast and require timing tools to help coordinate the pump-probe experiment [99–102].

Light-driven strategies to initiate catalysis in systems that are not naturally light sensitive include: (a) caged compounds, (b) caged proteins, (c) ligand exchange, or (d) temperature jump methods [68, 103–119]. Each of these must satisfy the requirements of high selectivity, high quantum yield, and temporal resolution. To these ends, *o*-nitrobenzyl moieties are among the most common photocaging groups for substrates and amino acids, but their decaging photochemistry may not be as fast nor as clean as *p*-hydroxyphenacyl or coumarylmethyl derivatives. Caged substrates are either co-crystallized with the target macromollecule or soaked into slurries of microcrystals. Incorporation of non-natural amino acids that convert a given protein into a caged protein is more difficult. Consequently, caged protein approaches typically include a computational evaluation stage, followed by protein translation using *amber* TAG codon-suppression methods, or post-translational modification strategies. The photoactive moiety is most often linked to thio, amino, carboxy, or hydroxy groups of proteins or substrate ligands, and is then cleaved by irradiation with UV to visible light. Photo-cleavage of the caging group varies in rate (ps–μs) and quantum yield (<0.2–1), which then generates: (i) authentic substrate in the active site vicinity, (ii) a rapid pH shift, (iii) a temperature jump, (iv) removes an active site barrier and enables substrate binding, or (v) eliminates a dynamic or conformational restraint required for catalysis. Ligand exchange methods include photolabile metal-CO or NO complexes that mimic metal-$O_2$ intermediates in a reaction cycle. Photodissociation of the blocking diatomic molecules then allows for $O_2$ binding and the ensuing reaction. All of these methods are experiencing a resurgence of research and development activity and provide important opportunities for time-resolved structural biology at XFELs. Some of these techniques will require careful coordination of the timing between the visible light pump laser and the XFEL probe pulse.

A remaining frontier challenge in structural biology is to determine time-resolved structures at atomic resolution directly from systems engaged in function at physiological temperatures and pressures. To this end, *Dynamic Structural Biology* is as much a philosophy as a set of tools to collect as much data as possible, from every sample and every X-ray pulse, at physiological temperature and pressure, with an aim to create time-resolved molecular movies of macromolecules engaged in function. XFEL methods coupled with pump-probe strategies are making important advances toward this goal. This chapter highlights recent results in time-resolved SFX. In the first two subsections, Nango, Iwata, Mous, Standfuss, and Nogly describe the structure, function, and dynamics of the light-driven proton pump, bacteriorhodopsin. Next Suga and Shen summarize results from photosystem II, perhaps the best benchmark system for time-resolved SFX studies. Kubo concludes the chapter with an example of a caged nitric oxide compound driving the P450nor enzyme reaction, which is an unusual member of the P450 superfamily of heme-dependent proteins.

## 15.2 A Three-Dimensional Movie of Structural Changes in Bacteriorhodopsin Captured by X-Ray Free-Electron Lasers

### *15.2.1 Introduction*

Bacteriorhodopsin (bR) is a light-driven proton pump found in the cell membrane of the salt-tolerant microorganism, *Halobacterium salinarum*. bR is one of the microbial rhodopsins, such as halorhodopsin and channelrhodopsin, which are used as a tool for optogenetics. It also serves as a simple model for G-protein coupled receptors, including visual rhodopsin, located in the retina of the eye. bR uses light as a trigger to transport protons from the cytoplasmic side to the extracellular side against the electrochemical potential. The resulting proton gradient across the cell membrane is converted into chemical energy by the activity of adenosine triphosphate synthase. How does bR pump protons out without the backflow of protons? The proton transfer mechanism has been investigated across various research fields, including biochemistry, biophysics, and structural biology, since its discovery in the 1960s and still attracts interest.

The bR structure is composed of seven transmembrane helices, with a buried all-*trans*-retinal chromophore that is covalently bound to Lys216 via a protonated Schiff-base linkage (Fig. 15.1a). Upon light absorption, the all-*trans*-retinal isomerizes to the 13-*cis* conformation, which initiates the photocycle, forming several intermediates (K, L, M, N, and O) with different maximum absorption wavelengths (Fig. 15.1b, c). Primary proton transfer occurs between the protonated Schiff-base and Asp85 [120]. In the resting state (bR structure before light exposure), the protonated Schiff-base forms a hydrogen bond with one water molecule (W402), which further forms a pentagonal hydrogen-bond cluster with Asp85, Asp212, and two other water molecules (Fig. 15.2). The hydrogen-bonded network results in a difference of 11 orders of magnitude between the proton affinities of the primary donor and acceptor [121, 122], which prevents the leakage of protons from the extracellular medium to the cytoplasm. After light excitation, these proton affinities close together to facilitate spontaneous proton exchange, raising the question of what brings about the change in proton affinities. It is also puzzling why primary proton transfer occurs in microseconds although the Schiff-base and Asp85 are separated by only 4 Å and the water-mediated proton exchange pathway between the Schiff-base and Asp85 is in the resting state. In addition, retinal isomerization redirects the Schiff-base proton away from the extracellular side and toward the cytoplasmic side, which seems to be an unfavorable orientation for proton transfer.

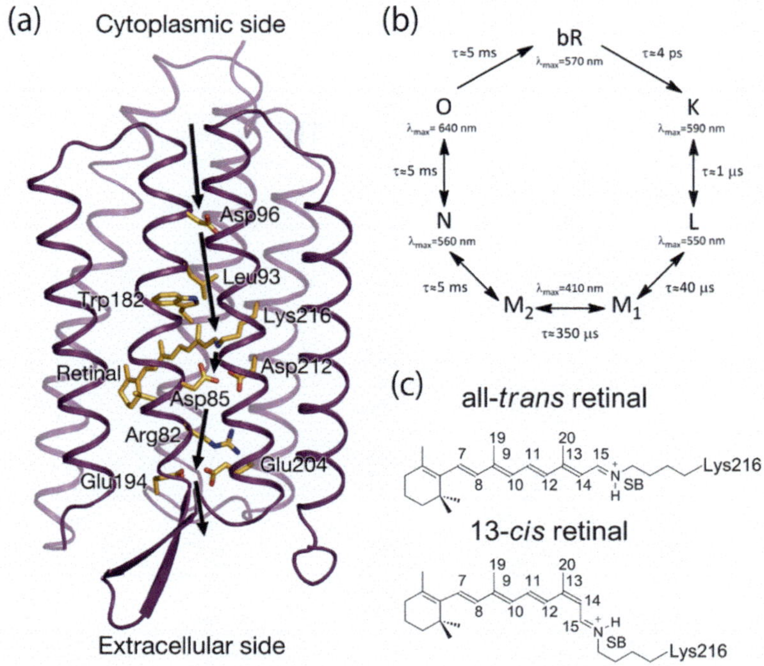


**Fig. 15.1** Structure and function of bacteriorhodopsin (bR). **a** An overall structure of bR. *Trans*-retinal is shown linked to Lys216. The black arrows indicate the proton transfer pathway. **b** A schematic representation of the bR photocycle. **c** A schematic illustration of retinal bound to Lys216 through a protonated Schiff-base (SB) in an all-*trans* and a 13-*cis* configuration. From [127]. Reprinted with permission from AAAS

Many research groups have reported X-ray crystal structures to reveal both light- and mutation-induced structural changes in bR that has been trapped at cryogenic temperature. However, these results have been controversial for the following reasons: Firstly, the reported structures show considerable variation. For instance, three crystal structures [123–125] of the K-intermediate of bR have been reported, which propose structural changes in the retinal and the periphery of the retinal; however, the three models are not in agreement with each other. A water molecule (W402) bound to the Schiff-base in the ground state is displaced in one K-structure [123], whereas the same water molecule is retained with different hydrogen-bond distances between W402 and the Schiff-base in the other two K-structures [124, 125]. Secondly, intermediate structures trapped at cryogenic temperature or by mutation do not correlate directly with time-dependent structural changes in wild-type bR. Thirdly, bR is very sensitive to X-ray induced radiation damage; even a very low dose of X-ray radiation—less than 0.06 MGy—may cause structural alterations around the retina [126]. Therefore, the mechanism of proton transfer in bR is debatable.

We overcame these barriers by using time-resolved serial femtosecond crystallography (TR-SFX) at an X-ray free-electron laser to record a three-dimensional

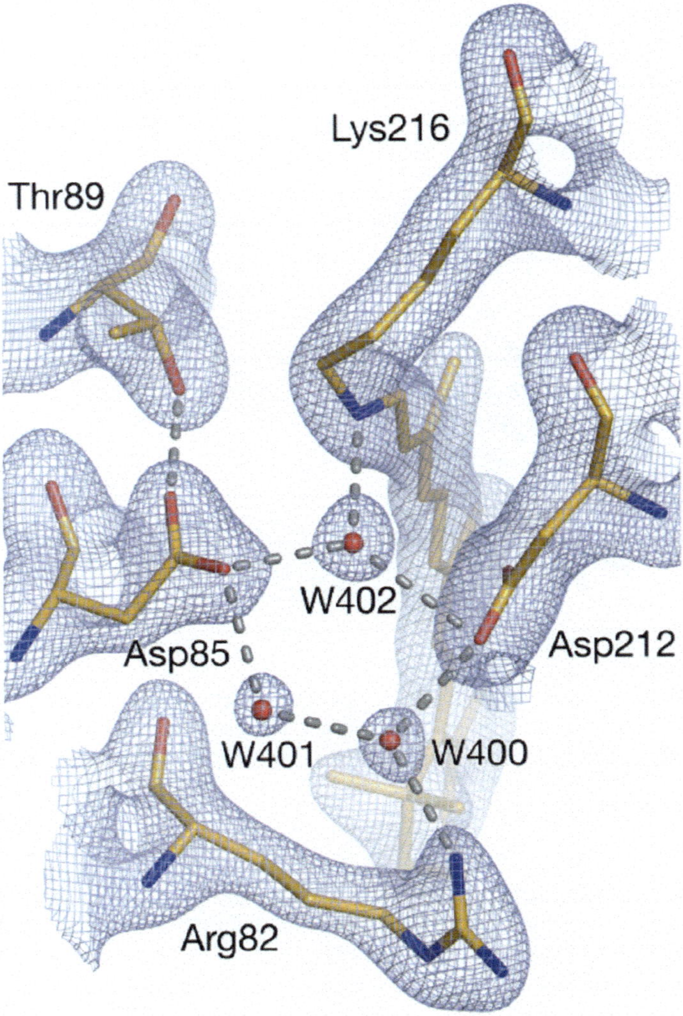


**Fig. 15.2** bR structure in resting state. The $2F_o - F_c$ electron density map (gray) is contoured at 1.3$\sigma$. W400, W401, and W402 denote water molecules. The dotted lines indicate hydrogen-bond interactions. From [127]. Reprinted with permission from AAAS

movie of structural changes in bR at room temperature at 2.1-Å resolution [127]. Extremely intense, ultra-short X-ray pulses from X-ray free-electron lasers (XFEL) enable the acquisition of diffraction data prior to the onset of radiation damage [128]. Furthermore, TR-SFX can help observe the structural dynamics of a protein molecule at atomic resolution due to femtosecond X-ray pulses. We successfully captured a series of structural changes during proton transfer in bR from 16 ns to 1.7 ms, which reveals a mechanism of how bR achieves unidirectional membrane transport.

### 15.2.2 TR-SFX Experiment

In serial femtosecond crystallography (SFX), microcrystals, which are suspended in a carrier media such as a buffer, are continuously delivered into an intersection point with XFEL, using an injector, at room temperature [16]. For SFX, several sample delivery methods have been used: a gas dynamic virtual nozzle utilizing gas flow for focusing the liquid-jet [27]; a microfluidic electrokinetic sample holder, based on the principle of electrospinning [129]; and a highly viscous carrier media, such as lipidic cubic phase (LCP), allowing sample extrusion with a slow flow rate [130, 131]. As bR crystals grown in LCP diffract over 2 Å [123, 132], sample injection using LCP crystals was applied to TR-SFX.

TR-SFX requires the trigger for a reaction in a protein before XFEL irradiation. The retinal in bR is excited by light and the first proton transfer occurs microseconds after light absorption [133]. Therefore, a nanosecond pump laser installed into an optical-fiber-based setup [134] was used for triggering the photocycle. In the TR-SFX experiment, light-adapted bR crystals in LCP were loaded into a high-viscosity cartridge-type injector [135] and injected from a nozzle, with an inner diameter of 75 μm, to an interacting point of an XFEL pulse, with a pulse duration of <10 fs and a repetition rate of 30 Hz, at the SPring-8 Angstrom Compact free-electron Laser (SACLA) (Fig. 15.3a). A nanosecond laser beam with a 532 nm wavelength was focused to 40 μm (full width at half maximum) and operated at 15 Hz to provide 'light' and 'dark' interleaved images (Fig. 15.3b). The delay between sample photoexcitation and the arrival of an XFEL pulse was controlled by a delay generator. Finally, TR-SFX data were collected from bR microcrystals after photoactivation by a pump laser pulse at $\Delta t = 16$ ns, 40 ns, 110 ns, 290 ns, 760 ns, 2 μs, 5.25 μs, 13.8 μs, 36.2 μs, 95.2 μs, 250 μs, 657 μs, and 1.725 ms, yielding 13 snapshots of the bR photoreaction.

### 15.2.3 Early Structural Changes in the Photocycle

Structural changes were clustered at the retinal and its binding site at the earliest time point, $\Delta t = 16$ ns, which corresponds to essentially the K-intermediate. The retinal is initially tilted toward helix G. By $\Delta t = 290$ ns, the retinal conformation straightens, which results in the displacement of Trp182 toward the cytoplasmic side (Fig. 15.4a). Lys216 also indicates a large movement associated with retinal isomerization.

A water molecule (W402), which forms a hydrogen bond with the Schiff base, is displaced in response to photoisomerization (Fig. 15.4a). Trapping intermediate studies by cryo-crystallography have suggested that retinal photoisomerization disorders W402. However, this is controversial because X-ray-induced radiation damage also shows similar results [125, 126, 136]. On the other hand, no effects of radiation damage are visible when using XFEL pulses with a duration of <10 fs. Therefore, the TR-SFX data provides a conclusion to this debate. Interestingly, one cryo-trapped

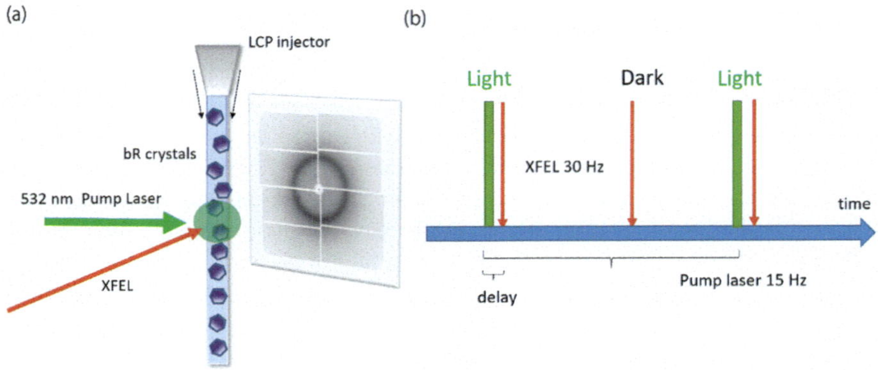


**Fig. 15.3** Time-resolved serial femtosecond crystallography (TR-SFX) of bR. **a** A schematic view of the experimental setup. A lipidic cubic phase microstream continuously transports microcrystals across the focused X-ray free-electron laser (XFEL) beam. X-ray diffraction is recorded on a detector for every XFEL exposure. A green ns laser is used to activate the bR microcrystals prior to the arrival of an XFEL pulse (red arrow). **b** A data collection sequence illustrating how X-ray diffraction data were collected at 30 Hz from photoactivated (green laser flash, 15 Hz) and dark (no laser flash) states in an interleaved fashion. From [127]. Reprinted with permission from AAAS

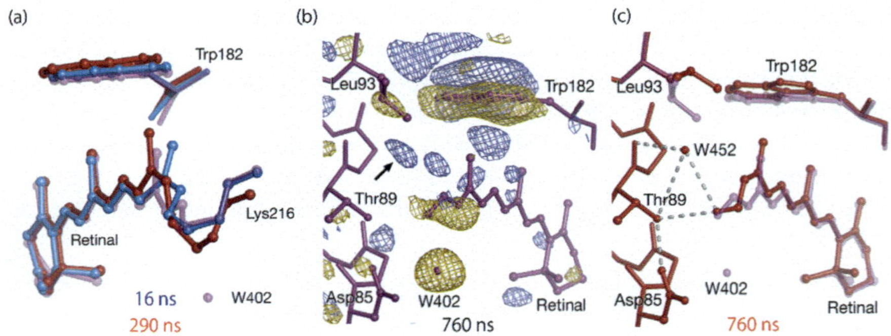


**Fig. 15.4** Structural changes in bR observed at $\Delta t = 16$, 290, and 760 ns. **a** Early structural changes in the bR photocycle. Crystallographic models for $\Delta t = 16$ ns (blue) and $\Delta t = 290$ ns (red) superimposed upon the resting bR structure (purple). **b** The retinal binding site of bR is shown as a $F_0(\text{light}) - F_0(\text{dark})$ difference Fourier electron density map (contoured at $\pm 3.1\sigma$) for $\Delta t = 760$ ns. Blue indicates a positive difference in electron density. Yellow denotes a negative difference in electron density. A positive difference feature (black arrow) suggests the ordering of a water molecule (W452). **c** Crystallographic model for bR at $\Delta t = 760$ ns (red) superimposed upon the resting bR structure (purple). W452 denotes the water molecule, Wat452. From [127]. Reprinted with permission from AAAS

K-intermediate structure [123] showed the disordering of W402 but did not indicate a twist of the retinal C20 methyl toward helix G, whereas another K-state model captured this twist but maintained W402. The third K-intermediate structure showed neither change [124].

### 15.2.4 Unidirectional Proton Transfer Mechanism

The key step in achieving unidirectional proton transport by bR is the primary transfer event from the Schiff-base to Asp85 [120], which occurs during the transition from the L-intermediate to the M-intermediate. For $\Delta t = 760$ ns, the side chain of Leu93, near Trp182, moves toward the cytoplasmic side and a positive difference in electron density arises between the retinal, Leu93, and Thr89 (Fig. 15.4b). This positive electron density is observed from $\Delta t = 40$ ns up to $\Delta \mathrm{t} = 13.8$ μs and is strongest at $\Delta t = 760$ ns. This timescale is consistent with the L-to-M transition shown in our time-resolved difference absorption spectra from bR crystals in LCP. We modeled this feature as a transient water molecule (W452 in Fig. 15.4c). W452 forms weak hydrogen-bond interactions with the Schiff-base and Thr89, which creates a pathway for proton transfer from the Schiff-base to Asp85. A proton from the Schiff-base accesses Asp85 via hydrogen-bond interactions even though the Schiff-base has turned toward the cytoplasmic side by photoisomerization. Previously, a water molecule has been observed at the W452 location in one cryo-trapped L-intermediate structure [137] but has not been captured in other L-state structures [138, 139].

On the extracellular side of the retinal, the disordering of W402 triggers water rearrangements between Asp85 and Asp212. From $\Delta t \geq 13.8$ μs, two water molecules (W400 and W401) were found to be disordered, and a new water molecule (W451) was found to be ordered (Fig. 15.5), which raises the pKa of Asp85 to the point such that it may spontaneously accept a proton from the Schiff-base. The collapse of the hydrogen-bond networks allows helix C to bend toward helix G, approximately 10 μs after photoactivation. The inward movement of helix C progresses until the Schiff-base is deprotonated (to $\Delta t \leq 36.2$ μs). Consequently, the time required for helix movement is the rate-limiting step that controls primary proton transfer.

Proton transfer from the Schiff-base to Asp85 effects a structural change within the retinal, which is modeled as a subtle displacement of the Schiff-base toward the cytoplasmic side and of the C20 methyl, over the 760 ns $\leq \Delta t \leq$ 36.2 μs interval. These changes may be due to a conformational change caused by deprotonation [140] or an electrostatic effect induced by proton transfer [141]. After primary proton transfer ($\Delta t \geq 36.2$ μs), the transient water molecule, W452, is not visible; a hydrogen-bond interaction connecting Thr89 to Asp85 is also lost with significant negative difference electron density appearing between Oγ of Thr89 and Oδ of Asp85. These structural changes hinder reverse proton transfer from Asp85 to the Schiff-base. Furthermore, Thr89 moves closer to the retinal to form a tighter hydrogen-bond interaction with the deprotonated Schiff-base, which allows the Asp85 side chain to rotate and break its linkage to Thr89, while forming a new hydrogen-bond interaction with W451 (Fig. 15.5).

The TR-SFX data, spanning five orders of magnitude in time, captured structural changes arising for a moment, including a transient water molecule, which reveals how bR controls proton transport from the extracellular side to the cytoplasmic side without reversing proton-flow. Nogly et al. have successfully recorded a sequence of snapshots for photoisomerization of the retinal in bR which occurs from fs to ps

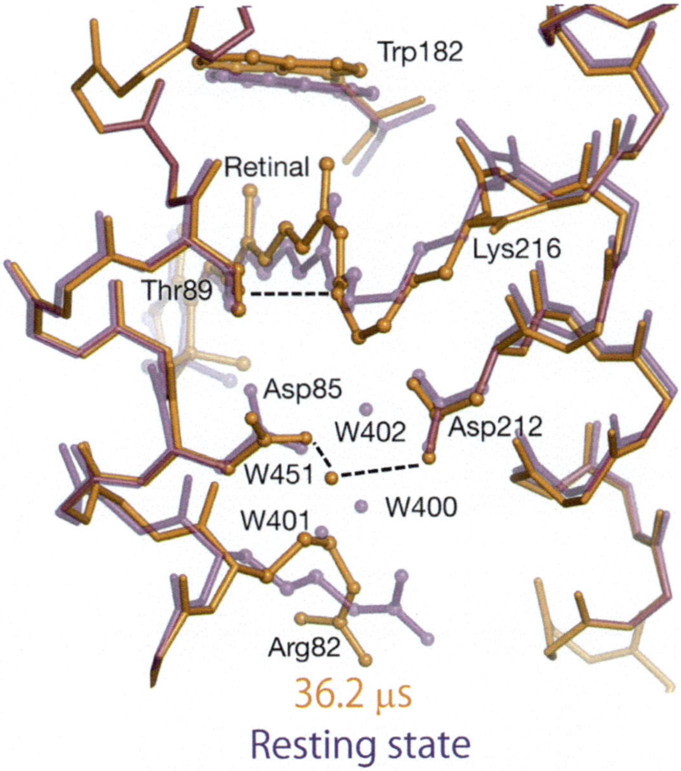


**Fig. 15.5** A crystallographic model for bR at $\Delta t = 36.2\ \mu s$ (orange) superimposed upon the resting bR structure (purple). The dotted lines indicate hydrogen-bond interactions

[84], as shown in the next section. Recently, Weinert et al. have captured structural changes in bR over 200 ms by serial millisecond crystallography using synchrotrons [86]. The data reveal large structural rearrangements, up to 9 Å, on the cytoplasmic side and reorganizations of water-mediated networks for proton uptake. Thus, TR crystallography has contributed to elucidating the mechanism of the bR photocycle, which has been debated for many years. These methods developed at the XFELs and synchrotrons should provide a powerful tool to solve enigmas in structural biology.

## 15.3 Ultrafast Structural Dynamics of Bacteriorhodopsin

### *15.3.1 Retinal Chromophore*

The light-induced isomerization of retinal is an ultrafast process implicated in fundamental biological processes such as light sensing, energy conversion, and ion pumping. Retinal binding proteins are activated through the absorption of a photon by the conjugated double-bond system of the retinal chromophore (Fig. 15.6a). Once the reaction is initiated, the proteins progress through characteristic photocycles that reflect different functional stages. High interest in the mechanism of retinal isomerization and the resulting molecular motions in the connected proteins led to many studies on bacteriorhodopsin (bR), an archetypical proton pump. The tight retinal binding pocket in bR (Fig. 15.6b, [142]) guides a stereoselective isomerization of the chromophore from all-*trans* to 13-*cis* conformation with a quantum yield of up to 64% [143], while in comparison illumination in solution results in a mixture of 7-*cis*, 9-*cis*, 11-*cis* and 13-*cis* isomers [144]. Further spectroscopic studies on bR in solution identified the rise and decay of the I excited state intermediate, forming 200 fs after photoexcitation [145], and the J 'tumbling state' intermediate, evolving after 500 fs, which then relaxes into the isomerized K-state within a few picoseconds [146]. Compared to the isomerization of the chromophore in a protein environment, photoisomerization of free retinal in solution has a slower excited state decay time of 2.0 ps (50%) and 7.2 ps (50%) [147]. Together, the high quantum yield, stereoselectivity, and altered dynamics suggest that the isomerization process is somewhat catalyzed by the opsin pocket [148].

### *15.3.2 Structural Dynamics of Retinal*

The development of time-resolved serial femtosecond crystallography (TR-SFX) at X-ray free-electron lasers (XFEL) made it possible to study such ultrafast structural changes in proteins [78, 96]. Applying this powerful method to bR has allowed to directly probe the structural dynamics of the primary photoresponse at 1.5 Å spatial and around 200 fs temporal resolution [84]. Using different time delays between an activating optical laser pulse and the probing X-ray pulse of the XFEL, diffraction data of bR crystals were collected for 19 different time intervals from the femtosecond to the picosecond temporal regime. Merging data from 6 consecutive time bins (time delay $\Delta t$ = 49–406 fs and 457–646 fs) enabled the refinement of two ultrafast structural bR intermediates [84], which approximately correlate to the accumulation of the spectroscopic I and J intermediates in solubilized bR [145, 146] and bR crystals [149]. In addition, two structures were solved from data collected at a time delay of $\Delta t$ = 10 ps and 8.33 ms [84], which correlate with the accumulation of the K and M intermediates.

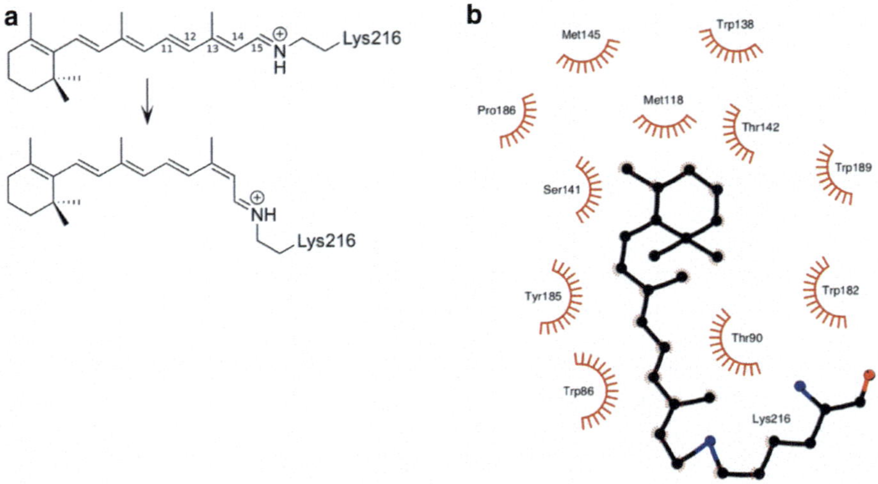


**Fig. 15.6** Retinal chromophore mediating bacteriorhodopsin activation. **a** Schematic structure of all-*trans* and 13-*cis* retinal (K intermediate after photoexcitation). The chromophore is covalently bound to the Lys216 residue of bR via a protonated Schiff base. **b** The schematic illustrates between retinal and its binding pocket within the protein. Amino acids are numbered and hydrophobic contacts within 2.9–3.9 Å distance to the retinal are indicated with red lines

The time-resolved X-ray diffraction data showed that changes in the electron density already occurred during the earliest measured time point ($\Delta t = 0$–142 fs) along the retinal polyene backbone. These changes were also observed in the first structural intermediate at $\Delta t = 49$–406 fs, which showed strong features in the difference Fourier electron density map ($F_{obs}{}^{light} - F_{obs}{}^{dark}$) at a contour level of 4.0σ (root mean square deviation of the electron density) [84]. The changes in the electron density along the backbone were interpreted as an out-of-plane distortion, which may accompany the inversion of the bond length alternation, as predicted in the excited state retinal by quantum chemical computations [150]. It appears that in the high energy state, the conjugated double-bond system undergoes sampling of potential isomerization pathways, most of which are sterically prevented by the tight binding pocket (Fig. 15.6b).

## *15.3.3 Response of the Counterion*

Interestingly, the ultrafast changes in bR upon photoactivation are not constrained to the chromophore. The protonated Schiff-base (PSB) and the associated counterion network (formed by the residues Asp85, Asp212, and Tyr57 together with the water molecules Wat402, Wat401, and Wat400) move even before retinal isomerization has begun (Fig. 15.7). This counterintuitive observation could be explained in terms of the response to the charge redistribution in the excited state retinal. It has been proposed

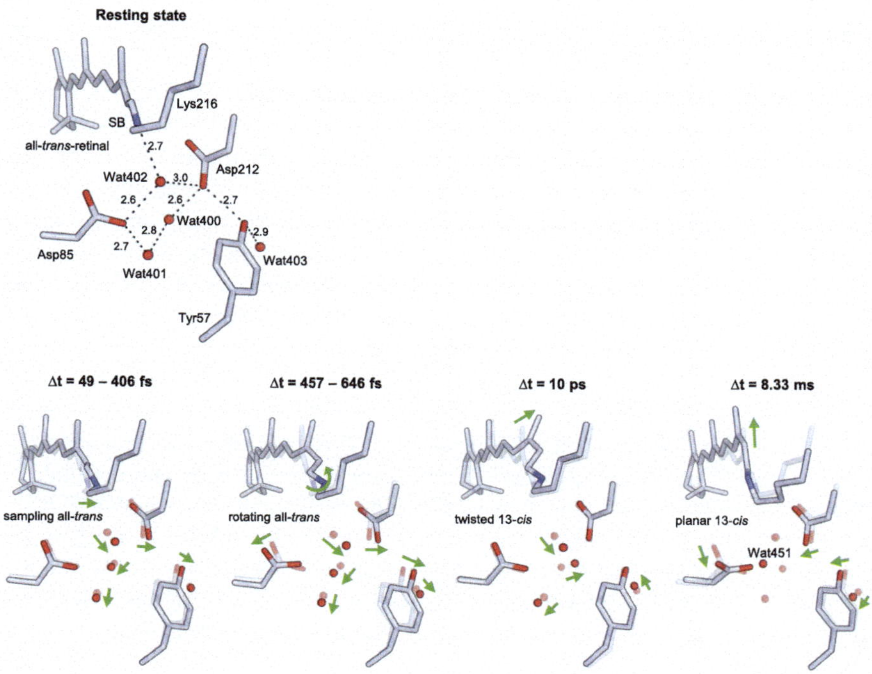


**Fig. 15.7** Structural intermediates of the bacteriorhodopsin photocycle. Refined structures at pump-probe delays of t = 49–406 fs, 457–646 fs, 10 ps and 8.33 ms (PDB accession codes 6G7I, 6G7J, 6G7K, and 6G7L, respectively) show rearrangements of the retinal chromophore and counterion network (indicated by the green arrows) as bacteriorhodopsin progresses through the photocycle. Distances of the hydrogen bonds in the resting state structure (PDB accession code 6G7H) are given in Å

that the positive charge from the PSB in the dark state relocates toward the β-ionone ring in the excited state [145]. The change in the retinal dipole moment weakens the interaction with the counterion cluster, resulting in the shift of water molecules and residues in this region. Quantum chemical calculation of the difference in electron density between the ground and excited state retinal validates that the changes in the electron density immediately after photoexcitation extend to Wat402 [84]. In addition, the ultrafast response of the protein environment to photoexcitation before isomerization was shown by Raman spectroscopy for bR [151] and IR spectroscopy for the homologous Channelrhodopsin-2 protein [152].

### *15.3.4 Trans-cis Retinal Transition*

The structural intermediate at $\Delta t = 457$–646 fs corresponds to the accumulation of the J intermediate, during which the isomerization reaction is initiated [84, 146]. This intermediate is modeled with a C13 = C14 torsion angle of −82°, representing an intermediate state between the 13-*trans* and 13-*cis* isomers (with a torsion angle of −165° and 2°, respectively). Analysis of the twist of the C15 = Nζ, C13 = C14, and C11 = C12 torsion angles in the $\Delta t$ = 49–406 fs and $\Delta t$ = 457–646 fs structures (−159° to 178° (Δ° = +23), −135° to −82° (Δ° = −53), and 179°–166° (Δ° = +13), respectively) reveal that the observed geometry is in agreement with the aborted bicycle-pedal model describing the evolution of the retinal in the I and J intermediates [153, 154]. This specific geometry minimizes the spatial requirements for the isomerization to take place. After the first ultrafast structural intermediates of bR were published [84], further insight into the ultrafast dynamics were provided by additional time-resolved crystallographic data obtained at the XFEL [149] as well as quantum chemical calculations [155]. These independent studies described similar structural changes of the retinal chromophore and counterion cluster and are in agreement with the initial ultrafast time-resolved crystallography experiment on bR. Moreover, the TR-SFX studies described oscillatory behavior upon photoexcitation, either in the difference electron density (showing peak fluctuations at W402, Y57, and the retinal C20) [84], the torsion angles of the retinal skeleton, Lys216 $\chi_4$, Asp212 $\chi_2$, Trp86 $\chi_2$, Tyr185 $\chi_1$, and Met118 $\chi_2$, and the distances PSB-Wat402 and Asp212-Wat402. The torsional oscillations in the residues around the retinal suggest that the chromophore and surrounding residues form a vibrationally coupled network [149].

The aforementioned sub-picosecond structural intermediates and complementary quantum chemical calculations show that the sudden polarization in the excited state eventually causes the hydrogen bond between Wat402 and the PSB to break [84]. Changes in the hydrogen bonding pattern in the sub-picosecond timescale have also been observed by independent spectroscopy studies [156, 157]. The loss of the hydrogen bond between the PSB and Wat402 is thought to lower the energy barrier for structural rearrangements in the vicinity of the PSB, allowing full *trans*-to-*cis* isomerization of the C13 = C14 bond at $\Delta t$ = 10 ps (Fig. 15.7). However, other studies [149, 155] suggest that the elongation of the hydrogen bond between Wat402 and the PSB may be less pronounced. Based on these results, breaking this hydrogen bond may be unnecessary to complete the isomerization of the retinal chromophore to 13-*cis* conformation. The refined structure at $\Delta t$ = 10 ps still displays an out-of-plane twisted C20 methyl group, which is likely the result of steric hindrance by residue Trp182, which did not have enough time to respond to the new retinal geometry. Finally, for comparison, at $\Delta t$ = 8.33 ms, bR features a fully planar 13-*cis* retinal [84], analogous to the retinal conformation observed in all the later photocycle intermediates [127].

### *15.3.5 Long-Range Protein Response to Light Absorption*

The fast movement observed in the retinal binding pocket was found to propagate through the entire structure [84], causing what has been previously described as a protein quake [158]. The earthquake-like motions are thought to be a mechanism to release the excess energy that is generated by the decay of the excited state of the retinal to the ground state. These changes start early at the retinal chromophore, the epicenter of the protein, and coherent structural changes gradually expand up to a distance of 12 Å away from the PSB within 600 fs, progressing with a speed that is on the same order of magnitude as the speed of sound in water [84]. Recently, concerns were raised about the high optical laser pump power density used for the activation of bR in both TR-SFX studies. It has been suggested that high photoexcitation intensities result in multiphoton excitation [149]. This mode of excitation could influence the dynamics of bR, result in ionization effects, and possibly cause the oscillatory behavior and quake-like motions in the protein as the system finds a way to quickly dissipate energy. It was therefore recommended to investigate ultrafast dynamics in bR further using a lower laser excitation intensity.

The proton pump bR has proven to be a useful model system for studying a fundamental photochemical reaction occurring in a protein environment. A better understanding of the ultrafast electronic and structural dynamics of the retinal chromophore advances the understanding of homologue retinal binding proteins, implicated in diverse processes such as future optogenetic tools or the photoreceptors that enable our vision. Furthermore, the recent studies on the retinal isomerization in bR are an excellent example of how time-resolved serial crystallography can be used to answer challenging questions in biology and the research paves the way for the study of other photochemical reactions at XFELs.

## 15.4 Time-Resolved Studies of Photosystem II Reveal Intermediate $S_i$-State Structures in Photosynthetic Water Oxidation

### *15.4.1 Photosystem II*

Plants and algae synthesize biomass in the form of sugars from water molecules and carbon dioxide by utilizing light energy from the sun. This process is called oxygenic photosynthesis, which is achieved by a chain of electron transport, a series of proton transfer events and enzymatic reactions cooperatively catalyzed by many proteins. The initial step of oxygenic photosynthesis is the light-driven water oxidation reaction of photosystem II (PSII), in which four electrons and protons are extracted from two water molecules, and dioxygen molecules are released into the atmosphere. This reaction is catalyzed at the catalytic center of PSII, namely, the oxygen-evolving complex (OEC), through five oxidation steps of the $S_i$-state ($i = 0$–$4$) where the $S_4$ state is the most oxidized transient state that releases dioxygen and successively turns back to the ground $S_0$ state. Unveiling the mechanism of nature's water oxidation is of considerable importance as it may have great potential in the application for developing a clean, renewable energy source; in other words, a framework for the

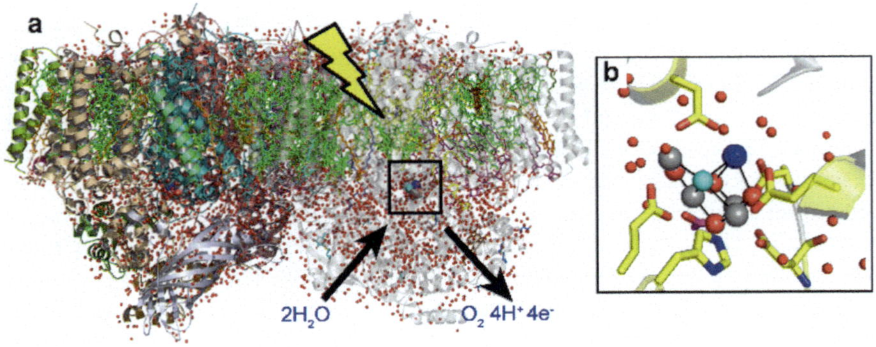


**Fig. 15.8** Structure of PSII (**a**) and structure of the OEC and its ligand environment (**b**). Reprinted from [68]

$S_i$-state may provide a structural basis for the bio-inspired catalysts capable of water oxidation by visible light.

The X-ray structure of PSII in the $S_1$ state under dark-adapted condition was analyzed at 1.9 Å resolution at a synchrotron source, which revealed the detailed organization of the OEC [159]. The OEC is a $Mn_4CaO_5$ cluster with a shape resembling a distorted chair, located at the lumenal surface of the thylakoid membrane and linked to four hydrogen-bonded water channels (Fig. 15.8). The first high-resolution structure of PSII advanced our understanding of the reaction mechanism of water oxidation greatly and served as the standard model for theoretical calculations and spectroscopy [160, 161]. However, two critical issues remained unresolved: First, irradiation by strong X-rays of synchrotron radiation (SR) could damage the OEC [162], and therefore it remained to be confirmed if the reported structure reflects the native structure of OEC or not. Second, the crystal structures in the intermediate $S_i$-states were not reported, but these structures are indispensable for elucidating the full mechanism of water splitting. Here, we review how XFEL has been utilized to bring solutions to these challenges [68].

### 15.4.2 *Radiation Damage-Free Structure of PSII in the $S_1$ State*

In the $Mn_4CaO_5$ cluster of the OEC, three Mn atoms and one Ca atom form the $Mn_3CaO_4$ cuboid cluster and the fourth Mn atom is connected at the outside of the cuboid by two oxo-bridges (Fig. 15.8). The Mn–Mn distances determined are slightly longer (0.1–0.2 Å) than those obtained by extended X-ray absorption fine structures (EXAFS) [163] or theoretical calculations based on the SR structure [164–166]. For instance, the three shortest Mn–Mn distances are 2.8, 2.9, and 3.0 Å in the SR structure (PDB 3WU2), whereas three short Mn–Mn distances of 2.7, 2.7, and 2.8 Å have been reported by EXAFS [163]. Since Mn ions are in the III and IV oxidation states in OEC, they are easily damaged and reduced by strong X-rays, resulting in elongations of the Mn–Mn distances. EXAFS indeed showed that the X-ray dose used for analyzing the SR structure reduces 25% of the Mn atoms of OEC to $Mn^{2+}$ ions [162]. Therefore, the discrepancy in the Mn–Mn distances between the SR structure and those determined by EXAFS could be due to radiation damage during the X-ray data collection for solving the SR structure. In order to eliminate the possible effects of radiation damage, we collected diffraction images using femtosecond XFEL pulses and analyzed the structure at 1.95 Å resolution [167]. Since PSII is a large membrane protein complex of 700-kDa in a dimeric form and its crystals diffract weakly, large crystals with a maximum length of 1 mm were used to ensure the collection of high-resolution data. For collecting a full dataset from such large crystals, a fixed-target serial femtosecond rotational crystallography (SF-ROX) approach [168] was used. In this approach, a large crystal was fixed at low temperature, and an XFEL pulse was irradiated at one point of the crystal to collect the first diffraction image, then the crystal was immediately moved to a next point separated by 50 μm with a simultaneous rotation of the crystal by 0.2° to collect the next diffraction image. This approach allowed the collection of several tens of diffraction images from one large crystal and 100–200 crystals were used to collect one full dataset. Two independent datasets were obtained from two different preparations of the crystals and the structure was analyzed at 1.95 Å resolution from both datasets (PDB 4UB6 and 4UB8). The structure of OEC was determined independently in two non-crystallographic symmetry-related PSII monomers, providing some indications of the agreement of OEC structure across different monomers and datasets.

The overall structure of PSII determined by XFEL is similar to the SR structure. The orientation of the side chains except for some side chains in the vicinity of metal-binding sites, as well as most of the water molecules found in each structure, are almost identical, indicating that X-ray diffraction using continuous X-rays of SR is sufficient to determine the overall protein environment. However, despite the similarity, clear differences are found in the interatomic distances of the Mn atoms within the $Mn_4CaO_5$ cluster (Fig. 15.9). All the interatomic distances between the Mn atoms are shorter by 0.1–0.2 Å compared with the SR structure, and they are mostly consistent with the results of EXAFS. This comparison suggests that the XFEL structure is radiation damage free and the previous SR structure has suffered some radiation damage, leading to a partial reduction of the Mn atoms and therefore causing elongations of the distances between the Mn atoms. The XFEL structure confirmed the presence of a unique central oxo-bridged oxygen (O5) located in between two manganese atoms, Mn1 and Mn4, which was originally found in the SR structure [159], with unusually long distances to the two Mn atoms. The bond distances between the Mn atoms and O5 are 2.7 Å (Mn1–O5), 2.2 Å (Mn3–O5), and 2.3 Å (Mn4–O5), respectively, whereas the other Mn–O distances are within 1.8–2.1 Å (Fig. 15.9) that are typical for Mn oxides. This feature suggests a weak binding of O5 with the nearby Mn ions and therefore possible participation of O5 in the dioxygen formation. Furthermore, the valances of the four Mn atoms were estimated to be (Mn1, Mn2, Mn3, Mn4) = (+3, +4, +4, +3) in the $S_1$ state based on

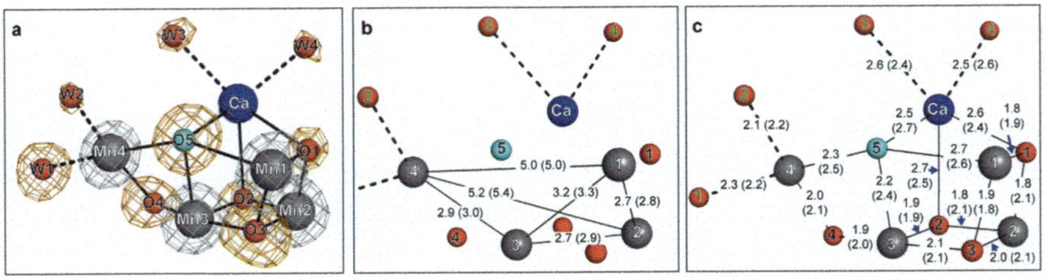


**Fig. 15.9** Radiation damage-free structure of PSII in the $S_1$ state determined by the SF-ROX method. **a** Structure of the $Mn_4CaO_5$ cluster shown with electron density maps of individual atoms. The $2mF_o - DF_c$ map and the oxygen-omit $mF_o - DF_c$ map are shown as gray and brown meshes and contoured at $7\sigma$ and $6\sigma$ levels, respectively. **b** and **c** Interatomic distances (Å) between Mn atoms and M–O (**b**) and Ca–O (**c**). Numbers in bracket indicate the corresponding distances in the SR structure. Reprinted from [68]

the metal–ligand distances as well as the existence of the Jahn–Teller distortion on the $Mn^{+3}$ atoms in the XFEL structure. This is consistent with estimations by previous electron-nuclear double resonance results and theoretical calculations [169, 170]. Thus, the high-resolution XFEL structure of PSII provides a vital understanding of the relationship between the structure of the OEC and its physical properties.

### *15.4.3 Structures of PSII in the Intermediate $S_i$-State*

Although the XFEL structure in the $S_1$ state provides important implications for water oxidation in PSII, detailed structural changes that occur during the reaction cycle remain unknown, but they are indispensable for fully resolving the mechanism of water oxidation. Such structural changes have been implicated by spectroscopy and electron paramagnetic resonance measurements showing that all Mn atoms of the OEC are six-coordinated in the $S_3$ state whereas one of the Mn atoms is five-coordinated in the $S_2$ state, suggesting insertion and binding of a possible substrate water molecule into the OEC during $S_2$ to $S_3$ transition [171]. Pump-probe time-resolved serial femtosecond crystallography (TR-SFX) using XFEL pulses is a robust method for analyzing the structural dynamics of enzymes at room temperature. While there have been several reports on the structural analysis of the intermediate $S_i$-states by using TR-SFX with simultaneous measurements of X-ray spectroscopy, showing that Mn ions remain intact during the data collection using the intense XFEL pulses [65, 74, 172], those works were not conclusive for the dioxygen formation mechanism due to the insufficient resolution of the diffraction data. For example, no insertion of new water during the $S_i$-state transition was observed; thus, we analyzed the structures of PSII in the intermediate state by using TR-SFX at higher resolutions.

The PSII microcrystals were excited by two pump laser flashes separated by 10 ms to advance the dark-adapted $S_1$ state to the $S_3$ state, followed by exposure to the probe XFEL pulses 10 ms after the second pump flash. The structures of the $S_1$ and doubly flashed $S_3$ dominant state were analyzed at 2.35 Å resolutions [67]. The results showed that the overall structure of PSII in the $S_3$ state is almost identical to the $S_1$ state structure. In such TR-SFX studies, an isomorphous Fourier difference map between the two datasets with a reasonably high isomorphicity is an excellent indicator to detect subtle structural changes. The $S_3$–$S_1$ Fourier difference map obtained showed pairs of positive and negative peaks covering the region around the OEC, indicating the occurrence of light-induced structural changes within the crystals. The difference map revealed six substantial structural changes around the OEC during the $S_1$–$S_3$ transition and provided important implications for the mechanism of dioxygen formation (Fig. 15.10). They are: (i) The Mn4 atom moves slightly outward from the OEC, resulting in an elongation of the interatomic distance between Mn4 and Mn1 by 0.1–0.2 Å. (ii) The Ca atom of the OEC moves slightly away from Mn4. (iii) A sharp electron density appeared at a position near the O5 atom, which was modeled as a new water molecule O6. (iv) The side chain of Glu189 flips away from the OEC, providing a space that can accommodate the O6 between Mn1 and O5. (v) The water molecule W665 that existed in the O4 channel becomes disordered. (vi) As a result of the above structural changes, other ligands to the OEC (Asp61, Asp170, His332, and Ala344) slightly moved from their original positions or changed their orientations. Among these structural changes, the insertion of O6 provides novel insights into the dioxygen formation, that is, the OEC changes to a $Mn_4CaO_6$ cluster in the $S_3$ state. The distance between O6 and O5 is around 1.5–2.0 Å, which is suitable for the O=O bond formation. The possible chemical structures of the O6 and O5 are (i) superoxide (1.3 Å), (ii) peroxo (1.5 Å), (iii) oxyl/oxo (2.0 Å), and (iv) hydro-oxo/oxo (2.4 Å), and the structure fits well with either peroxo and oxyl/oxo, but the assignment was still not conclusive due to the limited resolution.

To distinguish among these chemical species and determine the exact chemical structure of the species responsible for O=O bond formation, we further analyzed the XFEL structures of PSII in the $S_1$, $S_2$, and $S_3$ states by using the fixed-target SFX method with XFELs and PSII microcrystals. In this method, the $S_2$ and $S_3$ states were generated by excitations with one or two flashes at room temperature with PSII microcrystals evenly sprayed on a mesh, and trapped immediately at 77 K. Single-shot diffraction images were collected in a fixed-target data collection manner at cryogenic temperature. Compared with the previous TR-SFX approach, this method reduced the sample consumption by one order of magnitude and ensured low background images, allowing us to analyze the structure of multiple intermediate states at 2.15 Å resolutions [173]. The results showed that no insertion of water occurs in the $S_2$ state, but upon transition to the $S_3$ state, flipping of Glu189, the only monodentate carboxyl ligand of OEC, provides a space for incorporation of the additional oxygen O6, and the $Mn_4CaO_5$ cluster remains in the open-cubane form (Fig. 15.11). Four possible chemical species, namely, superoxo, peroxo, oxyl/oxo, and oxyl/hydroxo were examined to determine the exact chemical structure of the species of O5–O6 as described above (Fig. 15.11). By altering the O5–O6 distance and examining the residual densities in the difference map, a distance of 1.9 Å between O5 and O6 resulted in the weakest residual densities. The insertion of O6 at the $S_3$ state and a similar distance of 2.0 Å between O5 and O6 (Ox) has been reported by another

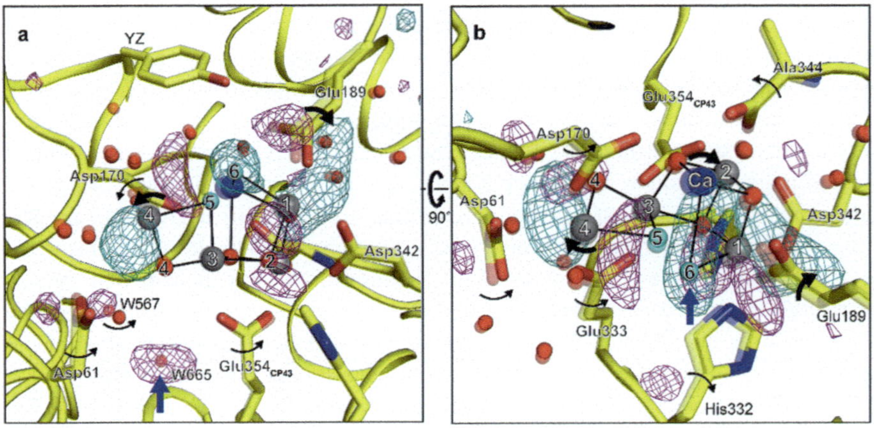


**Fig. 15.10** Structural changes around the OEC during the transition from the $S_1$ to $S_3$ state determined by the SFX method. **a, b** The structure in the $S_1$ and the $S_3$ states are shown with the isomorphous difference Fourier map contoured at $\pm 5\sigma$ levels in turquoise (positive) and magenta (negative). The $S_1$ state structure where changes occurred during the progress of the $S_i$-state cycle are shown in a transparent presentation and black arrows indicate corresponding structural changes. Blue arrows indicate the positions of O6 and W665. Reprinted from [68]

group [34]. These results suggest an oxyl/oxo coupling mechanism for the dioxygen formation in OEC. Moreover, the flipping of the Glu189 induces van der Waals repulsion between the carbonyl oxygen of Glu189 and CP43-Ala411, which moves a short loop of the CP43 subunit that restricts the size of the water channel connecting the O1 oxygen of the OEC (O1 channel), leading to the widening of this channel. The O1 channel thus likely functions in delivering the substrate water to the Mn-cluster during the $S_2$ to $S_3$ transition. These structural changes revealed the mechanism of photosynthetic water oxidation by the cooperative action of substrate water access, proton release, and O=O bond formation (Fig. 15.12).

### *15.4.4 Structural Insights into the Mechanism of Nature's Water Oxidation Revealed by XFEL*

Since the first demonstration of successful structural analysis of photosystem I using XFEL pulses [16], XFEL has provided the new opportunity to study the structural dynamics of PSII at room temperature [34, 67] as well as its radiation damage-free structure. The radiation damage-free structure in the $S_1$ state confirmed the distorted chair form of OEC with its unique, long $\mu$-oxo-bridged O5 atom, and the structure in the $S_3$ state analyzed by TR-SFX revealed the insertion of a new water molecule (O6) into the catalyst in a vicinity of the O5 atom. The high-resolution structures in the $S_1$, $S_2$, and $S_3$ states analyzed by fixed-target SFX identified the chemical structure of O5 and O6 and revealed an oxyl/oxo coupling mechanism between them

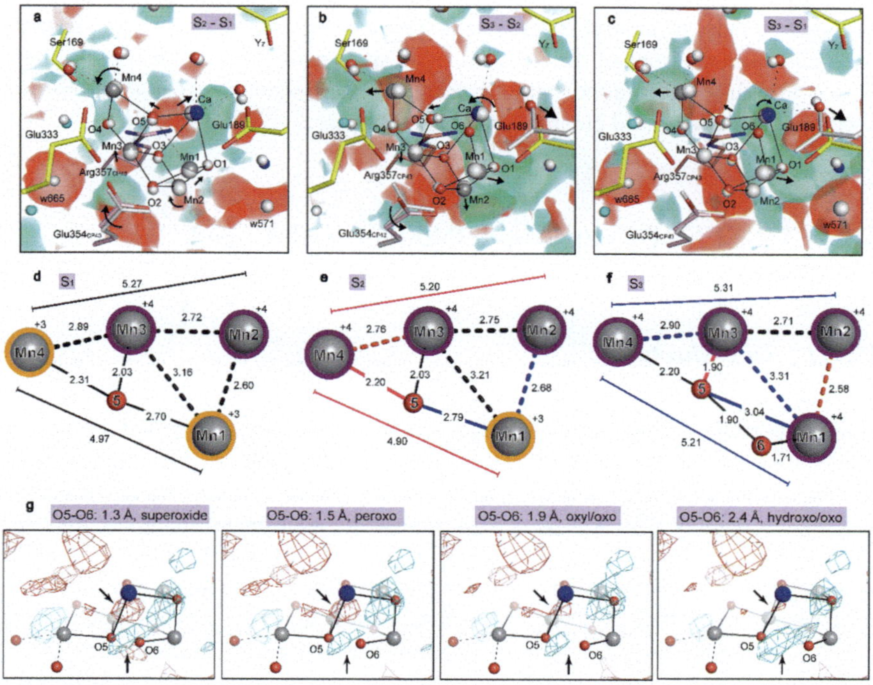

**Fig. 15.11** Structural changes of the OEC during the $S_i$-state determined by the SF-ROX method. **a, b** OEC structures superposed with $F_o - F_o$ isomorphous difference Fourier maps of **a** 1F minus dark, **b** 2F minus 1F, and **c** 2F minus dark datasets. Structures before and after $S_i$-state transition are shown in gray and color, respectively. Isomorphous difference Fourier maps are contoured at $\pm 3\sigma$ levels in cyan (positive) and red (negative). Structural changes consistent with isomorphous difference Fourier maps are represented by black arrows, where larger arrowheads represent the larger changes. **d–f** Interatomic distances (Å) of the OECs in the **d** $S_1$, **e** $S_2$, and **f** $S_3$ states. Blue and red lines indicate elongation and shorting of the interatomic distances compared with the structure in the precedent $S_i$ state. Presumed Mn (+4) and Mn (+3) cations are shown. **g** The $F_o - F_c$ difference Fourier maps contoured at $\pm 2.2\sigma$ levels in cyan (positive) and red (negative) after structural refinement by fixing the distance between O5 and O6 at 1.3 Å (superoxide), 1.5 Å (peroxide), 1.7 Å, 1.9 Å (oxyl/oxo), and 2.4 Å (hydroxo/oxo), respectively. Reprinted from [69] with permission from AAAS

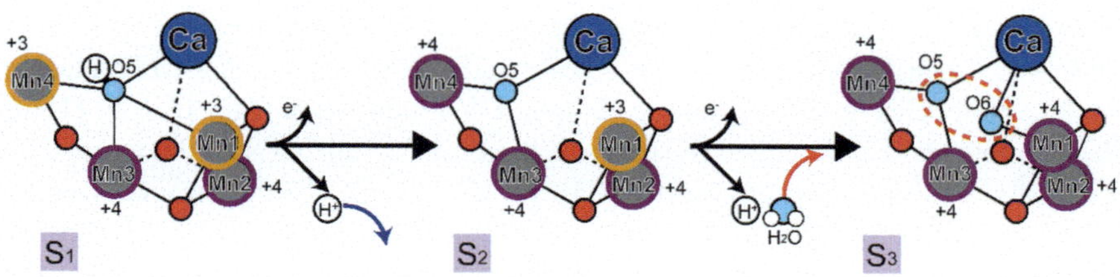

**Fig. 15.12** Schematic structures of OEC during the $S_1$–$S_3$ state transition. Reprinted from [69] with permission from AAAS

for the O=O bond formation. These results provide important bases for the catalytic mechanism of OEC, one of the essential catalysts in nature, that is, the OEC adjusts its structure finely during the catalytic cycle to ensure the water oxidation reaction to proceed under the mild environmental conditions.

## 15.5 TR-SFX of Cytochrome P450nor with Caged Substrate

### *15.5.1 Introduction*

The pump–probe technique for TR-SFX has been successfully applied for visualizing the structural dynamics of various photo-driven proteins, including bacteriorhodopsin and PSII. For further applications of TR-SFX, it is important to establish methods for dealing with non-photo-driven proteins, such as enzymes. Mixing-type TR-SFX is useful for this purpose [174–178]. On the other hand, another method is the pump-probe TR-SFX used with photosensitive caged compounds. Very recently, the application of TR-SFX using a photosensitive caged substrate was demonstrated at SACLA by employing cytochrome P450nor from *Fusarium oxysporum* [103].

P450nor is an unusual member of the P450 superfamily. This enzyme catalyzes the reduction of NO to $N_2O$ using NADH as part of the nitrogen cycle ($2NO + NADH + H^+ \rightarrow N_2O + H_2O$) [179]. Because the product $N_2O$ is the main ozone-depleting substance and a greenhouse gas [180], the NO reduction mechanism catalyzed by P450nor has received increasing attention. In the resting state of P450nor, a water molecule is bound to the ferric heme active site with Cys as a *trans* axial ligand (Fig. 15.13a). In the NO reduction reaction, one NO molecule first displaces the water molecule to form the ferric NO complex as an initial intermediate. Then, the ferric NO complex is two-electron reduced with hydride ($H^-$) from NADH to form intermediate *I*. Finally, *I* reacts with a second NO to generate $N_2O$. These reaction steps are known based on various spectroscopic and theoretical studies [181–185]; however, the structural basis for understanding the reaction mechanism of P450nor remains to be elucidated.

To initiate the reaction of P450nor in TR-SFX, [N,N′-bis-(carboxymethyl)-N,N′-dinitroso-p-phenylenediamine] was used as caged NO, which quantitatively releases two NO molecules on the μs time scale upon UV illumination (Fig. 15.13b) [186]. The quantum yield of the caged NO upon excitation at 308 nm is 1.4, as determined by UV–visible absorption spectroscopy [103]. TR UV–visible absorption spectroscopy and IR spectroscopy were successfully used in conjunction with TR-SFX for evaluating *in crystallo* reactivity of P450nor, which were key to the success of observing the intermediate state. In the following subchapters, we will review the TR-SFX study combined with *in crystallo* spectroscopies of P450nor, using caged NO as an enzymatic reaction trigger.

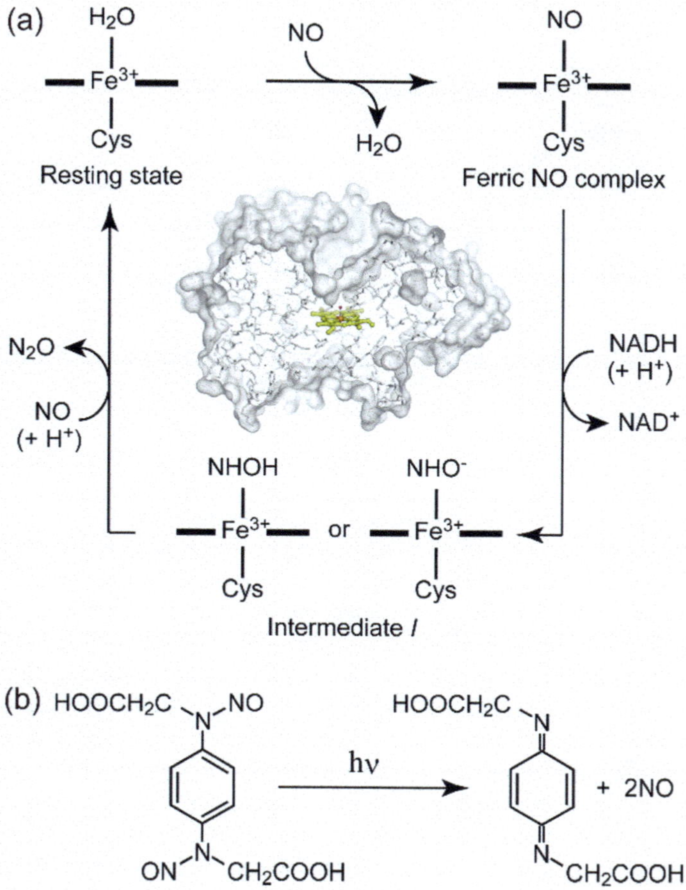


**Fig. 15.13** **a** Reaction cycle of P450nor. **b** Caged NO photolysis. This figure is modified from Tosha et al. *Nat. Commun.* **8**, 1585 (2017)

### *15.5.2 TR UV–Visible Microspectroscopy for P450nor Microcrystals*

The rate-limiting steps of enzymatic reactions, and thus the rates of intermediate formation and decomposition, generally depend on experimental conditions, such as temperature, pH, pressure, solvent viscosity, etc. Therefore, the kinetics and detectable reaction intermediates in microcrystals are not necessarily identical to those in solution. In the study of P450nor, a TR visible absorption microspectrometer was used to evaluate the *in crystallo* reaction of P450nor [103]. The slurries of P450nor microcrystals, pre-soaked with caged NO and NADH, were packed with an SFX carrying medium between two $CaF_2$ windows with a 100 μm spacer and mounted on the pump-probe focal spot in the spectrometer.

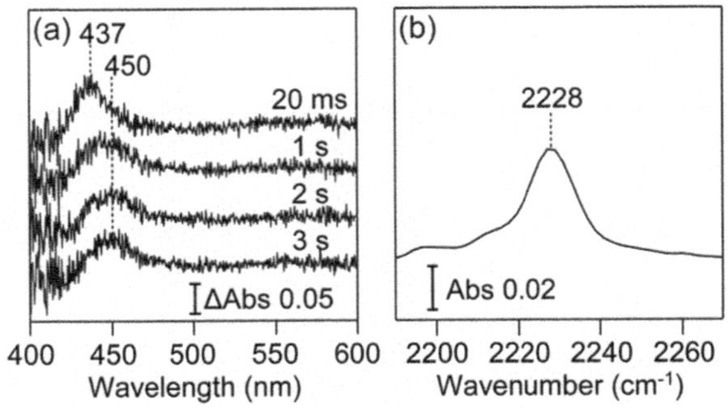


**Fig. 15.14** **a** *In crystallo* TR UV–visible absorption spectra of P450nor during the catalytic reaction. The difference was calculated by subtracting the spectrum recorded prior to photolysis. **b** *In crystallo* IR spectrum of P450nor after the catalytic reaction. This figure is modified from Tosha et al. *Nat. Commun.* **8**, 1585 (2017)

TR UV–visible absorption spectra of the P450nor microcrystal upon caged NO photolysis are shown in Fig. 15.14a. The Soret difference peak appears at 437 nm, which subsequently shifts to 450 nm, as previously reported for the solution sample [181], indicating that the ferric NO complex and *I* are formed following caged NO photolysis. This demonstrates that the reaction pathway in the crystalline phase is the same as in solution. However, the time scale of the intermediate formation is significantly different. Because the TR spectra of microcrystals were measured with a minimum delay time of 20 ms, the NO binding time (<20 ms) in the microcrystal remains unknown. However, the *I* formation in the microcrystal is significantly slower than in solution, by about three orders of magnitude [181]. This was attributed to the crystal packing contact occurring near the NADH channel in the microcrystals [103], which might have affected NADH access to the heme pocket from the solvent region. Based on the TR spectroscopic results on microcrystals, TR-SFX for P450nor was first performed with a delay time of 20 ms in order to analyze the ferric NO complex.

### 15.5.3 IR Microspectroscopy for P450nor Microcrystals

IR spectroscopy is more informative and useful for chemical analyses than UV–visible absorption spectroscopy. *In crystallo* TR-IR spectroscopy has been previously applied to photocyclic systems [98, 187], but is yet to be technically established for enzymatic systems due to the difficulties in single-shot measurements. For studying P450nor, static IR microspectroscopy was applied using a synchrotron-based IR microspectrometer at SPring-8/BL43IR [103]. First, a functional analysis was performed by detecting $N_2O$ production within microcrystals. To this end, the slurries of P450nor microcrystals, pre-soaked with caged NO and NADH, were packed with an SFX carrying medium between two $CaF_2$ windows with a 100 μm spacer, and mounted on the IR focal spot after UV illumination at 308 nm. The measured IR spectrum showed an N–N stretching IR band of $N_2O$ at 2228 $cm^{-1}$, indicating that P450nor was functional in microcrystals (Fig. 15.14b).

IR microspectroscopy was further performed for analyzing the electronic state of the ferric NO complex. The NO-bound sample was prepared by UV illumination to microcrystals soaked with caged NO, in the absence of NADH. A N–O stretching IR band corresponding to the heme-bound NO was observed at 1853 $cm^{-1}$, suggesting that the bound NO existed in almost a neutral radical state with a N–O bond length of 1.15 Å [188]. This IR result was used for the refinement of the TR-SFX result as described later.

### *15.5.4 TR-SFX for Capturing the Initial Intermediate of P450nor*

Based on the TR spectroscopic analysis of the *in crystallo* kinetics, TR-SFX for P450nor microcrystals, pre-soaked with caged NO and NADH, was performed with a delay time of 20 ms to capture the substrate NO binding. A structure of P450nor was thus obtained, shown in Fig. 15.15a. The difference Fourier map before and after caged NO photolysis shows a strong positive electron density at the heme moiety, which is assigned to the bound NO. During the refinement of the NO coordination structure, a restraint was imposed on the N–O bond length of 1.15 Å, which was based on the IR microspectroscopic result. This restraint helped in the refinement of other geometric parameters, such as the Fe–N bond length (1.67 Å) and the Fe–N–O angle (158°), leading to the NO coordination structure adopting a slightly bent form [103]. For comparison, the NO-bound structures determined at SPring-8 in the absence of NADH are also shown in Fig. 15.15b, c, where the Fe–N–O angle was reduced due to the inevitable radiation damage. It is evident that the TR-SFX experiment successfully clarified the intact ferric Fe–N–O coordination geometry of P450nor with no radiation damage, based on the diffraction-before-destruction principle.

The slightly bent form (with a radical character) of the bound NO as revealed by TR-SFX and IR spectroscopic analyses would be functionally important for P450nor. The ferric NO heme complexes with a linear Fe–N–O geometry have a $Fe^{2+}$–$NO^+$ electronic character and are easily decomposed by reductive nitrosylation ($Fe^{2+}$–$NO^+$ + $OH^- \rightarrow Fe^{2+} + NO_2^- + H^+$) [189]. The slight bending of the Fe–NO unit avoids such decomposition despite the fact that the heme pocket of P450nor is exposed to the solvent for NADH accommodation.

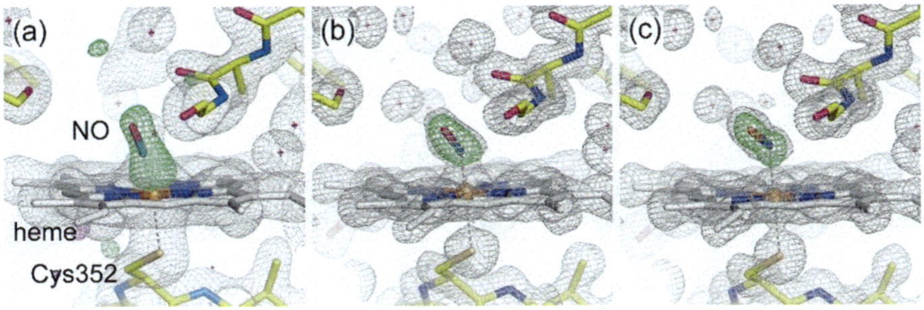


**Fig. 15.15** Structures of the ferric NO complex of P450nor obtained by **a** TR-SFX and **b**, **c** synchrotron X-ray crystallography. The X-ray doses in the synchrotron data collection were **b** 0.72 and **c** 5.7 MGy. The synchrotron data were taken at 100 K in the absence of NADH. This figure is modified from Tosha et al. *Nat. Commun.* **8**, 1585 (2017)

## *15.5.5 Future Directions*

Various caged compounds, such as caged ATP, caged $O_2$, and photosensitive electron donors are available for use [190–192]. An engineered enzyme that is covalently bound with a photosensitive electron donor has also been developed recently [193]. Therefore, TR-SFX with photosensitive caged compounds will be increasingly applied over the next decade. Here, it should be stressed that this technique should ideally be used in combination with *in crystallo* spectroscopy. The combination of TR visible absorption spectroscopy and static IR spectroscopy with TR-SFX in the case of P450nor was paramount to the success of TR-SFX measurements and analyses. So far, TR X-ray emission spectroscopy has also been combined to TR-SFX [34, 36], where the electronic state of metal sites was assigned using spectroscopy. A combination of TR X-ray crystallography and TR *in crystallo* spectroscopies realizes the tracking of catalytic reactions of enzymes at atomic and electronic levels at ambient temperature, which has been a long-desired feature in structural enzymology.

However, one technical limitation of TR-SFX became apparent during the course of the P450nor study. TR observations with a delay time ($\Delta t$) beyond tens of ms are difficult, because pump-illuminated crystals flow downstream from the XFEL spot during the period of $\Delta t$ under the sample flow speed (~5–10 mm/s) with a high-viscosity sample injector used at SACLA [135]. To overcome this problem, if the pump illumination area is set upstream and distant from the XFEL spot, the accuracy of the sample flow speed would be strictly required for positioning a pump-illuminated crystal at the XFEL spot at $t = \Delta t$. For this reason, TR X-ray crystallography for capturing the intermediate *I* of P450nor is not yet successful. Tape-based and/or chip-based "fixed-target" sample delivery systems would be more suitable for the observation of late intermediates of enzymes [36, 47, 53]. Such fixed-target systems may also be useful for various TR *in crystallo* spectroscopies as sample exchange schemes.

## References


1. R. Henderson, The potential and limitations of neutrons, electrons and X-rays for atomic resolution microscopy of unstained biological molecules. Q. Rev. Biophys. **28**(2), 171–193 (1995)
2. R. Henderson, Cryoprotection of protein crystals against radiation-damage in electron and X-ray-diffraction. Proc. R. Soc. B-Biol. Sci. **241**(1300), 6–8 (1990)
3. E. de la Mora, N. Coquelle, C.S. Bury, M. Rosenthal, J.M. Holton, I. Carmichael et al., Radiation damage and dose limits in serial synchrotron crystallography at cryo- and room temperatures. Proc. Natl. Acad. Sci. USA **117**(8), 4142–4151 (2020)
4. C.S. Bury, J.C. Brooks-Bartlett, S.P. Walsh, E.F. Garman, Estimate your dose: RADDOSE-3D. Protein Sci. **27**(1), 217–228 (2018)
5. E.F. Garman, M. Weik, Radiation damage in macromolecular crystallography. Methods Mol. Biol. **1607**, 467–489 (2017)
6. O.B. Zeldin, S. Brockhauser, J. Bremridge, J.M. Holton, E.F. Garman, Predicting the X-ray lifetime of protein crystals. Proc. Natl. Acad. Sci. USA **110**(51), 20551–20556 (2013)
7. J.M. Holton, K.A. Frankel, The minimum crystal size needed for a complete diffraction data set. Acta Crystallogr. D Biol. Crystallogr. **66**(Pt 4), 393–408 (2010)
8. J.M. Holton, A beginner's guide to radiation damage. J. Synchrotron Radiat. **16**(Pt 2), 133–142 (2009)
9. H.N. Chapman, X-ray free-electron lasers for the structure and dynamics of macromolecules. Annu. Rev. Biochem. **88**, 35–58 (2019)
10. I. Schlichting, Serial femtosecond crystallography: the first five years. IUCrJ **2**(Pt 2), 246–255 (2015)
11. P. Fromme, XFELs open a new era in structural chemical biology. Nat. Chem. Biol. **11**(12), 895–899 (2015)
12. J.C.H. Spence, XFELs for structure and dynamics in biology. IUCrJ **4**(Pt 4), 322–339 (2017)
13. C. Gati, D. Oberthuer, O. Yefanov, R.D. Bunker, F. Stellato, E. Chiu et al., Atomic structure of granulin determined from native nanocrystalline granulovirus using an X-ray free-electron laser. Proc. Natl. Acad. Sci. USA **114**(9), 2247–2252 (2017)
14. G. Tetreau, A.S. Banneville, E.A. Andreeva, A.S. Brewster, M.S. Hunter, R.G. Sierra et al., Serial femtosecond crystallography on in vivo-grown crystals drives elucidation of mosquitocidal Cyt1Aa bioactivation cascade. Nat. Commun. **11**(1), 1153 (2020)
15. S. Boutet, L. Lomb, G.J. Williams, T.R. Barends, A. Aquila, R.B. Doak et al., High-resolution protein structure determination by serial femtosecond crystallography. Science **337**(6092), 362–364 (2012)
16. H.N. Chapman, P. Fromme, A. Barty, T.A. White, R.A. Kirian, A. Aquila et al., Femtosecond X-ray protein nanocrystallography. Nature **470**(7332), 73–77 (2011)
17. Consortium PDB, Protein Data Bank: the single global archive for 3D macromolecular structure data. Nucl. Acids Res. **47**(D1), D520-D8 (2019)
18. J.L. Dickerson, P.T.N. McCubbin, E.F. Garman, RADDOSE-XFEL: femtosecond time-resolved dose estimates for macromolecular X-ray free-electron laser experiments. J. Appl. Crystallogr. **53**(2), 549–560 (2020)
19. K. Nass, A. Gorel, M.M. Abdullah, A.V. Martin, M. Kloos, A. Marinelli, et al., Structural dynamics in proteins induced by and probed with X-ray free-electron laser pulses. Nat. Commun. **11**(1), 1814 (2020)
20. K. Nass, L. Foucar, T.R. Barends, E. Hartmann, S. Botha, R.L. Shoeman et al., Indications of radiation damage in ferredoxin microcrystals using high-intensity X-FEL beams. J. Synchrotron Radiat. **22**(2), 225–238 (2015)
21. N.L. Opara, I. Mohacsi, M. Makita, D. Castano-Diez, A. Diaz, P. Juranic et al., Demonstration of femtosecond X-ray pump X-ray probe diffraction on protein crystals. Struct. Dyn. **5**(5), 054303 (2018)

22. S.C. Jensen, B. Sullivan, D.A. Hartzler, J.M. Aguilar, S. Awel, S. Bajt et al., X-ray emission spectroscopy at X-ray free electron lasers: limits to observation of the classical spectroscopic response for electronic structure analysis. J. Phys. Chem. Lett. **10**(3), 441–446 (2019)
23. C.A. Stan, D. Milathianaki, H. Laksmono, R.G. Sierra, T.A. McQueen, M. Messerschmidt et al., Liquid explosions induced by X-ray laser pulses. Nat. Phys. **12**(10), 966–971 (2016)
24. D. Kim, A. Echelmeier, J. Cruz Villarreal, S. Gandhi, S. Quintana, A. Egatz-Gomez, A. Ros, Electric triggering for enhanced control of droplet generation. Anal. Chem. **91**(15), 9792–9799 (2019)
25. J. Knoska, L. Adriano, S. Awel, K.R. Beyerlein, O. Yefanov, D. Oberthuer et al., Ultracompact 3D microfluidics for time-resolved structural biology. Nat. Commun. **11**(1), 657 (2020)
26. D. Oberthuer, J. Knoska, M.O. Wiedorn, K.R. Beyerlein, D.A. Bushnell, E.G. Kovaleva et al., Double-flow focused liquid injector for efficient serial femtosecond crystallography. Sci. Rep. **7**, 44628 (2017)
27. D.P. DePonte, U. Weierstall, K. Schmidt, J. Warner, D. Starodub, J.C.H. Spence, R.B. Doak, Gas dynamic virtual nozzle for generation of microscopic droplet streams. J. Phys. D Appl. Phys. **41**(19), 195505 (2008)
28. J. Zook, M. Shekhar, D. Hansen, C. Conrad, T. Grant, C. Gupta, et al., XFEL and NMR structures of francisella lipoprotein reveal conformational space of drug target against tularemia. Structure **28**(5), 54054–540547 e3 (2020)
29. C.E. Conrad, S. Basu, D. James, D. Wang, A. Schaffer, S. Roy-Chowdhury et al., A novel inert crystal delivery medium for serial femtosecond crystallography. IUCrJ **2**(Pt 4), 421–430 (2015)
30. U. Weierstall, D. James, C. Wang, T.A. White, D. Wang, W. Liu et al., Lipidic cubic phase injector facilitates membrane protein serial femtosecond crystallography. Nat. Commun. **5**, 3309 (2014)
31. G. Kovacsova, M.L. Grunbein, M. Kloos, T.R.M. Barends, R. Schlesinger, J. Heberle et al., Viscous hydrophilic injection matrices for serial crystallography. IUCrJ **4**(Pt 4), 400–410 (2017)
32. R.G. Sierra, C. Gati, H. Laksmono, E.H. Dao, S. Gul, F. Fuller et al., Concentric-flow electrokinetic injector enables serial crystallography of ribosome and photosystem II. Nat. Methods **13**(1), 59–62 (2016)
33. E.H. Dao, F. Poitevin, R.G. Sierra, C. Gati, Y. Rao, H.I. Ciftci et al., Structure of the 30S ribosomal decoding complex at ambient temperature. RNA **24**(12), 1667–1676 (2018)
34. J. Kern, R. Chatterjee, I.D. Young, F.D. Fuller, L. Lassalle, M. Ibrahim et al., Structures of the intermediates of Kok's photosynthetic water oxidation clock. Nature **563**(7731), 421–425 (2018)
35. A.M. Orville, CHAPTER 18 Acoustic Methods for On-demand Sample Injection into XFEL Beams. X-Ray Free Electron Lasers: Applications in Materials, Chemistry and Biology: The Royal Society of Chemistry (2017), pp. 348–364
36. F.D. Fuller, S. Gul, R. Chatterjee, E.S. Burgie, I.D. Young, H. Lebrette et al., Drop-on-demand sample delivery for studying biocatalysts in action at X-ray free-electron lasers. Nat. Methods **14**(4), 443–449 (2017)
37. C.G. Roessler, R. Agarwal, M. Allaire, R. Alonso-Mori, B. Andi, J.F.R. Bachega et al., Acoustic injectors for drop-on-demand serial femtosecond crystallography. Structure **24**(4), 631–640 (2016)
38. A.S. Soares, M.A. Engel, R. Stearns, S. Datwani, J. Olechno, R. Ellson et al., Acoustically mounted microcrystals yield high-resolution X-ray structures. Biochemistry **50**(21), 4399–4401 (2011)
39. B. Davy, D. Axford, J.H. Beale, A. Butryn, P. Docker, A. Ebrahim et al., Reducing sample consumption for serial crystallography using acoustic drop ejection. J. Synchrotron Radiat. **26**(5), 1820–1825 (2019)
40. P. Wu, C. Noland, M. Ultsch, B. Edwards, D. Harris, R. Mayer, S.F. Harris, Developments in the implementation of acoustic droplet ejection for protein crystallography. J. Lab. Autom. **21**(1), 97–106 (2016)

41. F. Mafune, K. Miyajima, K. Tono, Y. Takeda, J.Y. Kohno, N. Miyauchi et al., Microcrystal delivery by pulsed liquid droplet for serial femtosecond crystallography. Acta Crystallogr. D Struct. Biol. **72**(Pt 4), 520–523 (2016)
42. B. Hadimioglu, R. Stearns, R. Ellson, Moving liquids with sound: the physics of acoustic droplet ejection for robust laboratory automation in life sciences. J. Lab. Autom. **21**(1), 4–18 (2016)
43. M. Ibrahim, T. Fransson, R. Chatterjee, M.H. Cheah, R. Hussein, L. Lassalle, et al., Untangling the sequence of events during the S2 → S3 transition in photosystem II and implications for the water oxidation mechanism. Proc. Natl. Acad. Sci. USA (2020)
44. M.S. Hunter, B. Segelke, M. Messerschmidt, G.J. Williams, N.A. Zatsepin, A. Barty et al., Fixed-target protein serial microcrystallography with an x-ray free electron laser. Sci. Rep. **4**, 6026 (2014)
45. R.B. Doak, G.N. Kovacs, A. Gorel, L. Foucar, T.R.M. Barends, M.L. Grunbein et al., Crystallography on a chip—without the chip: sheet-on-sheet sandwich. Acta Crystallogr. D **74**, 1000–1007 (2018)
46. S. Oghbaey, A. Sarracini, H.M. Ginn, O. Pare-Labrosse, A. Kuo, A. Marx et al., Fixed target combined with spectral mapping: approaching 100% hit rates for serial crystallography. Acta Crystallogr. D Struct. Biol. **72**(Pt 8), 944–955 (2016)
47. D.A. Sherrell, A.J. Foster, L. Hudson, B. Nutter, J. O'Hea, S. Nelson et al., A modular and compact portable mini-endstation for high-precision, high-speed fixed target serial crystallography at FEL and synchrotron sources. J. Synchrotron Radiat. **22**(6), 1372–1378 (2015)
48. C. Mueller, A. Marx, S.W. Epp, Y. Zhong, A. Kuo, A.R. Balo et al., Fixed target matrix for femtosecond time-resolved and in situ serial micro-crystallography. Struct. Dyn. **2**(5), 054302 (2015)
49. P. Aller, J. Sanchez-Weatherby, J. Foadi, G. Winter, C.M. Lobley, D. Axford et al., Application of in situ diffraction in high-throughput structure determination platforms. Methods Mol. Biol. **1261**, 233–253 (2015)
50. J. Lieske, M. Cerv, S. Kreida, D. Komadina, J. Fischer, M. Barthelmess et al., On-chip crystallization for serial crystallography experiments and on-chip ligand-binding studies. IUCrJ **6**(Pt 4), 714–728 (2019)
51. P. Roedig, H.M. Ginn, T. Pakendorf, G. Sutton, K. Harlos, T.S. Walter et al., High-speed fixed-target serial virus crystallography. Nat. Methods **14**(8), 805–810 (2017)
52. P. Roedig, R. Duman, J. Sanchez-Weatherby, I. Vartiainen, A. Burkhardt, M. Warmer et al., Room-temperature macromolecular crystallography using a micro-patterned silicon chip with minimal background scattering. J. Appl. Crystallogr. **49**(Pt 3), 968–975 (2016)
53. P. Roedig, I. Vartiainen, R. Duman, S. Panneerselvam, N. Stube, O. Lorbeer et al., A micro-patterned silicon chip as sample holder for macromolecular crystallography experiments with minimal background scattering. Sci. Rep. **5**, 10451 (2015)
54. M.L. Shelby, D. Gilbile, T.D. Grant, C. Seuring, B.W. Segelke, W. He et al., A fixed-target platform for serial femtosecond crystallography in a hydrated environment. IUCrJ **7**(Pt 1), 30–41 (2020)
55. G. Chreifi, E.L. Baxter, T. Doukov, A.E. Cohen, S.E. McPhillips, J. Song et al., Crystal structure of the pristine peroxidase ferryl center and its relevance to proton-coupled electron transfer. Proc. Natl. Acad. Sci. USA **113**(5), 1226–1231 (2016)
56. E.L. Baxter, L. Aguila, R. Alonso-Mori, C.O. Barnes, C.A. Bonagura, W. Brehmer et al., High-density grids for efficient data collection from multiple crystals. Acta Crystallogr. D Struct. Biol. **72**(Pt 1), 2–11 (2016)
57. A.E. Cohen, S.M. Soltis, A. Gonzalez, L. Aguila, R. Alonso-Mori, C.O. Barnes et al., Goniometer-based femtosecond crystallography with X-ray free electron lasers. Proc. Natl. Acad. Sci. USA **111**(48), 17122–17127 (2014)
58. U. Zander, G. Bourenkov, A.N. Popov, D. de Sanctis, O. Svensson, A.A. McCarthy et al., MeshAndCollect: an automated multi-crystal data-collection workflow for synchrotron macromolecular crystallography beamlines. Acta Crystallogr. D Biol. Crystallogr. **71**(Pt 11), 2328–2343 (2015)

59. M.L. Grunbein, M. Stricker, G. Nass Kovacs, M. Kloos, R.B. Doak, R.L. Shoeman, et al., Illumination guidelines for ultrafast pump-probe experiments by serial femtosecond crystallography. Nat. Methods (2020)
60. A. Aquila, M.S. Hunter, R.B. Doak, R.A. Kirian, P. Fromme, T.A. White et al., Time-resolved protein nanocrystallography using an X-ray free-electron laser. Opt. Express **20**(3), 2706–2716 (2012)
61. L.C. Johansson, D. Arnlund, G. Katona, T.A. White, A. Barty, D.P. DePonte et al., Structure of a photosynthetic reaction centre determined by serial femtosecond crystallography. Nat. Commun. **4**, 2911 (2013)
62. M.O. Wiedorn, S. Awel, A.J. Morgan, K. Ayyer, Y. Gevorkov, H. Fleckenstein et al., Rapid sample delivery for megahertz serial crystallography at X-ray FELs. IUCrJ **5**(Pt 5), 574–584 (2018)
63. A. Echelmeier, D. Kim, J. Cruz Villarreal, J. Coe, S. Quintana, G. Brehm et al., 3D printed droplet generation devices for serial femtosecond crystallography enabled by surface coating. J. Appl. Crystallogr. **52**(Pt 5), 997–1008 (2019)
64. C. Gisriel, J. Coe, R. Letrun, O.M. Yefanov, C. Luna-Chavez, N.E. Stander et al., Membrane protein megahertz crystallography at the European XFEL. Nat. Commun. **10**(1), 5021 (2019)
65. C. Kupitz, S. Basu, I. Grotjohann, R. Fromme, N.A. Zatsepin, K.N. Rendek et al., Serial time-resolved crystallography of photosystem II using a femtosecond X-ray laser. Nature **513**(7517), 261–265 (2014)
66. K. Ayyer, O.M. Yefanov, D. Oberthur, S. Roy-Chowdhury, L. Galli, V. Mariani et al., Macromolecular diffractive imaging using imperfect crystals. Nature **530**(7589), 202–206 (2016)
67. M. Suga, F. Akita, M. Sugahara, M. Kubo, Y. Nakajima, T. Nakane et al., Light-induced structural changes and the site of O=O bond formation in PSII caught by XFEL. Nature **543**(7643), 131–135 (2017)
68. M. Suga, A. Shimada, F. Akita, J.R. Shen, T. Tosha, H. Sugimoto, Time-resolved studies of metalloproteins using X-ray free electron laser radiation at SACLA. Biochim. Biophys. Acta Gen. Subj. **1864**(2), 129466 (2020)
69. M. Suga, F. Akita, K. Yamashita, Y. Nakajima, G. Ueno, H. Li et al., An oxyl/oxo mechanism for oxygen-oxygen coupling in PSII revealed by an x-ray free-electron laser. Science **366**(6463), 334–338 (2019)
70. M. Suga, F. Akita, K. Hirata, G. Ueno, H. Murakami, Y. Nakajima, et al., Native structure of photosystem II at 1.95 A resolution viewed by femtosecond X-ray pulses. Nature **517**(7532), 99–103 (2015)
71. I.D. Young, M. Ibrahim, R. Chatterjee, S. Gul, F.D. Fuller, S. Koroidov et al., Structure of photosystem II and substrate binding at room temperature. Nature **540**(7633), 453–457 (2016)
72. J. Kern, V.K. Yachandra, J. Yano, Metalloprotein structures at ambient conditions and in real-time: biological crystallography and spectroscopy using X-ray free electron lasers. Curr. Opin. Struct. Biol. **34**, 87–98 (2015)
73. J. Kern, R. Tran, R. Alonso-Mori, S. Koroidov, N. Echols, J. Hattne et al., Taking snapshots of photosynthetic water oxidation using femtosecond X-ray diffraction and spectroscopy. Nat. Commun. **5**, 4371 (2014)
74. J. Kern, R. Alonso-Mori, R. Tran, J. Hattne, R.J. Gildea, N. Echols et al., Simultaneous femtosecond X-ray spectroscopy and diffraction of photosystem II at room temperature. Science **340**(6131), 491–495 (2013)
75. R. Alonso-Mori, J. Kern, R.J. Gildea, D. Sokaras, T.C. Weng, B. Lassalle-Kaiser et al., Energy-dispersive X-ray emission spectroscopy using an X-ray free-electron laser in a shot-by-shot mode. Proc. Natl. Acad. Sci. USA **109**(47), 19103–19107 (2012)
76. T. Fransson, R. Chatterjee, F.D. Fuller, S. Gul, C. Weninger, D. Sokaras et al., X-ray emission spectroscopy as an in situ diagnostic tool for X-ray crystallography of metalloproteins using an X-ray free-electron laser. Biochemistry **57**(31), 4629–4637 (2018)
77. J. Tenboer, S. Basu, N. Zatsepin, K. Pande, D. Milathianaki, M. Frank et al., Time-resolved serial crystallography captures high-resolution intermediates of photoactive yellow protein. Science **346**(6214), 1242–1246 (2014)

78. K. Pande, C.D. Hutchison, G. Groenhof, A. Aquila, J.S. Robinson, J. Tenboer et al., Femtosecond structural dynamics drives the trans/cis isomerization in photoactive yellow protein. Science **352**(6286), 725–729 (2016)
79. S. Pandey, R. Bean, T. Sato, I. Poudyal, J. Bielecki, J. Cruz Villarreal, et al., Time-resolved serial femtosecond crystallography at the European XFEL. Nat. Methods (2019)
80. Y. Kang, X.E. Zhou, X. Gao, Y. He, W. Liu, A. Ishchenko et al., Crystal structure of rhodopsin bound to arrestin by femtosecond X-ray laser. Nature **523**(7562), 561–567 (2015)
81. X.E. Zhou, X. Gao, A. Barty, Y. Kang, Y. He, W. Liu et al., X-ray laser diffraction for structure determination of the rhodopsin-arrestin complex. Sci. Data. **3**, 160021 (2016)
82. P. Nogly, V. Panneels, G. Nelson, C. Gati, T. Kimura, C. Milne et al., Lipidic cubic phase injector is a viable crystal delivery system for time-resolved serial crystallography. Nat. Commun. **7**, 12314 (2016)
83. T. Nakane, S. Hanashima, M. Suzuki, H. Saiki, T. Hayashi, K. Kakinouchi et al., Membrane protein structure determination by SAD, SIR, or SIRAS phasing in serial femtosecond crystallography using an iododetergent. Proc. Natl. Acad. Sci. USA **113**(46), 13039–13044 (2016)
84. P. Nogly, T. Weinert, D. James, S. Carbajo, D. Ozerov, A. Furrer, et al., Retinal isomerization in bacteriorhodopsin captured by a femtosecond x-ray laser. Science **361**(6398) (2018)
85. C. Wickstrand, P. Nogly, E. Nango, S. Iwata, J. Standfuss, R. Neutze, Bacteriorhodopsin: structural insights revealed using X-ray lasers and synchrotron radiation. Annu. Rev. Biochem. **88**(88), 59–83 (2019)
86. T. Weinert, P. Skopintsev, D. James, F. Dworkowski, E. Panepucci, D. Kekilli et al., Proton uptake mechanism in bacteriorhodopsin captured by serial synchrotron crystallography. Science **365**(6448), 61–65 (2019)
87. V. Panneels, W. Wu, C.J. Tsai, P. Nogly, J. Rheinberger, K. Jaeger et al., Time-resolved structural studies with serial crystallography: a new light on retinal proteins. Struct Dyn. **2**(4), 041718 (2015)
88. G. Nass Kovacs, J.P. Colletier, M.L. Grunbein, Y. Yang, T. Stensitzki, A. Batyuk et al., Three-dimensional view of ultrafast dynamics in photoexcited bacteriorhodopsin. Nat. Commun. **10**(1), 3177 (2019)
89. J.H. Yun, X. Li, J.H. Park, Y. Wang, M. Ohki, Z. Jin et al., Non-cryogenic structure of a chloride pump provides crucial clues to temperature-dependent channel transport efficiency. J. Biol. Chem. **294**(3), 794–804 (2019)
90. C.D.M. Hutchison, V. Cordon-Preciado, R.M.L. Morgan, T. Nakane, J. Ferreira, G. Dorlhiac, et al., X-ray free electron laser determination of crystal structures of dark and light states of a reversibly photoswitching fluorescent protein at room temperature. Int. J. Mol. Sci. **18**(9) (2017)
91. J. Woodhouse, G. Nass Kovacs, N. Coquelle, L.M. Uriarte, V. Adam, T.R.M. Barends et al., Photoswitching mechanism of a fluorescent protein revealed by time-resolved crystallography and transient absorption spectroscopy. Nat. Commun. **11**(1), 741 (2020)
92. J.P. Colletier, M. Sliwa, F.X. Gallat, M. Sugahara, V. Guillon, G. Schiro et al., Serial femtosecond crystallography and ultrafast absorption spectroscopy of the photoswitchable fluorescent protein IrisFP. J. Phys. Chem. Lett. **7**(5), 882–887 (2016)
93. N. Coquelle, M. Sliwa, J. Woodhouse, G. Schiro, V. Adam, A. Aquila et al., Chromophore twisting in the excited state of a photoswitchable fluorescent protein captured by time-resolved serial femtosecond crystallography. Nat. Chem. **10**(1), 31–37 (2018)
94. P. Edlund, H. Takala, E. Claesson, L. Henry, R. Dods, H. Lehtivuori et al., The room temperature crystal structure of a bacterial phytochrome determined by serial femtosecond crystallography. Sci. Rep. **6**, 35279 (2016)
95. E. Claesson, W.Y. Wahlgren, H. Takala, S. Pandey, L. Castillon, V. Kuznetsova et al., The primary structural photoresponse of phytochrome proteins captured by a femtosecond X-ray laser. Elife **9**, e53514 (2020)
96. T.R. Barends, L. Foucar, A. Ardevol, K. Nass, A. Aquila, S. Botha et al., Direct observation of ultrafast collective motions in CO myoglobin upon ligand dissociation. Science **350**(6259), 445–450 (2015)

97. I. Ishigami, N.A. Zatsepin, M. Hikita, C.E. Conrad, G. Nelson, J.D. Coe et al., Crystal structure of CO-bound cytochrome c oxidase determined by serial femtosecond X-ray crystallography at room temperature. Proc. Natl. Acad. Sci. USA **114**(30), 8011–8016 (2017)
98. A. Shimada, M. Kubo, S. Baba, K. Yamashita, K. Hirata, G. Ueno et al., A nanosecond time-resolved XFEL analysis of structural changes associated with CO release from cytochrome c oxidase. Sci. Adv. **3**(7), e1603042 (2017)
99. K. Nakajima, Y. Joti, T. Katayama, S. Owada, T. Togashi, T. Abe et al., Software for the data analysis of the arrival-timing monitor at SACLA. J. Synchrotron Radiat. **25**(Pt 2), 592–603 (2018)
100. T. Katayama, S. Owada, T. Togashi, K. Ogawa, P. Karvinen, I. Vartiainen et al., A beam branching method for timing and spectral characterization of hard X-ray free-electron lasers. Struct. Dyn. **3**(3), 034301 (2016)
101. A. Sanchez-Gonzalez, A.S. Johnson, A. Fitzpatrick, C.D.M. Hutchison, C. Fare, V. Cordon-Preciado, et al., Coincidence timing of femtosecond optical pulses in an X-ray free electron laser. J. Appl. Phys. **122**(20) (2017)
102. T. Yabuuchi, A. Kon, Y. Inubushi, T. Togahi, K. Sueda, T. Itoga et al., An experimental platform using high-power, high-intensity optical lasers with the hard X-ray free-electron laser at SACLA. J. Synchrotron Radiat. **26**(Pt 2), 585–594 (2019)
103. T. Tosha, T. Nomura, T. Nishida, N. Saeki, K. Okubayashi, R. Yamagiwa et al., Capturing an initial intermediate during the P450nor enzymatic reaction using time-resolved XFEL crystallography and caged-substrate. Nat. Commun. **8**(1), 1585 (2017)
104. A. Deiters, D. Groff, Y. Ryu, J. Xie, P.G. Schultz, A genetically encoded photocaged tyrosine. Angew. Chem. Int. Ed. Engl. **45**(17), 2728–2731 (2006)
105. J. Wang, Y. Liu, Y. Liu, S. Zheng, X. Wang, J. Zhao et al., Time-resolved protein activation by proximal decaging in living systems. Nature **569**(7757), 509–513 (2019)
106. R.S. Givens, M. Rubina, J. Wirz, Applications of p-hydroxyphenacyl (pHP) and coumarin-4-ylmethyl photoremovable protecting groups. Photochem. Photobiol. Sci. **11**(3), 472–488 (2012)
107. L.N. Johnson, Time-resolved protein crystallography. Protein Sci. **1**(10), 1237–1243 (1992)
108. R.H. Austin, K.W. Beeson, L. Eisenstein, H. Frauenfelder, I.C. Gunsalus, Dynamics of ligand binding to myoglobin. Biochemistry **14**(24), 5355–5373 (1975)
109. P. Mondal, M. Meuwly, Solvent composition drives the rebinding kinetics of nitric oxide to microperoxidase. Sci. Rep. **8**(1), 5281 (2018)
110. Y. Murakawa, M. Nagai, Y. Mizutani, Differences between protein dynamics of hemoglobin upon dissociation of oxygen and carbon monoxide. J. Am. Chem. Soc. **134**(3), 1434–1437 (2012)
111. D. Beece, L. Eisenstein, H. Frauenfelder, D. Good, M.C. Marden, L. Reinisch et al., Dioxygen replacement reaction in myoglobin. Biochemistry **18**(15), 3421–3423 (1979)
112. J.C. Flanagan, C.R. Baiz, Ultrafast pH-jump two-dimensional infrared spectroscopy. Opt. Lett. **44**(20), 4937–4940 (2019)
113. S. Abbruzzetti, S. Sottini, C. Viappiani, J.E. Corrie, Kinetics of proton release after flash photolysis of 1-(2-nitrophenyl)ethyl sulfate (caged sulfate) in aqueous solution. J. Am. Chem. Soc. **127**(27), 9865–9874 (2005)
114. M.C. Thompson, B.A. Barad, A.M. Wolff, H. Sun Cho, F. Schotte, D.M.C. Schwarz et al., Temperature-jump solution X-ray scattering reveals distinct motions in a dynamic enzyme. Nat. Chem. **11**(11), 1058–1066 (2019)
115. D.A. Keedy, L.R. Kenner, M. Warkentin, R.A. Woldeyes, J.B. Hopkins, M.C. Thompson, et al., Mapping the conformational landscape of a dynamic enzyme by multitemperature and XFEL crystallography. Elife **4** (2015)
116. M.J. Reddish, R. Callender, R.B. Dyer, Resolution of submillisecond kinetics of multiple reaction pathways for lactate dehydrogenase. Biophys. J. **112**(9), 1852–1862 (2017)
117. M.B. Winter, M.A. Herzik Jr., J. Kuriyan, M.A. Marletta, Tunnels modulate ligand flux in a heme nitric oxide/oxygen binding (H-NOX) domain. Proc. Natl. Acad. Sci. USA **108**(43), E881–E889 (2011)

118. N. Alberding, H. Frauenfelder, P. Hanggi, Stochastic theory of ligand migration in biomolecules. Proc. Natl. Acad. Sci. USA **75**(1), 26–29 (1978)
119. N. Alberding, R.H. Austin, S.S. Chan, L. Eisenstein, H. Frauenfelder, D. Good et al., Fast reactions in carbon monoxide binding to heme proteins. Biophys. J. **24**(1), 319–334 (1978)
120. H.G. Khorana, Two light-transducing membrane proteins: bacteriorhodopsin and the mammalian rhodopsin. Proc. Natl. Acad. Sci. USA **90**(4), 1166–1171 (1993)
121. M. Sheves, A. Albeck, N. Friedman, M. Ottolenghi, Controlling the pKa of the bacteriorhodopsin Schiff base by use of artificial retinal analogues. Proc. Natl. Acad. Sci. USA **83**(10), 3262–3266 (1986)
122. C.H. Chang, R. Jonas, R. Govindjee, T.G. Ebrey, Regeneration of blue and purple membranes from deionized bleached membranes of halobacterium-halobium. Photochem. Photobiol. **47**(2), 261–265 (1988)
123. K. Edman, P. Nollert, A. Royant, H. Belrhali, E. Pebay-Peyroula, J. Hajdu et al., High-resolution X-ray structure of an early intermediate in the bacteriorhodopsin photocycle. Nature **401**(6755), 822–826 (1999)
124. B. Schobert, J. Cupp-Vickery, V. Hornak, S.O. Smith, J.K. Lanyi, Crystallographic structure of the K intermediate of bacteriorhodopsin: conservation of free energy after photoisomerization of the retinal. J. Mol. Biol. **321**(4), 715–726 (2002)
125. Y. Matsui, K. Sakai, M. Murakami, Y. Shiro, S. Adachi, H. Okumura, T. Kouyama, Specific damage induced by X-ray radiation and structural changes in the primary photoreaction of bacteriorhodopsin. J. Mol. Biol. **324**(3), 469–481 (2002)
126. V. Borshchevskiy, E. Round, I. Erofeev, M. Weik, A. Ishchenko, I. Gushchin et al., Low-dose X-ray radiation induces structural alterations in proteins. Acta Crystallogr. D Biol. Crystallogr. **70**(Pt 10), 2675–2685 (2014)
127. E. Nango, A. Royant, M. Kubo, T. Nakane, C. Wickstrand, T. Kimura et al., A three-dimensional movie of structural changes in bacteriorhodopsin. Science **354**(6319), 1552–1557 (2016)
128. R. Neutze, R. Wouts, D. van der Spoel, E. Weckert, J. Hajdu, Potential for biomolecular imaging with femtosecond X-ray pulses. Nature **406**(6797), 752–757 (2000)
129. R.G. Sierra, H. Laksmono, J. Kern, R. Tran, J. Hattne, R. Alonso-Mori et al., Nanoflow electrospinning serial femtosecond crystallography. Acta Crystallogr. D **68**, 1584–1587 (2012)
130. U. Weierstall, D. James, C. Wang, T.A. White, D.J. Wang, W. Liu, et al., Lipidic cubic phase injector facilitates membrane protein serial femtosecond crystallography. Na. Commun. **5** (2014)
131. M. Sugahara, E. Mizohata, E. Nango, M. Suzuki, T. Tanaka, T. Masudala et al., Grease matrix as a versatile carrier of proteins for serial crystallography. Nat. Methods **12**(1), 61–63 (2015)
132. E.M. Landau, J.P. Rosenbusch, Lipidic cubic phases: a novel concept for the crystallization of membrane proteins. Proc. Natl. Acad. Sci. USA **93**(25), 14532–14535 (1996)
133. H. Luecke, H.T. Richter, J.K. Lanyi, Proton transfer pathways in bacteriorhodopsin at 2.3 Angstrom resolution. Science **280**(5371), 1934–1937 (1998)
134. M. Kubo, E. Nango, K. Tono, T. Kimura, S. Owada, C.Y. Song et al., Nanosecond pump-probe device for time-resolved serial femtosecond crystallography developed at SACLA. J. Synchrotron Radiat. **24**, 1086–1091 (2017)
135. Y. Shimazu, K. Tono, T. Tanaka, Y. Yamanaka, T. Nakane, C. Mori et al., High-viscosity sample-injection device for serial femtosecond crystallography at atmospheric pressure. J. Appl. Crystallogr. **52**, 1280–1288 (2019)
136. V.I. Borshchevskiy, E.S. Round, A.N. Popov, G. Buldt, V.I. Gordeliy, X-ray-radiation-induced changes in bacteriorhodopsin structure. J. Mol. Biol. **409**(5), 813–825 (2011)
137. T. Kouyama, T. Nishikawa, T. Tokuhisa, H. Okumura, Crystal structure of the L intermediate of bacteriorhodopsin: Evidence for vertical translocation of a water molecule during the proton pumping cycle. J. Mol. Biol. **335**(2), 531–546 (2004)
138. A. Maeda, S.P. Balashov, J. Lugtenburg, M.A. Verhoeven, J. Herzfeld, M. Belenky et al., Interaction of internal water molecules with the Schiff base in the L intermediate of the bacteriorhodopsin photocycle. Biochemistry **41**(11), 3803–3809 (2002)

139. A. Maeda, J. Herzfeld, M. Belenky, R. Needleman, R.B. Gennis, S.P. Balashov, T.G. Ebrey, Water-mediated hydrogen-bonded network on the cytoplasmic side of the Schiff base of the L photointermediate of bacteriorhodopsin. Biochemistry **42**(48), 14122–14129 (2003)
140. S. Subramaniam, R. Henderson, Molecular mechanism of vectorial proton translocation by bacteriorhodopsin. Nature **406**(6796), 653–657 (2000)
141. R. Neutze, E. Pebay-Peyroula, K. Edman, A. Royant, J. Navarro, E.M. Landau, Bacteriorhodopsin: a high-resolution structural view of vectorial proton transport. BBA-Biomembranes **1565**(2), 144–167 (2002)
142. R.A. Laskowski, M.B. Swindells, LigPlot+: multiple ligand-protein interaction diagrams for drug discovery. J. Chem. Inf. Model. **51**(10), 2778–2786 (2011)
143. O.P. Ernst, D.T. Lodowski, M. Elstner, P. Hegemann, L.S. Brown, H. Kandori, Microbial and animal rhodopsins: structures, functions, and molecular mechanisms. Chem. Rev. **114**(1), 126–163 (2014)
144. K.A. Freedman, R.S. Becker, Comparative investigation of the photoisomerization of the protonated and unprotonated n-butylamine Schiff bases of 9-cis-, 11-cis-, 13-cis-, and all-trans-retinals. J. Am. Chem. Soc. **108**(6), 1245–1251 (1986)
145. S. Schenkl, F. van Mourik, G. van der Zwan, S. Haacke, M. Chergui, Probing the ultrafast charge translocation of photoexcited retinal in bacteriorhodopsin. Science **309**(5736), 917–920 (2005)
146. J. Herbst, K. Heyne, R. Diller, Femtosecond infrared spectroscopy of bacteriorhodopsin chromophore isomerization. Science **297**(5582), 822–825 (2002)
147. P. Hamm, M. Zurek, T. Roschinger, H. Patzelt, D. Oesterhelt, W. Zinth, Femtosecond spectroscopy of the photoisomerisation of the protonated Schiff base of all-trans retinal. Chem. Phys. Lett. **263**(5), 613–621 (1996)
148. A. Aharoni, B. Hou, N. Friedman, M. Ottolenghi, I. Rousso, S. Ruhman, et al. Non-isomerizable artificial pigments: implications for the primary light-induced events in bacteriorhodopsin. Biochemistry-Moscow+ **66**(11), 1210–1219 (2001)
149. G.N. Kovacs, J.P. Colletier, M.L. Grunbein, Y. Yang, T. Stensitzki, A. Batyuk, et al., Three-dimensional view of ultrafast dynamics in photoexcited bacteriorhodopsin. Nat. Commun. **10** (2019)
150. R. Gonzalez-Luque, M. Garavelli, F. Bernardi, M. Merchan, M.A. Robb, M. Olivucci, Computational evidence in favor of a two-state, two-mode model of the retinal chromophore photoisomerization. Proc. Natl. Acad. Sci. USA **97**(17), 9379–9384 (2000)
151. S. Tahara, H. Kuramochi, S. Takeuchi, T. Tahara, Protein dynamics preceding photoisomerization of the retinal chromophore in bacteriorhodopsin revealed by deep-UV femtosecond stimulated raman spectroscopy. J. Phys. Chem. Lett. **10**(18), 5422–5427 (2019)
152. E. Buhl, P. Eberhardt, C. Bamann, E. Bamberg, M. Braun, J. Wachtveitl, Ultrafast protein response in channelrhodopsin-2 studied by time-resolved infrared spectroscopy. J. Phys. Chem. Lett. **9**(24), 7180–7184 (2018)
153. P. Altoe, A. Cembran, M. Olivucci, M. Garavelli, Aborted double bicycle-pedal isomerization with hydrogen bond breaking is the primary event of bacteriorhodopsin proton pumping. Proc. Natl. Acad. Sci. USA **107**(47), 20172–20177 (2010)
154. A. Warshel, Z.T. Chu, Nature of the surface crossing process in bacteriorhodopsin: computer simulations of the quantum dynamics of the primary photochemical event. J. Phys. Chem. B **105**(40), 9857–9871 (2001)
155. J.K. Yu, R.B. Liang, F. Liu, T.J. Martinez, First-principles characterization of the elusive I fluorescent state and the structural evolution of retinal protonated schiff base in bacteriorhodopsin. J. Am. Chem. Soc. **141**(45), 18193–18203 (2019)
156. N. Shibata, H. Kandori, FTIR studies of internal water molecules in the Schiff base region of bacteriorhodopsin. Biochemistry **44**(20), 7406–7413 (2005)
157. S. Shim, J. Dasgupta, R.A. Mathies, Femtosecond time-resolved stimulated raman reveals the birth of bacteriorhodopsin's J and K intermediates. J. Am. Chem. Soc. **131**(22), 7592–7597 (2009)

158. A. Ansari, J. Berendzen, S.F. Bowne, H. Frauenfelder, I.E. Iben, T.B. Sauke et al., Protein states and proteinquakes. Proc. Natl. Acad. Sci. USA **82**(15), 5000–5004 (1985)
159. Y. Umena, K. Kawakami, J.R. Shen, N. Kamiya, Crystal structure of oxygen-evolving photosystem II at a resolution of 1.9 angstrom. Nature **473**(7345), 55-U65 (2011)
160. M.M. Najafpour, G. Renger, M. Holynska, A.N. Moghaddam, E.M. Aro, R. Carpentier et al., Manganese compounds as water-oxidizing catalysts: from the natural water-oxidizing complex to nanosized manganese oxide structures. Chem. Rev. **116**(5), 2886–2936 (2016)
161. J.R. Shen, The structure of photosystem ii and the mechanism of water oxidation in photosynthesis. Annu. Rev. Plant Biol. **66**, 23–48 (2015)
162. J. Yano, J. Kern, K.D. Irrgang, M.J. Latimer, U. Bergmann, P. Glatzel et al., X-ray damage to the $Mn_4Ca$ complex in single crystals of photosystem II: a case study for metalloprotein crystallography. Proc. Natl. Acad. Sci. USA **102**(34), 12047–12052 (2005)
163. J. Yano, J. Kern, Y. Pushkar, K. Sauer, P. Glatzel, U. Bergmann et al., High-resolution structure of the photosynthetic $Mn_4Ca$ catalyst from X-ray spectroscopy. Philos. Trans. R. Soc. B **363**(1494), 1139–1147 (2008)
164. W. Ames, D.A. Pantazis, V. Krewald, N. Cox, J. Messinger, W. Lubitz, F. Neese, Theoretical evaluation of structural models of the S2 state in the oxygen evolving complex of photosystem II: protonation states and magnetic interactions. J. Am. Chem. Soc. **133**(49), 19743–19757 (2011)
165. H. Isobe, M. Shoji, S. Yamanaka, Y. Umena, K. Kawakami, N. Kamiya et al., Theoretical illumination of water-inserted structures of the $CaMn_4O_5$ cluster in the S-2 and S-3 states of oxygen-evolving complex of photosystem II: full geometry optimizations by B3LYP hybrid density functional. Dalton Trans. **41**(44), 13727–13740 (2012)
166. S. Luber, I. Rivalta, Y. Umena, K. Kawakami, J.R. Shen, N. Kamiya et al., S1-state model of the $O_2$-evolving complex of photosystem II. Biochemistry **50**(29), 6308–6311 (2011)
167. M. Suga, F. Akita, K. Hirata, G. Ueno, H. Murakami, Y. Nakajima, et al., Native structure of photosystem II at 1.95 angstrom resolution viewed by femtosecond X-ray pulses. Nature **517**(7532), 99-U265 (2015)
168. K. Hirata, K. Shinzawa-Itoh, N. Yano, S. Takemura, K. Kato, M. Hatanaka et al., Determination of damage-free crystal structure of an X-ray-sensitive protein using an XFEL. Nat. Methods **11**(7), 734-U174 (2014)
169. N. Cox, D.A. Pantazis, F. Neese, W. Lubitz, Biological water oxidation. Acc. Chem. Res. **46**(7), 1588–1596 (2013)
170. L.V. Kulik, B. Epel, W. Lubitz, J. Messinger, Electronic structure of the $Mn_4OxCa$ cluster in the S0 and S2 states of the oxygen-evolving complex of photosystem II based on pulse $^{55}$Mn-ENDOR and EPR spectroscopy. J. Am. Chem. Soc. **129**(44), 13421–13435 (2007)
171. N. Cox, M. Retegan, F. Neese, D.A. Pantazis, A. Boussac, W. Lubitz, Photosynthesis. Electronic structure of the oxygen-evolving complex in photosystem II prior to O–O bond formation. Science **345**(6198), 804–808 (2014)
172. I.D. Young, M. Ibrahim, R. Chatterjee, S. Gul, F.D. Fuller, S. Koroidov, et al., Structure of photosystem II and substrate binding at room temperature. Nature **540**(7633), 453–457 (2016)
173. M. Suga, F. Akita, K. Yamashita, Y. Nakajima, G. Ueno, H.J. Li, et al., An oxyl/oxo mechanism for oxygen-oxygen coupling in PSII revealed by an x-ray free-electron laser. Science **366**(6463), 334–338 (2019)
174. J.R. Stagno, Y. Liu, Y.R. Bhandari, C.E. Conrad, S. Panja, M. Swain, et al., Structures of riboswitch RNA reaction states by mix-and-inject XFEL serial crystallography. Nature **541**(7636), 242–246 (2017)
175. C. Kupitz, J.L. Olmos Jr., M. Holl, L. Tremblay, K. Pande, S. Pandey et al., Structural enzymology using X-ray free electron lasers. Struct. Dyn. **4**(4), 044003 (2017)
176. K.R. Beyerlein, D. Dierksmeyer, V. Mariani, M. Kuhn, I. Sarrou, A. Ottaviano et al., Mix-and-diffuse serial synchrotron crystallography. IUCrJ **4**, 769–777 (2017)
177. J.L. Olmos, S. Pandey, J.M. Martin-Garcia, G. Calvey, A. Katz, J. Knoska, et al., Enzyme intermediates captured "on the fly" by mix-and-inject serial crystallography. BMC Biol. **16** (2018)

178. I. Ishigami, A. Lewis-Ballester, A. Echelmeier, G. Brehm, N.A. Zatsepin, T.D. Grant et al., Snapshot of an oxygen intermediate in the catalytic reaction of cytochrome c oxidase. Proc. Natl. Acad. Sci. USA **116**(9), 3572–3577 (2019)
179. H. Shoun, S. Fushinobu, L. Jiang, S.W. Kim, T. Wakagi, Fungal denitrification and nitric oxide reductase cytochrome P450nor. Philos. Trans. R. Soc. B **367**(1593), 1186–1194 (2012)
180. A.R. Ravishankara, J.S. Daniel, R.W. Portmann, Nitrous oxide ($N_2O$): the dominant ozone-depleting substance emitted in the 21st century. Science **326**(5949), 123–125 (2009)
181. Y. Shiro, M. Fujii, T. Iizuka, S. Adachi, K. Tsukamoto, K. Nakahara, H. Shoun, Spectroscopic and kinetic-studies on reaction of cytochrome P450nor with nitric-oxide—implication for its nitric-oxide reduction-mechanism. J. Biol. Chem. **270**(4), 1617–1623 (1995)
182. E. Obayashi, S. Takahashi, Y. Shiro, Electronic structure of reaction intermediate of cytochrome P450nor in its nitric oxide reduction. J. Am. Chem. Soc. **120**(49), 12964–12965 (1998)
183. A. Daiber, T. Nauser, N. Takaya, T. Kudo, P. Weber, C. Hultschig et al., Isotope effects and intermediates in the reduction of NO by P450(NOR). J. Inorg. Biochem. **88**(3–4), 343–352 (2002)
184. L.E. Goodrich, F. Paulat, V.K.K. Praneeth, N. Lehnert, Electronic structure of heme-nitrosyls and its significance for nitric oxide reactivity, sensing, transport, and toxicity in biological systems. Inorg. Chem. **49**(14), 6293–6316 (2010)
185. C. Riplinger, E. Bill, A. Daiber, V. Ullrich, H. Shoun, F. Neese, New insights into the nature of observable reaction intermediates in cytochrome P450 NO reductase by using a combination of spectroscopy and quantum mechanics/molecular mechanics calculations. Chem.-Eur. J. **20**(6), 1602–1614 (2014)
186. S. Namiki, T. Arai, K. Fujimori, High-performance caged nitric oxide: a new molecular design, synthesis, and photochemical reaction. J. Am. Chem. Soc. **119**(16), 3840–3841 (1997)
187. Y. Kato, F. Akita, Y. Nakajima, M. Suga, Y. Umena, J.R. Shen, T. Noguchi, Fourier transform infrared analysis of the S-state cycle of water oxidation in the microcrystals of photosystem II. J. Phys. Chem. Lett. **9**(9), 2121–2126 (2018)
188. J.S. Stamler, D.J. Singel, J. Loscalzo, Biochemistry of nitric-oxide and its redox-activated forms. Science **258**(5090), 1898–1902 (1992)
189. D.P. Linder, K.R. Rodgers, Fe–N–O structure and bonding in six-coordinate {FeNO}(6) porphyrinates containing imidazole: implications for reactivity of coordinated NO. Inorg. Chem. **44**(5), 1367–1380 (2005)
190. G.C.R. Ellis-Davies, Caged compounds: photorelease technology for control of cellular chemistry and physiology. Nat. Methods **4**(8), 619–628 (2007)
191. C. Ludovici, R. Fröhlich, K. Vogtt, B. Mamat, M. Lübben, Caged $O(_2)$. Reaction of cytochrome $bo(_3)$ oxidase with photochemically released dioxygen from a cobalt peroxo complex. Eur. J. Biochem. **269**(10), 2630–2637 (2002)
192. M. Lübben, K. Gerwert, Redox FTIR difference spectroscopy using caged electrons reveals contributions of carboxyl groups to the catalytic mechanism of haem-copper oxidases. FEBS Lett. **397**(2–3), 303–307 (1996)
193. J. Spradlin, D. Lee, S. Mahadevan, M. Mahomed, L. Tang, Q. Lam et al., Insights into an efficient light-driven hybrid P450 BM3 enzyme from crystallographic, spectroscopic and biochemical studies. BBA-Proteins Proteom. **1864**(12), 1732–1738 (2016)

# Correction to: Attosecond Dynamics of Non-resonant Atomic Photoionization

**David Busto, Shiyang Zhong, Jan Marcus Dahlström, Anne L'Huillier, and Mathieu Gisselbrecht**

## Correction to:
## Chapter 1 in: K. Ueda (ed.), *Ultrafast Electronic and Structural Dynamics*, https://doi.org/10.1007/978-981-97-2914-2_1

In the original version of the book, the following correction has been incorporated in Chapter 1: The first name of Mathieu Gisselbrecht was unfortunately published with an error. The initially published version has now been corrected.

---

The updated version of this chapter can be found at
https://doi.org/10.1007/978-981-97-2914-2_1



K. Ueda (ed.), *Ultrafast Electronic and Structural Dynamics*,
https://doi.org/10.1007/978-981-97-2914-2_16

The manufacturer's authorised representative in the EU is Springer Nature Customer Service Centre GmbH, Europaplatz 3, 69115 Heidelberg, Germany. If you have any concerns regarding our products, please contact ProductSafety@springernature.com

Printed and bound by CPI Group (UK) Ltd, Croydon, CR0 4YY

30/09/2026

02225253-0017